AIRFRAME AND POWERPLANT

MECHANICS

AIRFRAME HANDBOOK

U.S. DEPARTMENT OF TRANSPORTATION

FEDERAL AVIATION ADMINISTRATION

FLIGHT STANDARDS SERVICE

First Edition 1972

First Revision 1976

JS312605B

PREFACE

This handbook was developed and first printed in 1972 as one of a series of three handbooks for persons preparing for certification as an airframe or powerplant mechanic. It is intended that this handbook will provide basic information on principles, fundamentals and technical procedures in the subject matter areas relating to the airframe rating. It is designed to aid students enrolled in a formal course of instruction as well as the individual who is studying on his own. Since the knowledge requirements for the airframe and powerplant ratings closely parallel each other in some subject areas, the chapters which discuss fire protection systems and electrical systems contain some material which is also duplicated in the Airframe and Powerplant Mechanics Powerplant Handbook, AC 65–12A.

This volume contains information on airframe construction features, assembly and rigging, fabric covering, structural repairs, and aircraft welding. The handbook also contains an explanation of the units which make up the various airframe systems.

Because there are so many different types of aircraft in use today, it is reasonable to expect that differences exist in airframe components and systems. To avoid undue repetition, the practice of using representative systems and units is carried out throughout the handbook. Subject matter treatment is from a generalized point of view, and should be supplemented by reference to manufacturers' manuals or other textbooks if more detail is desired. This handbook is not intended to replace, substitute for, or supersede official regulations or the manufacturers' instructions.

Grateful acknowledgement is extended to the manufacturers of engines, propellers, and powerplant accessories for their cooperation in making material available for inclusion in this handbook.

Copyright material is used by special permission of the following organizations and may not be extracted or reproduced without permission of the copyright owner.

	(R)
Monsanto Chemicals Co.	Skydrol ® Fluids
Townsend Corporation	Cherry Rivets
	Acres Sleeves
J. O. King, Inc.	Acres Sleeves
Gravines, Inc.	Fire Extinguishers
Walter Kidde	Fire Extinguishers
DuPont De Nemours	Fire Extinguishants
National Fire Protection Association	Fire Extinguisher and Extinguishant Specifications
National Association of Fire Extinguisher Distributors	Fire Extinguishers
Flight Safety Foundation	Refueling Data
American Petroleum Institute	Aviation Fuels
Exxon Corporation	Aviation Fuels
Parker Hannifin	Aircraft Fittings
Goodyear Tire and Rubber Co.	Aircraft Tires
	Aircraft Wheels
	Aircraft Brakes
Firestone	Aircraft Tires

Bendix Energy Controls	Aircraft Wheels
	Aircraft Brakes
Rohm and Haas	Plastics
Douglas Aircraft Company	Portable Oxygen Generators in the DC–10
Aviation Maintenance Foundation, Inc.	Air Conditioning
BF Goodrich	Aircraft Tires
	Aircraft Wheels
	Aircraft Brakes
Puritan Equipment, Inc.	Portable Oxygen Generators

The advancements in aeronautical technology dictate that an instructional handbook must be under continuous review and brought up to date periodically to be valid. Flight Standards requested comments, from the certificated mechanic schools on the three handbooks. As a result, the handbooks have been updated to this extent: indicated errors have been corrected, new material has been added in the areas which were indicated as being deficient, and some material has been rearranged to improve the usefulness of the handbooks.

We would appreciate having errors brought to our attention, as well as receiving suggestions for improving the usefulness of the handbooks. Your comments and suggestions will be retained in our files until such time as the next revision will be accomplished.

Address all correspondence relating to these handbooks to:

U.S. Department of Transportation
Federal Aviation Administration
Flight Standards National Field Office
P.O. Box 25082
Oklahoma City, Oklahoma 73125

The companion handbooks to AC 65–15A are the Airframe and Powerplant Mechanics General Handbook, AC 65–9A and the Airframe and Powerplant Mechanics Powerplant Handbook, AC 65–12A.

CONTENTS

PREFACE _____ iii

CONTENTS _____ v

CHAPTER 1. AIRCRAFT STRUCTURES

General _____ 1
Major Structural Stresses _____ 1
Fixed-Wing Aircraft _____ 2
Fuselage _____ 2
Wing Structure _____ 6
Nacelles or Pods _____ 13
Empennage _____ 16
Flight Control Surfaces _____ 18
Landing Gear _____ 23
Skin and Fairing _____ 24
Access and Inspection Doors _____ 24
Helicopter Structures _____ 24

CHAPTER 2. ASSEMBLY AND RIGGING

General _____ 27
Theory of Flight _____ 27
Aerodynamics _____ 27
The Atmosphere _____ 27
Pressure _____ 27
Density _____ 28
Humidity _____ 28
Bernoulli's Principle and Subsonic Flow _____ 29
Motion _____ 29
Airfoils _____ 30
Center of Gravity _____ 33
Thrust and Drag _____ 33
Axes of an Aircraft _____ 35
Stability and Control _____ 35
Control _____ 39
Flight Control Surfaces _____ 40
Control Around the Longitudinal Axis _____ 41
Control Around the Vertical Axis _____ 44
Control Around the Lateral Axis _____ 44
Tabs _____ 45
Boundary Layer Control Devices _____ 48
Forces Acting on a Helicopter _____ 49
Helicopter Axes of Flight _____ 55
High-Speed Aerodynamics _____ 56
Typical Supersonic Flow Patterns _____ 59
Aerodynamic Heating _____ 64
Flight Control Systems _____ 64
Hydraulic Operated Control Systems _____ 65
Cable Guides _____ 66
Mechanical Linkage _____ 68

CHAPTER 2. ASSEMBLY AND RIGGING—(Cont.)

Torque Tubes	68
Stops	68
Control Surface Snubbers and Locking Devices	69
Aircraft Rigging	70
Rigging Checks	72
Adjustment of Control Surfaces	75
Helicopter Rigging	77
Principles of Balancing or Re-balancing	80
Re-balancing Procedures	83
Methods	83

CHAPTER 3. AIRCRAFT FABRIC COVERING

Aircraft Fabrics	85
Miscellaneous Textile Materials	88
Seams	89
Applying Covering	91
Covering Wings	93
Covering Fuselages	95
Ventilation, Drain, and Inspection Openings	95
Repair of Fabric Covers	99
Replacing Panels in Wing Covers	103
Re-covering Aircraft Surface with Glass Cloth	104
Causes of Fabric Deterioration	104
Checking Condition of Doped Fabric	106
Testing Fabric Covering	106
Strength Criteria for Aircraft Fabric	107
Dopes and Doping	107
Dope Materials	108
Aluminum-Pigmented Dopes	109
Temperature and Humidity Effects on Dope	109
Common Troubles in Dope Application	109
Technique of Application	110
Number of Coats Required	111

CHAPTER 4. AIRCRAFT PAINTING AND FINISHING

General	113
Finishing Materials	113
Paint Touchup	117
Identification of Paint Finishes	117
Paint Removal	117
Restoration of Paint Finishes	118
Nitrocellulose Lacquer Finishes	118
Acrylic Nitrocellulose Lacquer Finish	119
Epoxy Finishes	120
Fluorescent Finishes	121
Enamel Finishes	121
Paint System Compatibility	122
Methods of Applying Finishes	122
Preparation of Paint	123

CHAPTER 4. AIRCRAFT PAINTING AND FINISHING—(Cont.)

Common Paint Troubles _____ 124
Painting Trim and Identification Numbers _____ 125
Decalcomanias (Decals) _____ 125

CHAPTER 5. AIRCRAFT STRUCTURAL REPAIRS

Basic Principles of Sheet Metal Repair _____ 127
General Structural Repair _____ 129
Inspection of Damage _____ 130
Classification of Damage _____ 131
Stresses in Structural Members _____ 131
Special Tools and Devices for Sheet Metal _____ 133
Metalworking Machines _____ 136
Forming Machines _____ 141
Forming Operations and Terms _____ 145
Making Straight Line Bends _____ 146
Setback _____ 148
Making Layouts _____ 151
Hand Forming _____ 155
Rivet Layout _____ 164
Rivet Installation _____ 166
Preparation of Rivet Holes _____ 168
Driving Rivets _____ 173
Rivet Failures _____ 175
Removing Rivets _____ 177
Special Rivets _____ 177
Self-Plugging (Friction Lock) Rivets _____ 178
Self-Plugging (Mechanical Lock) Rivets _____ 181
Pull-Thru Rivets _____ 184
Rivnuts _____ 184
Dill Lok-Skrus and Lok-Rivets _____ 186
Deutsch Rivets _____ 187
Hi-Shear Rivets _____ 187
Specific Repair Types _____ 189
Structural Sealing _____ 198
Metal Bonded Honeycomb _____ 200
Construction Features _____ 201
Damage _____ 201
Repairs _____ 202
Repair Materials _____ 205
Potted Compound Repair _____ 206
Glass Fabric Cloth Overlay Repairs _____ 208
One Skin and Core Repair Procedures _____ 209
Plastics _____ 213
Transparent Plastics _____ 213
Storage and Protection _____ 214
Forming Plastics _____ 215
Installation Procedures _____ 220
Laminated Plastics _____ 221
Fiber Glass Components _____ 221
Radomes _____ 223
Wooden Aircraft Structures _____ 224
Inspection of Wooden Structures _____ 224
Service and Repair of Wooden Structures _____ 228
Glues _____ 230
Gluing _____ 232

CHAPTER 5. AIRCRAFT STRUCTURAL REPAIRS—(Cont.)

Spliced Joints _____ 234
Plywood Skin Repairs _____ 235
Spar and Rib Repair _____ 242
Bolt and Bushing Holes _____ 244
Rib Repairs _____ 245

CHAPTER 6. AIRCRAFT WELDING

General _____ 247
Oxyacetylene Welding Equipment _____ 248
Welding Positions _____ 255
Welded Joints _____ 255
Expansion and Contraction of Metals _____ 257
Correct Forming of a Weld _____ 258
Oxyacetylene Welding of Ferrous Metals _____ 258
Welding Nonferrous Metals Using Oxyacetylene _____ 260
Titanium _____ 262
Cutting Metal Using Oxyacetylene _____ 263
Brazing Methods _____ 264
Soft Soldering _____ 265
Electric Arc Welding _____ 266
Welding Procedures and Techniques _____ 269
Welding of Aircraft Steel Structures _____ 276

CHAPTER 7. ICE AND RAIN PROTECTION

General _____ 285
Pneumatic Deicing Systems _____ 286
Deicer Boot Construction _____ 287
Deicing System Components _____ 288
Pneumatic Deicing System Maintenance _____ 291
Thermal Anti-Icing Systems _____ 293
Pneumatic System Ducting _____ 296
Ground Deicing of Aircraft _____ 299
Windshield Icing Control Systems _____ 300
Water and Toilet Drain Heaters _____ 303
Rain Eliminating Systems _____ 303
Maintenance of Rain Eliminating Systems _____ 308

CHAPTER 8. HYDRAULIC AND PNEUMATIC POWER SYSTEMS

Aircraft Hydraulic Systems _____ 309
Hydraulic Fluid _____ 309
Types of Hydraulic Fluids _____ 310
Phosphate Ester Base Fluids _____ 311
Filters _____ 313
Basic Hydraulic System _____ 315
Reservoirs _____ 316
Pressure Regulation _____ 323
Actuating Cylinders _____ 328
Selector Valves _____ 329
Aircraft Pneumatic Systems _____ 331
Pneumatic System Components _____ 334
Typical Pneumatic Power System _____ 338

CHAPTER 9. LANDING GEAR SYSTEMS

General _____ 341
Main Landing Gear Alignment, Support, Retraction _____ 348

CHAPTER 9. LANDING GEAR SYSTEMS—(Cont.)

Emergency Extension Systems _____ 351
Landing Gear Safety Devices _____ 351
Nosewheel Steering System _____ 354
Shimmy Dampers _____ 356
Brake Systems _____ 360
Brake Assemblies _____ 366
Inspection and Maintenance of Brake Systems _____ 372
Aircraft Landing Wheels _____ 373
Aircraft Tires _____ 377
Aircraft Tire Maintenance _____ 379
Tire Inspection—Mounted On Wheel _____ 381
Tire Inspection—Tire Demounted _____ 382
Tube Inspection _____ 384
Mounting and Demounting _____ 385
Causes of Air Pressure Loss in Tubeless Aircraft Tires _____ 389
The Wheel _____ 390
Good Pressure Gage Practice _____ 391
Repairing _____ 392
Operating and Handling Tips _____ 394
Tube Repair _____ 396
Sidewall-Inflated Aircraft Tires _____ 396
Tire Inspection Summary _____ 397
Antiskid System _____ 399
Landing Gear System Maintenance _____ 400

CHAPTER 10. FIRE PROTECTION SYSTEMS

General _____ 407
Fire Detection Systems _____ 408
Types of Fires _____ 411
Fire Zone Classification _____ 412
Extinguishing Agent Characteristics _____ 412
Fire Extinguishing Systems _____ 417
Reciprocating Engine Conventional CO_2 System _____ 417
Turbojet Fire Protection System _____ 419
Turbine Engine Fire Extinguishing System _____ 420
Turbine Engine Ground Fire Protection _____ 422
Fire Detection System Maintenance Practices _____ 423
Fire Detection System Troubleshooting _____ 425
Fire Extinguisher System Maintenance Practices _____ 425
Fire Prevention and Protection _____ 429
Cockpit and Cabin Interiors _____ 429
Smoke Detection Systems _____ 430

CHAPTER 11. AIRCRAFT ELECTRICAL SYSTEMS

General _____ 433
Lacing and Tying Wire Bundles _____ 444
Cutting Wire and Cable _____ 446
Emergency Splicing Repairs _____ 450
Connecting Terminal Lugs to Terminal Blocks _____ 451
Bonding and Grounding _____ 452
Connectors _____ 455
Conduit _____ 457
Electrical Equipment Installation _____ 457
Aircraft Lighting Systems _____ 459
Maintenance and Inspection of Lighting Systems _____ 464

CHAPTER 12. AIRCRAFT INSTRUMENT SYSTEMS

General _____ 469
Instrument Cases _____ 469
Dials _____ 469
Range Markings _____ 470
Instrument Panels _____ 470
Repair of Aircraft Instruments _____ 471
Aircraft Pressure Gages _____ 471
Pitot-Static System _____ 474
Maintenance of Pitot-Static Systems _____ 481
Turn-and-Bank Indicator _____ 482
Synchro-Type Remote Indicating Instruments _____ 483
Remote-Indicating Fuel and Oil Pressure Gages _____ 485
Capacitor-Type Fuel Quanity System _____ 485
Angle-of-Attack Indicator _____ 487
Tachometers _____ 488
Synchroscope _____ 491
Temperature Indicators _____ 491
Ratiometer Electrical Resistance Thermometer _____ 497
Fuel Flowmeter Systems _____ 497
Gyroscopic Instruments _____ 499
Sources of Power for Gyro Operation _____ 501
Vacuum-Driven Attitude Gyros _____ 504
Pressure-Operated Gyros _____ 506
Vacuum System Maintenance Practices _____ 506
Electric Attitude Indicator _____ 507
Autopilot System _____ 511
Basic Autopilot Components _____ 513
Flight Director Systems _____ 518
Autopilot System Maintenance _____ 516

CHAPTER 13. COMMUNICATIONS AND NAVIGATION SYSTEMS

General _____ 519
Basic Radio Principles _____ 519
Basic Equipment Components _____ 520
Power Supply _____ 522
Communication Systems _____ 522
Airborne Navigation Equipment _____ 524
VHF Omnirange System _____ 524
Instrument Landing System _____ 525
Distance-Measuring Equipment _____ 528
Automatic Direction Finders _____ 529
Radar Beacon Transponder _____ 530
Doppler Navigation Systems _____ 530
Inertial Navigation System _____ 531
Airborne Weather Radar System _____ 532
Radio Altimeter _____ 533
Emergency Locator Transmitter (ELT) _____ 533
Installation of Communication and Navigation Equipment _____ 534
Reducing Radio Interference _____ 536
Installatoin of Aircraft Antenna Systems _____ 537

CHAPTER 14. CABIN ATMOSPHERE CONTROL SYSTEM

Need for Oxygen _____ 539
Composition of the Atmosphere _____ 539
Pressurization _____ 541
Air Conditioning and Pressurization Systems _____ 543

CHAPTER 14.   CABIN ATMOSPHERE CONTROL SYSTEM—(Cont.)

Basic Requirements _____ 545
Sources of Cabin Pressure _____ 545
Supercharger Instruments _____ 549
Pressurization Valves _____ 550
Cabin Pressure Control System _____ 551
Air Distribution _____ 556
Air Conditioning System _____ 558
Heating Systems _____ 559
Combustion Heaters _____ 561
Maintenance of Combustion Heater Systems _____ 564
Cooling Systems _____ 565
Air Cycle Cooling System _____ 565
Air Cycle System Component Operation _____ 568
Electronic Cabin Temperature Control System _____ 574
Electronic Temperature Control Regulator _____ 575
Vapor Cycle System (Freon) _____ 576
Freon System Components _____ 577
Description of a Typical System _____ 580
Air Conditioning and Pressurization System Maintenance _____ 583
Cabin Pressurization Operational Checks _____ 585
Cabin Pressurization Troubleshooting _____ 586
Oxygen Systems General _____ 587
Portable Oxygen Equipment _____ 587
Smoke Protection Equipment _____ 588
Oxygen Cylinders _____ 588
Solid State Oxygen Systems _____ 589
Oxygen Plumbing _____ 591
Oxygen Valves _____ 592
Regulators _____ 594
Oxygen System Flow Indicators _____ 597
Pressure Gages _____ 597
Oygen Masks _____ 597
Servicing Gaseous Oxygen Systems _____ 598
Prevention of Oxygen Fires or Explosions _____ 600

GENERAL

The airframe of a fixed-wing aircraft is generally considered to consist of five principal units, the fuselage, wings, stabilizers, flight control surfaces, and landing gear. Helicopter airframes consist of the fuselage, main rotor and related gearbox, tail rotor (on helicopters with a single main rotor), and the landing gear.

The airframe components are constructed from a wide variety of materials and are joined by rivets, bolts, screws, and welding or adhesives. The aircraft components are composed of various parts called structural members (*i.e.*, stringers, longerons, ribs, bulkheads, etc.). Aircraft structural members are designed to carry a load or to resist stress. A single member of the structure may be subjected to a combination of stresses. In most cases the structural members are designed to carry end loads rather than side loads: that is, to be subjected to tension or compression rather than bending.

Strength may be the principal requirement in certain structures, while others need entirely different qualities. For example, cowling, fairing, and similar parts usually are not required to carry the stresses imposed by flight or the landing loads. However, these parts must have such properties as neat appearance and streamlined shapes.

MAJOR STRUCTURAL STRESSES

In designing an aircraft, every square inch of wing and fuselage, every rib, spar, and even each metal fitting must be considered in relation to the physical characteristics of the metal of which it is made. Every part of the aircraft must be planned to carry the load to be imposed upon it. The determination of such loads is called stress analysis. Although planning the design is not the function of the aviation mechanic, it is, nevertheless, important that he understand and appreciate the stresses involved in order to avoid changes in the original design through improper repairs.

There are five major stresses to which all aircraft are subjected (figure 1–1):

(1) Tension.
(2) Compression.
(3) Torsion.
(4) Shear.
(5) Bending.

The term "stress" is often used interchangeably with the word "strain." Stress is an internal force of a substance which opposes or resists deformation. Strain is the deformation of a material or substance. Stress, the internal force, can cause strain.

Tension (figure 1–1a) is the stress that resists a force that tends to pull apart. The engine pulls the aircraft forward, but air resistance tries to hold it back. The result is tension, which tries to stretch the aircraft. The tensile strength of a material is measured in p.s.i. (pounds per square inch) and is calculated by dividing the load (in pounds) required to pull the material apart by its cross-sectional area (in square inches).

Compression (figure 1–1b) is the stress that resists a crushing force. The compressive strength of a material is also measured in p.s.i. Compression is the stress that tends to shorten or squeeze aircraft parts.

Torsion is the stress that produces twisting (figure 1–1c). While moving the aircraft forward, the engine also tends to twist it to one side, but other aircraft components hold it on course. Thus, torsion is created. The torsional strength of a material is its resistance to twisting or torque.

Shear is the stress that resists the force tending to cause one layer of a material to slide over an adjacent layer. Two riveted plates in tension (figure 1–1d) subject the rivets to a shearing force. Usually, the shearing strength of a material is either equal to or less than its tensile or compressive strength. Aircraft parts, especially screws, bolts, and rivets, are often subject to a shearing force.

Bending stress is a combination of compression and tension. The rod in figure 1–1e has been shortened (compressed) on the inside of the bend and stretched on the outside of the bend.

1

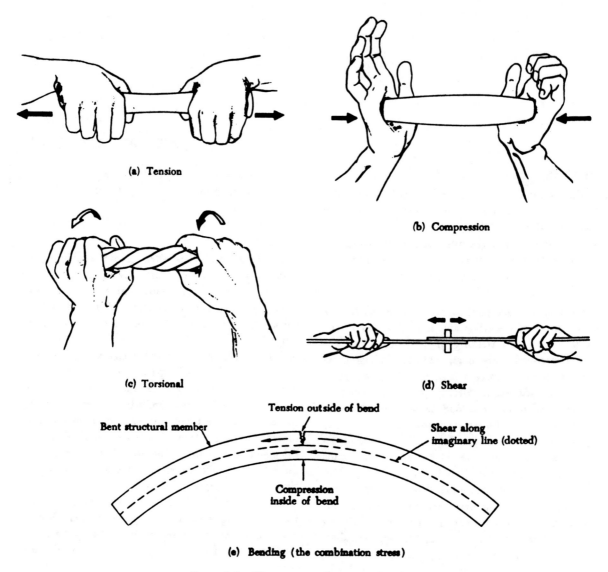

(a) Tension

(b) Compression

(c) Torsional

(d) Shear

Bent structural member

Tension outside of bend

Shear along imaginary line (dotted)

Compression inside of bend

(e) Bending (the combination stress)

FIGURE 1–1. Five stresses acting on an aircraft.

FIXED-WING AIRCRAFT

The principal components of a single-engine, propeller-driven aircraft are shown in figure 1–2.

Figure 1–3 illustrates the structural components of a typical turbine powered aircraft. One wing and the empennage assemblies are shown exploded into the many components which, when assembled, form major structural units.

FUSELAGE

The fuselage is the main structure or body of the aircraft. It provides space for cargo, controls, accessories, passengers, and other equipment. In single-engine aircraft, it also houses the powerplant. In

multi-engine aircraft the engines may either be in the fuselage, attached to the fuselage, or suspended from the wing structure. They vary principally in size and arrangement of the different compartments.

There are two general types of fuselage construction, the truss type, and the monocoque type. A truss is a rigid framework made up of members such as beams, struts, and bars to resist deformation by applied loads. The truss-framed fuselage is generally covered with fabric.

Truss Type

The truss type fuselage frame (figure 1–4) is usually constructed of steel tubing welded together in such a manner that all members of the truss can carry both tension and compression loads. In some

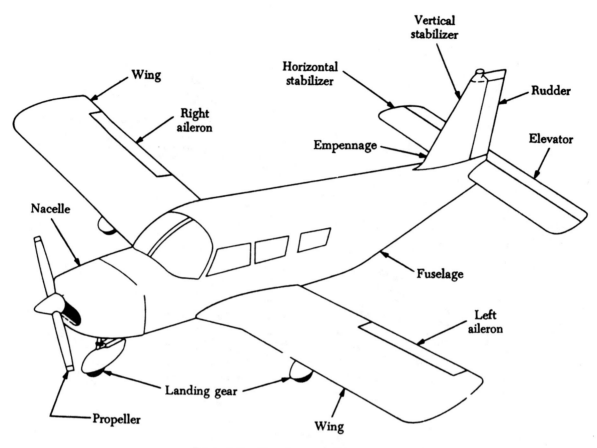

FIGURE 1-2. Aircraft structural components.

aircraft, principally the light, single-engine models, truss fuselage frames are constructed of aluminum alloy and may be riveted or bolted into one piece, with cross-bracing achieved by using solid rods or tubes.

Monocoque Type

The monocoque (single shell) fuselage relies largely on the strength of the skin or covering to carry the primary stresses. The design may be divided into three classes: (1) Monocoque, (2) semimonocoque, or (3) reinforced shell. The true monocoque construction (figure 1-5) uses formers, frame assemblies, and bulkheads to give shape to the fuselage, but the skin carries the primary stresses. Since no bracing members are present, the skin must be strong enough to keep the fuselage rigid. Thus, the biggest problem involved in monocoque construction is maintaining enough strength while keeping the weight within allowable limits.

To overcome the strength/weight problem of monocoque construction, a modification called semimonocoque construction (figure 1-6) was developed.

In addition to formers, frame assemblies, and bulkheads, the semimonocoque construction has the skin reinforced by longitudinal members. The reinforced shell has the skin reinforced by a complete framework of structural members. Different portions of the same fuselage may belong to any one of the three classes, but most aircraft are considered to be of semimonocoque type construction.

Semimonocoque Type

The semimonocoque fuselage is constructed primarily of the alloys of aluminum and magnesium, although steel and titanium are found in areas of high temperatures. Primary bending loads are taken by the *longerons*, which usually extend across several points of support. The longerons are supplemented by other longitudinal members, called *stringers*. Stringers are more numerous and lighter in weight than longerons. The vertical structural members are referred to as *bulkheads, frames, and formers*. The heaviest of these vertical members are located at intervals to carry concentrated loads and at points where fittings are used to attach other units, such as the wings, powerplants, and stabiliz-

3

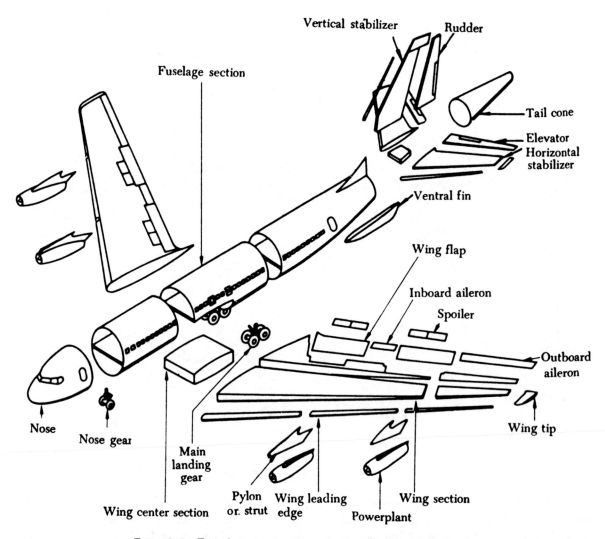

Vertical stabilizer
Rudder
Fuselage section
Tail cone
Elevator
Horizontal stabilizer
Ventral fin
Wing flap
Inboard aileron
Spoiler
Outboard aileron
Nose
Wing tip
Nose gear
Main landing gear
Wing center section
Pylon or strut
Wing leading edge
Wing section
Powerplant

FIGURE 1–3. Typical structural components of a turbine powered aircraft.

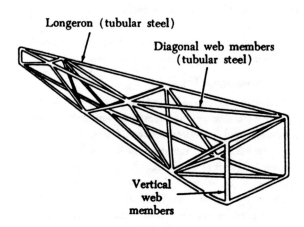

Longeron (tubular steel)

Diagonal web members (tubular steel)

Vertical web members

FIGURE 1–4. Warren truss of welded tubular steel.

ers. Figure 1–7 shows one form of the semi-monocoque design now in use.

The stringers are smaller and lighter than longe-

rons and serve as fill-ins. They have some rigidity, but are chiefly used for giving shape and for attachment of the skin. The strong, heavy longerons hold the bulkheads and formers, and these, in turn, hold the stringers. All of these joined together form a rigid fuselage framework.

There is often little difference between some rings, frames, and formers. One manufacturer may call a brace a former, whereas another may call the same type of brace a ring or frame. Manufacturers' instructions and specifications for a specific aircraft are the best guides.

Stringers and longerons prevent tension and compression from bending the fuselage. Stringers are usually of a one-piece aluminum alloy construction, and are manufactured in a variety of shapes by casting, extrusion, or forming. Longerons, like stringers, are usually made of aluminum alloy; how-

4

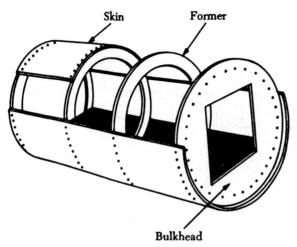

FIGURE 1-5. Monocoque construction.

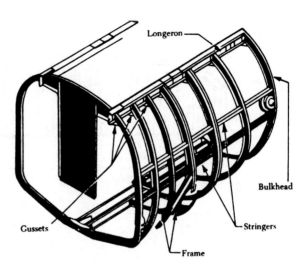

FIGURE 1-7. Fuselage structural members.

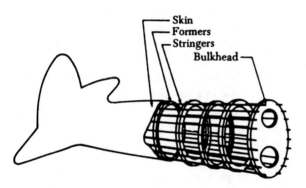

FIGURE 1-6. Semimonocoque construction.

ever, they may be of either a one-piece or a built-up construction.

By themselves, the structural members discussed do not give strength to a fuselage. They must first be joined together by such connective devices as gussets, rivets, nuts and bolts, or metal screws. A gusset (figure 1-7) is a type of connecting bracket. The bracing between longerons is often referred to as web members. They may be installed vertically or diagonally.

The metal skin or covering is riveted to the longerons, bulkheads, and other structural members and carries part of the load. The fuselage skin thickness will vary with the load carried and the stresses sustained at a particular location.

There are a number of advantages in the use of the semimonocoque fuselage. The bulkheads, frames, stringers, and longerons facilitate the design and construction of a streamlined fuselage, and add to the strength and rigidity of the structure. The main advantage, however, lies in the fact that it

does not depend on a few members for strength and rigidity. This means that a semimonocoque fuselage, because of its stressed-skin construction, may withstand considerable damage and still be strong enough to hold together.

Fuselages are generally constructed in two or more sections. On small aircraft, they are generally made in two or three sections, while larger aircraft may be made up of as many as six sections.

Quick access to the accessories and other equipment carried in the fuselage is provided for by numerous access doors, inspection plates, landing wheel wells, and other openings. Servicing diagrams showing the arrangement of equipment and location of access doors are supplied by the manufacturer in the aircraft maintenance manual.

Location Numbering Systems

There are various numbering systems in use to facilitate location of specific wing frames, fuselage bulkheads, or any other structural members on an aircraft. Most manufacturers use some system of station marking; for example, the nose of the aircraft may be designated zero station, and all other stations are located at measured distances in inches behind the zero station. Thus, when a blueprint reads "fuselage frame station 137," that particular frame station can be located 137 in. behind the nose of the aircraft. A typical station diagram is shown in figure 1-8.

To locate structures to the right or left of the center line of an aircraft, many manufacturers consider the center line as a zero station for structural member location to its right or left. With such a

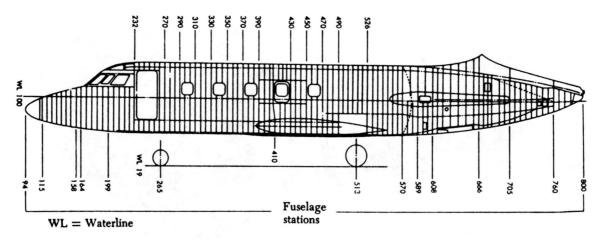

WL = Waterline

FIGURE 1–8. Fuselage stations.

system the stabilizer frames can be designated as being so many inches right or left of the aircraft center line.

The applicable manufacturer's numbering system and abbreviated designations or symbols should always be reviewed before attempting to locate a structural member. The following list includes location designations typical of those used by many manufacturers.

(1) **Fuselage stations** (Fus. Sta. or F.S.) are numbered in inches from a reference or zero point known as the reference datum. The reference datum is an imaginary vertical plane at or near the nose of the aircraft from which all horizontal distances are measured. The distance to a given point is measured in inches parallel to a center line extending through the aircraft from the nose through the center of the tail cone. Some manufacturers may call the fuselage station a body station, abbreviated B.S.

(2) **Buttock line or butt line** (B.L.) is a width measurement left or right of, and parallel to, the vertical center line.

(3) **Water line** (W.L.) is the measurement of height in inches perpendicular from a horizontal plane located a fixed number of inches below the bottom of the aircraft fuselage.

(4) **Aileron station** (A.S.) is measured outboard from, and parallel to, the inboard edge of the aileron, perpendicular to the rear beam of the wing.

(5) **Flap station** (F.S.) is measured perpen-

dicular to the rear beam of the wing and parallel to, and outboard from, the inboard edge of the flap.

(6) **Nacelle station** (N.C. or Nac. Sta.) is measured either forward of or behind the front spar of the wing and perpendicular to a designated water line.

In addition to the location stations listed above, other measurements are used, especially on large aircraft. Thus, there may be horizontal stabilizer stations (H.S.S.), vertical stabilizer stations (V.S.S.) or powerplant stations (P.P.S.). In every case the manufacturer's terminology and station location system should be consulted before locating a point on a particular aircraft.

WING STRUCTURE

The wings of an aircraft are surfaces which are designed to produce lift when moved rapidly through the air. The particular design for any given aircraft depends on a number of factors, such as size, weight, use of the aircraft, desired speed in flight and at landing, and desired rate of climb. The wings of a fixed-wing aircraft are designated left and right, corresponding to the left and right sides of the operator when seated in the cockpit.

The wings of some aircraft are of cantilever design; that is, they are built so that no external bracing is needed. The skin is part of the wing structure and carries part of the wing stresses. Other aircraft wings use external bracings (struts, wires, etc.) to assist in supporting the wing and carrying the aerodynamic and landing loads. Both aluminum alloy and magnesium alloy are used in wing construction. The internal structure is made up of spars and stringers running spanwise, and

6

ribs and formers running chordwise (leading edge to trailing edge). The spars are the principal structural members of the wing. The skin is attached to the internal members and may carry part of the wing stresses. During flight, applied loads which are imposed on the wing structure are primarily on the skin. From the skin they are transmitted to the ribs and from the ribs to the spars. The spars support all distributed loads as well as concentrated weights, such as fuselage, landing gear, and, on multi-engine aircraft, the nacelles or pylons.

The wing, like the fuselage, may be constructed in sections. One commonly used type is made up of a center section with outer panels and wing tips. Another arrangement may have wing stubs as an integral part of the fuselage in place of the center section.

Inspection openings and access doors are provided, usually on the lower surfaces of the wing. Drain holes are also placed in the lower surface to provide for drainage of accumulated moisture or fluids. On some aircraft built-in walkways are provided on the areas where it is safe to walk or step. On some aircraft jacking points are provided on the underside of each wing.

Various points on the wing are located by station number. Wing station 0 (zero) is located at the center line of the fuselage, and all wing stations are measured outboard from that point, in inches.

In general, wing construction is based on one of three fundamental designs: (1) Monospar, (2) multi-spar, or (3) box beam. Modifications of these basic designs may be adopted by various manufacturers.

The monospar wing incorporates only one main longitudinal member in its construction. Ribs or bulkheads supply the necessary contour or shape to the airfoil. Although the strict monospar wing is not common, this type of design, modified by the addition of false spars or light shear webs along the trailing edge as support for the control surfaces, is sometimes used.

The multi-spar wing incorporates more than one main longitudinal member in its construction. To give the wing contour, ribs or bulkheads are often included.

The box beam type of wing construction uses two main longitudinal members with connecting bulkheads to furnish additional strength and to give contour to the wing. A corrugated sheet may be

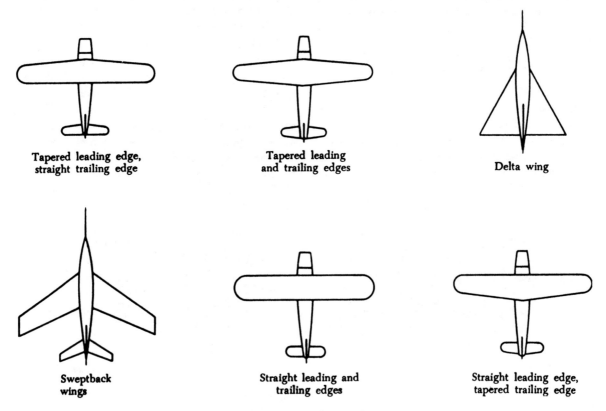

Tapered leading edge, straight trailing edge

Tapered leading and trailing edges

Delta wing

Sweptback wings

Straight leading and trailing edges

Straight leading edge, tapered trailing edge

FIGURE 1-9. Typical wing leading and trailing edge shapes.

placed between the bulkheads and the smooth outer skin so that the wing can better carry tension and compression loads. In some cases, heavy longitudinal stiffeners are substituted for the corrugated sheets. A combination of corrugated sheets on the upper surface of the wing and stiffeners on the lower surface is sometimes used.

Wing Configurations

Depending on the desired flight characteristics, wings are built in many shapes and sizes. Figure 1–9 shows a number of typical wing leading and trailing edge shapes.

In addition to the particular configuration of the leading and trailing edges, wings are also designed to provide certain desirable flight characteristics, such as greater lift, balance, or stability. Figure 1–10 shows some common wing forms.

Features of the wing will cause other variations in its design. The wing tip may be square, rounded, or even pointed. Both the leading edge and the trailing edge of the wing may be straight or curved, or one edge may be straight and the other curved. In addition, one or both edges may be tapered so that the wing is narrower at the tip than at the root where it joins the fuselage. Many types of modern aircraft employ sweptback wings (figure 1–9).

Wing Spars

The main structural parts of a wing are the spars, the ribs or bulkheads, and the stringers or stiffeners, as shown in figure 1–11.

Spars are the principal structural members of the wing. They correspond to the longerons of the fuselage. They run parallel to the lateral axis, or toward the tip of the wing, and are usually attached to the fuselage by wing fittings, plain beams, or a truss system.

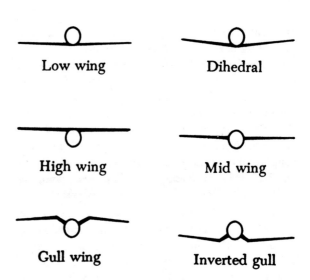

FIGURE 1–10. Common wing forms.

Wooden spars can be generally classified into four different types by their cross sectional configuration. As shown in figure 1–12, they may be partly hollow, in the shape of a box, solid or laminated, rectangular in shape, or in the form of an I-beam.

Spars may be made of metal or wood depending on the design criteria of a specific aircraft. Most aircraft recently manufactured use spars of solid extruded aluminum or short aluminum extrusions riveted together to form a spar.

The shape of most wooden spars is usually similar to one of the shapes shown in figure 1–12. The rectangular form, figure 1–12A, can be either solid or laminated. Figure 1–12B is an I-beam spar that has been externally routed on both sides to reduce weight while retaining adequate strength. A box spar, figure 1–12C, is built up from plywood and solid spruce. The I-beam spar, figure 1–12D, may be built up of wood or manufac-

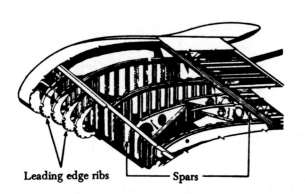

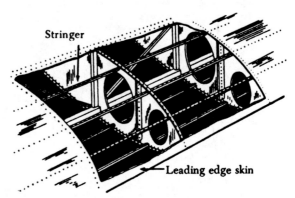

FIGURE 1–11. Internal wing construction.

8

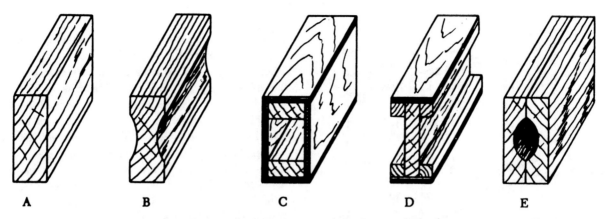

A   B   C   D   E

FIGURE 1–12. Typical spar cross sectional configurations.

tured by an aluminum extrusion process. The I-beam construction for a spar usually consists of a web (a deep wall plate) and cap strips, which are extrusions or formed angles. The web forms the principal depth portion of the spar. Cap strips are extrusions, formed angles, or milled sections to which the web is attached. These members carry the loads caused by the wing bending and also provide a foundation for attaching the skin. An example of a hollow or internally routed spar is represented in figure 1–12E.

Figure 1–13 shows the basic configuration of some typical metal spars. Most metal spars are built up from extruded aluminum alloy sections, with riveted aluminum alloy web sections to provide extra strength.

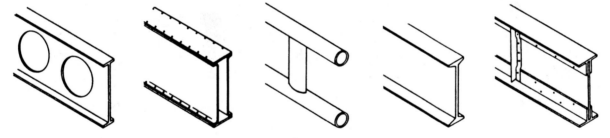

FIGURE 1–13. Metal spar shapes.

Although the spar shapes of figure 1–13 are typical of most basic shapes, the actual spar configuration may assume many forms. For example, a spar may have either a plate or truss type web. The plate web (figure 1–14) consists of a solid plate with vertical stiffeners which increase the strength of the web. Some spar plate webs are constructed differently. Some have no stiffeners; others contain flanged holes for reducing weight. Figure 1–15 shows a truss spar made up of an upper cap, a lower cap, and connecting vertical and diagonal tubes.

A structure may be designed so as to be considered "fail-safe." In other words, should one member of a complex structure fail, some other member would assume the load of the failed member.

A spar with "fail-safe" construction is shown in figure 1–16. This spar is made in two sections. The top section consists of a cap, riveted to the upper web plate. The lower section is a single extrusion, consisting of the lower cap and web plate. These two sections are spliced together to form the spar. If either section of this type of spar breaks, the other section can still carry the load, which is the "fail-safe" feature.

As a rule, a wing has two spars. One spar is usually located near the front of the wing, and the other about two-thirds of the distance toward the wing's trailing edge. Regardless of type, the spar is the most important part of the wing. When other structural members of the wing are placed under load, they pass most of the resulting stress on to the wing spars.

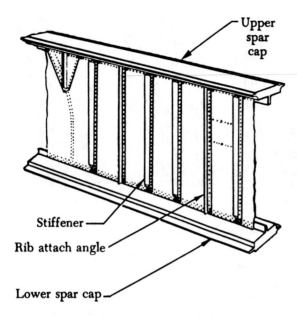

FIGURE 1-14.   Plate web wing spar.

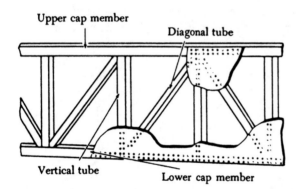

FIGURE 1-15.   Truss wing spar.

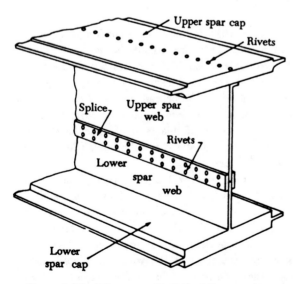

FIGURE 1-16.   Wing spar with "fail-safe" construction.

Wing Ribs

Ribs are the structural crosspieces that make up the framework of the wing. They usually extend from the wing leading edge to the rear spar or to the trailing edge of the wing. The ribs give the wing its cambered shape and transmit the load from the skin and stringers to the spars. Ribs are also used in ailerons, elevators, rudders, and stabilizers.

Ribs are manufactured from wood or metal. Either wood or metal ribs are used with wooden spars while metal ribs are usually used with metal spars. Some typical wooden ribs, usually manufactured from spruce, are shown in figure 1-17.

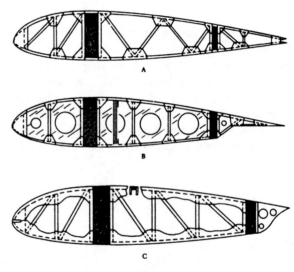

FIGURE 1-17.   Typical wooden ribs.

The most common types of wooden ribs are the plywood web, the lightened plywood web, and the truss types. Of these three types, the truss type is the most efficient, but it lacks the simplicity of the other types.

The wing rib shown in Figure 1-17A is a truss type, with plywood gussets on both sides of the rib and a continuous rib cap around the entire rib. Rib caps, often called cap strips, are usually made of the same material as the rib itself, especially when using wooden ribs. They stiffen and strengthen the rib and provide an attaching surface for the rib covering.

A lightened plywood web rib is illustrated in figure 1-17B. On this type the cap strip may be laminated, especially at the leading edge. Figure 1-17C shows a rib using a continuous gusset, which provides extra support throughout the entire rib with very little additional weight.

10

A continuous gusset stiffens cap strips in the plane of the rib. This aids in preventing buckling and helps to obtain better rib/skin glue joints where nail-gluing is used because such a rib can resist the driving force of nails better than the other types. Continuous gussets are more easily handled than the many small separate gussets otherwise required.

Figure 1–18 shows the basic rib and spar structure of a wooden wing frame, together with some of the other wing structural members. In addition to the front and rear spars, an aileron spar, or false spar, is shown in figure 1–18. This type of spar extends only part of the spanwise length of the wing and provides a hinge attachment point for the aileron.

Various types of ribs are also illustrated in figure 1–18. In addition to the wing rib, sometimes called "plain rib" or even "main rib," nose ribs and the butt rib are shown. A nose rib is also called a false rib, since it usually extends from the wing leading edge to the front spar or slightly beyond. The nose ribs give the wing leading edge area the necessary curvature and support. The wing rib, or plain rib, extends from the leading edge of the wing to the rear spar and in some cases to the trailing edge of the wing. The wing butt rib is normally the heavily stressed rib section at the inboard end of the wing

near the attachment point to the fuselage. Depending on its location and method of attachment, a butt rib may be called a bulkhead rib or a compression rib, if it is designed to receive compression loads that tend to force the wing spars together.

Since the ribs are laterally weak, they are strengthened in some wings by tapes that are woven above and below rib sections to prevent sidewise bending of the ribs.

Drag and antidrag wires (figure 1–18) are crisscrossed between the spars to form a truss to resist forces acting on the wing in the direction of the wing chord. These tension wires are also referred to as tie rods. The wire designed to resist the backward forces is called a drag wire; the antidrag wire resists the forward forces in the chord direction.

The wing attachment fittings, shown in figure 1–18, provide a means of attaching the wing to the aircraft fuselage.

The wing tip is often a removable unit, bolted to the outboard end of the wing panel. One reason for this is the vulnerability of the wing tips to damage, especially during ground handling and taxiing.

Figure 1–19 shows a removable wing tip for a large aircraft wing. The wing-tip assembly is of aluminum alloy construction. The wing-tip cap is secured to the tip with countersunk screws and is

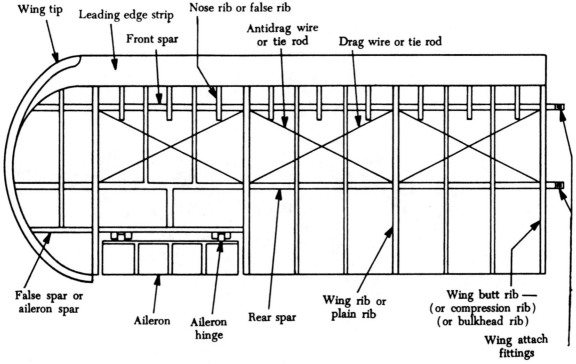

FIGURE 1–18. Basic rib and spar structure.

11

secured to the interspar structure at four points with ¼-in. bolts. The tip leading edge contains the heat anti-icing duct. Wing-heated air is exhausted through a louver on the top surface of the tip. Wing position lights are located at the center of the tip and are not directly visible from the cockpit. As an indication that the wing tip light is operating, some wing tips are equipped with a lucite rod to transmit the light to the leading edge.

Figure 1–20 shows a cross sectional view of an

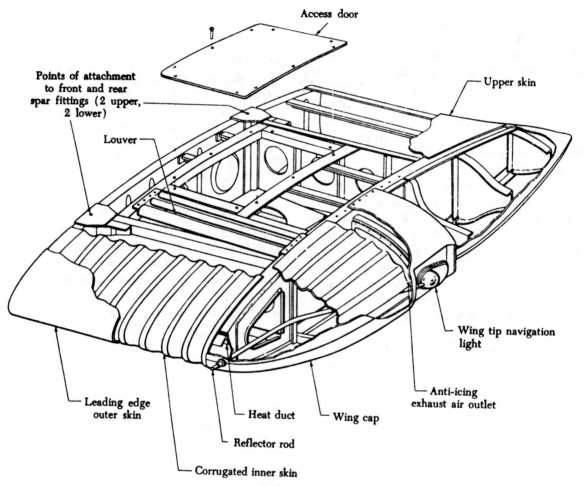

Access door

Points of attachment to front and rear spar fittings (2 upper, 2 lower)

Upper skin

Louver

Wing tip navigation light

Anti-icing exhaust air outlet

Leading edge outer skin

Heat duct

Wing cap

Reflector rod

Corrugated inner skin

FIGURE 1–19. Removable wing tip.

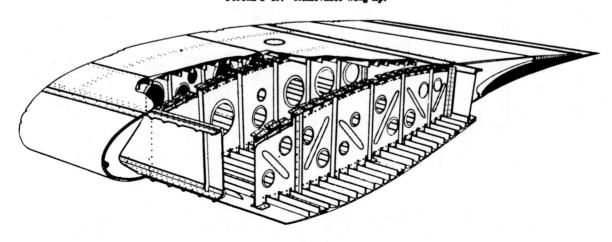

FIGURE 1–20. All-metal wing with chemically milled channels.

all-metal full cantilever (no external bracing) wing section. The wing is made up of spars, ribs, and lower and upper wing skin covering. With few exceptions, wings of this type are of the stressed-skin design (the skin is part of the wing structure and carries part of the wing stresses).

The top and bottom wing skin covers are made up of several integrally stiffened sections. This type of wing construction permits the installation of bladder-type fuel cells in the wings or is sealed to hold fuel without the usual fuel cells or tanks. A wing which is constructed to allow it to be used as a fuel cell or tank is referred to as a "wet-wing."

A wing that uses a box-beam design is shown in figure 1–21. This type of construction not only increases strength and reduces weight, but it also enables the wing to serve as a fuel tank when properly sealed.

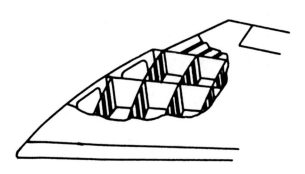

FIGURE 1–21. Box-beam milled wing.

Both aluminum honeycomb and fiber glass honeycomb sandwich material are commonly used in the construction of wing and stabilizer surfaces, bulkheads, floors, control surfaces, and trim tabs. Aluminum honeycomb material is made of aluminum foil honeycomb core, bonded between sheets of aluminum. Fiber glass honeycomb material consists of fiber glass honeycomb core bonded between layers of fiber glass cloth.

In the construction of large aircraft structures, and in some small aircraft as well, the honeycomb sandwich structure employs either aluminum or reinforced plastic materials. Honeycomb panels are usually a lightweight cellular core sandwiched between two thin skins or facing materials such as aluminum, wood, or plastic.

Aircraft honeycomb material is manufactured in various shapes, but is usually of the constant thickness or tapered core types. An example of each is shown in figure 1–22.

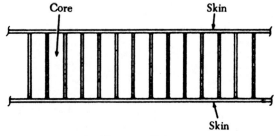

A. Constant thickness

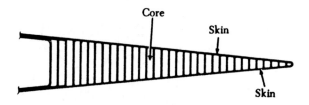

B. Tapered core

FIGURE 1–22. Constant-thickness and tapered-core honeycomb sections.

Figure 1–23 shows a view of the upper surface of a large jet transport wing. The various panels manufactured from honeycomb material are outlined by diagonal lines and labeled.

Still another type of construction is illustrated in figure 1–24. In this case the sandwich structure of the wing leading edge is bonded to the metal spar. Also shown is the integrally bonded deicer panel.

NACELLES OR PODS

Nacelles or pods are streamlined enclosures used on multi-engine aircraft primarily to house the engines. They are round or spherical in shape and are usually located above, below, or at the leading edge of the wing on multi-engine aircraft. If an aircraft has only one engine, it is usually mounted at the forward end of the fuselage, and the nacelle is the streamlined extension of the fuselage.

An engine nacelle or pod consists of skin, cowling, structural members, a firewall, and engine mounts. Skin and cowling cover the outside of the nacelle. Both are usually made of sheet aluminum alloy, stainless steel, magnesium, or titanium. Regardless of the material used, the skin is usually attached to the framework by rivets.

The framework usually consists of structural members similar to those of the fuselage. The framework includes lengthwise members, such as

13

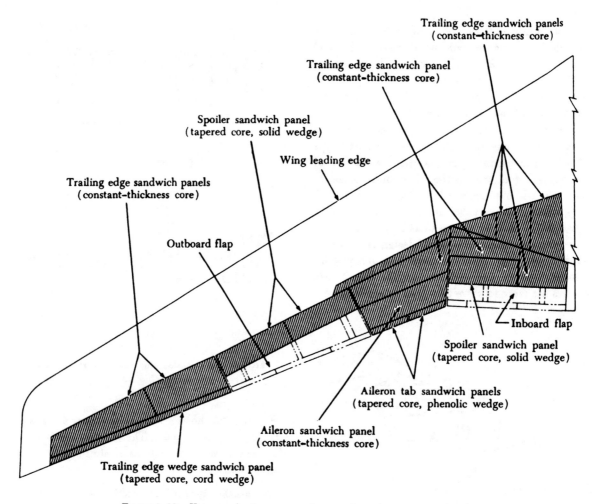

Trailing edge sandwich panels
(constant-thickness core)

Trailing edge sandwich panel
(constant-thickness core)

Spoiler sandwich panel
(tapered core, solid wedge)

Wing leading edge

Trailing edge sandwich panels
(constant-thickness core)

Outboard flap

Inboard flap

Spoiler sandwich panel
(tapered core, solid wedge)

Aileron tab sandwich panels
(tapered core, phenolic wedge)

Aileron sandwich panel
(constant-thickness core)

Trailing edge wedge sandwich panel
(tapered core, cord wedge)

FIGURE 1–23. Honeycomb wing construction on a large jet transport aircraft.

longerons and stringers, and widthwise/vertical members, such as bulkheads, rings, and formers.

A nacelle or pod also contains a firewall which separates the engine compartment from the rest of the aircraft. This bulkhead is usually made of stainless steel sheet metal, or as in some aircraft, of titanium.

Another nacelle or pod member is the engine mount. The mount is usually attached to the firewall, and the engine is attached to the mount by nuts, bolts, and vibration-absorbing rubber cushions or pads. Figure 1–25 shows examples of a semimonocoque and a welded tubular steel engine mount used with reciprocating engines.

Engine mounts are designed to meet particular conditions of installation, such as the location and the method of attachment of the engine mount and the size, type, and characteristics of the engine it is intended to support. An engine mount is usually constructed as a single unit which can be detached quickly and easily from the remaining structure. Engine mounts are commonly made of welded chrome/molybdenum steel tubing, and forgings of chrome/nickel/molybdenum are used for the highly stressed fittings.

To reduce wind resistance during flight, the landing gear of most high-speed or large aircraft is retracted (drawn up into streamlined enclosures). The part of the aircraft which receives or encloses the landing gear as it retracts is called a wheel well. In many instances, the wheel well is part of the nacelle; however, on some aircraft the landing gear retracts into the fuselage or wing.

Cowling

Cowling usually refers to the detachable covering of those areas into which access must be gained regularly, such as engines, accessory sections, and engine mount or firewall areas. Figure 1–26 shows an exploded view of the pieces of cowling for a horizontally opposed engine on a light aircraft.

14

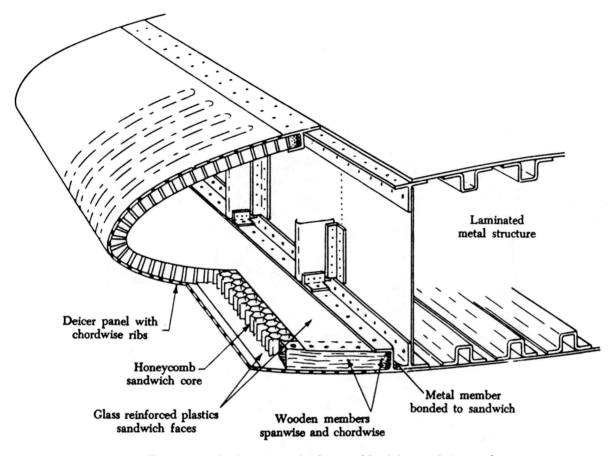

Figure 1–24. Leading edge sandwich material bonded to metal wing member.

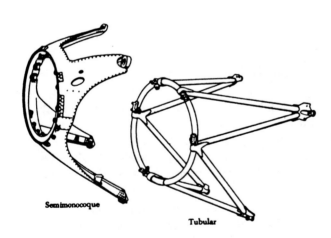

Figure 1–25. Semimonocoque and welded tubular steel engine mounts.

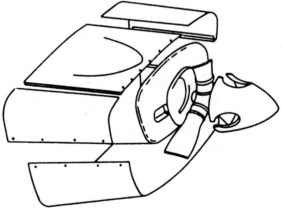

Figure 1–26. Cowling for horizontally opposed engine.

Some large reciprocating engines are enclosed by "orange-peel" cowl panels. The cowl panels are attached to the firewall by mounts which also serve as hinges when the cowl is opened (figure 1–27).

The lower cowl mounts are secured to the hinge brackets by pins which automatically lock in place,

but can be removed by simply pulling on a ring. The side panels are held open by short rods; the top panel is held open by a longer rod, and the lower panel is restrained in the "open" position by a spring and cable.

All four panels are locked in the "closed" position by over-center steel latches, which are secured

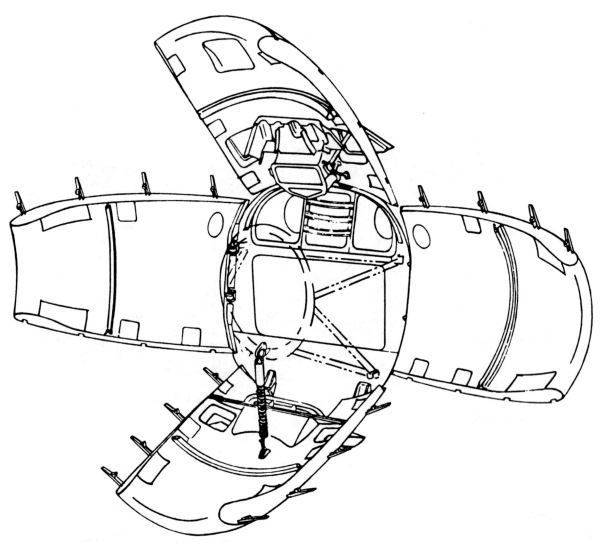

FIGURE 1-27. "Orange-peel" cowling opened.

in the closed position by spring-loaded safety catches. Cowl panels are generally of aluminum alloy construction; however, stainless steel is generally used as the inner skin aft of the power section, for cowl flaps and near the cowl flap openings, and for oil cooler ducts.

On turbojet engine installations, cowl panels are designed to provide a smooth airflow over the engines and to protect the engine from damage. The entire engine cowling system includes a nose cowl, upper and lower hinged removable cowl panels, and fixed cowl panel. Typical upper and lower hinged removable panels are shown in figure 1-28.

EMPENNAGE

The empennage is also called the tail section and most aircraft designs consist of a tail cone, fixed surfaces, and movable surfaces.

The tail cone serves to close and streamline the aft end of most fuselages. The cone is made up of structural members (figure 1-29) like those of the fuselage; however, cones are usually of lighter construction since they receive less stress than the fuselage.

Other components of the typical empennage are of heavier construction than the tail cone. These members include fixed surfaces that help steady the aircraft and movable surfaces that help to direct an aircraft's flight. The fixed surfaces are the horizontal and vertical stabilizers. The movable surfaces are usually a rudder and elevators.

Figure 1-30 shows how the vertical surfaces are braced, using spars, ribs, stringers, and skin in a similar manner to the systems used in a wing.

Stress in an empennage is also carried like stress

16

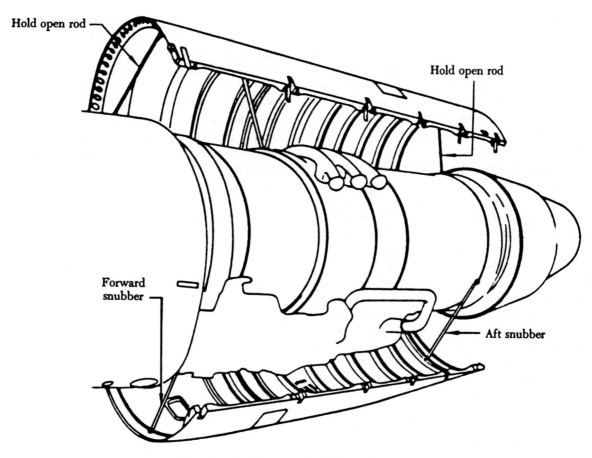

FIGURE 1–28. Side-mounted turbojet engine cowling.

in a wing. Bending, torsion, and shear, created by airloads, pass from one structural member to another. Each member absorbs some of the stress and passes the remainder to other members. The overload of stress eventually reaches the spars, which transmit it to the fuselage structure.

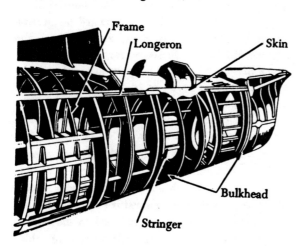

FIGURE 1–29. The fuselage terminates in a tail cone.

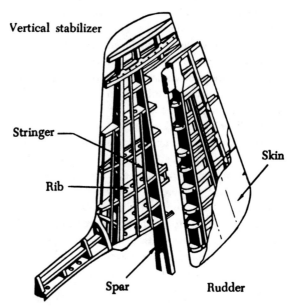

FIGURE 1–30. Construction features of rudder and vertical stabilizer.

17

FLIGHT CONTROL SURFACES

The directional control of a fixed-wing aircraft takes place around the lateral, longitudinal, and vertical axes by means of flight control surfaces. These control devices are hinged or movable surfaces through which the attitude of an aircraft is controlled during takeoff, flight, and landing. They are usually divided into two major groups, the primary or main, and the auxiliary control surfaces.

The primary group of flight control surfaces consists of ailerons, elevators, and rudders. Ailerons are attached to the trailing edge of both wings of an aircraft. Elevators are attached to the trailing edge of the horizontal stabilizer. The rudder is hinged to the trailing edge of the vertical stabilizer.

Primary control surfaces are similar in construction and vary only in size, shape, and methods of attachment. In construction, control surfaces are similar to the all-metal wing. They are usually made of an aluminum alloy structure built around a single spar member or torque tube. Ribs are fitted to the spar at the leading and trailing edges and are joined together with a metal strip. The ribs, in many cases, are formed from flat sheet stock. They are seldom solid; more often, the formed, stamped-out ribs are reduced in weight by holes which are punched in the metal.

The control surfaces of some aircraft are fabric covered. However, all turbojet powered aircraft have metal-covered surfaces for additional strength.

The control surfaces previously described can be considered conventional, but on some aircraft, a control surface may serve a dual purpose. For example, one set of control surfaces, the elevons, combines the functions of both ailerons and elevators. Flaperons are ailerons which can also act as flaps. A movable horizontal tail section is a control surface which supplies the action of both the horizontal stabilizer and the elevators.

The secondary or auxiliary group of control surfaces consists of such members as trim tabs, balance tabs, servo tabs, flaps, spoilers, and leading edge devices. Their purpose is to reduce the force required to actuate the primary controls, to trim and balance the aircraft in flight, to reduce landing speed or shorten the length of the landing roll, and to change the speed of the aircraft in flight. They are usually attached to, or recessed in, the main control surfaces.

Ailerons

Ailerons are primary control surfaces which make up part of the total wing area. They are movable through a pre-designed arc and are usually hinged to the aileron spar or rear wing spar. The ailerons are operated by a lateral (side-to-side) movement of the aircraft control stick, or a turning motion of the wheel on the yoke.

In a conventional configuration, one aileron is hinged to the outboard trailing edge of each wing. Figure 1–31 shows the shape and location of typical small-aircraft ailerons on various wing-tip designs.

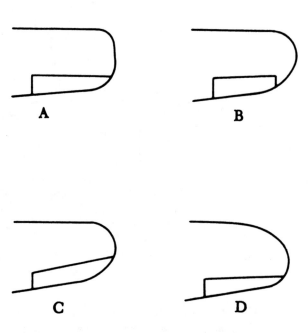

FIGURE 1–31. Aileron location on various wing-tip designs.

The ailerons are interconnected in the control system to operate simultaneously in opposite directions. As one aileron moves downward to increase lift on its side of the fuselage, the aileron on the opposite side of the fuselage moves upward to decrease lift on its side. This opposing action results in more lift being produced by the wing on one side of the fuselage than on the other, resulting in a controlled movement or roll due to unequal aerodynamic forces on the wings.

An end view of a typical metal rib in an aileron is shown in figure 1–32. The hinge point of this type of aileron is behind the leading edge of the aileron to provide a more sensitive response to control movements. The horns attached to the aileron spar are levers to which the aileron control cables are secured.

18

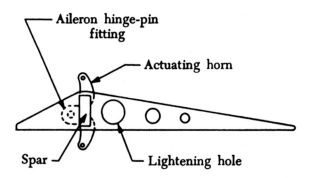

FIGURE 1–32. End view of aileron rib.

Large aircraft may use all-metal ailerons, except for fiber glass trailing edges, hinged to the rear wing spar in at least four places. Figure 1–33 shows several examples of aileron installation.

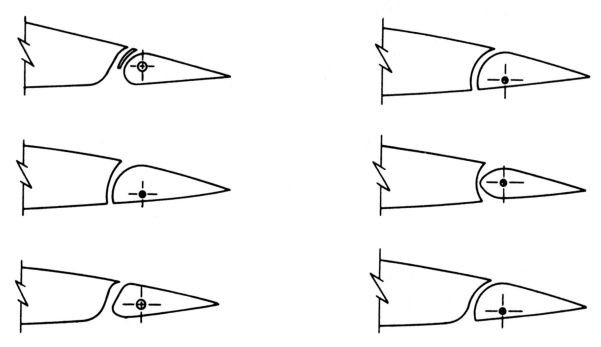

FIGURE 1–33. Aileron hinge locations.

All the control surfaces of a large turbojet aircraft are shown in figure 1–34. As illustrated, each wing has two ailerons, one in the conventional position at the outboard trailing edge of the wing and another hinged to the trailing edge of the wing center section.

The complex lateral control system in large turbojet aircraft is far more sophisticated than the type employed in a light airplane. During low-speed flight all lateral control surfaces operate to provide maximum stability. This includes all four ailerons, flaps, and spoilers. At high speeds, flaps are retracted and the outboard ailerons are locked out of the aileron control system.

The major part of the skin area of the inboard ailerons is aluminum honeycomb panels. Exposed honeycomb edges are covered with sealant and protective finish. The aileron nose tapers and extends forward of the aileron hinge line. Each inboard aileron is positioned between the inboard and out-

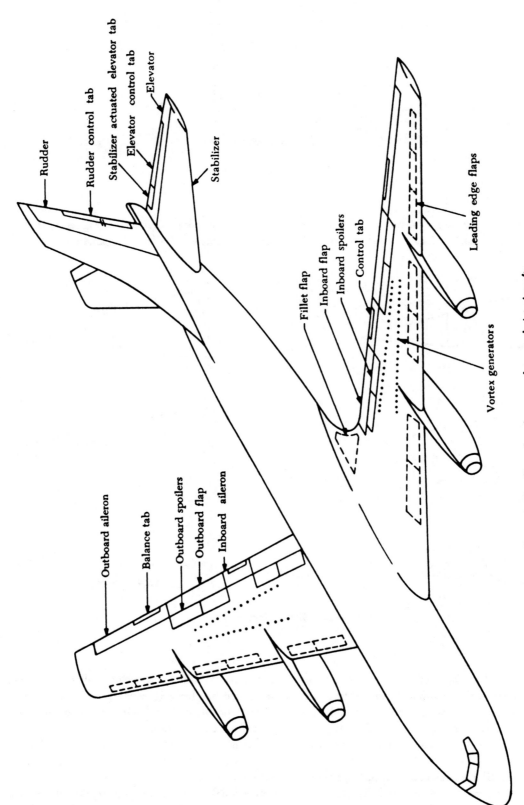

Rudder

Rudder control tab

Stabilizer actuated elevator tab

Elevator control tab

Elevator

Stabilizer

Fillet flap

Inboard flap

Inboard spoilers

Control tab

Leading edge flaps

Vortex generators

Outboard aileron

Balance tab

Outboard spoilers

Outboard flap

Inboard aileron

FIGURE 1-34   Control surfaces on a large turbojet aircraft.

board flaps at the trailing edge of the wing. The aileron hinge supports extend aft and are attached to aileron hinge bearings to support the aileron.

The outboard ailerons are made up of a nose spar and ribs covered with aluminum honeycomb panels. A continuous hinge attached to the forward edge of the nose is grooved to mate with the hem of a fabric seal.

The outboard ailerons are located in the trailing edge of each outboard wing section. Hinge supports extend aft from the wing and are attached to the aileron hinge bearing to support the aileron. The nose of the aileron extends into a balance chamber in the wing and is attached to balance panels.

Aileron balance panels (figure 1–35) reduce the force necessary to position and hold the ailerons. The balance panels may be made of aluminum honeycomb skin bonded to an aluminum frame, or of

aluminum skin-covered assemblies with hat-section stiffeners. Clearance between the aileron nose and wing structure provides a controlled airflow area necessary for balance panel action. Seals attached to the panels control air leakage.

Air loads on the balance panels (figure 1–35) depend on aileron position. When the ailerons are moved during flight to either side of the streamline position, differential pressure is created across the balance panels. This differential pressure acts on the balance panels in a direction that assists aileron movement. Full balance panel force is not required for small angles of aileron displacement because the manual force necessary to rotate the control tab through small angles is slight. A controlled air bleed is progressively decreased as the aileron displacement angle is increased. This action increases the differential air pressure on the balance panels as

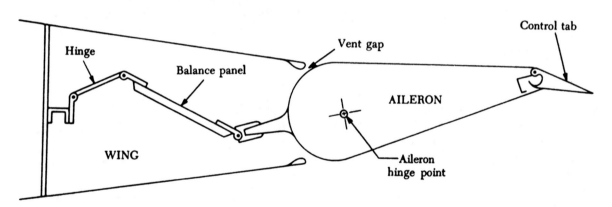

FIGURE 1–35. Aileron balance panel.

the ailerons rotate from the streamline position. The increasing load on the balance panel counteracts the increasing load on the ailerons.

Auxiliary Wing Flight Surfaces

The ailerons are the primary wing flight surfaces. Auxiliary wing flight surfaces include trailing edge flaps, leading edge flaps, speed brakes, spoilers, and leading edge slats. The number and type of auxiliary wing flap surfaces on an aircraft vary widely, depending on the type and size of aircraft.

Wing flaps are used to give the aircraft extra lift. They reduce the landing speed, thereby shortening the length of the landing rollout to facilitate landing in small or obstructed areas by permitting the gliding angle to be increased without greatly in-

creasing the approach speed. In addition, the use of flaps during takeoff reduces the length of the takeoff run.

Most flaps are hinged to the lower trailing edges of the wings, inboard of the ailerons. Leading edge flaps are also used, principally on large high-speed aircraft. When they are in the "up" (or retracted) position, they fair in with the wings and serve as part of the wing trailing edge. When in the "down" (or extended) position, the flaps pivot on the hinge points and drop to about a 45° or 50° angle with the wing chord line. This increases the wing camber and changes the airflow, providing greater lift.

Some common types of flaps are shown in figure 1–36. The plain flap (figure 1–36A) forms the trailing edge of the wing when the flap is in the up (or

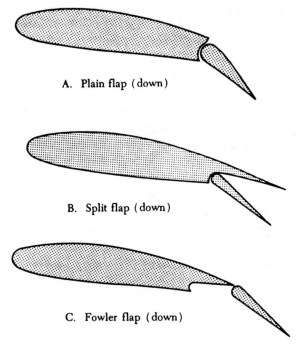

A. Plain flap (down)

B. Split flap (down)

C. Fowler flap (down)

FIGURE 1–36. Wing flaps.

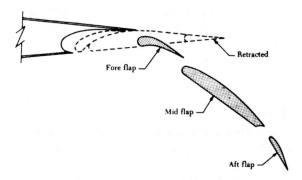

FIGURE 1–37. Triple-slotted trailing edge flaps.

flap area. The resulting slots between flaps prevents separation of the airflow over the flap area.

The leading edge flap (figure 1–38) is similar in operation to the plain flap; that is, it is hinged on the bottom side, and, when actuated, the leading edge of the wing extends in a downward direction to increase the camber of the wing. Leading edge flaps are used in conjunction with other types of flaps.

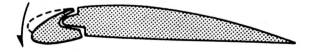

FIGURE 1–38. Cross section of a leading edge flap.

Figure 1–34 shows the location of the leading edge flaps on a large multi-engine turbine aircraft. Three Kruger-type flaps are installed on each wing. The flaps are machined magnesium castings with integral ribs and stiffeners. The magnesium casting of each flap is the principal structural component and consists of a straight section with a hollow core called the torque tube extending from the straight section at the forward end.

Each leading edge flap has three gooseneck hinges attached to fittings in the fixed wing leading edge, and a hinged fairing is installed on the trailing edge of each flap. Figure 1–39 shows a typical leading edge flap in retracted position, with an outline of the extended position.

Speed brakes, sometimes called dive flaps or dive brakes, serve to slow an aircraft in flight. These brakes are used when descending at a steep angle or when approaching the runway for a landing. The brakes themselves are manufactured in many

retracted) position. It contains both the upper and lower surface of the wing trailing edge.

The plain split flap (figure 1–36B) is normally housed flush with the undersurface of the wing. It is similar to a plain flap except that the upper surface of the wing extends to the flap trailing edge and does not droop with the flap. This flap is also called the split-edge flap. It is usually just a braced, flat metal plate hinged at several points along its leading edge.

Aircraft requiring extra wing area to aid lift often use Fowler flaps (figure 1–36C). This system houses the flaps flush under the wings much as does the plain split flap system. But, instead of the flaps hinging straight down from a stationary hinge line, worm-gear drives move the flaps leading edge rearward as the flaps droop. This action provides normal flap effect, and, at the same time, wing area is increased when the flaps are extended.

An example of a triple-slotted segmented flap used on some large turbine aircraft is shown in figure 1–37. This type of trailing edge flap system provides high lift for both takeoff and landing. Each flap consists of a foreflap, a mid-flap, and an aft-flap. The chord length of each flap expands as the flap is extended, providing greatly increased

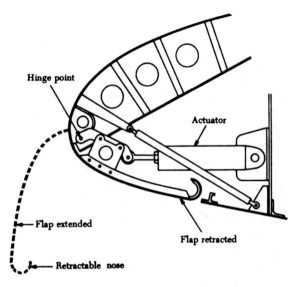

Figure 1-39. Leading edge flap.

shapes, and their location depends on the design of the aircraft and the purpose of the brakes.

The brake panels may be located on certain parts of the fuselage or on the wing surfaces. Brakes on the fuselage are small panels that can be extended into the smooth airflow to create turbulence and drag. Wing-type brakes may be multiple-finger channels extending above and below the wing surfaces to break up smooth airflow. Usually speed brakes are controlled by electrical switches and actuated by hydraulic pressure.

Another type of air brake is a combination of spoilers and speed brakes. A typical combination consists of spoiler flaps located in the upper wing surfaces ahead of the ailerons. When the operator wishes to use both air brakes and spoilers, he can slow the flight speed and maintain lateral control as well.

Spoilers are auxiliary wing flight control surfaces, mounted on the upper surface of each wing, which operate in conjunction with the ailerons to provide lateral control.

Most spoiler systems can also be extended symmetrically to serve a secondary function as speed brakes. Other systems are equipped with separate ground and flight spoilers. Most spoiler panels are bonded honeycomb structures with aluminum skin. They are attached to the wing structure by machined hinge fittings which are bonded into the spoiler panel.

Tabs

One of the simplest yet most important devices to aid the pilot of an aircraft is the tab attached to a control surface. Although a tab does not take the place of a control surface, it is mounted on or attached to a movable control surface and causes easier movement or better balance of the control surface.

All aircraft, except a few of the very lightest types, are equipped with tabs that can be controlled from the cockpit. Tabs on some of these aircraft are usually adjustable only when the aircraft is on the ground. Figure 1-40 shows the location of a typical rudder tab.

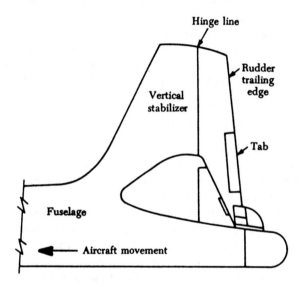

Figure 1-40. Typical location of rudder control tab.

LANDING GEAR

The landing gear is the assembly that supports the aircraft during landing or while it is resting or moving about on the ground. The landing gear has shock struts to absorb the shock of landing and taxiing. By means of a gear-retraction mechanism, the landing gear attaches to the aircraft structure and enables the gear to extend and retract. The landing gear arrangement either has a tailwheel or a nosewheel. Landing gear arrangements having a nosewheel are usually equipped for nosewheel steering. Nosewheel aircraft are protected at the fuselage tail section with a tail skid or bumper. By means of wheels and tires (or skis), the landing gear forms a stabilizing contact with the ground during landing and taxiing. Brakes installed in the wheels enable the aircraft to be slowed or stopped during movement on the ground.

SKIN AND FAIRING

The smooth outer cover of the aircraft is referred to as skin. The skin covers the fuselage, wings, empennage, nacelles, and pods. The material used for the skin covering is usually sheet aluminum alloy, treated so that it will not corrode. Magnesium and stainless steel may also be used to a limited extent. The thickness of the skin materials covering a structural unit may differ, depending on the load and stresses imposed within and throughout the structure. To smooth out the airflow over the angles formed by the wings and other structural units with the fuselage, shaped and rounded panels or metal skin are attached. This paneling or skin is called fairing. Fairing is sometimes referred to as a fillet. Some fairing is removable to provide access to aircraft components, whereas other fairing is riveted to the aircraft structure.

ACCESS AND INSPECTION DOORS

Access doors permit normal or emergency entrance into or exit from the aircraft. Also, they provide access to servicing points and manually operated drains. Inspection doors provide access to a particular part of the aircraft being inspected or maintained. Access or inspection doors are either hinged or removable. They are fastened in the closed position with catch and locking mechanisms, screws, quick-release devices, or cowling type fasteners. Access and inspection doors that are removable often have a stenciled identification number that is identical to a number stenciled near the opening that they cover. Other access and inspection doors have a stenciled nomenclature to identify the opening that they cover.

HELICOPTER STRUCTURES

Like the fuselages in fixed-wing aircraft, helicopter fuselages may be welded truss or some form of monocoque construction. Although their fuselage configurations may vary a great deal, most helicopter fuselages employ structural members similar to those used in fixed-wing aircraft. For example, most helicopters have such vertical/widthwise braces as bulkheads, formers, rings, and frames. They are also provided with such lengthwise braces as stringers and longerons. In addition, the gussets, joiners, and skin hold the other structural members together.

The basic body and tail boom sections of a typical helicopter are of conventional, all-metal, riveted structures incorporating formed aluminum alloy bulkheads, beams, channels, and stiffeners. Stressed skin panels may be either smooth or beaded. The firewall and engine deck are usually stainless steel. The tail boom is normally of semimonocoque construction, made up of formed aluminum bulkheads, extruded longerons, and skin panels or of welded tubular steel.

The major structural components of one type of helicopter are shown in figure 1–41. The members of a helicopter's tail group vary widely, depending on the individual type and design. In this case, a stabilizer is mounted on a pylon to make up the group. In other cases, the stabilizer may be mounted on the helicopter tail cone or fuselage. In either case, both the pylon and stabilizer usually contain aluminum alloy structural members covered with magnesium alloy skin. The types of structural members used, however, usually vary. A pylon usually has bulkheads, formers, frames, stringers, and beams, making it somewhat of a blend of aircraft wing and fuselage structural members. The stabilizer is usually built more like an aircraft wing, with ribs and spars.

In a typical helicopter, the tail, body, and tail boom are constructed of all-metal stressed skin and metal reinforcing members. The helicopter cabin is normally a plexiglass enclosure which is supported by aluminum tubing in some models.

A large single-rotor helicopter is shown in figure 1–42. It is all-metal and is basically composed of two major sections, the cabin and the tail cone. The cabin section is further divided into passenger or cargo compartments, which provide space for the crew, passengers, cargo, fuel and oil tanks, controls, and powerplant. In multi-engine helicopters, the powerplants are usually mounted in separate engine nacelles.

As shown in figure 1–42, the aft section of a typical single-rotor helicopter consists of the tail cone, the fin, the tail-cone housing, the tail-rotor pylon, and the tail-end fairing. The tail cone is bolted to the rear of the forward section and supports the tail rotor, tail-rotor drive shafts, stabilizers, tail-cone housing, and tail-rotor pylon. The tail cone is of magnesium alloy and aluminum alloy construction. The tail-cone housing is bolted to the aft end of the tail cone. Trim stabilizers extend out

24

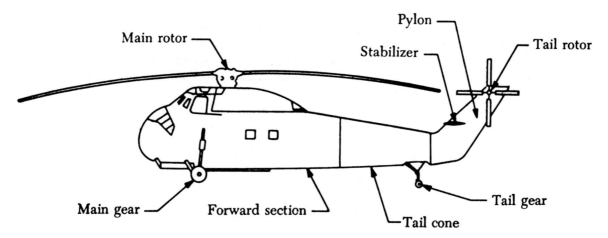

FIGURE 1-41. Typical helicopter structural components.

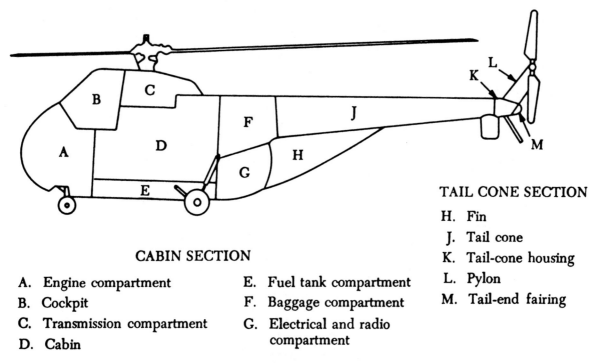

CABIN SECTION

A. Engine compartment
B. Cockpit
C. Transmission compartment
D. Cabin

E. Fuel tank compartment
F. Baggage compartment
G. Electrical and radio
   compartment

TAIL CONE SECTION

H. Fin
J. Tail cone
K. Tail-cone housing
L. Pylon
M. Tail-end fairing

FIGURE 1-42. Location of major helicopter components.

on both sides of the tail cone forward of the housing.

Helicopter structural members are designed to carry a load or, stated differently, to resist stress. A single member of the helicopter structure may be subjected to a combination of stresses. In most cases it is desirable for structural members to carry end loads rather than side loads; that is, to be subjected to tension or compression rather than bending. Structural members are usually combined into a truss to carry end loads. In a typical Pratt truss, the longitudinal and vertical members are tubes or rods capable of carrying compression loads.

Nonstructural members that are not removable from the helicopter are usually attached by riveting or spot welding. Riveting is the most common method of attaching aluminum alloy sheets together. Parts that can be removed from the helicopter structure are usually bolted together.

Transparent materials are used for windshields and windows and sometimes to cover parts requiring frequent visual inspection. Transparent plastic sheet and laminated glass are the materials most commonly used.

Some helicopter manufacturers use impregnated glass cloth laminate (fiber glass) as a lightweight substitute for certain metal parts, since fiber glass is simple to manufacture, has a high strength-weight ratio, and resists mildew, corrosion, and rot.

GENERAL

This chapter includes both assembly and rigging since the subjects are directly related. Assembly involves putting together the component sections of the aircraft, such as wing sections, empennage units, nacelles, and landing gear. Rigging is the final adjustment and alignment of the various component sections to provide the proper aerodynamic reaction.

Two important considerations in all assembly and rigging operations are: (1) Proper operation of the component in regard to its aerodynamic and mechanical function, and (2) maintaining the aircraft's structural integrity by the correct use of materials, hardware, and safetying devices. Improper assembly and rigging may result in certain members being subjected to loads greater than those for which they were designed.

Assembly and rigging must be done in accordance with the requirements prescribed by the aircraft manufacturer. These procedures are usually detailed in the applicable maintenance or service manuals. The Aircraft Specification or Type Certificate Data Sheets also provide valuable information regarding control surface travel.

The rigging of control systems varies with each type of aircraft, therefore, it would be impracticable to define a precise procedure. However, certain principles apply in all situations and these will be discussed in this chapter. It is essential that the aircraft manufacturer's instructions be followed when rigging an aircraft.

THEORY OF FLIGHT

Numerous comprehensive texts have been written about the aerodynamics involved in the flight of an aircraft. It is unnecessary that a mechanic be totally versed on the subject. However, he must understand the relationships between the atmosphere, the aircraft, and the forces acting on it in flight, in order to make intelligent decisions affecting the flight safety of both airplanes and helicopters.

Understanding why the aircraft is designed with a particular type of primary and secondary control system, and why the surfaces must be aerodynamically smooth, becomes essential when maintaining today's complex aircraft.

AERODYNAMICS

Theory of flight deals with aerodynamics. The term aerodynamics is derived from the combination of two Greek words—"aer" meaning air, and "dyne" meaning force of power. Thus, when aero is joined with dynamics, we have aerodynamics, meaning the study of objects in motion through the air and the forces that produce or change such motion.

Aerodynamics is the science of the action of air on an object. It is further defined as that branch of dynamics which deals with the motion of air and other gases, with the forces acting upon an object in motion through the air, or with an object which is stationary in a current of air. In effect, aerodynamics is concerned with three distinct parts. These parts may be defined as the aircraft, the relative wind, and the atmosphere.

THE ATMOSPHERE

Before discussing the fundamentals of the theory of flight, there are several basic ideas that must be considered. An aircraft operates in the air; therefore, the properties of air that affect aircraft control and performance must be understood.

Air is a mixture of gases composed principally of nitrogen and oxygen. Since air is a combination of gases, it follows the laws of gases. Air is considered a fluid because it answers the definition of a fluid, namely, a substance which may be made to flow or change its shape by the application of moderate pressure. Air has weight, since something lighter than air, such as a balloon filled with helium, will rise in the air.

PRESSURE

The deeper a diver goes beneath the surface of the ocean, the greater the pressure becomes on his body due to the weight of the water overhead. Since air also has weight, the greater the depth from the outer surface of the atmosphere, the greater the pressure. If a 1-in. square column of air extending from sea level to the "top" of the atmosphere could be weighed, it would be found to weigh about 14.7 lbs. Thus, atmospheric pressure at sea level is 14.7 p.s.i. (pounds per square inch). However, pounds per square inch is rather a crude unit for the measurement of a light substance such as air. Therefore, atmospheric pressure is usually measured in terms of inches of mercury.

The apparatus for measuring atmospheric pressure is shown in figure 2–1. A glass tube, 36 in. long, open at one end, and closed at the other, is filled with mercury. The open end is sealed temporarily and then submerged into a small container partly filled with mercury, after which the end is unsealed. This allows the mercury in the tube to descend, leaving a vacuum at the top of the tube. Some of the mercury flows into the container while a portion of it remains in the tube. The weight of the atmosphere pressing on the mercury in the open container exactly balances the weight of the mercury in the tube, which has no atmospheric pressure pushing down on it due to the vacuum in the top of the tube. As the pressure of the surrounding air decreases or increases, the mercury column lowers or rises correspondingly. At sea level the height of the mercury in the tube measures approximately 29.92 in., although it varies slightly with atmospheric conditions.

An important consideration is that atmospheric pressure varies with altitude. The higher an object rises above sea level, the lower the pressure. Various atmospheric conditions have a definite relation to flying. The effect of temperature, altitude, and density of the air on aircraft performance is discussed in the following paragraphs.

DENSITY

Density is a term that means weight per unit volume. Since air is a mixture of gases, it can be compressed. If the air in one container is under one-half as much pressure as the air in another identical container, the air under the greater pressure weighs twice as much as that in the container under lower pressure. The air under greater pressure is twice as dense as that in the other container. For equal weights of air, that which is under the greater pressure occupies only half the volume of that under half the pressure.

The density of gases is governed by the following rules:

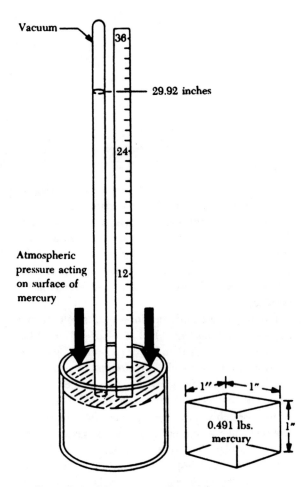

FIGURE 2–1. Measurement of atmospheric pressure.

(1) Density varies in direct proportion with the pressure.
(2) Density varies inversely with the temperature.

Thus, air at high altitudes is less dense than air at low altitudes, and a mass of hot air is less dense than a mass of cool air.

Changes in density affect the aerodynamic performance of aircraft. With the same horsepower, an aircraft can fly faster at a high altitude where the density is low than at a low altitude where the density is great. This is because air offers less resistance to the aircraft when it contains a smaller number of air particles per unit volume.

HUMIDITY

Humidity is the amount of water vapor in the air. The maximum amount of water vapor that air can hold varies with the temperature. The higher the temperature of the air, the more water vapor it can absorb. By itself, water vapor weighs approximately

28

five-eighths as much as an equal amount of perfectly dry air. Therefore, when air contains water vapor it is not as heavy as air containing no moisture.

Assuming that the temperature and pressure remain the same, the density of the air varies inversely with the humidity. On damp days the air density is less than on dry days. For this reason, an aircraft requires a longer runway for takeoff on damp days than it does on dry days.

BERNOULLI'S PRINCIPLE AND SUBSONIC FLOW

Bernoulli's principle states that when a fluid (air) flowing through a tube reaches a constriction, or narrowing of the tube, the speed of the fluid flowing through that constriction is increased and its pressure is decreased. The cambered (curved) surface of an airfoil (wing) affects the airflow exactly as a constriction in a tube affects airflow. This resemblance is illustrated in figure 2–2.

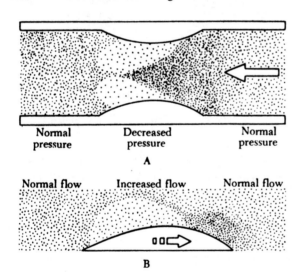

Normal pressure   Decreased pressure   Normal pressure

A

Normal flow   Increased flow   Normal flow

B

FIGURE 2–2. Bernoulli's principle.

Diagram A of figure 2–2 illustrates the effect of air passing through a constriction in a tube. In B, the air is flowing past a cambered surface, such as an airfoil, and the effect is similar to that of air passing through a restriction.

As the air flows over the upper surface of an airfoil, its speed or velocity increases and its pressure decreases. An area of low pressure is thus formed. There is an area of greater pressure on the lower surface of the airfoil, and this greater pressure tends to move the wing upward. This difference in pressure between the upper and lower surfaces of the wing is called lift. Three-fourths of the total lift

of an airfoil is the result of the decrease in pressure over the upper surface. The impact of air on the under surface of an airfoil produces the other one-fourth of the total lift.

An aircraft in flight is acted upon by four forces:
- (1) Gravity, or weight, the force that pulls the aircraft toward the earth.
- (2) Lift, the force that pushes the aircraft upward.
- (3) Thrust, the force that moves the aircraft forward.
- (4) Drag, the force that exerts a braking action.

MOTION

Motion is the act or process of changing place or position. An object may be in motion with respect to one object and motionless with respect to another. For example, a person sitting quietly in an aircraft flying at 200 knots is at rest or motionless with respect to the aircraft; however, the person is in motion with respect to the air or the earth, the same as is the aircraft.

Air has no force or power, except pressure, unless it is in motion. When it is moving, however, its force becomes apparent. A moving object in motionless air has a force exerted on it as a result of its own motion. It makes no difference in the effect then, whether an object is moving with respect to the air or the air is moving with respect to the object.

The flow of air around an object caused by the movement of either the air or the object, or both, is called the relative wind.

Velocity and Acceleration

The terms "speed" and "velocity" are often used interchangeably, but they do not mean the same. Speed is the rate of motion, and velocity is the rate of motion in a particular direction in relation to time.

An aircraft starts from New York City and flies 10 hrs. at an average speed of 260 m.p.h. At the end of this time the aircraft may be over the Atlantic Ocean, the Pacific Ocean, the Gulf of Mexico, or, if its flight were in a circular path, it may even be back over New York. If this same aircraft flew at a velocity of 260 m.p.h. in a southwestward direction, it would arrive in Los Angeles in about 10 hrs. Only the rate of motion is indicated in the first example and denotes the speed of the aircraft. In the last example, the particular direction is included with the rate of motion, thus, denoting the velocity of the aircraft.

Acceleration is defined as the rate of change of velocity. An aircraft increasing in velocity is an example of positive acceleration, while another aircraft reducing its velocity is an example of negative acceleration. (Positive acceleration is often referred to as acceleration and negative acceleration as deceleration.)

Newton's Laws of Motion

The fundamental laws governing the action of air about a wing are Newton's laws of motion.

Newton's first law is normally referred to as the law of inertia. It simply means that a body at rest will not move unless force is applied to it. If it is moving at uniform speed in a straight line, force must be applied to increase or decrease that speed.

Since air has mass, it is a "body" in the meaning of the law. When an aircraft is on the ground with its engines stopped, inertia keeps the aircraft at rest. An aircraft is moved from its state of rest by the thrust force created by the propeller, by the expanding exhaust gases, or both. When it is flying at uniform speed in a straight line, inertia tends to keep the aircraft moving. Some external force is required to change the aircraft from its path of flight.

Newton's second law, that of force, also applies to objects. This law states that if a body moving with uniform speed is acted upon by an external force, the change of motion will be proportional to the amount of the force, and motion will take place in the direction in which the force acts. This law may be stated mathematically as follows:

Force = mass × acceleration (F = ma).

If an aircraft is flying against a headwind, it is slowed down. If the wind is coming from either side of the aircraft's heading, the aircraft is pushed off course unless the pilot takes corrective action against the wind direction.

Newton's third law is the law of action and reaction. This law states that for every action (force) there is an equal and opposite reaction (force). This law is well illustrated by the action of a swimmer's hands. He pushes the water aft and thereby propels himself forward, since the water resists the action of his hands. When the force of lift on an aircraft's wing equals the force of gravity, the aircraft maintains level flight.

The three laws of motion which have been discussed are closely related and apply to the theory of flight. In many cases, all three laws may be operating on an aircraft at the same time.

AIRFOILS

An airfoil is a surface designed to obtain a desirable reaction from the air through which it moves. Thus, we can say that any part of the aircraft which converts air resistance into a force useful for flight is an airfoil. The blades of a propeller are so designed that when they rotate, their shape and position cause a higher pressure to be built up behind them than in front of them so that they will pull the aircraft forward. The profile of a conventional wing, shown in figure 2–3, is an excellent example of an airfoil. Notice that the top surface of the wing profile has greater curvature than the lower surface.

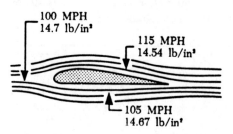

FIGURE 2-3. Airflow over a wing section.

The difference in curvature of the upper and lower surfaces of the wing builds up the lift force. Air flowing over the top surface of the wing must reach the trailing edge of the wing in the same amount of time as the air flowing under the wing. To do this, the air passing over the top surface moves at a greater velocity than the air passing below the wing because of the greater distance it must travel along the top surface. This increased velocity, according to Bernoulli's principle, means a corresponding decrease in pressure on the surface. Thus, a pressure differential is created between the upper and lower surfaces of the wing, forcing the wing upward in the direction of the lower pressure.

The theoretical amount of lift of the airfoil at a velocity of 100 m.p.h. can be determined by sampling the pressure above and below the airfoil at the point of greatest air velocity. As shown in figure 2–3, this pressure is 14.54 p.s.i. above the airfoil. Subtracting this pressure from the pressure below the airfoil, 14.67, gives a difference in pressure of 0.13 p.s.i. Multiplying 0.13 by 144 (number of square inches in a square foot) shows that each square foot of this wing will lift 18.72 pounds. Thus, it can be seen that a small pressure differential across an airfoil section can produce a large lifting force. Within limits, lift can be increased by increasing the angle of attack, the wing area, the freestream velocity, or the density of the air, or by changing the shape of the airfoil.

Angle of Attack

Before beginning the discussion on angle of attack and its effect on airfoils, we shall first consider the terms "chord" and "center of pressure."

The chord of an airfoil or wing section is an imaginary straight line which passes through the section from the leading edge to the trailing edge, as shown in figure 2–4. The chord line provides one side of an angle which ultimately forms the angle of attack. The other side of the angle is formed by a line indicating the direction of the relative airstream. Thus, angle of attack is defined as the angle between the chord line of the wing and the direction of the relative wind. This is not to be confused with the angle of incidence, which is the angle between the chord line of the wing and the longitudinal axis of the aircraft.

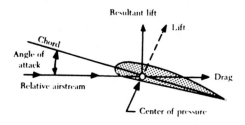

FIGURE 2–4. Positive angle of attack.

On each minute part of an airfoil or wing surface, a small force is present. This force is different in magnitude and direction from any forces acting on other areas forward or rearward from this point. It is possible to add all of these small forces mathematically, and the sum is called the resultant force (lift). This resultant force has magnitude, direction, and location, and can be represented as a vector, as shown in figure 2–4. The point of intersection of the resultant force line with the chord line of the airfoil is called the center of pressure. The center of pressure moves along the airfoil chord as the angle of attack changes. Throughout most of the flight range, the center of pressure moves forward with increasing angle of attack and rearward as the

angle of attack decreases. The effect of increasing angle of attack on the center of pressure is shown in figure 2–5.

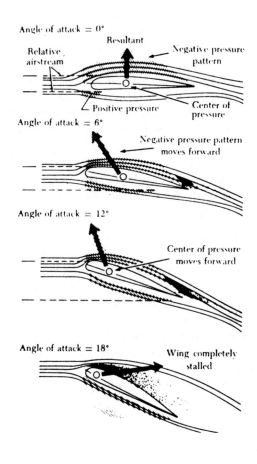

FIGURE 2–5. Effect of increasing angle of attack.

The angle of attack changes as the aircraft's attitude changes. Since the angle of attack has a great deal to do with determining lift, it is given primary consideration when designing airfoils. In a properly designed airfoil, the lift increases as the angle of attack is increased.

When the angle of attack is increased gradually toward a positive angle of attack, the lift component increases rapidly up to a certain point and then suddenly begins to drop off. During this action the drag component increases slowly at first and then rapidly as lift begins to drop off.

When the angle of attack increases to the angle of maximum lift, the burble point is reached. This is known as the critical angle. When the critical

31

angle is reached, the air ceases to flow smoothly over the top surface of the airfoil and begins to burble, or eddy. This means that air breaks away from the upper camber line of the wing. What was formerly the area of decreased pressure is now filled by this burbling air. When this occurs, the amount of lift drops and drag becomes excessive. The force of gravity exerts itself, and the nose of the aircraft drops. Thus we see that the burble point is the stalling angle.

As we have seen, the distribution of the pressure forces over the airfoil varies with the angle of attack. The application of the resultant force, that is, the center of pressure, varies correspondingly. As this angle increases, the center of pressure moves forward; and as the angle decreases, the center of pressure moves back. The unstable travel of the center of pressure is characteristic of practically all airfoils.

Angle of Incidence

The acute angle which the wing chord makes with the longitudinal axis of the aircraft is called the angle of incidence (figure 2–6), or the angle of wing setting. The angle of incidence in most cases is a fixed, built-in angle. When the leading edge of the wing is higher than the trailing edge, the angle of incidence is said to be positive. The angle of incidence is negative when the leading edge is lower than the trailing edge of the wing.

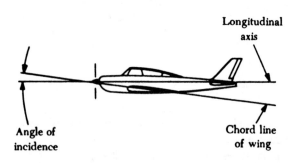

FIGURE 2–6. Angle of incidence.

Wing Area

Wing area is measured in square feet and includes the part blanked out by the fuselage. Wing area is adequately described as the area of the shadow cast by the wing at high noon. Tests show that lift and drag forces acting on a wing are roughly proportional to the wing area. This means that if the wing area is doubled, all other variables remaining the same, the lift and drag created by the wing is doubled. If the area is tripled, lift and drag are tripled.

Shape of the Airfoil

The shape of the airfoil determines the amount of turbulence or skin friction that it will produce. The shape of a wing consequently affects the efficiency of the wing.

Airfoil section properties differ from wing or aircraft properties because of the effect of the wing planform. A wing may have various airfoil sections from root to tip, with taper, twist, and sweepback. The resulting aerodynamic properties of the wing are determined by the action of each section along the span.

Turbulence and skin friction are controlled mainly by the fineness ratio, which is defined as the ratio of the chord of the airfoil to the maximum thickness. If the wing has a high fineness ratio, it is a very thin wing. A thick wing has a low fineness ratio. A wing with a high fineness ratio produces a large amount of skin friction. A wing with a low fineness ratio produces a large amount of turbulence. The best wing is a compromise between these two extremes to hold both turbulence and skin friction to a minimum.

Efficiency of a wing is measured in terms of the lift over drag (L/D) ratio. This ratio varies with the angle of attack but reaches a definite maximum value for a particular angle of attack. At this angle, the wing has reached its maximum efficiency. The shape of the airfoil is the factor which determines the angle of attack at which the wing is most efficient; it also determines the degree of efficiency. Research has shown that the most efficient airfoils for general use have the maximum thickness occurring about one-third of the way back from the leading edge of the wing.

High-lift wings and high-lift devices for wings have been developed by shaping the airfoils to produce the desired effect. The amount of lift produced by an airfoil will increase with an increase in wing camber. Camber refers to the curvature of an airfoil above and below the chord line surface. Upper camber refers to the upper surface, lower camber to the lower surface, and mean camber to the mean line of the section. Camber is positive when departure from the chord line is outward, and negative when it is inward. Thus, high-lift wings have a large positive camber on the upper surface and a slight negative camber on the lower surface. Wing flaps cause an ordinary wing to approximate this same condition by increasing the upper camber and by creating a negative lower camber.

It is also known that the larger the wingspan as compared to the chord, the greater the lift obtained. This comparison is called aspect ratio. The higher the aspect ratio, the greater the lift. In spite of the benefits from an increase in aspect ratio, it was found that definite limitations were of structural and drag considerations.

On the other hand, an airfoil that is perfectly streamlined and offers little wind resistance sometimes does not have enough lifting power to take the aircraft off the ground. Thus, modern aircraft have airfoils which strike a medium between extremes, with the shape varying according to the aircraft for which it is designed.

CENTER OF GRAVITY

Gravity is the pulling force that tends to draw all bodies in the earth's sphere to the center of the earth. The center of gravity may be considered as a point at which all the weight of the aircraft is concentrated. If the aircraft were supported at its exact center of gravity, it would balance in any position. Center of gravity is of major importance in an aircraft, for its position has a great bearing upon stability.

The center of gravity is determined by the general design of the aircraft. The designer estimates how far the center of pressure will travel. He then fixes the center of gravity in front of the center of pressure for the corresponding flight speed in order to provide an adequate restoring moment for flight equilibrium.

THRUST AND DRAG

An aircraft in flight is the center of a continuous battle of forces. Actually, this conflict is not as violent as it sounds, but it is the key to all maneuvers performed in the air. There is nothing mysterious about these forces; they are definite and known. The directions in which they act can be calculated; and the aircraft itself is designed to take advantage of each of them. In all types of flying, flight calculations are based on the magnitude and direction of four forces: weight, lift, drag, and thrust. (See fig. 2–7.)

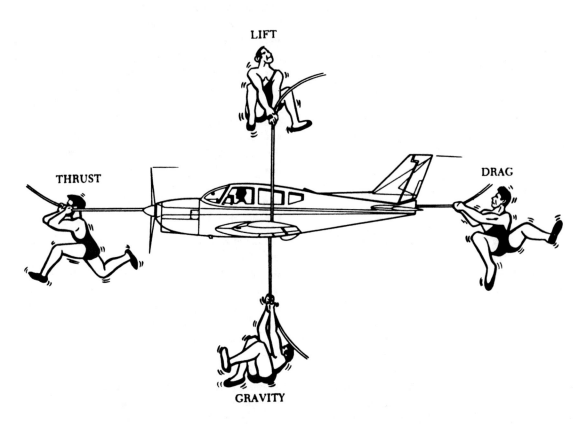

FIGURE 2–7. Forces in action in flight.

33

Weight is the force of gravity acting downward upon everything that goes into the aircraft, such as the aircraft itself, the crew, the fuel, and the cargo.

Lift acts vertically and by so doing counteracts the effects of weight.

Drag is a backward deterrent force and is caused by the disruption of the airflow by the wings, fuselage, and protruding objects.

Thrust produced by the powerplant is the forward force that overcomes the force of drag.

Notice that these four forces are only in perfect balance when the aircraft is in straight and level unaccelerated flight.

The force of lift and drag are the direct result of the relationship between the relative wind and the aircraft. The force of lift always acts perpendicular to the relative wind, and the force of drag always acts parallel to the relative wind and in the same direction. These forces are actually the components that produced a resultant lift force on the wing as shown in figure 2–8.

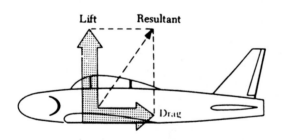

FIGURE 2–8. Resultant of lift and drag.

Weight has a definite relationship with lift, and thrust with drag. This relationship is quite simple, but very important in understanding the aerodynamics of flying. As stated previously, lift is the upward force on the wing acting perpendicular to the relative wind. Lift is required to counteract the aircraft's weight, caused by the force of gravity acting on the mass of the aircraft. This weight force acts downward through a point called the center of gravity which is the point at which all the weight of the aircraft is considered to be concentrated. When the lift force is in equilibrium with the weight force, the aircraft neither gains nor loses altitude.

If lift becomes less than weight, the aircraft loses altitude. When the lift is greater than weight, the aircraft gains altitude.

Drag must be overcome in order for the aircraft to move, and movement is essential to obtain lift. To overcome the drag and move the aircraft forward, another force is essential. This force is thrust. Thrust is derived from jet propulsion or from a propeller and engine combination. Jet propulsion theory is based on Newton's third law of motion which states that for every action there is an equal and opposite reaction. For example, in firing a gun the action is the bullet going forward while the reaction is the gun recoiling backwards. The turbine engine causes a mass of air to be moved backward at high velocity causing a reaction forward that moves the aircraft.

In a propeller/engine combination, the propeller is actually two or more revolving airfoils mounted on a horizontal shaft. The motion of the blades through the air produces lift similar to the lift on the wing, but acts in a horizontal direction, pulling the aircraft forward.

Before the aircraft begins to move, thrust must be exerted. It continues to move and gain speed until thrust and drag are equal. In order to maintain a steady speed, thrust and drag must remain equal, just as lift and weight must be equal for steady, horizontal flight. We have seen that increasing the lift means that the aircraft moves upward, whereas decreasing the lift so that it is less than the weight causes the aircraft to lose altitude. A similar rule applies to the two forces of thrust and drag. If the r.p.m. of the engine is reduced, the thrust is lessened, and the aircraft slows down. As long as the thrust is less than the drag, the aircraft travels more and more slowly until its speed is insufficient to support it in the air.

Likewise, if the r.p.m. of the engine is increased, thrust becomes greater than drag, and the speed of the aircraft increases. As long as the thrust continues to be greater than the drag, the aircraft continues to accelerate. When drag equals thrust, the aircraft flies at a steady speed.

The relative motion of the air over an object that produces lift also produces drag. Drag is the resistance of the air to objects moving through it. If an aircraft is flying on a level course, the lift force acts vertically to support it while the drag force acts horizontally to hold it back. The total amount of drag on an aircraft is made up of many drag forces, but for our purposes, we will only consider three—*parasite drag, profile drag, and induced drag.*

Parasite drag is made up of a combination of many different drag forces. Any exposed object on an aircraft offers some resistance to the air, and the more objects in the airstream, the more parasite drag. While parasite drag can be reduced by reducing the number of exposed parts to as few as practical and streamlining their shape, skin friction is the type of parasite drag most difficult to reduce. No surface is perfectly smooth. Even machined surfaces when inspected under magnification have a ragged uneven appearance. These ragged surfaces deflect the air near the surface causing resistance to smooth airflow. Skin friction can be reduced by using glossy flat finishes and eliminating protruding rivet heads, roughness, and other irregularities.

Profile drag may be considered the parasite drag of the airfoil. The various components of parasite drag are all of the same nature as profile drag.

The action of the airfoil that gives us lift also causes induced drag. Remember that the pressure above the wing is less than atmospheric, and the pressure below the wing is equal to or greater than atmospheric pressure. Since fluids always move from high pressure toward low pressure, there is a spanwise movement of air from the bottom of the wing outward from the fuselage and upward around the wing tip. This flow of air results in "spillage" over the wing tip, thereby setting up a whirlpool of air called a vortex (figure 2–9). The air on the upper surface has a tendency to move in toward the fuselage and off the trailing edge. This air current forms a similar vortex at the inner portion of the trailing edge of the wing. These vortices increase drag, because of the turbulence produced, and constitute induced drag.

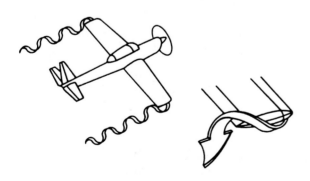

FIGURE 2–9. Wing tip vortices.

Just as lift increases with an increase in angle of attack, induced drag also increases as the angle of attack becomes greater. This occurs because as the angle of attack is increased, there is a greater pressure difference between the top and bottom of the wing. This causes more violent vortices to be set up, resulting in more turbulence and more induced drag.

AXES OF AN AIRCRAFT

Whenever an aircraft changes its attitude in flight, it must turn about one or more of three axes. Figure 2–10 shows the three axes, which are imaginary lines passing through the center of the aircraft. The axes of an aircraft can be considered as imaginary axles around which the aircraft turns like a wheel. At the center, where all three axes intersect, each is perpendicular to the other two. The axis which extends lengthwise through the fuselage from the nose to the tail is called the longitudinal axis. The axis which extends crosswise, from wing tip to wing tip, is the lateral axis. The axis which passes through the center, from top to bottom, is called the vertical axis.

Motion about the longitudinal axis resembles the roll of a ship from side to side. In fact, the names used in describing the motion about an aircraft's three axes were originally nautical terms. They have been adapted to aeronautical terminology because of the similarity of motion between an aircraft and a ship.

Thus, the motion about the longitudinal axis is called roll; motion along the lateral (crosswing) axis is called pitch. Finally, an aircraft moves about its vertical axis in a motion which is termed yaw. This is a horizontal movement of the nose of the aircraft.

Roll, pitch, and yaw—the motions an aircraft makes about its longitudinal, lateral, and vertical axes—are controlled by three control surfaces. Roll is produced by the ailerons, which are located at the trailing edges of the wings. Pitch is affected by the elevators, the rear portion of the horizontal tail assembly. Yaw is controlled by the rudder, the rear portion of the vertical tail assembly.

STABILITY AND CONTROL

An aircraft must have sufficient stability to maintain a uniform flight path and recover from the various upsetting forces. Also, to achieve the best performance, the aircraft must have the proper response to the movement of the controls.

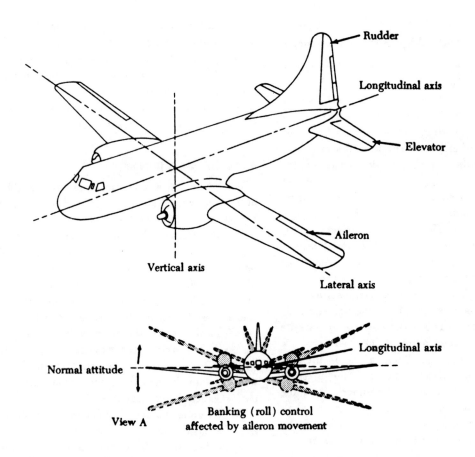

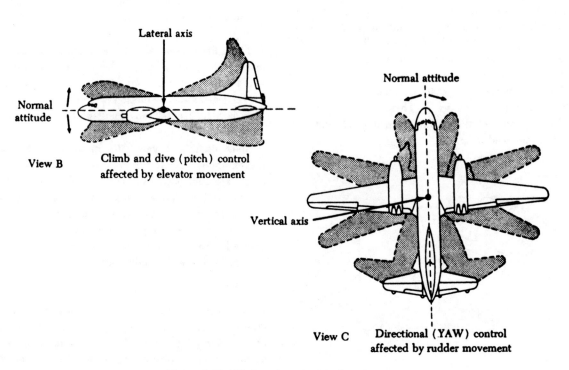

FIGURE 2-10. Motion of an aircraft about its axes.

36

Three terms that appear in any discussion of stability and control are: (1) Stability, (2) maneuverability, and (3) controllability. Stability is the characteristic of an aircraft which tends to cause it to fly (hands off) in a straight and level flight path. Maneuverability is the ability of an aircraft to be directed along a desired flight path and to withstand the stresses imposed. Controllability is the quality of the response of an aircraft to the pilot's commands while maneuvering the aircraft.

Static Stability

An aircraft is in a state of equilibrium when the sum of all the forces acting on the aircraft and all the moments is equal to zero. An aircraft in equilibrium experiences no accelerations, and the aircraft continues in a steady condition of flight. A gust of wind or a deflection of the controls disturbs the equilibrium, and the aircraft experiences acceleration due to the unbalance of moment or force.

The three types of static stability are defined by the character of movement following some disturbance from equilibrium. Positive static stability exists when the disturbed object tends to return to equilibrium. Negative static stability or static instability exists when the disturbed object tends to continue in the direction of disturbance. Neutral static stability exists when the disturbed object has neither the tendency to return nor continue in the displacement direction, but remains in equilibrium in the direction of disturbance. These three types of stability are illustrated in figure 2–11.

Dynamic Stability

While static stability deals with the tendency of a displaced body to return to equilibrium, dynamic stability deals with the resulting motion with time. If an object is disturbed from equilibrium, the time history of the resulting motion defines the dynamic stability of the object. In general, an object demonstrates positive dynamic stability if the amplitude of motion decreases with time. If the amplitude of motion increases with time, the object is said to possess dynamic instability.

Any aircraft must demonstrate the required degrees of static and dynamic stability. If an aircraft were designed with static instability and a rapid rate of dynamic instability, the aircraft would be very difficult, if not impossible, to fly. Usually, positive dynamic stability is required in an aircraft design to prevent objectionable continued oscillations of the aircraft.

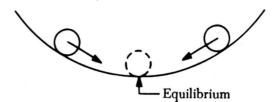

Tendency to return to equilibrium

Equilibrium

A. Positive Static Stability

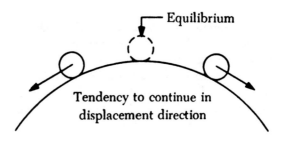

Equilibrium

Tendency to continue in displacement direction

B. Negative static stabiliy or static instability

Equilibrium encountered at any point of displacement

C. Neutral static stability

FIGURE 2–11. Static stability.

Longitudinal Stability

When an aircraft has a tendency to keep a constant angle of attack with reference to the relative wind—that is, when it does not tend to put its nose down and dive or lift its nose and stall—it is said to have longitudinal stability. Longitudinal stability refers to motion in pitch. The horizontal stabilizer is the primary surface which controls longitudinal stability. The action of the stabilizer depends upon the speed and angle of attack of the aircraft.

Figure 2–12 illustrates the contribution of tail lift to stability. If the aircraft changes its angle of attack, a change in lift takes place at the aerodynamic center (center of pressure) of the horizontal stabilizer.

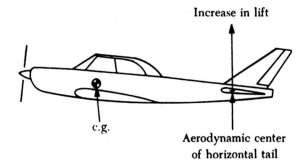

FIGURE 2-12. Producing tail lift.

Under certain conditions of speed, load, and angle of attack, the flow of air over the horizontal stabilizer creates a force which pushes the tail up or down. When conditions are such that the airflow creates equal forces up and down, the forces are said to be in equilibrium. This condition is usually found in level flight in calm air.

Directional Stability

Stability about the vertical axis is referred to as directional stability. The aircraft should be designed so that when it is in straight and level flight it remains on its course heading even though the pilot takes his hands and feet off the controls. If an aircraft recovers automatically from a skid, it has been well designed and possesses good directional balance. The vertical stabilizer is the primary surface which controls directional stability.

As shown in figure 2–13, when an aircraft is in a sideslip or yawing, the vertical tail experiences a change in angle of attack with a resulting change in lift [not to be confused with the lift created by the wing]. The change in lift, or side force, on the vertical tail creates a yawing moment about the center of gravity which tends to return the aircraft to its original flight path.

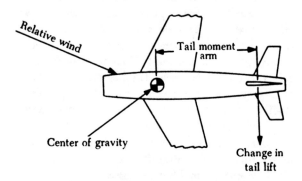

FIGURE 2–13. Contribution of vertical tail to directional stability.

Sweptback wings aid in directional stability. If the aircraft yaws from its direction of flight, the wing which is farther ahead offers more drag than the wing which is aft. The effect of this drag is to hold back the wing which is farther ahead, and to let the other wing catch up.

Directional stability is also aided by using a large dorsal fin and a long fuselage.

The high Mach numbers of supersonic flight reduce the contribution of the vertical tail to directional stability. To produce the required directional stability at high Mach numbers, a very large vertical tail area may be necessary. Ventral (belly) fins may be added as an additional contribution to directional stability.

Lateral Stability

We have seen that pitching is motion about the aircraft's lateral axis and yawing is motion about its vertical axis. Motion about its longitudinal (fore and aft) axis is a lateral or rolling motion. The tendency to return to the original attitude from such motion is called lateral stability.

The lateral stability of an airplane involves consideration of rolling moments due to sideslip. A sideslip tends to produce both a rolling and a yawing motion. If an airplane has a favorable rolling moment, a sideslip will tend to return the airplane to a level flight attitude.

The principal surface contributing to the lateral stability of an airplane is the wing. The effect of the geometric dihedral (figure 2–14) of a wing is a powerful contribution to lateral stability. As shown in figure 2–14, a wing with dihedral develops stable rolling moments with sideslip. With the relative wind from the side, the wing into the wind is subject to an increase in angle of attack and develops an increase in lift. The wing away from the wind is subject to a decrease in angle of attack and develops less lift. The changes in lift effect a rolling moment tending to raise the windward wing.

When a wing is swept back, the effective dihedral increases rapidly with a change in the lift coefficient of the wing. Sweepback is the angle between a line perpendicular to the fuslage center line and the quarter chord of each wing airfoil section. Sweepback in combination with dihedral causes the dihedral effect to be excessive. As shown in figure 2–15, the swept-wing aircraft in a sideslip has the wing that is into the wind operating with an effective decrease in sweepback, while the wing out of the wind is operating with an effective increase in

38

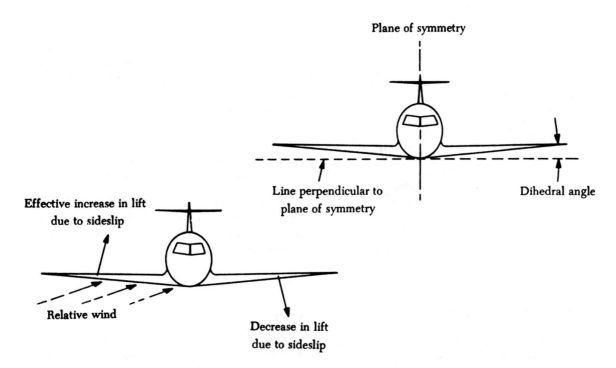

FIGURE 2–14. Contribution of dihedral to lateral stability.

sweepback. The wing into the wind develops more lift, and the wing out of the wind develops less. This tends to restore the aircraft to a level flight attitude.

The amount of effective dihedral necessary to produce satisfactory flying qualities varies greatly with the type and purpose of the aircraft. Generally, the effective dihedral is kept low, since high roll due to sideslip can create problems. Excessive dihedral effect can lead to Dutch Roll, difficult rudder coordination in rolling maneuvers, or place extreme demands for lateral control power during crosswind takeoff and landing.

CONTROL

Control is the action taken to make the aircraft follow any desired flight path. When an aircraft is said to be controllable, it means that the craft responds easily and promptly to movement of the controls. Different control surfaces are used to control the aircraft about each of the three axes. Moving the control surfaces on an aircraft changes the airflow over the aircraft's surface. This, in turn, creates changes in the balance of forces acting to keep the aircraft flying straight and level.

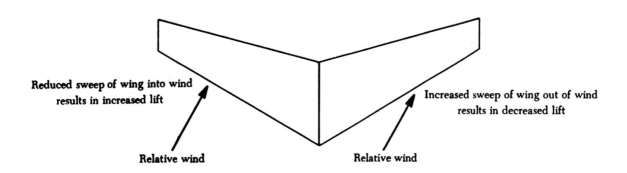

FIGURE 2–15. Effect of sweepback on lateral stability.

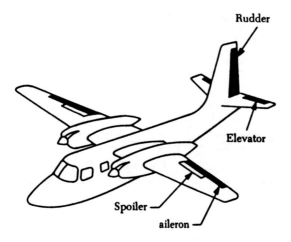

FIGURE 2-16. Primary flight controls.

FLIGHT CONTROL SURFACES

The flight control surfaces are hinged or movable airfoils designed to change the attitude of the aircraft during flight. These surfaces may be divided into three groups, usually referred to as the primary group, secondary group, and auxiliary group.

Primary Group

The primary group includes the ailerons, elevators, and rudder (figure 2–16). These surfaces are used for moving the aircraft about its three axes.

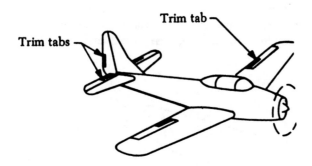

FIGURE 2-17. Trim tabs.

The ailerons and elevators are generally operated from the cockpit by a control stick on single-engine aircraft and by a wheel and yoke assembly on multi-engine aircraft. The rudder is operated by foot pedals on all types of aircraft.

Secondary Group

Included in the secondary group are the trim tabs and spring tabs. Trim tabs (figure 2–17) are small airfoils recessed into the trailing edges of the primary control surfaces. The purpose of trim tabs is to enable the pilot to trim out any unbalanced condition which may exist during flight, without exerting any pressure on the primary controls. Each trim tab is hinged to its parent primary control surface, but is operated by an independent control.

Spring tabs are similar in appearance to trim tabs, but serve an entirely different purpose. Spring tabs are used to aid the pilot in moving the primary control surfaces.

Auxiliary Group

Included in the auxiliary group of flight control surfaces are the wing flaps, spoilers, speed brakes, slats, leading edge flaps and slots.

The auxiliary groups may be divided into two sub-groups. Those whose primary purpose is lift augmenting and those whose primary purpose is lift decreasing. In the first group are the flaps, both trailing edge and leading edge (slats), and slots. The lift decreasing devices are speed brakes and spoilers.

The trailing edge airfoils (flaps) increase the wing area thereby increasing lift on takeoff and decrease the speed during landing. These airfoils are retractable and fair into the wing contour. Others are simply a portion of the lower skin which extends into the airstream thereby slowing the aircraft.

Leading edge flaps are airfoils extended from and retracted into the leading edge of the wing. Some installations create a slot (an opening between the extended airfoil and the leading edge). The flap (termed slat by some manufacturers) and slot create additional lift at the slower speeds of takeoff and landing. Other installations have permanent slots built in the leading edge of the wing. At cruising speeds, the trailing edge and leading edge flaps (slats) are retracted into the wing proper.

Lift decreasing devices are the speed brakes (spoilers). In some installations, there are two types of spoilers. The ground spoiler is extended only after the aircraft is on the ground thereby assisting in the braking action. The flight spoiler assists in lateral control by being extended whenever the aileron on that wing is rotated up. When actuated as speed brakes, the spoiler panels on both wings raise up—the panel on the "up" aileron wing raising more than the panel on the down aileron side. This provides speed brake operation and lateral control simultaneously.

Slats are movable control surfaces attached to the leading edges of the wings. When the slat is closed, it forms the leading edge of the wing. When in the open position (extended forward), a slot is created between the slat and the wing leading edge. At low airspeeds this increases lift and improves handling characteristics, allowing the aircraft to be controlled at airspeeds below the otherwise normal landing speed.

CONTROL AROUND THE LONGITUDINAL AXIS

The motion of the aircraft about the longitudinal axis is called rolling or banking. The ailerons (figure 2–18) are used to control this movement. The ailerons form a part of the wing and are located in the trailing edge of the wing toward the tips. Ailerons are the movable surfaces of an otherwise fixed-surface wing. The aileron is in neutral position when it is streamlined with the trailing edge of the wing.

Ailerons respond to side pressure applied to the control stick. Pressure applied to move the stick toward the right raises the right aileron and lowers

the left aileron, causing the aircraft to bank to the right. Ailerons are linked together by control cables so that when one aileron is down, the opposite aileron is up. The function of the lowered aileron is to increase the lift by increasing the wing camber. At the same time, the down aileron also creates some additional drag since it is in the area of high pressure below the wing. The up aileron, on the opposite end of the wing, decreases lift on that end of the wing. The increased lift on the wing whose aileron is down, raises this wing. This causes the aircraft to roll about its longitudinal axis as shown in figure 2–19.

As a result of the increased lift on the wing with the lowered aileron, drag is also increased. This drag attempts to pull the nose in the direction of the high wing. Since the ailerons are used with the rudder when making turns, the increased drag tries to turn the aircraft in the direction opposite to that desired. To avoid this undesirable effect, aircraft are often designed with differential travel of the ailerons.

Differential aileron travel (figure 2–20) provides more aileron up travel than down travel for a given movement of the control stick or wheel in the cockpit.

The spoilers, or speed brakes as they are also called, are plates fitted to the upper surface of the wing. They are usually deflected upward by hydraulic actuators in response to control wheel movement in the cockpit. The purpose of the spoilers is to disturb the smooth airflow across the top of the airfoil thereby creating an increased amount of drag and a decreased amount of lift on that airfoil.

Spoilers are used primarily for lateral control. When banking the airplane, the spoilers function with the ailerons. The spoilers on the up aileron side raise with that aileron to further decrease the lift on that wing. The spoiler on the opposite side remains in the faired position. When the spoilers are used as a speed brake, they are all deflected upward simultaneously. A separate control lever is provided for operating the spoilers as speed brakes.

While we tend to think of a spoiler as being a fairly complicated, controlled device, we should keep in mind that some are not controllable. Some spoilers are automatic in operation in that they come into effect only at a high angle of attack. This

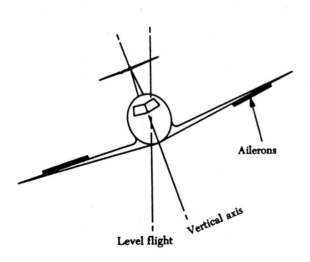

Ailerons

Level flight

Vertical axis

FIGURE 2–18. Aileron action.

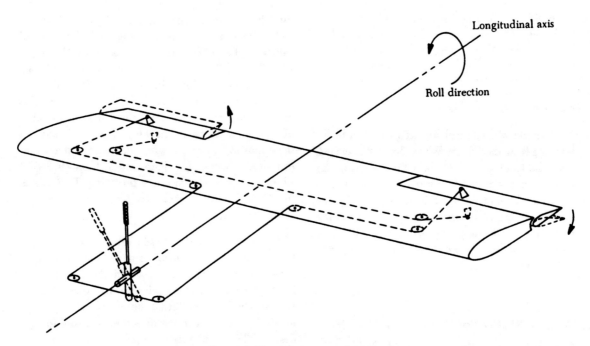

FIGURE 2-19. Aileron control system.

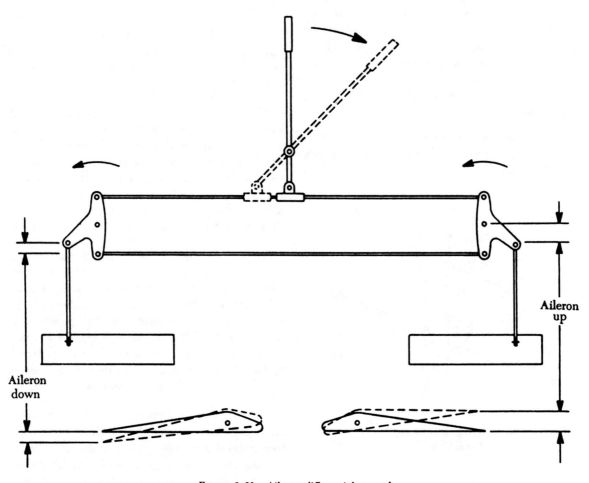

FIGURE 2-20. Aileron differential control.

42

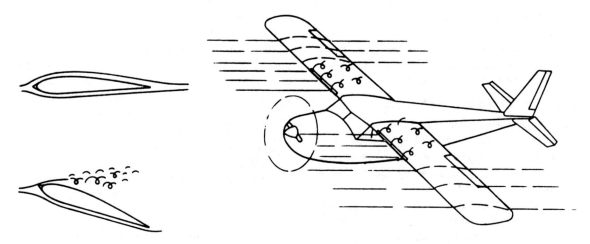

FIGURE 2-21. Fixed spoilers or stall strip.

arrangement keeps them out of the slipstream at cruise and high speeds.

A fixed spoiler may be a small wedge affixed to the leading edge of the airfoil as shown in figure 2-21. This type spoiler causes the inboard portion of the wing to stall ahead of the outboard portion which results in aileron control right up to the occurrence of complete wing stall.

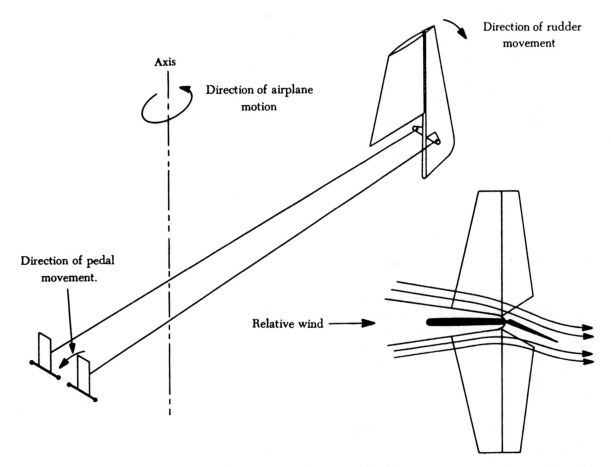

FIGURE 2-22. Rudder action.

Use extreme accuracy in positioning leading edge spoilers when re-installing them after they have been removed for maintenance. Improper positioning may result in adverse stall characteristics. Always follow the manufacturers' instructions regarding location and method of attachment.

CONTROL AROUND THE VERTICAL AXIS

Turning the nose of the aircraft causes the aircraft to rotate about its vertical axis. Rotation of the aircraft about the vertical axis is called yawing. This motion is controlled by using the rudder as illustrated in figure 2-22.

The rudder is a movable control surface attached to the trailing edge of the vertical stabilizer. To turn the aircraft to the right, the rudder is moved to the right. The rudder protrudes into the airstream, causing a force to act upon it. This is the force necessary to give a turning movement about the center of gravity which turns the aircraft to the right. If the rudder is moved to the left, it induces a counterclockwise rotation and the aircraft similarly turns to the left. The rudder can also be used in controlling a bank or turn in flight.

The main function of the rudder is to turn the aircraft in flight. This turn is maintained by the side pressure of the air moving past the vertical surfaces. When an aircraft begins to slip or skid, rudder pressure is applied to keep the aircraft headed in the desired direction (balanced).

Slip or sideslipping refers to any motion of the aircraft to the side and downward toward the inside of a turn. Skid or skidding refers to any movement upward and outward away from the center of a turn.

CONTROL AROUND THE LATERAL AXIS

When the nose of an aircraft is raised or lowered, it is rotated about its lateral axis. Elevators are the movable control surfaces that cause this rotation (figure 2-23). They are normally hinged to the trailing edge of the horizontal stabilizer.

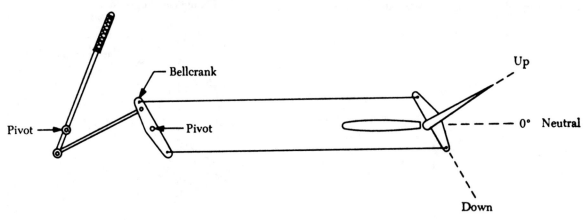

FIGURE 2-23. Elevator action.

The elevators are used to make the aircraft climb or dive and also to obtain sufficient lift from the wings to keep the aircraft in level flight at various speeds.

The elevators can be moved either up or down. If the elevator is rotated up, it decreases the lift force on the tail causing the tail to lower and the nose to rise. If the elevator is rotated downward, it increases the lift force on the tail causing it to rise and the nose to lower. Lowering the aircraft's nose increases forward speed, and raising the nose decreases forward speed.

Some aircraft use a movable horizontal surface called a stabilator (figure 2-24). The stabilator serves the same purpose as the horizontal stabilizer and elevator combined. When the cockpit control is moved, the complete stabilator is moved to raise or lower the leading edge, thus changing the angle of attack and the amount of lift on the tail surfaces.

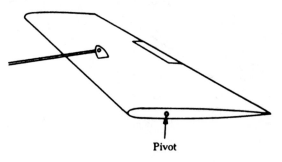

FIGURE 2-24. Movable horizontal stabilator.

44

Aircraft empennages have been designed which combine the vertical and horizontal stabilizers. Such empennages have the stabilizers set at an angle as shown in figure 2–25. This arrangement is referred to as a butterfly or vee tail.

The control surfaces are hinged to the stabilizers at the trailing edges. The stabilizing portion of this arrangement is called a stabilator, and the control portion is called the ruddervator. The ruddervators can be operated both up or both down at the same time. When used in this manner, the result is the same as with any other type of elevator. This action is controlled by the stick or control column.

The ruddervators can be made to move opposite each other by pushing the left or right rudder pedal (figure 2–26). If the right rudder padel is pushed, the right ruddervator moves down and the left ruddervator moves up. This produces turning moments to move the nose of the aircraft to the right.

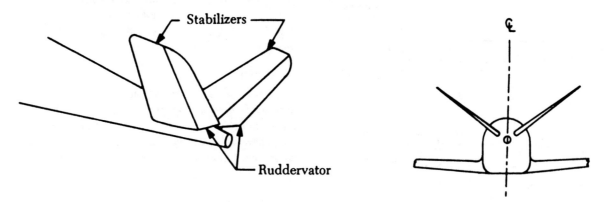

FIGURE 2–25. A butterfly or vee tail.

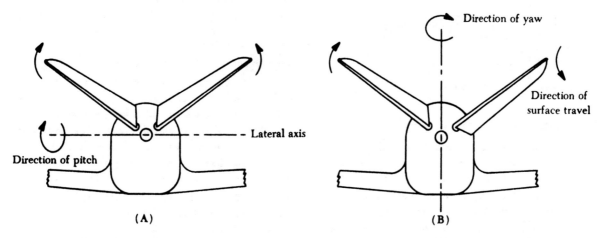

(A)　　　　　　　　　　(B)

FIGURE 2–26. Ruddervator action. (A) functioning as an elevator; (B) functioning as a rudder.

TABS

Even though an aircraft has inherent stability, it does not always tend to fly straight and level. The weight of the load and its distribution affect stability. Various speeds also affect its flight characteristics. If the fuel in one wing tank is used before that in the other wing tank, the aircraft tends to roll toward the full tank. All of these variations require constant exertion of pressure on the controls for correction. While climbing or gliding, it is necessary to apply pressure on the controls to keep the aircraft in the desired attitude.

To offset the forces that tend to unbalance an aircraft in flight, ailerons, elevators, and rudders are provided with auxiliary controls known as tabs. These are small, hinged control surfaces (figure 2–27) attached to the trailing edge of the primary control surfaces. Tabs can be moved up or down by means of a crank or moved electrically from the cockpit. These tabs can be used to balance the forces on the controls so that the aircraft flies straight and level, or may be set so that the aircraft maintains either a climbing or gliding attitude.

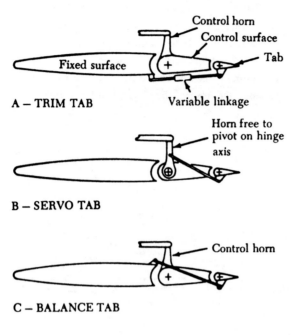

A – TRIM TAB

B – SERVO TAB

C – BALANCE TAB

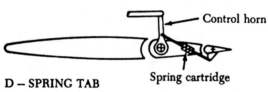

D – SPRING TAB

FIGURE 2–27. Flight control trim tab types.

Trim Tabs

Trim tabs trim the aircraft in flight. To trim means to correct any tendency of the aircraft to move toward an undesirable flight attitude. Trim tabs control the balance of an aircraft so that it maintains straight and level flight without pressure on the control column, control wheel, or rudder pedals. Figure 2–27A illustrates a trim tab. Note that the tab has a variable linkage which is adjustable from the cockpit. Movement of the tab in one direction causes a deflection of the control surface in the opposite direction. Most of the trim tabs installed on aircraft are mechanically operated from the cockpit through an individual cable system. However, some aircraft have trim tabs that are operated by an electrical actuator. Trim tabs are either controlled from the cockpit or adjusted on the ground before taking off. Trim tabs are installed on elevators, rudders, and ailerons.

Servo Tabs

Servo tabs (figure 2–27B) are very similar in operation and appearance to the trim tabs just discussed. Servo tabs, sometimes referred to as flight tabs, are used primarily on the large main control surfaces. They aid in moving the control surface and holding it in the desired position. Only the servo tab moves in response to movement of the cockpit control. (The servo tab horn is free to pivot to the main control surface hinge axis.) The force of the airflow on the servo tab then moves the primary control surface. With the use of a servo tab less force is needed to move the main control surface.

Balance Tabs

A balance tab is shown in figure 2–27C. The linkage is designed in such a way that when the main control surface is moved, the tab moves in the opposite direction. Thus, aerodynamic forces, acting on the tab, assist in moving the main control surface.

Spring Tabs

Spring tabs (figure 2–27D) are similar in appearance to trim tabs, but serve an entirely different purpose. Spring tabs are used for the same purpose as hydraulic actuators; that is, to aid in moving a primary control surface. There are various spring arrangements used in the linkage of the spring tab.

On some aircraft, a spring tab is hinged to the trailing edge of each aileron and is actuated by a spring-loaded push-pull rod assembly which is also linked to the aileron control linkage. The linkage is connected in such a way that movement of the aileron in one direction causes the spring tab to be deflected in the opposite direction. This provides a balanced condition, thus reducing the amount of force required to move the ailerons.

The deflection of the spring tabs is directly proportional to the aerodynamic load imposed upon the aileron; therefore, at low speeds the spring tab remains in a neutral position and the aileron is a direct manually controlled surface. At high speeds, however, where the aerodynamic load is great, the tab functions as an aid in moving the primary control surface.

To lessen the force required to operate the control surfaces they are usually balanced statically and aerodynamically. Aerodynamic balance is usually achieved by extending a portion of the con-

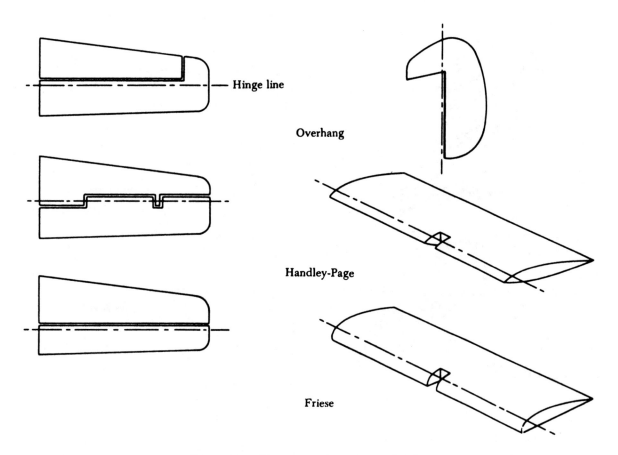

FIGURE 2-28. Three forms of aerodynamic balance.

trol surface ahead of the hinge line. This utilizes the airflow about the aircraft to aid in moving the surface. The various methods of achieving aerodynamic balance are shown in figure 2–28.

Static balance is accomplished by adding weight to the section forward of the hinge line until it weighs the same as the section aft of it. When repairing a control surface use care to prevent upsetting or disturbing the static balance. An unbalanced surface has a tendency to flutter as air passes over it.

High-Lift Devices

High-lift devices are used in combination with airfoils in order to reduce the takeoff or landing speed by changing the lift characteristics of an airfoil during the landing or takeoff phases. When these devices are no longer needed they are returned to a position within the wing to regain the normal characteristics of the airfoil.

Two high-lift devices commonly used on aircraft are shown in figure 2–29. One of these is known as a slot, and is used as a passageway through the leading edge of the wing. At high angles of attack the air flows through the slot and smooths out the airflow over the top surface of the wing. This enables the wing to pass beyond its normal stalling point without stalling. Greater lift is obtained with the wing operating at the higher angle of attack.

The other high-lift device is known as a flap. It is a hinged surface on the trailing edge of the wing. The flap is controlled from the cockpit, and when not in use fits smoothly into the lower surface of each wing. The use of flaps increases the camber of a wing and therefore the lift of the wing, making it possible for the speed of the aircraft to be decreased without stalling. This also permits a steeper gliding angle to be obtained as in the landing approach. Flaps are primarily used during takeoff and landing.

The types of flaps in use on aircraft include: (1) Plain, (2) split, (3) Fowler, and (4) slotted. The plain (figure 2–30) is simply hinged to the wing and forms a part of the wing surface when raised.

The split flap (figure 2–30) gets its name from the hinge at the bottom part of the wing near the trailing edge permitting it to be lowered from the

47

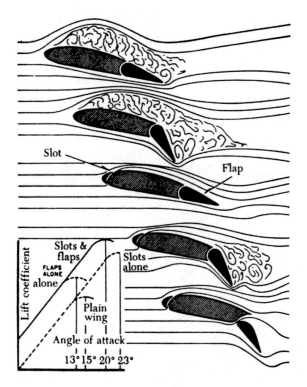

FIGURE 2-29. High-lift devices.

fixed top surface. The Fowler flap (figure 2–30) fits into the lower part of the wing so that it is flush with the surface. When the flap is operated, it slides backward on tracks and tilts downward at the same time. This increases wing camber, as do the other types of flaps. However, Fowler flaps also increase the wing area; thus, they provide added lift without unduly increasing drag.

The slotted flap (figure 2–30) is like the Fowler flap in operation, but in appearance it is similar to the plain flap. This flap is equipped with either tracks and rollers or hinges of a special design. During operation, the flap moves downward and rearward away from the position of the wing. The "slot" thus opened allows a flow of air over the upper surface of the flap. The effect is to streamline the airflow and to improve the efficiency of the flap.

BOUNDARY LAYER CONTROL DEVICES

The layer of air over the surface which is slower moving in relation to the rest of the slipstream is called the boundary layer. The initial airflow on a smooth surface (figure 2–31) gives evidence of a very thin boundary layer with the flow occurring in smooth laminations of air sliding smoothly over one another. Therefore, the term for this type of flow is the laminar boundary layer.

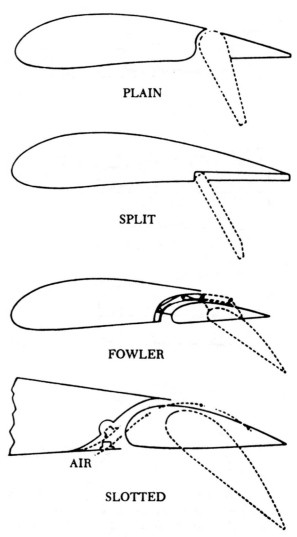

FIGURE 2-30. Types of wing flaps.

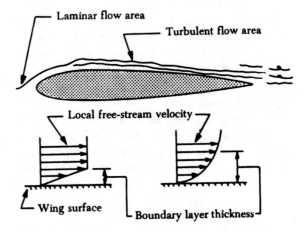

FIGURE 2-31. Boundary layer characteristics.

As the flow continues back from the leading edge, friction forces in the boundary layer continue to

48

dissipate the energy of the airstream, slowing it down. The laminar boundary layer increases in thickness with increased distance from the wing leading edge. Some distance back from the leading edge, the laminar flow begins an oscillatory disturbance which is unstable. A waviness occurs in the laminar boundary layer which ultimately grows larger and more severe and destroys the smooth laminar flow. Thus, a transition takes place in which the laminar boundary layer decays into a turbulent boundary layer. The same sort of transition can be noticed in the smoke from a cigarette in still air. At first, the smoke ribbon is smooth and laminar, then develops a definite waviness, and decays into a random turbulent smoke pattern.

Boundary layer control devices are additional means of increasing the maximum lift coefficient of a section. The thin layer of air adjacent to the surface of an airfoil shows reduced local velocities from the effect of skin friction. At high angles of attack, the boundary layer on the upper surface tends to stagnate (come to a stop). When this happens, the airflow separates from the surface and stall occurs.

Boundary layer control devices for high-lift applications feature various devices to maintain high velocity in the boundary layer and delay separation of the airflow. Control of the boundary layer's kinetic energy can be accomplished using slats and the application of suction to draw off the stagnant air and replace it with high-velocity air from outside the boundary layer.

Slats (figure 2–32) are movable control surfaces attached to the leading edge of the wing. When the slat is closed, it forms the leading edge of the wing. When in the open position (extended forward), a slot is created between the slat and the wing leading edge. Thus, high-energy air is introduced into the boundary layer over the top of the wing. This is known as "boundary layer control." At low airspeeds this improves handling characteristics, allowing the aircraft to be controlled laterally at airspeeds below the otherwise normal landing speed.

Controlling boundary layer air by surface suction allows the wing to operate at higher angles of attack. The effect on lift characteristics is similar to that of a slot, because the slot is essentially a boundary layer control device ducting high-energy air to the upper surface.

Boundary layer control can also be accomplished by directing high-pressure engine bleed air through a narrow orifice located just forward of the wing

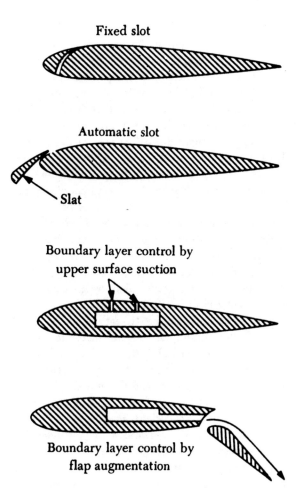

Fixed slot

Automatic slot

Slat

Boundary layer control by upper surface suction

Boundary layer control by flap augmentation

FIGURE 2–32. Methods of controlling boundary layer air.

flap leading edge. This directs a laminar flow (air in layers) of air over the wing and flaps when the flaps have opened sufficiently to expose the orifice. The high-temperature, high-velocity laminar air passing over the wing and flaps delays flow separation (the airstream over an airfoil no longer follows the contour of the airfoil), hence reduces turbulence and drag (see figure 2–29). This results in a lower stall speed and allows slower landing speeds.

FORCES ACTING ON A HELICOPTER

One of the differences between a helicopter and a fixed-wing aircraft is the main source of lift. The fixed-wing aircraft derives its lift from a fixed airfoil surface while the helicopter derives lift from a rotating airfoil called the rotor. Aircraft are classified as either fixed-wing or rotating wing. The word helicopter comes from a Greek word meaning "helical wing" or "rotating wing."

During any kind of horizontal or vertical flight, there are four forces acting on the helicopter—lift, thrust, weight, and drag. Lift is the force required to support the weight of the helicopter. Thrust is the force required to overcome the drag on the fuselage and other helicopter components.

During hovering flight in a no-wind condition, the tip-path plane is horizontal, that is, parallel to the ground. Lift and thrust act straight up; weight and drag act straight down. The sum of the lift and thrust forces must equal the sum of the weight and drag forces in order for the helicopter to hover.

During vertical flight in a no-wind condition, the lift and thrust forces both act vertically upward. Weight and drag both act vertically downward. When lift and thrust equal weight and drag, the helicopter hovers; if lift and thrust are less than weight and drag, the helicopter descends vertically; if lift and thrust are greater than weight and drag, the helicopter rises vertically.

For forward flight, the tip-path plane is tilted forward, thus tilting the total lift-thrust force forward from the vertical. This resultant lift-thrust force can be resolved into two components—lift acting vertically upward, and thrust acting horizontally in the direction of flight. In addition to lift and thrust, there are weight, the downward acting force, and drag, the rearward acting or retarding force of inertia and wind resistance.

In straight-and-level, unaccelerated forward flight, lift equals weight and thrust equals drag. (Straight-and-level flight is flight with a constant heading and at a constant altitude.) If lift exceeds weight, the helicopter climbs; if the lift is less than weight, the helicopter descends. If thrust exceeds drag, the helicopter speeds up; if thrust is less than drag, it slows down.

In sideward flight, the tip-path plane is tilted sideward in the direction that flight is desired thus tilting the total lift-thrust vector sideward. In this case, the vertical or lift component is still straight up, weight straight down, but the horizontal or thrust component now acts sideward with drag acting to the opposite side.

For rearward flight, the tip-path plane is tilted rearward tilting the lift-thrust vector rearward. The thrust component is rearward and drag forward, just the opposite to forward flight. The lift component is straight up and weight straight down.

Torque

Newton's third law of motion states, "To every action there is an equal and opposite reaction." As the main rotor of a helicopter turns in one direc-

tion, the fuselage tends to rotate in the opposite direction. This tendency for the fuselage to rotate is called torque. Since torque effect on the fuselage is a direct result of engine power supplied to the main rotor, any change in engine power brings about a corresponding change in torque effect. The greater the engine power, the greater the torque effect. Since there is no engine power being supplied to the main rotor during autorotation, there is no torque reaction during autorotation.

The force that compensates for torque and provides for directional control can be produced by means of an auxiliary rotor located on the end of the tail boom. This auxiliary rotor, generally referred to as a tail rotor, or anti-torque rotor, produces thrust in the direction opposite to torque reaction developed by the main rotor (figure 2–33). Foot pedals in the cockpit permit the pilot to increase or decrease tail-rotor thrust, as needed, to neutralize torque effect.

Other methods of compensating for torque and providing directional control are illustrated in figure 2–33.

The spinning main rotor of a helicopter acts like a gyroscope. As such, it has the properties of gyroscopic action, one of which is precession. Gyroscopic precession is the resultant action or deflection of a spinning object when a force is applied to this object. This action occurs approximately 90° in the direction of rotation from the point where the force is applied (figure 2–34). Through the use of this principle, the tip-path plane of the main rotor may be tilted from the horizontal.

The movement of the cyclic pitch control in a two-bladed rotor system increases the angle of attack of one rotor blade with the result that a greater lifting force is applied at this point in the plane of rotation. This same control movement simultaneously decreases the angle of attack of the other blade a like amount thus decreasing the lifting force applied at this point in the plane of rotation. The blade with the increased angle of attack tends to rise; the blade with the decreased angle of attack tends to lower. However, because of the gyroscopic precession property, the blades do not rise or lower to maximum deflection until a point approximately 90° later in the plane of rotation.

As shown in figure 2–35, the retreating blade angle of attack is increased and the advancing blade angle of attack is decreased resulting in a tipping forward of the tip-path plane, since maxi-

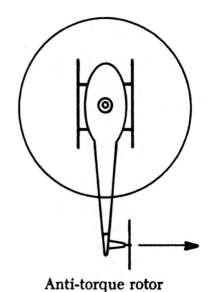

Anti-torque rotor

Vane in rotor slipstream

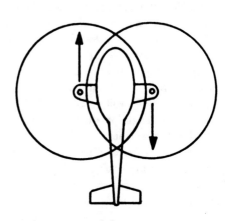

Differential tilt of rotor thrusts

Differential torque between two rotors

FIGURE 2–33. Methods for achieving directional control.

mum deflection takes place 90° later when the blades are at the rear and front respectively.

In a three-bladed rotor, the movement of the cyclic pitch control changes the angle of attack of each blade an appropriate amount so that the end result is the same, a tipping forward of the tip-path plane when the maximum change in angle of attack is made as each blade passes the same points at which the maximum increase and decrease are made for the two-bladed rotor as shown in figure 2–35. As each blade passes the 90° position on the left, the maximum increase in angle of attack occurs. As each blade passes the 90° position to the right, the maximum decrease in angle of attack occurs. Maximum deflection takes place 90° later, maximum up-

ward deflection at the rear and maximum downward deflection at the front, and the tip-path plane tips forward.

Dissymmetry of Lift

The area within the tip-path plane of the main rotor is known as the disk area or rotor disk. When hovering in still air, lift created by the rotor blades at all corresponding positions around the rotor disk is equal. Dissymmetry of lift is created by horizontal flight or by wind during hovering flight, and is the difference in lift that exists between the advancing blade half of the disk area and the retreating blade half.

At normal rotor operating r.p.m. and zero air-

51

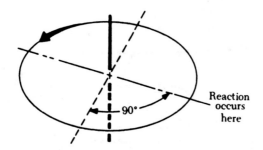

Force applied here

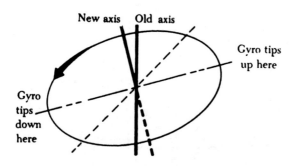

FIGURE 2-34. Gyroscopic precession principle.

speed, the rotating blade tip speed of most helicopter main rotors is approximately 400 m.p.h. When hovering in a no-wind condition, the speed of the relative wind at the blade tips and at any specific point along the blade is the same throughout the tip-path plane (figure 2-36). However, the speed is reduced as this point moves closer to the rotor hub as indicated in figure 2-36 by the two inner circles.

As the helicopter moves into forward flight, the relative wind moving over each rotor blade becomes a combination of the rotational speed of the rotor and the forward movement of the helicopter. As shown in figure 2-37, the advancing blade has the combined speed of the blade velocity plus the speed of the helicopter. On the opposite side, the retreating blade speed is the blade velocity less the speed of the helicopter.

It is apparent that the lift over the advancing blade half of the rotor disk will be greater than the lift over the retreating blade half during horizontal flight or when hovering in a wind.

Due to the greater lift of the advancing blade the helicopter would roll to the left unless something is done to equalize the lift of the blades on both sides of the helicopter.

Blade Flapping

In a three-bladed rotor system, the rotor blades are attached to the rotor hub by a horizontal hinge which permits the blades to move in a vertical plane, i.e., flap up or down, as they rotate (figure 2-38). In forward flight and assuming that the blade-pitch angle remains constant, the increased lift on the advancing blade will cause the blade to flap up decreasing the angle of attack because the relative wind will change from a horizontal direction to more of a downward direction. The decreased lift on the retreating blade will cause the blade to flap down increasing the angle of attack because the relative wind changes from a horizontal direction to more of an upward direction. The com-

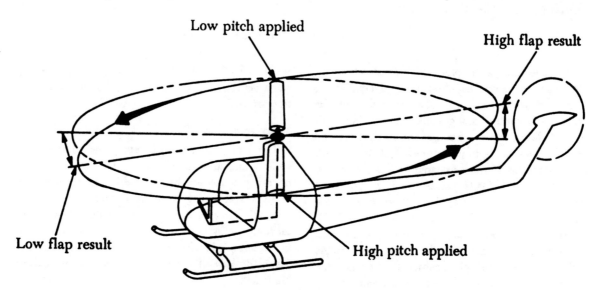

FIGURE 2-35. Rotor disk acts like a gyro.

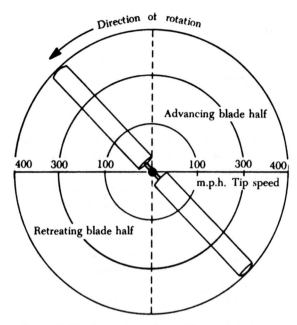

FIGURE 2-36. Comparison of rotor blade speeds for the **advancing blade and retreating blade during hover.**

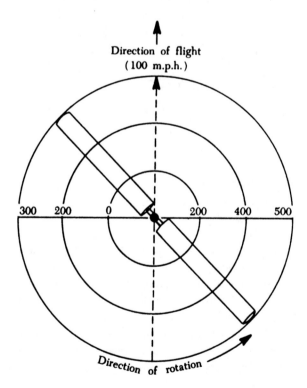

FIGURE 2-37. Comparison of rotor blade speeds for the **advancing and retreating blade during forward flight.**

bination of decreased angle of attack on the advancing blade and increased angle of attack on the retreating blade through blade flapping action tends

FIGURE 2-38. Blade flapping action (vertical plane).

to equalize the lift over the two halves of the rotor disk.

The amount that a blade flaps up is a compromise between centrifugal force, which tends to hold the blade straight out from the hub, and lift forces which tend to raise the blade on its flapping hinge. As the blades flap up they leave their normal tip-path plane momentarily. As a result, the tip of the blade which is flapping must travel a greater distance. Therefore, it has to travel at a greater speed for a fraction of a second, in order to keep up with the other blades.

The blade flapping action creates an unbalance condition with resulting vibration. To prevent this vibration, a drag hinge (figure 2-39) is incorporated which permits the blade to move back and forth in a horizontal plane.

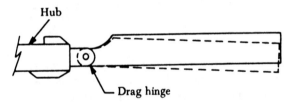

FIGURE 2-39. Action of drag hinge (horizontal plane).

With the blades free to move back and forth on the drag hinges, an unbalanced condition is created since the c.g. (center of gravity) will not remain fixed but moves around the mast. This c.g. movement causes excessive vibration. To dampen out the vibrations, hydraulic dampers limit the movement of the blades on the drag hinge. These dampers also tend to maintain the geometric relationship of the blades.

A main rotor which permits individual movement of the blades from the hub in both a vertical and horizontal plane is called an articulated rotor. The hinge points and direction of motion around each hinge are illustrated in figure 2-40.

In a two-bladed system, the blades flap as a unit. As the advancing blade flaps up due to the increased lift, the retreating blade flaps down due to the decreased lift. The change in angle of attack on

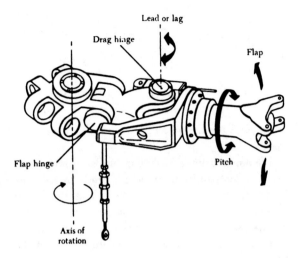

FIGURE 2-40. Articulated rotor head.

each blade brought about by this flapping action tends to equalize the lift over the two halves of the rotor disk.

The position of the cyclic pitch control in forward flight also causes a decrease in angle of attack on the advancing blade and an increase in angle of attack on the retreating blade. This together with blade flapping equalizes lift over the two halves of the rotor disk.

Coning

Coning (figure 2-41) is the upward bending of the blades caused by the combined forces of lift and centrifugal force. Before takeoff, the blades rotate in a plane nearly perpendicular to the rotor mast, since centrifugal force is the major force acting on them.

As a vertical takeoff is made, two major forces are acting at the same time—centrifugal force acting outward perpendicular to the rotor mast, and lift acting upward and parallel to the mast. The result of these two forces is that the blades assume

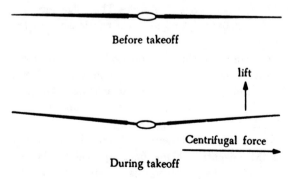

FIGURE 2-41. Blade coning.

a conical path instead of remaining in the plane perpendicular to the mast.

Coning results in blade bending in a semirigid rotor; in an articulated rotor, the blades assume an upward angle through movement about the flapping hinges.

Ground Effect

When a helicopter is in a hovering position close to the ground, the rotor blades will be displacing air downward through the disk faster than it can escape from beneath the helicopter. This builds up a cushion of dense air between the ground and the helicopter (figure 2-42). This cushion of more dense air, referred to as ground effect, aids in supporting the helicopter while hovering. It is usually effective to a height of approximately one-half the rotor disk diameter. At approximately 3 to 5 m.p.h. groundspeed, the helicopter will leave its ground cushion.

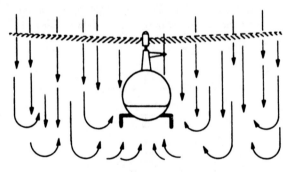

FIGURE 2-42. Ground effect.

Autorotation

Autorotation is the term used for the flight condition during which no engine power is supplied and the main rotor is driven only by the action of the relative wind. The helicopter transmission or power train is designed so that the engine, when it stops, is automatically disengaged from the main rotor system to allow the main rotor to rotate freely in its original direction.

When engine power is being supplied to the main rotor, the flow of air is downward through the rotor. When engine power is not being supplied to the main rotor, that is, when the helicopter is in autorotation, the flow of air is upward through the rotor. It is this upward flow of air that causes the rotor to continue turning after engine failure.

The portion of the rotor blade that produces the forces that cause the rotor to turn when the engine is no longer supplying power to the rotor, is that portion between approximately 25% and 70% of the radius outward from the center. This portion is often referred to as the "autorotative or driving region." Aerodynamic forces along this portion of the blade tend to speed up the blade rotation.

The inner 25% of the rotor blade, referred to as the "stall region," operates above its maximum angle of attack (stall angle), thereby contributing little lift but considerable drag which tends to slow the blade rotation.

The outer 30% of the rotor blade is known as the "propeller or driven region." Aerodynamic forces here result in a small drag force which tends to slow the tip portion of the blade.

The aerodynamic regions as described above are for vertical autorotations. During forward flight autorotations, these regions are displaced across the rotor disk to the left.

Rotor r.p.m. stabilizes when the autorotative forces (thrust) of the "driving region" and the autorotative forces (drag) of the "driven region" and the "stall region" are equal.

Forward speed during autorotative descent permits a pilot to incline the rotor disk rearward, thus causing a flare. The additional induced lift created by the greater volume of air momentarily checks forward speed as well as descent. The greater volume of air acting on the rotor disk will normally increase rotor r.p.m. during the flare. As the forward speed and descent rate near zero, the upward flow of air has practically ceased and rotor r.p.m. again decreases; the helicopter settles at a slightly increased rate but with reduced forward speed. The flare enables the pilot to make an emergency landing on a definite spot with little or no landing roll or skid.

HELICOPTER AXES OF FLIGHT

As a helicopter maneuvers through the air, its attitude in relation to the ground changes. These changes are described with reference to three axes (figure 2–43) of flight: (1) Vertical, (2) longitudinal, and (3) lateral.

Movement about the vertical axis produces yaw, a nose swing (or change in direction) to the right or left. This is controlled by the directional-control pedals. The various methods of achieving directional control were discussed earlier in this section.

Movement about the longitudinal axis is called roll. This movement is effected by moving the cyclic pitch control to the right or left. The cyclic pitch control is similar to the control stick of a conven-

tional aircraft. It acts through a mechanical linkage (figure 2–44) to change the pitch of each main-rotor blade during a cycle of rotation.

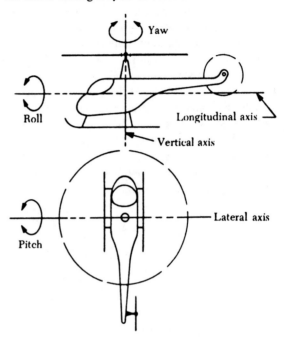

FIGURE 2–43. Axes of flight.

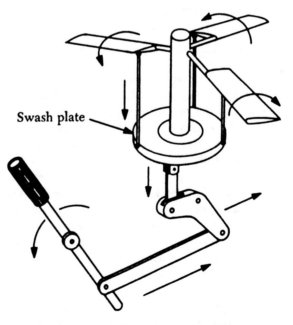

FIGURE 2–44. Cyclic pitch control mechanism.

The rapidly rotating rotor blades create a disk area that can be tilted in any direction with respect to the supporting rotor mast. Horizontal movement is controlled by changing the direction of tilt of the main rotor to produce a force in the desired direction.

Movement about the lateral axis produces a nose-up or nosedown attitude. This movement is effected by moving the cyclic pitch control fore and aft.

The collective pitch control (figure 2–45) varies the lift of the main rotor by increasing or decreasing the pitch of all blades at the same time. Raising the collective pitch control increases the pitch of the blades, thereby increasing the lift. Lowering the control decreases the pitch of the blades, causing a loss of lift. Collective pitch control is also used in coordination with cyclic pitch control to regulate the airspeed of the helicopter.

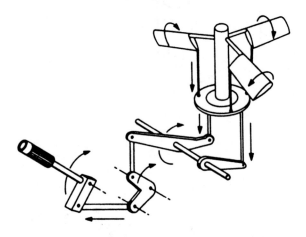

FIGURE 2–45. Collective pitch control mechanism.

Many factors determine the amount of lift available in helicopter operation. Generally speaking, the pilot has control of two of these. One is the pitch angle of the rotor blades; the other is the power delivered to the rotor, represented by r.p.m. and manifold pressure. By controlling the pitch angle of the rotor blades, the pilot can establish the vertical flight of the helicopter. By manipulating the throttle control, a constant engine speed can be maintained regardless of the increase or decrease in blade pitch. The throttle is mounted on the collective pitch grip and is operated by rotating the grip. The throttle is synchronized with the main-rotor pitch control in such a manner that an increase of pitch increases power and a decrease in pitch decreases power. A complete control system of a conventional helicopter is shown in figure 2–46.

HIGH-SPEED AERODYNAMICS

Developments in aircraft and powerplants are yielding high-performance transports with capabilities for very high speed flight. Many significant differences arise in the study of high-speed aerodynamics when compared with low-speed aerodynamics. It is quite necessary, therefore, that persons associated with commercial aviation be familiar with the nature of high-speed airflow and the peculiarities of high-performance airplanes.

General Concepts of Supersonic Flow Patterns

At low flight speeds, air experiences small changes in pressure which cause negligible variations in density, greatly simplifying the study of low-speed aerodynamics. The flow is called incompressible since the air undergoes small changes in pressure without significant changes in density. At high flight speeds, however, the pressure changes that take place are quite large and significant changes in air density occur. The study of airflow at high speeds must account for these changes in air density and must consider that the air is compressible, or that there are compressibility effects.

The speed of sound is very important in the study of high-speed airflow. The speed of sound varies with the ambient temperature. At sea level, on a standard day, the speed of sound is about 661.7 knots (760 m.p.h.).

As a wing moves through the air, local velocity changes occur which create pressure disturbances in the airflow around the wing. These pressure disturbances are transmitted through the air at the speed of sound. If the wing is traveling at low speed, the pressure disturbances are transmitted and extend indefinitely in all directions. Evidence of these pressure disturbances is seen in the typical subsonic flow pattern illustrated in figure 2–47 where upwash and flow direction change well ahead of the wing leading edge.

If the wing is traveling above the speed of sound, the airflow ahead of the wing is not influenced by the pressure field of the wing, since pressure disturbances cannot be propagated faster than the speed of sound. As the flight speed nears the speed of sound, a compression wave forms at the leading edge and all changes in velocity and pressure take place quite sharply and suddenly. The airflow ahead of the wing is not influenced until the air molecules are suddenly forced out of the way by the wing. Evidence of this phenomenon is seen in the typical supersonic flow pattern shown in figure 2–48.

Compressibility effects depend not on airspeed, but rather on the relationship of airspeed to the

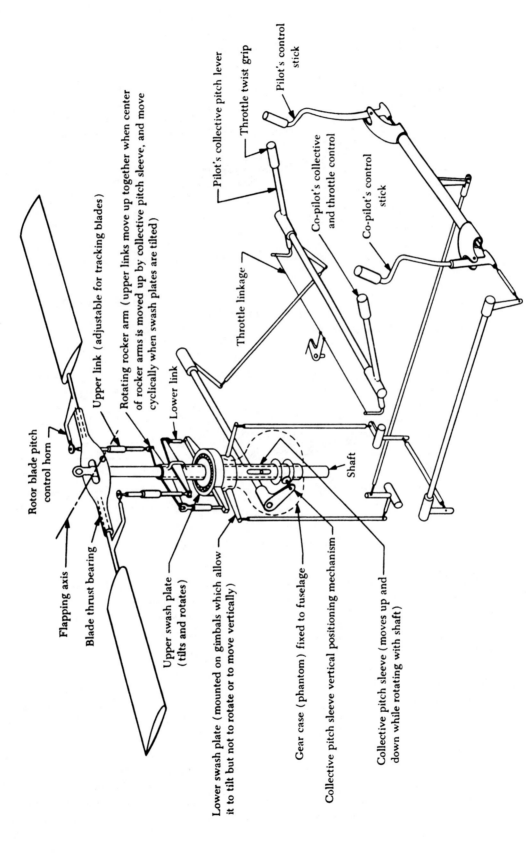

Rotor blade pitch
control horn

Flapping axis

Blade thrust bearing

Upper swash plate
(tilts and rotates)

Lower swash plate (mounted on gimbals which allow
it to tilt but not to rotate or to move vertically)

Gear case (phantom) fixed to fuselage

Collective pitch sleeve vertical positioning mechanism

Collective pitch sleeve (moves up and
down while rotating with shaft)

Upper link (adjustable for tracking blades)

Rotating rocker arm (upper links move up together when center
of rocker arms is moved up by collective pitch sleeve, and move
cyclically when swash plates are tilted)

Lower link

Pilot's collective pitch lever

Throttle twist grip

Pilot's control
stick

Co-pilot's collective
and throttle control

Co-pilot's control
stick

Throttle linkage

Shaft

FIGURE 2–46. Control system of conventional helicopter.

57

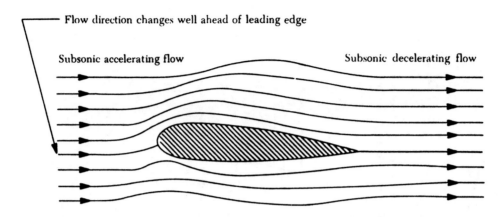

FIGURE 2-47. Typical subsonic flow pattern, subsonic wing.

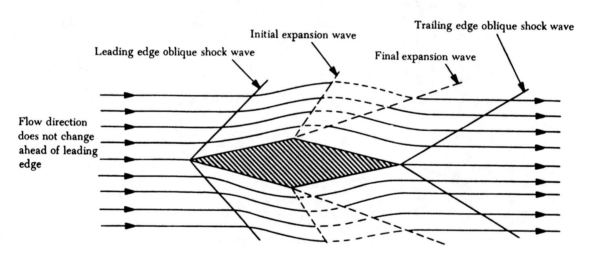

FIGURE 2-48. Typical supersonic flow pattern, supersonic wing.

speed of sound. This relationship is called Mach number, and is the ratio of true airspeed to the speed of sound at a particular altitude.

Compressibility effects are not limited to flight speeds at and above the speed of sound. Since any airplane is made up of aerodynamic shapes, air accelerates and decelerates around these shapes and attains local speeds above the flight speed. Thus, an aircraft can experience compressibility effects at flight speeds well below the speed of sound. Since it is possible to have both subsonic and supersonic flows on the airplane at the same time, it is best to define certain regimes of flight. These approximate regimes are defined as follows:

(1) Subsonic—flight Mach numbers below 0.75.

(2) Transonic—flight Mach numbers from 0.75 to 1.20.

(3) Supersonic—flight Mach numbers from 1.20 to 5.00.

(4) Hypersonic—flight Mach numbers above 5.00.

While the flight Mach numbers used to define these regimes are approximate, it is important to appreciate the types of flow existing in each area. In the subsonic regime, subsonic airflow exists on all parts of the aircraft. In the transonic regime, the flow over the aircraft components is partly subsonic and partly supersonic. In the supersonic and hypersonic regimes, supersonic flow exists over all parts of the aircraft. Of course, in supersonic and hypersonic flight some portions of the boundary layer are

58

subsonic, but the predominating flow is still supersonic.

Difference Between Subsonic and Supersonic Flow

In a subsonic flow every molecule is affected more or less by the motion of every other molecule in the whole field of flow. At supersonic speeds, an air molecule can influence only that part of the flow contained in the Mach cone formed behind that molecule.

The peculiar differences between subsonic flow and supersonic flow can best be seen by considering airflow in a closed contracting/expanding tube, as depicted in figure 2–48.

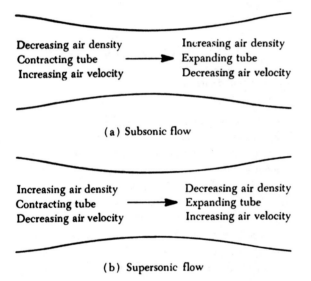

(a) Subsonic flow

(b) Supersonic flow

FIGURE 2–49. Comparison of subsonic and supersonic airflow through a closed tube.

Unlike subsonic flow, a supersonic airstream accelerates along an expanding tube, causing the air density to decrease rapidly to compensate for the combined effects of increased speed and increased cross sectional area.

Unlike subsonic flow, a supersonic airstream decelerates along a contracting tube, causing the air density to increase rapidly to compensate for the combined effects of decreased speed and decreased cross sectional area.

In order to clarify these fundamental points, figure 2–50 lists the nature of the two types of tubes. An understanding of figures 2–49 and 2–50 is essential if one is to grasp the fundamentals of supersonic flow.

	Contracting tube	Expanding tube
Subsonic flow	Accelerates and rarefies slightly	Decelerates and compresses slightly
Supersonic flow	Decelerates and compresses greatly	Accelerates and rarefies greatly

FIGURE 2–50. High-speed flows.

TYPICAL SUPERSONIC FLOW PATTERNS

With supersonic flow, all changes in velocity, pressure, temperature, density, and flow direction take place suddenly and over a short distance. The areas of flow change are distinct and the phenomena causing the flow change are called wave formations. All compression waves occur abruptly and are wasteful of energy. Compression waves are more familiarly known as shock waves. Expansion waves result in smoother flow transition and are not wasteful of energy like shock waves. Three types of waves can take place in supersonic flow: (1) The oblique (inclined angle) shock wave (compression), (2) the normal (right angle) shock wave (compression), and (3) the expansion wave (no shock). The nature of the wave depends on the Mach number, the shape of the object causing the flow change, and the direction of flow.

A supersonic airstream passing through the oblique shock wave experiences these changes:

(1) The airstream is slowed down. Both the velocity and the Mach number behind the wave are reduced, but the flow is still supersonic.

(2) The flow direction is changed so that the airstream runs parallel to the new surface.

(3) The static pressure behind the wave is increased.

(4) The static temperature behind the wave is increased (and hence the local speed of sound is increased).

(5) The density of the airstream behind the wave is increased.

(6) Some of the available energy of the airstream (indicated by the sum of dynamic and static pressure) is dissipated by conversion into unavailable heat energy. Hence, the shock wave is wasteful of energy.

The Normal Shock Wave

If a blunt-nosed object is placed in a supersonic airstream, the shock wave which is formed is detached from the leading edge. The detached wave

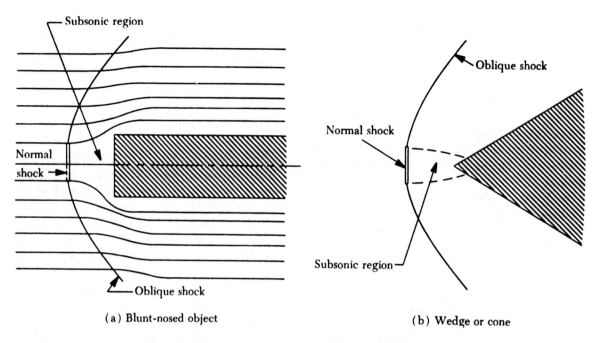

Subsonic region

Oblique shock

Normal shock

Normal shock

Subsonic region

Oblique shock

(a) Blunt-nosed object

(b) Wedge or cone

FIGURE 2–51. Normal shock-wave formation with a detached wave.

also occurs when a wedge or cone half angle exceeds some critical value. Figure 2–51 shows the formation of a normal shock wave in the above two cases. Whenever a shock wave forms perpendicular to the free stream flow, the shock wave is termed a normal (right angle) shock wave and the flow immediately behind the wave is subsonic. No matter how high the free stream Mach number may be, the flow directly behind a normal shock is always subsonic. In fact, the higher the supersonic free stream Mach number (M) is in front of the normal shock wave, the lower the subsonic Mach number is aft of the wave. For example, if M_1 is 1.5, M_2 is 0.7; while if M_1 is 2.6, M_2 is only 0.5. A normal shock wave forms immediately in front of any relatively blunt object in a supersonic airstream, slowing the airstream to subsonic so that the airstream may feel the presence of the blunt object and thus flow around it. Once past the blunt nose, the airstream may remain subsonic or it may accelerate back to supersonic, depending on the shape of the nose and the Mach number of the free stream.

A normal wave may also be formed when there is no object in the supersonic airstream. It so happens that whenever a supersonic airstream is slowed to subsonic without a change in direction, a normal shock wave forms as the boundary between the supersonic and subsonic regions. This is why airplanes encounter compressibility effects before the

flight speed is sonic. Figure 2–52 illustrates the manner in which an airfoil at a high subsonic speed has local flow velocities which are supersonic. As the local supersonic flow moves aft, a normal shock wave forms so that the flow may return to subsonic and rejoin the subsonic free stream at the trailing edge without discontinuity. The transition of flow from subsonic to supersonic is smooth and is not accompanied by shock waves if the transition is made gradually with a smooth surface. The transition of flow from supersonic to subsonic without direction change always forms a normal shock wave.

A supersonic airstream passing through a normal shock wave experiences these changes:

(1) The airstream is slowed to subsonic. The local Mach number behind the wave is approximately equal to the reciprocal of the Mach number ahead of the wave. For example, if the Mach number ahead of the wave is 1.25, the Mach number of the flow behind the wave is about 0.8 (more ex-0.81264).

airflow direction immediately behind ave is unchanged.

(3) The static pressure behind the wave is greatly increased.

(4) The static temperature behind the wave is

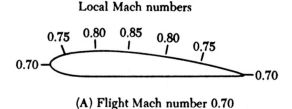

Local Mach numbers

(A) Flight Mach number 0.70

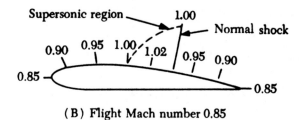

(B) Flight Mach number 0.85

FIGURE 2–52. Normal shock-wave formation on an airfoil in a subsonic airstream.

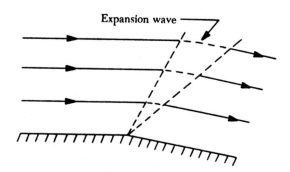

(a) Supersonic flow around a sharp corner

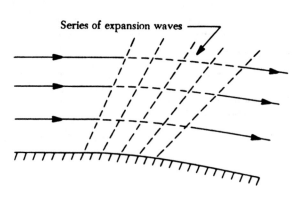

(b) Supersonic flow around a rounded corner

FIGURE 2–53. Expansion wave formation.

greatly increased (and hence the local speed of sound is increased).

(5) The density of the airstream behind the wave is greatly increased.

(6) The available energy of the airstream (indicated by the sum of dynamic and static pressure) is greatly reduced. The normal shock wave is very wasteful of energy.

The Expansion Wave

If a supersonic airstream is turned away from the preceding flow, an expansion wave is formed. The flow around a corner shown in figure 2–53 does not cause sharp, sudden changes in the airflow except at the corner itself and thus is not actually a shock wave. A supersonic airstream passing through an expansion wave experiences these changes:

(1) The supersonic airstream is accelerated. The velocity and Mach number behind the wave are greater.

(2) The flow direction is changed so that the airstream runs parallel to the new surface, provided separation does not occur.

(3) The static pressure behind the wave is decreased.

(4) The static temperature behind the wave is decreased (and hence the local speed of sound is decreased).

(5) The density of the airstream behind the wave is decreased.

(6) Since the flow changes in a rather gradual

manner, there is no shock and no loss of energy in the airstream. The expansion wave does not dissipate airstream energy.

A summary of the characteristics of the three principal wave forms encountered with supersonic flow is shown in figure 2–54.

Figure 2–55 shows the wave pattern for a conventional blunt-nosed subsonic airfoil in a supersonic stream. When the nose is blunt, the wave must detach and become a normal shock wave immediately ahead of the leading edge. Since the flow just behind a normal shock wave is always subsonic, the airfoil's leading edge is in a subsonic region of very high static pressure, static temperature, and density.

In supersonic flight, the zero lift of an airfoil of some finite thickness includes a wave drag. Wave drag is separate and distinct from drag due to lift. The thickness of the airfoil has an extremely powerful effect on the wave drag. The wave drag varies as the square of the thickness ratio (maximum thickness divided by the chord). For example, if the thickness is cut by one-half, the wave drag is cut by three-fourths. The leading edges of supersonic

61

Type of wave	Flow direction change	Effect on velocity and Mach number	Effect on static pressure, static temperature and density	Effect on available energy
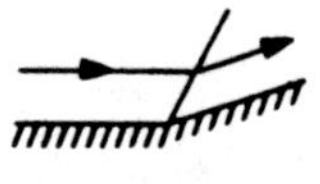 Oblique shock wave	Flow into a corner	Decreased, but still supersonic	Increase	Decrease
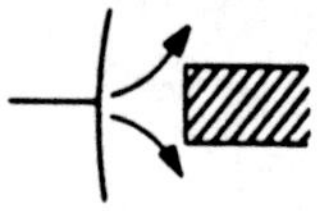 Normal shock wave	No change	Decreased to subsonic	Great increase	Great decrease
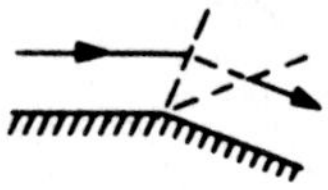 Expansion wave	Flow around a corner	Increased to higher supersonic	Decrease	No change (no shock)

FIGURE 2-54. Supersonic wave characteristics.

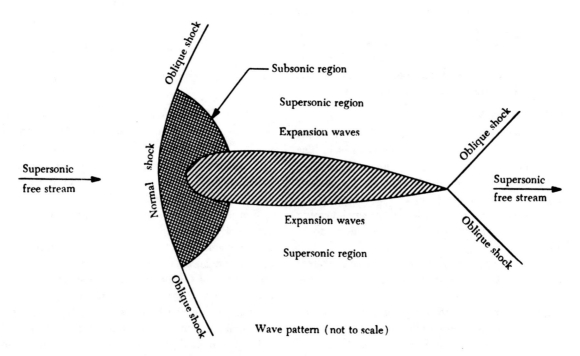

FIGURE 2-55. The conventional subsonic airfoil in supersonic flow.

shapes must be sharp. If they are not, the wave formed near the leading edge is a strong detached normal shock wave.

Once the flow over the airfoil is supersonic, the aerodynamic center of the surface is located ap-proximately at the 50% chord position. This con-trasts with the subsonic location of the aerodynamic center, which is near the 25% chord position.

During supersonic flow all changes in velocity, Mach number, static pressure, static temperature, density, and flow direction take place quite suddenly through the various wave forms. The shape of the object, the Mach number, and the required flow direction change dictate the type and strength of the wave formed.

Any object in subsonic flight which has some finite thickness or is producing lift must have local velocities on the surface which are greater than the free stream velocity. Hence, compressibility effects can be expected to occur at flight speeds which are less than the speed of sound. The transonic regime of flight provides the opportunity for mixed subsonic and supersonic local velocities and accounts for the first significant effects of compressibility.

As the flight speed approaches the speed of sound, the areas of supersonic flow enlarge and the shock waves move nearer the trailing edge. The boundary layer may remain separated or may re-attach, depending much upon the airfoil shape and angle of attack. When the flight speed exceeds the speed of sound, a bow wave suddenly appears in front of the leading edge with a subsonic region behind the wave. The normal shock waves move to the trailing edge. If the flight speed is increased to some higher supersonic value, the bow wave moves closer to the leading edge and inclines more downstream, and the trailing edge normal shock waves become oblique shock waves.

Of course, all components of the aircraft are affected by compressibility in a manner somewhat similar to that of the basic airfoil (the empennage, fuselage, nacelles, and so forth).

Since most of the difficulties of transonic flight are associated with shock-wave-induced flow separation, any means of delaying or lessening the shock-induced separation improves the aerodynamic characteristics. An aircraft configuration can make use of thin surfaces of low aspect ratio with sweepback to delay and reduce the magnitude of transonic force divergence. In addition, various methods of boundary layer control, high-lift devices, vortex generators, and so forth, may be applied to improve transonic characteristics. For example, the mounting of vortex generators on a surface can produce higher local surface velocities and increase the kinetic energy of the boundary layer. Thus, a more severe pressure gradient (stronger shock wave) would be necessary to produce the unwanted airflow separation.

A vortex generator is a complementary pair of small, low aspect ratio (short span in relation to chord) airfoils mounted at opposite angles of attack

to each other and perpendicular to the aerodynamic surface they serve. Figure 2–56 shows the airfoils and the airflow characteristics of a vortex generator. Like any airfoil, those of the generator develop lift. In addition, like any airfoil of especially low aspect ratio, the airfoils of the generator also develop very strong tip vortices. These tip vortices cause air to flow outward and inward in circular paths around the ends of the airfoils. The vortices generated have the effect of drawing high-energy air from outside the boundary layer into the slower moving air close to the skin. The strength of the vortices is proportional to the lift developed by the airfoils of the generator.

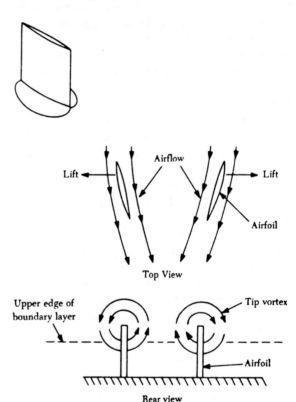

FIGURE 2–56. Wing vortex generator.

Vortex generators serve two distinctly different purposes, depending on the aerodynamic surface upon which they are mounted. Rows of vortex generators located on the upper surface of the wing just upstream of the ailerons delay the onset of drag divergence at high speeds and also aid in maintaining aileron effectiveness at high speeds. In contrast, rows of vortex generators mounted on both sides of the vertical fin just upstream of the rudder prevent flow separation over the rudder during extreme angles of yaw which are attained only when rudder application is delayed after an engine loss at very low speeds. In addition, rows of vortex generators placed on the underside (and occasionally on the upper surface) of the horizontal stabilizer just upstream of the elevators prevent flow separation over the elevators at very low speeds.

In summary, vortex generators on wing surfaces improve high-speed characteristics, while vortex generators on tail surfaces, in general, improve low-speed characteristics.

Control Surfaces

The control surfaces used on aircraft operating at transonic and supersonic flight speeds involves some important considerations. Trailing edge control surfaces can be affected adversely by the shock waves formed in flight above the control surface critical Mach number. If the airflow is separated by the shock wave, the resulting buffet of the control surface can be very objectionable. Installation of vortex generators can reduce buffet caused by shock-induced flow separation. In addition to the buffet of the surface, the change in the pressure distribution due to separation and shock-wave location can create very large changes in control surface hinge moments. Such large changes in hinge moments produce undesirable control forces which may require the use of an irreversible control system. An irreversible control system employs powerful hydraulic or electric actuators to move the control surfaces, hence the airloads developed on the surfaces cannot be felt by the pilot. Suitable feedback must be synthesized by bungees, "q" springs, bobweights, and so forth.

AERODYNAMIC HEATING

When air flows over any aerodynamic surface, certain reductions in velocity take place which produce corresponding increases in temperature. The greatest reduction in velocity and increase in temperature occur at the various stagnation points on the aircraft. Of course, smaller changes occur at other points on the aircraft, but these lower temperatures can be related to the ram temperature rise at the stagnation point. While subsonic flight does not produce temperatures of any real concern, supersonic flight can create temperatures high enough to be of major importance to the airframe, fuel system, and powerplant.

Higher temperatures produce definite reductions in the strength of aluminum alloys and require the use of titanium alloys and stainless steels. Continued exposure at elevated temperatures further reduces strength and magnifies the problems of creep failure and structural stiffness.

The effect of aerodynamic heating on the fuel system must be considered in the design of a supersonic airplane. If the fuel temperature is raised to the spontaneous ignition temperature, the fuel vapors will burn in the presence of air without the need of an initial spark or flame.

Turbojet engine performance is adversely affected by high compressor inlet air temperature. The thrust output of the turbojet, obviously, is some function of the fuel flow. But the maximum allowable fuel flow depends on the maximum permissible turbine operating temperature. If the air entering the engine is already hot, less fuel can be added in order to avoid exceeding turbine temperature limits.

FLIGHT CONTROL SYSTEMS

Three types of control systems commonly used are: (1) The cable, (2) push-pull, and (3) the torque tube system. The cable system is the most widely used because deflections of the structure to which it is attached do not affect its operation. Many aircraft incorporate control systems that are combinations of all three types.

Flight Control System Hardware, Mechanical Linkage, and Mechanisms

The systems which operate the control surfaces, tabs, and flaps include flight control system hardware, linkage, and mechanisms. These items connect the control surfaces to the cockpit controls. Included in these systems are cable assemblies, cable guides, linkage, adjustable stops, control surface snubber or locking devices, surface control booster units, actuators operated by electric motors, and actuators operated by hydraulic motors.

Cable Assembly

The conventional cable assembly consists of flexible cable, terminals (end fittings) for attaching to other units, and turnbuckles. Information concerning conventional cable construction and end fittings is contained in Chapter 6 of the Airframe and Powerplant Mechanics General Handbook, AC 65–9A.

At each regular inspection period, cables should be inspected for broken wires by passing a cloth along their length and observing points where the cloth snags. To thoroughly inspect the cable, move the surface control to its extreme travel limits. This will reveal the cable in pulley, fairlead, and drum areas. If the surface of the cable is corroded, relieve cable tension. Then carefully force the cable open by reverse twisting, and visually inspect the interior for corrosion. Corrosion on the interior strands of the cable indicates failure of the cable and requires replacement of the cable. If there is no internal corrosion, remove external corrosion with a coarse-weave rag or fiber brush. Never use metallic wools or solvents to clean flexible cable. Metallic wools imbed dissimilar metal particles, which cause further corrosion. Solvents remove the internal cable lubricant, which also results in further corrosion. After thoroughly cleaning the flexible cable, apply corrosion-preventive compound. This compound preserves and lubricates the cable.

Breakage of wires occurs most frequently where cables pass over pulleys and through fairleads. Typical breakage points are shown in figure 2–57. Control cables and wires should be replaced if worn, distorted, corroded, or otherwise damaged.

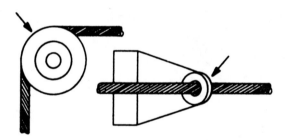

FIGURE 2–57. Typical breakage points.

Lockclad cable is used on some large aircraft for all long, straight runs. It consists of the conventional flexible steel cable with aluminum tubing swaged to it to lock the cable inside the tubing. Lockclad cable construction has certain advantages. Changes in tension due to temperature are less than with conventional cable. Furthermore, the amount of stretch at a given load is less than with conventional cable.

Lockclad cables should be replaced when the covering is worn through, exposing worn wire strands; is broken; or shows worn spots which cause the cable to bump when passing over fairlead rollers.

Turnbuckles

The turnbuckle is a device used in cable control systems to adjust cable tension. The turnbuckle bar-

rel is threaded with left-hand threads inside one end and right-hand threads inside the other. When adjusting cable tension, the cable terminals are screwed into either end of the barrel an equal distance by turning the barrel. After a turnbuckle is adjusted, it must be safetied. The methods of safetying turnbuckles is discussed in Chapter 6 of the Airframe and Powerplant Mechanics General Handbook, AC 65–9.

Cable Connectors

In addition to turnbuckles, cable connectors are used in some systems. These connectors enable a cable length to be quickly connected or disconnected from a system. Figure 2–58 illustrates one type of cable connector in use. This type is connected or disconnected by compressing the spring.

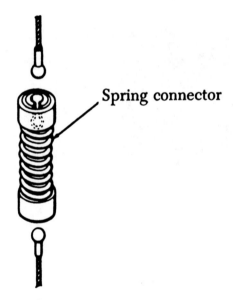

Spring connector

FIGURE 2–58. Spring type of cable connector.

HYDRAULIC OPERATED CONTROL SYSTEMS

As the airspeed of late model aircraft increased, actuation of controls in flight became more difficult. It soon became apparent that the pilot needed assistance to overcome the airflow resistance to control movement. Spring tabs which were operated by the conventional control system were moved so that the airflow over them actually moved the primary control surface. This was sufficient for the aircraft operating in the lowest of the high speed ranges (250–300 mph).

For high speeds a power assist (hydraulic) control system was designed.

Conventional cable or push pull rod systems are installed and are tied into a power transmission quadrant. With the system activated, the pilot's effort is used to open valves thereby directing hydraulic fluid to actuators, which are connected to the control surfaces by control rods. The actuators move the control surface to the desired flight condition. Reversing the input effort moves the control surface in the opposite direction.

Manual Control

The control system from the cockpit is connected by a rod across the power transmission quadrant to the control actuating system. During manual operation, the pilot's effort is transmitted from the control wheel through this direct linkage to the control surface. Those aircraft which do not have the manual reversion system may have as many as three sources of hydraulic power-primary, back-up and auxiliary. Any or all of the primary controls may be operated by these systems.

Gust Lock

A cam on the control quadrant shaft engages a spring-loaded roller for the purpose of centering and neutralizing the controls with the hydraulic system off (aircraft parked). Pressure is trapped in the actuators and since the controls are neutralized by the cam and roller, no movement of the control surfaces is permitted.

CABLE GUIDES

Cable guides (figure 2–59) consist primarily of fairleads, pressure seals, and pulleys.

A fairlead may be made from a nonmetallic material, such as phenolic or a metallic material such as soft aluminum. The fairlead completely encircles the cable where it passes through holes in bulkheads or other metal parts. Fairleads are used to guide cables

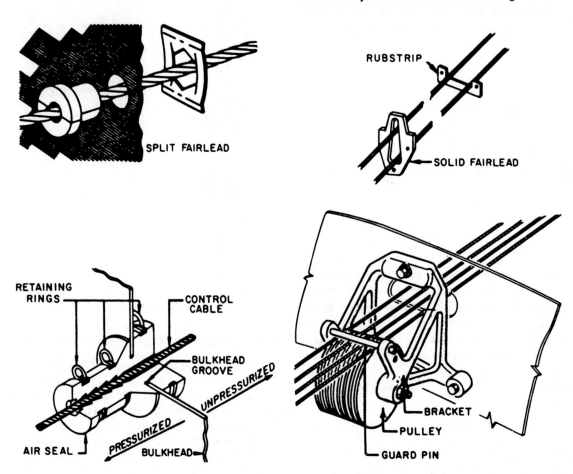

FIGURE 2-59. Cable guides.

in a straight line through or between structural members of the aircraft. Fairleads should never deflect the alignment of a cable more than 3° from a straight line.

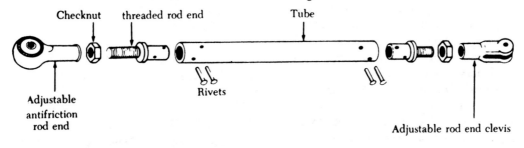

A   Control or push-pull rod

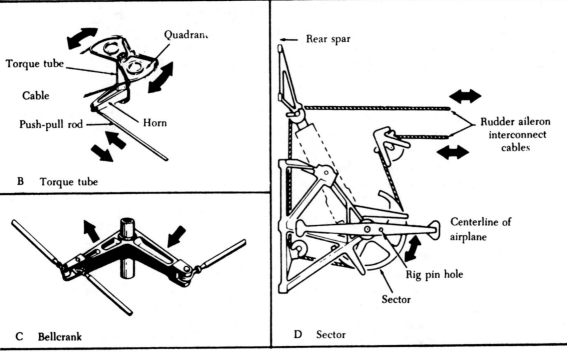

B   Torque tube

C   Bellcrank

D   Sector

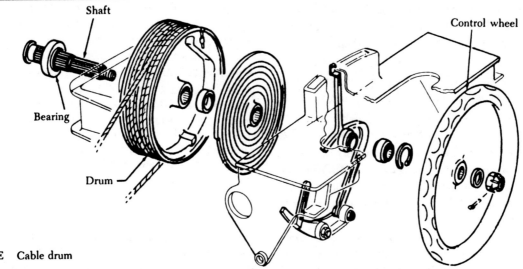

E   Cable drum

FIGURE 2–60. Flight control system mechanical linkages.

Pressure seals are installed where cables (or rods) move through pressure bulkheads. The seal grips tightly enough to prevent excess air pressure loss but not enough to hinder movement of the cable. Pressure seals should be inspected at regular intervals to determine that the retaining rings are in place. If a retaining ring comes off, it may slide along the cable and cause jamming of a pulley.

Pulleys are used to guide cables and also to change the direction of cable movement. Pulley bearings are sealed, and need no lubrication other than the lubrication done at the factory. Brackets fastened to the structure of the aircraft support the pulleys. Cables passing over pulleys are kept in place by guards. The guards are close-fitting to prevent jamming or to prevent the cables from slipping off when they slacken due to temperature variations.

MECHANICAL LINKAGE

Various mechanical linkages connect the cockpit controls to control cables and surface controls. These devices either transmit motion or change the direction of motion of the control system. The linkage consists primarily of control (push-pull) rods, torque tubes, quadrants, sectors, bellcranks, and cable drums.

Control rods are used as links in flight control systems to give a push-pull motion. They may be adjusted at one or both ends. View A of figure 2–60 shows the parts of a control rod. Notice that it consists of a tube having threaded rod ends. An adjustable antifriction rod end, or rod end clevis, attaches at each end of the tube. The rod end, or clevis, permits attachment of the tube to flight control system parts. The checknut, when tightened, prevents the rod end or clevis from loosening.

Control rods should be perfectly straight, unless designed to be otherwise, when they are installed. The bellcrank to which they are attached should be checked for freedom of movement before and after attaching the control rods. The assembly as a whole should be checked for correct alignment. When the rod is fitted with self-aligning bearings, free rotational movement of the rods must be obtained in all positions.

It is possible for control rods fitted with bearings to become disconnected because of failure of the peening that retains the ball races in the rod end. This can be avoided by installing the control rods

so that the flange of the rod end is interposed between the ball race and the anchored end of the attaching pin or bolt as shown in figure 2–61.

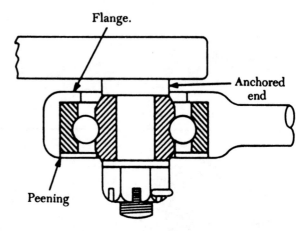

FIGURE 2–61. Rod end flange interposed between the bearing race and the end of the attaching bolt.

Another alternative is to place a washer, having a larger diameter than the hole in the flange, under the retaining nut on the end of the attaching pin or bolt.

TORQUE TUBES

Where an angular or twisting motion is needed in a control system, a torque tube is installed. View B of figure 2–60 shows how a torque tube is used to transmit motion in opposite directions.

Quadrants, bellcranks, sectors, and drums change direction of motion and transmit motion to parts such as control rods, cables, and torque tubes. The quadrant shown in figure 2–60B is typical of flight control system linkages used by various manufacturers. Figures 2–60C and 2–60D illustrate a bellcrank and a sector. View E illustrates a cable drum. Cable drums are used primarily in trim tab systems. As the trim tab control wheel is moved clockwise or counterclockwise, the cable drum winds or unwinds to actuate the trim tab cables.

STOPS

Adjustable and nonadjustable stops (whichever the case requires) are used to limit the throw-range or travel movement of the ailerons, elevator, and rudder. Usually there are two sets of stops for each of the three main control surfaces, one set being located at the control surface, either in the snubber cylinders or as structural stops (figure 2–62), and the other at the cockpit control. Either of these may serve as the actual limit stop. However, those situated at the control surface usually perform this function. The other stops do not normally contact each other, but are adjusted to a definite clearance when the control surface is at the full extent of its travel. These work as over-ride stops to prevent stretching of cables and damage to the control system during violent maneuvers. When rigging control systems, refer to the applicable maintenance manual for the sequence of steps for adjusting these stops to limit the control surface travel.

through a cable system by a spring-loaded plunger (pin) that engages a hole in the control surface mechanical linkage to lock the surface. A spring connected to the pin forces it back to the unlock position when the cockpit control lever is placed in the "unlock" position. An over-center toggle linkage is used on some other type aircraft to lock the control surfaces.

Control surface locking systems are usually so designed that the throttles cannot be advanced until the control surfaces are unlocked. This prevents taking off with the control surfaces in the locked position.

A typical control lock for small aircraft consists of a metal tube that is installed to lock the control wheel and rudder pedals to an attachment in the cockpit. Such a system is illustrated in figure 2–63.

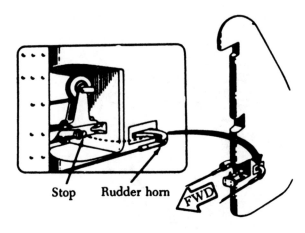

FIGURE 2–62. Adjustable rudder stops.

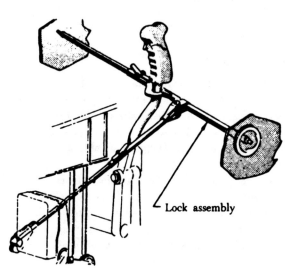

FIGURE 2–63. Typical control lock assembly for small aircraft.

CONTROL SURFACE SNUBBERS AND LOCKING DEVICES

Various types of devices are in use to lock the control surfaces when the aircraft is parked or moored. Locking devices prevent damage to the control surfaces and their linkages from gusts and high-velocity winds. Common devices that are in use are the internal locking brake (sector brake) spring-loaded plunger, and external control surface locks.

Internal Locking Devices

The internal locking device is used to secure the ailerons, rudder, and elevator in their neutral positions. The locking device is usually operated

Control Surface Snubbers

Hydraulic booster units are used on some aircraft to move the control surfaces. The surfaces are usually protected from wind gusts by snubbers incorporated into the booster unit. On some aircraft an auxiliary snubber cylinder is connected directly to the surface to provide protection. The snubbers hydraulically check or cushion control surface movement when the aircraft is parked. This prevents wind gusts from slamming the control surfaces into their stops and possibly causing damage.

External Control Surface Locks

External control surface locks are in the form of channeled wood blocks. The channeled wood blocks slide into the openings between the ends of the movable surfaces and the aircraft structure. This locks the surfaces in neutral. When not in use, these locks are stowed within the aircraft.

Tension Regulators

Cable tension regulators are used in some flight control systems because there is considerable difference in temperature expansion of the aluminum aircraft structure and the steel control cables. Some large aircraft incorporate tension regulators in the control cable systems to automatically maintain a given cable tension. The unit consists of a compression spring and a locking mechanism which allows the spring to make correction in the system only when the cable system is in neutral.

AIRCRAFT RIGGING

Control surfaces should move a certain distance in either direction from the neutral position. These movements must be synchronized with the movement of the cockpit controls. The flight control system must be adjusted (rigged) to obtain these requirements.

Generally speaking, the rigging consists of the following: (1) Positioning the flight control system in neutral and temporarily locking it there with rig pins or blocks, and (2) adjusting surface travel, system cable tension, linkages, and adjustable stops to the aircraft manufacturer's specifications.

When rigging flight control systems, certain items of rigging equipment are needed. Primarily, this equipment consists of tensiometers, cable rigging tension charts, protractors, rigging fixtures, contour templates, and rulers.

Measuring Cable Tension

To determine the amount of tension on a cable, a tensiometer is used. When properly maintained, a tensiometer is 98% accurate. Cable tension is determined by measuring the amount of force needed to make an offset in the cable between two hardened steel blocks, called anvils. A riser or plunger is pressed against the cable to form the offset. Several manufacturers make a variety of tensiometers, each type designed for different kinds of cable, cable sizes, and cable tensions.

One type of tensiometer is illustrated in figure 2–64. With the trigger lowered, place the cable to be tested under the two anvils. Then close the trigger (move it up). Movement of the trigger pushes up the riser, which pushes the cable at right angles to the two clamping points under the anvils. The force that is required to do this is indicated by the dial pointer. As the sample chart beneath the illustration shows, different numbered risers are used with different size cables. Each riser has an identifying number and is easily inserted into the tensiometer.

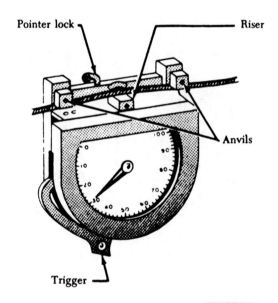

	No. 1			Riser	No. 2		No. 3	
Dia. 1/16	3/32	1/8	Tension Lb.	5/32	3/16	7/32	1/4
12	16	21	30	12	20		
19	23	29	40	17	26		
25	30	36	50	22	32		
31	36	43	60	26	37		
36	42	50	70	30	42		
41	48	57	80	34	47		
46	54	63	90	38	52		
51	60	69	100	42	56		
			110	46	60		
			120	50	64		

Sample only   Example

Figure 2–64. Tensiometer.

70

In addition, each tensiometer has a calibration table (figure 2–64) which is used to convert the dial reading to pounds. (The calibration table is very similar to the sample chart shown below the illustration.) The dial reading is converted to pounds of tension as follows. Using a No. 2 riser (figure 2–64) to measure the tension of a 5/32-in. diameter cable a reading of "30" is obtained. The actual tension (see calibration table) of the cable is 70 lbs. Observing the chart, also notice that a No. 1 riser is used with 1/16-, 3/32-, and 1/8-in. cable. Since the tensiometer is not designed for use in measuring 7/32- or 1/4-in. cable, no values are shown in the No. 3 riser column of the chart.

When taking a reading, it may be difficult to see the dial. Therefore, a pointer lock is present on the tensiometer. Push it in to lock the pointer. Then remove the tensiometer from the cable and observe the reading. After observing the reading, pull the lock out and the pointer will return to zero.

Cable rigging tension charts (figure 2–65) are graphic tools used to compensate for temperature variations. They are used when establishing cable tensions in flight control systems, landing gear systems, or any other cable-operated systems.

To use the chart, determine the size of the cable that is to be adjusted and the ambient air temperature. For example, assume that the cable size is 1/8-in. in diameter, that it is a 7 x 19 cable, and the ambient air temperature is 85° F. Follow the 85° F. line upward to where it intersects the curve for 1/8-in. cable. Extend a horizontal line from the point of intersection to the right edge of the chart. The value at this point indicates the tension (rigging load in pounds) to establish on the cable. The tension for this example is 70 lbs.

Surface Travel Measurement

The tools for measuring surface travel primarily include protractors, rigging fixtures, contour templates, and rulers. These tools are used when rigging flight control systems to assure that the desired travel has been obtained.

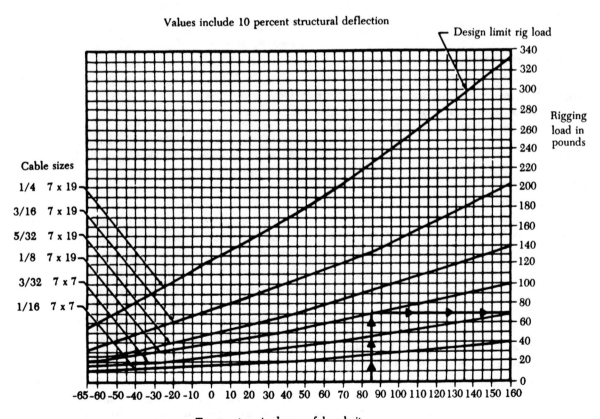

FIGURE 2–65. Typical cable rigging chart.

Protractors are tools for measuring angles in degrees. Various types of protractors are used to determine the travel of flight control surfaces. One protractor that can be used to measure aileron, elevator, or wing flap travel is the universal propeller protractor. Notice that this protractor (figure 2–66) is made up of a frame, a disk, a ring, and two spirit levels. The disk and ring turn independently of each other and of the frame. (The corner spirit level is used to position the frame vertically when measuring propeller blade angle.) The center spirit level is used to position the disk when measuring control surface travel. A disk-to-ring lock is provided to secure the disk and ring together when the zero on the ring vernier scale and the zero on the disk degree scale align. The ring-to-frame lock prevents the ring from moving when the disk is moved. Note that they start at the same point and advance in opposite directions. A double 10-part vernier is marked on the ring.

The procedure to use for operating the protractor to measure control surface travel is shown at the bottom of figure 2–66.

Rigging Fixtures and Contour Templates

Rigging fixtures and templates are special tools (gages) designed by the manufacturer to measure control surface travel. Markings on the fixture or template indicate desired control surface travel.

Rulers

In many instances the aircraft manufacturer gives the travel of a particular control surface in degrees and inches. If the travel in inches is provided, a ruler can be used to measure surface travel in inches.

RIGGING CHECKS

The purpose of this section is to explain the methods of checking the relative alignment and adjustment of an aircraft's main structural components. It is not intended to imply that the procedures are exactly as they may be in a particular aircraft. When rigging an aircraft, always follow the procedures and methods specified by the aircraft manufacturer.

Structural Alignment

The position or angle of the main structural components is related to a longitudinal datum line parallel to the aircraft center line and a lateral datum line parallel to a line joining the wing tips. Before checking the position or angle of the main components, the aircraft should be leveled.

Small aircraft usually have fixed pegs or blocks attached to the fuselage parallel to or coincident with the datum lines. A spirit level and a straight edge are rested across the pegs or blocks to check the level of the aircraft. This method of checking aircraft level also applies to many of the larger types of aircraft. However, the grid method is sometimes used on large aircraft. The grid plate (figure 2–67) is a permanent fixture installed on the aircraft floor or supporting structure. When the aircraft is to be leveled, a plumb bob is suspended from a predetermined position in the ceiling of the aircraft over the grid plate. The adjustments to the jacks necessary to level the aircraft are indicated on the grid scale. The aircraft is level when the plumb bob is suspended over the center point of the grid.

Certain precautions must be observed in all instances. Normally, rigging and alignment checks should not be undertaken in the open. If this cannot be avoided, the aircraft should be positioned with the nose into the wind.

The weight and loading of the aircraft should be exactly as described in the manufacturer's manual. In all cases, the aircraft should not be jacked until it is ensured that the maximum jacking weight (if any) specified by the manufacturer is not exceeded.

With a few exceptions, the dihedral and incidence angles of conventional modern aircraft cannot be adjusted. Some manufacturers permit adjusting the wing angle of incidence to correct for a wing-heavy condition. The dihedral and incidence angles should be checked after hard landings or after experiencing abnormal flight loads to ensure that the components are not distorted and that the angles are within the specified limits.

There are several methods for checking structural alignment and rigging angles. Special rigging boards which incorporate, or on which can be placed, a special instrument (spirit level or clinometer) for determining the angle are used on some aircraft. On a number of aircraft the alignment is checked using a transit and plumb bobs or a theo-

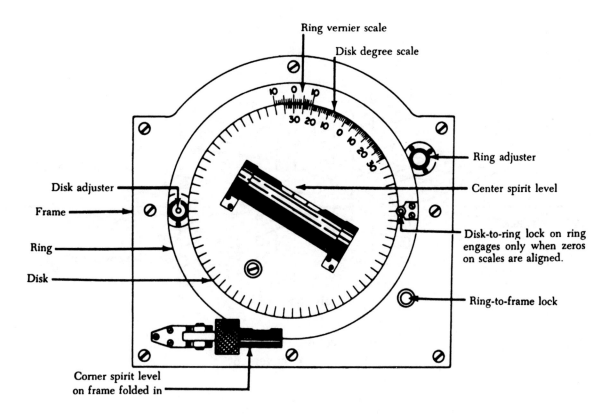

Ring vernier scale

Disk degree scale

Ring adjuster

Center spirit level

Disk adjuster

Frame

Ring

Disk

Disk-to-ring lock on ring engages only when zeros on scales are aligned.

Ring-to-frame lock

Corner spirit level on frame folded in

1 With disk-to-ring lock in the deep slot, turn disk adjuster to lock disk to ring.

2 Move control surface to neutral. Place protractor on control surface and turn ring adjuster to center bubble in center spirit level (ring must be unlocked from frame).

3 Lock ring to frame with ring-to-frame lock.

4 Move control surface to extreme limit of movement

5 Unlock disk from ring with disk-to-ring lock.

6 Turn disk adjuster to center bubble in center spirit level.

7 Read surface travel in degrees on disk and tenths of a degree on vernier scale.

FIGURE 2–66. Using the universal propeller protractor to measure control surface travel.

dolite and sighting rods. The particular equipment to use is usually specified in the manufacturer's manuals.

When checking alignment, a suitable sequence should be developed and followed to be certain that the checks are made at all the positions specified. The alignment checks specified usually include:

(1) Wing dihedral angle.

(2) Wing incidence angle.

(3) Engine alignment.

(4) Horizontal stabilizer incidence.

(5) Horizontal stabilizer dihedral.

(6) Verticality of the fin.

(7) A symmetry check.

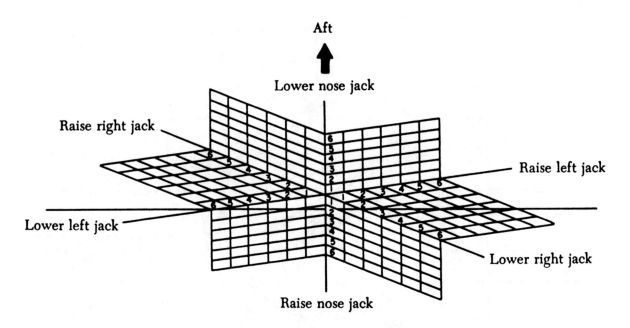

Aft

Lower nose jack

Raise right jack

Raise left jack

Lower left jack

Lower right jack

Raise nose jack

Scale reading one unit per
inch of jack movement

FIGURE 2–67. Typical grid plate.

Checking Dihedral

The dihedral angle should be checked in the spec-ified positions using the special boards provided by the aircraft manufacturer. If no such boards are available, a straight edge and a clinometer can be used. The methods for checking dihedral are shown in figure 2–68.

It is important that the dihedral be checked at the positions specified by the manufacturer. Certain portions of the wings or horizontal stabilizer may sometimes be horizontal or, on rare occasions, an-hedral angles may be present.

Checking Incidence

Incidence is usually checked in at least two speci-fied positions on the surface of the wing to ensure

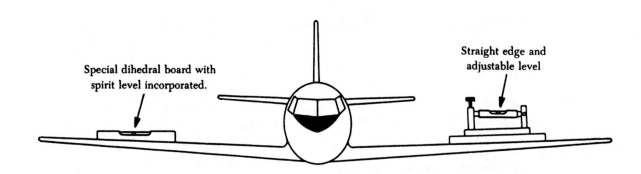

Special dihedral board with
spirit level incorporated.

Straight edge and
adjustable level

FIGURE 2–68. Checking dihedral.

that the wing is free from twist. A variety of incidence boards are used to check the incidence angle. Some have stops at the forward edge which must be placed in contact with the leading edge of the wing. Others are equipped with location pegs which fit into some specified part of the structure. The purpose in either case is to ensure that the board is fitted in exactly the position intended. In most instances, the boards are kept clear of the wing contour by short extensions attached to the board. A typical incidence board is shown in figure 2–69.

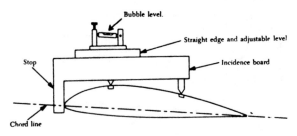

FIGURE 2–69. A typical incidence board.

When used, the board is placed at the specified locations on the surface being checked. If the incidence angle is correct, a clinometer on top of the board will read zero, or within a specified tolerance of zero. Modifications to the areas where incidence boards are located can affect the reading. For example, if leading-edge deicer boots have been installed, this will affect the position taken by a board having a leading edge stop.

Checking Fin Verticality

After the rigging of the horizontal stabilizer has been checked, the verticality of the vertical stabilizer relative to the lateral datum can be checked. The measurements are taken from a given point on either side of the top of the fin to a given point on the left and right horizontal stabilizers (figure 2–70). The measurements should be similar within prescribed limits. When it is necessary to check the

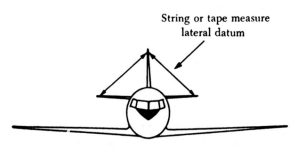

FIGURE 2–70. Checking fin verticality.

alignment of the rudder hinges, remove the rudder and pass a plumb bob line through the rudder hinge attachment holes. The line should pass centrally through all the holes. It should be noted that some aircraft have the leading edge of the vertical fin offset to the longitudinal center line to counteract engine torque.

Checking Engine Alignment

Engines are usually mounted with the thrust line parallel to the horizontal longitudinal plane of symmetry. However, this is not always true when the engines are mounted on the wings. Checking to ensure that the position of the engines, including any degree of offset, is correct depends largely on the type of mounting. Generally, the check entails a measurement from the center line of the mounting to the longitudinal center line of the fuselage (figure 2–71) at the point specified in the applicable manual.

Symmetry Check

The principle of a typical symmetry check is illustrated in figure 2–71. The precise figures, tolerances and checkpoints for a particular aircraft will be found in the applicable service or maintenance manual.

On small aircraft the measurements between points are usually taken using a steel tape. When measuring long distances, it is suggested that a spring scale be used with the tape to obtain equal tension. A 5-lb. pull is usually sufficient.

Where large aircraft are concerned, the positions where the dimensions are to be taken are usually chalked on the floor. This is done by suspending a plumb bob from the checkpoints, and marking the floor immediately under the point of each plumb bob. The measurements are then taken between the center of each marking.

ADJUSTMENT OF CONTROL SURFACES

In order for a control system to function properly, it must be correctly adjusted. Correctly rigged control surfaces will move through a prescribed arc (surface-throw) and be synchronized with the movement of the cockpit controls.

Rigging any system requires that the step-by-step procedures be followed as outlined in the aircraft maintenance manual. Although the complete rigging procedure for most aircraft is of a detailed nature that requires several adjustments, the basic method follows three steps:

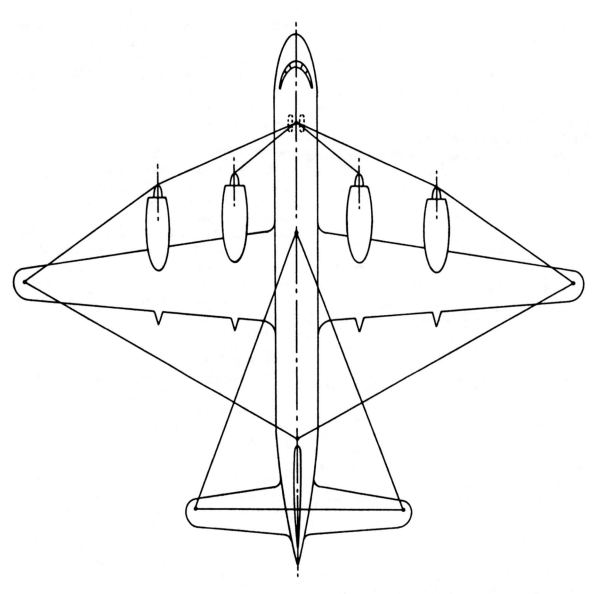

FIGURE 2-71. A typical method of checking aircraft symmetry.

(1) Lock the cockpit control, bellcranks, and the control surfaces in the neutral position.

(2) Adjust the cable tension, maintaining the rudder, elevators, or ailerons in the neutral position.

(3) Adjust the control stops to limit the control surface travel to the dimensions given for the aircraft being rigged.

The range of movement of the controls and control surfaces should be checked in both directions from neutral.

The rigging of the trim tab systems is performed in a similar manner. The trim tab control is set to the neutral (no trim) position, and the surface tab is usually adjusted to streamline with the control surface. However, on some aircraft the trim tabs may be offset a degree or two from streamline when in the "neutral" position. After the tab and tab control are in the neutral position, adjust the control cable tension.

Pins, usually called rig pins, are sometimes used to simplify the setting of pulleys, levers, bellcranks, etc., in their neutral positions. A rig pin is a small metallic pin or clip. When rig pins are not provided, the neutral positions can be established by means of alignment marks, by special templates, or by taking linear measurements.

If the final alignment and adjustment of a system are correct, it should be possible to withdraw the rigging pins easily. Any undue tightness of the pins in the rigging holes indicates incorrect tensioning or misalignment of the system.

After a system has been adjusted, the full and synchronized movement of the controls should be checked. When checking the range of movement of the control surface, the controls must be operated from the cockpit and not by moving the control surfaces. During the checking of control surface travel, ensure that chains, cables, etc., have not reached the limit of their travel when the controls are against their respective stops. Where dual controls are installed, they must be synchronized and function satisfactorily when operated from both positions.

Trim tabs and other tabs should be checked in a manner similar to the main control surfaces. The tab position indicator must be checked to see that it functions correctly. If jackscrews are used to actuate the trim tab, check to see that they are not extended beyond the specified limits when the tab is in its extreme positions.

After determining that the control system functions properly and is correctly rigged, it should be thoroughly inspected to determine that the system is correctly assembled, and will operate freely over the specified range of movement. Make certain that all turnbuckles, rod ends, and attaching nuts and bolts are correctly safetied.

HELICOPTER RIGGING

The flight control units located in the cockpit (figure 2–72) of all helicopters are very nearly the same. All helicopters have either one or two of each of the following: (1) Collective pitch control, (2) cyclic pitch control, and (3) directional control pedals. Basically, these units do the same things, regardless of the type of helicopter on which they are installed. But this is where most of the similarity ends. The operation of the systems in which these units are installed varies greatly according to the helicopter model.

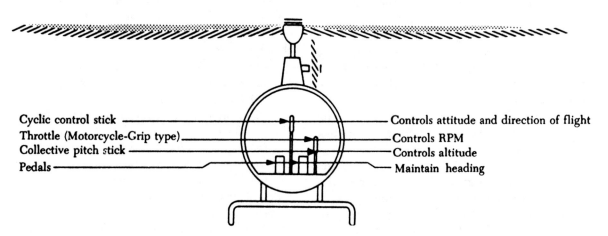

Cyclic control stick —————— Controls attitude and direction of flight
Throttle (Motorcycle-Grip type) —————— Controls RPM
Collective pitch stick —————— Controls altitude
Pedals —————— Maintain heading

FIGURE 2–72. Controls of the helicopter and the principal function of each.

Rigging the helicopter coordinates the movements of the flight controls and establishes the relationship between the main rotor and its controls and between the tail rotor and its controls. Rigging is not a difficult job, but it requires great precision and attention to detail. Strict adherence to rigging procedures is a must. Adjustments, clearances, and tolerances must be exact.

Rigging of the various flight control systems can be broken down into three major steps.

(1) Step one consists of placing the control system in a particular position; holding it in position with pins, clamps, or jigs; and adjusting the various linkages to fit the immobilized control component.

(2) Step two consists of placing the control surfaces in a specific reference position; using a rigging jig (figure 2–73), a precision bubble protractor, or a spirit level to check the angular difference between the control surface and some fixed surface on the aircraft.

(3) Step three consists of setting the maximum range of travel of the various components.

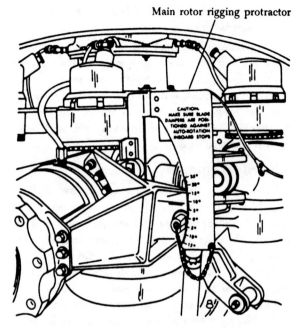

Main rotor rigging protractor

FIGURE 2-73. A typical rigging protractor.

This adjustment limits the physical movement of the control system.

After completion of the static rigging, a functional check of the flight control system must be accomplished. The nature of the functional check will vary with the type of rotorcraft and system concerned, but usually includes determining that:

(1) The direction of movement of the main and tail rotor blades is correct in relation to movement of the pilot's controls.

(2) The operation of interconnected control systems (engine throttle and collective pitch) are properly coordinated.

(3) The range of movement and neutral position of the pilot's controls are correct.

(4) The maximum and minimum pitch angles of the main rotor blades are within specified limits. This includes checking the fore-and-aft and lateral cyclic pitch and collective pitch blade angles.

(5) The tracking of the main rotor blades is correct.

(6) In the case of multi-rotor aircraft, the rigging and movement of the rotor blades are synchronized.

(7) When tabs are provided on main rotor blades, they are correctly set.

(8) The neutral, maximum and minimum pitch angles and coning angles of the tail rotor blades are correct.

(9) When dual controls are provided, they function correctly and in synchronization.

Upon completion of rigging, a thorough check should be made of all attaching, securing, and pivot points. All bolts, nuts, and rod ends should be properly secured and safetied.

Blade Tracking

When the main rotor blades do not "cone" by the same amount during rotation, it is referred to as "out of track." This may result in excessive vibration at the control column. Blade tracking is the process of determining the positions of the tips of the rotor blade relative to each other while the rotor head is turning, and of determining the corrections necessary to hold these positions within certain tolerances. Tracking shows only the relative position of the blades, not their path of flight. The blades should all track one another as closely as possible. The purpose of blade tracking is to bring the tips of all blades into the same tip path throughout their entire cycle of rotation.

In order to track rotor blades with minimum time and maximum accuracy, the correct equipment must be available. The equipment generally used to track blades includes:

(1) Tracking flag with flag material.

(2) Grease pencils or colored chalk.

(3) Suitable marking material.

(4) Reflectors and tracking lights (figure 2-74).

(5) Tracking stick.

(6) Trim-tab bending tool.

(7) Trim-tab angle indicator.

Before starting a blade tracking operation, new or recently overhauled blades should be checked for proper incidence. Trim tabs should be set at zero on new or overhauled blades. Trim tabs of blades in service should not be altered until blade track has been determined.

One means of tracking is the flag tracking method (figure 2-75). The blade tips are marked with chalk or grease pencil. Each blade tip should be marked with a different color so that it will be easy to determine the relationship of the tips of the rotor blades to each other. This method can be used on all types of helicopters that do not have jet propulsion at the blade tips. The man holding the flag faces in the direction of blade rotation, watching the retreating blades. Facing away from the oncoming blades permits the flag holder to observe the blades as they come in contact with the flag.

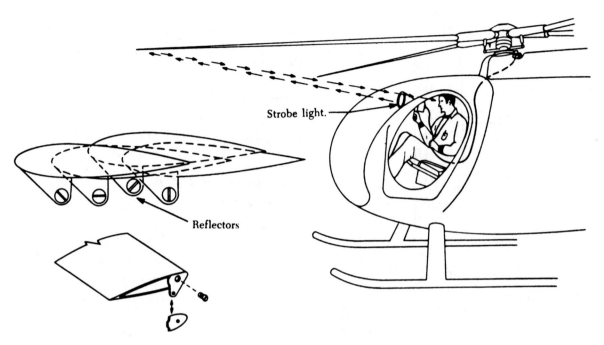

FIGURE 2-74. Blade tracking with strobe light.

The angle of the flag to the chord of the blade is important. If the angle is too great, the marks will be long and the flag will flutter excessively. If the angle is too straight, the blade may cut the flag. The most satisfactory angle is about 80° to the chord line of the blade. The marks on the flag will then be approximately 3/16 in. to 1/4 in. long. The flag method of tracking can be used not only to ascertain the relative positions of the blades but also the flight characteristics of the blades at different r.p.m. and power settings.

In order to plot the flight characteristics of a set of blades, it is necessary to take a trace at different r.p.m. settings and record the results. A minimum of three traces is necessary to produce a satisfactory plot. Four traces are desirable to produce a plot on heads having three or more rotor blades. When the tracking plot is completed, one blade is chosen as a reference blade. Usually, the reference blade is the center blade of a plot on a multi-blade rotor system and the lower blade on a two-blade rotor system. If the center or lower blade of a plot shows unusual flight characteristics, another blade may be chosen as the reference blade. A blade track that rises with an increase in r.p.m. is a climbing blade; one that lowers with increase of r.p.m. or power is a diving blade. When a climbing blade and a diving blade cross, it is termed a crossover. Because of the climb-

ing and diving tendencies of imporperly rigged blades, it is possible to have all blades at a common point at certain r.p.m. and power settings, but out of track at other r.p.m. or power settings.

The most common error in blade tracking is to bring the blades into track with trim tabs at cruise r.p.m. only. The blades may then be at the meeting point of a crossover and will spread at different r.p.m. power settings, or forward speed; an out-of-track condition will result. The correct tracking procedure is to maintain a constant blade spread at all r.p.m. power settings and flight speeds. A constant spread can be held only by proper adjustment of trim tabs. After a constant spread has been established with trim tabs, it is necessary to bring the blade tips into a single path of rotation with the pitch links. Bending the trim tabs up will raise the blade and bending them down will lower the blade. Bending of trim tabs should be kept to a minimum because tab angle produces excessive drag on the blades. The setting of the tabs on main rotor blades (if provided) should be checked to eliminate out-of-balance moments which will apply torque to the rotor blades. The tab setting is checked for correctness by running the rotor at the prescribed speed and ensuring that the cyclic-pitch control column remains stationary. Out-of-balance moments impart a stirring motion to the column.

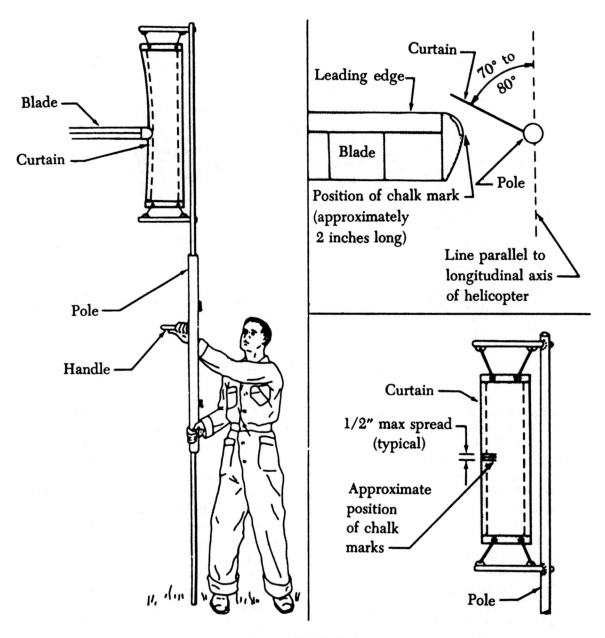

FIGURE 2–75. Tracking.

PRINCIPLES OF BALANCING OR RE-BALANCING

The principles that are essential in the balancing or re-balancing of the control surfaces are not too difficult to understand if some simple comparison is used. For example, a seesaw that is out of balance may be compared to a control surface that does not have balance weights installed, as in figure 2–76. From this illustration, it is easy to understand how a control surface is naturally tail (trailing edge) heavy.

It is this out-of-balance condition that can cause a damaging flutter or buffeting of an aircraft and therefore must be eliminated. This is best accomplished by adding weights either inside or on the leading edge of the tabs, ailerons, or in the proper location on the balance panels. When this is done properly, a balanced condition exists and can be compared to the seesaw with a child sitting on the short end of the plank.

The effects of moments on control surfaces can be easily understood by a closer observation and study of a seesaw and two children of different weights

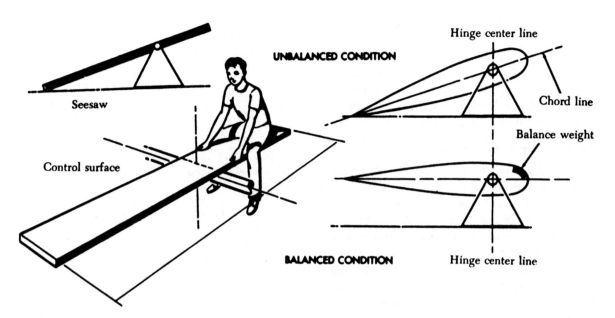

FIGURE 2-76. (A) Unbalanced, (B) Balanced conditions.

seated in different positions thereon. Figure 2-77 illustrates a seesaw with an 80-lb. child seated at a distance of 6 ft. from the fulcrum point of the seesaw. The weight of the child tends to rotate the seesaw in a clockwise direction until it touches the ground. To bring the seesaw into a level or balanced condition, a child is placed on the opposite end of the seesaw. The child must be placed at a point equal to the moment of the child on the left side of the seesaw.

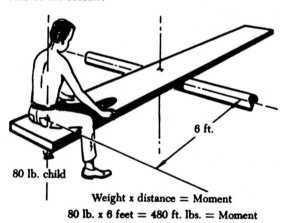

Weight x distance = Moment
80 lb. x 6 feet = 480 ft. lbs. = Moment

FIGURE 2-77. Moment.

Assume that the child is placed a distance of 8 ft. to the right of the fulcrum point. A simple formula can be used to determine the exact weight that the child must have to balance or bring the seesaw into a leveled condition.

To produce a balanced condition of the seesaw (or control surface), the counterclockwise moment must equal the clockwise moment. Moment is found by multiplying weight times distance. Therefore, the formula to balance the seesaw is:

$$W_2 \times D_2 = W_1 \times D_1.$$

W_2 would be the unknown weight of the second child. D_2 would be the distance (in feet) from the fulcrum that the second child is seated (8). W_1 would be the weight of the first child (80 lbs.). D_1 would be the distance (in feet) from the fulcrum that the first child is seated (6).

Finding the weight of the second child is now a simple matter of substitution and solving the formula as follows:

$$W_2 \times D_2 = W_1 \times D_1$$
$$W_2 \times 8 = 80 \text{ lbs.} \times 6$$
$$W_2 = \frac{480 \text{ lbs.}}{8}$$
$$W_2 = 60 \text{ lbs.}$$

So the weight of the second child would have to be 60 lbs. To prove the formula:

$$60 \text{ lbs.} \times 8 \text{ ft.} = 80 \text{ lbs.} \times 6 \text{ ft.}$$
$$480 \text{ ft. lbs.} = 480 \text{ ft. lbs.}$$

This would result in a balanced condition of the seesaw since the counterclockwise moments around the fulcrum are equal to the clockwise moments around the fulcrum.

81

The same effect is obtained in a control surface by the addition of weight. Since most of the repairs to control surfaces are aft of the hinge center line, resulting in a trailing-edge-heavy condition, the weight is added forward of the hinge center line. The correct re-balance weight must be calculated and installed in the proper position.

Re-balancing of Movable Surfaces

The material in this section is presented for familiarization purposes only, and should not be used when re-balancing a control surface. Explicit instructions for the balancing of control surfaces are given in the service and overhaul manuals for the specific aircraft and must be followed closely.

Any time repairs on a control surface add weight fore or aft of the hinge center line, the control surface must be re-balanced. Any control surface that is out of balance will be unstable and will not remain in a streamlined position during normal flight. For example, an aileron that is trailing-edge heavy will move down when the wing deflects upward, and up when the wing deflects downward. Such a condition can cause unexpected and violent maneuvers of the aircraft. In extreme cases, fluttering and buffeting may develop to a degree that could cause the complete loss of the aircraft.

Re-balancing a control surface concerns both static and dynamic balance. A control surface that is statically balanced will also be dynamically balanced.

Static Balance

Static balance is the tendency of an object to remain stationary when supported from its own center of gravity. There are two ways in which a control surface may be out of static balance. They are called underbalance and overbalance.

When a control surface is mounted on a balance stand, a downward travel of the trailing edge below the horizontal position indicates underbalance. Some manufacturers indicate this condition with a plus (+) sign. Figure 2–78A illustrates the underbalance condition of a control surface.

An upward movement of the trailing edge, above the horizontal position (figure 2–78B), indicates overbalance. This is designated by a minus (—) sign. These signs show the need for more or less weight in the correct area to achieve a balanced control surface as shown in figure 2–78C.

A tail-heavy condition (static underbalance) causes undesirable flight performance and is not

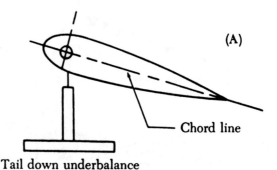

Tail down underbalance

Plus (+) condition

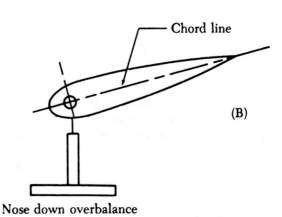

Nose down overbalance

Minus (—) condition

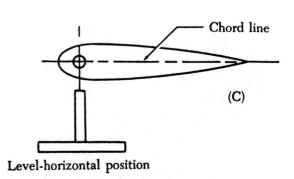

Level-horizontal position

Balanced condition

FIGURE 2–78. Control surface static balance.

usually allowed. Better flight operations are gained by nose heaviness static overbalance. Most manufacturers advocate the existence of nose-heavy control surfaces.

Dynamic Balance

Dynamic balance is that condition in a rotating body wherein all rotating forces are balanced within themselves so that no vibration is produced while the body is in motion. Dynamic balance as related to control surfaces is an effort to maintain balance when the control surface is submitted to movement on the aircraft in flight. It involves the placing of weights in the correct location along the span of the surfaces. The location of the weights will, in most cases, be forward of the hinge center line.

RE-BALANCING PROCEDURES

Requirements

Repairs to a control surface or its tabs generally increase the weight aft of the hinge center line, requiring static re-balancing of the control surface system as well as the tabs. Control surfaces to be re-balanced should be removed from the aircraft and supported, from their own points, on a suitable stand, jig, or fixture (figure 2–79).

Trim tabs on the surface should be secured in the neutral position when the control surface is mounted on the stand. The stand must be level and be located in an area free of air currents. The control surface must be permitted to rotate freely about the hinge points without binding. Balance condition is determined by the behavior of the trailing edge when the surface is suspended from its hinge points. Any excessive friction would result in a false reaction as to the overbalance or underbalance of the surface.

When installing the control surface in the stand or jig, a neutral position should be established with the chord line of the surface in a horizontal position (figure 2–80). Use a bubble protractor to determine the neutral position before continuing balancing procedures. Sometimes a visual check is all that is needed to determine whether the surface is balanced or unbalanced.

Any trim tabs or other assemblies that are to remain on the surface during balancing procedures should be in place. If any assemblies or parts must be removed before balancing, they should be removed.

METHODS

At the present time, four methods of balancing

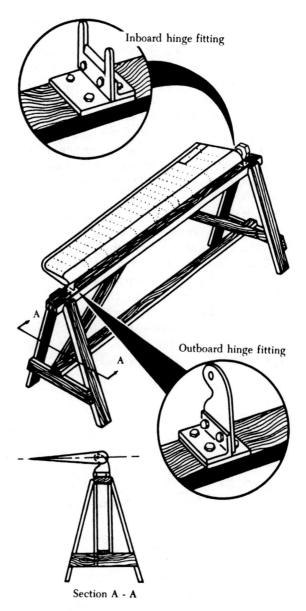

Inboard hinge fitting

Outboard hinge fitting

Section A - A

FIGURE 2–79. Field type balancing jigs.

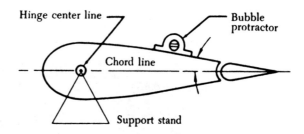

Hinge center line — Bubble protractor

Chord line

Support stand

FIGURE 2–80. Establishing a neutral position.

(re-balancing) control surfaces are in use by the various manufacturers of aircraft. The four methods are commonly called the calculation method,

83

scale method, trial weight (trial and error) method, and component method.

The calculation method of balancing a control surface is directly related to the principles of balancing discussed previously. It has one advantage over the other methods in that it can be performed without removing the surface from the aircraft.

In using the calculation method, the weight of the material from the repair area and the weight of the materials used to accomplish the repair must be known. Subtracting the weight removed from the weight added will give the resulting net gain in the amount added to the surface.

The distance from the hinge center line to the center of the repair area is then measured in inches. This distance must be determined to the nearest one-hundredth of an inch (figure 2–81).

The next step is to multiply the distance times the net weight of the repair. This will result in an in.-lbs. (inch-pounds) answer. If the in.-lbs. result of the calculations is within specified tolerances, the control surface will be considered balanced. If it is

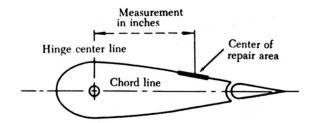

FIGURE 2–81. Calculation method measurements.

not within specified limits, consult the manufacturer's service manuals for the needed weights, material to use for weights, design for manufacture, and installation locations for addition of the weights.

The scale method of balancing a control surface requires the use of a scale that is graduated in hundredths of a pound. A support stand and balancing jigs for the surface are also required. Figure 2–82 illustrates a control surface mounted for rebalancing purposes. Use of the scale method requires the removal of the control surface from the aircraft.

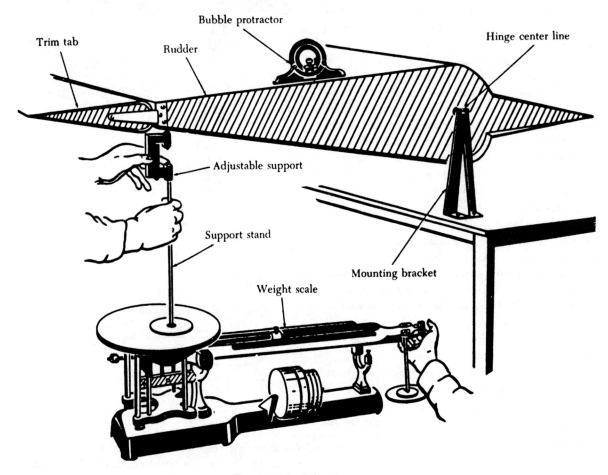

FIGURE 2–82. Balancing setup.

AIRCRAFT FABRICS

General

Most aircraft in production today are of all-metal construction. However, many aircraft in service still use fabric for covering wings, fuselages, and control surfaces. In the United States cotton fabrics have long been the standard material for covering aircraft. Cotton fabrics are still used, but other fabrics, such as linen, Dacron, and fiber glass, are gaining in popularity.

Organic and synthetic fibers are used in the manufacture of fabrics or cloth for covering aircraft. The organic fibers include cotton and linen; the synthetic fibers include fiber glass and heat-shrinkable synthetic fibers.

Three of the most common heat-shrinkable synthetic fibers available are a polyamide, manufactured and marketed under the trade name Nylon; an acrylic fiber called Orlon; and a polyester fiber known as Dacron.

Fabric Quality and Strength Requirements

In the original manufacture of a fabric-covered aircraft, the quality and strength of the fabric, surface tape, lacing cord, thread, etc., are determined by the aircraft's never-exceed speed and the pounds per square foot of wing loading. The never-exceed speed for a particular aircraft is that safe speed beyond which it should never be operated. The aircraft wing loading is determined by dividing its total wing planform area (in square feet) into the maximum allowable gross weight.

All fabric, surface tape, reinforcing tape, machine thread, lacing cord, etc., used for re-covering or repairing an aircraft's cover should be of high-grade aircraft textile material. The materials must also be at least as good a quality and of equivalent strength as those originally used by the aircraft manufacturer.

Acceptable fabrics for covering wings, control surfaces, and fuselages are listed in figures 3-1 and 3-2. Fabrics conforming to AMS (Aeronautical Material Specifications), incorporate a continuous

marking of specification numbers along the selvage edges to permit identification of the fabric in the field.

The following definitions are presented to simplify the discussion of fabrics. Some of these terms are shown graphically in figure 3-3.

(1) **Warp**—The direction along the length of fabric.

(2) **Warp ends**—The woven threads that run the length of the fabric.

(3) **Filling, woof, or weft**—The direction across the width of the fabric.

(4) **Count**—The number of threads per inch in warp or filling.

(5) **Ply**—The number of yards making up a thread.

(6) **Bias**—A cut, fold, or seam made diagonally to the warp or fill threads.

(7) **Calendering**—The process of ironing fabric by threading it wet between a series of hot and cold rollers to produce a smooth finish.

(8) **Mercerization**—The process of dipping cotton yarn or fabric in a hot solution of diluted caustic soda. This treatment causes the material to shrink and acquire greater strength and luster.

(9) **Sizing**—Material, such as starch, used to stiffen the yarns for ease in weaving the cloth.

(10) **Pinked edge**—An edge which has been cut by machine or shears in a continuous series of V's to prevent raveling.

(11) **Selvage edge**—An edge of cloth, tape, or webbing woven to prevent raveling.

Cotton Fabrics

Grade A airplane cloth is a 4-oz. mercerized fabric made of high-grade, long-staple cotton. It is calendered to reduce the thickness and lay the nap so that the surface will be smooth. There are from 80 to 84 threads per in., warp and fill. The minimum tensile strength is 80 lbs. per in. of width,

Materials	Specification	Minimum tensile strength new (undoped)	Minimum tearing strength new (undoped)	Minimum tensile strength deteriorated (undoped)	Thread count per inch	Use and remarks
Airplane cloth mercerized cotton (Grade "A").	Society Automotive Engineers AMS 3806 (TSO-C15 references this spec.).	80 pounds per inch warp and fill.	5 pounds warp and fill.	56 pounds per inch.	80 minimum, 84 maximum warp and fill.	For use on all aircraft. Required on aircraft with wing loadings greater than 9 p.s.f. Required on aircraft with placarded never-exceed speed greater than 160 m.p.h.
"	MIL-C-5646	"	"	"	"	Alternate to AMS 3806.
Airplane cloth cellulose nitrate pre-doped.	MIL-C-5643	"	"	"	"	Alternate to MIL-C-5646 or AMS 3806 (undoped). Finish with cellulose nitrate dope.
Airplane cloth cellulose acetate butyrate, predoped.	MIL-C-5642	"	"	"	"	Alternate to MIL-C-5646 or AMS 3806 (undoped). Finish with cellulose acetate butyrate dope.
Airplane cloth mercerized cotton.	Society Automotive Engineers AMS 3804 (TSO-C14 references this spec.).	65 pounds per inch warp and fill.	4 pounds warp and fill.	46 pounds per inch.	80 minimum, 94 maximum warp and fill.	For use on aircraft with wing loadings of 9 p.s.f. or less, provided never-exceed speed is 160 m.p.h. or less.
Airplane cloth mercerized cotton.	Society Automotive Engineers AMS 3802.	50 pounds per inch warp and fill.	3 pounds warp and fill	35 pounds per inch.	110 maximum warp and fill.	For use on gliders with wing loading of 8 p.s.f. or less, provided the placarded never-exceed speed is 135 m.p.h. or less.
Glider fabric cotton.	A.A.F. No. 16128. AMS 3802.	55 pounds per inch warp and fill.	4 pounds warp and fill.	39 pounds per inch.	80 minimum warp and fill.	Alternate to AMS 3802-A.
Aircraft linen..........	British 7F1					This material meets the minimum strength requirements of TSO-C15.

Figure 3–1. Textile fabrics used in covering aircraft.

Materials	Specification	Yarn Size	Minimum tensile strength	Yards per pound	Use and remarks
Reinforcing tape, cotton.	MIL-T-5661		150 pounds per one-half-inch width		Used as reinforcing tape on fabric and under rib lacing cord. Strength of other widths approx. in proportion.
Lacing cord, pre-waxed braided cotton.	MIL-C-5649		80 pounds double.	310 minimum.	Lacing fabric to structures. Unless already waxed, must be lightly waxed before using.
Lacing cord, special cotton.	U.S. Army No. 6-27.	20/3/3/3	85 pounds double.		"
Lacing cord, braided cotton.	MIL-C-5648		80 pounds single.	170 minimum.	"
Lacing cord thread; linen and linen hemp.	MIL-T-6779	9 ply	59 pounds single.	620 minimum.	"
		11 ply	70 pounds single.	510 minimum.	
Lacing cord thread; high-tenacity cotton.	MIL-T-5660	Ticket No. 10.	62 pounds single.	480 minimum.	"
Machine thread cotton.	Federal V-T-276b.	20/4 ply	5 pounds single	5,000 normal.	Use for all machine sewing.
Hand sewing thread cotton.	V-T-276b. Type III B.	8/4 ply	14 pounds single.	1,650 normal.	Use for all hand sewing. Use fully waxed thread.
Surface tape cotton (made from AN-C-121).	MIL-T-5083		80 lbs/in.		Use over seams, leading edges, trailing edges, outer edges and ribs, pinked, scalloped or straight edges.
Surface tape cotton.	Same as fabric used.		Same as fabric used.		Alternate to MIL-T-5083.

FIGURE 3-2. Miscellaneous textile materials.

warp and fill. The term "4 ounce" indicates that the normal weight of the finished cloth is 4 oz./sq. yd. for 36- and 42-in. widths. Fabric of this grade and weight is acceptable for covering any aircraft fabric surface.

Linen Fabrics

Unbleached linen fabric is used extensively in England and to a limited degree in the United States. This fabric is practically identical to Grade

87

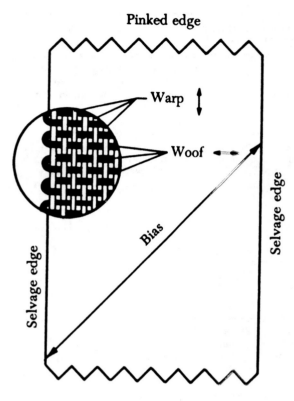

Pinked edge

Warp

Woof

Bias

Selvage edge

Selvage edge

FIGURE 3 3. Fabric terms

A cotton fabric insofar as weight, strength, and threads per inch are concerned.

Dacron Fabrics

Dacron is a very smooth monofilament, polyester fiber manufactured by the condensation of dimethyl terephthalate and ethylene glycol. A generally standard style and weight of Dacron cloth has evolved for use as aircraft covering. It is a plain weave with a weight of about 3.7 oz./sq. yd. This heavy-duty fabric has a tensile strength of approximately 148 lbs./in. and can be used to replace Grade A cotton and linen fabrics.

A fine weave, medium weight Dacron fabric is used when a minimum covering weight and a very smooth finish are desired. The medium weight fabric has a tensile strength of approximately 96 lbs./in., weighs about 2.7 oz./sq. yd., and can also be used as a replacement for Grade A cotton fabric.

Glass Cloth

Glass cloth, or fiber glass cloth, is made from fine-spun glass filaments which are woven into a strong, tough fabric. Glass cloth used for aircraft covering has a plain weave and weighs about 4.5 oz./sq. yd.

Glass cloth is not affected by moisture, mildew, chemicals, or acids. It is also fire resistant.

Glass cloth applications generally fall into the following classes:

(1) Class A is a complete or partial reinforcement of a serviceable fabric covering. No direct structural attachment of the glass cloth is provided. This composite covering is considered airworthy until the underlying conventional fabric deteriorates below the values listed in figure 3-1.

(2) Class B is a reinforcement of a fabric covering wherein the glass cloth is provided with the same direct structural attachment as that used with the original covering. This composite covering is considered airworthy until the underlying conventional fabric has deteriorated to a strength less than 50% of the minimum tensile strength values for new fabric listed in figure 3-1.

(3) Class C is replacement coverings applied either independently or over a conventional covering. The glass covering should possess all the necessary characteristics for airworthiness and is in no way dependent upon the underlying covering if one is present.

MISCELLANEOUS TEXTILE MATERIALS

Surface Tape

Surface tape is the finishing tape that is doped over each rib or seam to cover the stitching and provides a neat, smooth, finished appearance. It can be obtained with serrated or pinked edges, or with a straightedge impregnated with a sealing compound. The compound-impregnated edges or pinked edges provide better adhesion to the fabric covering.

Surface tape is made from Grade A fabric in various widths from 1¼ to 5″ and from glider fabric in 1½ and 6″ widths. Cotton surface tape may be used with Grade A cotton, linen, or Dacron fabric. Surface tape is also available in Dacron fabric, which should be the first choice when covering an aircraft with Dacron fabric. Linen surface tape frequently is used with fiber glass covering, especially for covering screwheads. If fiber glass tape is used, it is difficult to remove the irregularities caused by the screwheads. Using linen tape to cover screwheads gives a smooth, finished appearance.

Surface tape or finishing tape should be placed over all lacing, seams (both machine- and hand-sewn), corners, edges, and places where wear is likely to occur. Two-inch tape generally is used for this purpose. Pinked surface tape is sometimes applied over the trailing edges of control surfaces and airfoils. For such application the tape must be at least 3 inches in width and if the aircraft "never-exceed speed" is greater than 200 mph, notch the tape at equal intervals not to exceed 18″ between notches. Notching of trailing edge is unnecessary if the never exceed speed is under 200 mph. If the tape begins to separate from the trailing edge, it will tear at a notched section and thereby prevent loosening of the entire strip.

Tape is applied over a second wet coat of dope which is applied after the first coat has dried. Another coat of dope is applied immediately over the tape. The tape adheres firmly to the covering because both surfaces of the tape are impregnated with dope.

Reinforcing Tape

Reinforcing tape is used over ribs between the fabric covering and the rib stitching to prevent the stitching cord from cutting through the fabric. It is also used for cross-bracing ribs and for binding. Reinforcing tape is fabricated from cotton, Dacron, fiber glass, or linen materials. A tape made from fiber glass on acetate with a pressure-sensitive adhesive is also available.

Reinforcing tape is available in a variety of widths conforming to the different widths of ribs or rib capstrips. The tape should be slightly wider than the member it covers. A double width is sometimes necessary for very wide members.

Reinforcing tape is used under all lacing to protect the fabric from cuts. This tape should be under a slight tension and secured at both ends. For wings with plywood or metal leading edge covering, the reinforcing tape is extended only to the front spar on the upper and lower surfaces.

Sewing Thread

Thread is made with a right or left twist that is identified by various terms. Machine, machine twist, left twist, or Z-twist indicates a left-twist thread; S-twist indicates a right-twist thread.

An unbleached silk-finish, left-twist cotton thread is used to machine sew cotton fabrics. Silk-finish refers to a thread which has been sized to produce a hard, glazed surface. This finish prevents the thread from fraying and becoming weak. A thread having a tensile strength of at least 5 lbs. per single strand should be used. An unbleached white cotton, silk-finish thread is used in hand sewing cotton fabrics. This thread must have a strength of at least 14 lbs. per single strand.

Dacron fabrics are sewn with Dacron sewing thread. Glass fabrics, when sewn, are sewn with special synthetic threads.

Thread for hand sewing and lacing cord should be waxed lightly before using. The wax should not exceed 20% of the weight of the finished cord. A beeswax free from paraffin should be used for waxing.

Rib Lacing Cord

Rib lacing cord is used to sew the fabric to the ribs. The cord must be strong to transmit the suction on the upper surface of the wing from the fabric to the ribs, which, in turn, carry the load into the main wing structure. The cord must also resist fraying caused by the flexing action of the fabric and wing ribs. Dacron, linen, glass, or cotton cords are used for rib lacing cord.

Special Fasteners

When repairs are made to fabric surfaces attached by special mechanical methods, the original type of fastening should be duplicated. Screws and washers are used on several models of aircraft, and wire clips are used on other models. Screws or clips may not be used unless they were used by the manufacturer of the aircraft. When self-tapping screws are used to attach fabric to metal rib structure, the following procedure should be observed. Worn or distorted holes should be re-drilled, and a screw one size larger than the original should be used as a replacement. The length of the screw should be sufficient so that at least two threads of the grip (threaded part) extend through and beyond the rib capstrip. A thin washer, preferably celluloid, should be used under the heads of screws, and pinked-edge tape should be doped over each screwhead.

SEAMS

A seam consists of a series of stitches joining two or more pieces of material. Properly formed seam stitches possess the following characteristics:

(1) **Strength.** A seam must have sufficient strength to withstand the strain to which it will be subjected. The strength of a seam is affected by the type of stitch and thread used, number of stitches per inch of seam, tightness of the seam, construction of the seam, and the size and type of needle used.

(2) **Elasticity.** The elasticity of the material to be sewed determines the degree of elasticity desirable in a seam. Elasticity is affected by the quality of thread used, tension of the thread, length of stitch, and type of seam.

(3) **Durability.** The durability of a seam is determined by the durability of the material. Tightly woven fabrics are more durable than loosely woven fabrics which tend to work or slide upon each other. For this reason, the stitches must be tight and the thread well set into the material to minimize abrasion and wear caused by contact with external objects.

(4) **Good appearance.** The appearance of a seam is largely controlled by its construction. However, appearance should not be the principal factor when constructing covers. Due consideration must be given to strength, elasticity, and durability.

Sewed Seams

Machine-sewed seams (figure 3–4) should be of the folded-fell or French-fell types. A plain lapped seam is satisfactory where selvage edges or pinked edges are joined.

All machine sewing should have two rows of stitches with 8 to 10 stitches per inch. A lockstitch is preferred. All seams should be as smooth as possible and provide adequate strength. Stitches should be approximately 1/16 in. from the edge of the seam, and from 1/4 to 3/8 in. from the adjacent row of stitches.

Hand sewing is necessary to close the final openings in the covering. Final openings in wooden wing coverings are sometimes closed by tacking, but sewing is preferable. A 1/2-in. hem should be turned under on all seams to be hand sewn. Preparatory to hand sewing, the fabric on wooden wings can be held under tension by tacks; fabric on metal wings can be pinned to adhesive tape pasted to the trailing edge of the wings.

Hand sewing or tacking should begin where machine sewing stops and should continue to the point where machine sewing or uncut fabric is again reached. Hand sewing should be locked at 6-in. intervals and the seams should be properly finished with a lockstitch and a knot (figure 3–5). Where hand sewing or permanent tacking is necessary, the fabric should be so cut that it can be doubled under before it is sewed or permanently tacked. After hand sewing has been completed, the temporary tacks should be removed. In hand sewing there should be a minimum of four stitches per inch.

A double-stitched lap joint should be covered with pinked-edge surface tape at least 4 in. wide.

Spanwise seams on the upper or lower surface should be sewed with a minimum protuberance. The seam should be covered with pinked-edge tape at least 3 in. wide.

A spanwise seam sewed at the trailing edge should be covered with pinked-edge surface tape at least 3 in. wide. Notches (V-shaped) at least 1 in. deep and 1 in. wide should be cut into both edges of the surface tape if it is used to cover spanwise seams on trailing edges, especially the trailing edges of control surfaces. For application on aircraft with never-exceed speed of over 200 mph the tape should be notched at equal intervals not to exceed 18″ between notches. If the tape begins to separate because of poor adhesion or other causes, it will tear at a notched section, thus preventing progressive loosening of the entire length of tape.

Sewed seams parallel to the line of flight (chordwise) may be placed over a rib, but the seams should be placed so that the lacing will not be through them.

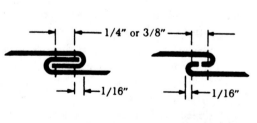

French fell seam   Folded fell seam

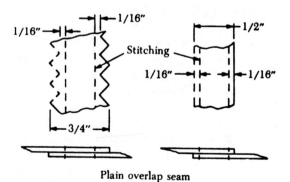

Plain overlap seam

FIGURE 3–4. Machine-sewed seams.

90

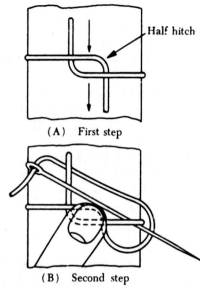

(A) First step — Half hitch

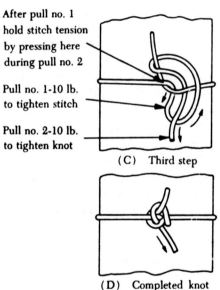

(B) Second step

After pull no. 1
hold stitch tension
by pressing here
during pull no. 2

Pull no. 1-10 lb.
to tighten stitch

Pull no. 2-10 lb.
to tighten knot

(C) Third step

(D) Completed knot

FIGURE 3-5. Standard knot for rib lacing
(modified seine knot).

Doped Seams

(1) For a lapped and doped spanwise seam on a metal-or-wood-covered leading edge, lap the fabric at least 4 inches and cover with pinked-edge surface tape at least 4 inches wide.

(2) For a lapped and doped spanwise seam at the trailing edge, lap the fabric at least 4 inches and cover with pinked-edge surface tape at least 3 inches wide.

APPLYING COVERING

General

The proper application of cloth on the surfaces is essential if a good appearance and the greatest strength are to be obtained from the material selected. A good covering job is important not only from a strength and appearance standpoint, but also because it affects the performance of the airplane. All covering must be taut and smooth for best performance.

All fabric materials to be used in covering should be stored in a dry place and protected from direct sunlight until needed. The room in which the sewing and application of the covering is done should be clean and well ventilated.

Preparation of the Structure for Covering

One of the most important items in covering aircraft is proper preparation of the structure. Dope proofing, covering edges which are likely to wear the fabric, preparing plywood surfaces, and similar operations, if properly done, will do much toward ensuring an attractive and long-lasting job.

Dope Proofing

Treat all parts of the structure that come in contact with doped fabric with a protective coating, such as aluminum foil, dope-proof paint, or cellulose tape. Clad aluminum and stainless steel parts need not be dope proofed.

Chafe Points

All points of the structure, such as sharp edges or boltheads, that are likely to chafe or wear the covering should be covered with doped-on fabric strips or taped with cellophane or other nonhygroscopic adhesive tape. After the cover has been installed, the chafe points of the fabric should be reinforced by the doping-on of fabric patches. Where a stronger reinforcement is required, a cotton duck or leather patch should be sewed to the fabric patch and then doped in place. All portions of the fabric pierced by wires, bolts, or other projections should be reinforced. Patches should fit the protruding part as closely as possible to prevent the entrance of moisture or dirt.

Inter-Rib Bracing

A continuous line of reinforcing tape may be used to successively tie the rib sections between the spars together at equally spaced intervals to hold the ribs in correct alignment and prevent their warping. Wing ribs that do not have permanent inter-rib bracing should be tied in position with reinforcing tape. Approximately half way between the front and rear spar, apply the tape diagonally between the top and bottom capstrip of each suc-

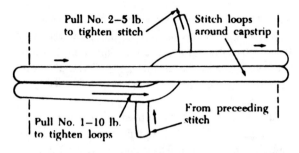

Pull No. 2–5 lb. to tighten stitch

Stitch loops around capstrip

Pull No. 1–10 lb. to tighten loops

From preceeding stitch

Operation No. 1

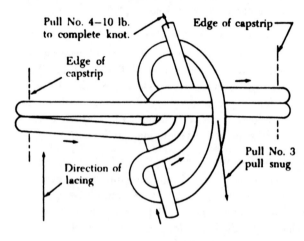

Pull No. 4–10 lb. to complete knot.

Edge of capstrip

Edge of capstrip

Pull No. 3 pull snug

Direction of lacing

Operation No. 2

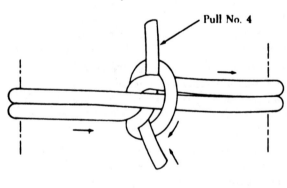

Pull No. 4

Completed knot

FIGURE 3-6. Standard knot for double loop lacing.

cessive rib from the wing butt rib to the tip rib. Tape is to be continuous and will be anchored with one turn around each individual rib cap strip.

Preparing Plywood Surfaces for Covering

Before covering plywood surfaces with fabric, prepare the surface by cleaning and applying sealer and dope.

Sand all surface areas that have been smeared with glue to expose a clean wood surface. Remove loose deposits such as wood chips and sawdust. Remove oil or grease spots by carefully washing with naphtha.

After cleaning the surfaces, apply one brush coat or two dip coats (wiped) of a dope-proof sealer, such as that conforming to Military Specification MIL-V-6894 thinned to 30% nonvolatile content, and allow to dry 2 to 4 hrs. Finally, before covering, apply two brush coats of clear dope, allowing the first coat of dope to dry approximately 45 min. before applying the second coat.

Covering Practices

The method of fabric attachment should be identical, as far as strength and reliability are concerned, to the method used by the manufacturer of the airplane to be re-covered or repaired. Fabric may be applied so that either the warp or fill threads are parallel to the line of flight. However, it is usually preferable for the warp threads to be parallel to the line of flight. Either the envelope method or blanket method of covering is acceptable.

The envelope method of covering consists of sewing together widths of fabric cut to specified dimensions and machine sewn to form an envelope that can be drawn over the frame. The trailing and outer edges of the covering should be machine sewn unless the component is not favorably shaped for sewing, in which case the fabric should be joined by hand sewing.

In the blanket method of covering, widths of fabrics of sufficient lengths are sewn together to form a blanket over the surfaces of the frame. The trailing and outer edges of the covering should be joined by a plain overthrow or baseball stitch. For airplanes with a placarded never-exceed speed of 150 m.p.h. or less, the blanket may be lapped at least 1 in. and doped to the frame or the blanket; it may be lapped at least 4 in. at the nose of metal- or wood-covered leading edges, doped, and finished with pinked-edge surface tape at least 4 in. wide. In both the envelope and blanket coverings, the fabric should be cut in lengths sufficient to pass completely around the frame, starting at the trailing edge and returning to the trailing edge. Seams parallel to the line of flight are preferable; however, spanwise seams are acceptable.

Before applying cotton or linen fabrics, brush on several coats of clear, full-bodied nitrate dope on all points to which the fabric edges will be cemented. If the structure is not doped, the dope used to cement the fabric edges will be absorbed by the surface as well as by the fabric. This will result in a poor bond to the structure after the dope has dried. Dacron fabric can be attached to the structure by using either nitrate dope or specially formulated cements.

After securing the cover, cotton and linen fabrics may be water-shrunk to remove wrinkles and excess slack. The fabric must be dried thoroughly before doping begins. Dacron may be heat-shrunk by using an electric iron set at 225° F. or by using a reflector heater. Do not apply excessive heat, because the Dacron, as well as the understructures of wood, may be damaged.

Shrinking should be done in several stages on opposite sides to shrink the entire area uniformly. Remove the excess slack with the initial application of heat. The second pass will then shrink the fabric to the desired tautness und remove most of the remaining wrinkles. Nonshrinking nitrate and butyrate dopes are available and produce no further shrinking or tightening. Regular dopes will pull the fibers and strands together and can damage light structures. A nonshrinking dope must be used when Dacron is heat-shrunk to its final tautness.

Taping

Sewed seams, lapped edges, and rib stitching or screws must be covered with pinked-edge surface tape. Use surface tape having the same properties as the fabric used for covering.

Apply the tape by first laying down a wet coat of dope, followed immediately by the tape. Press the tape into the dope. Work out any trapped air and apply a coat of dope over the surface of the tape.

COVERING WINGS

Wings may be covered with fabric by the envelope, blanket, or combination method. The envelope method is preferable and should be used whenever possible.

The envelope method of covering wings consists of sewing together several widths of fabric of definite dimensions and then running a transverse (spanwise) seam to make an envelope or sleeve. The advantage of the envelope method is that practically all sewing is by machine, and there is an enormous saving of labor in fitting the covering.

The envelope is pulled over the wing through its open end, which is then closed over the butt by hand sewing. When the envelope is used in repairing a portion of a surface the open end is fitted to extend 3 inches beyond the adjacent rib. If the envelope is of the proper dimensions, it will fit the wing snugly. When possible, the spanwise seam should be placed along the trailing edge.

In the blanket method several widths of fabric are machine sewed together and placed over the wing with a hand-sewed, spanwise seam along the trailing edge. Care must be taken to apply equal tension over the whole surface.

The combination method uses the envelope method as much as possible. and the blanket method on the remainder of the covering. This method is applicable to wings with obstructions or recesses that prevent full application of an envelope.

After the cover is sewn in place and shrunk, reinforcing tape of at least the width of the capstrip is placed over each rib and the fabric is laced to each rib. Except on very thick wings, the rib lacing passes completely around the rib. On thick wings the rib top and bottom cap strips are individually laced. In lacing any covering to a wing, the lacing is held as near as possible to the capstrip, by inserting the needle immediately adjacent to the capstrip. The rib should not have any rough or sharp edges in contact with the lacing, or it will fray and break. Each time the lacing cord goes around the rib it is tied, and the next stitch is made at the specified distance.

In order not to overstress the lacing, it is necessary to space the stitches a definite distance apart, depending on the speed of the airplane. Because of the additional buffeting caused by the propeller slipstream, the stitching must be spaced closer on all ribs included within the propeller slipstream. It is customary to use this closer spacing on the rib just outboard of the propeller diameter as well.

The stitch spacing should not exceed the approved spacing on the original covering of the aircraft. If the spacing cannot be ascertained because of destruction of the covering, acceptable rib stitch spacing may be found in figure 3–7. The lacing holes should be placed as close as possible to the capstrip to minimize the tendency of the cord to tear the fabric. All lacing cord should be lightly waxed with beeswax for protection.

Anti-tear Strips

In very high speed airplanes difficulty is often experienced with rib lacing breaking or with fabric tearing in the slipstream.

On aircraft with never-exceed speeds in excess of 250 m.p.h., anti-tear strips are recommended under reinforcing tape on the upper surface of wings and on the bottom surface of that part of the wing in the slipstream. Where the anti-tear strip is used on both the top and the bottom surfaces, extend it continuously up to and around the leading edges and back to the trailing edge. Where the strip is used only on the top surface, carry it up to and around the leading edge and back on the lower surface as far aft as the front beam or spar. For this purpose the slipstream should be considered as being equal to the propeller diameter, plus one extra rib space on each side. Cut anti-tear strips from the same material as that used for covering, and cut them wide enough to extend beyond the reinforcing tape on each side to engage the lacing cord. Attach the strips by applying dope to that part of the fabric to be covered by the strip and applying dope freely over the strip.

Single-Loop Wing Lacing

Both surfaces of fabric covering on wings and control surfaces should be securely fastened to the ribs by lacing cord or any other method originally approved for the aircraft.

All sharp edges against which the lacing cord may bear must be protected by tape to prevent abrasion of the cord. Separate lengths of lacing cord should be joined by the splice knot shown in figure 3–8. The common square knot, which has a very low slippage resistance, should not be used to splice lengths of cord. The utmost care should be used to assure uniform tension and security of all stitches.

Rib stitching usually is started at the leading edge of the rib and continued to the trailing edge. If the leading edge is covered with plywood or metal, start the lacing immediately aft of these coverings. The first or starting stitch is made with a

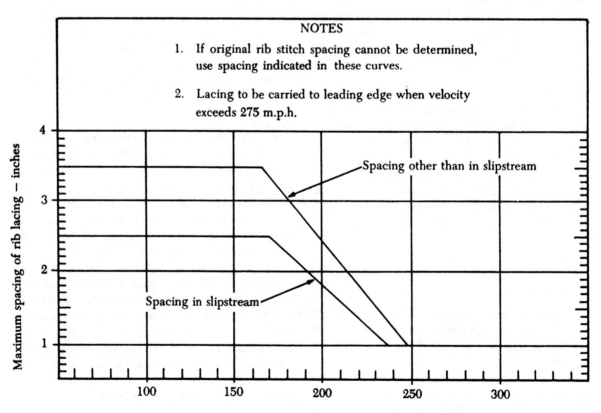

NOTES

1. If original rib stitch spacing cannot be determined, use spacing indicated in these curves.

2. Lacing to be carried to leading edge when velocity exceeds 275 m.p.h.

Spacing other than in slipstream

Spacing in slipstream

Maximum spacing of rib lacing — inches

Placard never exceeds speeds — m.p.h. (indicated).
(Curves presume leading edge support reinforcement such as plywood, metal)

FIGURE 3–7. Rib stitch spacing chart.

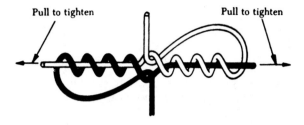

Knot formed but not tightened.

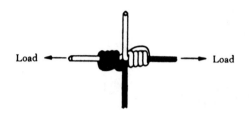

Knot completed.

FIGURE 3-8. Splice knot.

double loop, using the method illustrated in figure 3-9. All subsequent stitches can be made with a single loop. The spacing between the starting stitch and the next stitch should be one-half the normal stitch spacing. Where stitching ends, such as at the rear spar and the trailing edge, the last two stitches should be spaced at one-half normal spacing.

Double-Loop Wing Lacing

The double-loop lacing illustrated in figures 3-6 and 3-10 represents a method for obtaining higher strengths than are possible with the standard single lacing. When using the double-loop lacing, make the tie-off knot by the method shown in figure 3-6.

Tie-off Knots

All stitches other than the starting stitch must be tied off using the standard knot (modified seine) for rib lacing Figure 3-5. This knot is placed at the edge of the reinforcing tape Figure 3-9. Knots installed on top of the reinforcing tape are subject to increased wear and also have an adverse effect on the aerodynamics of the airfoil.

Tie-off knots usually are placed on the lower surface of low-wing aircraft and on the top surface of high-wing aircraft, to improve the final appearance of the surfaces.

Final location of the knot depends upon the original location selected by the manufacturer. If

such information is not available, consider positioning the knot where it will have the least effect on the aerodynamics of the airfoil.

The seine knot admits a possibility of improper tightening, resulting in a false (slip) form with greatly reduced efficiency and must not be used for last stitch tie-offs. Lock the tie-off knot for the last stitch by an additional half-hitch. Under no circumstances pull tie-off knots back through the lacing holes.

COVERING FUSELAGES

Fuselages are covered by either the sleeve or blanket method, similar to the methods described for covering wings. In the sleeve method several widths of fabric are joined by machine-sewed seams to form a sleeve that will fit snugly when drawn over the end of the fuselage. When the sleeve is in place, all seams should be as nearly parallel as possible to the longitudinal members of the fuselage.

In the blanket method all seams are machine sewed, except one final longitudinal seam along the bottom center of the fuselage. In some cases the blanket is put on it two or three sections and hand sewed on the fuselage. All seams should run fore and aft.

Fuselage Lacing

Fabric lacing is also necessary on deep fuselages and on those where former strips and ribs shape the fabric to a curvature. In the latter case the fabric should be laced to the formers at intervals. The method of attaching the fabric to the fuselage should be at least the equivalent in strength and reliability to that used by the manufacturer of the airplane.

VENTILATION, DRAIN, AND INSPECTION OPENINGS

The interior of covered sections is ventilated and drained to prevent moisture from accumulating and damaging the structure. Ventilation and drainage holes are provided and the edges reinforced with plastic, aluminum, or brass grommets.

Grommets are doped to the underside of fabric

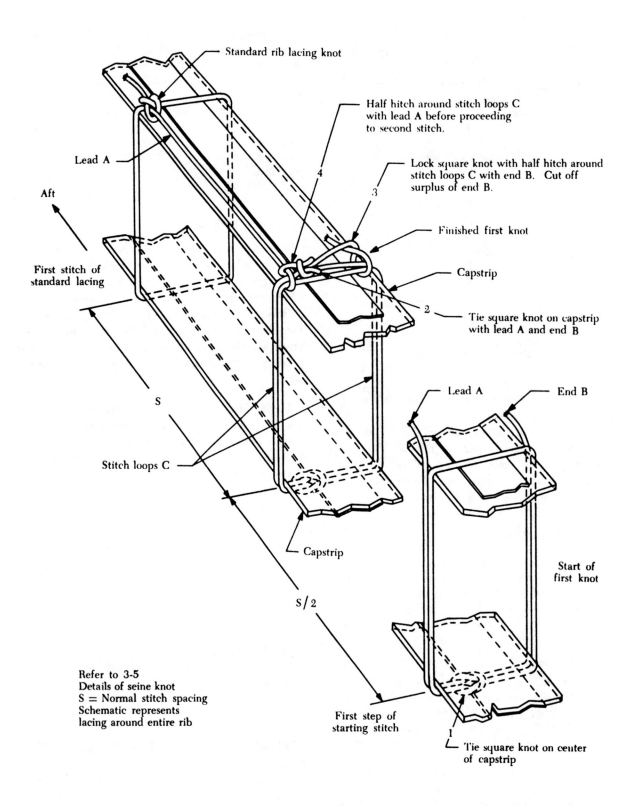

Standard rib lacing knot

Half hitch around stitch loops C with lead A before proceeding to second stitch.

Lead A

Aft

Lock square knot with half hitch around stitch loops C with end B. Cut off surplus of end B.

Finished first knot

First stitch of standard lacing

Capstrip

S

Tie square knot on capstrip with lead A and end B

Stitch loops C

Lead A   End B

Capstrip

Start of first knot

S/2

Refer to 3-5
Details of seine knot
S = Normal stitch spacing
Schematic represents
lacing around entire rib

First step of starting stitch

Tie square knot on center of capstrip

FIGURE 3–9. Starting stitch for rib lacing.

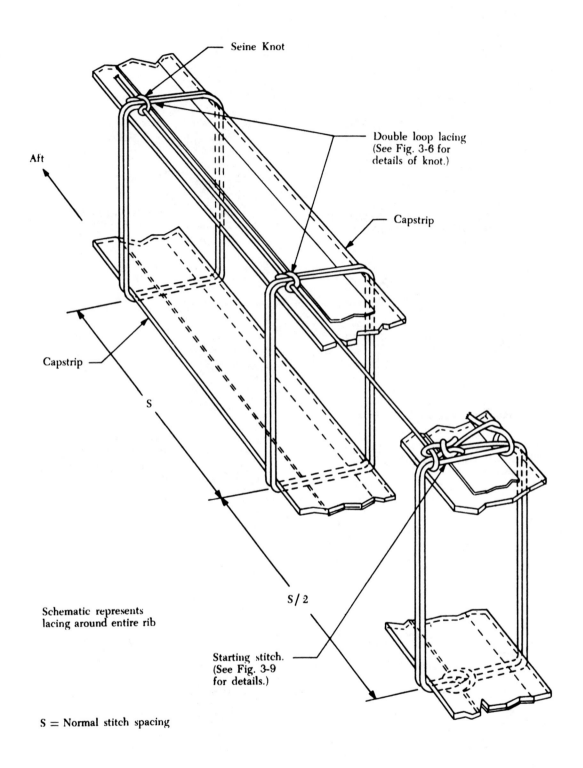

Seine Knot

Double loop lacing
(See Fig. 3-6 for
details of knot.)

Aft

Capstrip

Capstrip

S

S / 2

Schematic represents
lacing around entire rib

Starting stitch.
(See Fig. 3-9
for details.)

S = Normal stitch spacing

FIGURE 3–10. Standard double-loop lacing.

97

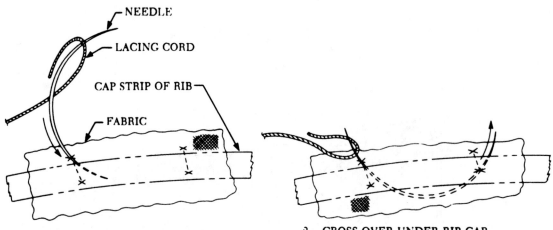

1 — START NEEDLE THROUGH FABRIC CLOSE TO SIDE OF RIB CAP STRIP

2 — CROSS OVER UNDER RIB CAP STRIP AND THROUGH FABRIC

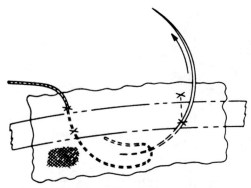

3 — PULL EYE END OF NEEDLE THROUGH THE FIRST HOLE IN FABRIC.

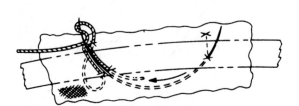

4 — BRING EYE END OF NEEDLE UP THROUGH FABRIC OPPOSITE FIRST HOLE AND FORM CORD ON END OF NEEDLE, AS SHOWN, TO MAKE HALF HITCH.

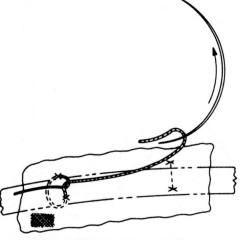

5 — PULL NEEDLE COMPLETELY OUT AND TIGHTEN HALF HITCH AS SHOWN.

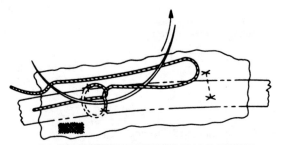

6 — PUT NEEDLE UNDER HALF HITCH AND THROUGH LOOP "K" AS SHOWN — THEN PULL NEEDLE THROUGH AND TIGHTEN HALF HITCH — THEN HOLD THUMB AT "J" TO KEEP HALF HITCH TIGHT, AND TIGHTEN LOOP "K", BACK OF HALF HITCH TO FORM A SEINE KNOT.

FIGURE 3–11. Rib lacing around capstrip.

surfaces wherever moisture may be trapped. It is customary to place one of these grommets on each side of a rib on the underside at the trailing edge. The grommets are also placed at the lowest drainage points of wings, ailerons, fuselage, and empennage components to provide complete drainage.

Plastic grommets (figure 3–12) are either in the shape of a thin, circular washer or are streamlined. Plastic grommets are doped to the fabric cover immediately after the surface tape is applied. Streamlined grommets usually are installed with the opening located toward the trailing edge of the surface.

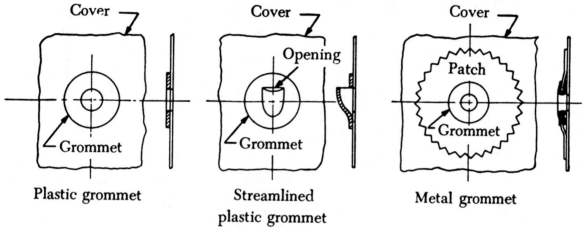

FIGURE 3–12. Typical grommets.

Brass and aluminum grommets, also shown in figure 3–12, are mounted on either circular or square fabric patches. The edges of the patch are pinked to provide better adhesion. The patch assembly is doped to the cover after the surface tape is applied.

Inspection doors and access holes are provided in all surfaces, whether fabric or metal covered. One way to provide these openings on fabric-covered surfaces is to dope a zipper-equipped patch in the desired place. Another inspection method for cloth or metal surfaces is to install a framework inside the wing to which a cover plate can be attached by screws. These frameworks are built into the structure wherever access or inspection holes are necessary.

REPAIR OF FABRIC COVERS

General

Repair fabric-covered surfaces so that the original strength and tautness are returned to the fabric. Repair all tears or punctures immediately to prevent the entry of moisture or foreign objects into the interior of the structure. Sewn and unsewn repairs are permitted. The type of repair technique to be used depends on the size and location of the damage as well as the never-exceed speed of the aircraft.

When re-covering or repairing control surface fabric, especially on high-performance airplanes, the repairs must not involve the addition of weight

aft of the hinge line. The addition of weight disturbs the dynamic and static balance of the surface to a degree that will induce flutter.

Repair of Tears

Small cuts or tears are repaired by sewing the edges together and doping a pinked- or frayed-edge patch over the area. The baseball stitch is used in repairing tears. The type illustrated in figure 3–13 enables the damaged edges to be drawn to their original location, thus permitting a tighter repair to be made. The first stitch is started by inserting the needle from the underneath side. All remaining stitches are made by inserting the needle from the top instead of from the bottom so that the points for making the stitch can be more accurately located. The edges are sewn together using the proper thread. The last stitch is anchored with a modified seine knot. Stitches should not be more than 1/4 in. apart and should extend 1/4 in. into the untorn cover.

Cut two patches of sufficient size to cover the tear and extend at least 1-1/2 in. beyond the tear in all directions (figure 3–14). The fabric used should be at least as good as the original fabric. The edges of the patch should be either pinked or frayed about 1/4 in. on all sides. One patch is saturated with nitrate thinner or acetone and laid over the sewn tear to remove the old finish. The patch is occasionally moistened with a brush until all but the clear

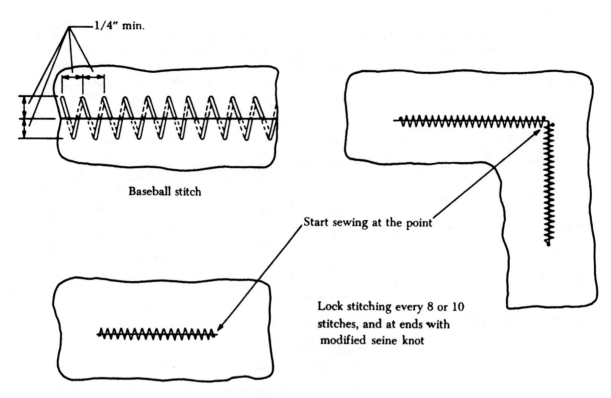

1/4" min.

Baseball stitch

Start sewing at the point

Lock stitching every 8 or 10
stitches, and at ends with
modified seine knot

FIGURE 3–13.   Repair of tears in fabric.

undercoats are soft enough to be removed with a putty knife. Since only the finish under the patch is removed, a neat, smooth repair can be made. A coat of clear tautening dope is applied to the second patch and to the area of the cover from which the finish has been removed. While still wet, this patch is applied to the cover and rubbed so it is smooth and free of air bubbles. Successive coats of clear, pigmented dope are applied until the patched surface has attained the same tension and appearance as the original surrounding surface.

Sewed Patch Repair

Damage to covers where the edges of the tear are tattered beyond joining or where a piece has been completely torn away is repaired by sewing a fabric patch into the damaged area and doping a surface patch over the sewed insert. A sewed-in repair patch may be used if the damage is not longer than 16 in. in any one direction.

The damaged area is trimmed in the form of a circle (figure 3–15) or oval-shaped opening. A fabric insert is cut large enough to extend 1/2 in. beyond the diameter of the opening. The 1/2-in. allowance is folded under as reinforcement. Before sewing, fasten the patch at several points with a few temporary stitches to aid in sewing the seams. The edges of the insert are sewed with a baseball stitch.

After the sewing is completed, clean the area of the old fabric to be doped as indicated for repair of tears and then dope the patch in the regular manner. Apply surface tape over any seams that have a second coat of dope. If the opening extends over or closer than 1 in. to a rib or other laced member, the patch should be cut to extend 3 in. beyond the member. After sewing has been completed, the patch should be laced to the rib over a new section of reinforcing tape. The old rib lacing and reinforcing tape should not be removed.

If the fabric covering is damaged at the trailing edge or part of it is torn away as shown in figure 3–16A, it can be repaired as follows. The damaged portion of the panel is removed, and a rectangular- or square-shaped opening is made as shown in figure 3–16B. A patch is cut of sufficient size to extend 3/4 in. beyond both sides and the bottom edge of the opening, and 1/2 in. beyond the top. The edges of the patch are reinforced by being folded under 1/2 in. before being sewed, and each corner is stretched and temporarily held in place with T-pins. Two sides and the trailing edge, as shown in figure 3–16C, are sewed to the old cover with the folded edge extending 1/4 in. beyond both

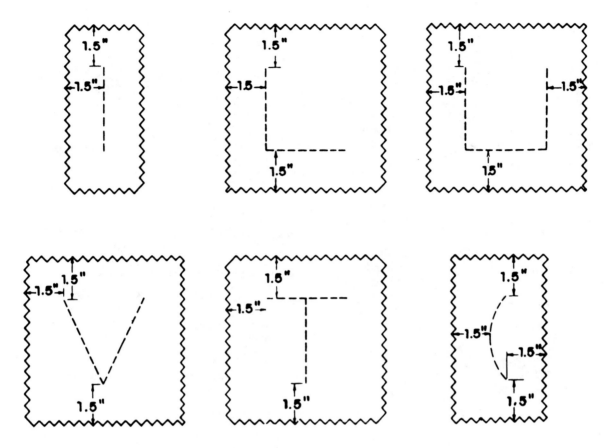

FIGURE 3-14. Patching over tears. Dash line represents a stitched tear.

ribs. The top of the opening is then sewed. Taping and doping, as shown in figure 3–16D, completes the repair.

Sewed-In Panel Repair

When the damaged area exceeds 16 in. in any direction, a new panel should be installed. Remove the surface tape from the ribs adjacent to the damaged area and from the trailing and leading edges of the section being repaired. Leave the old reinforcing tape in place.

Cut the old fabric along a line approximately 1 in. from the center of the ribs on the sides nearest to the damage, and continue the cuts to remove the damaged section completely. The old fabric should not be removed from the leading and trailing edges unless both upper and lower surfaces are being re-covered. Do not remove the reinforcing tape and lacing at the ribs.

Cut a patch to extend from the trailing edge up to and around the leading edge and back approximately to the front beam. The patch should extend approximately 3 in. beyond the ribs adjacent to the damage.

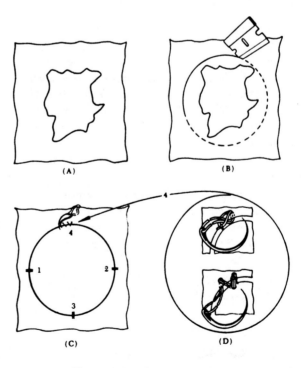

FIGURE 3–15. Sewed patch repair.

101

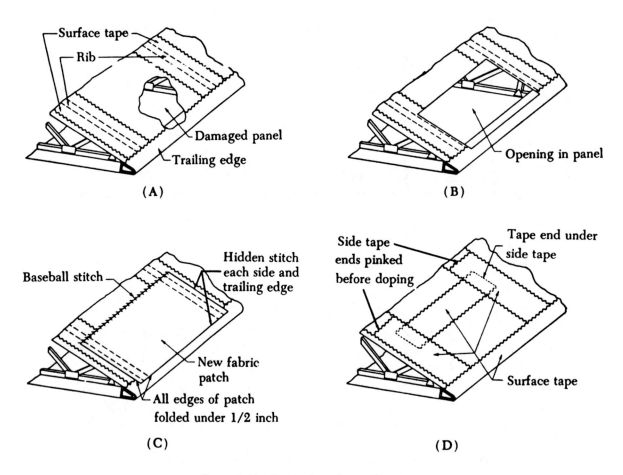

FIGURE 3-16. Repair of panel at trailing edge.

Clean the area of the old fabric to be covered by the patch, put the patch in place, stretch taut and pin. After the patch is pinned in place, fold under the trailing and leading edges of the patch 1/2 in. and sew to the old fabric. Fold the side edges under 1/2 in. and sew to the old cover. After completing the sewing, place reinforcing tape over the ribs under moderate tension and lace down (rib lace or stitch) to the ribs. Remove the temporary pinning.

Give the panel a coat of clear dope and allow to dry. Apply surface tape with the second coat of dope over the reinforcing tape and over the edges of the panel. Finish the doping using regular doping procedures.

This type of repair may be extended to cover both the upper and lower surfaces and to cover several rib bays if necessary. The panel must be laced to all ribs covered.

Unsewed (Doped-On) Fabric Repairs

Unsewed (doped-on) repairs may be made on all aircraft fabric-covered surfaces, provided the aircraft never-exceed speed is not greater than 150 m.p.h. A doped-on patch repair may be used if the damage does not exceed 16 in. in any direction. Cut out the damaged section, making a round or oval-shaped opening trimmed to a smooth contour. Use a grease solvent to clean the edges of the opening to be covered by the patch. Sand off the dope from the area around the patch or wash it off with a dope thinner. Support the fabric from underneath while sanding.

For holes up to 8 in. in size, make the fabric patch of sufficient size to provide a lap of at least 2 in. around the hole. For holes over 8 in. in size, make the overlap of the fabric around the hole at least one-fourth the hole diameter with a maximum lap limit of 4 in. If the hole extends over a rib or closer than the required overlap to a rib or other laced member, the patch should be extended at least 3 in. beyond the rib. In this case, after the edges of the patch have been doped in place and the dope has dried, the patch should be laced to the rib over a new section of reinforcing tape in the usual manner. The old rib lacing and reinforcing should not be removed. All patches should have pinked edges or, if smooth, should be finished with pinked-edge surface tape.

Doped-In Panel Repair

When the damage exceeds 16 in. in any direction, make the repair by doping in a new panel. This type of repair may be extended to cover both the upper and lower surfaces and to cover several rib bays if necessary. The panel should be laced to all ribs covered, and it should be doped or sewed as in the blanket method.

Remove the surface tape from the ribs adjacent to the damaged area and from the trailing and leading edges of the section being repaired. Leave the old reinforcing tape and lacing in place. Next cut the fabric along a line approximately 1 in. from the ribs on the sides nearest to the damage and continue cutting to remove the damaged section completely. The old fabric should not be removed from the leading and trailing edges unless both upper and lower surfaces are being re-covered.

Cut a patch to run around the trailing edge 1 in. and to extend from the trailing edge up to and around the leading edge and back approximately to the front beam. The patch should extend approximately 3 in. beyond the ribs adjacent to the damage. As an alternative attachment on metal or wood-covered leading edges, the patch may be lapped over the old fabric at least 4 in. at the nose of the leading edge, doped, and finished with at least 8 in. of pinked-edge surface tape.

Clean the area of the old fabric that is to be covered by the patch and apply a generous coat of dope to this area. Put the new panel in place, pull as taut as possible, and apply a coat of dope to the portion of the panel that overlaps the old fabric. After this coat has dried, apply a second coat of dope to the overlapped area and allow to dry.

Place reinforcing tape, under moderate tension, over the ribs and lace the fabric to the ribs.

Give the panel a coat of clear dope and allow to dry. Apply surface tape with the second coat of dope over the reinforcing tape and over the edges of the panel. Finish the doping process using the regular doping procedure.

REPLACING PANELS IN WING COVERS

Repairs to structural parts require opening of the fabric. The surface tape is removed from the damaged rib, the ribs on either side of the damaged rib, and along the leading and trailing edges where the fabric is to be cut. The rib lacing is removed from the damaged rib. The cover is cut along the top of the damaged rib and along the leading and trailing edges as shown in figure 3–17.

To close an opening of this size, the cut edges are joined over the rib, the leading edge, and the trailing edge with the baseball stitch and a new fabric panel is sewn over all the repaired area. The new panel extends between the adjacent ribs and from the trailing edge to the leading edge (figure 3–18). The new fabric is cut so that it can be folded under 1/2 in. and carried 1/4 in. beyond the adjacent ribs where it is sewed. The leading and trailing edges are folded and sewed in the same manner. After the panel has been sewed in place, new reinforcing tape is laced over the repaired rib. The new fabric is laced at each of the adjacent ribs without using any additional reinforcing tape. Finally, all surface tapes are replaced, and the new surface is finished to correspond to the original covering.

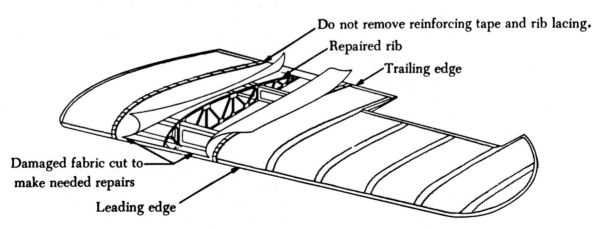

Do not remove reinforcing tape and rib lacing.
Repaired rib
Trailing edge
Damaged fabric cut to make needed repairs
Leading edge

FIGURE 3–17. Opening cover for internal structural repair.

RE-COVERING AIRCRAFT SURFACE WITH GLASS CLOTH

Fiber glass fabrics are acceptable for re-covering or reinforcing an aircraft surface, provided the material meets the requirements of Military Specifications MIL–C–9084, MIL–Y–1140, and MIL–G–1140. The tensile strength of the glass cloth should be at least equivalent to the tensile strength of the fabric originally installed on the aircraft. The chemical finish of the glass cloth should be chemically compatible with the dope or resin to be used.

Either the blanket or envelope method of reinforcement should be used on treated fabrics that can be sewn. Untreated fabric that cannot be sewn may be applied in overlapping sections. The practices recommended for doped seams should be used. Where the glass cloth is applied only to the upper surface of the wings for hail protection, it should wrap around the trailing edge at least 1 in. and extend from the trailing edge up to and around the leading edge and back approximately to the front spar. Before starting the work, make certain that the bonding agents used will be satisfactory. Blistering or poor adhesion can occur when using bonding agents which are not chemically compatible with the present finish on the aircraft, or which have already deteriorated because of age. A simple means of determining this is to apply a small piece of the reinforcement cloth to the original cover, using the proposed finishing process. The test sample should be visually checked the next day for blistering or poor adhesion.

When butyrate dope is used to bond glass cloth, the finishing can be accomplished in the following manner:

(1) Thoroughly clean the surface and allow to dry. If the surface has been waxed or previously covered with other protective coatings, thoroughly remove at least the top finish coat. After placing the glass cloth on the surface, brush out smoothly and thoroughly with butyrate dope thinner and 10% (by volume) retarder.

(2) Apply a heavy coat of butyrate dope between all glass cloth overlaps. When dry, brush in butyrate rejuvenator and allow to set until the surface has again drawn tight.

(3) Install reinforcing tape and structural attachments (Class B) and dope on finishing tape (cotton is recommended); then brush in one coat of 50% thinner and 50% butyrate dope.

(4) Follow by conventional finishing schedules, which call for application of one or more coats of full-bodied clear butyrate dope, two spray coats of aluminum pigmented butyrate dope, light surface sanding, and two spray coats of pigmented butyrate dope.

When resin is used to bond the glass cloth, after surface cleaning, the finishing may be done in the following manner:

(1) Rejuvenate the doped surface. After placing the glass cloth on the surface, brush in thoroughly a coat of resin. Saturate overlapped areas thoroughly and allow to cure.

(2) Brush in a second coat of resin smoothly and evenly and allow to cure. The finished surface should not be considered completed until all the holes between the weave of the cloth are filled flush with resin.

(3) After water sanding, paint the surface with one coat of primer surfacer and finish as desired.

Install drain grommets and inspection holes as provided in the original cover.

When using glass fabric to reinforce movable control surfaces, check to ascertain that no change has been made in their static and dynamic balance.

CAUSES OF FABRIC DETERIORATION

Aircraft fabrics deteriorate more rapidly in areas of heavy industrialization than in areas that have cleaner air. The greatest single cause of aircraft fabric deterioration is sulfur dioxide. This toxic compound is present in variable amounts in the

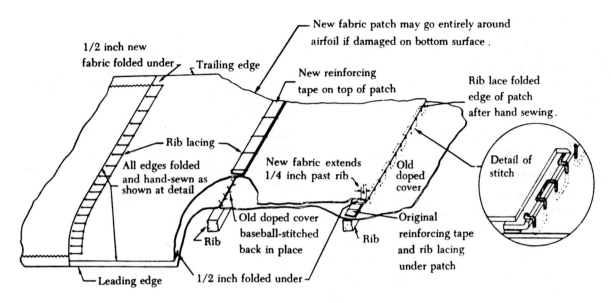

FIGURE 3-18. Method of replacing cover.

atmosphere. It occurs in large concentrations in industrial areas. Sulfur dioxide combines with oxygen, sunlight, and moisture to form sulfuric acid, which, readily attacks cotton fabrics. Linen fabrics are also affected, but to a lesser degree than cotton. Dacron fabric is far more resistant to sulfur dioxide and other chemicals than any other fabric except fiber glass fabric. Fiber glass fabric is not affected by moisture, mildew, chemicals, or most acids.

Mildew

Mildew spores attack fabrics when they are damp. All natural cellulose fibers provide nourishment for mildew growth when conditions are right. Mildew spores, also known as fungus or mold, can be controlled by using a fungus inhibitor. The inhibitor is usually mixed with dope and applied with the first coat of dope. Dope containing fungicides should not be sprayed because they contain poisons.

Re-covering should be done in dry, clean buildings. Damp, dirty buildings encourage mildew growth. The spores grow on damp rags, paper, etc., and are deposited directly on the fabric surfaces by any movement which stirs the air in the area. Spores are always present in the atmosphere in varying degrees and are induced into the airframe enclosures by air movement. An aircraft should be flown frequently to circulate dry air into the wings and fuselage so that moisture, which supports mildew, will not accumulate.

Acid Dopes and Thinners

The use of dopes or thinners whose acidity has increased beyond safe limits can cause rapid deterioration of aircraft fabrics. When dope is stored under extremes of heat or cold, chemical reactions increase the acidity beyond safe limits.

Stocks of Military dope compound are sold as surplus when periodic tests indicate that the dope has developed an acid content in excess of a safe value. Using surplus dope can lead to early fabric deterioration.

General-purpose thinners should not be used to thin aircraft dope. Such thinners are usually acidic and are not formulated for use with dope.

Insufficient Dope Film

A thin dope film does not provide sufficient protection of the fabric from the elements, and early deterioration of the fabric may result. Ultraviolet light, which is invisible, combines with oxygen to form an oxidizing agent that attacks organic materials. The ultraviolet rays can be screened by adding pigments to the dope film and by adequately covering the fabric with the dope. Aluminum powder usually is added to two of the dope coats to stop any ultraviolet light from reaching the fabric. Undoped fabric or fabric covering that is not protected by coats of aluminum-pigmented dope should not be exposed to sunlight for long periods.

Adequate protection of the fabric usually is achieved if the dope film hides the weave of the fabric, leaving a smooth surface. This cannot be determined by the number of coats of dope applied, but rather by the dope film thickness. This varies with application technique, temperature, dope consistency, and equipment.

Cracks in the dope film admit moisture and light, causing localized deterioration of the fabric.

Storage Conditions

It is generally assumed that a hangared aircraft is protected from fabric deterioration. However, premature deterioration can occur, especially on aircraft stored in an unheated hangar that has a dirt floor. During the day, sun shining on the roof raises the air temperature in the hangar. This warm air absorbs moisture from the ground. When the air cools, the absorbed moisture condenses and settles on the aircraft. Atmospheric pressure changes draw the damp air into the airframe enclosures. These conditions provide an ideal situation for promoting mildew growth.

When storing fabric-covered aircraft, all openings large enough for rodents to enter should be taped: Uric acid from mice can rot fabric. It can also corrode metal parts, such as ribs, spars, and fittings.

CHECKING CONDITION OF DOPED FABRIC

The condition of doped fabric should be checked at intervals sufficient to determine that the strength of the fabric has not deteriorated to the point where airworthiness of the aircraft is affected.

The areas selected for test should be those known to deteriorate most rapidly. The top surfaces generally deteriorate more rapidly than the side or bottom surfaces. When contrasting colors are used on an aircraft, the fabric will deteriorate more rapidly under the darker colors. The dark colors absorb more heat than the lighter colors. The warmer inner surface of the fabric under the dark color absorbs more moisture from the air inside the wing or fuselage. When the surface cools, this moisture condenses and the fabric under the dark area becomes moist and promotes mildew growth in a localized area. When checking cloth fabric that has been reinforced by applying fiber glass, peel back the glass cloth in the areas to be tested. Test the underlying cloth in the conventional manner.

Checking fabric surfaces is made easier by using a fabric punch tester. There are several acceptable

fabric punch testers on the market; one such tester incorporates a penetrating cone (figure 3–19). Fabric punch testers are designed for use on the dope-finished-fabric surface of the aircraft and provide only a general indication of the degree of deterioration in the strength of the fabric covering. Their advantage is that they may be used easily and quickly to test the fabric surfaces without cutting samples from the airplane's fabric. If a fabric punch tester indicates that the aircraft fabric strength is marginal, a laboratory test should be performed to determine the actual fabric strength.

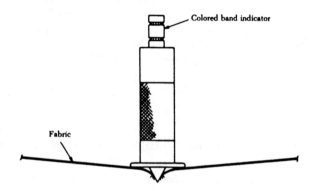

FIGURE 3–19. Fabric punch tester.

When using a punch tester similar to the one illustrated in figure 3–19, place the tip on the doped fabric. With the tester held perpendicular to the surface, apply pressure with a slight rotary action until the flange of the tester contacts the fabric. The condition of the fabric is indicated by a color-banded plunger that projects from the top of the tester. The last exposed band is compared to a chart supplied by the manufacturer of the tester to determine fabric condition.

The test should be repeated at various positions on the fabric. The lowest reading obtained, other than on an isolated reparable area, should be considered representative of the fabric condition as a whole. Fabrics that test just within the acceptable range should be checked frequently thereafter to ensure continued serviceability.

The punch tester makes only a small hole (approximately 1/2-in. diameter) or a depression in the fabric that can be repaired quickly by doping-on a 2-in. or 3-in. patch.

TESTING FABRIC COVERING

Tensile Testing of Undoped Fabric

Tensile testing of fabric is a practical means of

106

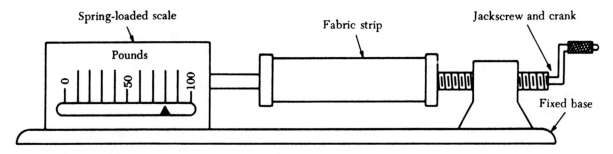

FIGURE 3-20. Fabric tensile tester.

determining whether a fabric covering has deteriorated to the point where re-covering is necessary. Figure 3-20 illustrates a typical fabric tensile tester.

A sample of the undoped fabric to be tested is cut to exactly 1-1/2-in. wide and to a sufficient length (usually 6 in.) to allow insertion in the fabric tester. Usually, each edge of the strip is frayed 1/4 in., reducing the woven width to 1 in. The ends of the fabric strip are fastened securely in the clamps of the tester. As the crank of the tester is turned, the threaded jackscrew is backed out, thus gradually increasing the tension (pull) on the fabric against the resistance of the spring-loaded scale until the fabric strip breaks. The scale reading, taken at the moment the fabric strip breaks, indicates the strength of the fabric in pounds per inch.

Fabric specimens must be tensile tested in the undoped condition. Use acetone dope thinner or other appropriate thinning agents to remove the finishing materials from the test specimen.

STRENGTH CRITERIA FOR AIRCRAFT FABRIC

Present minimum strength values for new aircraft fabric covering are provided in figure 3-1.

The maximum permissible deterioration for used aircraft fabric, based on a large number of tests, is 30%. Fabric that has less than 70% of the originally required tensile strength is not considered airworthy. Figure 3-1 contains the minimum tensile strength values for deteriorated fabric tested in the undoped condition.

Some light aircraft operators use the Grade A type fabric, but are only required to use intermediate grade fabric. In this case, the Grade A material is still considered airworthy, provided it has not deteriorated, as tested in the undoped condition, below 46 pounds, i.e., 70% of the originally required tensile strength value for new intermediate fabric.

DOPES AND DOPING

To tighten fabric covering and to make it airtight and watertight, brush or spray the cloth with dope. A tight fabric cover is essential to securing and holding the cross-sectional shape of the airfoil to the form given it by the ribs. This dope also protects the fabric from deterioration by weather or sunlight and, when polished, imparts a smooth surface to the fabric and reduces skin friction. Dopes must be applied under ideal conditions to obtain satisfactory and consistent results. A clean, fresh, dry atmosphere with a temperature above 70° F. and a relative humidity below 60%, combined with good ventilation, is necessary in the dope room. The dope must be of the proper consistency and be applied uniformly over the entire surface.

Dopes will deteriorate seriously if stored in too warm a place for a long period. The temperature should not exceed 60° F. for long-time storage and must not exceed 80° F. for periods up to 4 months. Precautions against fire should be taken wherever dope is stored or used because of its flammable nature. Dope and paint rooms that are not located in a separate building should be isolated from the rest of the building by metal partitions and fireproof doors.

As stated previously, the most desirable condition in a dope room is a temperature above 70° F. and a relative humidity below 60%. At lower temperatures the dope will not flow freely without the addition of excessive thinners. The relative humidity can be lowered by raising the temperature if the dope shop is not equipped with humidity control. To condition fabric surfaces to the desired temperature and moisture conditions, allow them to stand about 4 hrs. in the dope room after covering and prior to doping.

The number of coats of dope applied to a fabric surface depends on the finish desired. It is customary to apply two to four coats of clear dope, followed by two coats of pigmented dope. Sufficient clear dope should be applied to increase the weight of the fabric by 2.25 to 2.50 oz./sq. yd. The clear-dope film should weigh this amount after drying for 72 hrs. With fabric weighing 4 oz., the total weight of fabric and dope is approximately 9.5 oz./sq. yd.

Pigmented dopes must be applied over the clear dopes to protect the fabric from the sunlight. Sufficient pigment must be added to the dope to form an opaque surface. Pigmented dopes consist of the properly colored pigment added to the clear dope. When an aluminum finish is desired, 1 gal. of the clear nitrocellulose dope is mixed with 12 oz. of aluminum powder and an equal additional amount of glycol sebacate plasticizer. Sufficient thinner is then added so that two coats of this dope will give a film weight of about 2 oz./sq. yd.

Panels should be doped in a horizontal position, whenever possible, to prevent the dope from running to the bottom of the panel. Hand brush the first coat of dope and work it uniformly into the fabric. A minimum of 30 min. under good atmospheric conditions should be allowed for drying between coats. Surface tape and patches should be applied just prior to the second coat of dope. This second coat should also be brushed on as smoothly as possible. A third and fourth coat of clear dope can be applied by either brushing or spraying. These coats of clear dope provide a taut and rigid surface to the fabric covering. If desired, this surface may be smoothed by lightly rubbing with number 280 or 320 wet or dry sandpaper, or a similar abrasive. When being rubbed, all surfaces should be electrically grounded to dissipate static electricity. The doping is completed by spraying two or more coats of the properly colored pigmented dope on the surface.

Under certain unfavorable atmospheric conditions, a freshly doped surface will blush. Blushing is caused by the precipitation of cellulose ester, which is caused largely by a high rate of evaporation and/or high humidity. High temperatures or currents of air blowing over the work increase the evaporation rate and increase blushing tendencies. Blushing seriously reduces the strength of the dope film and the necessary precautions should be taken to guard against blushing. When a doped surface blushes, it becomes dull in spots, or white in extreme cases.

The surface under the doped fabric must be protected to prevent the dope from "lifting" the paint on the surface. A common method is to apply dope-proof paint or zinc chromate primer over all parts of the surface that come in contact with doped fabric. Another excellent method is to cover this surface with aluminum foil 0.0005 in. thick. This foil is glued to the surface and prevents the penetration of dope. It is applied over the regular finish. Other materials, such as a cellophane tape, have also been used successfully in place of aluminum foil.

DOPE MATERIALS

Aircraft dope is any liquid applied to a fabric surface to produce tautness by shrinkage, to increase strength, to protect the fabric, to waterproof, and to make the fabric airtight. Aircraft dopes are also used extensively in the repair and rejuvenation of aircraft fabric surfaces.

Aircraft dope is technically a colloidal solution of cellulose acetate butyrate or cellulose nitrate. If nitric acid was used in the chemical manufacturing of the dope, it is known as cellulose nitrate dope. If acetic and butyric acids were used, the dope is known as cellulose acetate butyrate dope.

Cellulose-Nitrate Dope

Nitrocellulose dope is a solution of nitrocellulose and a plasticizer, such as glycol sebacate, ethyl acetate, butyl acetate, or butyl alcohol, or toluene. The nitrocellulose base is made by treating cotton in nitric acid. The plasticizer aids in producing a flexible film. Both the plasticizer and the solvents are responsible for the tautening action of dope. Thinners such as benzol or ethyl alcohol are sometimes added to the dope to obtain the proper consistency. These thinners evaporate with the volatile solvents.

Nitrate dope flows more freely and is more easily applied to fabric than butyrate dope. It burns readily and rapidly and is difficult to extenguish, whereas butyrate dope burns slowly and is easily extinguished. The tautening (shrinking) effect of nitrate is not quite so great as that of butyrate, but it is sufficient to tighten the fabric the desired amount.

Cellulose-Acetate-Butyrate Dope

This type of dope is composed of cellulose acetate butyrate and a plasticizer, triphenyl phosphate, which are nonvolatile when mixed with ethyl acetate, butyl acetate, diacetone alcohol or methyl ethyl ketone, all of which are volatile.

Butyrate dope has a greater tautening effect on fabric and is more fire resistant than nitrate dope.

The solvents of butyrate dope are more penetrating than those of nitrate dope, and butyrate dope can be applied successfully over dried nitrate dope on a fabric surface.

Both the cellulose nitrate and cellulose acetate butyrate dopes, without the addition of color pigments, are a clear, transparent solution. Both are used on aircraft fabric covering to shrink and tighten the fabric to a drum-like surface, to impregnate and fill the fabric-mesh, and to waterproof, airproof, strengthen, and preserve the fabric.

Pigments of the desired color may be added to the final two or three coats of dope applied to the fabric to attain the desired color and trim on the aircraft.

ALUMINUM-PIGMENTED DOPES

When at leat two or more coats of aluminum-pigmented dope (brushed or sprayed) have been applied over the first two or three coats of clear dope after they have dried and have been sanded, a thin film of aluminum is formed over the fabric and the undercoats of clear dope. This aluminum film insulates the fabric from the sun's heat and reflects the heat and ultraviolet rays away from the fabric surfaces of the aircraft.

Aluminum-pigmented dopes may be purchased already mixed and ready for application by brush or spray. However, it is often more economical and desirable to mix the powdered aluminum into the clear dope in the shop.

The aluminum for mixing into the clear dope may be obtained in either the powdered form or the paste form. In the powdered form it is nothing more than finely ground (pulverized) aluminum metal. In the paste form the powdered aluminum metal has been mixed with an adhesive agent to form a putty-like paste.

Recommended mixing proportions are 1-1/2 lbs. of aluminum powder to 5 gal. of clear dope, or 1-3/4 lbs. of aluminum paste to 5 gal. of clear dope. First, thoroughly mix and dissolve the powder or paste in a small amount of alcohol thinner and then add to the clear dope.

TEMPERATURE AND HUMIDITY EFFECTS ON DOPE

The successful application of dope finishes on fabric depends on many things, including the method of application, temperature, humidity, correct mixture of anti-blush reducers and thinners, sanding, and preparation of the fabric. In addition to the special methods necessary in the application of dope, further precautions are required in the handling, storage, and use of dope because it is highly flammable and its fumes are harmful if breathed in excess. For the best and safest results, doping is usually done in a special dope room where many of these factors can be controlled.

Cold Effects on Dope

In cold weather, dopes left in unheated rooms or outside become quite viscous (thick). Cold dopes should be kept in a warm room between 75° F. and 80° F. at least 24 hrs. before being used. Dope in large drum containers (55 gal.) will require 48 hrs. to reach this temperature. Cold dopes will pull and rope under the brush and, if thinned sufficiently to spray or brush, will use extra thinner needlessly and will lack body when the thinner evaporates.

COMMON TROUBLES IN DOPE APPLICATION

Bubbles and Blisters

A heavy coat of lacquer applied over a doped surface that is not thoroughly dry will tend to form bubbles. To prevent this condition, allow the surface to dry for 10 to 12 hrs. Bubbles may be removed by washing the surface with dope thinner until smooth, allowing the surface to dry, and then sanding before refinishing. Blisters are caused by dope dripping through to the opposite fabric during application of the priming coat, as a result of excessive brushing over spars, ribs, or other parts. Dope may also seep through fittings, inspection openings, or patches, and form blisters. Extreme care should be taken to avoid blisters inasmuch as they can be removed only by cutting the fabric at the blister, and patching.

Slack Panels

Slack panels are caused by loose application of the fabric, or the fabric may have been applied with proper tension but permitted to remain undoped for too long a period, thus losing its tension. Fabric slackened by remaining undoped may be tightened by the application of acetone if it is applied as soon as the slackening is noticeable.

Extremes of temperature or humidity may cause dope to dry in such a condition that the fabric becomes slack. This can be remedied by spraying on another coat of dope containing either a slow dryer, such as butyl alcohol, or a rapid dryer, such as acetone, as conditions may require.

Inconsistent Coloring

Inconsistent coloring of enamels, paints, and pigmented dope, is caused by the pigments settling to the bottom of the container, thus depriving the upper portion of the vehicle of its proper percentage of pigment. If shaking the container does not distribute the pigment satisfactorily, a broad paddle or an agitator should be used to stir the mixture thoroughly.

Pinholes

Pinholes in the dope film can be caused by the temperature of the dope room being too high, by not brushing the priming coat well into fabric to seal it completely, by a heavy spray coat of a mixture containing too much thinner, or by water, oil, or dirt in the air supply of the spray gun.

Blushing

Blushing in dopes or lacquers is common in humid weather. This condition in cellulose nitrate and cellulose acetate dopes is caused by rapid evaporation of thinners and solvents. The evaporation lowers the temperature on the surface of the freshly doped fabric, causing condensation of moisture from the atmosphere. This moisture on the surface of the wet dope or lacquer precipitates the cellulose nitrate or cellulose acetate out of solution, thus giving the thick milky-white appearance known as blush. Of course, such a decomposed finish is of no value either in tautening or protecting the surface for any period of time. Therefore, the blush must be eliminated if the finish is to endure.

The common causes of blushing are:
(1) Temperature too low.
(2) Relative humidity too high.
(3) Drafts over freshly doped surface.
(4) Use of acetone as a thinner instead of nitrate thinner.

If causes (1) and (2) cannot be corrected, blushing may be avoided by adding butyl alcohol to the dope in sufficient quantity to correct the condition. Dope films that have blushed may be restored by applying another coat of dope, thinned with butyl alcohol, over the blushed film. This coat will dissolve the precipitation on the previous coat. If an additional coat is not desired, the blushed film may be removed by saturating a rag with butyl alcohol and rubbing it rapidly and lightly over the blushed film. If butyl alcohol does not remove the blushing, acetone may be applied in the same manner to accomplish this purpose.

Brittleness

Brittleness is caused by applying the fabric too tightly or by the aging of the doped surface. Overtight panels may be loosened by spraying a 50% solution of fast-evaporating solvent (acetone) and dope over the surface to soak into the dope layers, allowing the fabric to slacken. If the age of the doped surface causes brittleness, the only remedy is to re-cover the structure.

Peeling

Peeling is caused by failure to remove moisture, oil, or grease from the fabric before the surface is coated. Fabric areas so affected should be treated with acetone before the priming coat is applied.

Runs and Sags

Runs and sags in the finish are caused either by applying the dope too heavily or by allowing the dope to run over the sides and ends of the surface. Immediately after a surface is finished, the opposite and adjacent surfaces should be inspected for sags and runs.

TECHNIQUE OF APPLICATION

Apply the first two coats of dope by brush, spread on the surface as uniformly as possible, and thoroughly work into the fabric. Be careful not to work the dope through the fabric so that an excessive film is formed on the reverse side. The first coat should produce a thorough and uniform wetting of the fabric. To do so, work the dope with the warp and the fill threads for three or four brush strokes and stroke away any excess material to avoid piling up or dripping. Apply succeeding brush or spray coats with only sufficient brushing to spread the dope smoothly and evenly.

When doping fabric over plywood or metal-covered leading edges, care should be taken to ensure that an adequate bond is obtained between the fabric and the leading edge. Care should also be taken when using predoped fabric to use a thinned dope to obtain a good bond between the fabric and the leading edge of wings.

Applying Surface Tape and Reinforcing Patches

Apply surface tape and reinforcing patches with the second coat of dope. Apply surface tape over all rib lacing and all other points of the structure where tape reinforcements are required.

Installation of Drain Grommets

With the second coat of dope, install drain grommets on the underside of airfoils at the trailing edge and as close to the rib as practicable. On fuselages, install drain grommets at the center of the underside in each fuselage bay, located to ensure the best possible drainage. Special shielded grommets, sometimes called marine or suction grommets, are recommended for seaplanes to prevent the entry of spray. Also use this type of grommet on landplanes in the part of the structure that is subject to splash from the landing gear when operating from wet and muddy fields. Plastic type grommets are doped directly to the covering. Where brass grommets are used, mount them on fabric patches and then dope them to the covering. After the doping scheme is completed, open the drainholes by cutting out the fabric with a small-bladed knife. Do not open drain grommets by punching.

Use of Fungicidal Dopes

Fungicidal dope normally is used as the first coat for fabrics to prevent rotting. While it may be more advisable to purchase dope in which fungicide has already been incorporated, it is feasible to mix the fungicide with dope. Military Specification MIL–D–7850 requires that cellulose acetate butyrate dope incorporate a fungicide for the first coat used on aircraft. The fungicide designated in this specification is zinc dimethyldithiocarbonate, which forms a suspension with the dope. This material is a fine powder, and if it is mixed with the dope, it should be made into a paste, using dope, and then diluted to the proper consistency according to the manufacturer's instructions. It is not practicable to mix the powder with a large quantity of dope.

Copper naphtonate is also used as a fungicide and forms a solution with dope. However, this material has a tendency to bleed out, especially on light-colored fabric. It is considered satisfactory from a fungicidal standpoint.

The first coat of fungicidal dope should be applied extremely thin so that the dope can thoroughly saturate both sides of the fabric. Once the fabric is thoroughly saturated, subsequent coats may be applied at any satisfactory working consistency.

NUMBER OF COATS REQUIRED

Regulations require that the total number of coats of dope should be not less than that necessary to result in a taut and well-filled finish job. A guide for finishing fabric-covered aircraft is:

(1) Two coats of clear dope, brushed on and sanded after the second coat. To prevent damaging the rib stitch lacing cords and fabric, do not sand heavily over the center portion of pinked tape over ribs and spars.

(2) One coat of clear dope, either brushed or sprayed on, and sanded.

(3) Two coats of aluminum-pigmented dope, brushed or sprayed on, and sanded after each coat.

(4) Three coats of pigmented dope (the color desired), sanded and rubbed to give a smooth glossy finish when completed.

GENERAL

Metal- or wood-covered aircraft frequently are painted to protect their surfaces from deterioration and to provide a desirable finish. Many types of finishes are used on aircraft structures. Wood structures may be varnished, whereas aluminum and steel frequently are protected and preserved by applying paint. The term "paint" is used in a general sense and includes primers, enamels, lacquers, and epoxies.

Aircraft finishes can be separated into three general classes: (1) Protective, (2) appearance, and (3) decorative. Internal and unexposed parts are finished to protect them from deterioration. All exposed parts and surfaces are finished to provide protection and to present a pleasing appearance. Decorative finishing includes trim striping, the painting of emblems, the application of decals, and identification numbers and letters.

FINISHING MATERIALS

A wide variety of materials are used in aircraft finishing. Some of the more common materials and their uses are described in the following paragraphs.

Acetone

Acetone is a fast-evaporating dope solvent that is suitable for removing grease from fabric prior to doping, for cleaning spray paint guns, and as an ingredient in paint and varnish removers. It should not be used as a thinner in dope since its rapid drying action causes the doped area to cool and collect moisture. The absorbed moisture prevents uniform drying and results in blushing.

Alcohol

Butyl alcohol is a slow-drying solvent that can be mixed with dope to retard drying of the dope film on humid days, thus preventing blushing. Generally, 5 to 10% of butyl alcohol is sufficient for this purpose.

Butyl alcohol and ethyl alcohol are used together as a mixture to dilute wash coat primer as necessary for spray application. The percentage of butyl alcohol used will depend on the temperature and humidity. The butyl alcohol retards the evaporation rate. In some cases a 25% butyl/75% ethyl alcohol mixture may be satisfactory; in others, a 50/50 mixture may be required.

Denatured alcohol is used for thinning shellac to spray gun consistency, and as a constituent of paint and varnish remover.

Isopropyl alcohol is used as a diluent in the formulation of oxygen system cleaning solutions. It is also used in preparing nonionic detergent mixtures.

Benzene

Benzene is used for cleaning equipment in which enamel, paint, or varnish has been used. It is also used as a constituent of paint and varnish remover.

Thinner

Dopes, enamels, paints, etc., are thinned for use in spray guns, for more efficient brushing consistency, and for reducing the thickness of coats. The correct thinner must be used with a specific finishing material.

Several materials required for thinning specific paints and lacquers are also available for solvent cleaning, but they must be used with care. Most of these materials have very low flash points and, in addition, will damage existing paint finishes. Some of the more common paint thinners are briefly discussed in the following paragraphs.

Acrylic Nitrocellulose Lacquer Thinner

Acrylic nitrocellulose lacquer thinner may be effectively used to wipe small areas prior to paint touchup. It will soften the edges of the base paint film, which in turn will assure improved adhesion of the touchup coating. However, the thinner contains toluene and ketones and should never be used indiscriminately for cleaning painted surfaces.

Cellulose Nitrate Dope and Lacquer Thinner

This thinner is both explosive and toxic as well as damaging to most paint finishes. It may be used for hand removal of lacquer or primer overspray. It is the approved thinner for nitrocellulose lacquers and is a mixture of ketones, alcohols, and hydrocarbons.

Volatile Mineral Spirits

This material is very similar to dry-cleaning solvent but evaporates somewhat faster and leaves less residue after evaporation. It can be effectively used in wiping stripped metal surfaces just before the re-application of paint finishes. It is also used as a carrier for solvent-emulsion compounds in general cleaning.

Toluene

Toluene (toluol) may be used as a paint remover in softening fluorescent finish clear topcoat sealing materials. It is also an acceptable thinner for zinc chromate primer.

Turpentine

Turpentine is used as a thinner and quick-drier for varnishes, enamels, and other oil-base paints. Turpentine is a solvent for these types of materials and can be used to remove paint spots and to clean paint brushes.

Dope

Aircraft dope is essentially a colloidal solution of cellulose acetate or nitrate, combined with sufficient plasticizers to produce a smooth, flexible, homogeneous film. The dope imparts to the fabric cover additional qualities of increased tensile strength, airtightness, weather-proofing, and tautness of the fabric cover. Dope must possess maximum durability, flexibility, blush resistance, and adhesion, while adding the least additional weight. Each coating of dope applied over undercoats must penetrate and soften them, and build up a smooth, united surface without lessening the degree of fabric tautness.

The six essential constituents of dope are:

(1) Film-base compounds, which are either cellulose acetate or cellulose nitrate.

(2) Plasticizers, such as camphor oil and castor oil, used to produce a durable, flexible film.

(3) Solvents, used to dissolve the cellulose-base materials.

(4) Diluents, used for thinning the mixture. Toxic diluents, such as benzol, are never used.

(5) Slow dryers, such as butyl alcohol, used to prevent too rapid drying, which tends to

over-cool the surface, thus causing condensation of moisture and resultant blushing.

(6) Colors or pigments, which are finely ground particles of inorganic material added to clear dope to give a desired color.

The three types of dope used for aircraft finishes are: (1) Clear, (2) semi-pigmented, and (3) pigmented. Their characteristics and uses are:

(1) There are two clear nitrate dopes. One is used to produce a gloss finish over semi-pigmented finishes, and as a vehicle for bronze/aluminum doped finishes. The other is a specially prepared, quick-drying material to be used only for patching.

(2) Semi-pigmented nitrate dope contains a limited quantity of pigment. It is used for finishing fabric-covered surfaces.

(3) Pigmented nitrate dope contains a greater quantity of pigment than does semi-pigmented dope, and is normally used for code marking and finishing insignia. One or two coats applied over semi-pigmented dope will produce the desired color effect.

Dope should not be applied over paint or enamel because it tends to remove such material.

Nitrocellulose Lacquer

Nitrocellulose lacquers are available in both glossy and flat finishes. They are also available in either clear or pigmented form. These materials can be applied over either old type zinc chromate or the newer modified zinc chromate primer. The lacquer finish is applied in two coats; a mist coat first, with a full, wet cross coat applied within 20 to 30 min. afterward. The lacquer finishes should be thinned as necessary, using cellulose nitrate dope and lacquer thinner. Clear lacquer may be substituted for spar varnish over doped fabric and is also used with bronze/aluminum powder to produce aluminized lacquer. Clear lacquer should never be applied over paint, enamel, or varnish because it tends to remove such material.

Acrylic Nitrocellulose Lacquer

This is the most common topcoat in use today, available either in flat or glossy finish. Both types of material are required in refinishing conventional aircraft. Anti-glare areas generally require the use of flat finishes. The remaining surfaces usually are finished with glossy materials to reduce heat absorption. The base materials should be thinned as necessary for spray application with acrylic nitrocellulose thinner.

Paint Drier

Paint drier is added to paint when improved drying properties are desired. Excessive drier in paint will result in a brittle film, causing cracking and peeling.

Linseed Oil

Linseed oil is used to reduce semi-paste colors, such as dull black stenciling paint and insignia colors, to brushing consistency. It is also used as a protective coating on the interior of metal tubing.

Zinc Chromate Primer

Zinc chromate primer is applied to metallic surfaces, before the application of enamel or lacquer, as a corrosion-resistant covering and as a base for protective topcoats. Older type zinc chromate primer is distinguishable by its bright yellow color compared to the green cast of the modified primers currently used. The old type primer will adhere well to bare metal. It is still specified as an acceptable coating for internal surfaces, and it forms a part of the old type nitrocellulose system finishes. It can be applied by brush or spray and should be thinned for spraying as necessary with toluene. When this material is to be applied by brush, it should be thinned to brushing consistency with xylene to give better wet-edge retention. It dries adequately for overcoating within an hour. Zinc chromate primer is satisfactory for use under oil-base enamels or nitrocellulose lacquers. It is also an excellent dope proof paint.

Standard Wash Primer

Some paint finishes in general use include a standard wash primer undercoat, also termed a metal pre-treatment coating compound. It is a two-part material consisting of resin and alcoholic phosphoric acid, which is added just prior to application. The two components should be mixed very slowly and carefully and allowed to stand at least 30 min. before use. The primer should be used within a total time of 4 hrs. Any necessary thinning is done with a 25/75 to 50/50 mixture of butyl alcohol and ethyl alcohol, respectively. The percentage of butyl alcohol used will be determined by the evaporation rate. The percentage of butyl alcohol should be kept to the minimum possible under local temperature and humidity conditions. It is particu-larly important that the ratio of acid to resin in the wash primer be maintained. Any decrease in acid will result in poor coat formation. An excess of acid will cause serious brittleness.

Acrylic Cellulose Nitrate Modified Primer

The lacquer primer currently applied over the wash coat base is a modified alkyd-type zinc chromate developed for its adherence to the wash primer. It does not adhere well to bare metal, but works effectively as a sandwich between the wash coat primer and the acrylic nitrocellulose topcoating. It can be thinned as necessary for spray application with cellulose nitrate thinner. In areas where the relative humidity is high, it may be more desirable to use acrylic nitrocellulose thinner. It should be topcoated within 30 to 45 min. after its application for best results.

Under no condition should it dry more than an hour and a half before finish coats of acrylic lacquer are applied. If primer coats are exposed to atmospheric conditions for longer than this maximum drying period, a re-application of both the wash primer and modified primer is necessary, followed immediately by the application of the acrylic lacquer topcoat. Otherwise, complete stripping of the coatings and refinishing is required.

In general, freshly applied coatings can be removed with either acrylic lacquer thinner or methyl ethyl ketone. However, once the coat is dry, paint stripper is required for complete removal of the coating.

The finish coatings are usually applied in two coats over the modified zinc chromate primer; the first a light mist coat, and the second a wet cross-coat with 20 to 30 min. drying time allowed between the two coatings. On amphibians or sea-planes, where maximum protection is required, the finish is increased to two coats of primer and three coats of lacquer. Once the paint finish has set, paint stripper is necessary for its removal.

Enamel

Enamels are special types of varnish having either an oil base or nitrocellulose base as the solvent. Enamel finishes are generally glossy, although flat enamel finishes are available. Enameled surfaces are hard, resist scratching and the action of oils or water, and certain grades resist high temperatures. Enamel can be applied by spraying or brushing and is suitable for either interior or exterior application.

Varnish

Spar varnish is used for finishing interior or exterior wood surfaces. It produces a transparent, durable coating for use where high gloss and hardness are not the principal requirements.

Asphalt varnish is a black coating used for the protection of surfaces around lead acid batteries, and where acid or water are present.

Oil Stain

Oil stain is used to stain wood for decorative purposes. It is available in light and dark shades, simulating mahogany, oak, walnut, or other wood.

Color

Various coloring materials are used for special applications, such as insignia and signs. These colors are obtainable as pastes (powder ground in oil) to be mixed with the proper solvent.

Paint

Paint is a mechanical mixture of a vehicle and a pigment. The vehicle is a liquid that cements the pigment together and strengthens it after drying. The pigment gives solidity, color, and hardness to the paint. Among the commonly used pigments are: iron oxide, zinc chromate, titanium oxide, iron blue lead chromate, carbon black, and chrome green.

The vehicles used for paint can be divided into two general classes: (1) Solidifying oils, and (2) volatile oils. The solidifying oils dry and become tough leathery solids upon exposure to the air. China wood oil, tung oil, or linseed oil are the most common solidifying oils used in aircraft paint. Volatile oils, or spirits are those which evaporate when exposed. These oils are used to dilute paint to the proper consistency and to dissolve varnish resins. The most common volatile vehicles are: Alcohol, turpentine, benzine, toluene, ethyl acetate, and butyl acetate. Paints, varnishes, and enamels are usually composed of a pigment and a mixture of both solidifying and volatile oils. Lacquer, which is noted for its rapid drying, is composed of pigments, resins, and volatile oils.

Paint Remover

General-purpose paint and enamel remover is a good, nonflammable, water-rinsable paint remover. It is used for stripping lacquer and enamel coatings from metal surfaces, and it consists of active solvents, amines, ammonia, thinners, emulsifiers, a stable chlorinated solvent, and a cresol mixture that can be applied by fluid spray or brush. The cresol

additive swells the resins in the paint coatings, while the chlorinated constituents penetrate through and lift the softened resins by evaporation. This material is water-rinsable after application and can be applied several times on stubborn coatings. It should never be permitted to contact acrylic windows, plastic surfaces, or rubber products. This material should be stored indoors or in an area well protected against weather conditions. Goggle-type eyeglasses and protective clothing should be worn when using it. Paint stripping procedures, discussed later in this chapter, are the same for touchup as for a complete repainting.

Epoxy Coating Remover

Strong acid solutions or alkaline tank stripping agents are the most effective materials for removal of certain well-cured epoxies at the present time, but these stripping agents may not be used on aluminum surfaces. General-purpose paint and enamel remover can remove most epoxy finishes. Several applications or extended dwell times may be necessary for effective results.

Fluorescent Paint Remover

Fluorescent paint remover, water-rinsable type, is a paint stripper designed to remove fluorescent paint finishes from exterior surfaces of aircraft. This material is used for stripping the high-visibility coatings without affecting the permanent acrylic or cellulose nitrate coatings underneath. A permanent base coating of cellulose nitrate lacquer may be softened by this material if allowed to remain too long.

Work with paint remover should be done out of doors in shaded areas whenever practicable, or with adequate ventilation when used indoors. Rubber, plastic, and acrylic surfaces require adequate masking. Goggle type glasses, rubber gloves, aprons, and boots should be worn during any extensive application of this stripper. Hand stripping of small areas requires no special precautions.

Masking Material

Masks are used to exclude areas to which dope or lacquer, etc., is not to be applied. Masks are made of thin metal, fiberboard, paper, or masking tape. Metal and fiberboard masks are usually held in place by weights, and paper masks by masking tape.

Liquid spray shield is a solution applied to protect areas, thus serving as a liquid mask. The liquid shield and the finish deposited upon it are easily washed off with water when the design is dry.

116

Storage of Finishing Material

Dope, paint, enamel, and other finishing material should be stored in a dry place away from direct sunlight and heat. Each container is assigned a code and color number identifying the material contained therein.

Stored paint, enamel, and other finishing material that has separated from the vehicle must be mixed to regain usefulness. If the pigment is caked, pour most of the liquid into another container and mix the caked pigment until it is free of lumps. A broad paddle or an agitator may be used for this purpose. When the pigment is smooth and free from lumps, the liquid is added slowly and the stirring is continued to ensure complete mixing.

PAINT TOUCHUP

A good intact paint finish is one of the most effective barriers available for placement between metal surfaces and corrosive media. Touching up the existing paint finish and keeping it in good condition will eliminate most general corrosion problems.

When touching up paint, confine paint coverage to the smallest area possible. Acrylic primer or lacquer may be used, but adhesion is usually poor. Epoxy coatings, as well as the older type of zinc chromate primer, may be used for touchup on bare metal.

When a paint surface has deteriorated badly, it is best to strip and repaint the entire panel rather than attempt to touchup the area. Touchup materials should be the same as the original finish. Surfaces to be painted should be thoroughly cleaned and free from grease, oil, or moisture. Where conditions are not suitable for painting, preservatives may be used as temporary coatings until good painting conditions are restored. Paint finishes should not be too thick since thickness promotes cracking in service.

Much of the effectiveness of a paint finish and its adherence depends on the careful preparation of the surface prior to touchup and repair. It is imperative that surfaces be clean and that all soils, lubricants, or preservatives be removed.

Cleaning procedures for paint touchup are much the same as the procedures for cleaning before inspection. Many types of cleaning compounds are available. Chapter 6, "Hardware, Materials, and Processes," in the Airframe and Powerplant Mechanics General Handbook, AC 65–9A, describes many of these compounds.

IDENTIFICATION OF PAINT FINISHES

Existing finishes on current aircraft may be any one of several types, combinations of two or more types, or combinations of general finishes with special proprietary coatings.

Any of the finishes may be present at any given time, and repairs may have been made using materials from several different types. Some detailed information for the identification of each finish is necessary to assure adequate repair procedures. A simple test is valuable in confirming the nature of the coatings present. The following tests will aid in paint finish identification.

Apply a coating of engine oil (Military Specification MIL–L–7808, or equal) to a small area of the surface to be checked. Old nitrocellulose finishes will soften within a period of a few minutes. Acrylic and epoxy finishes will show no effects.

If not identified, next wipe down a small area of the surface in question with a rag wet with MEK (methyl ethyl ketone). MEK will pick up the pigment from an acrylic finish, but will have no effect on an epoxy coating. Wipe the surface; do not rub. Heavy rubbing will pick up even epoxy pigment from coatings that are not thoroughly cured. Do not use MEK on nitrocellulose finishes. No test of fluorescent finishes should be necessary other than visual examination.

PAINT REMOVAL

One of the most important jobs is the stripping of old paint finishes preparatory to applying a new surface cover coat. An original finish may have to be removed in any of the following cases:

(1) If a panel or other area on the aircraft has badly deteriorated paint surfaces.

(2) If repair materials are not compatible with the existing finish, thereby precluding touchup repair.

(3) If corrosion is evident or suspected under an apparently good paint coating.

The area to be stripped must be cleaned of grease, oil, dirt, or preservatives to assure maximum efficiency of the stripping compound. The selection of the type of cleaning materials to be used will depend on the nature of the matter to be removed. Dry-cleaning solvent may be used for removing oil, grease, and soft preservative compounds. For heavy-duty removal of thick or dried preservatives, other compounds of the solvent-emulsion type are available.

In general, paint stripping materials are toxic and must be used with care. The use of a general-purpose, water-rinsable stripper is recommended for most field applications. Wherever practicable, paint removal from any large area should be done out of doors and preferably in shaded areas. If indoor removal is necessary, adequate ventilation must be assured. Synthetic rubber surfaces, including aircraft tires, fabric, and acrylics, must be thoroughly protected against possible contact with paint remover. Care must be taken when using paint remover around gas- or water-tight seam sealants, since this material will soften and destroy the integrity of the sealants.

Mask any opening that would permit stripper to get into aircraft interiors or critical cavities. Paint stripper is toxic and contains ingredients harmful to both skin and eyes. Rubber gloves, aprons of acid repellent material, and goggle type eyeglasses should be worn if any extensive paint removal is to be done. A general stripping procedure is discussed in the following paragraphs.

No prepared paint remover should be used on aircraft fabric or be allowed to come in contact with any fiberglass reinforced parts such as radomes, radio antenna, or any component such as fiberglass reinforced wheel pants or wing tips. The active agents will attack and soften the binder in these parts.

CAUTION: Any time you use a paint stripper, always wear protective goggles and rubber gloves. If any stripper is splashed on your skin, wash it off immediately with water; and if any comes in contact with your eyes, flood them repeatedly with water and CALL A PHYSICIAN.

Brush the entire area to be stripped with a cover of stripper to a depth of 1/32 in. to 1/16 in. Any paint brush makes a satisfactory applicator, except that the bristles will be loosened by the effect of paint remover on the binder. The brush should not be used for other purposes after being exposed to paint remover.

After applying the stripping compound, it may be covered with an inexpensive polyethane drop cloth. Covering prevents rapid evaporation of the solvents and facilitates penetration of the paint film.

Allow the stripper to remain on the surface for a sufficient length of time to wrinkle and lift the paint. This may vary from 10 min. to several hours, depending on the temperature, humidity, and the condition of the paint coat being removed. Scrub the paint-remover-wet surface with a bristle brush saturated with paint remover to further loosen any finish that may still be adhering to the metal.

Re-apply the stripper as necessary in areas that remain tight or where the material has dried, and repeat the above process. Nonmetallic scrapers may be used to assist in removing persistent paint finishes.

Remove the loosened paint and residual stripper by washing and scrubbing the surface with water. If water spray is available, use a low-to-medium pressure stream of water directly on the scrubbing broom. If steam cleaning equipment is available and the area is sufficiently large, this equipment, together with a solution of steam cleaning compound, may be used for cleaning. On small areas, any method may be used that will assure complete rinsing of the cleaned area.

RESTORATION OF PAINT FINISHES

The primary objective of any paint finish is the protection of exposed surfaces against deterioration. Other reasons for particular paint schemes are:

(1) The reduction of glare by nonspecular coatings.
(2) The use of white or light-colored, high-gloss finishes to reduce heat absorption.
(3) High visibility requirements.
(4) Identification markings.

All of these are of secondary importance to the protection offered by a paint finish in good condition. A faded or stained, but well-bonded paint finish is better than a fresh touchup treatment improperly applied over dirt, corrosion products, or other contaminants.

NITROCELLULOSE LACQUER FINISHES

Nitrocellulose finishes ordinarily consist of a wash primer coat and a coat of zinc chromate primer. A nitrocellulose lacquer topcoat is applied over the prime coats.

Replacement of Existing Finish

When an existing nitrocellulose finish is extensively deteriorated, the entire aircraft may have to be stripped of paint and a complete new paint finish applied. When such damage is confined to one or more panels, the stripping and application of the new finish may be limited to such areas by masking to the nearest seam line.

The complete nitrocellulose lacquer finish is begun with the application of standard wash primer undercoat. The wash primer should be applied in a thin coat, with the texture of the metal still visible through the coating. If absorption of water results and the coat shows evidence of blushing, successive coatings will not adhere. The area should be re-sprayed with butyl alcohol to re-deposit the wash primer. If blushing is still evident, it should be stripped and re-sprayed. After 20 min. drying time, adherence of the film should be checked with a thumbnail test. A moderate thumbnail scratch should not remove the prime coat.

The wash primer must be applied over a pre-cleaned surface that has been wiped with a volatile solvent such as MEK, naphtha, or paint and lacquer thinners just before paint application. Evaporation of the solvent should be complete before the prime coat is added. Better results will be obtained if the solvent wipe-down is followed by a detergent wash.

Lacquer primer is a modified alkyd-type zinc chromate developed for its adherence to the wash primer. Lacquer primer does not adhere well to bare metal, but works effectively as a sandwich between the wash coat primer and the nitrocellulose opcoating, and can be thinned as necessary for spray application with cellulose nitrate thinner. In areas where the relative humidity is high, it may be more desirable to use acrylic nitrocellulose thinner. For best results, lacquer primer should be topcoated within 30 to 45 min. after its application.

The old type primer will adhere well to bare metal and is still specified as an acceptable coating for internal surfaces as well as a part of the nitrocellulose finishes. Apply by brush or spray; thin for spraying with toluene. When this material is to be applied by brush, thin to brushing consistency with xylene to give better wet-edge retention. Over-coating may be applied within an hour.

Nitrocellulose lacquers are available in both glossy and nonspecular finishes. The lacquer finish is applied in two coats: a mist coat first, with a full wet crosscoat applied within 20 to 30 min. The lacquer should be thinned as necessary, using cellulose nitrate dope and lacquer thinner.

Cellulose nitrate dope and lacquer thinner (Federal Specification TT-T-266) is both explosive and toxic, as well as damaging to most paint finishes. Dope and lacquer thinner may be used for hand removal of lacquer or primer overspray, is an approved thinner for nitrocellulose lacquers, and is a mixture of ketones, alcohols, and hydrocarbons.

The surface areas of damaged paint must be clean prior to touchup repair, and all soils, lubri-cants and preservatives must be removed. Cleaning procedures for paint touchup are much the same as those for paint removal.

If the old finish is not to be completely stripped, the existing surface must be prepared to receive the new cover coat after cleaning. If good adhesion is to be obtained, all loose paint should be brushed off, giving particular attention to overpaint usually found in wheel wells and wing butt areas. Curled or flaky edges must be removed and feathered to provide about 1/2 in. of overlap. A fine abrasive approved for aircraft use should be used and extreme care taken to ensure that existing surface treatments are not damaged.

After sanding, sanded areas and bare metal should be wiped with either mineral spirits, alcohol, aliphatic naphtha, or dry-cleaning solvent. Following complete evaporation of these solvents, a detergent wash using a nonionic detergent/isopropyl alcohol mixture should be applied just prior to painting. This will improve paint adhesion.

ACRYLIC NITROCELLULOSE LACQUER FINISH

Acrylic nitrocellulose lacquer is one of the most common topcoats in use today, available either as nonspecular material or glossy finish. Both types of material are required in refinishing conventional aircraft. Surfaces visible from above and other anti-glare areas generally require the use of nonspecular finishes. The remaining surfaces are usually finished with glossy materials to reduce heat absorption. The base materials should be thinned as necessary for spray application with acrylic nitrocellulose thinner.

Replacement of Existing Acrylic Nitrocellulose Lacquer Finish

This finish includes a wash primer coat, modified zinc chromate primer coat, and an acrylic nitrocellulose lacquer topcoat. This finish may be applied only in the sequence specified in the manufacturer's instructions and will not adhere to either the old nitrocellulose coatings or the new epoxy finishes. Even when finishes are applied over old acrylic coatings during touchup, a softening of the old film with a compatible thinner is required.

When a finish is being rebuilt from bare metal, the steps through the application of the modified primer are the same as for nitrocellulose finishes, except that old type zinc chromate primer may not be used. As with the nitrocellulose finish, the acrylic nitrocellulose topcoat should be applied within 30 to 45 min. The finish coatings are usually applied in two coats over the modified primer: the first a mist coat, and the second a wet, full-hiding crosscoat, with 20 to 30 min. drying time allowed between the two coatings. Once the paint finish has set, paint stripper is necessary for its removal.

Acrylic nitrocellulose lacquer thinner is used in thinning acrylic nitrocellulose lacquers to spray consistency.

When rebuilding acrylic finishes, use two separate thinners: (1) Cellulose nitrate dope and lacquer thinner to thin the modified primer, and (2) acrylic nitrocellulose lacquer thinner to reduce the topcoat material. Make sure that the thinner materials are used properly and that the two are not mixed.

Touchup of Acrylic Nitrocellulose Finishes

After removal of damaged paint, the first step before application of touchup acrylic nitrocellulose lacquer is preparing the old coat to receive the new. Acrylic nitrocellulose lacquer thinner may be effectively used to wipe small areas prior to painting. This will soften the edges of the base paint film around damaged areas, which in turn will assure improved adhesion of the touchup coating. However, the thinner contains toluene and ketones and should never be used indiscriminately for cleaning painted surfaces.

When softening old, good-condition acrylic nitrocellulose finishes with thinner, care should be taken to avoid penetration and separation of the old primer coats. The new acrylic lacquer coat should be applied directly over the softened surface without the use of primers between the old and new coats.

EPOXY FINISHES

Another type of paint becoming increasingly common is a Military Specification epoxy finish or proprietary epoxy primer and topcoats. These finishes ordinarily consist of a conventional wash primer coat and two coats of epoxy material. However, in some cases it may consist of a three-coat finish that includes wash primer plus epoxy-polyamide primer with an epoxy-polyamide topcoat.

The high gloss inherent with this system is primarily due to the slow flowing resins used. The thinners flash off quickly but the resins continue to flow for three to five days. It is this long flow-out time and the even cure throughout the film that gives the pigment and the film time to form a truly flat surface, one that reflects light and has the glossy "wet" look which makes them so popular.

Polyurethane finish is used on agricultural aircraft and seaplanes because of its abrasion resistance and resistance to chemical attack. Phosphateesten (Skydrol) hydraulic fluid, which quite actively attacks and softens other finishes, has only minimal effect on polyurethanes. Even acetone will not dull the finish. Paint strippers must be held to the surface for a good while to give the active ingredients time to break through the film and attack the primer.

The epoxy material presently in use is a two-package system that consists of a resin and a converter which must be mixed in definite proportions just before application. Since the proportions will vary between colors used and also with sources of procurement, it is important that instructions on the specific container be observed carefully. The converter should always be added to the resin, *never resin to the converter*. Also, do not mix materials from two different manufacturers. The mixture should be allowed to stand at least 15 min. before initial use.

In this time the curing action is started. The primary purpose of this waiting period is to aid in the application and actually has little to do with the results of the finish itself. After this induction period, the material is stirred and mixed with reducer to the proper viscosity for spraying. When you have the proper viscosity, spray on a very light tack coat, lighter than with a conventional enamel. Allow it to set for about 15 minutes so the thinner can flash off, or evaporate, and spray on a full wet cross coat. The main problem with the application of polyurethane lies in getting it on too thick. A film thickness of about 1.5 mils (one and a half thousandths of an inch) is about maximum for all areas except for those subject to excessive erosion, such as leading edges. Too thick a film which might build up in the faying strips can crack because of loss of flexibility. A good practical way to tell when you have enough material is to spray until you feel that one more pass will be just right, then quit right there, before you make that "one more pass." The high solids content of polyurethane, its slow drying, and low surface tension allows the finish to crawl for an hour or so after it has been put on. If you can still see the metal when you think you have almost enough, don't worry; it will flow out and cover it. Almost no polyurethane job will look good until the next day, because it is still flowing. It will actually flow for about 3 to 5 days. It will be hard in this time, and the airplane may be flown in good weather, but the paint below the surface is still moving.

Masking tape may be applied after 5 hours under the most ideal conditions, but it is far better if you can wait 24 hours after application of the finish; it should be removed as soon after the trim is sprayed as possible. If it is left on the surface for a day or so, it will be almost impossible to remove.

Both the polyurethane enamel and the epoxy primer which bonds the film to the surface are catalytic materials. They should be mixed and used within 6 hours. If they are not applied within this time, they will not have the full gloss because of the reduced flow time. If it is impossible to spray all of the polyurethane within the 6 hour time period, careful addition of reducer can add a couple of hours to the useful life of the material.

The catalysts used for these primers and finishes are highly reactive to moisture, and the cans should be recapped immediately after using. If a can of the catalyst is allowed to remain open for a period of time, and is then resealed, the moisture in the can will activate it, and swell it up so much there is danger of the can bursting. High humidity and/or heat accelerates the cure.

All catalyzed material must be removed from the pressure pot, the hose, and the gun, immediately upon completion of the spraying operation, and the equipment thoroughly washed. If any of this material is allowed to remain overnight, it will solidify and ruin the equipment.

Precautions must be taken to assure respiratory and eye protection when mixing the two-part resin and activator. Gloves and aprons should also be used to prevent skin contact. Smoking or eating in the mixing area should be specifically prohibited, and mixing should be accomplished in a well-ventilated area. The uncured resins and catalysts contained in these mixtures can cause skin sensitivity similar to a poison ivy reaction.

Touchup of Epoxy Finishes

Epoxy coatings may be applied directly over bare metal in small areas. Minor damage such as scratches and abrasions should be repaired by applying the epoxy topcoat directly to the damaged area, whether or not the damage extends through to the bare metal. The area should be thoroughly cleaned and the edges of the old coating roughened to assure adherence. This material builds up very rapidly. Coats that are too heavy are easily produced and are particularly subject to poor adhesion and cracking.

Larger areas of damage should be repaired by stripping to the nearest seam line and building a complete epoxy finish.

FLUORESCENT FINISHES

Fluorescent paint finishes are available in two types of equal fade- and weather-resistant qualities: (1) A removable finish which is designed for ease of removal and (2) a permanent finish which ordinarily may not be removed without stripping the entire paint finish down to bare metal. These fluorescent paints are applied over full-hiding, clean, white undercoats for maximum reflectance.

Replacement of Existing Finish

For optimum weather resistance and film properties, the dry film thickness must be at least 3 mils for the fluorescent body coat and 1 mil for the clear topcoat. A clear topcoat of from 1 to 1-1/2 mils is necessary to screen out ultraviolet rays from the sun and prevent early or spotty fading of the fluorescent finish. The use of clear lacquers, other than those provided with the fluorescent paint, may also promote fading.

When the permanent finish is white and a fluorescent finish is needed, the permanent white finish may serve as the undercoat. If the permanent finish is any other color, a white lacquer should be used under the fluorescent paint.

When applying fluorescent paint to epoxy finishes, first coat the epoxy surface with white nitrocellulose lacquer, since the fluorescent finish does not adhere too well to the epoxy films. These high-visibility finishes are effective for periods of 6 to 8 months.

Touchup of Fluorescent Finishes

Touchup of fluorescent finishes is difficult to control and should seldom be attempted. Any touchup will be noticeable because of the variations in shading.

Minor damage in fluorescent coatings is repaired by masking, stripping with toluene down to the white undercoat, and repainting with fluorescent paint. This should include one or more touchup coatings of fluorescent paint finish and then overcoated with a clear topcoat sealant.

ENAMEL FINISHES

Enamels frequently are used for the topcoats in finishing aircraft. Practically all aircraft enamels are made by mixing a pigment with spar varnish or glycerol phthalate varnish.

Most enamel finishes used on aircraft components are baked finishes that cannot be duplicated under field conditions. Some are proprietary (patented) materials that are not available in standard stock.

However, for touchup purposes on any enameled surface, standard air-drying, glossy enamel or quick-drying enamel may be used.

High-gloss enamel is thinned with mineral spirits, can be applied by brushing, and should ordinarily be used over a zinc chromate primer coat base. Quick-drying enamel is best thinned with aromatic naphtha. In situations where a primer is not available, either of these enamels may be applied directly to bare metal.

If no enamel is available for touchup purposes, epoxy topcoat material may be substitued. The use of acrylic nitrocellulose lacquer for enamel repairs usually is not satisfactory.

PAINT SYSTEM COMPATIBILITY

The use of several different types of paint, coupled with several proprietary coatings, makes repair of damaged and deteriorated areas particularly difficult, since paint finishes are not necessarily compatible with each other. The following general rules for constituent compatibility are included for information and are not necessarily listed in the order of importance:

(1) Old type zinc chromate primer may be used directly for touchup of bare metal surfaces and for use on interior finishes. It may be overcoated with wash primers if it is in good condition. Acrylic lacquer finishes will not adhere to this material.

(2) Modified zinc chromate primer will not adhere satisfactorily to bare metal. It must never be used over a dried film of acrylic nitrocellulose lacquer.

(3) Nitrocellulose coatings will adhere to acrylic finishes, but the reverse is not true. Acrylic nitrocellulose lacquers may not be used over old nitrocellulose finishes.

(4) Acrylic nitrocellulose lacquers will adhere poorly to both nitrocellulose and epoxy finishes and to bare metal generally. For best results the lacquers must be applied over fresh, successive coatings of wash primer and modified zinc chromate. They will also adhere to freshly applied epoxy coatings (dried less than 6 hrs.).

(5) Epoxy topcoats will adhere to all the paint systems that are in good condition and may be used for general touchup, including touchup of defects in baked enamel coatings.

(6) Old wash primer coats may be overcoated directly with epoxy finishes. A new second coat of wash primer must be applied if an acrylic finish is to be applied.

(7) Old acrylic finishes may be refinished with new acrylic if the old coating is thoroughly softened using acrylic nitrocellulose thinner before paint touchup.

(8) Damage to epoxy finishes can best be repaired by using more epoxy, since neither of the lacquer finishes will stick to the epoxy surface. In some instances, air-drying enamels may be used for touchup of epoxy coatings if edges of damaged areas are first roughened with abrasive paper.

METHODS OF APPLYING FINISHES

There are several methods of applying aircraft finishes. Among the most common are dipping, brushing, and spraying.

Dipping

The application of finishes by dipping is generally confined to factories or large repair stations. The process consists of dipping the part to be finished in a tank filled with the finishing material. Prime coats are frequently applied in this manner.

Brushing

Brushing has long been a satisfactory method of applying finishes to all types of surfaces. Brushing is generally used for small repair work and on surfaces where it is not practicable to spray paint.

The material to be applied should be thinned to the proper consistency for brushing. A material that is too thick has a tendency to pull or rope under the brush. If the materials are too thin, they are likely to run or will not cover the surface adequately.

Spray Painting

All spray systems have several basic similarities. There must be an adequate source of compressed air, a reservoir or feed tank to hold a supply of the finishing material, and a device for controlling the combination of air and finishing material ejected in an atomized cloud or spray against the surface to be coated.

There are two main types of spray equipment. A spray gun with integral paint container is satisfactory when painting small areas. When large areas are painted, pressure-feed equipment is usually preferred, since a large supply of finishing material can be provided under constant pressure to a pressure-feed type of spray gun.

The air-pressure supply must be entirely free from water or oil to obtain good spray painting. Oil and water traps as well as suitable filters must be incorporated in the air pressure supply line. These filters and traps must be serviced on a regular basis.

The spray gun can be adjusted to give a circular or fan type of spray pattern. Figure 4–1 shows the spray pattern at various dial settings. When covering large surfaces, set the gun just below maximum width of the fan spray. The circular spray is suitable for "spotting-in" small areas.

The gun should be held 6 to 10 in. away from the surface and the contour of the work carefully followed. It is important that the gun be kept at right angles to the surface. Each stroke of the spray gun should be straight and the trigger released just before completing the stroke, as shown in figure 4–2. The speed of movement should be regulated to deposit an even, wet, but not too heavy, coat.

Each stroke of the gun should be overlapped to keep a wet film, thus absorbing the dry edges of the previous stroke.

The spray should be applied as an even, wet coat that will flow out smoothly and be free from "spray dust." Inadequate coverage results from spraying too lightly and "runs" and "sags" from spraying too heavily.

To aid in obtaining good results, make sure the air pressure to the spray gun is between 40 and 80 p.s.i., depending on the material being used. With air pressures below 40 p.s.i. spraying is slow and tedious. Also, with viscous materials, full atomization is not obtained. Above 80 p.s.i. "dust" and blowback become troublesome.

When using pressure-feed equipment, adjust the air pressure in the container according to the viscosity of the paint and the length of the fluid hose used. The pressure must be such that the material reaches the spray gun head in a gentle and continuous flow. Generally, a pressure between 5 and 15 p.s.i. should be used. Higher pressures lead to runs and sags caused by the delivery of too much paint.

PREPARATION OF PAINT

Before paint is used, it must be stirred thoroughly so that any pigment which may have settled to the bottom of the container is brought into suspension and distributed evenly throughout the paint. If a film, called "skinning," has formed over the paint, the skin must be completely removed before stirring. Mechanical stirring is preferable to hand stirring. A mechanical agitator or tumbler may be used. However, as tumbling does not always remove pigment caked at the bottom of the container, a test with a stirrer should be made to ensure that the pigment is completely held in suspension. For hand stirring, a flat-bladed, nonferrous stirrer should be used.

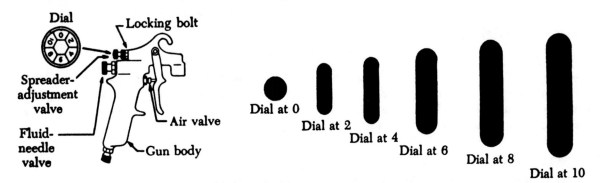

FIGURE 4–1. Spray patterns at various dial settings.

123

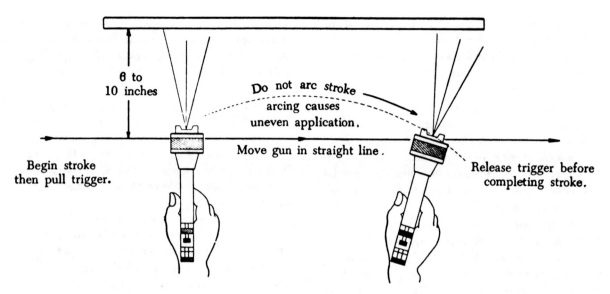

6 to 10 inches

Do not arc stroke arcing causes uneven application.

Move gun in straight line.

Begin stroke then pull trigger.

Release trigger before completing stroke.

FIGURE 4–2. Spray gun stroke.

The degree of thinning depends on the type of spray equipment, air pressure, atmospheric conditions and the type of paint being used. No hard and fast rule for thinning ratios can be applied. Because of the importance of accurate thinning, some manufacturers recommend the use of viscosity control. This is usually accomplished by using a viscosity (flow) cup. When the right proportion of thinner is mixed into the material, a cupful of material will flow out completely in a designated number of seconds. The finishing manufacturer can specify the number of seconds required for a given material. Material thinned using this method will be of the correct viscosity for best application.

In many cases manufacturers recommend that all materials should be strained before use. A 60- to 90-mesh strainer is suitable for this purpose. Strainers are available in metal gauze, paper, or nylon mesh.

COMMON PAINT TROUBLES

Poor Adhesion

Paint properly applied to correctly pretreated surfaces should adhere satisfactorily, and when it is thoroughly dry, it should not be possible to remove it easily, even by firm scratching with the fingernail. Poor adhesion may result from one of the following:

(1) Inadequate cleaning and pretreatment.
(2) Inadequate stirring of paint or primer.

(3) Coating at incorrect time intervals.
(4) Application under adverse conditions.
(5) Bad application.

Spray Dust

Spray dust is caused by the atomized particles becoming dry before reaching the surface being painted and thus failing to flow into a continuous film. The usual causes are incorrect air pressure or the distance the gun is held from the work.

Sags and Runs

Sags and runs result from too much paint being applied causing the film of wet paint to move by gravity and presenting a sagging appearance. Incorrect viscosity, air pressure, and gun handling are frequent causes. However, inadequate surface preparation may also be responsible.

Spray Mottle

Sometimes known as "orange peel" or "pebble," spray mottle is usually caused by incorrect paint viscosity, air pressure, spray gun setting, or the distance the gun is held from the work.

Blushing

Blushing is one of the most common troubles experienced and appears as a "clouding" or

124

"blooming" of the paint film. It is more common with the cellulose than synthetic materials. It may be caused by moisture in the air supply line, adverse humidity, drafts, or sudden changes in temperature.

PAINTING TRIM AND IDENTIFICATION NUMBERS

When an aircraft is being painted, the predominate color usually is applied first over the entire surface. The trim colors are applied over the base color after it dries. When the top of the fuselage is to be painted white with a dark color adjoining it, the light color is applied and feathered into the area to be painted with the dark color. When the light color has dried, masking tape and paper are placed along the line of separation and the dark color is then sprayed on.

Allow the paint to dry for several hours before removing the masking tape. Remove the tape by pulling slowly parallel to the surface. This will reduce the possibility of peeling off the finish with the tape.

All aircraft are required to display nationality and registration marks. These marks may be painted on or affixed using self-adhering plastic figures. The marks must be formed of solid lines using a color that contrasts with the background. No ornamentation may be used with the markings, and they must be affixed with a material or paint that produces a degree of permanence. Aircraft scheduled for immediate delivery to a foreign purchaser may display marks that can be easily removed. Aircraft manufactured in the United States for delivery outside the U. S. may display identification marks required by the State of registry of the aircraft. The aircraft may be operated only for test and demonstration flights for a limited period of time or for delivery to the purchaser.

Aircraft registered in the United States must display the Roman capital letter "N" followed by the registration number of the aircraft. The location and size of the identification marks vary according to the type of aircraft. The location and size are prescribed in the Federal Aviation Regulations.

DECALCOMANIAS (DECALS)

Markings are placed on aircraft surfaces to provide servicing instructions, fuel and oil specifications, tank capacities, and to identify lifting and leveling points, walkways, battery locations, cr any areas that should be identified. These markings can be applied by stenciling or by using decalcomanias.

Decalcomanias are used instead of painted instructions because they are usually cheaper and easier to apply. Decals used on aircraft are usually of three types: (1) Paper, (2) metal, or (3) vinyl film. These decals are suitable for exterior and interior surface application.

To assure proper adhesion of decals, clean all surfaces thoroughly with aliphatic naphtha to remove grease, oil, wax, or foreign matter. Porous surfaces should be sealed and rough surfaces sanded, followed by cleaning to remove any residue.

The instructions for applying decals are usually printed on the reverse side of each decal and should be followed. A general application procedure for each type of decal is presented in the following paragraphs to provide familiarization with the techniques involved.

Paper Decals

Immerse paper decals in clean water for 1 to 3 min. Allowing decals to soak longer than 3 min. will cause the backing to separate from the decal while immersed. If decals are allowed to soak less than 1 min., the backing will not separate from the decal.

Place one edge of the decal on the prepared receiving surface and press lightly, and then slide the paper backing from beneath the decal. Perform minor alignment with the fingers. Remove water by gently blotting the decal and adjacent area with a soft, absorbent cloth. Remove air or water bubbles trapped under the decal by wiping carefully toward the nearest edge of the decal with a cloth. Allow the decal to dry.

After the decal has dried, coat it with clear varnish to protect it from deterioration and peeling.

Metal Decals with Cellophane Backing

Apply metal decals with cellophane backing adhesive as follows:

(1) Immerse the decal in clean, warm water for 1 to 3 min.

(2) Remove it from the water and dry carefully with a clean cloth.

(3) Remove the cellophane backing but do not touch adhesive.

(4) Position one edge of the decal on the prepared receiving surface. On large foil decals, place the center on the receiving

surface and work outward from the center to the edges.

(5) Remove all air pockets by rolling firmly with a rubber roller, and press all edges tightly against the receiving surface to assure good adhesion.

Metal Decals with Paper Backing

Metal decals with a paper backing are applied similarly to those having a cellophane backing. However, it is not necessary to immerse the decal in water to remove the backing. It may be peeled from the decal without moistening. After removing the backing, apply a very light coat of cyclohexanone, or equivalent, to the adhesive. The decal should be positioned and smoothed out following the procedures given for cellophane-backed decals.

Metal Decals with No Adhesive

Apply decals with no adhesive in the following manner:

(1) Apply one coat of cement, Military Specification MIL–A–5092, to the decal and prepared receiving surface.

(2) Allow cement to dry until both surfaces are tacky.

(3) Apply the decal and smooth it down to remove air pockets.

(4) Remove excess adhesive with a cloth dampened with aliphatic naphtha.

Vinyl Film Decals

To apply vinyl film decals, separate the paper backing from the plastic film. Remove any paper backing adhering to the adhesive by rubbing the area gently with a clean cloth saturated with water; remove small pieces of remaining paper with masking tape.

Place the vinyl film, adhesive side up, on a clean porous surface, such as wood or blotter paper.

Apply cyclohexanone, or equivalent, in firm, even strokes to the adhesive side of decal.

Position the decal in the proper location, while adhesive is still tacky, with only one edge contacting the prepared surface.

Work a roller across the decal with overlapping strokes until all air bubbles are removed.

Removal of Decals

Paper decals can be removed by rubbing the decal with a cloth dampened with lacquer thinner. If the decals are applied over painted or doped surfaces, use lacquer thinner sparingly to prevent removing the paint or dope.

Remove the metal decals by moistening the edge of the foil with aliphatic naphtha and peeling the decal from the adhering surface.

Vinyl film decals are removed by placing a cloth saturated with cyclohexanone or MEK on the decal and scraping with a Micarta scraper. Remove the remaining adhesive by wiping with a cloth dampened with dry-cleaning solvent.

Methods of repairing structural portions of an aircraft are numerous and varied, and no set of specific repair patterns has been found which will apply in all cases. Since design loads acting in various structural parts of an aircraft are not always available, the problem of repairing a damaged section must usually be solved by duplicating the original part in strength, kind of material, and dimensions. Some general rules concerning the selection of material and the forming of parts which may be applied universally by the airframe mechanic will be considered in this chapter.

The repairs discussed are typical of those used in aircraft maintenance and are included to introduce some of the operations involved. For exact information about specific repairs, consult the manufacturer's maintenance or service manuals.

BASIC PRINCIPLES OF SHEET METAL REPAIR

The first and one of the most important steps in repairing structural damage is "sizing up" the job and making an accurate estimate of what is to be done. This sizing up includes an estimate of the best type and shape of patch to use; the type, size, and number of rivets needed; and the strength, thickness, and kind of material required to make the repaired member no heavier (or only slightly heavier) and just as strong as the original. Also inspect the surrounding members for evidence of corrosion and load damage so that the required extent of the "cleanout" of the old damage can be estimated accurately. After completing the cleanout, first make the layout of the patch on paper, then transfer it to the sheet stock selected. Then, cut and chamfer the patch, form it so that it matches the contour of that particular area, and apply it.

Maintaining Original Strength

In making any repair, certain fundamental rules must be observed if the original strength of the structure is to be maintained. The patch plate should have a cross-sectional area equal to, or greater than, that of the original damaged section. If the member is subjected to compression or to bending loads, place the splice on the outside of the member to secure a higher resistance to such loads. If the splice cannot be placed on the outside of the member, use material that is stronger than the material used in the original member.

To reduce the possibility of cracks starting from the corners of cutouts, try to make cutouts either circular or oval in shape. Where it is necessary to use a rectangular cutout, make the radius of curvature at each corner no smaller than 1/2 in. Either replace buckled or bent members or reinforce them by attaching a splice over the affected area.

Be sure the material used in all replacements or reinforcements is similar to the material used in the original structure. If it is necessary to substitute an alloy weaker than the original, use material of a heavier gage to give equivalent cross-sectional strength. But never practice the reverse; that is, never substitute a lighter gage stronger material for the original. This apparent inconsistency is because one material can have greater tensile strength than another, but less compressive strength, or vice versa. As an example, the mechanical properties of alloys 2024–T and 2024–T80 are compared in the following paragraph.

If alloy 2024–T were substituted for alloy 2024–T80, the substitute material would have to be thicker unless the reduction in compressive strength was known to be acceptable. On the other hand, if 2024–T80 material were substitued for 2024–T stock, the substitute material would have to be thicker unless the reduction in tensile strength was known to be acceptable. Similarly, the buckling and torsional strength of many sheet-metal and tubular parts are dependent primarily upon the thickness rather than the allowable compressive and shear strengths.

When forming is necessary, be particularly careful, for heat-treated and cold-worked alloys will stand very little bending without cracking. Soft alloys, on the other hand, are easily formed but are not strong enough for primary structures. Strong alloys can be formed in their annealed condition and heat treated to develop their strength before assembling.

In some cases, if the annealed metal is not available, heat the metal, quench it according to regular heat-treating practices, and form it before age-hardening sets in. The forming should be completed in about half an hour after quenching, or the material will become too hard to work.

The size of rivets for any repair can be determined by referring to the rivets used by the manufacturer in the next parallel rivet row inboard on the wing, or forward on the fuselage. Another method of determining the size of rivets to be used is to multiply the thickness of the skin by three and use the next larger size rivet corresponding to that figure. For example, if the skin thickness is 0.040-in., multiply 0.040 by 3, which equals 0.120; use the next larger size rivet, 1/8 in. (0.125 in.).

All repairs made on structural parts of aircraft require a definite number of rivets on each side of the break to restore the original strength. This number varies according to the thickness of the material being repaired and the size of the damage. The number of rivets or bolts required can be determined by referring to a similar splice made by the manufacturer, or by using the following rivet formula:

$$\frac{\text{Number of rivets required}}{\text{on each side of the break}} = \frac{L \times T \times 75{,}000}{S \text{ or } B}.$$

The number of rivets to be used on each side of the break is equal to the length of the break (L) times the thickness of the material (T) times 75,000, divided by the shear strength or bearing strength (S or B) of the material being repaired, whichever is the smaller of the two.

The length of the break is measured perpendicu-

lar to the direction of the general stress running through the damaged area.

The thickness of the material is the actual thickness of the piece of material being repaired and is measured in thousandths of an inch.

The 75,000 used in the formula is an assumed stress load value of 60,000 p.s.i. increased by a safety factor of 25%. It is a constant value.

Shear strength is taken from the charts shown in figure 5–1. It is the amount of force required to cut a rivet holding together two or more sheets of material. If the rivet is holding two parts, it is under single shear; if it is holding three sheets or parts, it is under double shear. To determine the shear strength, the diameter of the rivet to be used must be known. This is determined by multiplying the thickness of the material by three. For example, material thickness 0.040 multiplied by 3 equals 0.120; the rivet selected would be 1/8 in. (0.125 in.) in diameter.

Bearing strength is a value taken from the chart shown in figure 5–2 and is the amount of tension required to pull a rivet through the edge of two sheets riveted together, or to elongate the hole. The diameter of the rivet to be used and the thickness of material being riveted must be known to use the bearing strength chart. The diameter of the rivet would be the same as that used when determining the shear strength value. Thickness of material would be that of the material being repaired.

Example:

Using the formula, determine the number of 2117–T rivets needed to repair a break 2-1/4 in. long in material 0.040-in. thick:

$$\frac{\text{Number rivets}}{\text{per side}} = \frac{L \times T \times 75{,}000}{S \text{ or } B}.$$

*Single-Shear Strength of Aluminum-Alloy Rivets (Pounds)									
Composition of Rivet (Alloy)	Ultimate Strength of Rivet Metal (Pounds Per Square Inch)	Diameter of Rivet (Inches)							
		1/16	3/32	1/8	5/32	3/16	1/4	5/16	3/8
2117 T	27,000	83	186	331	518	745	1,325	2,071	2,981
2017 T	30,000	92	206	368	573	828	1,472	2,300	3,313
2024 T	35,000	107	241	429	670	966	1,718	2,684	3,865
*Double-shear strength is found by multiplying the above values by 2.									

FIGURE 5–1. Single shear strength chart.

Thickness of Sheet (Inches)	Diameter of Rivet (Inches)							
	1/16	3/32	1/8	5/32	3/16	1/4	5/16	3/8
0.014	71	107	143	179	215	287	358	430
.016	82	123	164	204	246	328	410	492
.018	92	138	184	230	276	369	461	553
.020	102	153	205	256	307	410	412	615
.025	128	192	256	320	284	512	640	768
.032	164	245	328	409	492	656	820	984
.036	184	276	369	461	553	738	922	1,107
.040	205	307	410	512	615	820	1,025	1,230
.045	230	345	461	576	691	922	1,153	1,383
.051	261	391	522	653	784	1,045	1,306	1,568
.064		492	656	820	984	1,312	1,640	1,968
.072		553	738	922	1,107	1,476	1,845	2,214
.081		622	830	1,037	1,245	1,660	2,075	2,490
.091		699	932	1,167	1,398	1,864	2,330	2,796
.102		784	1,046	1,307	1,569	2,092	2,615	3,138
.125		961	1,281	1,602	1,922	2,563	3,203	3,844
.156		1,198	1,598	1,997	2,397	3,196	3,995	4,794
.188		1,445	1,927	2,409	2,891	3,854	4,818	5,781
.250		1,921	2,562	3,202	3,843	5,125	6,405	7,686
.313		2,405	3,208	4,009	4,811	6,417	7,568	9,623
.375		2,882	3,843	4,803	5,765	7,688	9,068	11,529
.500		3,842	5,124	6,404	7,686	10,250	12,090	15,372

FIGURE 5-2. Bearing strength chart (pounds).

Given:

L = 2-1/4 (2.25) in.

T = 0.040 in.

Size of rivet: 0.040 x 3 = 0.120, so rivet must be 1/8 in. or 0.125.

S = 331 (from the shear strength chart).

B = 410 (from the bearing strength chart).

(Use S to find number of rivets per side as it is smaller than B.)

Substituting in the formula:

$$\frac{2.25 \times 0.040 \times 75,000}{331} = \frac{6,750}{331}$$
$$= 20.39 \text{ (or 21)}$$
$$\text{rivets/side.}$$

Since any fraction must be considered as a whole number, the actual number of rivets required would be 21 for each side, or 42 rivets for the entire repair.

Maintaining Original Contour

Form all repairs in such a manner that they will fit the original contour perfectly. A smooth contour is especially desirable when making patches on the smooth external skin of high-speed aircraft.

Keeping Weight to a Minimum

Keep the weight of all repairs to a minimum. Make the size of the patches as small as practicable and use no more rivets than are necessary. In many cases, repairs disturb the original balance of the structure. The addition of excessive weight in each repair may unbalance the aircraft so much that it will require adjustment of the trim-and-balance tabs. In areas such as the spinner on the propeller, a repair will require application of balancing patches so that a perfect balance of the propeller assembly can be maintained.

GENERAL STRUCTURAL REPAIR

Aircraft structural members are designed to perform a specific function or to serve a definite purpose. The prime objective of aircraft repair is to restore damaged parts to their original condition. Very often, replacement is the only way in which this can be done effectively. When repair of a damaged part is possible, first study the part carefully so that its purpose or function is fully understood.

Strength may be the principal requirement in the repair of certain structures, while others may need entirely different qualities. For example, fuel tanks and floats must be protected against leakage; but cowlings, fairings, and similar parts must have such properties as neat appearance, streamlined shape, and accessibility. The function of any damaged part must be carefully determined so that the repair will meet the requirements.

INSPECTION OF DAMAGE

When visually inspecting damage, remember that there may be other kinds of damage than that caused by impact from foreign objects or collision. A rough landing may overload one of the landing gear, causing it to become sprung; this would be classified as load damage. During inspection and "sizing up of the repair job," consider how far the damage caused by the sprung shock strut extends to supporting structural members.

A shock occurring at one end of a member will be transmitted throughout its length; therefore, inspect closely all rivets, bolts, and attaching structures along the complete member for any evidence of damage. Make a close examination for rivets that have partially failed and for holes which have been elongated.

Another kind of damage to watch for is that caused by weathering or corrosion. This is known as corrosion damage. Corrosion damage of aluminum material is usually detected by the white crystalline deposits that form around loose rivets, scratches, or any portion of the structure that may be a natural spot for moisture to settle.

Definition of Defects

Types of damage and defects which may be observed on parts of this assembly are defined as follows:

Brinelling—Occurrence of shallow, spherical depressions in a surface, usually produced by a part having a small radius in contact with the surface under high load.

Burnishing—Polishing of one surface by sliding contact with a smooth, harder surface. Usually no displacement nor removal of metal.

Burr—A small, thin section of metal extending beyond a regular surface, usually located at a corner or on the edge of a bore or hole.

Corrosion—Loss of metal from the surface by chemical or electrochemical action. The corrosion products generally are easily removed by mechanical means. Iron rust is an example of corrosion.

Crack—A physical separation of two adjacent portions of metal, evidenced by a fine or thin line across the surface, caused by excessive stress at that point. It may extend inward from the surface from a few thousandths inch to completely through the section thickness.

Cut—Loss of metal, usually to an appreciable depth over a relatively long and narrow area, by mechanical means, as would occur with the use of a saw blade, chisel or sharp-edged stone striking a glancing blow.

Dent—Indentation in a metal surface produced by an object striking with force. The surface surrounding the indentation will usually be slightly upset.

Erosion—Loss of metal from the surface by mechanical action of foreign objects, such as grit or fine sand. The eroded area will be rough and may be lined in the direction in which the foreign material moved relative to the surface.

Chattering—Breakdown or deterioration of metal surface by vibratory or "chattering" action. Usually no loss of metal or cracking of surface but generally showing similar appearance.

Galling—Breakdown (or build-up) of metal surfaces due to excessive friction between two parts having relative motion. Particles of the softer metal are torn loose and "welded" to the harder.

Gouge—Grooves in, or breakdown of, a metal surface from contact with foreign material under heavy pressure. Usually indicates metal loss but may be largely displacement of material.

Inclusion—Presence of foreign or extraneous material wholly within a portion of metal. Such material is introduced during the manufacture of rod, bar or tubing by rolling or forging.

Nick—Local break or notch on edge. Usually displacement of metal rather than loss.

Pitting—Sharp, localized breakdown (small, deep cavity) of metal surface, usually with defined edges.

Scratch—Slight tear or break in metal surface from light, momentary contact by foreign material.

Score—Deeper (than scratch) tear or break in metal surface from contact under pressure. May show discoloration from temperature produced by friction.

Stain—A change in color, locally causing a noticeably different appearance from the surrounding area.

Upsetting—A displacement of material beyond the normal contour or surface (a local bulge or bump). Usually indicates no metal loss.

CLASSIFICATION OF DAMAGE

Damages may be grouped into four general classes. In many cases, the availability or lack of repair materials and time are the most important factors in determining whether a part should be repaired or replaced.

Negligible Damage

Damage which does not affect the structural integrity of the member involved, or damage which can be corrected by a simple procedure without placing flight restrictions on the aircraft, is classified as negligible damage. Small dents, scratches, cracks, or holes that can be repaired by smoothing, sanding, stop drilling, or hammering out, or otherwise repaired without the use of additional materials, fall in this classification.

Damage Repairable by Patching

Damage repairable by patching is any damage exceeding negligible damage limits which can be repaired by bridging the damaged area of a component with a material splice. The splice or patch material used in internal riveted and bolted repairs is normally the same type of material as the damaged part, but one gage heavier. In a patch repair, filler plates of the same gage and type of material as that in the damaged component may be used for bearing purposes or to return the damaged part to its original contour.

Damage Repairable by Insertion

Damage which can be repaired by cutting away the damaged section and replacing it with a like section, then securing the insertion with splices at each end is classified as damage repairable by insertion.

Damage Necessitating Replacement of Parts

Replacement of an entire part is considered when one or more of the following conditions exist:

(1) When a complicated part has been extensively damaged.

(2) When surrounding structure or inaccessibility makes repair impractical.

(3) When damaged part is relatively easy to replace.

(4) When forged or cast fittings are damaged beyond the negligible limits.

STRESSES IN STRUCTURAL MEMBERS

Forces acting on an aircraft, whether it is on the ground or in flight, cause pulling, pushing, or twisting within the various members of the aircraft structure. While the aircraft is on the ground, the weight of the wings, fuselage, engines, and empennage causes forces to act downward on the wing and stabilizer tips, along the spars and stringers, and on the bulkheads and formers. These forces are passed on from member to member causing bending, twisting, pulling, compression, and shearing.

As the aircraft takes off, most of the forces in the fuselage continue to act in the same direction; but because of the motion of the aircraft, they increase in intensity. The forces on the wingtips and the wing surfaces, however, reverse direction and instead of being downward forces of weight, they become upward forces of lift. The forces of lift are exerted first against the skin and stringers, then are passed on to the ribs, and finally are transmitted through the spars to be distributed through the fuselage.

The wings bend upward at their ends and may flutter slightly during flight. This wing bending cannot be ignored by the manufacturer in the original design and construction, and cannot be ignored during maintenance. It is surprising how an aircraft structure composed of structural members and skin rigidly riveted or bolted together, such as a wing, can bend or act so much like a leaf spring.

The five types of stresses (figure 5–3) in an aircraft are described as tension, compression, shear, bending, and torsion (or twisting). The first three are commonly called basic stresses, the last two, combination stresses. Stresses usually act in combinations rather than singly.

Tension

Tension (or tensile stress) is the force per unit area tending to stretch a structural member. The strength of a member in tension is determined on the basis of its gross area (or total area), but calculations involving tension must take into consideration the net area of the member. Net area is defined as the gross area minus that removed by drilling holes or by making other changes in the section. Placing rivets or bolts in holes makes no appreciable difference in added strength, as the rivets or bolts will not transfer tensional loads across holes in which they are inserted.

Compression

Compression (or compressive stress) is the force per unit area which tends to shorten (or compress) a structural member at any cross section. Under a compressive load, an undrilled member will be stronger than an identical member with holes drilled through it. However, if a plug of equivalent or stronger material is fitted tightly in a drilled member, it will transfer compressive loads across the hole, and the member will carry approximately as large a load as if the hole were not there. Thus, for compressive loads, the gross or total area may be used in determining the stress in a member if all holes are tightly plugged with equivalent or stronger material.

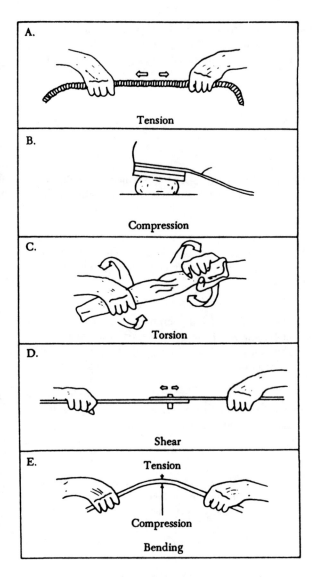

FIGURE 5–3. Five stresses acting on an aircraft.

Shear

Shear is the force per unit area which causes adjacent particles of material to slide past each other. The term "shear" is used because it is a sideways stress of the type that is put on a piece of paper or a sheet of metal when it is cut with a pair of shears. Shear stress concerns the aviation mechanic chiefly from the standpoint of rivet and bolt applications, particularly when attaching sheet stock, because if a rivet used in a shear application gives way, the riveted or bolted parts are pushed sideways.

Bending

Bending (or beam stress) is actually a combination of two forces acting upon a structural member at one or more points. In figure 5–3 note that the bending stress causes a tensile stress to act on the upper half of the beam and a compressive stress on the lower half. These stresses act oppositely on the two sides of the center line of the member, which is called the neutral axis. Since these forces acting in opposite directions are next to each other at the neutral axis, the greatest shear stress occurs along this line, and none exists at the extreme upper or lower surfaces of the beam.

Torsion

Torsion (or twisting stress) is the force which tends to twist a structural member. The stresses arising from this action are shear stresses caused by the rotation of adjacent planes past each other around a common reference axis at right angles to these planes. This action may be illustrated by a rod fixed solidly at one end and twisted by a weight placed on a lever arm at the other, producing the equivalent of two equal and opposite forces acting on the rod at some distance from each other. A shearing action is set up all along the rod, with the center line of the rod representing the neutral axis.

SPECIAL TOOLS AND DEVICES FOR SHEET METAL

The airframe mechanic does a lot of work with special tools and devices that have been developed to make his work faster, simpler, and better. These special tools and devices include dollies and stakes and various types of blocks and sandbags used as support in the bumping process.

Dollies and Stakes

Sheet metal is often formed or finished (planished) over variously shaped anvils called dollies and stakes. These are used for forming small, odd-shaped parts, or for putting on finishing touches for which a large machine may not be suited. Dollies are meant to be held in the hand, whereas stakes are designed to be supported by a flat cast iron bench plate fastened to the workbench (figure 5–4).

Most stakes have machined, polished surfaces which have been hardened. Do not use stakes to back up material when chiseling, or when using any similar cutting tool because this will deface the surface of the stake and make it useless for finish work.

V-Blocks

V-blocks made of hardwood are widely used in airframe metalwork for shrinking and stretching metal, particularly angles and flanges. The size of

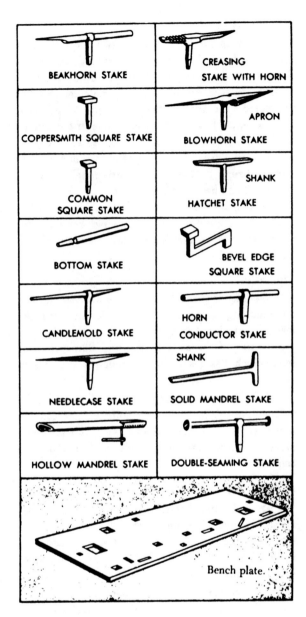

FIGURE 5–4. Bench plate and stakes.

the block depends on the work being done and on personal preference. Although any type of hardwood is suitable, maple and ash are recommended for best results when working with aluminum alloys.

Hardwood Form Blocks

Hardwood form blocks can be constructed to duplicate practically any aircraft structural or nonstructural part. The wooden block or form is shaped to the exact dimensions and contour of the part to be formed.

133

Shrinking Blocks

A shrinking block consists of two metal blocks and some device for clamping them together. One block forms the base, and the other is cut away to provide space where the crimped material can be hammered. The legs of the upper jaw clamp the material to the base block on each side of the crimp so that the material will not creep away but will remain stationary while the crimp is hammered flat (being shrunk). This type of crimping block is designed to be held in a bench vise.

Shrinking blocks can be made to fit any specific need. The basic form and principle remain the same, even though the blocks may vary considerably in size and shape.

Sandbags

A sandbag is generally used as a support during the bumping process. A serviceable bag can be made by sewing heavy canvas or soft leather to form a bag of the desired size, and filling it with sand which has been sifted through a fine mesh screen.

Before filling canvas bags with sand, use a brush to coat the inside of it with softened paraffin or beeswax, which forms a sealing layer and prevents the sand from working through the pores of the canvas.

Holding Devices

Vises and clamps are tools used for holding materials of various kinds on which some type of operation is being performed. The type of operation being performed and the type of metal being used determine the holding device to be used.

The most commonly used vises are shown in figure 5–5; the machinist's vise has flat jaws and usually a swivel base, whereas the utility bench vise has scored, removable jaws and an anvil-faced back jaw. This vise will hold heavier material than the machinist's vise and will also grip pipe or rod firmly. The back jaw can be used for an anvil if the work being done is light.

The carriage clamp, or C-clamp, as it is commonly called, is shaped like a large C and has three main parts: (1) The threaded screw, (2) the jaw, and (3) the swivel head. The swivel plate, which is at the bottom of the screw, prevents the end from turning directly against the material being clamped. Although C-clamps vary in size from 2 in. upward, their function is always that of clamping or holding

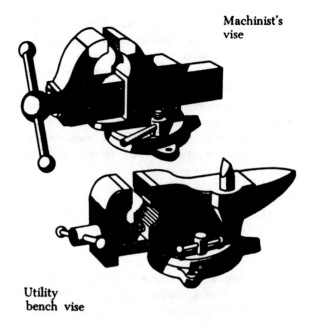

Machinist's vise

Utility bench vise

FIGURE 5–5. Vises.

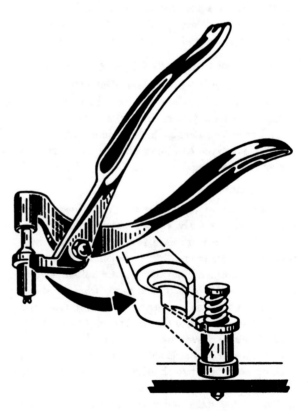

FIGURE 5–6. Cleco fastener.

The shape of the C-clamp allows it to span obstructions near the edge of a piece of work. The greatest limitation in the use of the carriage clamp is its tendency to spring out of shape. It should never be tightened more than hand-tight.

The most commonly used sheet-metal holder is the Cleco fastener (figure 5–6). It is used to keep drilled parts made from sheet stock pressed tightly together. Unless parts are held tightly together they will separate while being riveted.

This type of fastener is available in six different sizes: 3/32-, 1/8-, 5/32-, 3/16-, 1/4-, and 3/8-in.

The size is stamped on the fastener. Special pliers are used to insert the fastener in a drilled hole. One pair of pliers will fit the six different sizes.

Sheet-metal screws are sometimes used as temporary holders. The metal sheets must be held tightly together before installing these screws, since the self-tapping action of the threads tends to force the sheets apart. Washers placed under the heads of the screws keep them from marring or scratching the metal.

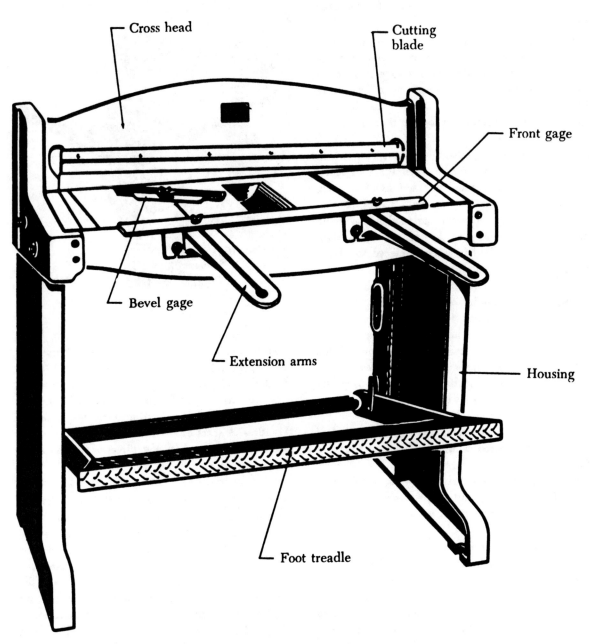

FIGURE 5–7. Squaring shears.

135

METALWORKING MACHINES

Without metalworking machines a job would be more difficult and tiresome, and the time required to finish a task would be much longer. Some of the machines used are discussed here; these include the powered and nonpowered metal-cutting machines, such as the various types of saws, powered and nonpowered shears, and nibblers. Also included is the forming equipment (both power driven and nonpowered), such as brakes and forming rolls, the bar folder, and shrinking and stretching machines.

Metal Cutting Manually Operated Tools—Lever Type

Squaring shears provide a convenient means of cutting and squaring metal. These shears consist of a stationary lower blade attached to a bed and a movable upper blade attached to a crosshead (figure 5–7). To make the cut, the upper blade is moved down by placing the foot on the treadle and pushing downward.

The shears are equipped with a spring which raises the blade and treadle when the foot is removed. A scale, graduated in fractions of an inch, is scribed on the bed. Two squaring fences, consisting of thick strips of metal and used for squaring metal sheets, are placed on the bed, one on the right side and one on the left. Each is placed so that it forms a 90° angle with the blades.

Three distinctly different operations can be performed on the squaring shears: (1) Cutting to a line, (2) squaring, and (3) multiple cutting to a specific size. When cutting to a line, the sheet is placed on the bed of the shears in front of the cutting blade with the cutting line directly even with the cutting edge of the bed. The sheet is cut by stepping on the treadle while the sheet is held securely in place by the holddown clamp.

Squaring requires several steps. First, one end of the sheet is squared with an edge (the squaring fence is usually used on the edge). Then the remaining edges are squared by holding one squared end of the sheet against the squaring fence and making the cut, one edge at a time, until all edges have been squared.

When several pieces must be cut to the same dimensions, use the gage which is on most squaring shears. The supporting rods are graduated in fractions of an inch, and the gage bar may be set at any point on the rods. Set the gage at the desired distance from the cutting blade of the shears and push each piece to be cut against the gage bar. All

the pieces can then be cut to the same dimensions without measuring and marking each one separately.

Scroll shears (figure 5–8) are used for cutting irregular lines on the inside of a sheet without cutting through to the edge. The upper cutting blade is stationary while the lower blade is movable. The machine is operated by a handle connected to the lower blade.

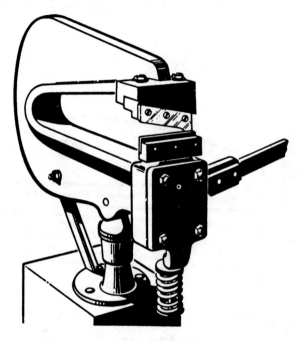

FIGURE 5–8. Scroll shears.

Throatless shears (figure 5–9) are best used to cut 10-gage mild carbon sheet metal and 12-gage stainless steel. The shear gets its name from its construction; it actually has no throat. There are no obstructions during cutting since the frame is throatless. A sheet of any length can be cut, and the metal can be turned in any direction to allow for cutting irregular shapes. The cutting blade (top blade) is operated by a hand lever.

The rotary punch (figure 5–10) is used in the airframe repair shop to punch holes in metal parts. This machine can be used for cutting radii in corners, for making washers, and for many other jobs where holes are required. The machine is composed of two cylindrical turrets, one mounted over the other and supported by the frame. Both turrets are synchronized so that they rotate together, and index pins assure correct alignment at all times. The index pins may be released from their locking position by

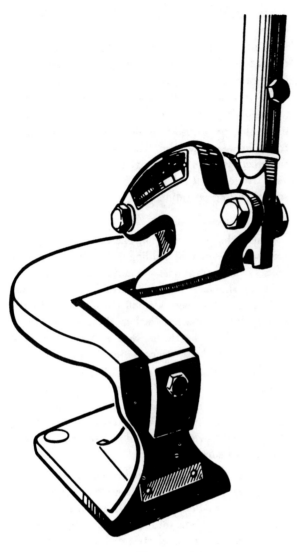

FIGURE 5–9. Throatless shears.

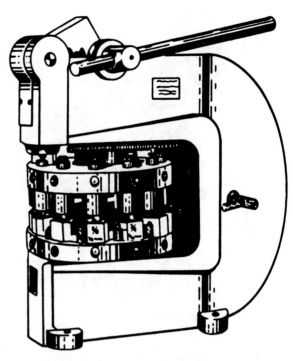

FIGURE 5–10. Rotary punch.

rotating a lever on the right side of the machine. This action withdraws the index pins from the tapered holes and allows an operator to turn the turrets to any size punch desired.

When rotating the turret to change punches, release the index lever when the desired die is within 1 in. of the ram, and continue to rotate the turret slowly until the top of the punch holder slides into the grooved end of the ram. The tapered index locking pins will then seat themselves in the holes provided and, at the same time, release the mechanical locking device, which prevents punching until the turrets are aligned.

To operate the machine, place the metal to be worked between the die and punch. Pull the lever on the top side of the machine toward you. This will actuate the pinion shaft, gear segment, toggle link, and the ram, forcing the punch through the metal. When the lever is returned to its original position, the metal is removed from the punch.

The diameter of the punch is stamped on the front of each die holder. Each punch has a point in its center which is placed in the centerpunch mark to punch the hole in the correct location.

Metal-Cutting Power-Operated Tools

The electrically operated portable circular-cutting Ketts saw (figure 5–11) uses blades of various diameters. The head of this saw can be turned to any desired angle, and is very handy for removing damaged sections on a stringer. Advantages of a Ketts saw are:

(1) The ability to cut metal up to 3/16 in. thick.
(2) No starting hole is required.
(3) A cut can be started anywhere on a sheet of metal.
(4) The capability of cutting an inside or outside radius.

To prevent grabbing, keep a firm grip on the saw handle at all times. Before installing a blade, it should be checked carefully for cracks. A cracked blade can fly apart and perhaps result in serious injury.

FIGURE 5–11. Ketts saw.

The portable, air powered reciprocating saw (figure 5–12) has a gun-type shape for balancing and ease of handling and operates most effectively at an air pressure of from 85 to 100 p.s.i. The reciprocating saw uses a standard hacksaw blade and can cut a 360° circle or a square or rectangular hole. This saw is easy to handle and safe to use.

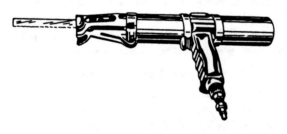

FIGURE 5–12. Reciprocating saw.

A reciprocating saw should be used in such a way that at least two teeth of the saw blade are cutting at all times. Avoid applying too much downward pressure on the saw handle because the blade may break.

Nibblers

Stationary and portable nibblers are used to cut metal by a high-speed blanking action. The cutting or blanking action is caused by the lower die moving up and down and meeting the upper stationary die. The shape of the lower die permits small pieces of metal approximately 1/16-in. wide to be cut out.

The cutting speed of the nibbler is controlled by the thickness of the metal being cut. Sheets of metal with a maximum thickness of 1/16 in. can be cut satisfactorily. Too much force applied to the metal during the cutting operation will clog the dies, causing the die to fail or the motor to overheat.

The spring-loaded screw on the base of the lower die should be adjusted to allow the metal to move freely between the dies. This adjustment must be sufficient to hold the material firmly enough to prevent irregular cuts. The dies may be shimmed for special cutting operations.

Portable Power Drills

One of the most common operations in airframe metalwork is that of drilling holes for rivets and bolts. This operation is not difficult, especially on light metal. Once the fundamentals of drills and their uses are learned, a small portable power drill is usually the most practical machine to use. However, there will be times when a drill press may prove to be the better machine for the job.

Some portable power drills will be encountered which are operated by electricity and others which are operated by compressed air. Some of the electrically operated drills work on either alternating or direct current, whereas others will operate on only one kind of current.

Portable power drills are available in various shapes and sizes to satisfy almost any requirement (figure 5–13). Pneumatic drills are recommended for use on projects around flammable materials where sparks from an electric drill might become a fire hazard.

When access to a place where a hole is to be drilled is difficult or impossible with a straight drill, various types of drill extensions and adapters are used. A straight extension can be made from an ordinary piece of drill rod. The twist drill is attached to the drill rod by shrink fit, brazing, or silver soldering. Angle adapters can be attached to either an electric or pneumatic drill when the location of the hole is inaccessible to a straight drill. Angle adapters have an extended shank fastened to the chuck of the drill. In use, the drill is held in one hand and the adapter in the other to prevent the adapter from spinning around the drill chuck.

A flexible extension can be used for drilling in places which are inaccessible to ordinary drills. Its flexibility permits drilling around obstructions with a minimum of effort.

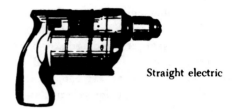

Straight electric

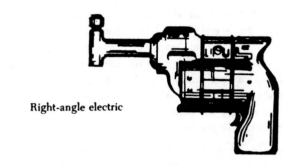

Right-angle electric

Straight air

Right-angle air

360° air

FIGURE 5-13. Portable power drills.

When using the portable power drill, hold it firmly with both hands. Before drilling, be sure to place a backup block of wood under the hole to be drilled to add support to the metal.

The twist drill should be inserted in the chuck and tested for trueness or vibration. This may be visibly checked by running the motor freely. A drill that wobbles or is slightly bent should not be used since such a condition will cause enlarged holes.

The drill should always be held at right angles to the work regardless of the position or curvatures. Tilting the drill at any time when drilling into or withdrawing from the material may cause elongation (egg shape) of the hole.

Always wear safety goggles while drilling.

When drilling through sheet metal, small burrs are formed around the edge of the hole. Burrs must be removed to allow rivets or bolts to fit snugly and to prevent scratching. Burrs may be removed with a bearing scraper, a countersink, or a twist drill larger than the hole. If a drill or countersink is used, it should be rotated by hand.

Drill Press

The drill press is a precision machine used for drilling holes that require a high degree of accuracy. It serves as an accurate means of locating and maintaining the direction of a hole that is to be drilled and provides the operator with a feed lever that makes the task of feeding the drill into the work an easy one.

A variety of drill presses are available; the most common type is the upright drill press (figure 5-14).

When using a drill press, the height of the drill press table is adjusted to accommodate the height of the part to be drilled. When the height of the part is greater than the distance between the drill and the table, the table is lowered. When the height of the part is less than the distance between the drill and the table, the table is raised.

After the table is properly adjusted, the part is placed on the table and the drill is brought down to aid in positioning the metal so that the hole to be drilled is directly beneath the point of the drill. The part is then clamped to the drill press table to prevent it from slipping during the drilling operation. Parts not properly clamped may bind on the drill and start spinning, causing the loss of fingers or hands or serious cuts on the operator's arms or body. Always make sure the part to be drilled is properly clamped to the drill press table before starting the drilling operation.

The degree of accuracy that it is possible to attain when using the drill press will depend to a certain extent on the condition of the spindle hole, sleeves, and drill shank. Therefore, special care must be exercised to keep these parts clean and free from nicks, dents, or warpage. Always be sure that the sleeve is securely pressed into the spindle hole. Never insert a broken drill in a sleeve or spindle hole. Be careful never to use the sleeve-clamping

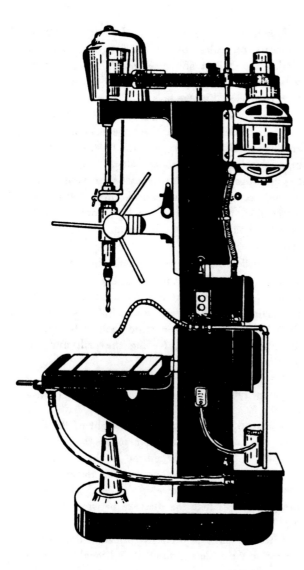

FIGURE 5–14. Drill press.

vise to remove a drill since this may cause the sleeve to warp.

Grinders

The term grinder applies to all forms of grinding machines. To be specific, it is a machine having an abrasive wheel which removes excess material while producing a suitable surface. There are many kinds of grinding machines, but only those which are helpful to the airframe mechanic will be discussed here.

Grinding Wheels

A grinding wheel is a cutting tool with a large number of cutting edges arranged so that when they become dull they break off and new cutting edges take their place.

Silicon carbide and aluminum oxide are the kinds of abrasives used in most grinding wheels. Silicon carbide is the cutting agent for grinding hard, brittle material, such as cast iron. It is also used in grinding aluminum, brass, bronze, and copper. Aluminum oxide is the cutting agent for grinding steel and other metals of high tensile strength.

The size of the abrasive particles used in grinding wheels is indicated by a number which corresponds to the number of meshes per linear inch in the screen through which the particles will pass. As an example, a number 30 abrasive will pass through a screen having 30 holes per linear inch, but will be retained by a smaller screen having more than 30 holes per linear inch.

The bond is the material which holds the abrasive particles together in forming the wheel. The kind and amount of bond used determines the hardness or softness of the wheel. The commonly used bonds are vitrified, silicate, resinoid, rubber, and shellac. Vitrified and silicate are the bonds used most frequently, vitrified bond being used in approximately three-fourths of all grinding wheels made. This bonding material forms a very uniform wheel and is not affected by oils, acids, water, heat, or cold. The silicate bond, however, is best suited for grinding edged tools.

Resinoid bonded wheels are better for heavy-duty grinding. Rubber bonded wheels are used where a high polish is required. Shellac bonded wheels are used for grinding materials where a buffed or burnished surface is needed.

A pedestal or floor type grinder usually has a grinding wheel on each end of a shaft which runs through an electric motor or a pulley operated by a belt. This grinder is used for sharpening tools and other general grinding jobs.

The wet grinder, although similar to the pedestal grinder, differs from it in that the wet grinder has a pump to supply a flow of water on a single grinding wheel. The water reduces the heat produced by material being ground against the wheel. It also washes away any bits of metal or abrasive removed during the grinding operation. The water returns to a tank and can be re-used.

A common type bench grinder found in most metalworking shops is shown in figure 5–15. This grinder can be used to dress mushroomed heads on chisels, and points on chisels, screwdrivers, and drills. It can be used for removing excess metal from work and smoothing metal surfaces.

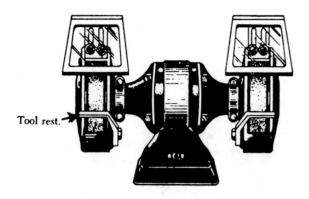

FIGURE 5-15. Bench grinder.

This type grinder is generally equipped with one medium-grain and one fine-grain abrasive wheel. The medium-grain wheel is usually used for rough grinding where a considerable quantity of material is to be removed or where a smooth finish is unimportant. The fine-grain wheel is usually used for sharpening tools and grinding to close limits because it removes metal more slowly, gives the work a smooth finish, and does not generate enough heat to anneal the edges of cutting tools. When it is necessary to make a deep cut on work or to remove a large amount of metal, it is usually good practice to grind with the medium-grain wheel first and then finish up with the fine-grain wheel.

The grinding wheels are removable, and the grinders are usually designed so that wire brushes, polishing wheels, or buffing wheels can be substituted for the abrasive wheels.

As a rule, it is not good practice to grind work on the side of an abrasive wheel. When an abrasive wheel becomes worn, its cutting efficiency is reduced because of a decrease in surface speed. When a wheel becomes worn in this manner, it should be discarded and a new one installed.

Before using a bench grinder, make sure the abrasive wheels are firmly held on the spindles by the flange nuts. If an abrasive wheel should come off or become loose, it could seriously injure the operator in addition to ruining the grinder.

Another hazard is loose tool rests. A loose tool rest could cause the tool or piece of work to be "grabbed" by the abrasive wheel and cause the operator's hand to come in contact with the wheel. If this should happen, severe wounds may result.

Always wear goggles when using a grinder, even if eyeshields are attached to the grinder. Goggles should fit firmly against your face and nose. This is the only way to protect your eyes from the fine pieces of steel. Goggles that do not fit properly should be exchanged for ones that do fit.

Be sure to check the abrasive wheel for cracks before using the grinder. A cracked abrasive wheel is likely to fly apart when turning at high speeds. Never use a grinder unless it is equipped with wheel guards.

FORMING MACHINES

Forming machines can be either hand operated or power driven. Small machines are usually hand operated, whereas the larger ones are power driven. Straight line machines include such equipment as the bar folder, cornice brake, and box and pan brake. Rotary machines include the slip roll former and combination machine. Power-driven machines are those that require a motor of some description for power. These include such equipment as the power-driven slip roll former, and power flanging machine.

Bar Folder

The bar folder (figure 5-16) is designed for use in making bends or folds along edges of sheets. This machine is best suited for folding small hems, flanges, seams, and edges to be wired. Most bar folders have a capacity for metal up to 22 gage in thickness and 42 inches in length.

Before using the bar folder, several adjustments must be made for thickness of material, width of fold, sharpness of fold, and angle of fold.

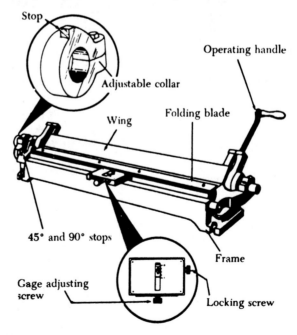

FIGURE 5-16. Bar folder.

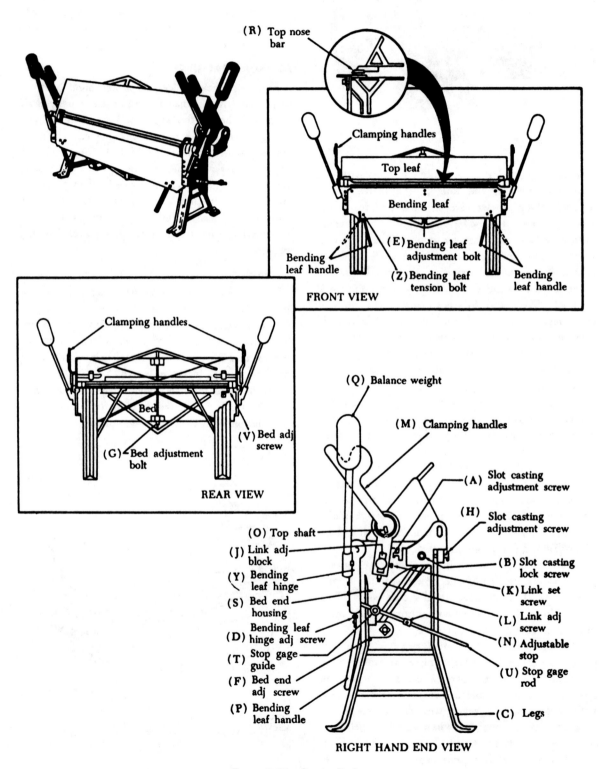

(R) Top nose bar

Clamping handles

Top leaf

Bending leaf

(E) Bending leaf adjustment bolt

(Z) Bending leaf tension bolt

Bending leaf handle

Bending leaf handle

FRONT VIEW

Clamping handles

Bed

(G) Bed adjustment bolt

(V) Bed adj screw

REAR VIEW

(Q) Balance weight

(M) Clamping handles

(A) Slot casting adjustment screw

(H) Slot casting adjustment screw

(O) Top shaft

(J) Link adj block

(Y) Bending leaf hinge

(S) Bed end housing

(D) Bending leaf hinge adj screw

(T) Stop gage guide

(F) Bed end adj screw

(P) Bending leaf handle

(B) Slot casting lock screw

(K) Link set screw

(L) Link adj screw

(N) Adjustable stop

(U) Stop gage rod

(C) Legs

RIGHT HAND END VIEW

Figure 5–17. Cornice Brake.

142

The adjustment for thickness of material is made by adjusting the screws at each end of the folder. As this adjustment is made, place a piece of metal of the desired thickness in the folder and raise the operating handle until the small roller rests on the cam. Hold the folding blade in this position and adjust the setscrews so that the metal is clamped securely and evenly the full length of the folding blade. After the folder has been adjusted, test each end of the machine separately with a small piece of metal by actually folding it.

There are two positive stops on the folder, one for 45° folds or bends and the other for 90° folds or bends. An additional feature (a collar) is provided and can be adjusted to any degree of bend within the capacity of the machine.

For forming angles of 45° or 90°, the correct stop is moved into place. This will allow the handle to be moved forward to the correct angle. For forming other angles, the adjustable collar shown in the inset view of figure 5–16 is used. This is accomplished by loosening the setscrew and setting the stop at the desired angle. After setting the stop, tighten the setscrew and complete the bend.

To make the fold, adjust the machine correctly and then insert the metal. The metal goes between the folding blade and the jaw. Hold the metal firmly against the gage and pull the operating handle toward the body. As the handle is brought forward, the jaw automatically raises and holds the metal until the desired fold is made. When the handle is returned to its original position, the jaw and blade will return to their original positions and release the metal.

Cornice Brake

The cornice brake (fig. 5–17) has a much greater range of usefulness than the bar folder. Any bend formed on a bar folder can be made on the cornice brake. The bar folder can form a bend or edge only as wide as the depth of the jaws. In comparison, the cornice brake allows the sheet that is to be folded or formed to pass through the jaws from front to rear without obstruction.

In making ordinary bends with the cornice brake, the sheet is placed on the bed with the sight line (mark indicating line of bend) directly under the edge of the clamping bar. The clamping bar is then brought down to hold the sheet firmly in place. The stop at the right side of the brake is set for the proper angle or amount of bend, and the bending leaf is raised until it strikes the stop. If other bends

are to be made, the clamping bar is lifted and the sheet is moved to the correct position for bending.

The bending capacity of a cornice brake is determined by the manufacturer. Standard capacities of this machine are from 12- to 22-gage sheet metal, and bending lengths are from 3 to 12 ft. The bending capacity of the brake is determined by the bending edge thickness of the various bending leaf bars.

Most metals have a tendency to return to their normal shape—a characteristic known as springback. If the cornice brake is set for a 90° bend, the metal bent will probably form an angle of about 87° to 88°. Therefore, if a bend of 90° is desired, set the cornice brake to bend an angle of about 93° to allow for springback.

Slip Roll Former

The slip roll former (figure 5–18) is manually operated and consists of three rolls, two housings, a base, and a handle. The handle turns the two front rolls through a system of gears enclosed in the housing.

FIGURE 5–18. Slip roll former.

The front rolls serve as feeding or gripping rolls. The rear roll gives the proper curvature to the work. The front rolls are adjusted by two front adjusting screws on each end of the machine. The rear roll is adjusted by two screws at the rear of each housing. The front and rear rolls are grooved to permit forming of objects with wired edges. The upper roll is equipped with a release which permits easy removal of the metal after it has been formed.

When using the slip roll former, the lower front roll must be raised or lowered so that the sheet of metal can be inserted. If the object has a folded edge, there must be enough clearance between the rolls to prevent flattening the fold. If a metal requiring special care (such as aluminum) is being formed, the rolls must be clean and free of imperfections.

The rear roll must be adjusted to give the proper curvature to the part being formed. There are no gages that indicate settings for a specific diameter; therefore, trial-and-error settings must be used to obtain the desired curvature.

The metal should be inserted between the rolls from the front of the machine. Start the metal between the rolls by rotating the operating handle in a clockwise direction.

A starting edge is formed by holding the operating handle firmly with the right hand and raising the metal with the left hand. The bend of the starting edge is determined by the diameter of the part being formed. If the edge of the part is to be flat or nearly flat, a starting edge should not be formed.

Be sure that fingers or loose clothing are clear of the rolls before the actual forming operation is started. Rotate the operating handle until the metal is partly through the rolls and change the left hand from the front edge of the sheet to the upper edge of the sheet. Then roll the remainder of the sheet through the machine.

If the desired curvature is not obtained, return the metal to its starting position by rotating the handle counterclockwise. Raise or lower the rear roll and roll the metal through the rolls again. Repeat this procedure until the desired curvature is obtained, then release the upper roll and remove the metal.

If the part to be formed has a tapered shape, the rear roll should be set so that the rolls are closer together on one end than on the opposite end. The amount of this adjustment will have to be determined by experiment.

If the job being formed has a wired edge, the distance between the upper and lower rolls and the distance between the lower front roll and the rear roll should be slightly greater at the wired end than at the opposite end.

Forming Processes

Before a part is attached to the aircraft during either manufacture or repair, it has to be shaped to fit into place. This shaping process is called forming. Forming may be a very simple process, such as making one or two holes for attaching, or it may be exceedingly complex, requiring shapes with complex curvatures.

Parts are formed at the factory on large presses or by drop hammers equipped with dies of the correct shape. Every part is planned by factory engineers, who set up specifications for the materials to be used so that the finished part will have the correct temper when it leaves the machines. A layout for each part is prepared by factory draftsmen.

Forming processes used on the flight line and those practiced in the maintenance or repair shop are almost directly opposite in the method of procedure. They have much in common, however, and many of the facts and techniques learned in the one process can be applied to the other.

Forming is of major concern to the airframe mechanic and requires the best of his knowledge and skill. This is especially true since forming usually involves the use of extremely light-gage alloys of a delicate nature which can be readily made useless by coarse and careless workmanship. A formed part may seem outwardly perfect, yet a wrong step in the forming procedure may leave the part in a strained condition. Such a defect may hasten fatigue or may cause sudden structural failure.

Of all the aircraft metals, pure aluminum is the most easily formed. In aluminum alloys, ease of forming varies with the temper condition. Since modern aircraft are constructed chiefly of aluminum and aluminum alloys, this section will deal with the procedures for forming aluminum or aluminum alloy parts.

Most parts can be formed without annealing the metal, but if extensive forming operations, such as deep draws (large folds) or complex curves are planned, the metal should be in the dead soft or annealed condition. During the forming of some complex parts, operations may have to be stopped and the metal annealed before the process can be continued or completed. Alloy 2024 in the "O" condition can be formed into almost any shape by the common forming operations, but it must be heat-treated afterward.

When forming, use hammers and mallets as sparingly as practicable, and make straight bends on bar folders or cornice brakes. Use rotary machines whenever possible. If a part fits poorly or not at all, do not straighten a bend or a curve and try to re-form it, discard the piece of metal and start with a new one.

When making layouts, be careful not to scratch aluminum or aluminum alloys. A pencil, if kept sharp, will be satisfactory for marking purposes. Scribers make scratches which induce fatigue failure; but they may be used if the marking lines fall outside the finished part, that is, if the scribed line will be part of the waste material. Keep bench tops covered with material hard enough to prevent chips and other foreign material from becoming imbedded in them. Be sure also to keep bench tops clean and free from chips, filings, and the like. For the protection of the metals being worked, keep vise jaws covered with soft metal jaw caps.

Stainless steel can be formed by any of the usual methods but requires considerably more skill than is required for forming aluminum or aluminum alloys. Since stainless steel work-hardens very readily, it requires frequent annealing during the forming operations. Always try to press out stainless steel parts in one operation. Use dies, if possible.

FORMING OPERATIONS AND TERMS

The methods used in forming operations include such sheetmetal work processes as shrinking, stretching, bumping, crimping, and folding.

Bumping

Shaping or forming malleable metal by hammering or pounding is called bumping. During this process, the metal is supported by a dolly, a sandbag, or a die. Each contains a depression into which hammered portions of the metal can sink. Bumping can be done by hand or by machine.

Crimping

Folding, pleating, or corrugating a piece of sheet metal in a way that shortens it is called crimping. Crimping is often used to make one end of a piece of stovepipe slightly smaller so that one section may be slipped into another. Turning down a flange on a seam is also called crimping. Crimping one side of a straight piece of angle iron with crimping pliers will cause it to curve, as shown in figure 5–19.

Stretching

Hammering a flat piece of metal in an area such as that indicated in figure 5–19 will cause the material in that area to become thinner. However, since the amount of metal will not have been decreased, it will cover a greater area because the metal will have been stretched.

Stretching one portion of a piece of metal affects the surrounding material, especially in the case of formed and extruded angles. For example, hammering the metal in the horizontal flange of the angle strip over a metal block, as shown in figure 5–19, would cause its length to be increased (stretched); therefore, that section would become longer than the section near the bend. To allow for this difference in length, the vertical flange, which tends to keep the material near the bend from stretching, would be forced to curve away from the greater length.

Shrinking

During the shrinking process, material is forced or compressed into a smaller area. The shrinking process is used when the length of a piece of metal, especially on the inside of a bend, is to be reduced. Sheet metal can be shrunk in two ways: (1) By hammering on a V-block (figure 5–20), or (2) by crimping and then shrinking on a shrinking block.

To curve the formed angle by the V-block method, place the angle on the V-block and gently

A.   B.

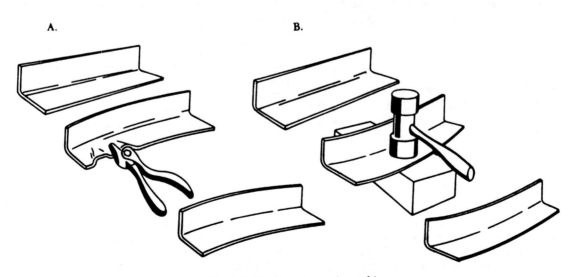

FIGURE 5–19. Crimping and stretching.

145

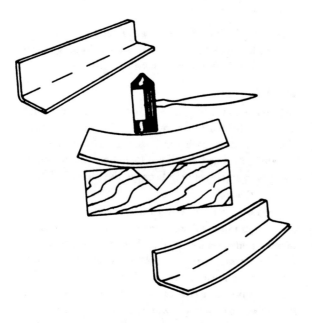

FIGURE 5–20. Shrinking on a V-block.

hammer downward against the upper edge directly over the "V" (figure 5–20). While hammering, move the angle back and forth across the V-block to compress the material along the upper edge. Compression of the material along the upper edge of the vertical flange will cause the formed angle to take on a curved shape. The material in the horizontal flange will merely bend down at the center, and the length of that flange will remain the same.

To make a sharp curve or a sharply bent flanged angle, crimping and a shrinking block can be used. In this process, crimps are placed in the one flange, and then by hammering the metal on a shrinking block, the crimps will be driven out (shrunk out) one at a time.

Folding

Making bends in sheets, plates, or leaves is called folding. Folds are usually thought of as sharp, angular bends; they are generally made on folding machines.

MAKING STRAIGHT LINE BENDS

When forming straight bends, the thickness of the material, its alloy composition, and its temper condition must be considered. Generally speaking, the thinner the material, the sharper it can be bent (the smaller the radius of bend), and the softer the material, the sharper the bend. Other factors that must be considered when making straight line bends are bend allowance, setback, and brake or sight line.

The radius of bend of a sheet of material is the radius of the bend as measured on the inside of the curved material. The minimum radius of bend of a sheet of material is the sharpest curve, or bend, to which the sheet can be bent without critically weakening the metal at the bend. If the radius of bend is too small, stresses and strains will weaken the metal and may result in cracking.

A minimum radius of bend is specified for each type of aircraft sheet metal. The kind of material, thickness, and temper condition of the sheet are factors affecting it. Annealed sheet can be bent to a radius approximately equal to its thickness. Stainless steel and 2024–T aluminum alloy require a fairly large bend radius (see fig. 5–28).

Bend Allowance

When making a bend or fold in a sheet of metal, the bend allowance must be calculated. Bend allowance is the length of material required for the bend. This amount of metal must be added to the overall length of the layout pattern to assure adequate metal for the bend.

Bend allowance depends on four factors: (1) The degree of bend, (2) The radius of the bend, (3) The thickness of the metal, and (4) The type of metal used. The radius of the bend is generally proportional to the thickness of the material. Furthermore, the sharper the radius of bend, the less the material that will be needed for the bend. The type of material is also important. If the material is soft it can be bent very sharply; but if it is hard, the radius of bend will be greater, and the bend allowance will be greater. The degree of bend will affect the overall length of the metal, whereas the thickness influences the radius of bend.

Bending a strip compresses the material on the inside of the curve and stretches the material on the outside of the curve. However, at some distance between these two extremes lies a space which is not affected by either force. This is known as the neutral line or neutral axis and occurs at a distance approximately 0.445 times the metal thickness $(0.445 \times T)$ from the inside of the radius of the bend (figure 5–21).

When bending metal to exact dimensions, the length of the neutral line must be determined so that sufficient material can be allowed for the bend. To save time in calculation of the bend allowance, formulas and charts for various angles, radii of bends, material thicknesses, and other factors have been established. The bend allowance formula for a 90° bend is discussed in the following paragraphs.

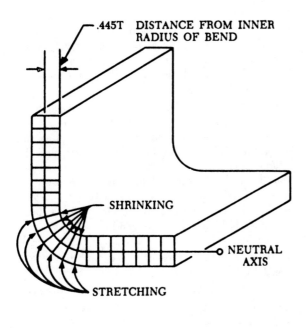

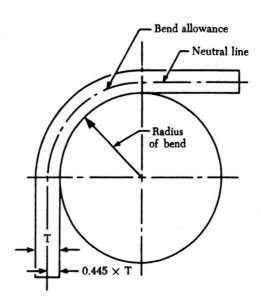

FIGURE 5-21. Neutral axis.

Method #1 Formula #1

To the radius of bend (R) add one-half the thickness of the metal, (½ T). This gives R + ½ T, or the radius of the circle of approximately the neutral axis.

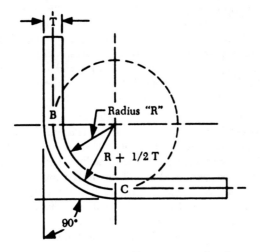

FIGURE 5-22. Bend allowance, 90° bend.

Compute the circumference of this circle by multiplying the radius of curvature of the neutral line (R + ½ T in figure 5-22) by 2π:

$$2\pi(R + \tfrac{1}{2} T).$$
Note: π = 3.1416.

Since a 90° bend is a quarter of the circle, divide the circumference by 4. This gives:

$$\frac{2\pi(R + \tfrac{1}{2} T)}{4}.$$

Therefore, bend allowance for a 90° bend is

$$\frac{2\pi(R + \tfrac{1}{2} T)}{4}.$$

To use the formula in finding the bend allowance for a 90° bend having a radius of ¼ in. for material 0.051-in. thick, substitute in the formula as follows:

Bend allowance

$$= \frac{2 \times 3.1416\ (0.250 + 1/2 \times 0.051)}{4}$$

$$= \frac{6.2832\ (0.250 + 0.02555)}{4}$$

$$= \frac{6.2832\ (0.2755)}{4}$$

$$= 0.4323.$$

Thus, if necessary, bend allowance or the length of material required for the bend is 0.4323, or 7/16 in.

The formula is slightly in error because actually the neutral line is not exactly in the center of the sheet being bent. (See figure 5-22.) However, the amount of error incurred in any given problem

147

is so slight that, for most work, since the material used is thin, the formula is satisfactory.

Method #2 Formula #2

This formula uses two constant values which have evolved over a period of years as being the relationship of the degrees in the bend to the thickness of the metal when determining the bend allowance for a particular application.

By experimentation with actual bends in metals, aircraft engineers have found that accurate bending results could be obtained by using the following formula for any degree of bend from 1° to 180°.

Bend allowance

$$= (0.01743 \times R + 0.0078 \times T) \times N$$

where:

R = The desired bend radius,

T = Thickness of the material, and

N = Number of degrees of bend.

BA = Bend allowance

BA = 0.01743 × 20°

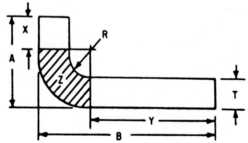

Bend allowance (Z) = (0.01743R + 0.0078T) x (No. of degrees of bend)

T = thickness of metal

R = radius of bend

Z = bend allowance

X = A - (R + T)

Y = B - (R + T)

Total developed length = X + Y + Z

FIGURE 5–23. Computing bend allowance.

Method #3 Use of 90° Bend Chart

Either formula may be used in the absence of a bend allowance chart. To determine bend allowance for any degree of bend by use of the chart (figure 5–24), find the allowance per degree for the number of degrees in the bend.

Radius of bend is given as a decimal fraction on the top line of the chart. Bend allowance is given

directly below the radius figures. The top number in each case is the bend allowance for a 90° angle, whereas the lower placed number is for a 1° angle. Material thickness is given in the left column of the chart.

To find the bend allowance when the sheet thickness is 0.051 in., the radius of bend is ¼ in. (0.250-in.), and the bend is to be 90°. Reading across the top of the bend allowance chart, find the column for a radius of bend of 0.250 in. Now find the block in this column that is opposite the gage of 0.051 in the column at left. The upper number in the block is 0.428, the correct bend allowance in inches for a 90° bend.

Method #4 use of chart for other than a 90° Bend

If the bend is to be other than 90°, use the lower number in the block (the bend allowance for 1°) and compute the bend allowance. The lower number in this case is 0.004756. Therefore, if the bend is to be 120°, the total bend allowance in inches will be 120X0.004756, or 0.5707 in.

SETBACK

When bending a piece of sheet stock, it is necessary to know the starting and ending points of the bend so that the length of the "flat" of the stock can be determined. Two factors are important in determining this, the radius of bend and the thickness of the material.

In figure 5–27, note that *setback is the distance from the bend tangent line to the mold point*. The mold point is the point of intersection of the lines extending from the outside surfaces, whereas the bend tangent lines are the starting and end points of the bend. Also note that setback is the same for the vertical flat and the horizontal flat.

Another way to look at setback is this: If the mandrel in a cornice brake is adjusted to the edge of the bed, a piece of metal is inserted, and a 90° bend is to be made, when the bending leaf is raised to 90°, the metal will be cut due to the compressing action of the leaf. The mandrel must be "set back" from the edge of the bed one thickness of the metal for a 90° bend. This permits the metal to flow thereby forming a correct bend.

Calculating Setback, Formula #1

To calculate the setback for a 90° bend, merely add the inside radius of the bend to the thickness of the sheet stock, i.e.

Setback = R + T.

Example:

Calculate the setback for a 90° bend, if the

RADIUS GAGE	1/32 .031	1/16 .063	3/32 .094	1/8 .125	5/32 .156	3/16 .188	7/32 .219	1/4 .250	9/32 .281	5/16 .313	11/32 .344	3/8 .375	7/16 .438	1/2 .500
.020	.062 .000693	.113 .001251	.161 .001792	.210 .002333	.259 .002874	.309 .003433	.358 .003974	.406 .004515	.455 .005056	.505 .005614	.554 .006155	.603 .006695	.702 .007795	.799 .008877
.025	.066 .000736	.116 .001294	.165 .001835	.214 .002376	.263 .002917	.313 .003476	.362 .004017	.410 .004558	.459 .005098	.509 .005657	.558 .006198	.607 .006739	.705 .007838	.803 .008920
.028	.068 .000759	.119 .001318	.167 .001859	.216 .002400	.265 .002941	.315 .003499	.364 .004040	.412 .004581	.461 .005122	.511 .005680	.560 .006221	.609 .006762	.708 .007862	.805 .007862
.032	.071 .000787	.121 .001345	.170 .001886	.218 .002427	.267 .002968	.317 .003526	.366 .004067	.415 .004608	.463 .005149	.514 .005708	.562 .006249	.611 .006789	.710 .007889	.807 .008971
.038	.075 .000837	.126 .001396	.174 .001937	.223 .002478	.272 .003019	.322 .003577	.371 .004118	.419 .004659	.468 .005200	.518 .005758	.567 .006299	.616 .006840	.715 .007940	.812 .009021
.040	.077 .000853	.127 .001411	.176 .001952	.224 .002493	.273 .003034	.323 .003593	.372 .004134	.421 .004675	.469 .005215	.520 .005774	.568 .006315	.617 .006856	.716 .007955	.813 .009037
.051		.134 .001413	.183 .002034	.232 .002575	.280 .003116	.331 .003675	.379 .004215	.428 .004756	.477 .005297	.527 .005855	.576 .006397	.624 .006934	.723 .008037	.821 009119
.064		.144 .001595	.192 .002136	.241 .002676	.290 .003218	.340 .003776	.389 .004317	.437 .004858	.486 .005399	.536 .005957	.585 .006498	.634 .007039	.732 .008138	.830 009220
.072			.198 .002202	.247 .002743	.296 .003284	.436 .003842	.394 .004283	.443 .004924	.492 .005465	.542 .006023	.591 .006564	.639 .007105	.738 .008205	.836 009287
.078			.202 .002249	.251 .002790	.300 .003331	.350 .003889	.399 .004430	.447 .004963	.496 .005512	.546 .006070	.595 .006611	.644 .007152	.745 .008252	.840 009333
.081			.204 .002272	.253 .002813	.302 .003354	.352 .003912	.401 .004453	.449 .004969	.498 .005535	.548 .006094	.598 .006635	.646 .007176	.745 .008275	.842 009357
.091			.212 .002350	.260 .002891	.309 .003432	.359 .003990	.408 .004531	.456 .005072	.505 .005613	.555 .006172	.604 .006713	.653 .0C7254	.752 .008353	.849 009435
.094			.214 .002374	.262 .002914	.311 .003455	.361 .004014	.410 .004555	.459 .005096	.507 .005637	.558 .006195	.606 .006736	.655 .007277	.754 .008376	.851 009458
.102				.268 .002977	.317 .003518	.367 .004076	.416 .004617	.464 .005158	.513 .005699	.563 .006257	.612 .006798	.661 .007339	.760 .008439	.857 009521
.109				.273 .003031	.321 .003572	.372 .004131	.420 .004672	.469 .005213	.518 .005754	.568 .006312	.617 .006853	.665 .008394	.764 .008493	.862 009575
.125				.284 .003156	.333 .003697	.383 .004256	.432 .004797	.480 .005338	.529 .005678	.579 .006437	.628 .006978	.677 .007519	.776 .008618	.873 009700
.156					.355 .003939	.405 .004497	.453 .005038	.502 .005579	.551 .006120	.601 .006679	.650 .007220	.698 .007761	.797 .008860	.895 009942
.188						.417 .004747	.476 .005288	.525 .005829	.573 .006370	.624 .006928	.672 .007469	.721 .008010	.820 009109	.917 010191
.250								.568 .006313	.617 .006853	.667 .007412	.716 .007953	.764 .008494	.863 009593	.961 010675

FIGURE 5–24. Bend allowance chart.

material is 0.051-in. thick and the radius of bend is specified to be 1/8 in. (0.125).

$$\text{Setback} = R + T.$$
$$= 0.125 + 0.051$$
$$= 0.176 \text{ in.}$$

Calculating Setback, Formula #2

To calculate setback for angles larger or smaller than 90°, consult standard setback charts (figure 5–25), or "K" chart, for a value called "K", and then substitute this value in the formula.

$$\text{Setback} = K \ (R + T).$$

The value of K varies with the number of degrees in the bend.

Example:

Calculate the setback for a 120° bend with a radius of bend of 0.125 in. in a sheet 0.032-in. thick.

$$\text{Setback} = K \ (R + T).$$
$$= 1.7320 \ (0.125 + 0.032)$$
$$= 0.272 \text{ in.}$$

Brake or Sight Line

The brake or sight line is the mark on a flat sheet which is set even with the nose of the radius bar of the cornice brake and serves as a guide in bending. The brake line can be located by measuring out one radius from the bend tangent line closest to the end which is to be inserted under the nose of the brake or against the radius form block. The nose of the brake or radius bar should fall directly over the brake or sight line as shown in figure 5–26.

Bend Allowance Terms

Familiarity with the following terms is necessary for an understanding of bend allowance and its application to an actual bending job. Figure 5–27 illustrates most of these terms.

Leg. The longer part of a formed angle.

Flange. The shorter part of a formed angle— the opposite of leg. If each side of the angle is the same length, then each is known as a leg.

149

A	K	A	K	A	K
1°	.00873	61°	.58904	121°	1.7675
2°	.01745	62°	.60086	122°	1.8040
3°	.02618	63°	.61280	123°	1.8418
4°	.03492	64°	.62487	124°	1.8807
5°	.04366	65°	.63707	125°	1.9210
6°	.05241	66°	.64941	126°	1.9626
7°	.06116	67°	.66188	127°	2.0057
8°	.06993	68°	.67451	128°	2.0503
9°	.07870	69°	.68728	129°	2.0965
10°	.08749	70°	.70021	130°	2.1445
11°	.09629	71°	.71329	131°	2.1943
12°	.10510	72°	.72654	132°	2.2460
13°	.11393	73°	.73996	133°	2.2998
14°	.12278	74°	.75355	134°	2.3558
15°	.13165	75°	.76733	135°	2.4142
16°	.14054	76°	.78128	136°	2.4751
17°	.14945	77°	.79543	137°	2.5386
18°	.15838	78°	.80978	138°	2.6051
19°	.16734	79°	.82434	139°	2.6746
20°	.17633	80°	.83910	140°	2.7475
21°	.18534	81°	.85408	141°	2.8239
22°	.19438	82°	.86929	142°	2.9042
23°	.20345	83°	.88472	143°	2.9887
24°	.21256	84°	.90040	144°	3.0777
25°	.22169	85°	.91633	145°	3.1716
26°	.23087	86°	.93251	146°	3.2708
27°	.24008	87°	.80978	147°	3.3759
28°	.24933	88°	.96569	148°	3.4874
29°	.25862	89°	.98270	149°	3.6059
30°	.26795	90°	1.0000C	150°	3.7320
31°	.27732	91°	1.0176	151°	3.8667
32°	.28674	92°	1.0355	152°	4.0108
33°	.29621	93°	1.0538	153°	4.1653
34°	.30573	94°	1.0724	154°	4.3315
35°	.31530	95°	1.0913	155°	4.5107
36°	.32492	96°	1.1106	156°	4.7046
37°	.33459	97°	1.1303	157°	4.9151
38°	.34433	98°	1.1504	158°	5.1455
39°	.35412	99°	1.1708	159°	5.3995
40°	.36397	100°	1.1917	160°	5.6713
41°	.37388	101°	1.2131	161°	5.9758
42°	.38386	102°	1.2349	162°	6.3137
43°	.39391	103°	1.2572	163°	6.6911
44°	.40403	104°	1.2799	164°	7.1154
45°	.41421	105°	1.3032	165°	7.5957
46°	.42447	106°	1.3270	166°	8.1443
47°	.43481	107°	1.3514	167°	8.7769
48°	.44523	108°	1.3764	168°	9.5144
49°	.45573	109°	1.4019	169°	10.385
50°	.46631	110°	1.4281	170°	11.430
51°	.47697	111°	1.4550	171°	12.706
52°	.48773	112°	1.4826	172°	14.301
53°	.49858	113°	1.5108	173°	16.350
54°	.50952	114°	1.5399	174°	19.081
55°	.52057	115°	1.5697	175°	22.904
56°	.53171	116°	1.6003	176°	26.636
57°	.54295	117°	1.6318	177°	38.188
58°	.55431	118°	1.6643	178°	57.290
59°	.56577	119°	1.6977	179°	114.590
60°	.57735	120°	1.7320	180°	infinite

Figure 5–25A Setback (K) chart.

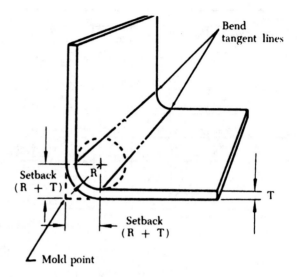

Figure 5–25 B. Setback, 90° bend.

Mold Line (ML). The line formed by extending the outside surfaces of the leg and flange. (An imaginary point from which real base measurements are provided on drawings.)

Bend Tangent Line (BL). The line at which the metal starts to bend and the line at which the metal stops curving. All the space between the band tangent lines is the bend allowance.

Bend Allowance (BA). The amount of material consumed in making a bend (figure 5–12).

Radius (R). The radius of the bend—always to the inside of the metal being formed unless otherwise stated. (The minimum allowable radius for bending a given type and thickness of material should always be ascertained before proceeding with any bend allowance calculations.)

Setback (SB). The setback is the distance from the bend tangent line to the mold point. In a 90-degree bend SB=R+T (radius of the bend plus thickness of the metal). The setback dimension must be determined prior to making the bend as it (setback) is used in determining the location of the beginning bend tangent line (figure 5–27).

Bend Line (also called Brake or Sight Line). The layout line on the metal being formed which is set even with the nose of the brake and serves as a guide in bending the work. (Before forming a bend, it must be decided which end of the material can be most conveniently inserted in the brake. The bend line is then measured and marked off with a soft-lead pencil from the bend tangent line closest to the end which is to be placed under the brake. This measurement should be equal to the radius of the bend. The metal is then inserted

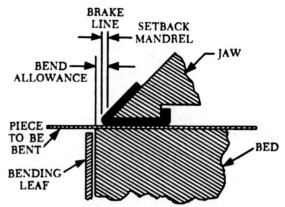

FIGURE 5-26. Setback – locating bend line in brake.

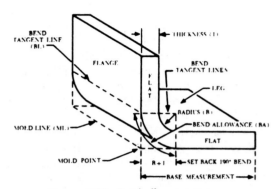

FIGURE 5-27. Bend allowance terms.

in the brake so that the nose of the brake will fall directly over the bend line, as shown in figure 5–26.)

Flat (short for flat portion). The flat portion or flat of a part is that portion not included in the bend. It is equal to the base measurement minus the setback.

Base Measurement. The outside dimensions of a formed part. Base measurement will be given on the drawing or blueprint, or may be obtained from the original part.

Closed Angle. An angle that is less than 90° when measured between legs, or more than 90° when the amount of bend is measured.

Open Angle. An angle that is more than 90° when measured between legs, or less than 90° when the amount of bend is measured.

"K" No. One of 179 numbers on the "K" chart corresponding to one of the angles between 0 and 180° to which metal can be bent. Whenever metal is to be bent to any angle other than 90° ("K" No. of 1.0), the corresponding "K" No. is selected from the chart and is multiplied by the sum of the radius and the thickness of the metal. The product is the amount of setback for the bend.

MAKING LAYOUTS

It is wise to make a layout or pattern of the part before forming it to prevent any waste of material and to get a greater degree of accuracy in the finished part. Where straight angle bends are concerned, correct allowances must be made for setback and bend allowance. If the shrinking or stretching processes are to be used, allowances must

Designation	Gage							
	0.020	0.025	0.032	0.040	0.050	0.063	0.071	0.080
2024-O	1/32	1/16	1/16	1/16	1/16	3/32	1/8	1/8
2024-T4	1/16	1/16	3/32	3/32	1/8	5/32	7/32	1/4
5052-O	1/32	1/32	1/16	1/16	1/16	1/16	1/8	1/8
5052-H34	1/32	1/16	1/16	1/16	3/32	3/32	1/8	1/8
6061-O	1/32	1/32	1/32	1/16	1/16	1/16	3/32	3/32
6061-T4	1/32	1/32	1/32	1/16	1/16	3/32	5/32	5/32
6061-T6	1/16	1/16	1/16	3/32	3/32	1/8	3/16	3/16
7075-O	1/16	1/16	1/16	1/16	3/32	3/32	5/32	3/16
7075-W	3/32	1/32	1/8	5/32	3/16	1/4	9/32	5/16
7075-T6	1/8	1/8	1/8	3/16	1/4	5/16	3/8	7/16

FIGURE 5-28. Minimum bend radii for aluminum alloys.

be made so that the part can be turned out with a minimum amount of forming.

The layout procedures can be put into three general groups: (1) Flat layout, (2) Duplication of pattern, and (3) Projection through a set of points. All three processes require a good working knowledge of arithmetic and geometry. This presentation will discuss only two processes, flat layout and duplication of pattern.

Referring to the "K" chart, figure 5–27, it is noted that the "K" value for 90° is equal to 1T (thickness of metal). Further observation will show that for an angle of less than 90° the setback is less than 1T, for an angle of more than 90° the setback is more than 1T.

The use of 1T setback in a bend of less than 90° (open angle) would result in the flange of the bend being too long. Conversely in an angle of over 90° with less than 1T setback the flange would be too short.

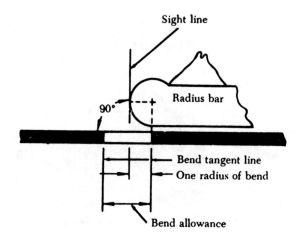

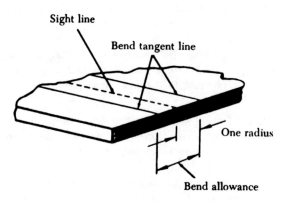

FIGURE 5–30. Brake or sight line.

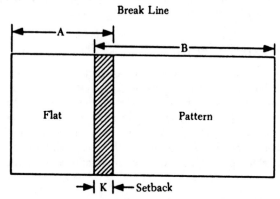

Developed Length of Pattern = A+B. To determine the developed Length of the flat pattern, deduct the "K" dimension from the sum of the dimensions A+B.

FIGURE 5–29. Setback development.

Flat Layout

Assume that it is necessary to lay out a flat pattern of a channel (figure 5–31) in which the left-hand flat, A, is to be 1 in. high, the right-hand flat, C, is to be 1–1/4 in. high, and the distance between the outside surface of the two flats, B, is to be 2 in. The material is 0.051-in. thick, and the radius of bend is to be 3/16 in. (0.188). The angles are to be 90°. Proceed as follows:

(1) Determine the setback to establish the distance of the flats.

(a) The setback for the first bend:
Setback $= R + T$
$= 0.188 + 0.051$
$= 0.239.$

(b) The first flat A is equal to the overall dimension less setback:
Flat A $= 1.000 - 0.239$
$= 0.761$ in.

(2) Calculate the bend allowance for the first bend by using the bend allowance chart (figure 5–24). (BA = 0.3307 or 0.331.)

(3) Now lay off the second flat, B. This is equal to the overall dimension less the setback at each end, or B minus two setbacks: (See figure 5–31.)
Flat B $= 2.00 - (0.239 + 0.239)$
$= 2.000 - .478$
$= 1.522$ in.

152

(4) The bend allowance for the second bend is the same as that for the first bend (0.331). Mark off this distance. (See figure 5–31.)

(5) The third flat, C, is equal to the overall dimension less the setback. Lay off this distance. (See figure 5–31.)

Flat C = 1.250 — 0.239
= 1.011 in.

(6) Adding the measurements of flats A, B, and C, and both bend allowances, (0.761 + 0.331 + 1.522 + 0.331 + 1.011), the sum is 3.956, or approximately 4.00 inches. Totaling the three flats, A, B, and C, 1 in., 2 in., and 1-1/4 in., respectively, the sum is 4.250 in. of material length. This illustrates how setback and bend allowance affect material lengths in forming straight line bends. In this case, the reduction is approximately 1/4 in.

After all measurements are calculated, cut the material and mark off the brake or sight lines as shown in figure 5–31.

Duplication of Pattern

When it is necessary to duplicate an aircraft part and blueprints are not available, take measurements directly from the original or from a duplicate part. In studying the following steps for laying out a part to be duplicated, refer to the illustrations in figure 5–32.

Draw a reference (datum) line, AB, on the sample part and a corresponding line on the template material (example 1, figure 5–32).

Next, with point A on the sample part as a center, draw an arc having a radius of approximately 1/2 in. and extending to the flanges (example 2, figure 5–32).

Draw similar arcs each with a radius 1/2 in. greater than the previous one until the entire part is marked. In case there is an extremely sharp curve in the object, decrease the distance between the arcs to increase the number of arcs. This procedure will increase the accuracy of the layout. An arc must

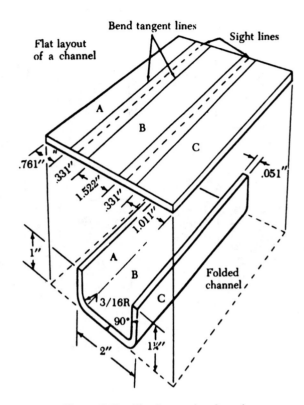

FIGURE 5–31. Flat layout of a channel.

pass through every corner of the part; one arc may pass through more than one corner (example 3, figure 5–32).

Locate the coordinate point on the layout by measuring on the part with dividers. Always measure the distance from the reference point to the beginning of the bend line on the flange of the part.

After locating all points, draw a line through them, using a French curve to ensure a smooth pattern (example 4, figure 5–32).

Allow for additional material for forming the flange and locate the inside bend tangent line by measuring, inside the sight line, a distance equal to the radius of bend of the part.

Using the intersection of the lines as a center, locate the required relief holes. Then cut out and form as necessary.

Relief Holes

Wherever two bends intersect, material must be removed to make room for the material contained in the flanges. Holes are therefore drilled at the intersection. These holes, called relief holes, prevent strains from being set up at the intersection of the

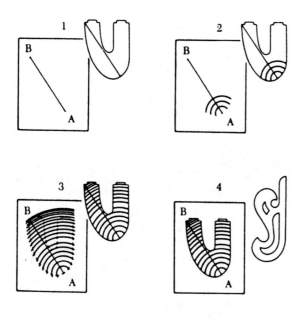

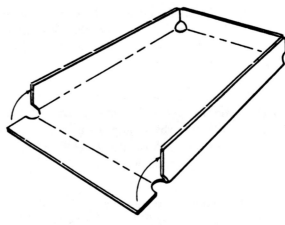

FIGURE 5–32. Duplicating a pattern.

inside bend tangent lines which would cause the metal to crack. Relief holes also provide a neatly trimmed corner from which excess material may be trimmed.

The size of relief holes varies with thickness of the material. They should be not less than 1/8 in. in diameter for aluminum alloy sheet stock up to and including 0.064-in. thick, or 3/16 in. for stock ranging from 0.072 in. to 0.128 in. in thickness. The most common method of determining the diameter of a relief hole is to use the radius of bend for this dimension, provided it is not less than the minimum allowance (1/8 in.).

Relief holes must touch the intersection of the inside bend tangent lines. To allow for possible error in bending, make the relief holes so they will extend 1/32 to 1/16 in. behind the inside bend tangent lines. It is good practice to use the intersection of these lines as the center for the holes (figure 5–33). The line on the inside of the curve is cut at an angle toward the relief holes to allow for the stretching of the inside flange.

Lightening Holes

Lightening holes are cut in rib sections, fuselage frames, and other structural parts to decrease weight. To keep from weakening the member by

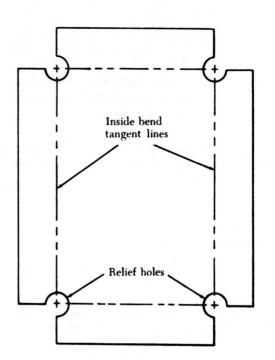

FIGURE 5–33. Locating relief holes.

removal of the material, flanges are often pressed around the holes to strengthen the area from which the material was removed.

Lightening holes should never be cut in any structural part unless authorized. The size of the lightening hole and the width of the flange formed around the hole are determined by design specifications. Margins of safety are considered in the specifications so that the weight of the part can be decreased and still retain the necessary strength.

154

Lightening holes may be cut by any one of the following methods:

(1) Punching out, if the correct size punch die is available.

(2) Cutting out with a fly cutter mounted on a drill.

(3) Scribing the circumference of a hole with dividers and drilling around the entire circumference with a small drill, allowing enough clearance to file smooth.

(4) Scribing the circumference of the hole with dividers, drilling the hole inside the circumference large enough to insert aviation snips, cutting out excess metal, and filing smooth.

Form the flange by using a flanging die, or hardwood or metal form blocks. Flanging dies consist of two matching parts, a female and a male die. For flanging soft metal, dies can be of hardwood, such as maple. For hard metal or for more permanent use, they should be made of steel. The pilot guide should be the same size as the hole to be flanged, and the shoulder should be the same width and angle as the desired flange.

When flanging lightening holes, place the material between the mating parts of the die and form it by hammering or squeezing the dies together in a

vise or in an arbor press. The dies will work more smoothly if they are coated with light machine oil.

Note that in the two form blocks shown on the left side of figure 5–34, the hole in the upper block is the same size as the hole to be flanged and is chamfered to the width of the flange and the angle desired, whereas in the lower block, the hole is the same diameter as that of the flange. Either type may be used. When using the upper block, center the material to be flanged and hammer it with a stretching mallet, around and around, until the flange conforms to the chamfer. When using the lower block, center the lightening hole over the hole in the block, then stretch the edges, hammering the material into the hole, around and around, until the desired flange is obtained. Occasionally, the chamfer is formed with a cone-shaped male die used in conjunction with the form block with which the part was formed.

HAND FORMING

All forming revolves around the process of shrinking and stretching, and hand forming processes are no exception. If a formed or extruded angle is to be curved, either stretch one leg or shrink the other, whichever will make the part fit. In bumping, the material is stretched in the bulge to make it "balloon," and in joggling, the material is stretched between the joggles. Material in the edge of lightening holes is often stretched to form a beveled reinforcing ridge around them.

Straight Line Bends

The cornice brake and bar folder are ordinarily used to make straight bends. Whenever such machines are not available, comparatively short sections can be bent by hand with the aid of wooden or metal bending blocks by proceeding as explained in the following paragraphs.

After laying out and cutting a blank to size, clamp it rigidly along the bending line between two wooden blocks held in a vise. The wooden forming block should have one edge rounded for the desired radius of bend. It should also be curved slightly beyond the 90° point to allow for springback.

By tapping lightly with a rubber, plastic, or rawhide mallet, bend the metal protruding beyond the bending blocks to the desired angle. Start tapping at one end and work back and forth along the edge to make a gradual and even bend.

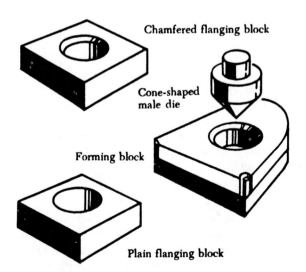

FIGURE 5–34. Flanging form blocks.

Continue this process until the protruding metal is forced down to the desired angle against the forming block. Allow for springback by driving the material slightly farther than the actual bend. If a large amount of metal extends beyond the bending blocks, maintain hand pressure against the protruding sheet to prevent "bouncing."

Remove any irregularities by holding a straight block of hardwood edgewise against the bend and striking it with heavy blows of a mallet or hammer. If the amount of metal protruding beyond the bending blocks is small, make the entire bend by using the hardwood block and hammer.

Formed or Extruded Angles

Both formed and extruded types of angles can be curved (not bent sharply) by stretching or shrinking either of the flanges. Curving by stretching the one flange is usually preferred since this process requires only a V-block and a mallet and is easily accomplished.

In the stretching process, place the flange to be stretched in the groove of the V-block. Using a stretching mallet, strike the flange directly over the V portion with light, even blows while gradually forcing it downward into the V. Too heavy a blow will buckle the angle strip. Keep moving the angle strip across the V-block, but always strike the spot directly above the V. Form the curve gradually and evenly by moving the strip slowly back and forth, distributing the hammer blows at equal spaces on the flange.

Lay out a full-sized, accurate pattern on a sheet of paper or plywood and periodically check the accuracy of the curve. Comparing the angle with the pattern will determine exactly how the curve is progressing and just where it needs to be increased or decreased. It is better to get the curve to conform roughly to the desired shape before attempting to finish any one portion, because the finishing or smoothing of the angle may cause some other portion of the angle to change shape. If any part of the angle strip is curved too much, reduce the curve by reversing the angle strip on the V-block, placing the bottom flange up, and striking it with light blows of the mallet.

Try to form the curve with a minimum amount of hammering, for excessive hammering will work-harden the metal. Work-hardening can be recognized by a lack of bending response or by springiness in the metal. It can be recognized very readily by an experienced worker. In some cases, the part

may have to be annealed during the curving operation. If so, be sure to heat treat the part again before installing it on the aircraft.

Curving an extruded or formed angle strip by shrinking may be accomplished by either of two methods, the V-block method or the shrinking block method. Of the two, the V-block is, in general, more satisfactory because it is faster, easier, and affects the metal less. However, very good results can be obtained by the shrinking block method.

In the V-block method, place one flange of the angle strip flat on the V-block with the other flange extending upward, as shown in figure 5–35. Hold it firmly so that it does not bounce when hammered, and strike the edge of the upper flange with light blows of a round, soft-faced mallet. Begin at one end of the angle strip and, working back and forth, strike light blows directly over the V-portion of the block. Strike the edge of the flange at a slight angle as this tends to keep the vertical flange from bending outward.

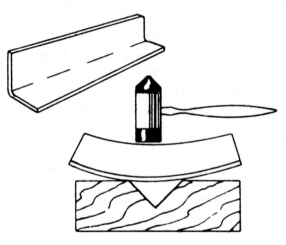

FIGURE 5–35. V-blocks.

Occasionally, check the curve for accuracy with the pattern. If a sharp curve is made, the angle (cross section of the formed angle) will close slightly. To avoid such closing of the angle, clamp the angle strip to a hardwood board with the hammered flange facing upward using small C-clamps. The jaws of the C-clamps should be covered with masking tape. If the angle has already closed, bring the flange back to the correct angle with a few blows of a mallet or with the aid of a small hardwood block. If any portion of the angle strip is curved too much, reduce it by reversing the angle on the V-block and hammering with a suitable mallet, as explained in the previous paragraph on stretching. After obtaining the proper curve, smooth the entire angle by planishing with a soft-faced mallet.

If the curve in a formed angle is to be quite sharp or if the flanges of the angle are rather broad, the shrinking block method is generally used. In this process, crimp the flange which is to form the inside of the curve.

When making a crimp, hold the crimping pliers so that the jaws are about 1/8 in. apart. By rotating the wrist back and forth, bring the upper jaw of the pliers into contact with the flange, first on one side and then on the other side of the lower jaw. Complete the crimp by working a raised portion into the flange, gradually increasing the twisting motion of the pliers. Do not make the crimp too large because it will be difficult to work out. The size of the crimp depends upon the thickness and softness of the material, but usually about 1/4 in. is sufficient. Place several crimps spaced evenly along the desired curve with enough space left between each crimp so that jaws of the shrinking block can easily be attached.

After completing the crimping, place the crimped flange in the shrinking block so that one crimp at a time is located between the jaws. Flatten each crimp with light blows of a soft-faced mallet, starting at the apex (the closed end) of the crimp and gradually working toward the edge of the flange. Check the curve of the angle with the pattern periodically during the forming process and again after all the crimps have been worked out. If it is necessary to increase the curve, add more crimps and repeat the process. Space the additional crimps between the original ones so that the metal will not become unduly work-hardened at any one point. If the curve needs to be increased or decreased slightly at any point, use the V-block.

After obtaining the desired curve, planish the angle strip over a stake or a wooden form.

Flanged Angles

The forming process for the following two flanged angles is slightly more complicated than that just discussed in that the bend is shorter (not gradually curved) and necessitates shrinking or stretching in a small or concentrated area. If the

flange is to point toward the inside of the bend, the material must be shrunk. If it is to point toward the outside, it must be stretched.

In forming a flanged angle by shrinking, use wooden forming blocks similar to those shown in figure 5-36 and proceed as follows:

(1) Cut the metal to size, allowing for trimming after forming. Determine the bend allowance for a 90° bend and round the edge of the forming block accordingly.

(2) Clamp the material in the form blocks as shown in figure 5-36, and bend the exposed flange against the block. After bending, tap the blocks slightly. This induces a setting process in the bend.

(3) Using a soft-faced shrinking mallet, start hammering near the center and work the flange down gradually toward both ends. The flange will tend to buckle at the bend because the material is made to occupy less space. Work the material into several small buckles instead of one large one and work each buckle out gradually by hammering lightly and gradually compressing the material in each buckle. The use of a small hardwood wedge block, as shown in figure 5-36, will aid in working out the buckles.

(4) Planish the flange after it is flattened against the form block and remove small irregularities. If the form blocks are made of hardwood, use a metal planishing hammer. If the forms are made of metal, use a soft-faced mallet. Trim the excess material away and file and polish.

Forming by Stretching

To form a flanged angle by stretching, use the same forming blocks, wooden wedge block, and mallet as in the shrinking process and proceed as follows:

(1) Cut the material to size (allowing for trim), determine bend allowance for a 90° bend, and round off the edge of the block to conform to the desired radius of bend.

(2) Clamp the material in the form blocks as shown in figure 5–36.

(3) Using a soft-faced stretching mallet, start hammering near the ends and work the flange down smoothly and gradually to prevent cracking and splitting. Planish the flange and angle as described in the previous procedure, and trim and smooth the edges, if necessary.

Curved Flanged Parts

Curved flanged parts are usually hand formed. Of the types shown in figure 5–37, the one with relief holes is probably the simplest to form. It has a concave flange (the inside flange) and a convex flange (the outside flange).

The concave flange is formed by stretching, the convex flange by shrinking. Such parts may be formed with the aid of hardwood or metal forming blocks. These blocks are made in pairs similar to those used for straight angle bends and are identified in the same manner. They differ in that they are made specifically for the particular part to be formed, they fit each other exactly, and they conform to the actual dimensions and contour of the finished article.

The mating parts may be equipped with aligning pins to aid in lining up the blocks and holding the metal in place. The blocks may be held together by C-clamps or a vise. They also may be held together with bolts by drilling through both forms and the metal, provided the holes do not affect the strength of the finished part. The edges of the forming block are rounded to give the correct radius of bend to the part, and are undercut to allow for springback of the metal. The undercut is especially necessary if the material is hard or if the bend must be highly accurate.

Note the various types of forming represented in figure 5–37. In the plain nose rib, only one large convex flange is used; but, because of the great distance around the part and the likelihood of buckles in forming, it is rather difficult to form. The flange and the beaded portion of this rib provide sufficient strength to make this a very good type to use. In the type with relief holes, the concave flange gives difficulty in forming; however, the outside flange is broken up into smaller sections by relief holes (notches inserted to prevent strains in a

Shrinking

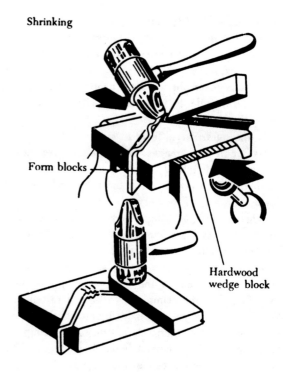

Form blocks

Hardwood wedge block

Stretching

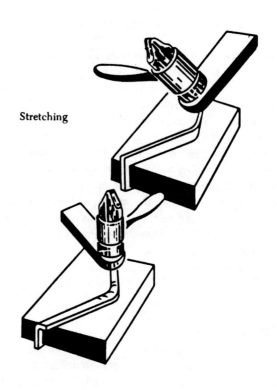

FIGURE 5–36. Forming a flanged angle.

158

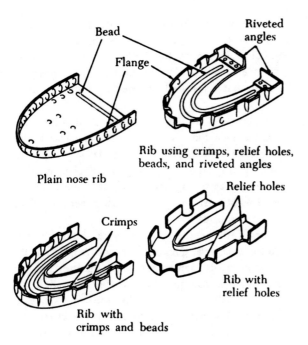

Bead

Flange

Riveted angles

Rib using crimps, relief holes, beads, and riveted angles

Plain nose rib

Crimps

Relief holes

Rib with relief holes

Rib with crimps and beads

FIGURE 5–37. Nose ribs.

bend). In the type with crimps and beads, note that crimps are inserted at equally spaced intervals. The crimps are placed to absorb material and cause curving, while also giving strength to the part.

In the other nose rib illustrated, note that a combination of the four common forming methods is applied. They are crimping, beading, putting in relief holes, and using a formed angle riveted on at each end. The beads and the formed angles supply strength to the part.

The major steps in forming a curved flange part are explained in the following paragraphs.

Cut the material to size (allowing for trim), locate and drill holes for alignment pins, and remove all burrs (jagged edges). Place the material between the wooden blocks. Clamp blocks tightly in a vise so that the material will not move or shift. Clamp the work as closely as possible to the particular area being hammered to prevent strain on the form blocks and to keep the metal from slipping (figure 5–38).

Bend the flange on the concave curve first. This practice may keep the flange from splitting open or cracking when the metal is stretched. (Should this occur, a new piece will have to be made.) Using a

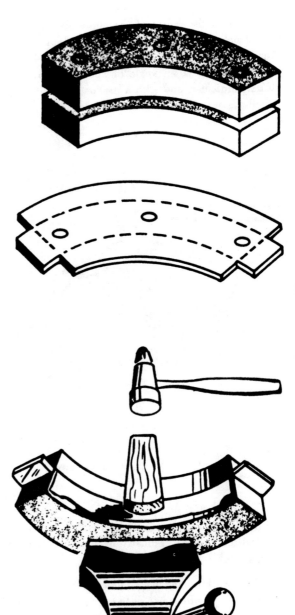

FIGURE 5–38. Forming a concave curve.

soft mallet or wooden wedge block, start hammering at a point a short distance away from the beginning of the concave bend and continue toward the center of the bend. This procedure permits some of the excess metal along the tapered portion of the flange to be worked into the curve where it will be needed. Continue hammering until the metal is gradually worked down over the entire flange, flush with the form block.

159

Starting at the center of the curve and working toward both ends, hammer the convex flange down over the form (figure 5–39). Strike the metal with glancing blows, at an angle of approximately 30° off perpendicular, and with a motion that will tend to pull the part away from the block.

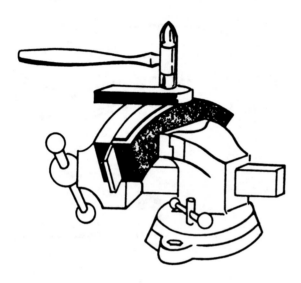

FIGURE 5–39. Forming a convex curve.

Stretch the metal around the radius bend and remove the buckles gradually by hammering on a wedge block.

While working the metal down over the form, keep the edges of the flange as nearly perpendicular to the block as possible. The wedge block helps keep the edge of the metal perpendicular to the block, lessens the possibility of buckles and of splitting or cracking the metal, and aids in removing buckles.

Finally, trim the flanges of excess metal, planish, remove burrs, round the corners (if any), and check the part for accuracy.

Bumping

Bumping on a form block or female die and bumping on a sandbag are the two common types practiced. In either method only one form is re-

quired, a wooden block, lead die, or sandbag. A good example of a part made by the block or die type of bumping is the "blister" or streamlined cover plate. Wing fillets constitute a good example of parts that are usually formed by bumping on a sandbag.

The lead die or the wooden block designed for bumping must have the same dimensions and contour as the outside of the blister. To provide sufficient bumping weight, and to give sufficient bearing surface for fastening the metal, the block or die should be at least 1 in. larger in all dimensions than the form requires.

When forming the wooden block, hollow it out with saws, chisels, gouges, files, and rasps. Smooth and finish it with sandpaper. Make the inside of the form as smooth as possible, because any slight irregularity will show up on the finished part. Prepare several templates (patterns of the cross section), such as those shown with the form block for the blister in figure 5–40, so that the form can be checked for accuracy.

Shape the contour of the form at points 2, 3, and 4. Shape the areas between the template check points to conform to the remaining contour and to template 4. Shaping of the form block requires particular care because the more nearly accurate it is, the less time it will take to produce a smooth, finished part.

Correct clamping of the material to the form block is an important part of the block-forming operation. Several methods are possible. For parts such as the blister, one of the best means of clamping the material is to use a full metal cutout or steel holddown plate as shown in figure 5–40.

In this process, place the holddown plate directly over the material to be formed and clamp it in position with bolts or C-clamps. Tighten the C-clamps or bolts just tight enough to hold the material flat against the face of the form block, but not so tight that the metal cannot be drawn into the form. If the material is not held flat against the face of the form, it will bend up or buckle away from the block. If it is not permitted to slip into the concave depression a little, the blister portion will become very thin in places.

Holddown plates should be of heavy steel, 1/8 in. for small forms and 1/4 in. or heavier for large forms.

If the material for making an all-metal holddown plate is not available, use a hardwood cutout. Make the cutout and use it in the same manner as the

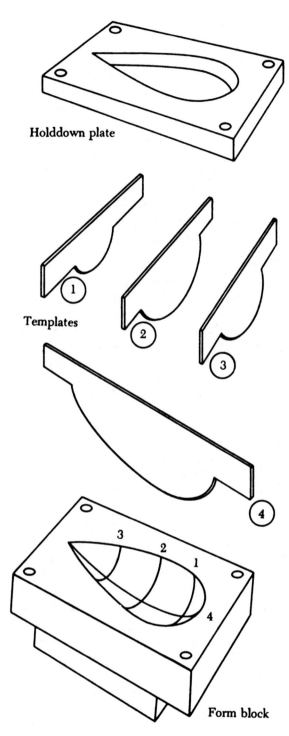

Holddown plate

Templates

①

②

③

④

3
2
1
4

Form block

FIGURE 5-40. Form blocks and templates.

steel plate, but take greater precautions to make sure that the material is held as desired.

Pieced form clamps can be used if an all-metal holddown plate or hardwood cutout is not available or if a full cutout cannot be used. Be careful to clamp them properly and locate them so that they align with the edge of the form. If they are not aligned accurately, the material will bulge.

After preparing and checking the form, perform the bumping process according to the following general steps:

(1) Cut a metal blank to size, allowing an extra 1/2 to 1 in. to permit "drawing."

(2) Apply a thin coat of light oil to the block and to the aluminum to prevent galling (scraping on rough spots).

(3) Clamp the material between the block and steel plate, as previously described, so that it will be firmly supported yet able to slip a little toward the inside of the form.

(4) Clamp the bumping block in a bench vise. With a soft-faced mallet or with a hardwood drive block and suitable mallet, start the bumping near the edges of the form.

(5) With light blows of the mallet, work the material down gradually from the edges. Remember that the object of the bumping process is to work the material into shape by stretching it, rather than by forcing it into the form with heavy blows. Always start bumping near the edge of the form; never start near the center of the blister.

(6) Smooth the work as much as possible before removing it from the form. This can be done by rubbing the work with the rounded end of a maple block or with the round end of a stretching mallet.

(7) Remove the blister from the bumping block and trim it, leaving a 1/2-in. flange.

(8) Finally, drill the rivet holes, chamfer the edges 45°, and clean and polish the part.

Bumping on a sandbag is one of the most difficult types of sheet-metal hand forming, because there is no exact form block to serve as a guide. In this type of forming operation, a depression is made into a sandbag to take the shape of the hammered portion of the metal. The depression or pit has a tendency to shift from the hammering. This necessitates readjusting from time to time during the bumping process. The degree of shifting depends largely on the contour or shape of the piece being formed, and whether glancing blows must be struck to stretch, draw, or shrink the metal.

When forming by this method, prepare a contour template or some sort of a pattern to serve as a working guide and to ensure accuracy of the finished part. Make the pattern from ordinary kraft or similar paper, folding it over the part to be duplicated. Cut the paper cover at the points where it would have to be stretched to fit, and attach additional pieces of paper with masking tape to cover the exposed portions. After completely covering the part, trim the pattern to exact size.

Open the pattern and spread it out on the metal from which the part is to be formed. Although the pattern will not lie flat, it will give a fairly accurate idea of the approximate shape of the metal to be cut, and the pieced-in sections will indicate where the metal is to be stretched. When the pattern has been placed on the material, outline the part and the portions to be stretched using a pencil. Add at least 1 in. of excess metal when cutting the material to size. Trim off the excess metal after bumping the part into shape.

If the part to be formed is radially symmetrical, it will be fairly easy to shape since a simple contour template can be used as a working guide, making a pattern to indicate the portions of unequal stretching unnecessary. However, the procedure for bumping sheet metal parts on a sandbag follows certain basic rules which can be applied to any part, regardless of its contour or shape.

(1) Lay out and cut the contour template. This can be made of sheet metal, medium-heavy cardboard, or thin plywood.

(2) Determine the amount of metal needed, lay it out, and cut it to size, allowing at least 1/2 in. excess.

(3) Place a sandbag on a solid foundation capable of supporting heavy blows and, with the aid of a smooth-faced mallet, make a pit in the bag. Analyze the part to determine the correct radius of the pit for the forming operation. The pit will change with the hammering it receives and must be re-adjusted occasionally.

(4) Select a soft round-faced or bell-shaped mallet having a contour slightly smaller than the contour desired on the sheet-metal part. Holding one edge of the metal in the left hand, place the portion to be bumped near the edge of the pit on the sandbag. Strike the metal with light glancing blows, about 1/2 to 1 in. from the edge.

(5) Continue bumping toward the center, revolving the metal and working gradually inward until the desired shape is obtained. Shape the entire part as a unit.

(6) At frequent intervals during the bumping process, check the part for accuracy of shape by applying the template. If wrinkles are formed, work them out before they become too large.

(7) Finally, with a suitable stake and planishing hammer, or with a hand dolly and planishing hammer, remove small dents and hammer marks.

(8) With a scribe, mark around the outside of the object. Trim the edge and file until it is smooth. Clean and polish the part.

Joggling

A joggle is an offset formed on an angle strip to allow clearance for a sheet or an extrusion. Joggles are often found at the intersection of stringers and formers. One of these members, usually the former, has the flange joggled to fit flush over the flange of the stringer. The amount of offset is usually small; therefore, the depth of the joggle is generally specified in thousandths of an inch. The thickness of the material to be cleared governs the depth of the joggle. In determining the necessary length of the joggle, it is common practice to allow an extra 1/16 in. to give enough added clearance to assure a fit between the joggled, overlapped part.

There are a number of different methods by which joggles can be formed. If the joggle is to be made on a straight flange or flat piece of metal, form it on a cornice brake by inserting and bending up along the line of the joggle. Hold a piece of metal of the correct thickness to give the desired offset under the bent-up portion, and pound the flange down while the metal is still in the same position in the brake.

Where a joggle is necessary on a curved flange, forming blocks or dies made of hardwood, steel, or aluminum alloy may be used. If the die is to be used only a few times, hardwood is satisfactory as it is easily worked. If a number of similar joggles are to be produced, then use steel or aluminum alloy dies. Dies of aluminum alloy are preferred since they are easier to fabricate than those of steel and will wear about as long. These dies are sufficiently soft and resilient to permit forming aluminum alloy parts on them without marring, and nicks and scratches are easily removed from their surfaces.

When using joggling dies for the first time, test

them for accuracy on a piece of waste stock. In this way you will avoid the possibility of ruining already fabricated parts. Always keep the surfaces of the blocks free from dirt, filings, and the like, so that the work will not be marred.

Working Stainless Steel

When working with stainless steel, make sure that the metal does not become unduly scratched or marred. Also take special precautions when shearing, punching, or drilling this metal. It takes about twice as much pressure to shear or punch stainless steel as it does mild steel. Keep the shear or punch and die adjusted very closely. Too much clearance will permit the metal to be drawn over the edge of the die and cause it to become work-hardened, resulting in excessive strain on the machine.

When drilling stainless steel, use a high-speed drill ground to an included angle of 140°. Some special drills have an offset point, whereas others have a chip curler in the flutes. When using an ordinary twist drill, grind its point to a stubbier angle than the standard drill point. Keep the drill speed about one-half that required for drilling mild steel, but never exceed 750 r.p.m. Keep a uniform pressure on the drill so the feed is constant at all times. Drill the material on a backing plate, such as cast iron, which is hard enough to permit the drill to cut all the way through the stock without pushing the metal away from the drill point. Spot the drill before turning on the power and also make sure that when the power is turned on, pressure is being exerted.

To avoid overheating, dip the drill in water after drilling each hole. When it is necessary to drill several deep holes in stainless steel, use a liquid coolant. A compound made up of 1 lb. of sulfur added to 1 gal. of lard oil will serve the purpose. Apply the coolant to the material immediately upon starting the drill. High-speed portable hand drills have a tendency to burn the drill points and excessively work-harden the material at the point of contact; thus high-speed portable hand drills should not be used because of the temperatures developed. A drill press adjustable to speeds under 750 r.p.m. is recommended.

Working Magnesium

Magnesium, in the pure state, does not have sufficient strength to be used for structural purposes; but, as an alloy, it has a high strength-to-weight ratio. Its strength is not affected by subzero temperatures, and this increases its adaptability to aircraft use. The nonmagnetic property of magnesium alloys makes it valuable for instrument cases and parts.

While magnesium alloys can usually be fabricated by methods similar to those used on other metals, it must be remembered that many of the details of shop practice cannot be applied. Magnesium alloys are difficult to fabricate at room temperature; therefore, operations other than the most simple ones must be performed at high temperatures. This requires preheating of the metal, or dies, or both.

Magnesium alloy sheets may be cut by blade shears, blanking dies, routers, or saws. Hand or circular saws are usually used for cutting extrusions to length. Conventional shears and nibblers should never be used for cutting magnesium alloy sheet because they produce a rough, cracked edge.

Shearing and blanking of magnesium alloys require close tool tolerances. A maximum clearance of from 3 to 5% of the sheet thickness is recommended. The top blade of the shears should be ground with an included angle of from 45° to 60°. The shear angle on a punch should be from 2° to 3°, with a 1° clearance angle on the die. For blanking, the shear angle on the die should be from 2° to 3° with a 1° clearance angle on the punch. Holddown pressures should be used when possible. Cold shearing should not be accomplished on hard-rolled sheet thicker than 0.064 in. or annealed sheet thicker than 1/8 in. Shaving is used to smooth the rough, flaky edges of magnesium sheet which has been sheared. This operation consists of removing approximately 1/32 in. by a second shearing.

Hot shearing is sometimes used to obtain an improved sheared edge. This is necessary for heavy sheet and plate stock. Annealed sheet may be heated to 600° F., but hard-rolled sheet must be held under 400° F., depending on the alloy used. Thermal expansion makes it necessary to allow for shrinkage after cooling, which entails adding a small amount of material to the cold metal dimensions before fabrication.

Sawing is the only method used in cutting plate stock more than 1/2 in.-thick. Bandsaw raker-set blades of 4- to 6-tooth pitch are recommended for cutting plate stock or heavy extrusions. Small and medium extrusions are more easily cut on a circular cutoff saw having six teeth per inch. Sheet stock can be cut on bandsaws having raker-set or straight-set teeth with an 8-tooth pitch. Bandsaws should be equipped with nonsparking blade guides to eliminate the danger of sparks igniting the magnesium alloy filings.

Cold-working most magnesium alloys at room temperature is very limited because they work-harden very rapidly and do not lend themselves to any severe cold-forming. Some simple bending operations may be performed on sheet material, but the radius of bend must be at least seven times the thickness of the sheet for soft material and 12 times the thickness of the sheet for hard material. A radius of two or three times the thickness of the sheet can be used if the material is heated for the forming operation.

Wrought magnesium alloys tend to crack after they are cold-worked. Therefore, the best results are obtained if the metal is heated to 450° F. before any forming operations are attempted. Parts formed at the lower temperature range are stronger because the higher temperature range has an annealing effect on the metal.

There are some disadvantages to hot-working. First, heating the dies and the material is expensive and troublesome. Second, there are problems in lubricating and handling materials at these temperatures. However, there are some advantages to hot-working magnesium in that it is more easily formed when hot than are other metals and springback is reduced, resulting in greater dimensional accuracy.

When heating magnesium and its alloys, watch the temperature carefully as the metal is easily burned. Overheating also causes small molten pools to form within the metal. In either case, the metal is ruined. To prevent burning, magnesium must be protected with a sulfur dioxide atmosphere while being heated.

Proper bending around a short radius requires the removal of sharp corners and burrs near the bend line. Layouts should be made with a carpenter's soft pencil because any marring of the surface may result in fatigue cracks.

It is permissible to heat small pieces of magnesium with a blowtorch, provided proper precautions are exercised. It must be remembered that magnesium will ignite when it is heated to a temperature near its boiling point in the presence of oxygen.

Press or leaf brakes can be used for making bends with short radii. Die and rubber methods should be used where bends are to be made at right angles, which complicate the use of a brake. Roll forming may be accomplished cold on equipment designed for forming aluminum. The most common method of forming and shallow drawing magnesium

is an operation in which a rubber pad is used as the female die. This rubber pad is held in an inverted steel pan which is lowered by a hydraulic press ram. The press exerts pressure on the metal and bends it to the shape of the male die.

The machining characteristics of magnesium alloys are excellent, making possible the use of maximum speeds of the machine tools with heavy cuts and high feed rates. Power requirements for machining magnesium alloys are about one-sixth of those for mild steel.

Filings, shavings, and chips from machining operations should be kept in covered metal containers because of the danger of combustion. To repeat a previous reminder, in case of a magnesium fire, do not try to extinguish it with water. The oxygen in the water supports the combustion and increases the intensity of the fire. Dry powder (sodium bicarbonate) is the recommended extinguishing agent for magnesium fires.

RIVET LAYOUT

Rivet layout consists of determining (1) the number of rivets required; (2) the size and style of rivet to use; (3) its material, temper condition, and strength; (4) the size of the rivet holes; (5) distance of the rivet holes and rivets from the edges of the patch; and (6) the spacing of the rivets throughout the repair. Since distances are measured in terms of rivet diameters, application of the measurements is simple once the correct rivet diameter is determined.

Single-row, two-row, and three-row layouts designed for small repair jobs are discussed in this section. More complicated layouts for large repairs, which require the application of rivet formulas, are discussed later in this chapter.

The type of head, size, and strength required in a rivet are governed by such factors as the kind of forces present at the point riveted, the kind and thickness of the material to be riveted, and location of the riveted part on the aircraft.

The type of head required for a particular job is determined by its installation location. Where a smooth aerodynamic surface is required, countersunk head rivets should be used. Universal head rivets may be used in most other locations. If extra strength is required and clearance permits, roundhead rivets may be used; if the necessary clearance is not available, flathead rivets may be used.

The size (or diameter) of the selected rivet shank should correspond in general to the thickness of the material being riveted. If too large a rivet is used in a thin material, the force necessary to drive the rivet properly will cause an undesirable bulging around the rivet head. On the other hand, if too small a rivet diameter is selected for thick material the shear strength of the rivet will not be great enough to carry the load of the joint. As a general rule, the rivet diameter should be not less than three times the thickness of the thicker sheet. Rivets most commonly chosen in the assembly and repair of aircraft range from 3/32-in. to 3/8-in. diameter. Ordinarily, rivets smaller than 3/32-in. diameter are never used on any structural parts which carry stresses.

When rivets are to pass completely through tubular members, select a rivet diameter equivalent to at least one-eighth the outside diameter of the tube. If one tube "sleeves" or fits over another, take the outside diameter of the outside tube and use one-eighth of that distance as the minimum rivet diameter. A good practice is to calculate the minimum rivet diameter and then use the next larger size rivet.

When determining the total length of a rivet for installation, the combined thickness of the materials to be joined must be known. This measurement is known as grip length (B of figure 5–41). The total length of the rivet (A of figure 5–41) should be equal to grip length plus the amount of rivet shank necessary to form a proper shop head. The length of rivet required to form a shop head is 1-1/2 times the diameter of the rivet shank (C of figure 5–41).

A — Total rivet length

B — Grip length

C — Amount of rivet length needed for proper shop head (1½ × rivet dia.)

D — Installed rivets

FIGURE 5–41. Determining length of rivet.

Using figure 5–41 and the above information, the formula A = B + C was developed. (A, total rivet length; B, grip length; C, material needed to form a shop head.)

Properly installed rivets are shown in D of figure 5–41. Note carefully the method used to measure total rivet lengths for countersunk rivets and the other types of heads.

Whenever possible, select rivets of the same alloy number as the material being riveted. For example, use 1100 and 3003 rivets on parts fabricated from 1100 and 3003 alloys, and 2117–T and 2017–T rivets on parts fabricated from 2017 and 2024 alloys.

The 2117–T rivet is usually used for general repair work, since it requires no heat treatment, is fairly soft and strong, and is highly corrosion resistant when used with most types of alloys. The 2024–T rivet is the strongest of the aluminum alloy rivets and is used in highly stressed parts. However, it must be soft when driven. Never replace 2024–T rivets with 2117–T rivets.

The type of rivet head to select for a particular repair job can be determined by referring to the type used within the surrounding area by the manufacturer. A general rule to follow on a flush-riveted aircraft is to apply flush rivets on the upper surface of the wing and stabilizers, on the lower leading edge back to the spar, and on the fuselage back to the high point of the wing. Use universal head rivets in all other surface areas.

In general, try to make the spacing of the rivets on a repair conform to that used by the manufacturer in the area surrounding the damage. Aside from this fundamental rule, there is no specific set of rules which governs spacing of rivets in all cases. However, there are certain minimum requirements which must be observed.

The edge distance, or distance from the center of the first rivet to the edge of the sheet, should be not less than two rivet diameters nor more than four. The recommended edge distance is about two and one-half rivet diameters. If rivets are placed too close to the edge of the sheet, the sheet is likely to crack or pull away from the rivets; and if they are spaced too far from the edge, the sheet is apt to turn up at the edges.

Rivet pitch is the distance between the centers of adjacent rivets in the same row. The smallest allowable rivet pitch is three rivet diameters. The average rivet pitch usually ranges from six to eight rivet diameters, although rivet pitch may range from four to 10 rivet diameters. Transverse pitch is the perpendicular distance between rivet rows; it is usually equal to 75% of the rivet pitch. The smallest allowable transverse pitch is two and one-half rivet diameters.

When splicing a damaged tube and the rivets pass completely through the tube, space the rivets four to seven rivet diameters apart if adjacent rivets are at right angles to each other, and space them five to seven rivet diameters apart if the rivets are in line (parallel to each other). The first rivet on each side of the joint should be not less than two and one-half rivet diameters from the end of the sleeve.

The general rules of rivet spacing, as applied to straight-row layout, are quite simple. In a single-row layout, first determine the edge distance at each end of the row then lay off the rivet pitch (distance between rivets) as shown in figure 5–42. In the two-row layout, lay off the first row as just described, place the second row a distance equal to the transverse pitch from the first row, and then lay off rivet spots in the second row so that they fall midway between those in the first row. In the three-row layout, first lay off the first and third rows, then determine the second row rivet spots by using a straightedge. (See figure 5–42.)

RIVET INSTALLATION

The various tools needed in the normal course of driving and upsetting rivets include drills, reamers, rivet cutters or nippers, bucking bars, riveting hammers, draw sets, dimpling dies or other types of countersinking equipment, rivet guns, and squeeze riveters. Self-tapping screws, C-clamps, and fasteners are riveting accessories commonly used to hold sheets together when riveting.

Several of these tools were discussed earlier in this chapter. Other tools and equipment needed in the installation of rivets are discussed in the following paragraphs.

Hole Duplicators

When sections of skin are replaced with new sections, the holes in the replacement sheet or in the patch must be drilled to match existing holes in the structure. These holes can be located with a hole duplicator. The peg on the bottom leg of the duplicator fits into the existing rivet hole. The hole in the new part is made by drilling through the bushing on the top leg. If the duplicator is properly made, holes drilled in this manner will be in perfect alignment. A separate duplicator must be used for each diameter of rivet.

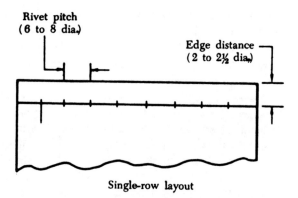

Single-row layout

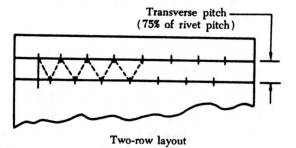

Two-row layout

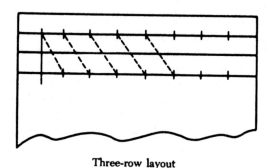

Three-row layout

FIGURE 5–42. Rivet spacing.

Rivet Cutters

In cases where rivets of the required length are unavailable, rivet cutters can be used to cut rivets to the desired length. When using the rotary rivet cutter, insert the rivet in the correct hole, place the required number of shims under the rivet head, and squeeze as though it were a pair of pliers. Rotation of the disks will cut the rivet to give the right length, which is determined by the number of shims inserted under the head. When using a large rivet cutter, place it in a vise, insert the rivet in the proper hole, and cut by pulling the handle, thus shearing off the rivet. If regular rivet cutters are not available, diagonal cutting pliers can be used as a substitute cutter.

166

Bucking Bars

A bucking bar is a tool which is held against the shank end of a rivet while the shop head is being formed. Most bucking bars are made of alloy bar stock, but those made of better grades of steel last longer and require less reconditioning. Bucking bars are made in a number of different shapes and sizes to facilitate rivet bucking in all places where rivets are used. Some of the various bucking bars are shown in figure 5–43.

FIGURE 5–43. Bucking bars.

The bars must be kept clean, smooth, and well polished. Their edges should be slightly rounded to prevent marring the material surrounding the riveting operation.

Hand Rivet and Draw Sets

A hand rivet set is a tool equipped with a die for driving a particular type rivet. Rivet sets are available to fit every size and shape of rivet head. The ordinary set is made of 1/2-in. carbon tool steel about 6 in. long and is knurled to prevent slipping in the hand. Only the face of the set is hardened and polished.

Sets for round and brazier head rivets are recessed (or cupped) to fit the rivet head. In selecting the correct set, be sure that it will provide the proper clearance between the set and the sides of the rivet head and between the surfaces of the metal and the set. Flush or flat sets are used for countersunk and flathead rivets. To seat flush rivets properly, be sure that the flush sets are at least 1 in. in diameter.

Special draw sets are used to "draw up" the sheets to eliminate any opening between them before the rivet is bucked. Each draw set has a hole 1/32 in. larger than the diameter of the rivet shank for which it is made. Occasionally, the draw set and rivet header are incorporated into one tool. The header part consists of a hole sufficiently shallow so that the set will expand the rivet and head it when struck with a hammer.

Countersinks

The countersink is a tool which cuts a cone-shaped depression around the rivet hole to allow the rivet to set flush with the surface of the skin. Countersinks are made with various angles to correspond to the various angles of the countersunk rivet heads.

Special stop countersinks are available. Stop countersinks are adjustable to any desired depth, and the cutters are interchangeable so that holes of various countersunk angles can be made. Some stop countersinks have a micrometer set arrangement, in increments of 0.001 in., for adjusting the cutting depths.

Dimpling Dies

The process of making an indentation or a dimple around a rivet hole so that the top of the head of a countersunk rivet will be flush with the surface of the metal is called dimpling. Dimpling is done with a male and female die, or forms, often called punch and die set. The male die has a guide the size of the rivet hole and is beveled to correspond to the degree of countersink of the rivet head. The female die has a hole into which the male guide fits, and is beveled to a corresponding degree of countersink.

When dimpling, rest the female die on a solid surface then place the material to be dimpled on the female die. Insert the male die in the hole to be dimpled and with a hammer strike the male die until the dimple is formed. Two or three solid hammer blows should be sufficient. A separate set of dies is necessary for each size of rivet and shape of rivet head.

An alternate method is to use a countersunk head rivet instead of the regular male punch die, and a draw set instead of the female die, and hammer the rivet until the dimple is formed.

Dimpling dies for light work can be used in portable pneumatic or hand squeezers. If the dies are used with a squeezer, they must, of course, be adjusted accurately to the thickness of the sheet being dimpled.

Pneumatic Rivet Guns

The most common upsetting tool used in airframe repair work is the slow-hitting pneumatic hammer called a rivet gun. Pneumatic guns are available in various sizes and shapes (figure 5–44). The capacity of each gun, as recommended by the manufacturer, is usually stamped on the barrel; pneumatic guns operate on air pressures of from 90 to 100 p.s.i.

Slow-hitting (long stroke) riveting hammers

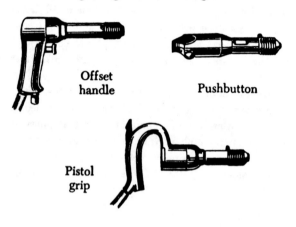

Offset handle

Pushbutton

Pistol grip

Fast-hitting (light) riveting hammers

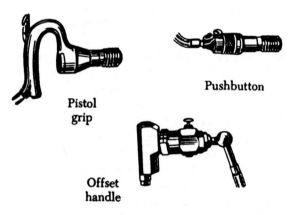

Pistol grip

Pushbutton

Offset handle

FIGURE 5–44. Types of rivet guns.

Pneumatic guns are used in conjunction with interchangeable rivet sets. Each set is designed to fit the type of rivet and location of the work. The shank of the set is designed to fit into the rivet gun. Force to buck the rivet is supplied by an air-driven hammer inside the barrel of the gun (figure 5–45). The sets are made of high-grade carbon tool steel and are heat treated to give them strength and wear resistance.

Some precautions to be observed when using a rivet gun are:

(1) Never point a rivet gun at anyone at any time. A rivet gun should be used for one purpose only—to drive or install rivets.

(2) Never depress the trigger mechanism unless the set is held tightly against a block of wood or a rivet.

(3) Always disconnect the air hose from the rivet gun when it will not be in use for any appreciable length of time.

Squeeze Riveters

The squeeze method of riveting is limited since it can be used only over the edges of sheets or assemblies where conditions permit, and where the reach of the squeeze riveter is deep enough. There are three types of rivet squeezers—hand, pneumatic, and pneudraulic. They are basically alike except that in the hand rivet squeezer, compression is supplied by hand pressure; in the pneumatic rivet squeezer, by air pressure; and in the pneudraulic, by a combination of air and hydraulic pressure. One jaw is stationary and serves as a bucking bar, the other jaw is movable and does the upsetting. Riveting with a squeezer is a quick method and requires only one operator.

Squeeze riveters are usually equipped with either a C-yoke or an alligator yoke. Yokes are available in various sizes to accommodate any size of rivet. The working capacity of a yoke is measured by its gap and its reach. The gap is the distance between the movable jaw and the stationary jaw; the reach is the inside length of the throat measured from the center of the end sets.

End sets for squeeze riveters serve the same purpose as rivet sets for pneumatic rivet guns and are available with the same type heads. They are interchangeable to suit any type of rivet head. One part of each set is inserted in the stationary jaw, while the other part is placed in the movable jaws. The manufactured head end set is placed on the stationary jaw whenever possible. However, during some operations, it may be necessary to reverse the end sets, placing the manufactured head end set on the movable jaw.

PREPARATION OF RIVET HOLES

It is very important that the rivet hole be of the correct size and shape and free from burrs. If the hole is too small, the protective coating will be scratched from the rivet when the rivet is driven

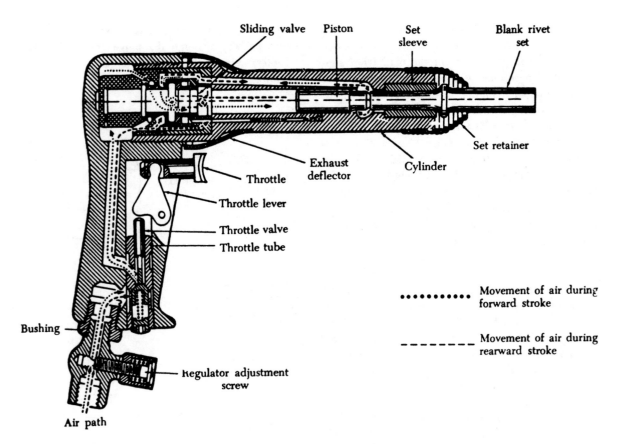

FIGURE 5–45. Rivet gun nomenclature.

through the hole. If the hole is too large, the rivet will not fill the hole completely. When it is bucked, the joint will not develop its full strength, and structural failure may occur at that spot.

If countersinking is required, consider the thickness of the metal and adopt the countersinking method recommended for that thickness. If dimpling is required, keep hammer blows or dimpling pressures to a minimum so that no undue work-hardening occurs in the surrounding area.

Drilling

To make a rivet hole of the correct size, first drill a hole slightly undersize. This is known as predrilling, and the hole is called a pilot hole. Ream the pilot hole with a twist drill of the correct size to get the required dimension. Pilot and reaming drill sizes are shown in figure 5–46. The recommended clearance for rivet holes is from 0.002 to 0.004 in.

When drilling hard metals the twist drill should have an included angle of 118° and should be operated at low speeds; but for soft metals, use a twist drill with an included angle of 90° and it should be operated at higher speeds. Thin sheets of aluminum

alloy are drilled with greater accuracy by a drill having an included angle of 118° because the large angle of the drill has less tendency to tear or elongate the hole.

Center punch locations for rivet holes before beginning the actual drilling. The center punch mark

Rivet Diameter	Pilot Size	Ream Size
3/32	3/32 (.0937)*	40 (.098)
1/8	1/8 (.125)	30 (.1285)
5/32	5/32 (.1562)	21 (.159)
3/16	3/16 (.1875)	11 (.191)
1/4	1/4 (.250)	F (.257)
5/16	5/16 (.3125)	O (.316)
3/8	3/8 (.375)	V (.377)

*Note that ream size exceeds the maximum tolerance of .004 inch. This is permissible only if the next larger drill size happens to be so much larger than the tolerance of .004 inch.

FIGURE 5–46. Pilot and reaming twist drill sizes.

169

acts as a guide and lets the drill grip or bite into the metal with greater ease. Make the center punch mark large enough to prevent the drill from slipping out of position, but punch lightly enough not to dent the surrounding material. Hold a hard, smooth, wooden backing block securely in position behind the hole locations when drilling.

Drilling is usually done with a hand drill or with a light power drill. Hold the power drill firmly with both hands. Extend the index and middle fingers of the left hand against the metal to act as a guide in starting a hole, and as a snubber or brake when the drill goes through the material. Before beginning to drill, always test the inserted twist drill for trueness and vibration by spinning the hand drill or running the motor freely and watching the drill end. If the drill wobbles, it may be because of burrs on its shank or because the drill is bent or incorrectly chucked. A drill that wobbles or is slightly bent must not be used because it causes enlarged holes.

Always hold the drill at right angles to the work, regardless of the position of the hole or the curvature of the material. Use an angle drill or drill extensions and adapters when access is difficult with a straight drill. Never tip the drill sideways when drilling or when withdrawing from the material because this causes elongation of the hole.

When holes are drilled through sheet metal, small burrs are formed around the edge of the hole. This is especially true when using a hand drill since the drill speed is slow and there is a tendency to apply more pressure per drill revolution. Remove all burrs with a burr remover before riveting.

Countersinking and Dimpling

An improperly made countersink reduces the strength of a flush-riveted joint and may even cause failure of the sheet or the rivet head. The two methods of countersinking commonly used for flush riveting in aircraft construction and repair are the machine or drill countersinking, and dimpling or press countersinking. The proper method for any particular application depends on the thickness of the parts to be riveted, the height and angle of the countersunk head, the tools available, and accessibility.

As a general rule, use the drill countersink method when the thickness of the material is greater than the thickness of the rivet head, and use the dimpling method on thinner material. Figure 5–47 illustrates general rules for countersinking. Note in figure 5–47A that the material is quite thick and

the head of the countersunk rivet extends only about halfway through the upper layer of metal. Countersinking will leave plenty of material for gripping.

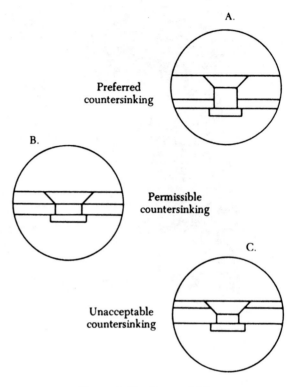

FIGURE 5–47. Countersinking.

In figure 5–47B, the countersunk head reaches completely through the upper layer. This condition is permissible but should be avoided.

In figure 5–47C, the head extends well into the second layer of material. This indicates that the material is thin and that most of it would be ground away by drill countersinking; therefore, dimpling is preferred. Dimpling will work best if the material is not over 0.040-in. thick.

Machine or drill countersinking is accomplished by a suitable cutting tool machined to the desired angle. The edge of the hole is cut away so that the countersunk rivet head fits snugly into the recess. The resulting recess is referred to as the "well" or "nest."

During the process of machine countersinking, first drill the original rivet hole to the exact rivet size, as recommended in the table in figure 5–46. The limits within which the head of the rivet may extend either above or below the surface of the metal are close, 0.006 in. in most cases. Therefore, perform the countersinking accurately, using equipment which is capable of producing results within the specified tolerance.

Hold the countersinking tool firmly at right angles to the material. Do not tip it. Tipping elongates the well and prevents the countersunk rivet head from fitting properly. Oversized rivet holes, undersized countersink pilots (in the case of the stop countersink), chattering caused by improper use of the countersink or by a countersink in poor condition, and a countersink not running true in the chuck of the drill are some of the causes of elongated wells.

Press countersinking or dimpling can be accomplished by either of two methods. Male and female die sets can be used, or using the rivet as the male die and the draw die as the female die is acceptable. In either case, the metal immediately surrounding the rivet hole is pressed to the proper shape to fit the rivet head. The depression thus formed, as in machine countersinking, is known as the "well" or "nest."

The rivet must fit the well snugly to obtain maximum strength. The number of sheets which can be dimpled simultaneously is limited by the capacity of the equipment used. The dimpling process may be accomplished by the use of hand tools, by dies placed in a pneumatic squeeze or single shot riveter, or by using a pneumatic riveting hammer.

Dimpling dies are made to correspond to any size and degree of countersunk rivet head available. The dies are usually numbered, and the correct combination of punch and die to use is indicated on charts specified by the manufacturer. Both male and female dies are machined accurately and have highly polished surfaces. The male die or punch is cone shaped to conform to the rivet head and has a small concentric pilot shaft that fits into the rivet hole and female die. The female die has a corresponding degree of countersink into which the male guide fits.

When dimpling a hole, rest the female die on some solid surface, place the material on the female die, insert the male die in the hole to be dimpled, and then hammer the male die. Strike with several solid blows until the dimple is formed.

In some cases, the face of the male die is convex to allow for springback in the metal. Dies of this type are used to advantage when the sheet to be dimpled is curved. Some dies have flat faces and are principally used for flat work. Dimpling dies are usually made so that their included angle is 5° less than that of the rivet. This arrangement allows for springback of the metal.

In die dimpling, the pilot hole of the female die should be smaller than the diameter of the rivet to be used. Therefore, the rivet hole must be reamed to the exact diameter after the dimpling operation has been completed so that the rivet fits snugly.

When using a countersink rivet as the male dimpling die, place the female die in the usual position and back it with a bucking bar. Place the rivet of the required type into the hole and strike the rivet with a pneumatic riveting hammer. This method of countersinking is often called "coin pressing." It should be used only when the regular male die is broken or not available.

Coin pressing has a distinct disadvantage in that the rivet hole must be drilled to correct rivet size before the dimpling operation is accomplished. Since the metal stretches during the dimpling operation, the hole becomes enlarged and the rivet must be swelled slightly before driving to produce a close fit. Because the rivet head will cause slight distortions in the recess, and these are characteristic only to that particular rivet head, it is wise to drive the same rivet that was used as the male die during the dimpling process. Do not substitute another rivet, either of the same size or a size larger.

Thermo-Dimpling

This type of dimpling consists of two processes, radius dimpling and coin dimpling. The major difference between radius and coin dimpling is in the construction of the female die. In radius dimpling a solid female die is used. Coin dimpling uses a sliding ram female die (figure 5–48) that makes this process superior.

During the coin dimpling process, the metal is coined (made to flow) into the contours of the dies so that the dimple assumes the true shape of the die. The pressure exerted by the coining ram prevents the metal from compressing and thereby assures uniform cross sectional thickness of the sides of the dimple and a true conical shape.

Coin dimpling offers several advantages. It improves the configuration of the dimple, produces a more satisfactory aerodynamic skin surface, eliminates radial and circumferential cracking, ensures a stronger and safer joint, and allows identical dies to be used for both skin and understructure dimpling.

The material being used is a very important factor to consider in any dimpling operation. Materials such as corrosion-resistant steel, magnesium,

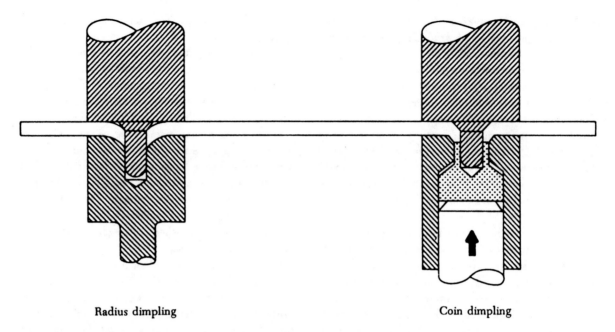

Radius dimpling   Coin dimpling

FIGURE 5-48. Radius and coin dimpling dies.

and titanium each present different dimpling problems.

The 2024–T aluminum alloy can be satisfactorily coin dimpled either hot or cold. However, cracking in the vicinity of the dimple may result from cold dimpling because of hard spots in the metal. Hot dimpling will prevent such cracking.

The 7075–T6 and 2024–T81 aluminum alloys are always hot dimpled. Magnesium alloys also must be hot dimpled because, like 7075–T6, they have low formability qualities. Titanium is another metal that must be hot dimpled because it is tough and resists forming. The same temperature and dwell time used to hot dimple 7075–T6 is used for titanium.

Corrosion-resistant steel is cold dimpled because the temperature range of the heating unit is not high enough to affect dimpling.

The coin ram dimpling dies are designed with a number of built-in features. The faces of both the male and female dies are dished (the male concave and the female convex) at an angle of 2° on the pilot. This facilitates removal of the metal after the dimple has been made.

The female dimpling set has two parts: (1) The body, which is merely a counterpart of the male die; and (2) the coining ram, which extends up through the center of the conical recess of the body. In forming a dimple, the metal is forced down into the female die by the male die. The metal first contacts the coining ram, and this supports the

metal as it is forced down into the conical recess. When the two dies close to the point where the forces of both are squeezing the material, the coining ram forces the metal back into the sharp corners of the dies.

When cold dimpling, the dies are used alone. When hot dimpling, a strap or block heater is slipped over either or both dies and connected to an electric current.

The dies should be kept clean at all times and in good working order. It is advisable to clean them regularly with steel wool. Special precautions must be taken when the dies are in the machine. If the machine is operated with the dies in place but without material between them, the male die will enlarge and ruin the coining ram.

When possible, coin dimpling should be performed on stationary equipment and before the assembly of parts. However, many instances arise in which dimpling must be done after parts are assembled to other structures. In such cases, dimpling operations are performed by portable squeeze dimplers. Most squeezers may be used either for cold dimpling or, combined with a junction box, for hot dimpling.

There are dimpling applications in which it is not possible to accommodate any squeezer- or yoke-type equipment. Under these circumstances, it is necessary to use a pneumatic hammer and a bucking bar type of tool to hold the dimpling dies.

DRIVING RIVETS

The methods of driving solid shank rivets can be classified into two types, depending on whether the riveting equipment is portable or stationary. Since stationary riveting equipment is seldom used in airframe repair work, only portable equipment that is used in hand, pneumatic, or squeezer methods is discussed here.

Before driving any rivets, be sure that all holes line up perfectly, all shavings and burrs have been removed, and that the parts to be riveted are securely fastened together.

Two men, a "gunner" and a "bucker," usually work as a team when installing rivets. However, on some jobs the riveter holds a bucking bar with one hand and operates a riveting gun with the other. When team riveting, an efficient signal system can be employed to develop the necessary teamwork. The code usually consists of tapping the bucking bar against the work; one tap may mean "not fully seated, hit it again"; two taps may mean "good rivet"; three taps may mean "bad rivet, remove and drive another"; and so on.

Bucking

Selection of the appropriate bucking bar is one of the most important factors in bucking rivets. If the bar does not have the correct shape, it will deform the rivet head; if the bar is too light, it will not give the necessary bucking weight, and the material may become bulged toward the shop head; and, if the bar is too heavy, its weight and the bucking force may cause the material to bulge away from the shop head. Weights of bucking bars range from a few ounces to 8 or 10 lbs., depending upon the nature of the work. Recommended weights of bucking bars to be used with various rivet sizes are given in figure 5–49.

Rivet Diameter (In Inches)	Approx. Weight (In Pounds)
3/32	2 to 3
1/8	3 to 4
5/32	3 to 4½
3/16	4 to 5
1/4	5 to 6½

FIGURE 5–49. Recommended bucking bar weights.

Always hold the face of the bucking bar at right angles to the rivet shank. Failure to do this will cause the rivet shank to bend with the first blows of the rivet gun, and will cause the material to become marred with the final blows. The bucker must hold the bucking bar in place until the rivet is completely driven. If the bucking bar is removed while the gun is in operation, the rivet set may be driven through the material. Do not bear down too heavily on the shank of the rivet. Allow the weight of the bucking bar to do most of the work. The hands merely guide the bar and supply the necessary tension and rebound action.

Allow the bucking bar to vibrate in unison with the gun set. This process is called coordinated bucking. Coordinated bucking can be developed through pressure and stiffness applied at the wrists; with experience, a high degree of deftness can be obtained.

Lack of proper vibrating action, the use of a bucking bar that is too light or too heavy, and failure to hold the bucking bar at right angles to the rivet can all cause defective rivet heads. A rivet going "clubhead" (malforming) can be corrected by rapidly moving the bucking bar across the rivet head in a direction opposite that of clubhead travel. This corrective action can be accomplished only while the gun is in action and the rivet is partly driven. If a rivet shank bends at the beginning of the bucking operation, place the bar in the corrective position only long enough to straighten the shank.

Hand Driving

Under certain conditions, it may be necessary to rivet by hand driving. Either of two methods can be used depending upon the location and accessibility of the work. In the one method, the manufactured head end of the rivet is driven with a hand set and hammer, the shank end is bucked with a suitable bucking bar. In the other method, the shank end of the rivet is driven with a hand set and a hammer, and the manufactured head is bucked with a hand set held in a vise or a bottle bar (a special bucking bar recessed to hold a rivet set). This method is known as reverse riveting. It is commonly used in hand riveting but is not considered good practice in pneumatic riveting.

When using either of the described methods, keep hammer strokes to a minimum. Too much hammering will change the crystalline structure of the rivet or the material around it, causing the joint to lose some of its strength. Hold the bucking bar and rivet set square with the rivet at all times. Misuse of the rivet set and bucking bar will result in marring or scratching the rivet head or material, and may cause undue corrosion. This, in turn, will weaken the structure of the aircraft.

The diameter of a correctly formed shop head should be one and one-half times the diameter of the rivet shank, and the height should be about one-half the diameter.

Pneumatic Driving

The procedure for pneumatic riveting is practically the same as for hand riveting. Preparation of the sheet, selection of rivets, and drilling of rivet holes are the same. In hand riveting, however, the pressure for bucking the rivet is applied using a hand set and hammer. In pneumatic riveting, the pressure is applied with a set and an air-driven hammer or gun.

To get good riveting results with a pneumatic rivet gun, follow these basic pointers:

(1) Select the right type and size of rivet gun and the correct rivet set for the size of rivet to be driven. Install the rivet set firmly, as shown in figure 5–50.

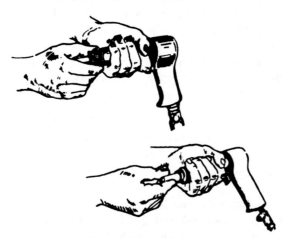

FIGURE 5–50. Installing rivet set.

(2) Adjust the speed of the riveting gun (vibrations per minute). Always press set firmly against a block of wood before pressing the trigger. Never operate the gun without resistance against the set because the vibrating action may cause the retaining spring to break, allowing the gun set to fly out of the gun. Also, free vibration may flare or mushroom the gun

end of the set, causing it to bind in the barrel of the gun.

(3) Hold the rivet set at right angles to the work to prevent damage to the rivet head or the surrounding material as shown in figure 5–51. Upset the rivet with a medium burst from the rivet gun.

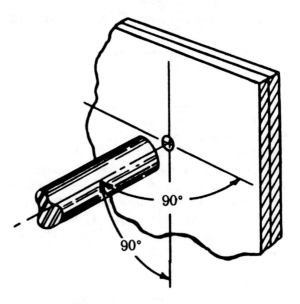

FIGURE 5–51. Position of the set.

(4) Remove the bucking bar and check the shop head of the rivet. It should be one and one-half times the diameter of the rivet in width and one-half times the rivet diameter in height. If the rivet needs further driving, repeat the necessary procedures to complete the job.

A small piece of adhesive tape applied to the cupped end of the rivet set often corrects an unsatisfactory cupped condition, which occasionally gives trouble in forming uniformly shaped rivet heads.

Squeeze Riveting

The squeeze method of driving a rivet produces the most uniform and balanced type of shop head. Each rivet is upset in a single operation; all rivets are headed over with uniform pressure; all heads are formed alike; and each rivet shank is sufficiently and uniformly expanded to completely fill each rivet hole. Squeeze riveters come equipped with pairs of end sets, each pair being designed for a particular job. Once the correct end set is selected

174

and the squeezer adjusted for a particular application, all the rivets will be driven uniformly, thus providing an efficient method of riveting.

Portable squeezers are particularly suited for riveting large assemblies where the tool must be moved in relation to the work. They are not too heavy and can easily be operated by one person. The preparation of the material for riveting with the squeeze riveter is the same as for hand or pneumatic riveting. For better results when using the squeeze riveter, observe these rules:

(1) Carefully select and insert suitable end sets to match the rivet being used. The importance of using the right end sets cannot be overemphasized. It is impossible to buck the rivet properly unless the correct pairs are used. Always be sure that the air is shut off or the squeezer is disconnected when inserting end sets.

(2) Adjust the squeezer cylinder pressure to obtain the correct pressure for the diameter of the rivet being used. Most squeezers are equipped with a blowoff valve to regulate the cylinder pressure. This unit governs the amount of air pressure allowed in the cylinder.

(3) Carefully regulate the gap to conform to the length of the rivet being used. Some squeezers are equipped with a gap regulator which controls the stroke of the plunger of a C-yoke squeezer, or the movement of the movable jaw of an alligator-yoke squeezer. For squeezers not equipped with a gap regulator, the gap can be adjusted by placing metal shims under the end sets of both jaws, or by using end sets of different lengths. On some types of squeeze riveters, the end set on the stationary jaw is held in place by an Allen screw, which allows regulation of the gap.

(4) Before using the squeezer on the work, test the cylinder pressure and gap for accuracy of adjustment on a piece of scrap material. The scrap material must be the same thickness as the material being used, and the rivets the same length and diameter.

(5) If the parts to be riveted are small and easily handled, mount the squeeze riveter in a bench vise or in a special clamp, and hold the part to be riveted in your hand.

Microshaving

Sometimes it is necessary to use a microshaver when making a repair involving the use of countersunk rivets. If the smoothness of the material (such as skin) requires that all countersunk rivets be driven within a specific tolerance, a microshaver is used. This tool has a cutter, stop, and two legs or stabilizers, as shown in figure 5–52.

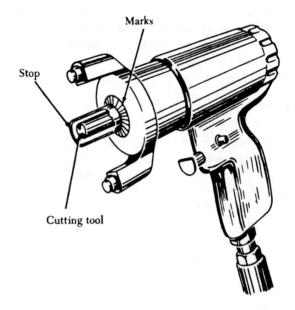

FIGURE 5–52. Microshaver.

The cutting portion of the microshaver is located inside the stop. The depth of cut can be adjusted by pulling outward on the stop and turning it in either direction (clockwise for deeper cuts). The marks on the stop permit adjustments of 0.001 in.

If the microshaver is adjusted and held correctly, it will cut the head of a countersunk rivet to within 0.002 in. without damaging the surrounding material. Adjustments should always be made on scrap material. When correctly adjusted, the shaver will leave a small round dot about the size of a pinhead on the microshaved rivet.

RIVET FAILURES

Generally speaking, the design of riveted joints is based on the theory that the total joint strength is simply the sum of the individual strengths of a whole group of rivets. It is then obvious that, if any one rivet fails, its load must immediately be carried by others of the group; if they are unable to carry this added load, progressive joint failure then occurs. Stress concentrations will usually cause one rivet to fail first; and careful analysis of such a rivet in a joint will indicate that it has been too highly loaded, with the possibility that neighboring rivets may have partially failed.

Shear Failure

Shear failure is perhaps the most common of rivet failures. It is simply a breakdown of the rivet shank by forces acting along the plane of two adjacent sheets, causing a slipping action which may be severe enough to cut the rivet shank in two. If the shank becomes loaded beyond the yield point of the material and remains overloaded, a permanent shift is established in the sheets and the rivet shank may become joggled.

Bearing Failure

If the rivet is excessively strong in shear, bearing failure occurs in the sheet at the edge of the rivet hole. The application of large rivets in thin sheets brings about such a failure. In that case, the sheet is locally crushed or buckled, and the buckling destroys the rigidity of the joint. Vibrations, set up by engine operation or by air currents in flight, may cause the buckled portion to flutter and the material to break off close to the rivet head. If buckling occurs at the end of the sheet, a tear-out may result. In either case, replacement of the sheet is necessary.

Head Failure

Head failure may result from complex loadings occuring at a joint, causing stresses of tension to be applied to the rivet head. The head may fail by shearing through the area corresponding to the rivet shank, or, in thicker sheets, it may fail through a prying action which causes failure of the head itself. Any visible head distortion is cause for replacement. This latter type of head failure is especially common in blind rivets.

Rivet Inspection

To obtain high structural efficiency in the manufacture and repair of aircraft, an inspection must be made of all rivets before the part is put in service. This inspection consists of examining both the shop and manufactured heads and the surrounding skin and structural parts for deformities. A scale or rivet gage can be used to check the condition of the upset rivet head to see that it conforms to the proper requirements. Deformities in the manufactured head can be detected by the trained eye alone. However, on flush rivets, a straightedge can be used as shown in figure 5–53.

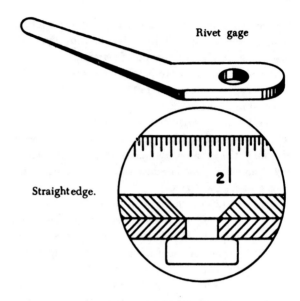

FIGURE 5–53. Tools used to gage rivets.

Some common causes of unsatisfactory riveting are improper bucking, rivet set slipping off or being held at the wrong angle, and rivet holes or rivets of the wrong size. Additional causes for unsatisfactory riveting are countersunk rivets not flush with the well; work not properly fastened together during riveting; the presence of burrs, rivets too hard, too much or too little driving; and rivets out of line.

Occasionally, during an aircraft structural repair, it is wise to examine adjacent parts to determine the true condition of neighboring rivets. In doing so, it may be necessary to remove the paint. The presence of chipped or cracked paint around the heads may indicate shifted or loose rivets. Look for tipped or loose rivet heads. If the heads are tipped or if rivets are loose, they will show up in groups of several consecutive rivets and will probably be tipped in the same direction. If heads which appear to be tipped are not in groups and are not tipped in the same direction, tipping may have occurred during some previous installation.

Inspect rivets known to have been critically loaded, but which show no visible distortion, by drilling off the head and carefully punching out the shank. If, upon examination, the shank appears joggled and the holes in the sheet misaligned, the rivet has failed in shear. In that case, try to determine what is causing the shearing stress and take the necessary corrective action. Flush rivets that show head slippage within the countersink or dimple, indicating either sheet bearing failure or rivet shear failure, must be removed for inspection and replacement.

Joggles in removed rivet shanks indicate partial shear failure. Replace these rivets with the next larger size. Also, if the rivet holes show elongation, replace the rivets with the next larger size. Sheet failures (such as tear-outs, cracks between rivets, and the like) usually indicate damaged rivets, and the complete repair of the joint may require replacement of the rivets with the next larger size.

The general practice of replacing a rivet with the next larger size (1/32 in. greater diameter) is necessary to obtain the proper joint strength of rivet and sheet when the original rivet hole is enlarged. If the rivet in an elongated hole is replaced by a rivet of the same size, its ability to carry its share of the shear load is impaired and joint weakness results.

REMOVING RIVETS

When removing a rivet for replacement, be very careful so that the rivet hole will retain its original size and shape and replacement with a larger size rivet will not be necessary. If the rivet is not removed properly, the strength of the joint may be weakened and the replacement of rivets made more difficult.

When removing a rivet, work on the manufactured head. It is more symmetrical about the shank than the shop head, and there will be less chance of damaging the rivet hole or the material around it. To remove rivets, use hand tools, a power drill, or a combination of both. The preferred method is to drill through the rivet head and drive out the remainder of the rivet with a drift punch. First, file a flat area on the head of any round or brazier head rivet, and center punch the flat surface for drilling. On thin metal, back up the rivet on the upset head when center punching to avoid depressing the metal. The dimple in 2117–T rivets usually eliminates the necessity of filing and center punching the rivet head.

Select a drill one size smaller than the rivet shank and drill out the rivet head. When using a power drill, set the drill on the rivet and rotate the chuck several revolutions by hand before turning on the power. This procedure helps the drill cut a good starting spot and eliminates the chance of the drill slipping off and tracking across the metal. Drill the rivet to the depth of its head, while holding the drill at a 90° angle. Be careful not to drill too deep because the rivet shank will turn with the drill and

cause a tear. The rivet head will often break away and climb the drill, which is a good signal to withdraw the drill. If the rivet head does not come loose of its own accord, insert a drift punch into the hole and twist slightly to either side until the head comes off.

Drive out the shank of the rivet with a drift punch slightly smaller than the diameter of the shank. On thin metal or unsupported structures, support the sheet with a bucking bar while driving out the shank. If the shank is exceptionally tight after the rivet head is removed, drill the rivet about two-thirds of the way through the thickness of the material and then drive out the remainder of the rivet with a drift punch.

The procedure for the removal of flush rivets is the same as that just described except that no filing is necessary. Be very careful to avoid elongation of the dimpled or the countersunk holes. The rivet head should be drilled to approximately one-half the thickness of the top sheet.

SPECIAL RIVETS

There are many places on an aircraft where access to both sides of a riveted structure or structural part is impossible, or where limited space will not permit the use of a bucking bar. Too, in the attachment of many nonstructural parts, such as aircraft interior furnishings, flooring, deicing boots, and the like, the full strength of solid shank rivets is not necessary.

For use in such places, special rivets have been designed which can be bucked from the front. They are sometimes lighter than solid shank rivets, yet amply strong for their intended use. These rivets are manufactured by several corporations and have unique characteristics that require special installation tools, special installation procedures, and special removal procedures. Because these rivets are often inserted in locations where one head (usually the shop head) cannot be seen, they are also called blind rivets.

The various types of mechanically expanded rivets, their fabrication, composition, uses, selection, and identification were discussed in Chapter 6, Hardware, Materials, and Processes, in the Airframe and Powerplant Mechanics General Handbook, AC 65–9A. The installation techniques will be covered in this section.

Installation Tools

The tools used to install self-plugging (friction lock) rivets depend upon the manufacturer of the rivet being installed. Each company has designed special tools which should always be used to ensure satisfactory results with its product. Hand tools as well as pneumatic tools are available.

After selection or determination of the rivet to be used in any installation, the proper size twist drill must be determined. Generally, manufacturers recommend the following finish drill sizes for the common shank diameters (figure 5–54).

Be very careful when drilling the material. Hold the drill at right angles to the work at all times to keep from drilling an elongated hole. The blind rivet will not expand as much as a solid shank rivet. If the hole is too large or elongated, the shank will not properly fill the drilled hole. Common hand or pneumatic powered drills can be used to drill the holes. Some manufacturers recommend predrilling the holes; others do not.

Equipment used to pull the stem of the rivet, as previously stated, will depend upon the manufacturer of the rivet. Both manually operated and power-operated guns are manufactured for this purpose. Nomenclature for various tools and assemblies available depends upon the manufacturer. Application and use of the equipment is basically the same. Whether the equipment is called hand tool, air tool, hand gun, or pneumatic gun (figure 5–55), all of these are used with but one goal, the proper installation of a rivet.

The choice of installation tools is influenced by several factors: The quantity of rivets to be installed, the availability of an air supply, the accessibility of the work, and the size and type of rivet to be installed. In addition to a hand or power riveter, it is necessary to select the correct "pulling head" to complete the installation tool.

Selection of the proper pulling head is of primary importance since it compensates for the variables of head style and diameter. Since your selection will depend on the rivets to be installed, you should consult the applicable manufacturer's literature.

SELF-PLUGGING (FRICTION LOCK) RIVETS

Self-plugging (friction lock) rivets are fabricated in two common head styles: (1) A protruding head similar to the AN470 or universal head, and (2) a 100° countersunk head. Other head styles are available from some manufacturers.

The stem of the self-plugging (friction lock) rivet may have a knot or knob on the upper portion, or it may have a serrated portion as shown in figure 5–56.

The sequence of steps to follow in the installation of self-plugging (friction lock) rivets is basically the same as that for solid shank rivets, but the methods and equipment vary. The following steps are typical of any installation:

(1) Select the rivet to be installed—determined by thickness of material to be riveted, strength desired in assembly, and location of installation (protruding or countersunk head).

(2) Drill the hole(s)—determine size of twist drill to be used, do not elongate rivet hole, remove burrs, and use a stop countersink if necessary.

CHERRYLOCK			
Rivet Diam.	Drill Size	Minimum	Maximum
3/32	#40	.097	.100
1/8	#30	.129	.132
5/32	#20	.160	.164
3/16	#10	.192	.196
1/4	F	.256	.261
BULBED CHERRYLOCK			
1/8	#27	.143	.146
5/32	#16	.176	.180
3/16	#5	.205	.209

COUNTERSINKING DIMENSIONS					
100° MS20426 HEAD		100° NAS1097 HEAD		90° UNISUNK HEAD	
Rivet Diam.	C Max. / C Min.		C Max. / C Min.		C Max. / C Min.
Rivet Diam.	C Max.	C Min.	C Max.	C Min.	C Max.	C Min.
3/32	.182	.176	—	—	—	—
1/8	.228	.222	.195	.189	.173	.167
5/32	.289	.283	.246	.240	.216	.210
3/16	.356	.350	.302	.296	.258	.252
1/4	.479	.473	.395	.389	—	—

FIGURE 5–54. Cherry rivet installation data.

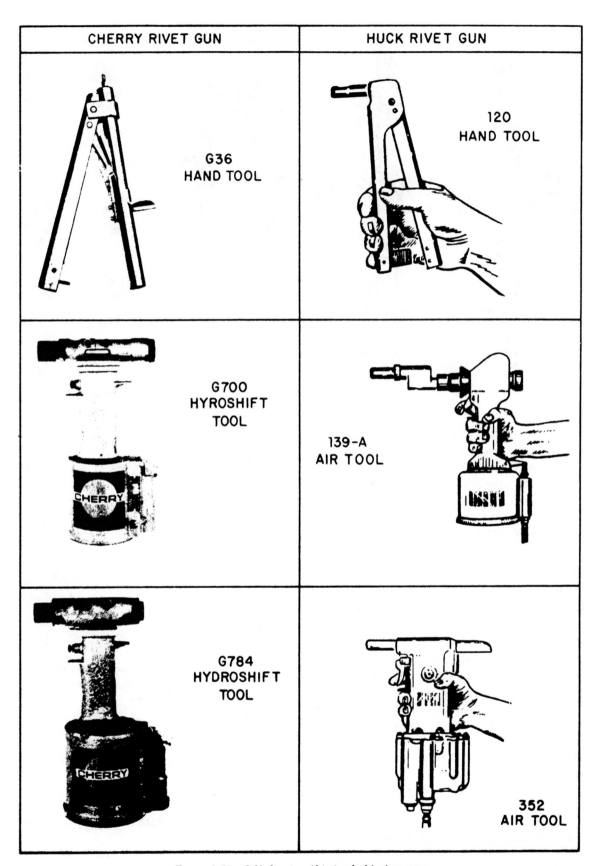

CHERRY RIVET GUN	HUCK RIVET GUN
G36 HAND TOOL	120 HAND TOOL
G700 HYROSHIFT TOOL	139-A AIR TOOL
G784 HYDROSHIFT TOOL	352 AIR TOOL

FIGURE 5–55. Self-plugging (friction lock) rivet guns.

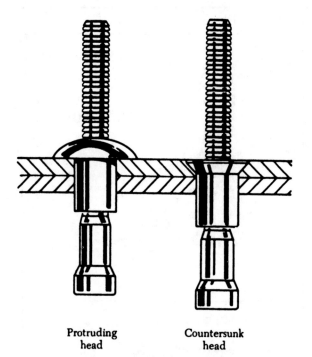

Protruding   Countersunk
head   head

FIGURE 5–56. Self-plugging (friction lock) rivets.

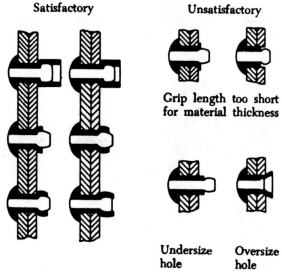

Satisfactory   Unsatisfactory

Grip length too short
for material thickness

Undersize   Oversize
hole   hole

FIGURE 5–57. Inspection of self-plugging (friction lock) rivets.

(3) Install the rivet—make certain the rivet head is seated firmly, position the selected tool on the rivet stem, pull rivet stem until the stem snaps, apply approximately 15 lbs. of pressure to the end of the stem, and trim the stem flush with the rivet head. If aerodynamic smoothness is a factor, the stem can be shaved with a rivet shaver.

Inspection

The inspection of installed self-plugging (friction lock) rivets is very limited. Often the only inspection that can be made is on the head of the rivet. It should fit tightly against the metal. The stem of the rivet should be trimmed flush with the head of the rivet whether it is a protruding head or a countersunk head.

If you can see the shop head side of the installed rivet, inspect it for the requirements illustrated in figure 5–57. When the rivet head is considered unsatisfactory, remove the rivet and install another in its place.

Removal Procedures

Self-plugging (friction lock) rivets are removed in the same manner as solid shank rivets except for the preliminary step of driving out the stem (figure

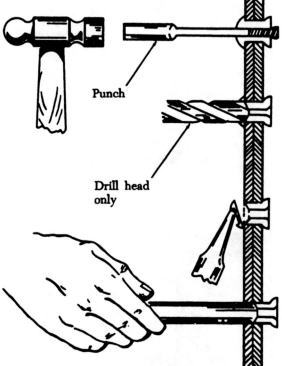

Punch

Drill head only

FIGURE 5–58. Removal of self-plugging (friction lock) rivets.

5–58). The following steps should be used in their proper sequence:

(1) Punch out the rivet stem with a pin punch.

(2) Drill out the rivet head, using a drill the same size as the rivet shank.

(3) Pry off the weakened rivet head with a pin punch.

(4) Push out the remainder of the rivet shank with a punch. If the shank will not push out, drill the shank, taking care not to enlarge the hole in the material.

SELF-PLUGGING (MECHANICAL LOCK) RIVETS

Self-plugging, mechanical lock rivets are similar to self-plugging, friction lock rivets, except for the manner in which they are retained in the material. This type of rivet has a positive mechanical locking collar that resists vibration that would cause the friction lock rivets to loosen and possibly fall out (figure 5–59). Also the mechanical locking type rivet stem breaks off flush with the head and usually does not require further stem trimming when properly installed. Self-plugging, mechanical lock rivets display all the strength characteristics of solid shank rivets and in almost all cases can be substituted rivet for rivet.

Huck Rivets

Self-plugging, mechanical lock rivets require special driving assemblies. It is best to use tools manufactured by the company that produces the rivet.

The Huck CKL rivet is installed by using the model CP350 blind rivet tool. The nose of the tool includes: (1) A set of chuck jaws which fit the serrated grooves on the rivet stem and pull it through the rivet shank to drive the rivet; (2) an outer anvil which bears against the outer portion of the manufactured head during the driving operation; and (3) an inner anvil which advances automatically to drive the locking collar into position after the blind head is formed (figure 5–60).

A change in rivet diameter requires a change in chuck jaws, outer anvil, and inner thrust bearing, and an adjustment of the shift operating pressure. Adjustment procedures are specified by the manufacturer.

Cherrylock Rivets

Cherrylock rivets are installed with a hydroshift or mechanical shift tooling system. The hydroshift system is a newer design and when available should be used in place of the mechanical system.

Cherrylock Mechanical Tooling

Most existing Cherry riveters, either hand or

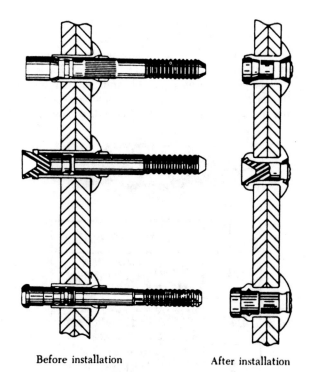

Before installation   After installation

FIGURE 5–59. Self-plugging, mechanical lock rivets.

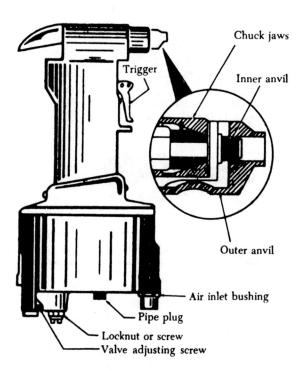

FIGURE 5–60. Huck model CP350 rivet pull tool.

power operated, may be used to install Cherrylock rivets when equipped with the proper mechanical pulling head.

181

Cherrylock mechanical pulling heads are of two types: the H615 (figure 5–61) and H640 (figure 5–62) series. They differ only in their method of attachment to the riveter. The H615 series is for the smaller screw-on type tools and the H640 is for the larger clip-on type. Both pulling heads will install bulbed and wiredraw Cherrylock rivets.

RIVET DIAMETER	PULLING HEAD NUMBER
1/8"	H615-4C { Universal Head / Countersunk Head H615-4S Uni-Sink Head
5/32"	H615-5C { Universal Head / Countersunk Head H615-5S Uni-Sink Head
3/16"	H615-6C { Universal Head / Countersunk Head H615-6S Uni-Sink Head

FIGURE 5–61. H615 Series pulling head tool.

A separate pulling head is required to install each diameter Cherrylock rivet. Separate pulling heads are recommended for universal and countersunk head rivets but countersunk pulling heads may be used for both styles.

RIVET DIAMETER	PULLING HEAD NUMBER
1/8"	H640-4C { Universal Head / Countersunk Head H640-4S Uni-Sink Head
5/32"	H640-5C { Universal Head / Countersunk Head H640-5S Uni-Sink Head
3/16"	H640-6C { Universal Head / Countersunk Head H640-6S Uni-Sink Head
1/4"	H640-8C { Universal Head / Countersunk Head

FIGURE 5–62. H640 Series pulling head tool.

RIVET DIAMETER	PULLING HEAD NUMBER		DIMENSIONS	
			A	B
3/32"	H681-3C	Universal Head	.188	.346
		Countersunk Head	.163	.331
1/8"	H681-4C	Universal Head	.250	.359
		Countersunk Head	.208	.341
	H681-B166-4	Uni-Sink Head	.250	.359
5/32"	H681-5C	Universal Head	.313	.377
		Countersunk Head	.269	.352
	H681-B166-5	Uni-Sink Head	.313	.377
3/16"	H681-6C	Universal Head	.375	.419
		Countersunk Head	.335	.385
	H681-B166-6	Uni-Sink Head	.375	.419
1/4"	H681-8C	Universal Head	.500	.452
		Countersunk Head	.458	.398

FIGURE 5–63. H681 Series pulling head tool.

Cherrylock Hydroshift Tooling

The hydroshift tooling system is an advanced design in which the sequence of operations necessary to install the rivet is accomplished hydraulically within the hydroshift tool rather than by means of a mechanical pulling head.

Cherrylock hydroshift pulling heads are of one type only, the H681 (figure 5–63).

A separate H681 pulling head is required to install each diameter Cherrylock rivet. Separate pulling heads are recommended for universal and countersunk head rivets but countersunk pulling heads may be used for both styles.

Hydroshift riveters are factory adjusted to break the rivet stem flush and set the collar properly. Fine adjustments to the shift point setting can be made by the operator. This adjustment determines the flushness of the break of the rivet stem (figures 5–64 and 5–65).

Installation Procedures

Procedures for installing self-plugging (mechanical lock) rivets are basically the same as those used for installing the friction lock type of rivets. Precautions to be observed are:

(1) Be sure the correct grip range is selected.

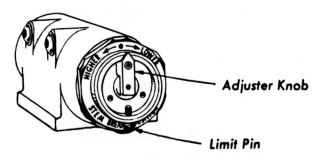

FIGURE 5–64. H681 pulling head adjuster.

182

FIGURE 5–65. Hydroshift pulling tool.

(2) Always use the correct nose assembly or pulling tool for the diameter rivet selected. (For the CKL rivet, check the tool air pressure for the correct setting.)

(3) When inserting the rivet in the tool and the material, hold a slight pressure against the head of the rivet.

(4) Determine that the rivet is completely driven before lifting the tool from the rivet head. (The stem should snap.)

(5) Check each rivet after the driving sequence has been completed for proper stem breakage. (The rivet stem should snap off even with the head of the rivet.)

Inspection

Visual inspection of the seating of the pin in the manufactured head is the most reliable and simplest means of inspection for mechanical lock rivets. If the proper grip range has been used and the locking collar and broken end of the stem are approximately flush with the manufactured head, the rivet has been properly upset and the lock formed. Insufficient grip length is indicated by the stem breaking below the surface of the manufactured head. Excessive grip length is indicated by the stem breaking

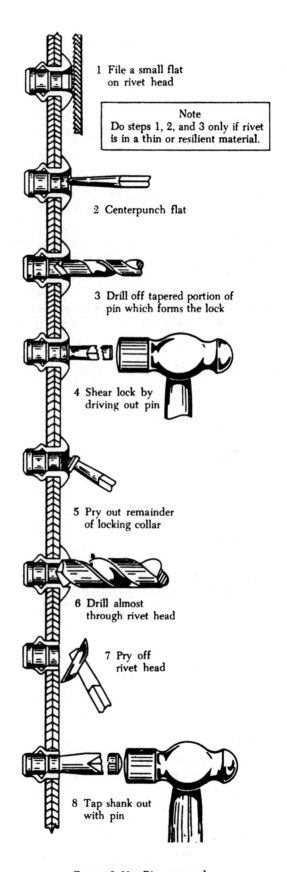

1 File a small flat on rivet head

Note
Do steps 1, 2, and 3 only if rivet is in a thin or resilient material.

2 Centerpunch flat

3 Drill off tapered portion of pin which forms the lock

4 Shear lock by driving out pin

5 Pry out remainder of locking collar

6 Drill almost through rivet head

7 Pry off rivet head

8 Tap shank out with pin

FIGURE 5–66. Rivet removal.

off well above the manufactured head. In either case, the locking collar might not be seated properly, thus forming an unsatisfactory lock.

Removal Procedures

The mechanical lock rivet can easily be removed by following the procedures illustrated in figure 5–66.

PULL-THRU RIVETS

This type of blind mechanically expanded rivet is used as a tacking rivet to attach assemblies to hollow tubes, and as a grommet. It differs from the two previously discussed rivets in that the stem pulls completely through the sleeve of the rivet during installation. Pull-thru rivets are structurally weak because of the hollow center after installation is completed. Methods and procedures for installation, inspection, and removal are not discussed here because of the limited use for this type rivet in the airframe field. Figure 5–67 illustrates a typical pull-thru rivet before and after installation.

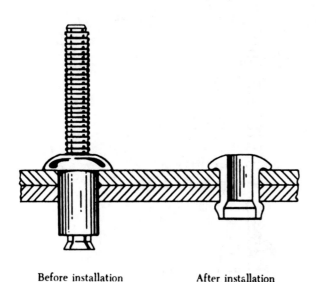

Before installation   After installation

FIGURE 5–67. Pull-thru rivet.

RIVNUTS

Rivnut is the trade name of a hollow blind rivet made of 6053 aluminum alloy, counterbored and threaded on the inside. Rivnuts are installed by one person using a special tool which heads the rivet on the blind side of the material (figure 5–68). The Rivnut is threaded on the mandrel of the heading tool and inserted in the rivet hole. The heading tool is held at right angles to the material; the handle is

squeezed, and the mandrel crank is turned clockwise after each stroke. Continue squeezing the handle and turning the mandrel crank of the heading tool until you feel a solid resistance, indicating that the rivet is set.

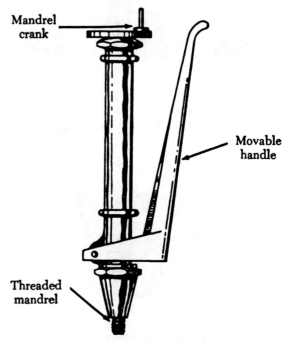

FIGURE 5–68. Rivnut heading tool.

All Rivnuts, except the thin head (0.048 in.) countersunk type, are available with or without small projections (keys) attached to the head to keep the Rivnut from turning. Keyed Rivnuts are used for service as a nut plate, whereas those without keys are used for straight blind riveting repairs where no torque loads are imposed. A keyway cutter is needed when installing Rivnuts which have keys (figure 5–69).

Tools used in the installation of Rivnuts include the hand-operated heading tools, the pneumatic power Rivnut driver, and the keyway cutter. All heading tools have a threaded mandrel onto which the Rivnut is threaded until the head of the Rivnut is against the anvil of the heading tool.

Hand-operated heading tools are made in three types: (1) Straight, (2) 45°, and (3) 90°. The pneumatic power driving tools are made in two types: (1) Lever throttle and (2) offset handle. With the power tool, the threading, upsetting, and withdrawal or unthreading, are accomplished by compressed air through the manipulation of finger-

Radial dash marks   **Keyed Rivnut**

FIGURE 5–69. Keyed Rivnut and keyway cutter.

tip controls. The keyway cutter is for cutting keyways only. In some instances, the keyway cutter cannot be used because the material may be too thick. If such is the case, use a small round file to form the keyway.

The important factors to be considered in selecting Rivnuts are grip range, style of head, condition of Rivnut end, and the presence or absence of a key.

Proper grip length is the most important of these conditions. The grip range of a Rivnut can be determined from its number. For example, a 6–45 has a maximum grip of 0.45 in. Note the procedure to follow when determining the grip range. The total thickness of the sheets shown in figure 5–70 is 0.052 in. By referring to the Rivnut data chart in figure 5–70, we see that 6–75 is the grip length to choose since the maximum grip length of the preceding size (6–45) is only 0.045 in. and would be too short. The grip length of the 6–75 Rivnut actually ranges from 0.045 in., the maximum length of the preceding size (6–45), to 0.075 in., which is the maximum grip length of the 6–75 Rivnut.

The objective when installing this type of rivet is to produce an ideal bulge on the blind side of the work without distorting any of the threads inside the Rivnut. In other words, be sure the bulge takes place between the first thread of the rivet and the lower edge of the riveted material. The space between the ideal bulge and the upper thread, where the gripping takes place, is considered the grip range.

When selecting head style, apply the same rules as for solid shank rivet application. Select key-type Rivnuts whenever screws are to be inserted, and use closed-end Rivnuts only in special places, such as sealed compartments of floats or pressurized compartments.

Drilling the holes for Rivnuts requires the same precision as for solid shank rivets. The shank of the Rivnut must fit snugly in the hole. To obtain the best results for a flathead installation, first drill a pilot hole smaller than the shank diameter of the Rivnut and then ream it to the correct size.

If keyed Rivnuts are used, cut the keyway after the hole has been reamed. In cutting the keyway, hold the keyway cutter so that it makes a 90° angle with the work. Also, cut the keyway on the side of the hole away from the edge of the sheet, especially when the Rivnut is used on the outside row. Operate the keyway cutter by inserting it in the hole and squeezing the handles.

The use of flush Rivnuts is limited. For metal which has a thickness greater than the minimum grip length of the first rivet of a series, use the machine countersink; for metal thinner than the minimum grip length of the first rivet, use the dimpling process. Do not use the countersunk Rivnut unless the metal is thick enough for machine countersinking, or unless the underside is accessible for the dimpling operation.

For a countersunk Rivnut the sheets to be joined can usually be machine countersunk. This method is preferred because the bearing surface in a dimpled hole in one sheet of average gage will normally occupy the entire gripping surface of the Rivnut, thus limiting its grip range to that of an anchored nut only.

When installing Rivnuts, among the things to check is the threaded mandrel of the heading tool to see that it is free from burrs and chips from the previous installation. Then screw the Rivnut on the mandrel until the head touches the anvil. Insert the Rivnut in the hole (with the key positioned in the keyway, if a key is used) and hold the heading tool at right angles to the work. Press the head of the

185

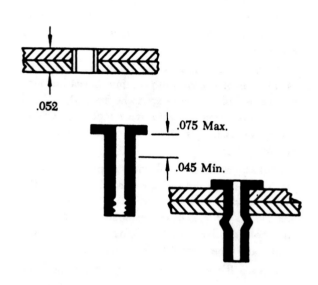

.052

.075 Max.

.045 Min.

Flat—0.32 Head Thickness		
6-45	6-75	6-100
8-45	8-75	8-100
10-45	10-75	10-100
6B45	6B75	6B100
8B45	8B75	8B100
10B45	10B75	10B100
6K45	6K75	6K100
8K45	8K75	8K100
10K45	10K75	10K100
6KB45	6KB75	6KB100
8KB45	8KB75	8KB100
10KB45	10KB75	10KB100
100°—0.48 Head Thickness		
6-91	6-121	6-146
8-91	8-121	8-146
10-91	10-121	10-146
6B91	6B121	6B146
8B91	8B121	8B146
10B91	10B121	10B146
100°—0.63 Head Thickness		
6-106	6-136	6-161
8-106	8-136	8-161
10-106	10-136	10-161
6B106	6B136	6B161
8B106	8B136	8B161
10B106	10B136	10B161
6K106	6K136	6K161
8K106	8K136	8K161
10K106	10K136	10K161
6KB106	6KB136	6KB161
8KB106	8KB136	8KB161
10KB106	10KB136	10KB161

FIGURE 5-70. Determining Rivnut grip length.

Rivnut tightly against the sheet while slowly squeezing the handles of the heading tool together until the Rivnut starts to head over. Then release the handle, and screw the stud further into the rivnut. This prevents stripping the threads of the Rivnut before it is properly headed. Again squeeze the handles together until the Rivnut heading is complete. Now remove the stud of the heading tool from the Rivnut by turning the crank counterclockwise.

The action of the heading tool draws the Rivnut against the anvil, causing a bulge to form in the counterbored portion of the Rivnut on the inaccessible side of the work. This bulge is comparable to the shop head on an ordinary solid shank rivet. The amount of squeeze required to head the Rivnut properly is best determined by practice. Avoid stripping the thread in the Rivnut.

The installation of a Rivnut is incomplete unless it is plugged either with one of the plugs designed for that purpose or with a screw used for attaching purposes. A Rivnut does not develop its full strength when left hollow.

Three types of screw plugs can be used: (1) The 100° countersunk screw plug, (2) the headless screw plug, and (3) the thin ovalhead screw plug.

The 100° countersunk and the headless screw plugs have either a Phillips or a Reed and Prince recess. The oval head either has a common screwdriver slot, a Phillips, or a Reed and Prince recess. All screw plugs are made of high tensile strength SAE steel and are cadmium plated.

The same tools are used for installation of the splined Rivnut as for installation of the standard types, but the pullup stud of the heading tool must be adjusted to accommodate the longer shank.

DILL LOK-SKRUS AND LOK-RIVETS

Dill Lok-Skru and Lok-Rivet are trade names for internally threaded rivets (two piece). They are used for blind attachment of such accessories as fairings, fillets, access door covers, door and window frames, floor panels, and the like. Lok-Skrus and Lok-Rivets are similar to the Rivnut in appearance and application. Lok-Skrus and Lok-Rivets, however, come in three parts and require more clearance on the blind side than the Rivnut to accommodate the barrel.

Special hand- and air-operated power tools are required for installation of Lok-Skrus. An interchangeable barrel blade fits into the blade handle and is held in place by a set screw. The barrel

186

blade has a flattened portion which fits into a slot in the end of the Lok-Skru barrel. The head driver has projections which fit into recesses in the Lok-Skru head. Head drivers and blades are interchangeable for use with various sizes and styles of Lok-Skrus.

The drilling procedure for Lok-Skrus is identical to that for common solid shank rivets. To install the Lok-Skru, insert the Lok-Skru tool so that the blade extends through the barrel slot and the driver sets firmly in the head slot. Insert the fastener in the drilled hole. Fit the ratchet handle assembly together and adjust the pawl lever for proper ratchet direction. Hold the ratchet handle stationary and turn the barrel blade handle to the left until the barrel is drawn firmly against the sheet on the opposite side. Press the tool firmly against the Lok-Skru to hold the tool blade and driver in the slots.

Stop turning the barrel handle when the Lok-Skru barrel has been drawn against the sheet. Finally, tighten by an additional quarter turn or less on the ratchet handle, drawing the head into the sheet. This time, hold the blade handle stationary while turning the ratchet handle. Test the tightness of the installation with an ordinary 8-in. screwdriver which has been ground round on the end. Attachments are made by using the attaching screw and a regular screwdriver.

DEUTSCH RIVETS

The Deutsch rivet is a high-strength blind rivet with a minimum shear strength of 75,000 p.s.i. and can be installed by one man. This rivet consists of two parts, a stainless steel sleeve and a hardened steel drive pin. The pin and sleeve are coated with a lubricant and a corrosion inhibitor.

A Deutsch rivet may be driven with an ordinary hammer or a pneumatic rivet gun and a flathead set. Seat the rivet in the previously drilled hole and then drive the pin into the sleeve. If the Deutsch rivet is driven into a tight hole, a hollow drift punch should be used to seat the rivet against the material. The punch should clear the drive pin and rest against the head of the rivet to prevent premature expansion of the sleeve and head.

The driving action causes the pin to exert pressure against the sleeve and forces the sides of the sleeve out. This stretching forms a shop head on the end of the rivet and provides a positive fastening action for the fastener. The ridge on the top of the rivet head locks the pin into the rivet as the last few blows are struck.

The head of the Deutsch rivet should never be shaved or milled. Milling or shaving will destroy the locking action of the ring on top of the rivet head.

Another feature of the Deutsch rivet is that it can be installed without going all the way through the second piece of material. However, this type of installation is not recommended unless the second piece is very thick.

One of the main restrictions to the use of the Deutsch rivet is that no bucking tool is used to take up the shock of driving. The structure where installation is made must be heavy and solid enough to support the driving forces.

If a Deutsch rivet that extends through the material is to be removed, use the same procedures used to remove a solid shank rivet. The head can be drilled off, and the pin can be driven out with a drift punch slightly smaller than the diameter of the drive pin. To drive the sleeve out of the material, use a drive punch slightly smaller than the diameter of the sleeve.

If the rivet does not extend through the material, drill out the drive pin to approximately one-half its depth. Then tap the hole and finish drilling out the remainder of the pin. Next, insert a screw through a spacer and tighten the screw into the sleeve. Continue tightening the screw until the sleeve is removed.

HI-SHEAR RIVETS

Hi-Shear pin rivets are essentially threadless bolts. The pin is headed at one end and is grooved about the circumference at the other. A metal collar is swaged onto the grooved end, effecting a firm tight fit.

The proper length rivet may be determined by part number or by trial. Part numbers for pin rivets can be interpreted to give the diameter and grip length of the individual rivets. A typical part number and an explanation of the terms are discussed in Chapter 6, Hardware, Materials, and Processes, in the Airframe and Powerplant Mechanics General Handbook, AC 65–9A.

To determine correct grip length by trial, insert the correct diameter rivet in the hole. The straight portion of the shank should not extend more than $1/16$ in. through the material, insert the correct diameter rivet in the hole. The straight portion of the shank should not extend more than $1/16$ in. through the material. Place a collar over the grooved end of the rivet. Check the position of the collar.

The collar should be positioned so that the shearing edge of the pin groove is just below the top of the collar. It is permissible to add a 0.032-in. (approximately) steel washer between the collar and the material to bring the collar to the desired location. The washer may be positioned on the rivet head side of the material when using a flathead rivet.

Hi-Shear rivets are installed with standard bucking bars and pneumatic riveting hammers. They require the use of a special gun set that incorporates collar swaging and trimming and a discharge port through which excess collar material is discharged. A separate size set is required for each shank diameter.

Prepare holes for pin rivets with the same care as for other close tolerance rivets or bolts. At times, it may be necessary to spot-face the area under the head of the pin so that the head of the rivet can fit tightly against the material. The spot-faced area should be $\frac{1}{16}$ in. larger in diameter than the head diameter.

Pin rivets may be driven from either end. Procedures for driving a pin rivet from the collar end are:

(1) Insert the rivet in the hole.

(2) Place a bucking bar against the rivet head.

(3) Slip the collar over the protruding rivet end.

(4) Place previously selected rivet set and gun over the collar. Align the gun so that it is perpendicular to the material.

(5) Depress the trigger on the gun, applying pressure to the rivet collar. This action will cause the rivet collar to swage into the groove on the rivet end.

(6) Continue the driving action until the collar is properly formed and excess collar material is trimmed off. (See figure 5-71.)

Procedures for driving a pin rivet from the head end are:

(1) Insert the rivet in the hole.

(2) Slip the collar over the protruding end of rivet.

(3) Insert the correct size gun rivet set in a bucking bar and place the set against the collar of the rivet.

(4) Apply pressure against the rivet head with a flush rivet set and pneumatic riveting hammer.

(5) Continue applying pressure until the

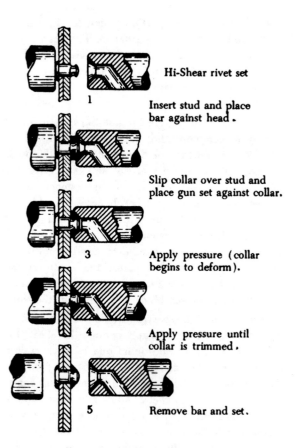

1 Hi-Shear rivet set

Insert stud and place bar against head.

2 Slip collar over stud and place gun set against collar.

3 Apply pressure (collar begins to deform).

4 Apply pressure until collar is trimmed.

5 Remove bar and set.

FIGURE 5-71. Using pin rivet set.

collar is formed in the groove and excess collar material is trimmed off.

Inspection

Pin rivets should be inspected on both sides of the material. The head of the rivet should not be marred and should fit tightly against the material. Figure 5-72 illustrates acceptable and unacceptable rivets.

Removal of Pin Rivets

The conventional method of removing rivets by drilling off the head may be utilized on either end of the pin rivet (figure 5-73). Center punching is recommended prior to applying drilling pressure. In some cases alternate methods may be more desirable for particular instances.

Grind a chisel edge on a small pin punch to a blade width of $\frac{1}{8}$ in. Place this tool at right angles to the collar and drive with a hammer to split the collar down one side. Repeat the operation on the opposite side. Then, with the chisel blade, pry the collar from the rivet. Tap the rivet out of the hole.

Use a special hollow punch having one or more

Acceptable rivets

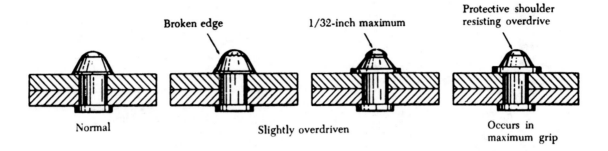

Broken edge

1/32-inch maximum

Protective shoulder resisting overdrive

Normal

Slightly overdriven

Occurs in maximum grip

Unacceptable rivets

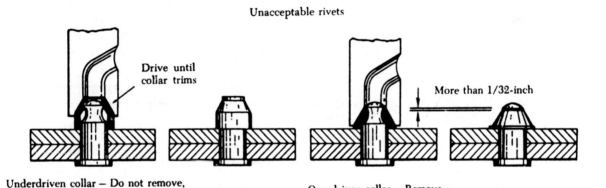

Drive until collar trims

More than 1/32-inch

Underdriven collar — Do not remove, but continue driving until collar trims.

Overdriven collar — Remove.

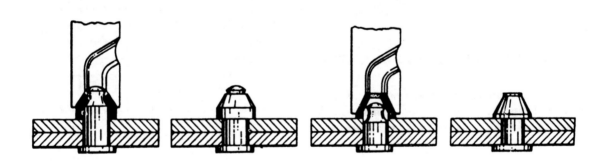

Pin too long — Remove.

Pin too short — Remove.

Figure 5-72. Pin rivet inspection.

blades placed to split the collar. Pry the collar from the groove and tap out the rivet.

Grind a pair of nippers so that cutting blades will cut the collar in two pieces, or use nippers at right angles to the rivet and cut through the small neck.

A hollow-mill collar cutter can be used in a power hand drill to cut away enough collar material to permit the rivet to be tapped out of the work.

SPECIFIC REPAIR TYPES

Before discussing any type of a specific repair

that could be made on an aircraft, remember that the methods, procedures, and materials mentioned in the following paragraphs are only typical and should not be used as the authority for the repair. When repairing a damaged component or part, consult the applicable section of the manufacturer's structural repair manual for the aircraft. Normally, a similar repair will be illustrated, and the types of material, rivets, and rivet spacing and the methods and procedures to be used will be listed. Any addi-

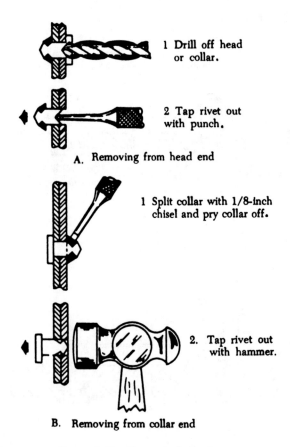

1 Drill off head
  or collar.

2 Tap rivet out
  with punch.

A. Removing from head end

1 Split collar with 1/8-inch
  chisel and pry collar off.

2. Tap rivet out
   with hammer.

B. Removing from collar end

FIGURE 5–73. Removal of pin rivets.

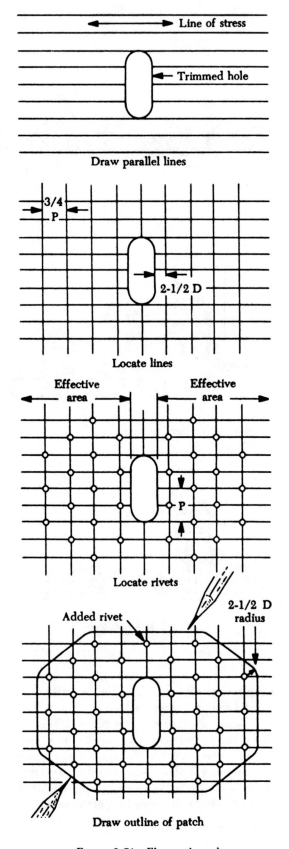

Line of stress

Trimmed hole

Draw parallel lines

3/4
P

2-1/2 D

Locate lines

Effective
area

Effective
area

P

Locate rivets

Added rivet

2-1/2 D
radius

Draw outline of patch

FIGURE 5–74. Elongated patch.

tional knowledge needed to make a repair will also be detailed.

If the necessary information is not found in the structural repair manual, attempt to find a similar repair or assembly installed by the manufacturer of the aircraft.

Smooth Skin Repair

Minor damage to the outside skin of an aircraft can be repaired by applying a patch to the inside of the damaged sheet. A filler plug must be installed in the hole made by the removal of the damaged skin area. It plugs the hole and forms a smooth outside surface necessary for aerodynamic smoothness of modern day aircraft.

The size and shape of the patch is determined in general by the number of rivets required in the repair. If not otherwise specified, calculate the required number of rivets by using the rivet formula. Make the patch plate of the same material as the original skin and of the same thickness or of the next greater thickness.

190

Elongated Octagonal Patch

Whenever possible, use an elongated octagonal patch for repairing the smooth skin. This type of patch provides a good concentration of rivets within the critical stress area, eliminates dangerous stress concentrations, and is very simple to lay out. This patch may vary in length according to the condition of the repair.

Follow the steps shown in the paper layout of this patch (figure 5–74). First, draw the outline of the trimmed-out damage. Then, using a spacing of three to four diameters of the rivet to be used, draw lines running parallel to the line of stress. Locate the lines for perpendicular rows two and one-half rivet diameters from each side of the cutout, and space the remaining lines three-fourths of the rivet pitch apart.

Locate the rivet spots on alternate lines perpendicular to the stress lines to produce a stagger between the rows and to establish a distance between rivets (in the same row) of about six to eight rivet diameters. After locating the proper number of rivets on each side of the cutout, add a few more if necessary so that the rivet distribution will be uniform. At each of the eight corners, swing an arc of two and one-half rivet diameters from each corner rivet. This locates the edge of the patch. Using straight lines, connect these arcs to complete the layout.

Round Patch

Use the round patch for flush repairs of small holes in smooth sheet sections. The uniform distribution of rivets around its circumference makes it an ideal patch for places where the direction of the stress is unknown or where it is known to change frequently.

If a two-row round patch is used (figure 5–75), first draw the outline of the trimmed area on paper. Draw two circles, one with a radius equal to the radius of the trimmed area plus the edge distance, and the other with a radius ¾-in. larger. Determine the number of rivets to be used and space two-thirds of them equally along the outer row. Using any two adjacent rivet marks as centers, draw intersecting arcs; then draw a line from the point of intersection of the arcs to the center of the patch. Do the same with each of the other pairs of rivet marks. This will give half as many lines as there are rivets in the outer row. Locate rivets where these lines intersect the inner circle. Then transfer the layout to the patch material, adding regular outer

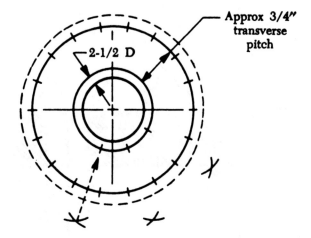

FIGURE 5–75. Layout of a two-row round patch.

edge material of two and one-half rivet diameters to the patch.

Use a three-row round patch (figure 5–76) if the total number of rivets is large enough to cause a pitch distance smaller than the minimum for a two-row patch. Draw the outline of the area on paper; then draw a circle with a radius equal to that of the trimmed area plus the edge distance. Equally space one-third of the required number of rivets in this row. Using each of these rivet locations as a center, draw arcs having a ¾-in. radius. Where they intersect, locate the second row rivets. Locate the third row in a similar manner. Then allow extra material of two and one-half rivet diameters around the outside rivet row. Transfer the layout to the patch material.

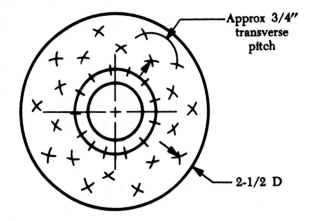

FIGURE 5–76. Layout of a three-row round patch.

After laying out and cutting the patch, remove the burrs from all edges. Chamfer the edges of all external patches to a 45° angle and turn them slightly downward so that they will fit close to the surface (figure 5–77).

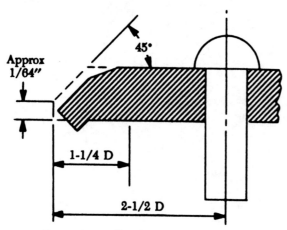

Approx 1/64″
45°
1-1/4 D
2-1/2 D

FIGURE 5–77. Chamfering and turning edge.

Panel Repair

In aircraft construction, a panel is any single sheet of metal covering. A panel section is the part of a panel between adjacent stringers and bulkheads. Where a section of skin is damaged to such an extent that it is impossible to install a standard skin repair, a special type of repair is necessary. The particular type of repair required depends on whether the damage is reparable outside the member, inside the member, or to the edges of the panel.

Damage which, after being trimmed, has less than eight and one-half manufacturer's rivet diameters of material inside the members requires a patch which extends over the members, plus an extra row of rivets along the outside of the members. For damage which, after being trimmed, has eight and one-half rivet diameters or more of material, extend the patch to include the manufacturer's row of rivets and add an extra row inside the members. Damage which extends to the edge of a panel requires only one row of rivets along the panel edge, unless the manufacturer used more than one row. The repair procedure for the other edges of the damage follows the previously explained methods.

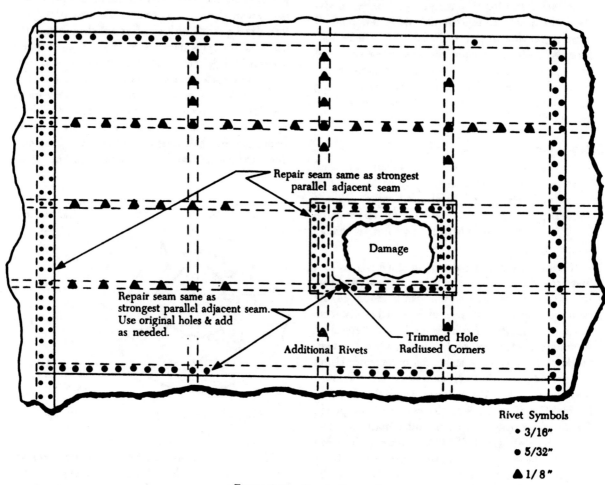

Repair seam same as strongest parallel adjacent seam

Damage

Repair seam same as strongest parallel adjacent seam. Use original holes & add as needed.

Additional Rivets

Trimmed Hole Radiused Corners

Rivet Symbols
• 3/16″
● 5/32″
▲ 1/8″

FIGURE 5–78. Panel skin patch.

The procedures for making all three types of panel repairs are similar. Trim out the damaged portion to the allowances mentioned in the preceding paragraph. For relief of stresses at the corners of the trim-out, round them to a minimum radius of ½ in. Lay out the new rivet row with a transverse pitch of approximately five rivet diameters and stagger the rivets with those put in by the manufacturer. See figure 5–78.

Cut the patch plate from material of the same thickness as the original or the next greater thickness, allowing an edge distance of two and one-half rivet diameters. At the corners, strike arcs having the radius equal to the edge distance. Chamfer the edges of the patch plate for a 45° angle and form the plate to fit the contour of the original structure. Turn the edges downward slightly so that the edges fit closely.

Place the patch plate in its correct position, drill one rivet hole, and temporarily fasten the plate in place with a fastener. Using a hole finder, locate the position of a second hole, drill it, and insert a second fastener. Then, from the back side and through the original holes, locate and drill the remaining holes. Remove the burrs from the rivet holes and apply corrosion protective material to the contacting surfaces before riveting the patch into place.

Stringer Repair

The fuselage stringers extend from the nose of the aircraft to the tail, and the wing stringers extend from the fuselage to the wing tip. Surface control stringers usually extend the length of the control surface. The skin of the fuselage, wing, or control surface is riveted to stringers.

Stringers may be damaged by vibration, corrosion, or collision. Damages are classified as negligible, damage reparable by patching, and damage necessitating replacement of parts. Usually the damage involves the skin and sometimes the bulkhead or formers. Such damage requires a combination of repairs involving each damaged member.

Because stringers are made in many different shapes repair procedures differ. The repair may require the use of preformed or extruded repair material, or it may require material formed by the airframe mechanic. Some repairs may need both kinds of repair material.

When repairing a stringer, first determine the extent of the damage and remove the rivets from the surrounding area. Then remove the damaged area by using a hacksaw, keyhole saw, drill, or file.

In most cases, a stringer repair will require the use of an insert and splice angle. When locating the splice angle on the stringer during repair, be sure to consult the applicable structural repair manual for the repair piece's position. Some stringers are repaired by placing the splice angle on the inside, whereas others are repaired by placing it on the outside.

Extrusions and preformed materials are commonly used to repair angles and insertions or fillers. If repair angles and fillers must be formed from flat sheet stock, use the brake. It may be necessary to use bend allowance and sight lines when making the layout and bends for these formed parts. For repairs to curved stringers, make the repair parts so that they will fit the original contour.

When calculating the number of rivets to be used in the repair, first determine the length of the break. In bulb-angle stringers, the length of the break is equal to the cross sectional length plus three times the thickness of the material in the standing leg (to allow for the bulb), plus the actual cross sectional length for the formed stringers and straight angles.

Substitute the value obtained, using the procedure above as the length of the break in the rivet formula, and calculate the number of rivets required. The rivet pitch should be the same as that used by the manufacturer for attaching the skin to the stringer. In case this pitch exceeds the maximum of 10 rivet diameters, locate additional rivets between the original rivets. Never make the spacing less than four rivet diameters.

When laying out this spacing, allow two and one-half rivet diameters for edge distance on each side of the break until all required rivets are located. At least five rivets must be inserted on each end of the splice section. If the stringer damage requires the use of an insertion or filler of a length great enough to justify more than 10 rivets, two splice angles should usually be used.

If the stringer damage occurs close to a bulkhead, cut the damaged stringer so that only the filler extends through the opening in the bulkhead. The bulkhead is weakened if the opening is enlarged to accommodate both the stringer and the splice angle. Two splice angles must be used to make such a repair.

Because the skin is fastened to the stringers, it is often impossible to drill the rivet holes for the repair splices with the common air drill. These holes can be drilled with an angle drill. When riveting a stringer, it may be necessary to use an offset rivet set and various shaped bucking bars.

Former or Bulkhead Repairs

Bulkheads are the oval-shaped members of the fuselage which give form to and maintain the shape of the structure. Bulkheads or formers are often called forming rings, body frames, circumferential rings, belt frames, and other similar names. They are designed to carry concentrated stressed loads.

There are various types of bulkheads. The most common type is a curved channel formed from sheet stock with stiffeners added. Others have a web made from sheet stock with extruded angles riveted in place as stiffeners and flanges. Most of these members are made from aluminum alloy. Corrosion-resistant steel formers are used in areas which are exposed to high temperatures.

Bulkhead damages are classified in the same manner as other damages. Specifications for each type of damage are established by the manufacturer and specific information is given in the maintenance manual or structural repair manual for the aircraft. Bulkheads are identified with station numbers, which are very helpful in locating repair information.

Repairs to these members are generally placed in one of two categories: (1) One-third or less of the cross sectional area damaged, or (2) more than one-third of the cross sectional area damaged. If one-third or less of the cross sectional area has been damaged, a patch plate, reinforcing angle, or both, may be used. First, clean out the damage and then use the rivet formula to determine the number of rivets required in order to establish the size of the patch plate. For the length of the break, use the depth of the cutout area plus the length of the flange.

If more than one-third of the cross sectional area is damaged, remove the entire section and make a splice repair (figure 5–79). When removing the damaged section, be careful not to damage the surrounding equipment, such as electrical lines, plumbing, instruments, and so forth. Use a hand file, rotary file, snips, or a drill to remove larger damages. To remove a complete section, use a hacksaw, keyhole saw, drill, or snips.

Measure the length of break as shown in figure 5–79 and determine the number of rivets required by substituting this value in the rivet formula. Use the double shear value of the rivet in the calculations. The result represents the number of rivets to be used in each end of the splice plate.

Most repairs to bulkheads are made from flat sheet stock if spare parts are not available. When fabricating the repair from flat sheet, remember that the substitute material must provide cross sectional tensile, compressive, shear, and bearing strength equal to the original material. Never substitute material which is thinner or has a cross sectional area less than the original material. Curved repair parts made from flat sheet must be in the "O" condition before forming, and then must be heat treated before installation.

Longeron Repair

Generally, longerons are comparatively heavy members which serve approximately the same function as stringers. Consequently, longeron repair is similar to stringer repair. Because the longeron is a heavy member and more strength is needed than with a stringer, heavy rivets will be used in the

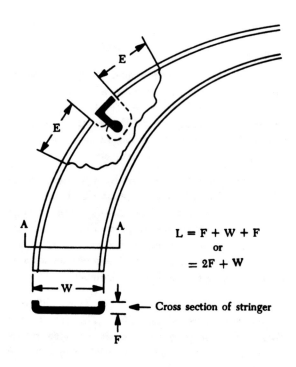

$$L = F + W + F$$
$$\text{or}$$
$$= 2F + W$$

FIGURE 5–79. Determining length of break.

repair. Sometimes bolts are used to install a longeron repair; but, because of the greater accuracy required, they are not as suitable as rivets. Also, bolts require more time for installation.

If the longeron consists of a formed section and an extruded angle section, consider each section separately. Make the longeron repair as you would a stringer repair. However, keep the rivet pitch between four- and six-rivet diameters. If bolts are used, drill the bolt holes for a light drive fit.

Spar Repair

The spar is the main supporting member of the wing. Other components may also have supporting members called spars which serve the same function as the spar does in the wing. Think of spars as the "hub" or "base" of the section in which they are located, even though they are not in the center. The spar is usually the first member located during the construction of the section, and the other components are fastened directly or indirectly to it.

Because of the load the spar carries, it is very important that particular care be taken when repairing this member to ensure that the original strength of the structure is not impaired. The spar is so constructed that two general classes of repairs, web repairs and cap strip repairs, are usually necessary.

For a spar web butt splice, first clean out the damage; then measure the full width of the web section. Determine the number of rivets to be placed in each side of the splice plate by substituting this value for the length of break in the rivet formula. Prepare an insert section of the same type material and thickness as that used in the original web. Make a paper pattern of the rivet layout for the splice plate using the same pitch as that used in the attachment of the web to the cap strip. Cut the splice plates from sheet stock having the same weight as that in the web, or one thickness heavier, and transfer the rivet layout from the paper pattern to the splice plates.

Give all contacting surfaces a corrosion-resistant treatment and rivet the component parts of the repair into place. The rivets used in attaching the insert section to the cap strips are in addition to those calculated for attaching the splice plates. Replace all web stiffeners removed during the repair. An exploded view of a spar web butt splice is shown in figure 5–80.

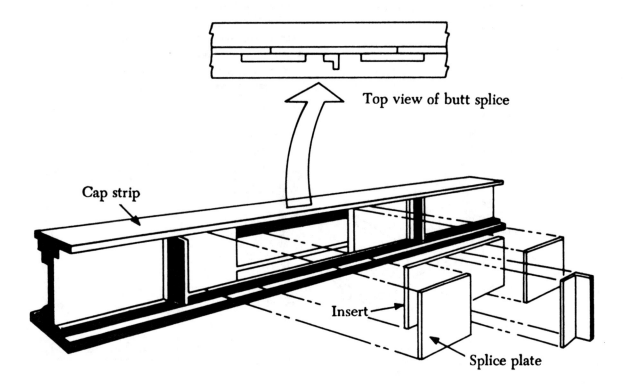

Top view of butt splice

Cap strip

Insert

Splice plate

FIGURE 5–80. Spar web butt splice.

When making a spar web joggle splice, no splice plates are needed. Instead, form the web repair section so that it overlaps the original web sufficiently to accommodate the required number of rivets. Make a joggle in each end of the repair section so that the repair piece contacts the cap strips to which it is riveted. Rivet calculation for this repair is similar to that described for butt splicing.

Many forms of cap strips are used in aircraft manufacturing, and each equires a distinct type of repair. In calculating the number of rivets required in an extruded T-spar cap strip repair, take the width of the base of the T, plus the length of the leg as the length of the break, and use double shear values.

Place one-fourth of the required number of rivets in each row of original rivets in the base of the T-section. Locate them midway between each pair of the original rivets. Locate the remainder of the rivets along the leg of the T-section in two rows. Consider all original rivets within the area of the splice as part of the required rivets.

Make the filler piece of a similar piece of T-section extrusion or of two pieces of flat stock. It is possible to make the splice pieces of extruded angle material or to form them from sheet stock; in either case, they must be the same thickness as the cap strip. Figure 5–81 shows an exploded view of a

T-spar cap strip repair. The rivets used in the leg of the cap strip may be either the round-, flat-, or brazier-head type; but the rivets used in the base must be the same type as those used in the skin.

The repair of milled cap strips is limited to damages occurring to flanges. Damages beyond flange areas require replacement of the entire cap strip. To make a typical flange repair, substitute the depth of the trimmed-out area as the length of break in the rivet formula and calculate the number of rivets required. Form a splice plate of the required length and drill it to match the original rivet layout. Cut an insert to fit the trimmed-out area and rivet the repair in place. If the trimmed-out area is more than 4 in. in length, use an angle splice plate to provide added strength.

Rib and Web Repair

Web repairs can be generally classified into two types: (1) Those made to web sections considered critical, such as those in the wing ribs, and (2) those considered less critical, such as those in elevators, rudders, flaps, and the like. Web sections must be repaired in such a way that the original strength of the member is restored.

In the construction of a member using a web (figure 5–82), the web member is usually a light-gage aluminum alloy sheet forming the principal depth of the member. The web is bounded by heavy aluminum alloy extrusions known as cap strips. These extrusions carry the loads caused by bending

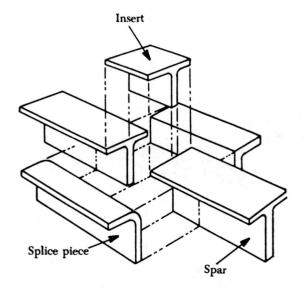

FIGURE 5–81. T-spar cap strip repair.

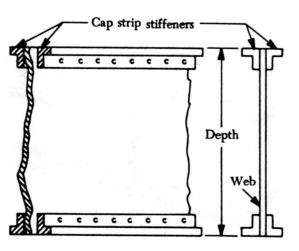

FIGURE 5–82. Construction of a web member.

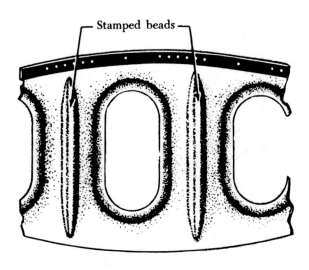

Stamped beads

FIGURE 5-83. Stamped beads in a web.

and also provide a foundation for attaching the skin. The web may be stiffened by stamped beads, formed angles, or extruded sections riveted at regular intervals along the web.

The stamped beads (figure 5-83) are a part of the web itself and are stamped in when the web is made. Stiffeners help to withstand the compressive loads exerted upon the critically stressed web members.

Often ribs are formed by stamping the entire piece from sheet stock. That is, the rib lacks a cap strip, but does have a flange around the entire piece, plus lightening holes in the web of the rib. Ribs may be formed with stamped beads for stiffeners, or they may have extruded angles riveted on the web for stiffeners.

Most damages involve two or more members; however, it may be that only one member is damaged and needs repairing. Generally, if the web is damaged, all that is required is cleaning out the damaged area and installing a patch plate.

The patch plate should be of sufficient size to ensure room for at least two rows of rivets around the perimeter of the damage; this will include proper edge distance, pitch, and transverse pitch for the rivets. The patch plate should be of material having the same thickness and composition as the original member. If any forming is necessary when making the patch plate, such as fitting the contour of a lightening hole, use material in the "O" condition and then heat treat it after forming.

Damage to ribs and webs which require a repair larger than a simple plate will probably need a patch plate, splice plates, or angles and an insertion. To repair such a damage by forming the necessary parts may take a great deal of time; therefore, if damaged parts which have the necessary areas intact are available from salvage, use them.

For example, if an identical rib can be located in salvage and it has a cracked web but the area in question is intact, clean out the damaged area; then cut the repair piece from the rib obtained from salvage. Be sure to allow plenty of material for correct rivet installation. Using a part from salvage will eliminate a great deal of hard work plus the heat-treating operation needed by a new repair piece.

Leading Edge Repair

The leading edge is the front section of a wing, stabilizer, or other airfoil. The purpose of the leading edge is to streamline the forward section of the wings or control surfaces so that the airflow is effective. The space within the leading edge is sometimes used to store fuel. This space may also house extra equipment such as landing lights, plumbing lines, or thermal anti-icing systems.

The construction of the leading edge section varies with the type of aircraft. Generally, it will consist of cap strips, nose ribs, stringers, and skin. The cap strips are the main lengthwise extrusions, and they stiffen the leading edges and furnish a base for the nose ribs and skin. They also fasten the leading edge to the front spar.

The nose ribs are stamped from aluminum alloy sheet. These ribs are U-shaped and may have their web sections stiffened. Regardless of their design, their purpose is to give contour to the leading edge.

Stiffeners are used to stiffen the leading edge and supply a base for fastening the nose skin. When fastening the nose skin, use only flush rivets.

Leading edges constructed with thermal anti-icing systems consist of two layers of skin separated by a thin air space. The inner skin, sometimes corrugated for strength, is perforated to conduct the hot air to the nose skin for anti-icing purposes.

Damage to leading edges are also classified in the same manner as other damages. Damage can be caused by contact with other objects, namely, pebbles, birds in flight, and hail. However, the major cause of damage is carelessness while the aircraft is on the ground.

A damaged leading edge will usually involve several structural parts. Flying-object damage will probably involve the nose skin, nose ribs, stringers, and possibly the cap strip. Damage involving all of these members will necessitate installing an access door to make the repair possible. First, the damaged area will have to be removed and repair procedures established. The repair will need insertions and splice pieces. If the damage is serious enough, it may require repair of the cap strip and stringer, a new nose rib, and a skin panel. When repairing a leading edge, follow the procedures prescribed in the appropriate repair manual for this type of repair.

Trailing Edge Repair

A trailing edge is the rearmost part of an airfoil, found on the wings, ailerons, rudders, elevators, and stabilizers. It is usually a metal strip which forms the shape of the edge by tying the ends of a rib section together and joining the upper and lower skins. Trailing edges are not structural members, but they are considered to be highly stressed in all cases.

Damage to a trailing edge may be limited to one point or extended over the entire length between two or more rib sections. Besides damage resulting from collision and careless handling, corrosion damage is often present. Trailing edges are particularly subject to corrosion because moisture collects or is trapped in them.

Thoroughly inspect the damaged area before starting repairs, and determine the extent of damage, the type of repair required, and the manner in which the repair should be performed. When making trailing edge repairs, remember that the repaired area must have the same contour and be made of material with the same composition and temper as the original section. The repair must also be made to retain the design characteristics of the airfoil.

Damage occurring in the trailing edge section between the ribs can be repaired as shown in figure 5–84. Cut out the damaged area and make a filler of either hardwood, fiber, or cast aluminum alloy to fit snugly inside the trailing edge. Then make an insert piece of the same material as the damaged section and shape it to match the trailing edge. Assemble the pieces as shown and rivet them into place using

countersunk rivets and forming countersunk shop heads to get a smooth contour.

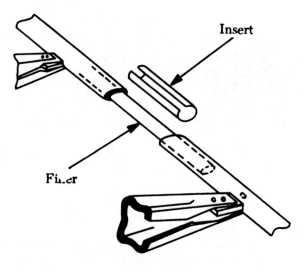

Insert

Filler

FIGURE 5–84. Trailing edge repair between ribs.

To repair damage occurring at or near a rib, first remove sufficient trailing edge material to allow a complete splice to fall between the ribs. This usually requires two splices joined by an insert piece of similar trailing edge material or of formed sheet stock. The repair procedure is similar to that for damage between ribs. Figure 5–85 shows this type of repair.

STRUCTURAL SEALING

Various areas of airframe structures are sealed compartments where fuels or air must be confined. Some of these areas contain fuel tanks; others consist of pressurized compartments such as the cabin. Because it is impossible to seal these areas completely airtight with a riveted joint alone, a sealing compound or sealant must be used. Sealants are also used to add aerodynamic smoothness to exposed surfaces such as seams and joints in the wings and fuselage.

Three types of seals are ordinarily used. Rubber seals are installed at all points where frequent breaking of the seal is necessary, such as emergency exits and entrance doors. Sealing compounds are used at points where the seal is seldom broken

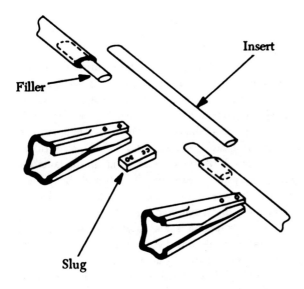

Figure 5–85. Trailing edge repair near a rib.

except for structural maintenance or part replacement, as with riveted lap and butt seams. Special seals are required for passing cables, tubing, mechanical linkages, or wires out of the pressurized or sealed areas.

Wires and tubes are passed through pressure bulkheads by using bulkhead fittings such as cannon plugs for wiring and couplings for tubing. These fittings are sealed to the bulkhead and the wires and tubes are fastened to them from each side. All seals of moving components such as flight controls are subject to wear and utmost care must be used when they are installed. Also, they must be checked regularly.

Determining Sealant Defects

Pressure tightness of an area or section is checked before and after a repair is made. Ground pressurization is accomplished by filling the section with air from an external source through ground pressure test fittings.

With the sections pressurized to a given pressure, locate leaks on the outside of the aircraft by applying a soapless bubble solution to all seams and joints in the suspected area. Air bubbles will locate the general area of leakage. A specific leak is then isolated on the inside of the aircraft by passing the free end of a stethoscope or similar listening device along the seams in the leakage area. The leak can

be detected by the change in sound when the instrument passes over it. After completing the test, remove the soapless bubble solution from the outside of the aircraft by washing with clear water to prevent corrosion.

Here are a few precautionary measures to follow during the testing procedure just discussed. With personnel inside, the area should never be pressurized to a pressure higher than has previously been established during testing with the section empty. No person who has a cold or who has recently had one, or whose sinuses are impaired in any way, should work in the pressurized section. A qualified operator should be present at the pressurization equipment control panel at all times while the section is being pressurized.

Pressurization may not always be necessary to determine defectively sealed areas. Sealants should be repaired when:

(1) Sealants have peeled away from the structure.

(2) Seams are exposed through the sealant fillet.

(3) Fillet- or hole-filling sealant is exposed through the smooth overcoating.

(4) Sealant is damaged by the removal and re-installation of fasteners, access doors, or other sealed parts.

(5) Cracks or abrasions exist in the sealant.

Sealant Repair

All surfaces which are to be sealed must be cleaned to ensure maximum adhesion between the sealant and the surface. Loose foreign material can be removed by using a vacuum cleaner on the affected area. Scrape all the old sealant from the repair area with a sharp plastic, phenolic, or hardwood block to prevent scratches, and apply a stripper and a cleaner.

The cleaner should not be allowed to dry on a metal surface, but should be wiped dry with clean rags. Do not remove the cleaner with soiled rags since the metal surface must be free of all dirt, grease, powder, and so forth. The surface can be checked for cleanliness by pouring water over it after being wiped dry of the cleanser. If the surface is not free of oily film, the water will separate into small droplets.

Be extremely careful to protect any undamaged sealant and acrylic plastics from the stripper compound. If artificial lighting is used when the repair is made, be sure the light is of the explosion proof type. Wear clothing which affords adequate protection from the stripper and cleaner so that these chemicals cannot contact the skin. Provide adequate ventilation in the work area. Personnel should wear a respirator when working in an enclosed area.

It may be necessary to replace rubber seals periodically to ensure tight door closure. Seals of this type should be replaced any time there is any degree of damage. Such a seal is usually not reparable because it must be continuous around the opening.

To remove the old seal, remove all the seal retainers from the frame and then pull off the old seal. Use aliphatic naphtha and clean rags to clean the frame on which the new seal is to be cemented. Cleaning should be done immediately before sealer installation. Then, using a clean paint brush, apply an even coat of rubber cement upon the metal parts and the seal surfaces which are to be joined.

Allow the rubber cement to dry until it becomes quite sticky. Then join the seal to the metal by pressing it firmly along all contact points. Install the seal retainers and allow the seal to set for at least 24 hrs. before using.

Toluene may be used for cleaning brushes and other equipment used in applying rubber cement. If the rubber cement needs thinning, use aliphatic naphtha.

Seals on pressurized sections must be able to withstand a certain amount of pressure. Therefore, damage to the seals in the compartment or section must be repaired with this question in mind: Can it withstand the pressures required? Pressure sealing must be performed on the pressurized side of the surface being sealed. Make sure that all areas are sealed before completing further assembly operations which would make the area inaccessible.

Sealing compounds should be applied only when the contacting surfaces are perfectly clean. The compound should be spread from the tube by a continuous forward movement to the pressure side of the joint. It is advisable to start the spreading of the compound 3 in. ahead of the repair area and continue 3 in. past it. If the compound is in bulk form, apply it with a pressure gun. Two coats or layers of compound are often required. If this is necessary, let the first application cure before the second is applied. Allow the compound to cure until it becomes tough and rubbery before joining the surfaces.

Curing time varies with temperature. High temperatures shorten the curing time and low temperatures lengthen it. Artificial heat may be used to speed up curing, but care must be used to avoid damaging the sealant with too high a temperature. Warm circulating air, not over 120°F., or infrared lamps placed 18 in. or more from the sealants are satisfactory heat sources. If infrared lamps are used, adequate ventilation must be provided to carry away the evaporated solvents.

Sealing compounds are most generally used on seams and joints, but they may also be used to fill holes and gaps up to $1/16$ in. wide.

Impregnated zinc chromate tape is sometimes used between seams and joints. Sealing tape is also used as a backing strip over holes and gaps which are $1/16$ to $1/2$ in. in width. The tape is applied over the opening on the pressure side, and a fillet of sealing compound is applied over the tape. Holes and gaps over $3/16$ in. in width are usually plugged with wood, metal caps, or metal plugs on the pressure side of the area; then, impregnated tape and sealing compound are applied over the repair.

Be sure that all forming, fitting, and drilling operations have been completed before applying the tape. With the repair surface area clean, unroll the tape with the white cloth innerleaf away from the metal surface. Leave the innerleaf on the tape until just before the parts are assembled. There must be no wrinkles in the tape, and the parts must be joined together with the least possible amount of sideways motion.

The application of putty sealant is similar to that of sealing compounds. A spatula or sharp-pointed plastic, phenolic, or hardwood block is sometimes used to force and pack the putty into the gaps or seams. Clean the gap or seam with compressed air before applying the putty to the pressure side.

Rivets, bolts, or screws do not always seal properly when used in these critical areas or sections. When pressure leaks occur around the fasteners, they should be removed and replaced. The holes should be filled with sealing compound and new fasteners installed. Remove excess sealant as soon as possible to avoid the difficulty encountered after it becomes cured.

METAL BONDED HONEYCOMB

The introduction of bonded honeycomb (sandwich construction) members in airframe design and manufacture came as a major breakthrough in the search for a more efficient type of structure. Because bonded honeycomb structures are manufactured and perform their jobs in a manner different from the previously used and more familiar conventional structures, new attitudes and methods of repair had to be developed with respect to the advantages, limitations, and physical peculiarities.

CONSTRUCTION FEATURES

Sandwich construction design is governed by the intended use of the panel or structure. It can be defined as a laminar construction consisting of a combination of alternating dissimilar materials, assembled and fixed in relation to each other so that the properties of each can be used to attain specific structural advantages for the whole assembly.

Sandwich-constructed assemblies can be found in a, variety of shapes and sizes on modern aircraft. They may consist of a whole section or a series of panels combined into an assembly. Sandwich-constructed assemblies are used for such areas as bulkheads, control surfaces, fuselage panels, wing panels, empennage skins, radomes, or shear webs.

Figure 5-86 illustrates a section of bonded honeycomb. The honeycomb stands on end and separates facings which are bonded to the core by means of an adhesive or resin. This type of construction has a superior strength/weight ratio over that of conventional structures. Also, it is better able to withstand sonic vibration, has relatively low cost when compared with fastener cost and installation of conventional structures, reduces the number of parts needed, and greatly reduces sealing problems while increasing aerodynamic smoothness.

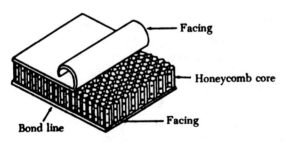

FIGURE 5-86. Bonded honeycomb section.

Special applications of metal-bonded honeycomb may employ stainless steel, titanium, magnesium, plywood, resin-impregnated paper, glass, nylon, or cotton cloth in various combinations.

DAMAGE

Causes of Damage

The majority of damages to bonded honeycomb assemblies result from flight loads or improper ground handling. Honeycomb structures may also be damaged by sonic vibrations. Such damage is usually a delamination or separation of the core and face along the bond line. (The bond line is the thin line of adhesive between the core and the face that holds the two together.) Occasionally the core may collapse.

Damage Inspection

Inspection for damage is more critical for honeycomb assemblies than for conventional structures. A honeycomb structure can suffer extensive damage without any observable indication. Sonic vibration, liquid leakage, internal condensation, or a misstep in manufacture or repair can cause or result in varied amounts of delamination.

The metallic ring test is the simplest way to inspect for delamination damage. When a coin (25-cent piece) is lightly bounced against a solid structure, a clear metallic ring should be heard. If delamination is present, a dull thud will be heard. A 1-oz. aluminum hammer makes an excellent tool for this type of inspection.

Occasionally, the delaminated skin will "oilcan" away from the core, making visual or thumb pressure detection possible. Punctures, dents, scratches, cracks, or other such damage may be inspected by conventional methods. Scratches should be given special attention since, with such thin material as that used in the metal bonded honeycomb, the scratch may actually be a crack.

A caustic soda solution can be used for testing scratches on aluminum surface panels. If the scratch area turns black after the application of a small amount of the solution, the scratch has penetrated through the clad surface. Caustic soda solutions are highly corrosive and must be handled with extreme care. Thoroughly neutralize the area after application of the solution.

Two additional instruments used in damage inspection of bonded panels are the panel inspection analyzer and the borescope.

Damage Evaluation

After inspections on metal bonded honeycomb structures are completed, any damage found must be evaluated to determine the type of repair needed to make the structure serviceable.

Damage to aluminum honeycomb structures can vary from minor dents or scratches to total panel destruction. Damage evaluation charts for honeycomb structures can be found in the applicable section of the structural repair manual for the specific aircraft. The charts specify types of damage, allowable damage, damage requiring repair, and figure numbers that illustrate similar repairs for each type of damage.

Once the type of repair is determined, procedures outlined in the structural repair manual should be rigidly followed.

REPAIRS

Recommendations for the type of repair to make and the methods or procedures to use vary among the different aircraft manufacturers. Tools, materials, equipment, and typical repairs that might be made on metal bonded honeycomb structures will be discussed in the following paragraphs.

Tools and Equipment

Effective repairs to bonded honeycomb assemblies depend largely on the knowledge and skill of the airframe mechanic in the proper use and maintenance of the tools and equipment used in making bonded honeycomb repairs. The design and high quality of workmanship built into these tools and equipment make them unique in the repair of bonded honeycomb assemblies. Therefore, it is essential that the techniques and procedures established for each tool or piece of equipment be known and practiced. Both personal injury and additional damage to the area being repaired can then be avoided.

Router

The primary tool used to prepare a damaged honeycomb area for repair is a pneumatically powered, hand-operated router with speeds ranging from 10,000 to 20,000 r.p.m. The router is used in conjunction with the support assembly, bit, and template as shown in figure 5–87.

The router support assembly threads onto the router body. It has provisions to adjust the desired depth of the cut with a locking (clamping nut) mechanism which secures the depth adjustment in place. One complete turn of the support adjustment changes the depth of cut approximately 0.083 in.

Metal-cutting, 1/4-in. mill bits are used with the router for removing the damaged areas. The router bits should be kept sharp, clean, and protected against nicks, breakouts, or other damages.

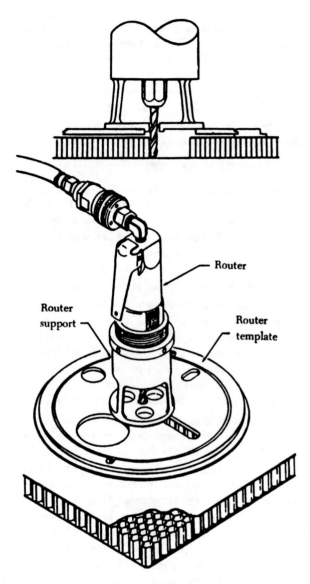

FIGURE 5–87. Router, support assembly, and template.

Router templates are used as guides when removing damaged honeycomb areas with a router. They can be designed and manufactured to the desired sizes, shapes, or contours of the repair. As an example, the multi-template (figure 5–88) can be used as a guide when cutting holes from 1/2 in. to 6 in. in diameter. For larger holes, a template can be manufactured locally from aluminum alloy 0.125-in. thick, whereas smaller holes can be cleaned out without the use of a template. The multi-templates should be kept clean and lightly oiled to prevent rusting and to maintain smooth operation during their use.

A routing template may be applied to a flat surface using the following procedures:

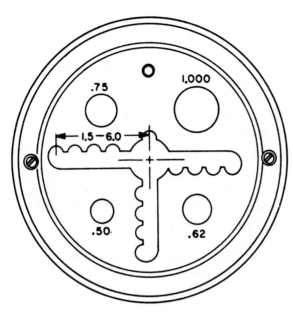

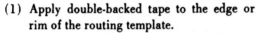

FIGURE 5-88. Multi-template.

(1) Apply double-backed tape to the edge or rim of the routing template.

(2) Place the template, centering the desired cutout guide hole directly over the damaged area.

(3) Press the template firmly down over the double-backed tape, making sure that it is secured in place; this will avoid any creeping or misalignment of the template during the routing operation.

A routing template may be applied to a tapered surface by using the follwoing procedures:

(1) Manufacture a bridge consisting of two wooden wedge blocks at least 6 in. long and with approximately the same degree of angle as that of the panel. (See figure 5-89.)

(2) Apply a strip of double-backed tape to one side of each wedge block.

(3) Place a wedge block on each side of the damaged area in a position that will bridge and support the template properly during the routing operation.

(4) Press the wedge blocks firmly in place.

(5) Place another strip of double-backed tape on the top side of each wooden wedge block.

(6) Place and align the template over the wedge blocks, thus avoiding any creeping or misalignment of the template during operation.

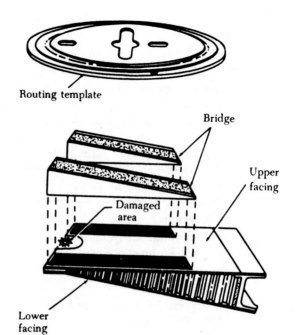

FIGURE 5-89. Wedge blocks and application.

Routing of Damaged Areas

When the extent of damage and type of repair have been determined, the proper size hole of the router template must be located around the damaged area in such a way that it will ensure that all the damage will be removed. The double-backed tape is used to secure the template to the surfaces around the damaged area, thus preventing creeping or misalignment of the template.

The router bit should be adjusted and set for the approximate depth required to remove the damaged area. During the routing operations, the router should be firmly gripped with both hands to prevent it from jumping or creeping. When the router is not in use, it should be disconnected from the air hose and stored properly until it is to be used again.

The following procedures for the removal of damaged bonded honeycomb areas are typical of those used by the various airframe manufacturers. Always follow the repair techniques specified by the applicable aircraft manufacturer.

(1) Determine the extent of the damage.

(2) Set up and adjust the router, router support assembly, and end-cutting mill bit for the removal of the damaged area.

(3) Select a routing template and position the template over the repair area according to procedures outlined in the discussion on templates.

(4) Attach the router's air intake plug in the socket of the air supply.

(5) Accomplish the routing operation.

    (a) Use face shield or goggles for eye protection against flying materials removed.

    (b) Place the air hose over the shoulder.

    (c) Holding the router firmly and at a 45° angle to the surface, place one edge of the router support assembly against the edge of the routing template.

    (d) Start the router by depressing the control lever.

    (e) Carefully, but firmly, lower the end-cutting mill bit into the material as close as possible to the center of the damaged area to be removed.

    (f) Straighten the router to be perpendicular to the surface.

    (g) Holding the router firmly, spiral it clockwise to the outer limits of the template's guide hole, removing all the damaged material.

    (h) Release the control lever, allow the router to stop, and remove it from the hole.

    (i) Disconnect the router from the air supply.

    (j) Check the damaged area removed. If additional removal is required, adjust the router's cutting depth and repeat routing operation.

(6) Upon completion of the routing operation, disconnect and clean the routing equipment.

During the routing operations, the aluminum core cells of a damaged honeycomb area tend to bend or close up. Therefore, they must be opened with tweezers and a pen knife before any further attempts to repair are made. At times, the core cells must be trimmed with a pen knife to the shape of the repair hole.

Pressure Jigs

Pressure jigs are used to apply pressure to repairs on the under surfaces of honeycomb panels or assemblies to hold the repair materials and resins in place. The pressure is maintained on the repair area until the repair material is cured.

C-clamps, locally manufactured jigs, or vacuum fixtures may be used to apply the necessary pressure to bonded honeycomb repairs.

The surfaces around the repair area must be absolutely clean and free of any foreign materials to ensure a good vacuum when vacuum fixtures or suction types of equipment are used. An application of water or glycerin to the surface areas will aid in obtaining a good vacuum. Normal cleaning, care, and corrosion prevention will maintain the above equipment in good working condition.

Infrared Heat Lamps

Infrared heat lamps are used to shorten the curing time of bonded honeycomb repairs from approximately 12 hrs. to 1 hr. A single-bulb lamp will adequately cure a repair up to 6 in. in diameter, but a large repair may require a battery of lamps to ensure uniform curing of the repair area.

The lamps should be centered directly over the repair at a distance of about 30 in. The setup is ideal to attain the recommended 130° F. curing temperature, provided the surrounding areas are at room temperature (70° F.). Warmer or colder surrounding areas will require that the heat lamps be adjusted to the prevailing condition. Caution must be used when working under extremely cold conditions, since a temperature differential of 150° or more will cause buckling of the surrounding skin surfaces because of thermal expansion.

As with any ordinary light bulbs, the infrared bulbs require little or no maintenance; however, the support stands, wiring, and switches should be handled carefully and maintained properly.

Fire Precautions

The potential of a fire hazard generally exists in the area of bonded honeycomb repairs because of the low flash point of the repair materials, such as cleaning solvents, primers, and resins. Therefore, it is necessary that all fire precautions be observed closely. Certain fire safety prevention equipment, such as utility cans, flammable-waste cans, and vapor- and explosion-proof lights, should be used.

With the potential of fire hazards in a honeycomb repair area, it is necessary to make sure that a suitable fire extinguisher is on hand or is located nearby and ready for use, if necessary. The extinguishing agents for all the materials used in bonded honeycomb repairs are dry chemicals or carbon dioxide; thus, the standard CO_2 fire extinguishers should be on hand for use in areas where bonded honeycomb structures will be repaired.

Handtools and Equipment

In addition to the tools and equipment described

in the preceding paragraphs, standard handtools and shop equipment are utilized in the repair of bonded honeycomb structures. Standard handtools and shop equipment used in the shop include an airframe mechanic's tool kit, face shields, scissors, power shear, drill press, horizontal and vertical belt sanders, contour metal-cutting saw, and pneumatic hand drills. The general uses and maintenance of these standard tools and equipment should be familiar to any airframe mechanic.

REPAIR MATERIALS

Cleaning Solvents

Before any repair is made to a bonded honeycomb structure, an area extending several inches away from the damage must be cleaned thoroughly of all paint or surface coating. This is best accomplished by the use of paint remover or MEK (methyl-ethyl-ketone) cleaning solvent. In some cases, Alconox, a powerful wetting and detergent agent, may be used for a final cleanup to remove any residue or oils remaining after application of the paint remover or the MEK cleaning solvent.

Paint removers are applied with a suitable size brush. When the paint or surface coating has loosened, it is either wiped off with a clean rag or removed with a nonabrasive scraper. Paint remover must not be allowed to enter the damaged area or be used along a bonded joint or seam because its chemical action will dissolve the bonding adhesive. These areas should be masked and final cleanup accomplished with the MEK cleaning solvent or emery cloth. The MEK cleaning solvent and the Alconox cleaning agent may be applied with a clean sponge.

After a damaged area has been completely removed, the surrounding surface areas must be thoroughly re-cleaned. This is accomplished by the use of the MEK cleaning solvent and gauze sponges. The MEK cleaning solvent is applied to the area with one sponge and immediately wiped off with another before it has had time to dry. This cleaning process should be continued until the surface area is lusterous in appearance and clean of any foreign matter.

To determine whether an area is completely and thoroughly clean, a water "break" test can be used. This test is a simple application of a thin film of distilled water to the cleaned area. Any "break" of the applied thin film of distilled water will indicate that the area has not been cleaned thoroughly enough and the cleaning process must be repeated.

Safety precautions must be closely observed when working with the above solvents, especially when the work is overhead or when working in confined areas. For personal protection, rubber gloves, face shields, adequate ventilation, and respirators should be worn. A CO_2 fire extinguisher should be on hand or nearby and ready for use if necessary.

Primers

Primers are applied to the cleaned surface areas primarily to ensure a good bond of the honeycomb repairs. The primer is applied to the cleaned surface areas with a clean gauze sponge or suitable brush. It is recommended that the primer be applied as rapidly as possible because it will become tacky in 10 to 15 sec., and it will pull and be ruined by any further brushing. The primer will cure in approximately 1 hr. at room temperature; however, this time may be reduced by the application of controlled heat.

Adhesives and Resins

Two types of adhesives presently used in the repairs of bonded honeycomb structures of some aircraft are known as the type 38 adhesive and the potting compound. The type 38 adhesive is applied to glass fabric overlay repairs, and the potting compound, as the name implies, to the potted compound (hole filling) repairs. In addition, the type 38 adhesive may be used as an alternate for the potting compound by adding micro-balloons (microscopic phenolic). The adhesives or potting compounds are prepared according to a batch mix (amount required for the repair) formula. The batch mix should be measured by weight.

Accurate mixing of the adhesive ingredients by batches is considered one of the more important steps in the repair of bonded honeycomb structures. The correct proportions of the epoxies, resins, and micro-balloons to be mixed into batches, both by weight and/or by volume, are given in the applicable section of the structural repair manual for the specific aircraft.

Core Material

Fiberglass honeycomb core materials ($3/16$ in. cell size) are usually used to replace the damaged aluminum cores of the bonded honeycomb structures. Aluminum core materials are not satisfactory for the repairs because of their flimsy and fragile structure. With this condition, it is impractical to cut the aluminum core materials accurately to the desired repair size. Fiberglass core materials are available in various thicknesses and are easily and accurately cut to size by the use of standard shop tools and equipment.

Glass Fabrics

Glass fabrics used in the overlay repairs to bonded honeycomb structures are manufactured from glass. The glass is spun into fibers which are in turn woven into a glass cloth with a variety of weaves.

Glass fabric cloth must be handled with care, stored properly, and be perfectly clean (free of any dirt, moisture, oil or other contaminants which may cause imperfect adhesion of the adhesives with which it is impregnated). Snags and sharp folds in the cloth will cause its strands to break, resulting in a local strength loss in the finished repair. Exposure to or contact with the glass fabric, dust, or particles may cause bodily itching or irritation.

Erosion and Corrosion Preventives

Two coatings of preventives are applied to the bonded honeycomb structure repairs to protect the areas against erosion and corrosion. The first is two layers of zinc chromate, preferably sprayed onto the repair area. The second is two layers of aluminum pigmented Corrogard (EC843), or equal, either sprayed or brushed on with a 30-min. drying period between each application. Both materials are flammable; therefore, the necessary fire precautions should be observed.

POTTED COMPOUND REPAIR

The following techniques, methods, and procedures are related to potted compound repairs and are typical of those used on most bonded honeycomb structures. For all repairs, consult the applicable section of the structural repair manual. The manufacturer's procedures should always be followed.

Bonded honeycomb structure damages up to 1 in. in diameter may be repaired by a hole-filling technique, using approved materials. The repair method is commonly known as the "potted compound repair." It is the easiest and fastest method of repairing a damaged area of a bonded honeycomb structure. However, be sure to follow the techniques, methods, or procedures established for potted compound repairs to avoid any further damage which might result in a more complicated repair.

Potted compound repairs may be applied both to single-face (one skin) and core damages, or to double-face (two skins) and core damages. (See figure 5–90.)

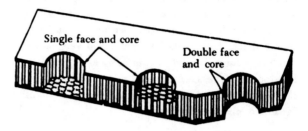

FIGURE 5–90. Typical potted compound repair areas.

Removal of Damage

Normally, no surface preparation is necessary when performing potted compound repairs. Oily or dirty surface areas to which multi-templates or pressure jigs are attached with adhesive tapes should be cleaned with any approved or recommended cleaning solvents, such as MEK.

Damages 1/4 in. or less in diameter can be satisfactorily removed with a twist drill. The multi-template and a high-speed router (10,000 to 20,000 r.p.m.) should be employed in the removal of damaged areas for potted compound repairs up to 1 in. in diameter. The amount of material removed by either method must be kept to a minimum to maintain as much of the original strength of the panel or structure as possible. Always use a face mask or protective glasses when using the router to remove damaged materials.

Repair Techniques

After a damaged area has been completely removed and cleaned, the necessary materials for the potted compound repair(s) are prepared.

Pieces of sheet plastic materials are prepared to provide a smooth surface effect of the potted compound repair, to provide part of the reservoir for the hole (cavity) filling operation of the repair, and to hold the potting compound in place until it is completely cured. The pieces of sheet plastic to be used for any or all of the above purposes should be at least 1/2 in. larger in diameter than the repair hole diameter.

A thinner piece of sheet plastic material (approximately 1/16 in. thick) is applied to the lower (bottom) surface of the double-face repair (figure 5–91). This is done not only to give the repaired surface a smooth effect, but mainly to hold the repair surface (potting compound) in place until it is cured. The same may be applied to a single-face

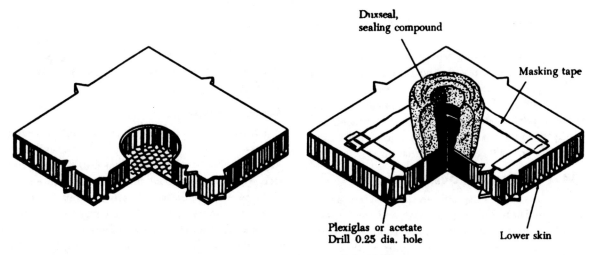

DAMAGE TO UPPER SKIN AND CORE

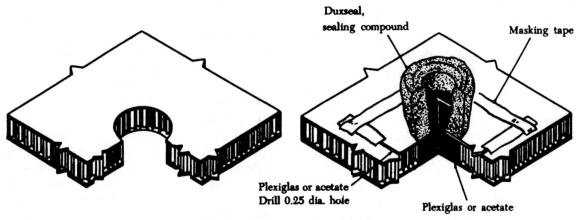

DAMAGE TO BOTH SKINS AND CORE

FIGURE 5–91. Potted compound repairs.

repair where the work must be accomplished in an overhead position.

Sheet plastic materials ranging from $\frac{1}{4}$ in. to $\frac{3}{16}$ in. in thickness are used on the upper (top side) surface of the repair during the hole (cavity) filling procedures. A $\frac{1}{4}$-in. hole is drilled directly in the center to permit easy application of the potting compounds to the repair cavity. The hole is also countersunk, allowing a buildup of the potting compound and thus assuring that the repair cavity has been completely filled.

This piece of plastic material is also a part of the "Duxseal" dam. After the prepared pieces of sheet plastic materials are properly located and taped in place over the repair area, the Duxseal (or equal) dam is built up around the hole. (See figure 5–91.) This dam is partially filled with the potting com-

pound during the hole-filling operation to ensure an adequately filled repair cavity. The dam also acts as the reservoir.

Next, a sufficient batch mix of the potting compound is prepared for the repair. The cavity is filled with the potting compound, and the air bubbles are removed with a toothpick or similar tool. The air bubbles are removed to ensure that the repair cavity is solidly filled.

When the potting compound within the repair cavity is completely cured, the pieces of plastic may then be removed. Generally, these pieces of plastic can be lifted off by hand; but, if necessary, they can be pried off easily with any dull straight hand-tool. When the drilled top piece of plastic is removed, it will leave a broken stem protruding above the repair surface. This stem may be filed, micro-

207

shaved, or routed down to make the repair surface area smooth.

The soundness of the repair can be tested by the metallic ring test. Pressure jigs may be used on the undersurface potted repairs as necessary. The repair is surface finished by the application of the recommended coatings of erosion or corrosion preventives, and a final coating of a finish of the same specifications as that of the original finish.

GLASS FABRIC CLOTH OVERLAY REPAIRS

Presently, two acceptable methods of repair are being applied to the damaged skin and core materials of some aircraft bonded honeycomb structures. One is the potted compound repair method previously discussed, and the other is the laminated glass fabric cloth overlay method applied to the various damages of honeycomb skin and core materials which exceed the repair limitations of the potted compound repair.

The differences between the two repair methods are in the techniques of removing the damaged area, preparing the damaged area for the repair, preparing and applying the repair materials, finishing and final inspection of the completed repair, and use and maintenance of the handtools and shop equipment.

Cleaning

Before repairing a damaged bonded honeycomb panel or section, thoroughly clean all paint or surface coatings from a surface area extending several inches away from the damage. Basically, this is necessary to attach and secure the templates or wedge blocks to the repair area with double-backed tape. Second, thoroughly clean the area of all foreign matter to ensure a perfect adhesion of the overlay repair materials.

Effective surface cleaning is of primary importance to the success of any repair. An area that is contaminated with paint, grease, oil, wax, oxides, or such, will not take a good bond. This cannot be emphasized too strongly since the quality of the repair will be no higher than the quality of the cleaning that precedes it. Even a fingerprint will prevent a good bond, because of natural oils in the skin.

Materials such as solvents, abrasives, alkaline detergents, and chemical etches can be used for effective cleaning. One of the easiest and most effective cleaning methods known is to apply MEK to the area with a sponge and immediately wipe it away

with another sponge. This procedure should be continued until a lusterous surface is obtained.

In removing paint, use caution, since paint remover will dissolve adhesives if allowed to enter the damaged area of a joint.

Removal of Damage

A high-speed router in conjunction with a router support assembly, metal-cutting mill bit, and template should be used in the removal of the damaged area. (Information about the uses and maintenance of the router was discussed earlier in this chapter.)

The techniques of removing damaged honeycomb skin and core material may differ from one repair to another. Their selection depends largely on the construction features of the bonded honeycomb panels, which are primarily of either flat, tapered, or combined (flat and tapered) surface design. Also, the location of the damaged area must be considered; that is, whether the damage occurred on the upper or lower side of the panel. Another factor that must be considered is that the honeycomb core is always installed within the panel with the cells perpendicular to the lower surface.

The techniques of preparing for and removing a damaged area on a tapered or upper surface of a panel are somewhat different from those for a flat or lower surface. Prior to the routing of a damaged area of an upper or a tapered surface, the routing template must be bridged over the repair area. This is done in such a manner that the routing template will be perpendicular to the core cells and parallel to the opposite (lower) facing. The bridge consists of two wedge blocks made of wood, at least 6 in. long, approximately 2 in. wide, and tapered to the same degree of angle as that of the panel. The method by which the bridge is attached to the damaged area is shown in figure 5–89.

Adhesives

Overlay repair adhesives consist of a type 38 batch mix. Micro-balloons are added to the resins and curing agent for "buttering" the fiber glass honeycomb core plug and cavity of the glass cloth overlay repair. The micro-balloons can also be used to control the consistency of the potted compound adhesive.

The type and location of the repair will determine the method of adhesive application. For example, a repair on an upper surface would use a low micro-balloon content and would be poured into the cavity, whereas the same repair on an under surface would use a high micro-balloon content and would have to be spooned into the cavity with a spatula or putty knife. Whichever method is used, the adhesive for all repairs should be applied evenly, without trapping any air bubbles.

The type 38 adhesive will set up and bond at room temperature. If a faster bond is required, the repair area should be preheated to 130° F., the repair parts and adhesives applied, and the whole repaired area heated at the same temperature for 1 hr. to effect a complete bond.

Upon completion of the repair, test it for any separation or other flaws, using the metallic ring test.

Core Plug

Core plugs are cut slightly larger than the desired thickness and shape from a glass fabric honeycomb core material (³⁄₁₆-in. cell size). They are sanded to the correct shape and thickness with a belt or disk sander to a tolerance of ±0.010 in. of the required size.

Before a core plug is inserted or assembled into the repair area, all contacting (faying) side surfaces of the core plug and the repair area must be "buttered" with an application of adhesive.

After the core plug has been properly installed into the repair area, the excessive potting compound is removed with a plastic scraper and the surface area thoroughly cleaned with a cleaning solvent. The core plug repair area should be cured for at least 30 min. to 1 hr. This is done to assure that the core plug is firmly in place before any further repair steps are accomplished.

Laminated Glass Cloth Overlay

A laminated glass cloth overlay consists of two layers of glass fabric cloth number 181 (three layers if number 128 is used) impregnated with type 38 adhesive and sandwiched between two sheets of polyethylene film. The glass cloth layers and sheets of polyethylene film are cut larger (approximately 4 in.) than the damage cutout. This is done to accommodate the cutting of the laminated overlay to correct size, allowing for the required minimum overlap of at least 1-½ in. beyond the edge of the damage cutout.

Before a laminated glass cloth overlay is applied to a repair area, the faying surface must be cleaned until no trace of foreign matter appears. After the area has been thoroughly cleaned and dried, a thin and continuous film of adhesive primer EC–776R (or equal) is applied to the faying surfaces of the area. The adhesive primer may be allowed to dry

at room temperature or may be accelerated by heat at a recommended temperature.

Protective coatings to prevent erosion and corrosion should be applied in accordance with the procedures outlined in the manufacturer's structural repair manual for the specific aircraft. A control surface repair should be checked to determine whether it is within balance limits or will require the surface to be re-balanced.

ONE SKIN AND CORE REPAIR PROCEDURES

Two typical methods of repairing damages to the honeycomb skin and core materials of aircraft are: (1) Single-face repair with damage extending through the core material and to the bond line of the opposite facing; and (2) transition area repair.

When the damage to the honeycomb structure is inspected and evaluated as damage to only one skin and the core (figure 5–92), the procedures

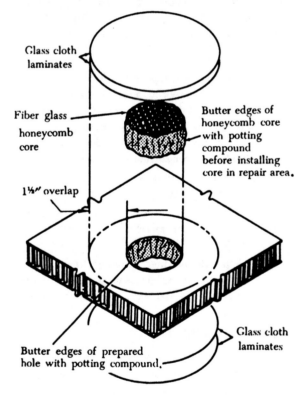

Glass cloth laminates

Fiber glass honeycomb core

Butter edges of honeycomb core with potting compound before installing core in repair area.

1½″ overlap

Butter edges of prepared hole with potting compound.

Glass cloth laminates

CORE REPLACEMENT

SECTIONAL VIEW OF REPAIR AREA

FIGURE 5–92. Skin and core repair.

discussed in the following paragraphs can be used. These procedures are typical but may not apply to all types of aircraft. Consult the manufacturer's repair manual for the specific aircraft, and follow the instructions for the particular type of repair.

Removing Damaged Area

A router and the applicable template should be used to remove the damaged material from the area. The depth of the router bit is determined by gradually increasing the depth of cut until it removes all of the damaged area. If the core is only partially damaged, remove only that portion. If the entire core is damaged, remove the core down to the opposite adhesive layer.

Preparing the Core Replacement

The core replacement must be fabricated from fiber glass core material. If the correct thickness is not available, the replacement section may be cut to size by hand sawing and/or sanding. The core plug should be flush or within ±0.010 in.

Potting Compound

Prepare the potting compound as follows:

(1) Select the desired mix for the repair. A stiff mix is desired when making overhead repairs or for core plug bonding. The average mix or thin mix is desirable when making upper surface repairs.

(2) Add micro-balloons to the resin and mix for 3 to 5 min.

(3) Add the curing agent to the resin and micro-balloon mixture. Mix for 3 to 5 min.

(4) Apply the potting compound to the edges of the core replacement and around the edges of the damaged area in the structure.

Insertion of Core Plug

Place the core plug in the repair area as follows:

(1) Insert the core plug into the repair area.

(2) Remove any excessive potting compound with a plastic scraper and clean the repair area thoroughly.

(3) Allow the core plug repair to cure for at least ½ to 1 hr. at room temperature (72° F) to assure that the core plug is firmly in place.

Application of Glass Cloth Laminates

The preparation for and application of the laminated sections of fiber glass cloth needed to complete the repair should be accomplished as follows:

(1) Remove surface coating from repair area.

(2) Wipe surface with clean cheesecloth moistened with MEK unil no trace of foreign material appears. Do not allow MEK to dry, but wipe it off with a clean cloth.

(3) Apply adhesive primer EC–776R (or equal) with a clean 1-in. varnish brush to faying surface area and allow to dry. Drying time is approximately 1 hr. at room temperature (72° F.). Drying may be speeded by the application of heat not to exceed 150° F. Primer should be applied in a thin and continuous film. Do not thin primer. Primer must be dry for proper adhesion of fiber glass cloth laminates.

(4) Prepare a clean work area, free of all foreign matter. This is generally accomplished by placing a clean piece of paper on a workbench.

(5) Select and cut two sheets of polyethylene film approximately 5 in. larger than the damage cutout.

(6) Prepare two disk templates of thin sheet metal to the correct size of the laminated overlays or 3 in. larger than the damage cutout.

(7) Prepare a batch mix of type 38 adhesive according to the procedures previously discussed in the text.

(8) Place one sheet of polyethylene film on a clean paper-covered work area. The corners of this sheet of polyethylene film may be taped to the work area.

(9) Pour a small amount of type 38 adhesive on this sheet of polyethylene film. With a plastic scraper, spread the adhesive evenly over the sheet of polyethylene film.

(10) Place and center one layer of the glass fabric cloth over the adhesive-covered area of the sheet of polyethylene film.

(11) Pour an adequate amount of type 38 adhesive over the first layer of glass fabric cloth to cover and penetrate its entire area. Spread the adhesive evenly over the area with a plastic scraper.

(12) Apply a second layer of glass fabric cloth in the same manner as the first layer.

(13) Apply a sufficient amount of type 38 adhesive over the second layer of glass fabric cloth in the same manner as over the first layer.

(14) Place and center the second sheet of polyethylene film over the layers of adhesive-impregnated glass fabric cloth.

(15) With a plastic scraper, work out all the air bubbles towards the edges of the laminated overlays. Turn the laminated overlay as necessary in working out the air bubbles.

(16) With a pair of scissors, cut the sandwiched polyethlyene-laminated overlay ½ in. to ¾ in. larger than the actual size of the laminated overlay.

(17) Place and center the sandwiched polyethylene-laminated overlay between the two disk templates that were previously prepared for the repair.

(18) With a pair of scissors, carefully cut the laminated overlay around the edge of the disk templates.

(19) Remove the disk templates from the sandwiched polyethylene-laminated overlay.

(20) Peel off one sheet of polyethylene film from the sandwiched laminated overlay. Discard the polyethylene film.

(21) Place the laminated overlay with the exposed adhesive side down and in position over the repair surface area.

(22) Remove the remaining sheet of polyethylene film from the top side of the laminated overlay. Discard the polyethylene film.

(23) Cut another sheet of polyethylene film ¾ in. to 1 in. larger than the laminated overlay.

(24) Place and center this sheet of polyethylene film over the laminated overlay positioned over the repair surface area.

(25) With a smooth plastic scraper, sweep out any excess resin or air bubbles that may be present within the laminated overlay. This step is of utmost importance to the overall quality of the repair. Therefore, this step should be accomplished with the greatest of care and patience.

(26) Clean the surrounding area of the repair

with MEK cleaning solvent. Take care to prevent any of the cleaning solvent from entering the bond of the repair area.

(27) Allow the laminated overlay repair to cure for at least 12 hrs. at room temperature (72° F.) before the final sheet of polyethylene film is removed.

Transition Area Repair

Some bonded honeycomb panels are constructed of a doubler separating an upper and lower skin and sectioned into bays of honeycomb core material. (See figure 5-93.) The edge section of a bay

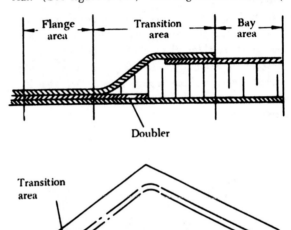

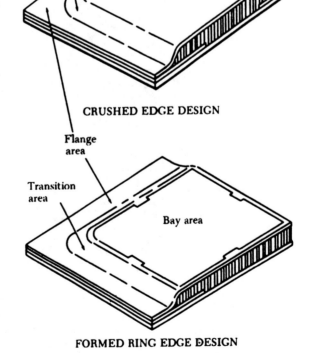

FIGURE 5-93. Typical bonded honeycomb panel bay construction.

211

area in which the honeycomb structure joins the laminated area of the panel or section is known as the transition area. Effective repairs to the transition areas are particularly essential because of the local transferring of the stresses.

The preparation of the repair materials and the assembly and curing of the core plug are basically the same as for the bonded honeycomb skin and core repairs. However, because of the shape and contour of a transition area, especially at the corners of a bay, give special attention to the cutting and shaping of the glass fabric honeycomb core material.

In this repair, four layers of impregnated glass cloth, number 181, are preferred for the overlays. The preference for glass cloth number 181 is because of its flexibility and ease of application, particularly when making repairs to a corner of a bay where a compound (double) contour is encountered.

Repair Procedures

The steps to be followed in the repair of a transition area are as follows:

(1) Outline the repair to a circular shape (not to exceed 2 in. in diameter) that will encircle the damaged area.

(2) Using a router, remove the damaged area down to the opposite adhesive layer. Depth of the router cut is determined by gradually increasing the depth of the cut until the adhesive layer is reached.

(3) Fabricate a fiber glass honeycomb core to replace the damaged core section. The correct thickness and contour of the transition area may be obtained by hand sawing and/or sanding. The core plug must be shaped to fit flush or within ±0.010 in.

(4) Prepare the potting compound.

(5) Butter the edges of the fiber glass honey-

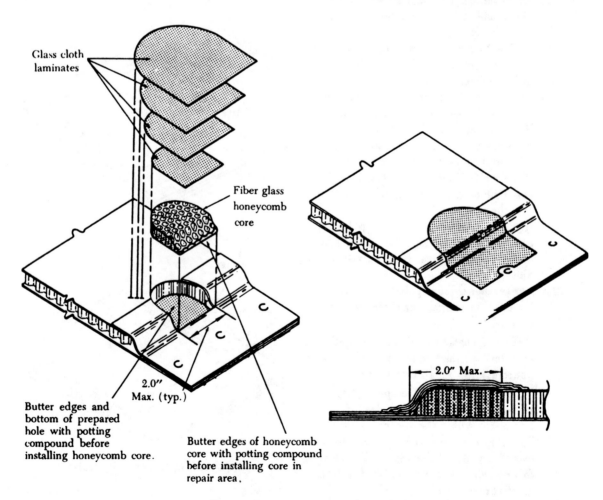

Glass cloth laminates

Fiber glass honeycomb core

2.0″ Max. (typ.)

Butter edges and bottom of prepared hole with potting compound before installing honeycomb core.

Butter edges of honeycomb core with potting compound before installing core in repair area.

2.0″ Max.

FIGURE 5–94. Transition area repair.

212

comb core plug with potting compound. (See figure 5–94.)

(6) Install the buttered core plug in the repair area.

(7) Prepare glass fabric adhesive.

(8) Prepare four glass fabric laminates—the first, sufficient in size to adequately cover the damaged area with no overlap; the others, 0.25 in. on all sides. The laminates should not extend over fasteners. If the laminates do extend over fasteners or fastener holes, cut around locations as shown in figure 5–94.

(9) Apply the four glass cloth laminates to the repair area as shown in the section view of the repair illustrated in figure 5–94.

(10) Allow the repair to cure properly.

(11) Apply the necessary erosion and corrosion preventives as specified.

PLASTICS

According to their chemical sources, plastics may be classified into four general groups: (1) Natural resins, (2) synthetic resins, (3) protein plastics, and (4) cellulose plastics.

Natural resins include such materials as shellac, pitch, amber, asphalt, and resin. These materials require fillers when molded.

Synthetic resins are made from petroleum, glycerol, indene, calcium-cyanamide, benzene, urea, ethylene, phenol, and formaldehyde. Products made from synthetic resins include acrylic plastics, nylon, vinyl, styrene, polyethylene, urea-formaldehyde, and others.

Protein plastics are manufactured from a variety of agricultural products. Sources included are peanuts, cashews, milk, coffee beans, and soy beans.

The cellulose plastics are the oldest of the group and include celluloid. Other plastics which fall into this class are acetate, nitrate, ethyl cellulose, butyrate, and propionate.

Nearly all of the early-day plastics were molded. However, today a large percentage of the plastics we know and use are cast, machined, rolled, laminated, or formed by other methods.

TRANSPARENT PLASTICS

Two types of transparent plastics used in aircraft windows, canopies, and similar transparent enclosures are thermoplastic and thermosetting materials.

Thermoplastic materials are originally hard but become soft and pliable when exposed to heat.

When pliable, the plastic can be molded; and, as it cools, it will retain the molded shape. When heated again and allowed to cool without being restrained, the plastic will return to its original shape. This process can be repeated many times without damage to the material unless the specified heat ranges are exceeded.

Thermosetting plastics are molded and allowed to cool and set in the desired shape. No amount of reheating will cause them to become pliable and workable. Once formed, they retain that shape and cannot be re-molded or re-shaped.

Each of these types of transparent plastics is available in monolithic or laminated forms. Monolithic plastic sheets are made in single solid uniform sheets. Laminated plastic sheets are made from transparent plastic face sheets bonded by an inner layer of material, usually polyvinyl butyral.

Optical Considerations

Optical qualities of the transparent material used in aircraft enclosures must be as good as those of the best quality plate glass. The ability to locate other aircraft in flight, positive depth perception necessary to land safely, all require a medium which can readily be molded into streamlined shapes and yet remain free from distortion of any kind. Such a medium must also be simple to maintain and repair.

In addition to their ease of fabrication and maintenance, plastics have other characteristics which make them better than glass for use in transparent enclosures. Plastics break in large dull-edged pieces; they have low water absorption and they do not readily fatigue-crack from vibration. But on the other hand, although they are nonconducters of electricity, they become highly electrostatic when polished.

Plastics do not possess the surface hardness of glass, so they are more easily scratched. Since scratches will impair vision, care must be used in servicing an aircraft. Specific procedures to avoid damaging the transparent plastic parts are discussed elsewhere in this chapter. Some general rules to follow are:

(1) Handle transparent plastic materials only with clean cotton gloves.

(2) Never use harmful liquids as cleaning agents, i.e., naphtha, gasoline, etc.

(3) Follow rigidly the applicable instructions for fabrication, repair, installation, and maintenance.

213

(4) Avoid operations which might scratch or distort the plastic surface. Be careful not to scratch the plastic with finger rings or other sharp objects.

Identification

The identity of transparent plastics used on aircraft can be determined by the MIL specification number on the part. Common MIL numbers and the type of material are as follows (fig. 5–95):

Specifications	Type Material	Edge Color
Thermoplastic		
MIL-P-6886	Regular Acrylic	Practically clear
MIL-P-5425	Heat Resistant	Practically clear
MIL-P-8184	Craze Resistant	Slightly yellow
Thermosetting		
MIL-P-8257	Polyester	Blue-green
Laminated		
MIL-P-7524	Laminated MIL-P–5425	Practically clear
MIL-P-25374	Laminated MIL-P–8184	Slightly yellow

Base	Name	Distinguishing Features
Acrylate	Plexiglas	Absence of color
	Lucite	Greater transparency
	Perspex (British)	Greater stiffness
Cellulose Acetate	Fibestos	Slightly yellow tint
	Lumarith	Greater flexibility
	Plastacele	Lower transparency
	Nixonite	Softer

FIGURE 5–95. Characteristics of plastics.

If the parts are not marked, the information in the following paragraphs will help to identify the material.

Transparent plastic enclosures and plate glass enclosures can be distinguished from each other by lightly tapping the material with a small blunt instrument. Plastic will resound with a dull or soft sound, whereas plate glass will resound with a metallic sound or ring.

Very few of the transparent plastics are color clear when viewed from the edge; some are practically clear, while others have a slight yellowish tint, or a bluish or blue-green tint.

The cellulose acetate plastics have a yellowish tint when viewed from the edge, and they are softer than the acrylic plastics.

Both acrylic and cellulose acetate base plastics have characteristic odors, especially when heated or burned. Burning a small sample and comparing its odor to that of a known sample is a very reliable method of identification. The acrylic odor is fairly pleasant, but acetate is very repugnant. Acrylic plastic burns with a steady, clear flame, whereas acetate burns with a sputtering flame and dark smoke.

These plastics can also be identified by the application of acetone and zinc chloride. Rub an area of the plastic with a solution of acetone, where it will not interfere with vision. Then blow on the area. If the plastic is acrylic, it will turn white; if it is acetate, it will soften but will not change color. A drop of zinc chloride placed on acetate base plastic will turn the plastic milky, but will have no effect on acrylic plastic.

STORAGE AND PROTECTION

Transparent plastics will soften and deform when heated sufficiently. Therefore, storage areas having high temperatures must be avoided. Plastic sheets should be kept away from heating coils, radiators, or hot water or steam pipes. Storage should be in a cool, dry location away from solvent fumes (such as may exist near paint spray and paint storage areas). Paper-masked transparent plastic sheets should be kept indoors. Direct rays of the sun will accelerate deterioration of the masking paper adhesive, causing it to cling to the plastic so that removal is difficult.

Plastic sheets should be stored, with the masking paper in place, in bins which are tilted at approximately a 10° angle from the vertical to prevent buckling. If it is necessary to store sheets horizontally, care should be taken to prevent chips and dirt from getting between the sheets. Stacks should not be over 18 in. high, and the smaller sheets should be stacked on the larger ones to avoid unsupported overhang.

Masking paper should be left on the plastic sheet as long as possible. Care should be used to avoid scratches and gouges which may be caused by sliding sheets against one another or across rough or dirty tables.

Formed sections should be stored so that they are amply supported and there is no tendency for them to lose their shape. Vertical nesting should be avoided. Protect formed parts from temperatures higher than 49° C. (120° F.). Protection from scratches can be provided by applying a protective coating, *i.e.*, masking paper, untreated builders' paper, posterboard, or similar material.

If masking paper adhesive deteriorates through long or improper storage, making removal of the paper difficult, moisten the paper with aliphatic naphtha. This will loosen the adhesive. Sheets so treated should be washed immediately with clear water.

Aliphatic naphtha is highly volatile and flammable. Use extreme care when applying this solvent.

Do not use gasoline, alcohol, kerosene, benzene, xylene, ketones (including acetone, carbon tetrachloride, fire extinguisher, or deicing fluids), lacquer thinners, aromatic hydrocarbons, ethers, glass cleaning compounds, or other unapproved solvents on transparent acrylic plastics to remove masking paper or other foreign material, because they will soften or craze the plastic surface.

When it is necessary to remove masking paper from the plastic sheet during fabrication, the surface should be re-masked as soon as possible. Either replace the original paper on relatively flat parts, or apply a protective coating on curved parts.

Certain protective spray coatings are available for formed parts. The thickness of the coating should be a minimum of 0.009 in. A layer of cheesecloth should be embedded in the coating at the time of application to assist in the removal of the masking spray. Coatings which remain on formed parts longer than 12 to 18 months become difficult to remove. Under no circumstances should transparent plastic, or formed parts coated with this material, be stored outdoors where it will be subject to direct sunlight for longer than 4 months.

To remove spray masking from the plastic, peel it off or lift a corner of the film and flow a jet of compressed air under it. If the film is too thin to be removed as a continuous film, apply a fresh coating of the compound, reinforced with a layer of cheesecloth, to obtain a thicker film. Allow to dry. Soaking the coated part, using a clean cloth saturated with water at room temperature, will help soften the film so that it can then be peeled off by hand. In no case should a solvent be used.

Extreme care must be used to avoid scratching the surface of the plastic. Tools must never be used in removing the film because of the danger of scratching the plastic.

FORMING PLASTICS

Transparent plastics become soft and pliable when heated to their respective forming temperatures. They can then be formed to almost any shape; and, on cooling, the material retains the shape to which it was formed, except for a small contraction. It is not desirable to cold form compound curvature transparent plastics (that is, to spring them into a curved frame without heating).

Transparent plastics may be cold bent (single curvature) if the material is thin and the radius of curvature to which it is cold bent is at least 180 times the thickness of the sheet. For example, an 18 in. length of transparent plastic, 0.250 in. thick, should not be deflected more than ¾ inch. Cold bending beyond these limits may eventually result in tiny fissures, called crazing, appearing on the surface of the plastic because of stresses being imposed beyond those recommended for continuous loading. For hot forming, transparent plastics should be maintained at the proper temperature recommended by the manufacturer.

Fabricating Processes

The fabrication of transparent plastics can be compared generally to that of wood or soft metal. Good craftsmanship, suitable equipment, and proper design are no less essential to the successful fabrication of transparent plastics than to that of other materials worked by similar methods. Light to medium woodworking equipment with minor modifications is satisfactory, but heavy-duty machines which are less apt to vibrate are better.

Where extreme accuracy is not required, the work can be laid out by penciling the cutting lines directly on the masking paper. For close tolerances, however, it is advisable to scribe layout lines directly on the surface of the plastic. Use straightedges or layout templates according to the requirements of the job. If the masking paper is removed before scribing, it should be replaced to within about ¼ in. of the scribed markings before the piece is cut.

Layout templates may be of plastic sheeting to which suitable handles can be cemented. Sharp edges or rough spots in such templates should be carefully rounded or smoothed. In the case of metal templates, it is good practice to cement thin flannel over the contact surfaces.

Cutting

Scribing and edge sanding is the cutting method most generally used on flat sections or two-dimensional curved pieces. The sheet is first cut to approximate shape on a band saw, using a scribed line as a guide and cutting approximately 1/16 in. oversize. Use disk sanders when removing material from straightedges and outside curves. Use drum or belt sanders for inside curved edges. When sanding irregular shapes or larger pieces which are awkward to manipulate around a fixed machine, use an air-driven sander or small electric hand sander.

Drilling

For the sake of both accuracy and safety, hold work in suitably designed clamps or fixtures. The twist drills commonly used for soft metals can be used successfully for transparent plastics if ordinary care is observed. However, the best results can be obtained if drills are re-pointed with the following in mind:

(1) The drill should be carefully ground free of nicks and burrs which would affect surface finish.

(2) It is particularly important that the cutting edge be dubbed off to zero rake angle.

(3) The length of the cutting edge (and hence the width of the lip) can be reduced by increasing the included angle of the drill. (See figure 5–96.)

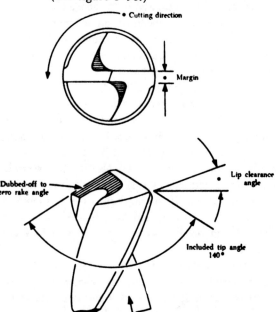

FIGURE 5–96. Drill for acrylic plastics.

Use drills with slow-spiral polished flutes. Flutes should be as wide as possible. The best lubricant and coolant for drilling plastics is a water-soluble cutting oil. For drilling shallow or medium depth holes, no coolant is needed. For deep holes, of course, a coolant is necessary.

Cleaner, more transparent deep holes can be produced by first drilling a pilot hole a little more than half the diameter of the final hole, filling this pilot hole with a wax stick, then redrilling to the final diameter. If the pilot hole is drilled all the way through, the Plexiglas must be backed with wood to close the hole and make the wax stick effective. The wax lubricates the cut and supports and expels the chips during drilling. In clear Plexiglas the resulting hole is cleaner, smoother, and more transparent than holes produced by other methods.

Large diameter holes can be cut with hollow-end mills, hole saws, fly cutters or trepanning tools. The cutters of the latter should be ground to zero rake angle and adequate back clearance, just as lathe tools are ground. All these tools can be used in the standard vertical spindle drill press or in flexible shaft or portable hand drills.

In general, the speed at which Plexiglas sheets can be drilled depends largely on the quality of the equipment used. Plexiglas can be drilled at the highest speed at which the drill will not "wobble" sufficiently to affect the finish of the hole. However, large diameter drills require slower rotative speeds for best results. Also, the Plexiglas should be backed with wood and the feed slowed as the drill point breaks through the underside of the sheet.

Whenever holes are drilled completely through Plexiglas, the standard twist drills should be modified to a 60° tip angle, the cutting edge to a zero rake angle, and the back lip clearance angle increased to 12–15°.

Drills specially modified for drilling Plexiglas are available from authorized distributors and dealers of Plexiglas.

For accuracy and safety, Plexiglas parts should be clamped or held rigidly during drilling.

SHALLOW HOLES—Hole depth/ hole diameter ratio of less than 1½ to 1, use slow spiral twist drills with wide flutes modified as for through drilling. Chip removal is no problem in drilling shallow holes and no coolant is needed.

MEDIUM DEEP HOLES—Hole depth/hole diameter ratio from 1½ to 1 up to 3 to 1.

Use slow spiral twist drills with polished flutes which should be as wide as possible to aid in removing a continuous ribbon of material. The opti-

216

mum included tip angle, between 60° and 140°, will depend on the size of the flute. Lip clearance angles should be ground to 12° to 15°. The feed of the drill should be controlled so that a continuous chip is cut and cleared without overheating the plastic at the tip of the drill. No coolant is needed for drilling holes up to 3 to 1 depth/diameter ratios although a jet of compressed air directed into the hole as it is being drilled is helpful. Drills with extra wide spirals and compressed air cooling can clear a continuous chip from holes with depth/diameter ratios up to 5 to 1.

DEEP HOLES—Hole depth/hole diameter ratio greater than 3 to 1.

Use slow spiral twist drills with wide polished flutes and an included tip angle of 140°. The wider tip angle results in a shorter cutting edge and narrow chip. The lip clearance angle should be ground to 12° to 15°. The feed should be slow—approximately 2½″ per minute—so that powder, rather than shavings or continuous chips, will be formed. A coolant is necessary for drilling deep holes to avoid scoring or burning the surface of the hole.

Compressed air can be used as a coolant for holes with depth diameter ratios up to 5 to 1. Water or a soluble oil-water coolant can also be used. When applied at the entry hole, however, a liquid coolant is actually pumped out of the hole by the drill and seldom reaches the drill point. A standard oil hole drill can be used to insure delivery of the coolant to the drill point. The collant can also be applied by filling a pilot hole drilled 95% of the way through the material or through a pilot hole drilled through from the opposite surface.

Cementing

With care and proper procedure, it is possible to obtain a cemented joint which approximates the original plastic in strength. Cementing of transparent acrylic plastics depends on the intermingling of the two surfaces of the joint so that actual cohesion exists. To effect cohesion, an organic liquid solvent is used to attack the plastic, forming a well-defined soft surface layer called a cushion, as shown in figure 5–97.

The most common method of cementing transparent plastics is the "soak method." This consists of dipping one of the two pieces to be cemented into the cement until a sufficient cushion is formed. When this surface is pressed against the opposite dry surface, the excess cement forms a second cushion—shallow, but enough to permit thorough intermingling of the two surfaces, as shown in figure 5–97.

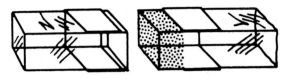
Before contact

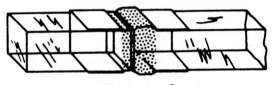
Contact only

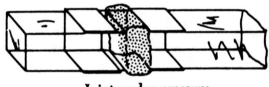
Joint under pressure

Joint "drying." Cushions harden.

FIGURE 5–97. Cementing with solvent cement.

Sometimes, for convenience in handling, clear transparent plastic shavings of the same type as the transparent plastic being cemented are dissolved in the cement to give it a thick, syrupy consistency so that it can be applied like glue. This viscous cement, however, works on exactly the same principle as a soak cement; for example, the excess solvent softens and swells both surfaces, permitting an intermingling of the cushions and the formation of a cohesive bond as shown in figure 5–98.

A solvent joint never dries completely; that is, it will never become entirely free of solvent. If the temperature is raised, the cushion will enlarge slowly until a new equilibrium is reached, as shown in figure 5–99.

Viscous cement

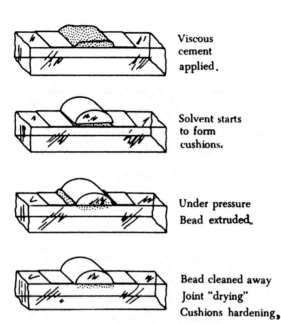

Viscous cement applied.

Solvent starts to form cushions.

Under pressure Bead extruded.

Bead cleaned away Joint "drying" Cushions hardening,

FIGURE 5–98. Cementing with syrup cement.

Room temperature equilibrium

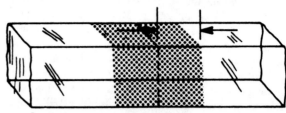

Joint dried at room temp., still contains solvent.

Equilibrium after heat treatment

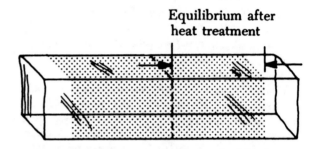

Heat treatment expands cushion, reduces concentration of solvent in joint.

FIGURE 5–99. Effect of heat treatment.

On cooling, the cushion will be larger and correspondingly harder since it contains less solvent per unit of volume. Heating a solvent joint long enough to expand its cushion, therefore, will produce a much stronger joint.

Cemented joints must be heat treated with caution. Heat first activates the solvent, which softens the cushion. The cushion then slowly expands as the solvent penetrates the material. In heat treating, it is important that the temperature does not approach the softening point of transparent plastics.

It is important that the joint be thoroughly hardened before machining, sanding, or polishing to remove the bead.

Care and Maintenance

Vision is so vital in aircraft that day-to-day maintenance of transparent enclosures is of the utmost importance. Proper maintenance methods should be carried out thoroughly whenever vision is impaired as a result of chemical or physical actions or defects, and every effort should be put forth to eliminate harmful action while servicing the aircraft.

The replacement of transparent plastic enclosures

has been necessitated by severe crazing, apparently caused by exposure to harmful solvents and improper maintenance handling. The crazing appears as a network of cracks running in all directions over the surface of the plastic. It can also occur within the plastic at or near cemented joints.

The use of improper cleaning fluids or compounds is one of the most common causes of these difficulties. The crazing action of a solvent is often delayed; that is, crazing may not appear for several weeks after the exposure to solvent or fumes. It is not always possible to determine immediately, by simple trial, whether a particular cleaner will be injurious or not. To minimize the damage, the precautions discussed in the following paragraphs should be observed.

Routine removal of film and other operational soils, where abrasive polishing for scratch removal is not required, can be accomplished by the use of aqueous detergent solutions. Two recommended solutions are: wetting agent, synthetic, nonionic, conforming to Military Specification MIL-D-16791; or wetting agent, alkyl aryl sulfonate, 40% active. These materials should be used in concentrations of 2 or 3 oz. per gal. of water. They should be applied with soft cloths or photographic cellulose sponges which have been used for no other purpose. Polish and cleaner conforming to Military Specification MIL–C–18767 will give satisfactory results for most cleaning requirements.

When cleaning exterior surfaces, always remove rings from the hands before washing the transparent plastic. The cleaning procedure is comprised of the following steps:

(1) Flush the plastic surface with plenty of water, using the bare hands to feel for and gently dislodge any dirt, sand, or mud.

(2) Wash with mild soap and water. Be sure the water is free of harmful abrasives. A soft cloth, sponge, or chamois may be used in washing, but only to carry the soapy water to the plastic. Go over the surface with bare hands to quickly detect and remove any remaining dirt before it scratches the plastic.

(3) Dry with a damp clean chamois, a clean soft cloth, or soft tissue. Do not continue rubbing the transparent plastic after it is dry. This not only scratches, but may build up an electrostatic charge which attracts dust particles. If the surface becomes charged, patting or gently blotting with a clean damp chamois will remove the charge as well as the dust.

(4) Never use a coarse or rough cloth for polishing. Cheesecloth is not acceptable.

The procedure for cleaning interior surfaces consists of three steps:

(1) Dust the plastic surface lightly with a clean soft cloth saturated with clean water. Do not use a dry cloth.

(2) Wipe carefully with a damp soft cloth or sponge. Keep the cloth or sponge free from grit by rinsing it frequently with clean water.

(3) Clean with an approved cleaner.

In hot weather the transparent enclosures of parked aircraft may absorb enough heat to soften and become distorted unless certain precautions are taken. Plastic enclosures installed on aircraft parked in the sun may receive heat directly from three sources.

Transparent plastic has a property of absorbing, selectively, the heat producing rays of the sun so that the platsic can become considerably hotter than the surrounding air inside or outside the aircraft.

Air inside an unshaded and unventilated aircraft will transfer the heat radiated by the metal members in the aircraft to the plastic enclosure by convection.

To prevent heat deformation of transparent plastic enclosures on aircraft parked exposed to the sun, the following precautions are recommended:

(1) If surrounding air temperature is below 100° F., no special precautions are necessary.

(2) If surrounding air temperature is between 100° and 120° F., enclosures should be opened sufficiently to permit free circulation of air through the aircraft and under the enclosure.

(3) If the surrounding air temperature is above 120° F., the enclosure must be opened and protected from the sun by a suitable cover which does not come into contact with the transparent plastic. If possible, the aircraft should be parked in the shade.

(4) To remove enclosure covers, lift them off; sliding may cause abrasion of the plastic surfaces.

Compounds for paint stripping, degreasing, and brightening, as well as most organic solvents, cause serious damage to transparent acrylic plastics. All such parts should be removed before starting paint stripping, and should not be replaced until the cleaning and painting is completed and the paint or lacquer is thoroughly dry, since paint and lacquer cause crazing of plastics. The plastic parts should be removed from the area where the stripping, degreasing, or painting is being done. The parts should be protected with soft cloth covers.

If it is impracticable to remove a plastic panel, cut a polyethylene sheet (minimum of 0.010-in. thick and containing no pinholes) to match as exactly as possible the size of the window. The polyethylene sheet should fit snugly over the surface of the plastic window, and the edges should be carefully taped with masking tape at least 2 in. wide to permit at least 1 in. of sealing width on both the plastic film and the aircraft. Make certain that no liquid or fumes can seep through to the window. It is important that the entire surface of the window be covered and that no cutting tools be used to remove the masking.

Aluminum foil is unsatisfactory as a protection from paint (and other sprays which contain solvents) because of its low resistance to tears, punctures, and pinholes. Protective coating conforming to Military Specification MIL-C-6799 is satisfactory as a protection from paint and other sprays which contain solvents.

Do not sand transparent plastics unless it is absolutely necessary. Hairline scratches of 0.001-in. maximum depth should be left as is, provided optical requirements are maintained.

INSTALLATION PROCEDURES

There are a number of methods for intalling transparent plastic panels in aircraft. The method the aircraft manufacturer utilizes depends on the position of the panel on the aircraft, the stresses to which it will be subjected, and a number of other factors. When installing a replacement panel, follow the same mounting method used by the aircraft manufacturer.

Where difficulty is encountered in rivet installation, bolts may be substituted when installing replacement panels, provided the manufacturer's original strength requirements are met and the bolts do not interfere with adjoining equipment.

In some instances replacement panels do not fit the installation exactly. Whenever adjustment of a replacement panel is necessary, the original design drawing, if available, should be consulted for proper clearances. The following principles should be considered in installing all replacement panels.

Fitting and handling should be done with masking material in place. Do not scribe plastic through masking material. On edges where transparent materials will be covered, or attached to, remove the masking material. When subject to large stresses, transparent plastics are apt to craze. It is of prime importance that plastics be mounted and installed so that such stresses are avoided.

Since transparent plastic is brittle at low temperatures, extra care must be taken to prevent cracking during maintenance operations. Transparent plastic parts should be installed at room temperature, if practicable.

Never force a transparent panel out of shape to make it fit a frame. If a replacement does not fit easily into the mounting, obtain a new replacement or sand the panel sufficiently to obtain the exact size that conforms with the mounting frame.

Do not heat and re-form areas of the panel, since local heating methods are likely to be only superficial and not thorough enough to reduce stress concentrations.

Since plastics expand and contract approximately three times as much as metal, suitable allowance for dimensional changes with temperature must be made. Use the values shown in figure 5–100 as minimum clearances between the frames and the plastics.

Dimension of Panel in Inches **	Dimensional Allowance in Inches *	
	Required for Expansion from 25°C (77°F) to 70°C (158°F)	Required for Contraction from 25°C (77°F) to −55°C (−67°F)
12	0.031	0.050
24	0.062	0.100
36	0.093	0.150
48	0.124	0.200
60	0.155	0.250
72	0.186	0.300

* Where the configuration of a curved part is such as to take up dimensional changes by change of contour, the allowances given may be reduced if it will not result in localized stress. Installations permitting linear change at both ends require half the listed clearances.

** For dimensions other than those given use proportional clearance.

FIGURE 5–100. Expansion and contraction allowances.

Bolt and Rivet Mountings

In bolt installations, spacers, collars, shoulders, or stop nuts should be used to prevent excessive tightening of the bolt. Whenever such devices are used by the aircraft manufacturer, they should be retained in the replacement installations.

To ensure long service, give special consideration to the following factors:

(1) Use as many bolts or rivets as practical.

(2) Distribute the total stresses as equally as possible along the bolts and rivets.

(3) Make sure the holes drilled in the plastic are sufficiently larger than the diameter of the bolt to permit expansion and contraction of the plastic relative to the frame.

(4) Make sure the holes in the plastic are concentric with the holes in the frame so that the greater relative expansion of the plastic will not cause binding at one edge of the hole. The hole should be smooth and free of any nicks or roughness.

(5) Use oversize tube spacers, shoulder bolts, rivets, cap nuts, or some other device to protect the plastic from direct pressure.

Synthetic Fiber Edge Attachment

Modern edge attachments to transparent plastic assemblies are made of synthetic fibers specially impregnated with plastic resins. The most commonly used fibers are glass, orlon, nylon, and dacron.

Reinforced laminated edge attachments are the preferred type, especially when mounting by bolts or rivets is necessary. The edges have the advantage of more efficiently distributing the load and reducing failures caused by differential thermal expansion.

Laminated edge attachments can be mounted by any of the foregoing methods, with any needed holes drilled through the edge attachment material and not the transparent plastic.

The most efficient method of mounting a laminated edge attachment is by the "slotted hole" method. The slotted holes are in the edge attachment and allow for differential thermal expansion.

Fabric loop attachments are sometimes attached to the plastic material with a cable or extrusion contained within the loop. A special extrusion is necessary to contain the loop and cable.

LAMINATED PLASTICS

Laminated plastic enclosures are made by bonding two layers of monolithic transparent sheets together with a soft plastic inner layer. They are installed in pressurized aircraft because of their superior shatter resistance and greater resistance to explosive decompression as compared to monolithic plastic enclosures.

Cellulose Acetate Base Plastics

In general, the methods used for fabrication, repair, and maintenance of cellulose acetate base plastics are similar to those used for acrylic plastics. In handling cellulose acetate base plastics, give attention to the following variations and additions to the recommendations already given for acrylic plastics.

Since the chemical composition of acetate base plastics differs greatly from that of acrylics, the cement used is of a different type. Generally, two types are used, solvent and dope.

Solvent type cement is generally used where transparency must be maintained in the joint. It is relatively quick drying and is well adapted for use in making emergency repairs. However, even though the cement is quick drying, the drying time will vary with the size of the joint and atmospheric conditions. Acetone may be used as a solvent type cement.

Dope type cement is preferred for use where the surfaces to be joined do not conform exactly. This cement softens the surfaces of a joint and, at the same time, creates a layer between the two pieces being cemented. However, it does not give a transparent joint and is slower drying than the solvent cement. It will take from 12 to 24 hrs. for the joint to reach full strength.

Since the expansion and contraction rates of acetate-base plastics are greater than those of acrylics, make greater allowances when mounting them. These plastics are affected by moisture and will swell as they absorb water. In general, allow $\frac{1}{8}$ in. per foot of panel length for expansion, and $\frac{3}{16}$ in. per foot for contraction.

FIBER GLASS COMPONENTS

Because of the unequaled strength/weight ratio, the ability to pass radio and radar waves, the ease of manufacture in contoured shapes, immunity to mildew and weather resistant characteristics, and adaptability to numerous places and shapes, fiber glass is a versatile material with numerous uses in modern aircraft construction. A few of the many applications are radomes, radio antenna covers, and junction boxes.

Fiber glass is manufactured from specially processed glass balls. By a fabrication process, the glass is turned into fibers which may result in an end product of cloth, molded mats, or yarn. The yarn is used to manufacture molded parts. The fiber glass cloth is used in making laminated shapes or in the repair of laminated assemblies. Another use is in the repair of metal structures.

Mat Molded Parts

Nonstructural parts, such as junction boxes, heater ducts, relay shields, and other electrical applications, are manufactured from mat-molded fiber glass. Molded mat fiber glass is short chopped fibers molded in a mat form. The assemblies are fabricated by a process wherein the chopped fibers are molded around a form, bonded together by use of a resin, and cured while under heat and pressure.

Carelessness in removing or handling mat-molded parts can cause the assembly to become damaged. Vibration may be another factor in the cause of cracks in the assemblies. Damage to mat-molded parts usually consists of holes or cracks (figure 5–101). Similar repair procedures are used for either type of damage.

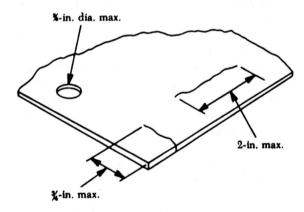

FIGURE 5–101. Typical damage to mat molded parts.

Repair Procedures

The following procedures are typical of those used in the repair of a mat-molded assembly. However, they are not to be construed as the only procedures that could be used. The correct section of the structural repair manual for the specific aircraft should be consulted and followed in all situations.

(1) Inspect the part for location of the crack.

(2) Remove the paint or protective coating from around the damaged area.

(3) Stop-drill the end of the crack. The size of the twist drill should be not smaller than $\frac{1}{8}$ in. and not larger than $\frac{3}{16}$ in.

(4) Lay out and sand the damaged area to the dimensions given in figure 5–102. Remove one-third of the material from both sides of the damaged area. Bevel the area 15° to 45°, as shown in figure 5–102, and sand $\frac{1}{2}$ in. beyond the beveled area.

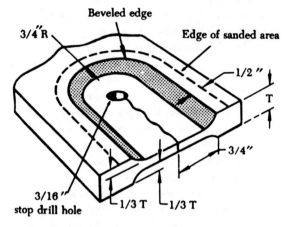

FIGURE 5–102. Mat molded repair.

(5) Prepare two pieces of PVA (polyvinyl alcohol) film large enough to cover the repair area.

(6) Prepare two pieces of metal large enough to cover the area. Use any piece of metal that will provide satisfactory holding strength.

(7) Check and start the air-circulating furnace. Set the temperature regulator at 220° F.

(8) Select and prepare the resin mixture.

(9) Cut the mat fiber glass material and saturate it in the prepared resin. Cut enough pieces of material to build up the beveled out area to its original contour.

(10) Insert the saturated mat fiber glass material into the repair area. (See figure 5–103.)

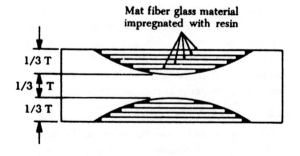

FIGURE 5–103. Insertion of precut saturated sections.

(a) Do one side at a time.

(b) Cover each side with the precut PVA film.

(c) Place the prepared metal plate on each side of the repair as it is completed.

(d) Secure the repair in place by use of C-clamps.

(11) Place the assembly in the preheated oven for at least 1 hr. (check applicable or manufacturer's instructions for the resin used).

(12) Remove the assembly from the oven and let it cool at room temperature.

(13) Disassemble the repair, removing C-clamps, metal plates, and PVA film.

(14) Sand both sides to a smooth finish and to the original contour of the part.

(15) Inspect the repair for soundness, using a metallic ring test. A good repair, when struck with a coin or light aluminum hammer, should resound with a metallic ring.

RADOMES

The protective dome or domelike covering for a radar antenna or other radar equipment is called a radome. It must be able to withstand the effects of hail, icing, wind, temperature changes, static electricity, supersonic speeds, and stratospheric altitudes. Also, it must have excellent dielectric qualities.

Handling, Installation, and Storage

Caution should always be used when handling, installing, or storing aircraft radomes. The necessity for the utmost care to prevent damage to sandwich parts cannot be overemphasized. Radomes (radar and radio antenna housings) are especially susceptible to damage. Damage is sometimes minute and invisible, but when exposed to vibration, stress, or liquids (water or oil), deterioration follows. Microwave distortion and energy losses occur as a result of cracks, punctures, and other physical damage, including moisture and oil contamination.

Take care also to avoid contamination with paint removers and stripping compounds normally used on the metal parts of the aircraft for removing finishes. Some of these materials have been found to penetrate the plastic facings of the radome and may have an adverse effect on its electric properties or its strength. Mild soap and water are used for general cleaning of radome surfaces. When a solvent cleaner is required for removing oils and greases on radome surfaces, use a clean cloth dampened with MEK.

Radomes must be handled with special care. Placing radomes upon rough surfaces and among metal parts must be avoided. Caution should always be used to avoid radome damage resulting from the radome striking against work stands, being dropped, or being drug across rough surfaces.

Correct radome installation begins with the uncrating procedure. Before uncrating a radome, provide a clean padded surface at least as large as that which the radome will occupy when uncrated. Adhere closely to the instructions for opening the radome crate. This will prevent damages that would be inflicted to the radome by protruding nails, bolts, staples, or other sharp objects. Installation instructions outlined in the applicable aircraft maintenance manual must be followed closely when installing radomes. Should sanding or grinding of the radome be required to fit a mounting frame, the sanded surfaces should be classified as a class I repair (discussed later in this chapter) and reworked accordingly.

Radomes should be stored where they are not subject to high humidity. They should be stored in suitable crates or padded racks and supported from the mounting holes. Avoid stacking radomes directly upon each other.

Detection and Removal of Oil and Moisture

All radomes are susceptible to moisture and oil contamination. Either can be the cause of very serious degradation of the performance of the aircraft's radar system. Contamination also causes weakening of the radome facing and the facing-to-core bonding strength.

Radomes should be inspected for moisture or oil contaimination prior to repairing or identifying as serviceable; they must be clean and dry prior to electrical testing. Radomes can be checked for moisture pockets using an electronic moisture meter. The probe unit of the meter should be held in contact with the inner surface of the radome and slowly moved over the surface. The presence of moisture will be indicated on the calibrated meter dial. Moisture detection and removal procedures should be accomplished on all radomes before performing repairs.

Inspection For Damage

Radomes should be visually inspected for delaminations, scars, scratches, or erosion of the protective coating that would affect only the outer ply. They should be inspected also for punctures, contamination, fractures of plies affecting either the plies on one side, the plies and core material, or damage extending completely through the outer plies, core material, and inner plies. Different aircraft have different limitations on damage that is reparable, the type of repairs allowed and on damage that is nonreparable. This information can usually be found in the maintenance manuals for the specific aircraft.

Damages to sandwich parts are divided into groups or classes according to the severity and possible effect upon the structure of the aircraft and upon electrical efficiency. Damages are classified in three basic classes: (1) Class I repairs—scars, scratches, or erosion affecting the outer ply only; (2) class II repairs—punctures, delaminations, contaminations, or fractures in one facing only, possibly accompanied by damage to the core; and (3) class III repairs—damage extending completely through the sandwich affecting both the facings and the core.

Radome Repairs

Repair procedures are developed with the objective of equaling as nearly as possible the electrical and strength properties of the original part with a minimum of increase in weight. This can be accomplished by repairing damaged parts with approved materials and working techniques. Therefore radome repairs should be accomplished in accordance with manufacturer's procedures by specially trained personnel of a shop which has proper facilities and adequate test equipment to ensure a satisfactory repair.

Testing of Repairs

Radomes must be repaired in a manner that will ensure not only the structural integrity of the radome, but the electrical characteristics as well. The type of electrical test required after a repair is completed depends on the purpose for which the radome was designed. Typical of the type of electrical tests conducted are:

(1) Transmissivity, which is the average one-way power transmission through the radome or the ratio of power transmitted through the radome to the same power transmitted with the radome removed.

(2) Incidence reflection, the power reflected into the radar system by the radome.

(3) Deflection or refraction to check for possible ghosts or false target returns.

WOODEN AIRCRAFT STRUCTURES

While the trend is undoubtedly toward all-metal aircraft, many airplanes still exist in which wood was used as the structural material. The inspection and repair of these wooden structures will continue to be the responsibility of the airframe mechanic. The ability to inspect wood structures and recognize defects such as dry rot, compression failures, etc., must be developed.

The information in this section is of a general nature and should not be regarded as a substitute for specific instructions contained in the aircraft manufacturer's maintenance and repair manuals. Methods of construction vary with different types of aircraft, as will the various repair and maintenance procedures.

INSPECTION OF WOODEN STRUCTURES

Whenever possible, the aircraft should be kept in a dry, well-ventilated hangar with all inspection covers, access panels, etc., removed for as long as possible prior to inspection. If the aircraft is thoroughly dried out, this will facilitate the inspection, especially when determining the condition of glued joints.

Before beginning a detailed inspection of the glued joints and the wood, a rough assessment of the general condition of the structure can sometimes be obtained from the external condition of the aircraft.

The wings, fuselage, and empennage should be checked for undulation, warping or any other departures from the original shape. In instances where the wings, fuselage, or empennage structures and skins form stressed structures (figure 5–104), no departure from the original contour or shape is permissible.

Where light structures using single plywood covering are concerned, some slight sectional undulation or bulging between panels may be permissible provided the wood and glue are sound. However, where such conditions exist, a careful check must be made of the attachment of the ply to its supporting structure. A typical example of a distorted single plywood structure is illustrated in figure 5–105.

The contours and alignment of leading and trailing edges are of particular importance, and a careful check should be made for departure from the original shape. Any distortion of these light ply-

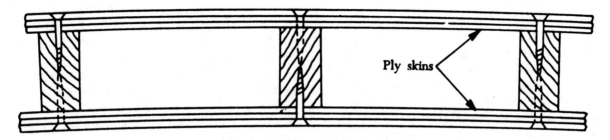

FIGURE 5–104. Cross sectional view of a stressed-skin structure.

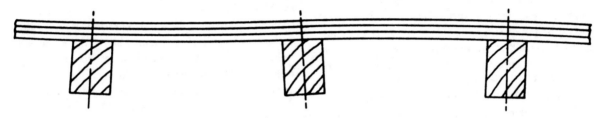

FIGURE 5–105. Single ply structure.

wood and spruce structures is indicative of deterioration, and careful internal inspection will have to be made for security of these parts to the main wing structure, and for general deterioration of the light plywood and spruce members. If deterioration is found in these components, the main wing structure may also be affected.

Splits in the fabric covering on plywood surfaces should not be repaired by doping on another piece of fabric over the affected area. In all cases, the defective fabric should be removed to ascertain whether the plywood skin beneath is serviceable, since it is common for a split in the plywood skin to be responsible for initiating a similar defect in the protective fabric covering.

Although a preliminary inspection of the external structure can be useful in assessing the general condition of the aircraft, it should be noted that wood and glue deterioration can often take place inside a structure without any external indications. Where moisture can enter a structure, it will seek the lowest point where it will stagnate and promote rapid deterioration. It should also be noted that glue deterioration can take place through other causes without the presence of water.

Glue failure and wood deterioration are often closely allied, and the inspection of glued joints must include an examination of the adjacent wood structure.

The inspection of a complete aircraft for glue or wood deterioration will necessitate checks on parts of the structure which may be known or suspected trouble spots and which are, in many instances, boxed in or otherwise inaccessible. In such instances, considerable dismantling is required, and it may be necessary to cut access holes in ply structures to facilitate the inspection. Such work must be done only in accordance with approved drawings or the repair manual for the aircraft concerned.

Glued Joint Inspection

The inspection of glued joints in aircraft structures presents considerable difficulties. Even where access to the joint exists, it is still difficult to positively assess the integrity of the joint. Keep this in mind when inspecting wooden structures.

Some of the more common factors which may cause glue deterioration are: (1) Chemical reactions of the glue caused by aging or moisture, extreme temperatures, or a combination of these factors, (2) mechanical forces caused mainly by wood shrinkage, and (3) development of fungus growths.

Aircraft exposed to large cyclic changes of temperature and humidity are especially prone to wood shrinkage which may lead to glue deterioration. The amount of movement of wooden members due to these changes varies with the size of each member, the rate of growth of the tree from which the wood was cut and the way in which the wood was converted. Thus, two major members in an aircraft structure, secured to each other by glue, are unlikely to have identical characteristics. Differential loads will, therefore, be transmitted across the glue film since the two members will not react in an identical manner relative to each other. This will impose stresses in the glued joint which can normally be accommodated when the aircraft is new and for some years afterwards. However, glue tends to deteriorate with age, and stresses at the glued joints may cause failure of the joints. This is true even when the aircraft is maintained under ideal conditions.

When checking a glue line (the edge of the glued joint) for condition, all protective coatings of paint should be removed by careful scraping. It is important to ensure that the wood is not damaged during the scraping operation. Scraping should cease immediately when the wood is revealed in its natural state and the glue line is clearly discernible.

The glue line is often inspected by the use of a magnifying glass. Where the glue line tends to part or where the presence of glue cannot be detected or is suspect, the glue line should be probed with a thin feeler gage. If any penetration is possible, the joint should be regarded as defective. It is important to ensure that the surrounding wood is dry; otherwise, a false impression of the glue line would be obtained due to closing of the joint by swelling. In instances where pressure is exerted on a joint, either by the surrounding structure or by metal

attachment devices such as bolts or screws, a false impression of the glue condition could be obtained unless the joint is relieved of this pressure before the glue line inspection is carried out.

The choice of feeler gage thickness will vary with the type of structure, but a rough guide is that the thinnest possible gage should be used. Figure 5–106 indicates the points where checks with a feeler gage should be made.

Wood Condition

Dry rot and wood decay are not usually difficult to detect. Dry rot is indicated by small patches of crumbling wood. A dark discoloration of the wood surface or gray stains running along the grain are indicative of water penetration. If such discoloration cannot be removed by light scraping, the part should be replaced. Local staining of the wood by the dye from a synthetic adhesive hardener can be disregarded.

In some instances where water penetration is suspected, the removal of a few screws from the area in question will reveal, by their degree of corrosion, the condition of the surrounding joint (figure 5–107).

The adhesive will cause slight corrosion of the screw following the original construction; therefore, the condition of the screw should be compared with that of a similar screw removed from another

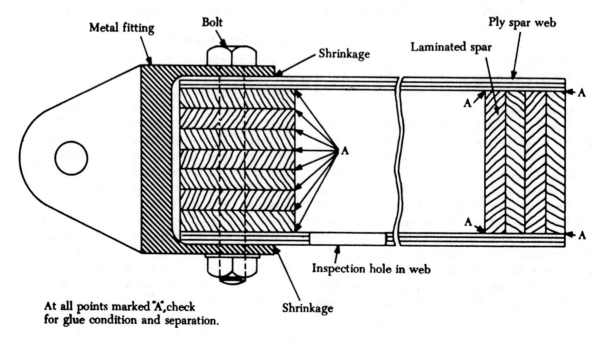

At all points marked "A", check for glue condition and separation.

FIGURE 5–106. Laminated joint.

226

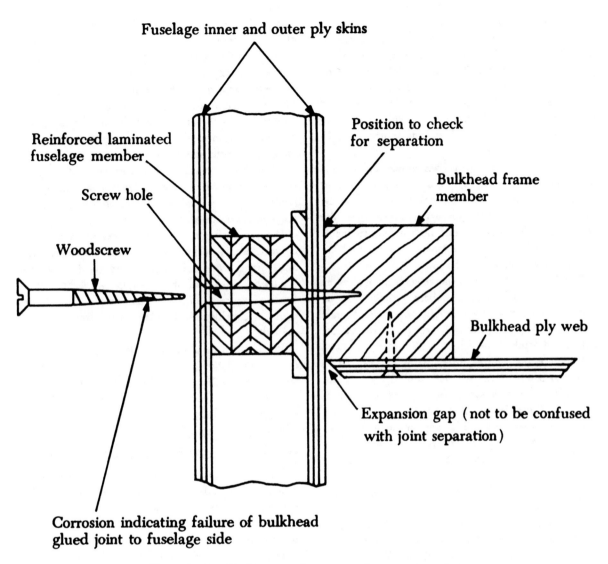

Fuselage inner and outer ply skins

Reinforced laminated fuselage member

Screw hole

Woodscrew

Position to check for separation

Bulkhead frame member

Bulkhead ply web

Expansion gap (not to be confused with joint separation)

Corrosion indicating failure of bulkhead glued joint to fuselage side

FIGURE 5–107. Checking a wooden structure for water penetration.

part of the structure known to be free from water soakage.

Plain brass screws are normally used for reinforcing glued wooden members, although zinc-coated brass is sometimes used. For hardwoods such as mahogany or ash, steel screws are sometimes used. Unless otherwise specified by the aircraft manufacturer, it is usual to replace screws with new screws of identical length but one gage larger.

Another means of detecting water penetration is to remove the bolts holding the fittings at spar root-end joints, aileron hinge brackets, etc. (figure 5–107). Corrosion on the surface of such bolts and wood discoloration will provide a useful indication of water penetration.

Experience with a particular aircraft will indicate those portions of the structure most prone to water penetration and moisture entrapment, e.g., at windows or the bottom lower structure of entrance doors. However, this is not necessarily indicative of the condition of the complete aircraft.

The condition of the fabric covering on ply surfaces is of great importance. If any doubt exists regarding its protective qualities or if there are any signs of poor adhesion, cracks, or other damage, it should be removed to reveal the ply skin.

The condition of the exposed ply surface should be examined. Water penetration will be shown by dark gray streaks along the grain and a dark discoloration at ply joints or screw holes. If these marks cannot be removed by light scraping or, in the case

227

of advanced deterioration, where there are small surface cracks or separation of the ply laminations, the ply should be replaced. Where evidence of water penetration is found, sufficient areas of the ply surface should be uncovered to determine its extent.

During the inspection, the structure should be examined for other defects of a mechanical nature. Information regarding such defects is presented in the following paragraphs.

Where bolts secure fittings which take load-carrying members, or where the bolts are subject to landing or shear loads, the bolt holes should be examined for elongation or surface crushing of the wood fibers. The bolts should be removed to facilitate the inspection. It is important to ensure that the bolts are a good fit in the holes.

Check for evidence of damage such as bruises or crushing of structural members which can be caused, for example, by overtightening bolts. Repair techniques for such damage are governed by the extent and depth of the defect.

Compression failures, often wrongly referred to as compression "shakes," are caused by rupture across the wood fibers. This is a serious defect which, at times, is difficult to detect. Special care is necessary when inspecting any wooden member which has been subjected to abnormal bending or compression loads during a hard landing. In the case of a member having been subjected to an excessive bending load, the failure will appear on the surface which has been compressed. The surface subjected to tension will normally show no defects. In the case of a member taking an excessive direct compression load, the failure will usually be apparent on all surfaces.

If a compression failure is suspected, a flashlight shone along the member, with the beam of light running parallel to the grain, will assist in revealing this type of failure.

A glued joint may fail in service as a result of an accident or because of excessive mechanical loads having been imposed upon it, either in tension or in shear. It is often difficult to decide the nature of the load which caused the failure, but remember that glued joints are generally designed to take shear loads.

If a glued joint is known to have failed in tension, it is difficult to assess the quality of the joint, since these joints may often show an apparent lack of adhesion. Tension failures often appear to strip the glue from one surface leaving the bare wood. In such cases, the glue should be examined with a magnifying glass, which should reveal a fine layer of wood fibers on the glued surface. The presence of fibers indicates that the joint itself is not at fault. If examination of the glue under magnification does not reveal any wood fibers but shows an imprint of the wood grain, this is caused by pre-drying of the glue before applying pressure during the manufacture of the joints. If the glue exhibits an irregular appearance with star-shaped patterns, this is an indication that pre-curing has occurred before pressure was applied or that pressure has been incorrectly applied or maintained. In all such instances other joints in the aircraft should be suspect.

If a joint is expected to take tension loads, it will be secured by a number of bolts or screws (or both) in the area of tension loading. If a failure occurs in this area, it is usually very difficult to form an opinion of the actual reasons for it, because of considerable breakup of the wood close to the bolts.

In all cases of glued joint failure, whatever the direction of loading, there should be a fine layer of wood fibers adhering to the glue, whether or not the glue has come away completely from one section of the wood member. If there is no evidence of fiber adhesion, this may indicate glue deterioration.

SERVICE AND REPAIR OF WOODEN STRUCTURES

Damage to wooden structures such as wing ribs, spars, and skin frequently requires repair. Whenever major wood parts have been damaged, a detailed inspection must be made. Secondary cracks sometimes start some distance away from the main damage and proceed in unrelated directions.

The purpose of repairing all wooden structural parts is to obtain a structure as strong as the original. Severe damage will require replacement of the entire damaged assembly, but minor damage can be repaired by cutting away the damaged members and replacing them with new sections. This replacement is accomplished by glued, or sometimes glued and nailed, or glued and screw-reinforced splicing.

Materials

Several forms of wood are commonly used in aircraft. Solid wood or the adjective "solid" used with such nouns as beam or spar refers to a member consisting of one piece of wood.

Laminated wood is an assembly of two or more layers of wood which have been glued together with the grain of all layers or laminations approximately parallel. Plywood is an assembled product of wood and glue that is usually made of an odd number of thin plies (veneers) with the grain of each layer at an angle of 90° with the adjacent ply or plies. High-density material includes compreg, impreg, or similar commercial products, heat stabilized wood, or any of the hardwood plywoods commonly used as bearing or reinforcement plates. The woods listed in figure 5–108 are those used for structural purposes. For interior trim, any of the decorative woods such as maple or walnut can be used since strength is of little consideration in this situation.

Species of wood	Strength properties as compared to spruce	Maximum permissible grain deviation (slope of grain)	Remarks
Spruce	100%	1:15	Excellent for all uses. Considered as standard for this table.
Douglas Fir	Exceeds spruce	1:15	May be used as substitute for spruce in same sizes or in slightly reduced sizes providing reductions are substantiated. Difficult to work with hand tools. Some tendency to split and splinter during fabrication. Large solid pieces should be avoided due to inspection difficulties. Gluing satisfactory.
Noble Fir	Slightly exceeds spruce except 8 percent deficient in shear.	1:15	Satisfactory characteristics with respect to workability, warping, and splitting. May be used as direct substitute for spruce in same sizes providing shear does not become critical. Hardness somewhat less than spruce. Gluing satisfactory.
Western Hemlock	Slightly exceeds spruce	1:15	Less uniform in texture than spruce. May be used as direct substitute for spruce. Gluing satisfactory.
Pine, Northern White	Properties between 85 percent and 96 percent those of spruce.	1:15	Excellent working qualities and uniform in properties but somewhat low in hardness and shock-resisting capacity. Cannot be used as substitute for spruce without increase in sizes to compensate for lesser strength. Gluing satisfactory.
White Cedar, Port Orford	Exceeds spruce	1:15	May be used as substitute for spruce in same sizes or in slightly reduced sizes providing reductions are substantiated. Easy to work with hand tools. Gluing difficult but satisfactory joints can be obtained if suitable precautions are taken.
Poplar, Yellow	Slightly less than spruce except in compression (crushing) and shear.	1:15	Excellent working qualities. Should not be used as a direct substitute for spruce without carefully accounting for slightly reduced strength properties. Somewhat low in shock-resisting capacity. Gluing satisfactory.

FIGURE 5–108. Woods for aircraft use.

All wood and plywood used in the repair of aircraft structures must be of aircraft quality. The species used to repair a part should be the same as that of the original whenever possible. If it is necessary to substitute a different species, follow the recommendations of the aircraft manufacturer in selecting a different species.

Defects Permitted.

a. Cross grain. Spiral grain, diagonal grain, or a combination of the two is acceptable providing the grain does not diverge from the longitudinal axis of the material more than specified in column 3 of figure 108. A check of all four faces of the board is necessary to determine the amount of divergence. The direction of free-flowing ink will frequently assist in determining grain direction.

b. Wavy, curly, and interlocked grain. Acceptable, if local irregularities do not exceed limitations specified for spiral and diagonal grain.

c. Hard knots. Sound hard knots up to ⅜-inch in maximum diameter are acceptable providing: (1) they are not in projecting portions of I-beams, along the edges of rectangular or beveled unrouted beams, or along the edges of flanges of box beams (except in lowly stressed portions); (2) they do not cause grain divergence at the edges of the board or in the flanges of a beam more than specified in column 3; and (3) they are in the center third of the beam and are not closer than 20 inches to another knot or other defect (pertains to ⅜-inch knots—smaller knots may be proportionately closer). Knots greater than ¼-inch must be used with caution.

d. Pin knot clusters. Small clusters are acceptable providing they produce only a small effect on grain direction.

e. Pitch pockets. Acceptable, in center portion of a beam providing they are at least 14 inches apart when they lie in the same growth ring and do not exceed 1½ inch length by ⅛-inch width by ⅛-inch depth and providing they are not along the projecting portions of I-beams, along the edges of rectangular or beveled unrouted beams, or along the edges of the flanges of box beams.

f. Mineral streaks. Acceptable, providing careful inspection fails to reveal any decay.

Defects Not Permitted.

g. Cross grain. Not acceptable, unless within limitations noted in **a.** above.

h. Wavy, curly, and interlocked grain. Not acceptable, unless within limitations noted in **b.** above.

i. Hard knots. Not acceptable, unless within limitations noted in **c.** above.

j. Pin knot clusters. Not acceptable, if they produce large effect on grain direction.

k. Spike knots. These are knots running completely through the depth of a beam perpendicular to the annual rings and appear most frequently in quartersawed lumber. Reject wood containing this defect.

l. Pitch pockets. Not acceptable, unless within limitations noted in **e.** above.

m. Mineral streaks. Not acceptable, if accompanied by decay (see **f.**).

n. Checks, shakes, and splits. Checks are longitudinal cracks extending, in general, across the annual rings. Shakes are longitudinal cracks usually between two annual rings. Splits are longitudinal cracks induced by artificially induced stress. Reject wood containing these defects.

o. Compression wood. This defect is very detrimental to strength and is difficult to recognize readily. It is characterized by high specific gravity; has the appearance of an excessive growth of summer wood; and in most species shows but little contrast in color between spring wood and summer wood. In doubtful cases reject the material, or subject samples to a toughness machine test to establish the quality of the wood. Reject all material containing compression wood.

p. Compression failures. This defect is caused from the wood being overstressed in compression due to natural forces during the growth of the tree, felling trees on rough or irregular ground, or rough handling of logs or lumber. Compression failures are characterized by a buckling of the fibers that appear as streaks on the surface of the piece substantially at right angles to the grain, and vary from pronounced failures to very fine hairlines that require close inspection to detect. Reject wood containing obvious failures. In doubtful cases reject the wood, or make a further inspection in the form of microscopic examination or toughness test, the latter means being the more reliable.

q. Decay. Examine all stains and discolorations carefully to determine whether or not they are harmless, or in a stage of preliminary or advanced decay. All pieces must be free from all forms of decay.

GLUES

Glues used in aircraft repair fall into two general groups: (1) Casein and (2) resin glues. Any glue that meets the performance requirements of applicable U.S. Military Specifications or has previously been accepted by the FAA is satisfactory for use in certificated civil aircraft. In all cases, glues are to be used strictly in accordance with the glue manufacturer's recommendations.

Casein glues have been widely used in wood aircraft repair work. The forms, characteristics, and properties of water-resistant casein glues have remained substantially the same for many years, except for the addition of preservatives. Casein glues for use in aircraft should contain suitable preservatives, such as the chlorinated phenols and their sodium salts, to increase their resistance to organic deterioration under high-humidity exposures. Most casein glues are sold in powder form ready to be mixed with water at ordinary room temperatures.

Synthetic resin glues for wood are outstanding in that they retain their strength and durability under moist conditions and after exposure to water. The best known and most commonly used synthetic resin glues are the phenol-formaldehyde, resorcinol-formaldehyde, and urea-formaldehyde types. The resorcinol-formaldehyde type glue is recommended for wood aircraft applications. Materials, such as walnut shell flour, are often added by the glue manufacturer to the resin glues to give better working characteristics and joint-forming properties. The suitable curing temperatures for both urea-formaldehyde and resorcinol glues are from 70° F. up. At the 70° F. minimum temperature, it may take as long as 1 week for the glue line in a spar splice to cure to full strength. Thinner pieces of wood and/or higher curing temperatures shorten curing time considerably. The strength of a joint cannot be depended upon if assembled and cured at temperatures below 70° F.

For those not familiar with the terms used relating to synthetic resin adhesives and their application, a glossary follows.

(1) **Cold setting adhesive.** An adhesive which sets and hardens satisfactorily at ordinary room temperature, *i.e.,* 50° F. to 86° F. (10° C. to 32° C.) within a reasonable period.

(2) **Close contact adhesive.** A nongap-filling adhesive suitable for use only in those joints where the surfaces to be joined can be brought into close contact by means of adequate pressure, and where glue lines exceeding 0.005 in. can be avoided with certainty.

(3) **Closed assembly time.** The time elapsing between the assembly of the joints and the application of pressure.

(4) **Double spread.** The spread of adhesive equally divided between the two surfaces to be joined.

(5) **Gap-filling adhesive.** An adhesive suitable for use in those joints where the surfaces to be joined may or may not be in close or continuous contact, owing either to the impossibility of applying adequate pressure or to slight inaccuracies of machining. Unless otherwise stated by the manufacturer, gap-filling adhesives are not suitable for use where the glue line exceeds 0.050 in. in thickness.

(6) **Glue line.** The resultant layer of adhesive joining any two adjacent wood layers in the assembly.

(7) **Hardener.** A material used to promote the setting of the glue. It may be supplied separately in either liquid or powder form, or it may have been incorporated with the resin by the manufacturer. It is an essential part of the adhesive, the properties of which depend upon using the resin and hardener as directed.

(8) **Open assembly time.** The period of time between the application of the adhesive and the assembly of the joint components.

(9) **Single spread.** The spread of adhesive to one surface only.

(10) **Spread of adhesive.** The amount of adhesive applied to join two surfaces.

(11) **Synthetic resin.** A synthetic resin (phenolic) is derived from the reaction of a phenol with an aldehyde. A synthetic resin (amino plastic) is derived from the reaction of urea, thiourea, melamine, or allied compounds with formaldehyde.

(12) **Synthetic resin adhesive.** A composition substantially consisting of a synthetic resin, either the phenolic or amino type, but including any hardening agent or modifier which may have been added by the manufacturer or which must be added before use, according to manufacturer's instructions.

Synthetic resin adhesives are used extensively for joining wooden structures to avoid the localized stresses and strains which may be set up by the use of mechanical methods of attachment. The strength of such structures depends largely on the efficiency of the glued joints, and cannot be verified by means other than the destruction of the joints. Acceptance must be governed by adequate precautions throughout the gluing process and by the results obtained by representative test pieces.

Synthetic resin adhesives usually consist of two

separate parts, the resin and the hardener. The resin develops its adhesive properties only as a result of a chemical reaction between it and the hardener. With some adhesives, an inert filler is added to increase viscosity and to improve the gap-filling properties of the mixed adhesive.

Synthetic resins can be obtained either in liquid or powder form. In general, powder resins have the longest storage life, since they are less susceptible to deterioration from high ambient temperatures.

Powder resins must be mixed with water in accordance with the manufacturer's instructions before they can be used in conjunction with a hardener. To obtain satisfactory results, it is essential that they be properly mixed. Once mixed, the adhesive must not be diluted unless this is permitted by the manufacturer's instructions. In many instances, manufacturers specify a definite period of time which must elapse between the mixing and the application of the adhesive. During this period, the adhesive should be covered to prevent contamination. When resins are supplied in liquid form, they are ready for immediate use in conjunction with the hardener. Liquid resin must not be diluted unless this is permitted by the manufacturer's instructions.

When mixing the hardener with the resin, the proportions must be in accordance with the manufacturer's instructions. Hardeners should not be permitted to come into contact with the resin except when the adhesive is mixed prior to use.

GLUING

The surface to be joined must be clean, dry, and free from grease, oil, wax, paint, etc. It is important that the parts to be joined have approximately the same moisture content, since variations will cause stresses to be set up because of swelling or shrinkage which may lead to the failure of the joint.

The moisture content of wood can be determined by taking a sample of the wood to be glued, weighing it, and then drying it in an oven at a temperature of 100° C. to 105° C. Calculate the moisture content by using the formula:

$$\frac{W_1 - W_2}{W_2} \times 100$$

where,

$W_1 =$ the weight of the sample prior to drying.

$W_2 =$ the weight of the sample after drying.

Example

Substitution and solution of above formula:

$$\frac{2 - 1.5}{1.5} \times 100 = .33 \text{ or } 33\%.$$

The approximate moisture content can also be determined by using a moisture meter. When this instrument is used, its accuracy should be checked periodically.

The wood to be glued should be at room temperature. The surfaces to be joined should not be overheated since this affects the surface of the wood and reduces the efficiency of most synthetic resin adhesives.

Synthetic resin adhesives are very sensitive to variations in temperature. The usable (pot) life of the adhesive, proportion of hardener to use, and clamping time all depend largely on the temperature of the glue room at the time of gluing.

It is generally desirable to apply adhesive to both surfaces of the material. This applies particularly where the glue line is likely to be variable or when it is not possible to apply uniform pressure.

The adhesive can be applied by a brush, glue spreaders, or rubber rollers that are slightly grooved. The amount of adhesive required depends largely on the type of wood and the accuracy of machining. Dense wood requires less adhesive than soft or porous types. Adhesive should be applied generously to an end grain. Smooth, side-grained surfaces may be satisfactorily glued with a thin spread. The general rule is that the adhesive should completely cover the surfaces to be glued and remain tacky until pressure is applied to the joint.

Difficult gluing conditions may occur when a soft wood is to be glued to one much denser because the adhesive tends to flow into the more porous wood. In such instances, unless otherwise specified by the manufacturer of the adhesive, precoating and partial drying of the softer surface, before normal spreading, is suggested.

Care should be taken before applying the adhesive to ensure that the surfaces will make good contact and that the joint will be positioned correctly. The interval between the application of the adhesive and assembly of the joint should be kept as short as possible. Some adhesives contain solvents which should be allowed to evaporate before the joint is assembled. If this is not done, bubbles may be created and result in a weak joint. For adhesives of this type, the manufacturer will specify a time interval which should elapse before the joint is closed.

To ensure that the two surfaces bind properly, pressure must be applied to the joint. The strength of the joint will depend to a great extent upon how evenly the force is applied. The results of evenly and unevenly applied pressure are illustrated in figure 5–109.

Pressure is used to squeeze the glue out into a thin continuous film between the wood layers, to force air from the joint, to bring the wood surfaces into intimate contact with the glue, and to hold them in this position during the setting of the glue.

Pressure should be applied to the joint before the glue becomes too thick to flow and is accomplished by means of clamps, presses, or other mechanical devices.

Non-uniform gluing pressure commonly results in weak and strong areas in the same joint. The amount of pressure required to produce strong joints in aircraft assembly operations may vary from 125 to 150 p.s.i. for softwoods and 150 to 200 p.s.i. for hardwoods. Insufficient pressure to poorly machined wood surfaces usually results in thick glue lines, which indicates a weak joint, and should be carefully guarded against.

The methods used in applying pressure to joints in aircraft gluing operations range from the use of brads, nails, screws, or clamps to the use of hydraulic and electrical power presses. Hand-nailing is used rather extensively in the gluing of ribs and in the application of plywood skins to the wing, control surfaces, and fuselage frames.

On small joints, such as those found in wood ribs, the pressure is usually applied only by nailing the joint gussets in place after spreading the glue. Since small nails must be used to avoid splitting, the gussets should be comparatively large in area to compensate for the relative lack of pressure. At least four nails (cement-coated or galvanized and barbed) per sq. in. are to be used, and in no event

must nails be more than ¾ in. apart. Small brass screws may also be used advantageoulsy when the particular parts to be glued are relatively small and do not allow application of pressure by means of clamps.

Apply pressure using cabinet maker clamps, parallel clamps, or similar types. Use handspring clamps in conjunction with softwood only. Because of their limited pressure area, they should be applied with a pressure-distributing strip or block at least twice as thick as the member to be pressed.

High clamping pressures are neither essential nor desirable, provided good contact between the surfaces being joined is obtained. When pressure is applied, a small quantity of glue should be squeezed from the joint. This should be wiped off before it dries. The pressure must be maintained during the full setting time. This is important since the adhesive will not reunite if disturbed before it is fully set.

The setting time depends on the temperature at which the operation is carried out. An increase in temperature results in a decrease in the setting period. Conversely, a decrease in temperature causes an increase in the setting time.

Full joint strength and resistance to moisture will develop only after conditioning for at least 2 days. Again, this depends on the ambient temperature and the type of hardener used. Usually when repairs are made, the joint will be of sufficient strength after 1 day.

Testing Glued Joints

Frequent tests should be made to ensure that the joints are satisfactory. Whenever possible, tests should be carried out on pieces cut from the actual component. The test samples should be 1 in. wide and at least 2 in. long. The pieces should be joined with one member overlapping the other by ⅜ in. The glued test sample should be placed in a vice and the joint broken by exerting pressure on the overlapping member. The fractured glue faces should show at least 75% of the wood fibers evenly distributed over the fractured glue surfaces. A typical broken test piece is shown in figure 5–110.

Where repairs are to be made on old aircraft in which the wooden structure is joined with a casein cement, all traces of the casein cement must be removed from the joint, since this material is alkaline and is liable to affect the setting of a synthetic resin adhesive. Local staining of the wood by the casein cement can, however, be disregarded.

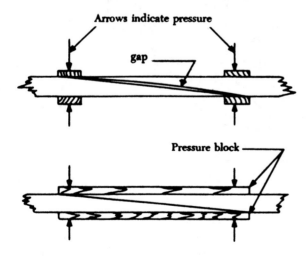

FIGURE 5–109. Results of uneven and even pressure.

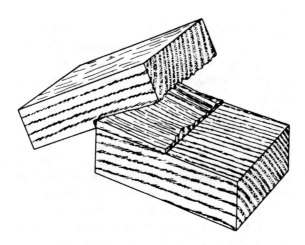

FIGURE 5–110. Typical broken test piece.

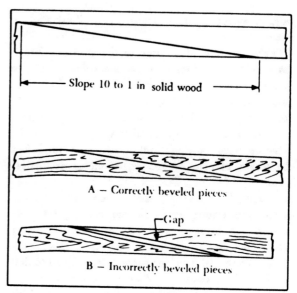

FIGURE 5–111. Beveling scarf joint.

SPLICED JOINTS

The scarf joint is generally used in splicing structural members in aircraft. As illustrated in figure 5–111, the two pieces to be joined are beveled and glued. The slope of the bevel should be not less than 10 to 1 in solid wood and 12 to 1 in plywood. Make the scarf cut in the general direction of the grain slope as shown in figure 5–111.

The chief difficulty encountered in making this type of joint is that of obtaining the same bevel on each piece. The strength of the joint will depend upon the accuracy of the two beveled surfaces, because an inaccurate bevel will reduce the amount of effective glue area. (See figure 5–111.)

One method of producing an accurate scarf joint is illustrated in figure 5–112. After the two bevels are cut, the pieces are clamped to a 2X4 backboard or some similar material. A fine-tooth saw is then run all the way through the joint. One of the pieces is then tapped on the end until it will move no farther, and the saw is again passed through the joint. This procedure is continued until the joint is perfect. A light cut of a plane is then used to smooth the surfaces of the joint.

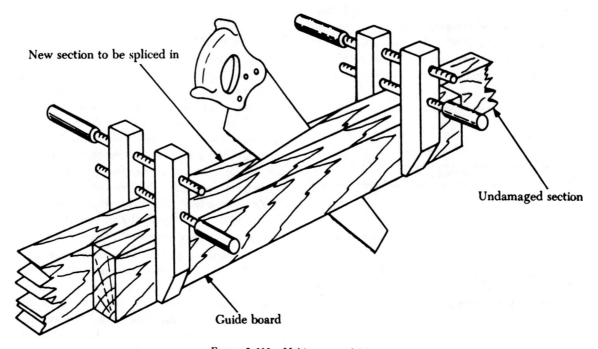

FIGURE 5–112. Making a scarf joint.

It is recommended that no more than 8 hrs. be permitted to elapse between final surfacing and gluing. The gluing surfaces should be machined smooth and true with planers, jointers, or special miter saws. Planer marks, chipped or loosened grain, and other surface irregularities are not permitted. Sandpaper must never be used to smooth softwood surfaces that are to be glued. Sawed surfaces must be similar to well-planed surfaces in uniformity, smoothness, and freedom from crushed fibers.

Tooth-planing, or other means of roughening smooth, well-planed surfaces of normal wood before gluing are not recommended. Such treatment of well-planed wood surfaces may result in local irregularities and objectionable rounding of edges. Although sanding of planed surfaces is not recommended for softwoods, sanding is a valuable aid in improving the gluing characteristics of some hard plywood surfaces, wood that has been compressed through exposure to high pressure and temperatures, resin-impregnated wood (impreg and compreg), or laminated paper plastic (papreg).

PLYWOOD SKIN REPAIRS

Most skin repairs can be made using either the surface or overlay patch, the splayed patch, the plug patch, or the scarf patch. Probably the easiest to make is the surface patch. Surface patches should not be used on skins over ⅛-in. thick.

To repair a hole by this method, trim the damaged skin to a rectangular or triangular shape depending upon the exact location of the damage relative to the framing members (figure 5–113). Where

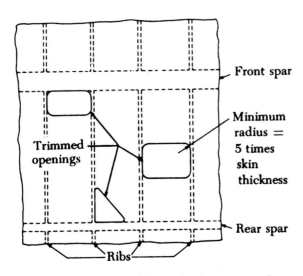

FIGURE 5–113. Typical shapes for damage removal.

the framing members form a square corner and the damage does not extend to the next parallel member, a triangular opening should be made. The corners of the cutouts should be rounded with a radius of at least five times the thickness of the skin.

Doublers, made of plywood at least as thick as the skin, are reinforcements placed under the edge of the hole inside the skin. The doublers are nailed and gluded in place. The doublers are extended from one framing member to another and are strengthened at the ends by saddle gussets attached to the framing members.

A patch large enough to extend at least 12 times the skin thickness beyond all edges of the opening is cut from material of the same kind and thickness as the original panel. The edges of the patch are then beveled (scarfed), as shown in figure 5–114.

It is usually impossible to use clamps when gluing an external patch; therefore, the pressure is applied by some other means. It is usually done by placing heavy weights on the patch until it is dry. Two or three small nails driven through the patch will prevent slipping during the drying.

After a surface patch has dried, it should be covered with fabric. The fabric should overlap the original plywood skin by at least 2 in.

Surface patches located entirely aft of the 10% chord line, or which wrap around the leading edge and terminate aft of the 10% chord line are permissible. The leading edge of a surface patch should be beleved with an angle of at least four times the skin thickness. Surface patches can have as much as a 50-in. perimeter and can extend from one rib to the next. The face-grain direction of the patch must be in the same direction as the original skin. Surface patches should not be used on skins over ⅛-in. thick.

Flush Patches

In places where an external patch would be objectionable, such as on wing coverings or fuselage skin, etc., a flush patch can be used.

Plug Patches

Two types of plug patches, oval and round, can be used on plywood skins. Since the plug patch is strictly a skin repair, it should be used only for damage that does not involve the supporting structure under the skin.

A plug patch has edges cut at right angles to the surface of the skin. The skin is cut out to a clean round or oval hole with square edges. The patch is cut to the exact size of the hole, and when installed,

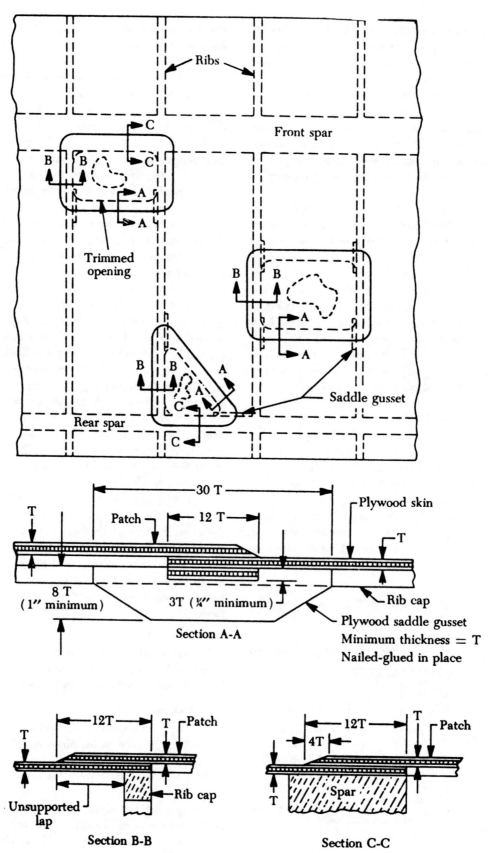

FIGURE 5–114. Surface patches.

236

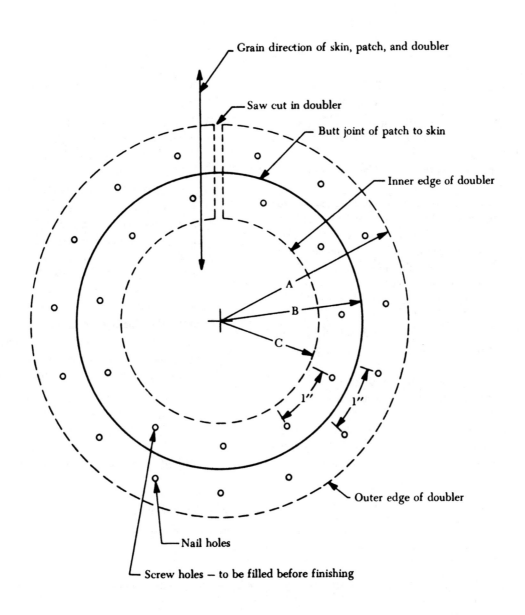

Grain direction of skin, patch, and doubler

Saw cut in doubler

Butt joint of patch to skin

Inner edge of doubler

A
B
C

1" 1"

Outer edge of doubler

Nail holes

Screw holes — to be filled before finishing

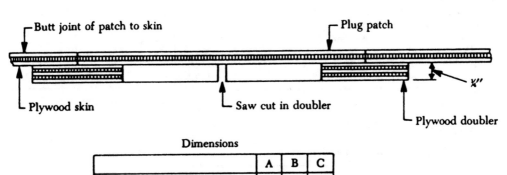

Butt joint of patch to skin

Plug patch

Plywood skin

Saw cut in doubler

Plywood doubler

X''

Dimensions

	A	B	C
Small circular plug patch	2X	2	1X
Large circular plug patch	3X	3	2X

(Two rows of screws and nails required for large patch)

FIGURE 5–115. A round plug patch.

237

the edge of the patch forms a butt joint with the edge of the hole.

A round plug patch, shown in figure 5–115, can be used where the cutout hole is no larger than 6 in. in diameter. Large and small circular patches have been designed for holes of 6 and 4 in. in diameter.

The steps for making a circular or round plug patch are:

(1) Cut a patch of the correct dimension for the repair. The patch must be of the same material and thickness as the original skin. Orientation of the face-grain direction of the round plug patch to that of the skin surface is no problem, since the round patch may be rotated until grain directions match.

(2) Lay the patch over the damaged spot and mark a circle of the same size as the patch.

(3) Cut the skin out so that the patch fits snugly into the hole around the entire perimeter.

(4) Cut a doubler of ¼-in. plywood so that its outside radius is ⅝ in. greater than the hole to be patched and the inside radius is

⅝ in. less. For a large patch, these dimensions would be ⅞ in. each. The doubler should be of a soft plywood such as poplar.

(5) Cut the doubler through one side so that it can be inserted through to the back of the skin. Apply a coat of glue to the outer half of the doubler surface where it will bear against the inner surface of the skin.

(6) Install the doubler by slipping it through the cutout hole and centering it so that it is concentric with the hole. Nail it in place with nailing strips, using a bucking bar or similar object for backing. Waxed paper must be placed between the nailing strips and the skin.

(7) After the glue has set in the installation of the doubler, apply glue to the surface of the doubler where the patch is to join and to the same area on the patch. Insert the patch in the hole.

(8) Apply pressure to the patch by means of a pressure plate and No. 4 wood screws placed at approximately 1-in. spacing. Waxed paper or cellophane placed be-

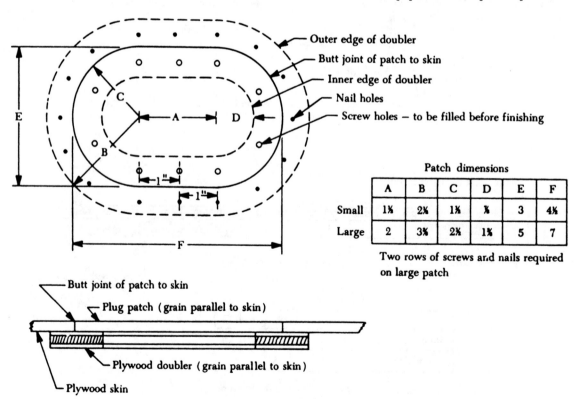

Patch dimensions

	A	B	C	D	E	F
Small	1⅛	2⅛	1⅜	⅞	3	4⅛
Large	2	3⅜	2⅛	1⅜	5	7

Two rows of screws and nails required on large patch

FIGURE 5–116. An oval patch.

238

tween the plate and patch will prevent glue from sealing the plate to the patch.

(9) After the glue has set, remove the nails and screws. Fill the nail and screw holes, sand, and finish to match the original surface.

The steps for making an oval plug patch, figure 5–116, are identical with those for making the round patch. The maximum dimensions for large oval patches are 7-in. long and 5-in. wide. Oval patches must be prepared with the face grain carefully oriented to the same direction as the original skin.

Splayed Patch

A splayed patch is a patch fitted into the plywood to provide a flush surface. The term "splayed" denotes that the edges of the patch are tapered, but the slope is steeper than is allowed in scarfing operations. The slope of the edges is cut at an angle of five times the thickness of the skin.

Splayed patches may be used where the largest dimension of the hole to be repaired is not more than 15 times the skin thickness and the skin thickness is not more than 1/10 in.

Lay out the patch as shown in figure 5–117. Tack a small piece of plywood over the hole for a center point. Draw two concentric circles around the damaged area. The difference between the radii of the circles is five times the skin thickness. The inner circle marks the limit of the hole and the outer one marks the limit of the taper.

Cut out the inner circle and taper the hole evenly to the outer mark with a chisel, knife, or rasp.

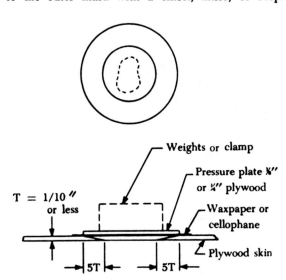

FIGURE 5–117. A splayed patch.

Prepare a circular patch, cut and tapered to match the hole. The patch is of the same type and thickness as the plywood being repaired.

Apply glue to the beveled surfaces and place the patch into place with the face-grain direction matching that of the original surface.

After the patch is in place, a pressure plate cut to the exact size of the patch is centered over the patch, with waxed paper between the two, and pressed firmly against the patch with a weight (sometimes a sandbag) or clamp. Since there is no reinforcement behind the splayed patch, use care to prevent excess pressure. After the glue has set, fill, sand, and finish the patch to match the original surface.

Scarf Patch

A properly prepared and inserted scarf patch is the best repair for damaged plywood and is preferred for most skin repairs. The scarf patch differs from the splayed patch in that the edges are scarfed to a 12 to 1 slope instead of the 5 to 1 used with the splayed patch. The scarf patch also uses reinforcements under the patch where the glue joints occur.

Much of the outside surface of plywood aircraft is curved. If the damaged plywood skin has a radius of curvature greater than 100 times the skin thickness, a scarf patch can be installed. Backing blocks or other reinforcements must be shaped to fit the skin curvature.

Figures 5–118 and 5–119 illustrate methods for making scarfed flush patches. Scarf cuts in plywood are made by hand plane, spoke shave, scraper, or accurate sandpaper block. Rasped surfaces, except at the corners of scarf patches and sawed surfaces, should be avoided since they are likely to be rough or inaccurate.

When the back of a damaged plywood skin is accessible (such as a fuselage skin), it should be repaired with scarf patches following the details shown in figure 5–118. Whenever possible, the edge of the patch should be supported as shown in section C–C. When the damage follows or extends to a framing member, the scarf should be supported as shown in section B–B.

Damages that do not exceed 25 times the skin thickness in diameter after being trimmed to a circular shape can be repaired as shown in figure 5–118, section D–D, provided the trimmed opening is not nearer than 15 times the skin thickness to a framing member. The backing block is carefully

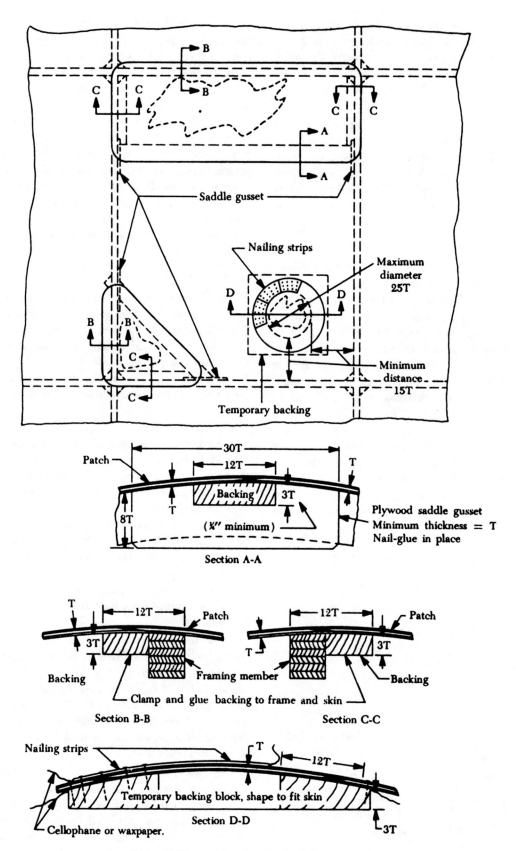

Saddle gusset

Nailing strips

Maximum diameter 25T

Minimum distance 15T

Temporary backing

Patch

30T

12T

T

Backing

3T

(⅛″ minimum)

8T

T

Plywood saddle gusset
Minimum thickness = T
Nail-glue in place

Section A-A

T

12T

Patch

3T

Backing

Framing member

Clamp and glue backing to frame and skin

Section B-B

12T

Patch

T

3T

Backing

Section C-C

Nailing strips

T

12T

Temporary backing block, shape to fit skin

Cellophane or waxpaper.

3T

Section D-D

FIGURE 5–118. Scarf patches, back of skin accessible.

240

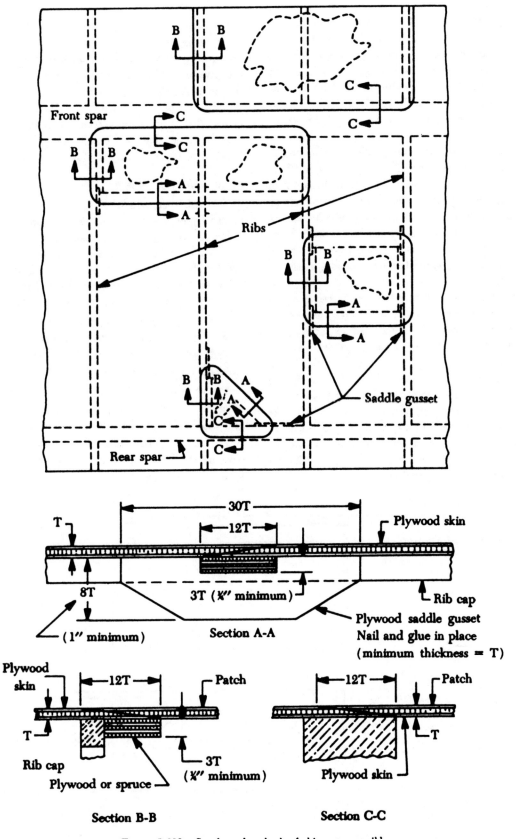

Front spar

B B

C C C

C C

Ribs

B B

Saddle gusset

A A

B B A A

C

Rear spar C

30T

T

12T

Plywood skin

8T

3T (¼″ minimum)

Rib cap

(1″ minimum)

Section A-A

Plywood saddle gusset
Nail and glue in place
(minimum thickness = T)

Plywood skin

12T

Patch

T

Rib cap

Plywood or spruce

3T (¼″ minimum)

Section B-B

12T

Patch

T

Plywood skin

Section C-C

FIGURE 5–119. Scarf patches, back of skin not accessible.

241

shaped from solid wood and fitted to the inside surface of the skin, and it is temporarily held in place with nails. A hole, the exact size of the inside circle of the scarf patch, is made in the block and is centered over the trimmed area of damage. The block is removed after the glue on the patch has set, leaving a flush surface to the repaired skin.

When the back of a damaged plywood skin is not accessible, it should be repaired as outlined in figure 5–119. After removing damaged sections, install backing strips along all edges that are not fully backed by a rib or a spar. To prevent warping of the skin, backing strips should be made of a soft-textured plywood, such as yellow poplar or spruce rather than solid wood. All junctions between backing strips and ribs or spars should have the end of the backing strip supported by a saddle gusset of plywood.

If needed, nail and glue the new gusset plate to the rib. It may be necessary to remove and replace the old gusset plate with a new saddle gusset, or it may be necessary to nail a saddle gusset over the original.

Attach nailing strips to hold backing strips in place while the glue sets. Use a bucking bar where necessary to provide support for nailing. Unlike the smaller patches made in a continuous process, work on the aircraft must wait while the glue, holding the backing strips, sets. After the glue sets, fill and finish to match the original skin.

Fabric Patch

Small holes that do not exceed 1 in. in diameter, after being trimmed to a smooth outline, can be repaired by doping a fabric patch on the outside of the plywood skin. The edges of the trimmed hole should first be sealed, and the fabric patch should overlap the plywood skin by at least 1 in. Holes nearer than 1 in. to any frame member, or in the wing leading edge or frontal area of the fuselage, should not be repaired with fabric patches.

SPAR AND RIB REPAIR

The web members of a spar or rib can be repaired by applying an external or flush patch, provided the damaged area is small. Plates of spruce or plywood of sufficient thickness to develop the longitudinal shear strength can be glued to both sides of the spar. Extend the plates well beyond the termination of the crack as shown in figure 5–120.

If more extensive damage has occurred, the web should be cut back to structural members and repaired with a scarf patch. Not more than two splices should be made in any one spar.

A spar may be spliced at any point except under wing attachment fittings, landing-gear fittings, engine-mount fittings, or lift-and-interplane strut fittings. Do not permit these fittings to overlap any part of the splice. Splicing under minor fittings such as drag wire, antidrag wire, or compression

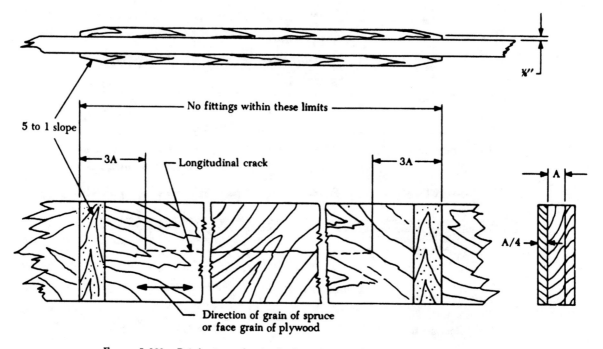

FIGURE 5–120. Reinforcing a longitudinal crack in a solid or internally routed spar.

strut fittings is acceptable under the following conditions:

    (1) The reinforcement plates of the splice should not interfere with the proper attachment or alignment of the fittings. The locations of pulley support brackets, bell-crank support brackets, or control surface support brackets should not be altered.

    (2) The reinforcement plate may overlap drag or antidrag wire or compression strut fittings if the reinforcement plates are on the front-face of the front spar or on the

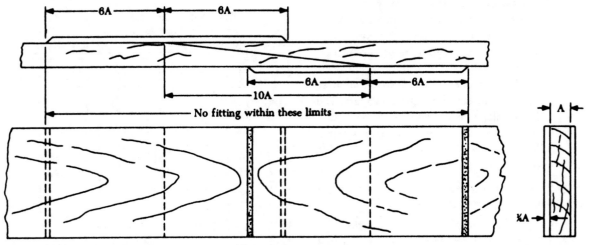

SPLICING OF SOLID RECTANGULAR SPAR

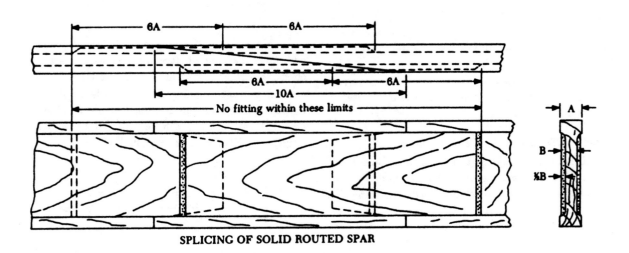

SPLICING OF SOLID ROUTED SPAR

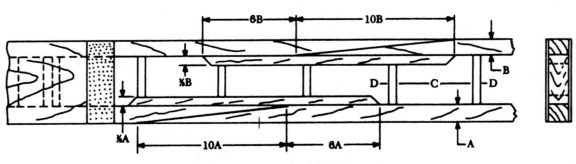

SPLICING OF BOX SPAR FLANGES

FIGURE 5–121. Spar splicing.

rear-face of the rear spar. In such cases it will be necessary to install slightly longer bolts. The inside reinforcement plate should not overlap drag strut fittings, unless such overlapping does not require sufficient shortening of compression struts or changes in drag-truss geometry to prevent adjustment for proper rigging. Even though takeup is sufficient, it may be necessary to change the angles on the fittings. Space the splices so that they do not overlap. Reinforcement plates must be used as indicated on all scarf repairs to spars. The desired slope for the scarf is 12 to 1, but a slope of not less than 10 to 1 is acceptable. The plates are held in place by glue and must not be nailed.

Figure 5–121 illustrates the general method for splicing common types of wooden spars.

Always splice and reinforce plywood webs with the same type of plywood as the original. Do not use solid wood to replace plywood webs. Plywood is stronger in shear than solid wood of the same thickness because of the variation in grain direction of the individual plies. The face-grain of plywood replacement webs and reinforcement plates must be in the same direction as that of the original member to ensure that the new web will have the required strength. One method of splicing plywood webs is shown in figure 5–122.

BOLT AND BUSHING HOLES

All bolts and bushings used in aircraft structures must fit snugly into the holes. Looseness allows the bolt or fitting to work back and forth, enlarging the hole. In cases of elongated bolt holes in a spar or cracks in the vicinity of bolt holes, splice in a new section of spar, or replace the spar entirely.

Holes drilled to receive bolts should be of such size that the bolt can be inserted by light tapping with a wood or rawhide mallet. If the hole is so tight that heavy blows are necessary to insert the bolt, deformation of the wood may cause splitting or unequal load distribution.

Rough holes are often caused by using dull bits or trying to bore too rapidly. Well-sharpened twist drills produce smooth holes in both solid wood and plywood. The twist drill should be sharpened to approximately a 60° angle. All holes bored for bolts which are to hold fittings in place should match exactly the hole in the fitting.

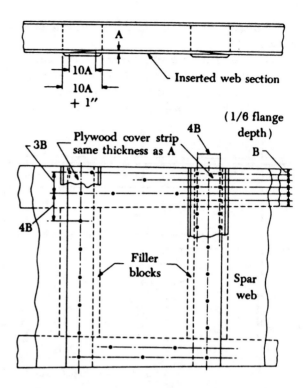

FIGURE 5–122. Method of splicing a box-spar web.

Bushings made of plastic or light metal provide additional bearing surface without any great increase in weight. Sometimes light steel bushings are used to prevent crushing the wood when bolts are tightened. The holes for bushings should be of such size that the bushing can be inserted by tapping lightly with a wood or rawhide mallet.

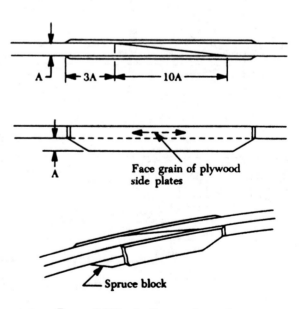

FIGURE 5–123. A rib cap-strip repair.

RIB REPAIRS

A cap strip of a wood rib can be repaired using a scarf splice. The repair is reinforced on the side opposite the wing covering by a spruce block which extends beyond the scarf joint not less than three times the thickness of the strips being repaired. The entire splice, including the reinforcing block, is reinforced on each side by a plywood side plate as shown in figure 5–123.

When the cap strip is to be repaired at a point where there is a joint between it and cross members of the rib, the repair is made by reinforcing the scarf joint with plywood gussets, as shown in figure 5–124.

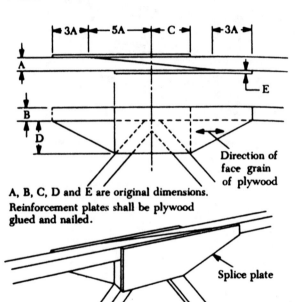

A, B, C, D and E are original dimensions.
Reinforcement plates shall be plywood glued and nailed.

FIGURE 5–124. A rib repair at a joint.

When it is necessary to repair a cap strip at a spar, the joint should be reinforced by a continuous gusset entending over the spar as shown in figure 5–125.

Edge damage, cracks, or other local damage to a spar can be repaired by removing the damaged portion and gluing in a properly fitted block, as shown in figure 5–126, reinforcing the joint by means of plywood or spruce blocks glued into place.

The trailing edge of a rib can be replaced and repaired by removing the damaged portion of the cap strip and inserting a softwood block of white pine, spruce, or basswood. The entire repair is then reinforced with plywood gussets and nailed and glued as shown in figure 5–127.

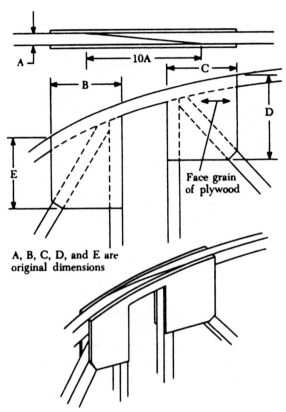

A, B, C, D, and E are original dimensions

FIGURE 5–125. A rib repair at a spar.

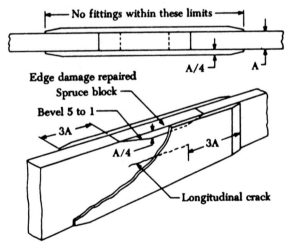

FIGURE 5–126. The repair of cracks and edge damage on a solid spar.

Compression ribs are of many different types, and the proper method of repairing any part of this type of rib is specified by the manufacturer. Figure 5–128 shows a typical repair made to a compression rib built up of a plywood web and three longitudinal members, the center one of which has been

245

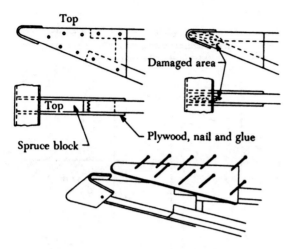

FIGURE 5–127. A rib trailing-edge repair.

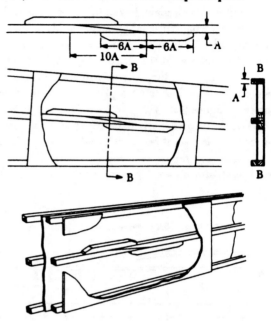

FIGURE 5–128. A compression-rib repair.

it is possible to replace a damaged member, it is always better to do so than to attempt a repair.

repaired by a properly reinforced scarf joint and an outside layer of plywood.

Such members as glue blocks, filler blocks, compression members, braces, and rib diagonals should not be repaired, but should be replaced. Wherever

GENERAL

Metals can be joined by mechanical means (bolting or riveting, or by welding, brazing, soldering or adhesive bonding). All of these methods are used in aircraft construction. This chapter will discuss the methods used to join metals by welding, brazing, and soldering.

Welding

Welding is the process of joining metal by fusing the materials while they are in a plastic or molten state. There are three general types of welding: (1) Gas, (2) electric arc, and (3) electric resistance welding. Each of these types of welding has several variations which are used in aircraft construction.

Welding is used extensively in the repair and manufacture of aircraft. Such parts as engine mounts and landing gear are often fabricated in this manner, and many fuselages, control surfaces, fittings, tanks, etc., are also of welded construction. Structures that have been welded in manufacture may generally be repaired economically by using the same welding process. Careful workmanship, both in preparation and actual welding, is of utmost importance.

Welding is one of the most practical of the many metal-joining processes available. The welded joint offers rigidity, simplicity, low weight, and high strength. Consequently, welding has been universally adopted in the manufacture and repair of all types of aircraft. Many structural parts as well as nonstructural parts are joined by some form of welding, and the repair of many of these parts is an indispensable part of aircraft maintenance.

It is equally important to know when not to weld, as it is to know when. Many of the alloy steels or high-carbon steel parts that have been hardened or strengthened by heat treatment cannot be restored to 100% of their former hardness and strength after they have been welded.

Gas Welding

Gas welding is accomplished by heating the ends or edges of metal parts to a molten state with a high-temperature flame. This flame is produced with a torch burning a special gas such as acetylene or hydrogen with pure oxygen. The metals, when in a molten state, flow together to form a union without the application of mechanical pressure or blows.

Aircraft parts fabricated from chrome-molybdenum or mild carbon steel are often gas welded. There are two types of gas welding in common use: (1) Oxyacetylene and (2) oxyhydrogen. Nearly all gas welding in aircraft construction is done with an oxyacetylene flame, although some manufacturers prefer an oxyhydrogen flame for welding aluminum alloys.

Electric Arc Welding

Electric arc welding is used extensively in both the manufacture and repair of aircraft, and can be satisfactorily used in the joining of all weldable metals. The process is based on using the heat generated by an electric arc. Variations of the process are: (1) Metallic arc welding, (2) carbon arc welding, (3) atomic hydrogen welding, (4) inert-gas (helium) welding, and (5) multi-arc welding. Metallic arc and inert-gas welding are the two electric arc welding processes most widely used in aircraft construction.

Electric Resistance Welding

Electric resistance welding is a welding process in which a low-voltage, high-amperage current is applied to the metals to be welded through a heavy, low-resistance copper conductor. The materials to be welded offer a high resistance to the flow of current, and the heat generated by this resistance fuses (welds) the parts together at their point of contact.

Three commonly used types of electric resistance welding are butt, spot, and seam welding. Butt welding is used in aircraft work to weld terminals to control rods. Spot welding is frequently used in airframe construction. It is the only welding method used for joining structural corrosion-resistant steel. Seam welding is similar to spot welding, except that power-driven rollers are used as electrodes. A continuous airtight weld can be obtained using seam welding.

OXYACETYLENE WELDING EQUIPMENT

Oxyacetylene welding equipment may be either stationary or portable. A portable equipment rig consists of the following:

(1) Two cylinders, one containing oxygen and one acetylene.
(2) Acetylene and oxygen pressure regulators, complete with pressure gages and connections.
(3) A welding torch, with a mixing head, extra tips and connections.
(4) Two lengths of colored hose, with adapter connections for the torch and regulators.
(5) A special wrench.
(6) A pair of welding goggles.
(7) A flint lighter.
(8) A fire extinguisher.

Figure 6–1 shows some of the equipment in a typical portable acetylene welding rig.

Oxygen pressure regulator

Acetylene pressure regulator

Oxygen cylinder

Acetylene cylinder

Torch

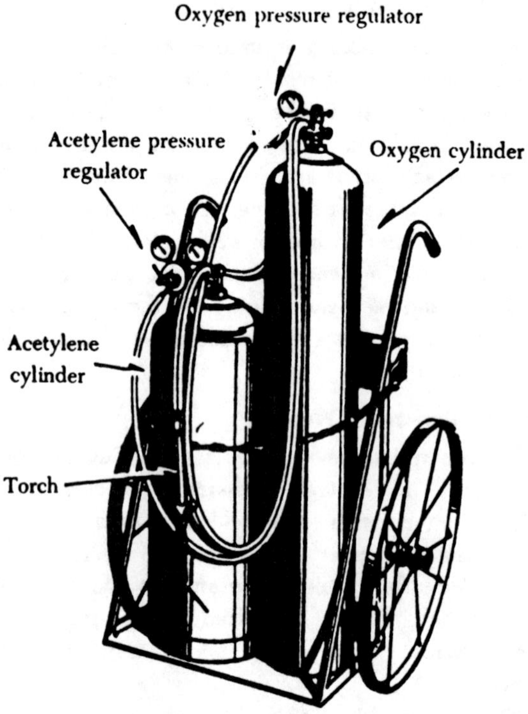

FIGURE 6–1. Typical portable acetylene welding rig.

Stationary oxyacetylene welding equipment is similar to portable equipment, except that acetylene and oxygen are piped to one or several welding stations from a central supply. The central supply usually consists of several cylinders connected to a common manifold. A master regulator controls the pressure in each manifold to ensure a constant pressure to the welding torch.

Acetylene Gas

Acetylene gas is a flammable, colorless gas which has a distinctive, disagreeable odor, readily detectable even when the gas is heavily diluted with air. Unlike oxygen, acetylene does not exist free in the atmosphere; it must be manufactured. The process is neither difficult nor expensive. Calcium carbide is made to react chemically with water to produce acetylene.

Acetylene is either used directly in a manifold system or stored in cylinders. If ignited, the result is a yellow, smoky flame with a low temperature. When the gas is mixed with oxygen in the proper proportions and ignited, the result is a blue-white flame with temperatures which range from approximately $5,700°$ to $6,300°$ F.

Under low pressure at normal temperatures, acetylene is a stable compound. But when compressed in a container to pressures greater than 15 p.s.i., it becomes dangerously unstable. For this reason, manufacturers fill the acetylene storage cylinders with a porous substance (generally a mixture of asbestos and charcoal) and saturate this substance with acetone. Since acetone is capable of absorbing approximately 25 times its own volume of acetylene gas, a cylinder containing the correct amount of acetone can be pressurized to 250 p.s.i.

Acetylene Cylinders

The acetylene cylinder is usually a seamless steel shell with welded ends, approximately 12 in. in diameter and 36 in. long. It is usually painted a distinctive color, and the name of the gas is stenciled or painted on the sides of the cylinder. A fully charged acetylene cylinder of this size contains approximately 225 cu. ft. of gas at pressures up to 250 p.s.i. In the event of fire or any excessive temperature rise, special safety fuse plugs installed in the cylinder will melt, allowing the excess gas to escape or burn, thus minimizing the chances of an explosion. The holes in the safety plugs are made small to prevent the flames from burning back into the cylinder. Acetylene cylinders should not be completely emptied, or a loss of filler material may result.

Oxygen Cylinders

The oxygen cylinders used in welding operations are made of seamless steel of different sizes. A typical small cylinder holds 200 cu. ft. of oxygen at 1,800 p.s.i. pressure. A large size holds 250 cu. ft. of oxygen at 2,265 p.s.i. pressure. Oxygen cylinders are usually painted green for identification The cylinder has a high-pressure valve located at the top of the cylinder. This valve is protected by a metal safety cap which should always be in place when the cylinder is not in use.

Oxygen should never come in contact with oil or grease. In the presence of pure oxygen, these substances become highly combustible. Oxygen hose and valve fittings should never be oiled or greased, or handled with oily or greasy hands. Even grease spots on clothing may flare up or explode if struck by a stream of oxygen. Beeswax is a commonly used lubricant for oxygen equipment and fittings.

Pressure Regulators

Acetylene and oxygen regulators reduce pressures and control the flow of gases from the cylinders to the torch. Acetylene and oxygen regulators are of the same general type, although those designed for acetylene are not made to withstand such high pressures as those designed for use with oxygen. To prevent interchange of oxygen and acetylene hoses, the regulators are built with different threads on the outlet fitting. The oxygen regulator has a right-hand thread, and the acetylene regulator has a left-hand thread.

On most portable welding units, each regulator is equipped with two pressure gages, a high-pressure gage which indicates the cylinder pressure and a low-pressure gage which indicates the pressure in the hose leading to the torch (working pressure).

In a stationary installation, where the gases are piped to individual welding stations, only one gage for oxygen and one for acetylene are required for each welding station, since it is necessary to indicate only the working pressure of the gases flowing through the hose to the welding torch.

A typical regulator, complete with pressure gages and connections, is shown in figure 6-2. The adjusting screw shown on the front of the regulator is for adjusting the working pressure. When this adjusting screw is turned to the left (counterclockwise) until it turns easily, the valve mechanism inside the regulator is closed. No gas can then flow to the torch. As the handle is turned to the right (clockwise), the screw presses against the regulating mechanism, the valve opens, and gas passes to the torch at the pressure shown on the working pressure gage. Changes in the working pressure can

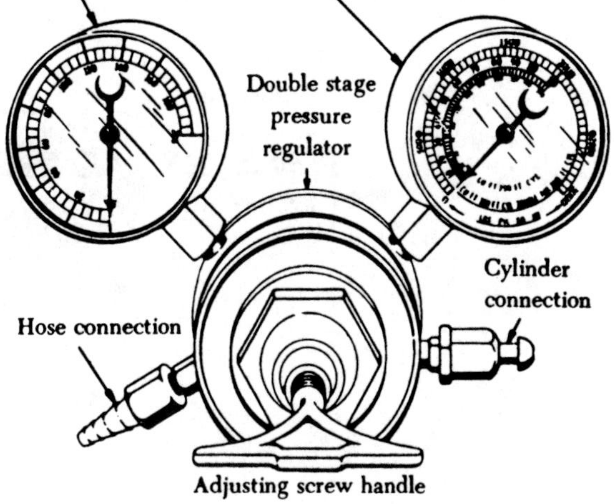

FIGURE 6-2. Typical oxygen pressure regulator.

be made by adjusting the handle until the desired pressure is registered.

Before opening the high-pressure valve on a cylinder, the adjusting screw on the regulator should be fully released by turning it counterclockwise. This closes the valve inside the regulator, protecting the mechanism against possible damage.

Welding Torch

The welding torch is the unit used to mix the oxygen and acetylene together in correct proportions. The torch also provides a means of directing and controlling the size and quality of the flame produced. The torches are designed with two needle valves, one for adjusting the flow of acetylene and the for other adjusting the flow of oxygen.

Welding torches are manufactured in different sizes and styles, thereby providing a suitable type for different applications. They are also available with several different sizes of interchangeable tips in order that a suitable amount of heat can be obtained for welding the various kinds and thicknesses of metals.

Welding torches can be divided into two classes: (1) The injector type and (2) the balanced-pressure type. The injector-type torch (figure 6-3A) is designed to operate with very low acetylene pressure as compared to the oxygen pressure.

A narrow passageway or nozzle within the torch, called the injector, through which the oxygen

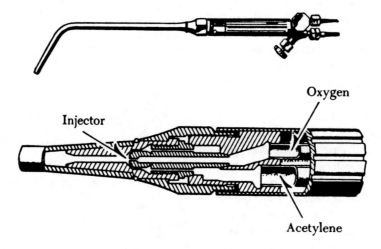

Injector

Oxygen

Acetylene

A   Injector-type welding torch.

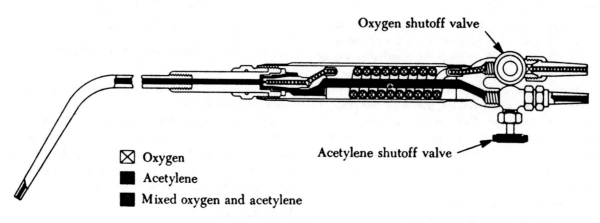

Oxygen shutoff valve

Acetylene shutoff valve

⊠ Oxygen
■ Acetylene
■ Mixed oxygen and acetylene

B   Balanced-pressure welding torch.

FIGURE 6–3.   Welding torches.

passes, causes the speed of oxygen flow to increase to a high velocity with a corresponding drop in pressure. This pressure drop across the injector creates a pressure differential which acts to draw the required amount of acetylene into the mixing chamber in the torch head.

In the balanced-pressure torch, the oxygen and acetylene are both fed to the torch at the same pressure (figure 6–3B). The openings to the mixing chamber for each gas are equal in size, and the delivery of each gas is independently controlled. This type of torch is generally better suited for aircraft welding than the injector type because of the ease of adjustment.

Welding Torch Tips

The torch tip delivers and controls the final flow of gases. It is important that the correct tip be selected and used with the proper gas pressures if a job is to be welded satisfactorily. The nature of the weld, the material, the experience of the welder, and the position in which the weld is to be made, all determine the correct size of the tip opening. The size of tip opening, in turn, determines the amount of heat (not the temperature) applied to the work. If a tip which is too small is used, the heat provided will be insufficient to produce penetration to the proper depth. If the tip is too large, the heat will be too great, and holes will be burned in the metal.

The torch tip sizes are designated by numbers, and each manufacturer has his own arrangement for classifying them. As an example, a number two tip is made with an orifice of approximately 0.040 in. diameter. The diameter of the tip orifice is related to the volume of heat it will deliver.

250

Torch tips are made of copper or copper alloy and are made so that they seat well when tightened handtight. Torch tips should not be rubbed across fire brick or used as tongs to position work.

With use, the torch tip will become clogged with carbon deposits and, if it is brought in contact with the molten pool, particles of slag may lodge in the opening. A split or distorted flame is an indication of a clogged tip. Tips should be cleaned with the proper size tip cleaners or with a piece of copper or soft brass wire. Fine steel wool may be used to remove oxides from the outside of the tip. These oxides hinder the heat dissipation and cause the tip to overheat.

A flint lighter is provided for igniting the torch. The lighter consists of a file-shaped piece of steel, usually recessed in a cuplike device, and a piece of flint that can be drawn across the steel, producing the sparks required to light the torch. Matches should never be used to ignite a torch since their length requires bringing the hand in close to the tip to ignite the gas. Accumulated gas may envelop the hand and, when ignited, cause a severe burn.

Goggles

Welding goggles, fitted with colored lenses, are worn to protect the eyes from heat, light rays, sparks, and molten metal. A shade or density of color that is best suited for the particular situation should be selected. The darkest shade of lens which will show a clear definition of the work without eyestrain is the most desirable. Goggles should fit closely around the eyes and should be worn at all times during welding and cutting operations.

Welding (Filler) Rods

The use of the proper type filler rod is very important in oxyacetylene welding operations. This material not only adds reinforcement to the weld area, but also adds desired properties to the finished weld. By selecting the proper rod, either tensile strength or ductility can be secured in a weld, or both can be secured to a reasonably high degree. Similarly, rods can be selected which will help retain the desired amount of corrosion resistance. In some cases, a suitable rod with a lower melting point will eliminate possible cracks caused by expansion and contraction.

Welding rods may be classified as ferrous or nonferrous. The ferrous rods include carbon and alloy steel rods as well as cast-iron rods. Nonferrous rods include brazing and bronze rods, aluminum and aluminum alloy rods, magnesium and magnesium alloy rods, copper rods, and silver rods.

Welding rods are manufactured in standard 36-in. lengths and in diameters from $\frac{1}{16}$ in. to $\frac{3}{8}$ in. The diameter of the rod to be used is governed by the thickness of the metals being joined. If the rod is too small, it will not conduct heat away from the puddle rapidly enough, and a burned weld will result. A rod that is too large will chill the puddle. As in selecting the proper size welding torch tip, experience enables the welder to select the proper diameter welding rod.

Setting Up Acetylene Welding Equipment

Setting up acetylene welding equipment and preparing for welding should be done systematically and in a definite order to avoid costly mistakes. The following procedures and instructions are typical of those used to assure safety of equipment and personnel:

(1) Secure the cylinders so they cannot be upset, and remove the protective caps from the cylinders.

(2) Open each cylinder shutoff valve for an instant to blow out any foreign matter that may be lodged in the outlet. Close the valves and wipe off the connections with a clean cloth.

(3) Connect the acetylene pressure regulator to the acetylene cylinder and the oxygen regulator to the oxygen cylinder. Use a regulator wrench and tighten the connecting nuts enough to prevent leakage.

(4) Connect the red (or maroon) hose to the acetylene pressure regulator and the green (or black) hose to the oxygen regulator. Tighten the connecting nuts enough to prevent leakage. Do not force these connections, since these threads are made of brass and are easily damaged.

(5) Release both pressure regulator adjusting screws by turning the adjusting screw handle on each regulator counterclockwise until it turns freely. This is to avoid damage to the regulators and pressure gages when the cylinder valves are opened.

(6) Open the cylinder valves slowly and read each of the cylinder pressure gages to check the contents in each cylinder. The oxygen cylinder shutoff valve should be opened fully and the acetylene cylinder

shutoff valve is opened approximately one and one-half turns.

(7) Blow out each hose by turning the pressure adjusting screw handle inward (clockwise) and then turning it out again. The acetylene hose should be blown out only in a well-ventilated space which is free from sparks, flame, or other sources of ignition.

(8) Connect both hoses to the torch and check the connections for leaks by turning the pressure regulator screws in, with the torch needle valves closed. When 20 p.s.i. shows on the oxygen working pressure gage and 5 p.s.i. on the acetylene gage, close the valves by turning the pressure regulator screws out. A drop in pressure on the working gage indicates a leak between the regulator and torch tip. A general tightening of all connections should remedy the situation. If it becomes necessary to locate a leak, use the soap suds method. Do this by painting all fittings and connections with a thick solution of the soapy water. **Never hunt for an acetylene leak with a flame**, since a serious explosion can occur in the hose or in the cylinder.

(9) Adjust the working pressure on both the oxygen and acetylene regulators by turning the pressure-adjusting screw on the regulator clockwise until the desired settings are obtained.

Oxyacetylene Flame Adjustment

To light the torch, open the torch acetylene valve a quarter to a half turn. Hold the torch to direct the flame away from the body and ignite the acetylene gas, using the flint lighter. The pure acetylene flame is long and bushy and has a yellowish color. Continue opening the acetylene valve until the flame leaves the tip approximately one-sixteenth of an inch. Open the torch oxygen valve. When the oxygen valve is opened, the acetylene flame is shortened, and the mixed gases burn in contact with the tip face. The flame changes to a bluish-white color and forms a bright inner cone surrounded by an outer flame envelope.

Oxyacetylene Welding Process

The oxyacetylene process of welding is a method in which acetylene and oxygen gases are used to produce the welding flame. The temperature of this flame is approximately 6,300° F., which is sufficiently high to melt any of the commercial metals to effect a weld. When the oxyacetylene flame is applied to the ends or edges of metal parts, they are quickly raised to a melting state and flow together to form one solid piece when solidified. Usually some additional metal is added to the weld, in the form of a wire or rod, to build up the weld seam to a greater thickness than the base metal.

There are three types of flames commonly used for welding. These are neutral, reducing or carburizing, and oxidizing. The characteristics of the different kinds of flames are shown in figure 6–4.

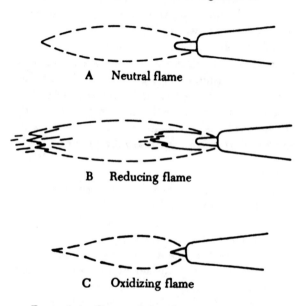

A   Neutral flame

B   Reducing flame

C   Oxidizing flame

Figure 6–4. Characteristics of oxyacetylene flames.

The neutral flame (figure 6–4A) is produced by burning acetylene with oxygen in such proportions as to oxidize all particles of carbon and hydrogen in the acetylene. This flame is distinguished by the well-rounded, smooth, clearly defined white central cone at the end of the tip. The envelope or outer flame is blue with a purple tinge at the point and edges. A neutral flame is generally used for welding and gives a thoroughly fused weld, free from burned metal or hard spots.

To obtain a neutral flame, gradually open the oxygen valve. This shortens the acetylene flame and causes a "feather" to appear in the flame envelope. Gradually increase the amount of oxygen until the "feather" disappears inside a clearly defined inner luminous cone.

The reducing or carburizing flame is shown in

figure 6–4B. Since the oxygen furnished through the torch is not sufficient to complete the combustion of the acetylene, carbon escapes unburned. This flame can be recognized by the greenish-white brushlike second cone at the tip of the first cone. The outer flame is slightly luminous and has about the same appearance as an acetylene flame burning freely in air alone. This type of flame introduces carbon into the steel.

To obtain a reducing flame, first adjust the flame to neutral; then open the acetylene valve slightly to produce a white streamer or "feather" of acetylene at the end of the inner cone.

An oxidizing flame (figure 6–4C) contains an excess of oxygen, which is the result of too much oxygen passing through the torch. The oxygen not consumed in the flame escapes to combine with the metal. This flame can be recognized by the short, pointed, bluish-white central cone. The envelope or outer flame is also shorter and of a lighter blue color than the neutral flame. It is accompanied by a harsh sound similar to high-pressure air escaping through a small nozzle. This flame oxidizes or burns most metals and results in a porous weld. It is used only when welding brass or bronze.

To obtain the oxidizing flame, first adjust the flame to neutral; then increase the flow of oxygen until the inner cone is shortened by about one-tenth of its length. The oxidizing flame has a pointed inner cone.

With each size of tip, a neutral, oxidizing or carburizing flame can be obtained. It is also possible to obtain a "harsh" or "soft" flame by increasing or decreasing the pressure of both gases.

For most regulator settings the gases are expelled from the torch tip at a relatively high velocity, and the flame is called "harsh." For some work it is desirable to have a "soft" or low-velocity flame without a reduction in thermal output. This may be achieved by using a larger tip and closing the gas needle valves until the neutral flame is quiet and steady. It is especially desirable to use a soft flame when welding aluminum, to avoid blowing holes in the metal when the puddle is formed.

Improper adjustment or handling of the torch may cause the flame to backfire or, in very rare cases, to flashback. A backfire is a momentary backward flow of the gases at the torch tip, which causes the flame to go out. A backfire may be causes by touching the tip against the work, by overheating the tip, by operating the torch at other than recommended pressures, by a loose tip or head, or by dirt or slag in the end of the tip. A backfire is rarely dangerous, but the molten metal may be splattered when the flame pops.

A flashback is the burning of the gases within the torch and is dangerous. It is usually caused by loose connections, improper pressures, or overheating of the torch. A shrill hissing or squealing noise accompanies a flashback; and unless the gases are turned off immediately, the flame may burn back through the hose and regulators and cause great damage. The cause of a flashback should always be determined and the trouble remedied before relighting the torch.

FIGURE 6–5. Holding the acetylene torch to weld light-gage metals.

Extinguishing The Torch

The torch can be shut off simply by closing both needle valves, but it is better practice to turn the acetylene off first and allow the gas remaining in the torch tip to burn out. The oxygen needle valve can then be turned off. If the torch is not to be used again for a long period, the pressure should be turned off at the cylinder. The hose lines should then be relieved of pressure by opening the torch needle valves and the working pressure regulator, one at a time, allowing the gas to escape. Again, it is a good practice to relieve the acetylene pressure and then the oxygen pressure. The hose should then be coiled or hung carefully to prevent damage or kinking.

Fundamental Oxyacetylene Welding Techniques

The proper method of holding the acetylene welding torch depends on the thickness of the metal being welded. When welding light-gage metal, the torch is usually held as illustrated in figure 6–5, with the hose draped over the wrist.

Figure 6–6 shows the method of holding the torch during the welding of heavy materials.

FIGURE 6–6.   Holding the acetylene torch to weld heavy materials.

The torch should be held so that the tip is in line with the joint to be welded, and inclined between 30° and 60° from the perpendicular. The best angle depends on the type of weld to be made, the amount of preheating necessary, and the thickness and type of metal. The thicker the metal, the more nearly vertical the torch must be for proper heat penetration. The white cone of the flame should be held about ⅛ in. from the surface of the base metal.

If the torch is held in the correct position, a small puddle of molten metal will form. The puddle should be composed of equal parts of the pieces being welded. After the puddle appears, movement of the tip in a semicircular or circular motion should be started. This movement ensures an even distribution of heat on both pieces of metal. The speed and motion of the torch movement are learned only by practice and experience.

Forehand welding is the technique of pointing the torch flame forward in the direction in which the weld is progressing, as illustrated in figure 6–7. The filler rod is added to the puddle as the edges of the joint melt before the flame. The forehand method is used in welding most of the lighter tubings and sheet metals.

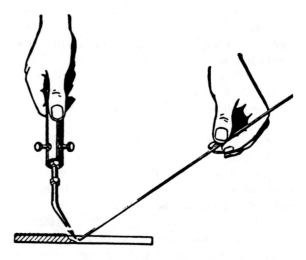

FIGURE 6–7.   Forehand welding.

Backhand welding is the technique of pointing the torch flame toward the finished weld and moving away in the direction of the unwelded area, melting the edges of the joint as it is moved (figure 6–8). The welding rod is added to the puddle between the flame and the finished weld.

Backhand welding is seldom used on sheet metal because the increased heat generated in this method is likely to cause overheating and burning. It is preferred for metals having a thick cross section. The large puddle of molten metal required for such welds is more easily controlled in backhand welding, and it is possible to examine the progress of the weld and determine if penetration is complete.

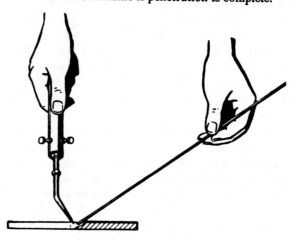

FIGURE 6–8.   Backhand welding.

WELDING POSITIONS

There are four general positions in which welds are made. These positions are shown in figure 6–9 and are designated as flat, overhead, horizontal, and vertical.

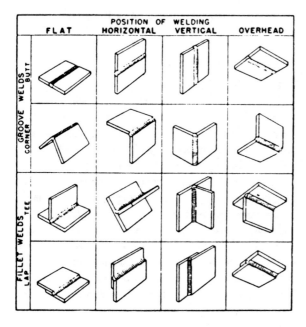

FIGURE 6–9. Four basic welding positions.

Welding is done in the flat position whenever possible, since the puddle is much easier to control in this position. Quite often, though, it is necessary to weld in the overhead, vertical, or horizontal position in aircraft repair.

The flat position is used when the material can be laid flat, or inclined at an angle of less than 45° and welded on the topside. The welding torch is pointed downward toward the work. This weld may be made by either the forehand or backhand technique, depending upon the thickness of the metal being welded.

The horizontal position is used when the line of the weld runs horizontally across a piece of work, and the torch is directed at the material in a horizontal or nearly horizontal position. The weld is made from right to left across the plate (for the right-handed welder). The flame is inclined upward at an angle of from 45° to 60°. The weld can be made using the forehand or backhand technique. Adding the filler rod to the top of the puddle will help prevent the molten metal from sagging to the lower edge of the bead.

The overhead position is used when the material is to be welded on the underside with the seam running horizontally or in a plane that requires the flame to point upward from below the work. In welding overhead, a large pool of molten metal should be avoided, as the metal will drip or run out of the joint. The rod is used to control the size of the molten puddle. The volume of flame used should not exceed that required to obtain good fusion of the base metal with the filler rod. The amount of heat needed to make the weld is best controlled by selecting the right tip size for the thickness of metal to be welded.

When the parts to be joined are inclined at an angle of more than 45°, with the seam running vertically, it is designated as a vertical weld. In a vertical weld, the pressure exerted by the torch flame must be relied upon to a great extent to support the puddle. It is highly important to keep the puddle from becoming too hot, to prevent the hot metal from running out of the puddle onto the finished weld. Vertical welds are begun at the bottom, and the puddle is carried upward using the forehand technique. The tip should be inclined from 45° to 60°, the exact angle depending upon the desired balance between correct penetration and control of the puddle. The rod is added from the top and in front of the flame.

WELDED JOINTS

The five fundamental types of welded joints (figure 6–10) are the butt joint, tee joint, lap joint, corner joint, and edge joint.

Butt Joints

A butt joint is made by placing two pieces of material edge to edge, so that there is no overlapping, and then welded. Some of the various types of butt joints are shown in figure 6–11. The flanged butt joint can be used in welding thin sheets, 1/16 in. or less. The edges are prepared for welding by turning up a flange equal to the thickness of the metal. This type of joint is usually made without the use of filler rod.

A plain butt joint is used for metals from 1/16 in. to 1/8 in. in thickness. A filler rod is used when making this joint to obtain a strong weld.

If the metal is thicker than 1/8 in., it is necessary to bevel the edges so that the heat from the torch can penetrate completely through the metal. These

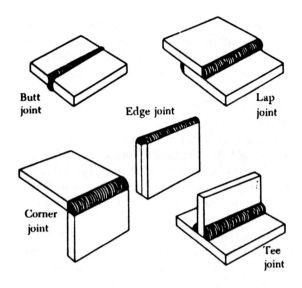

FIGURE 6–10. Basic welding joints.

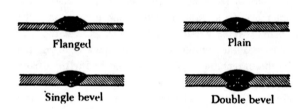

FIGURE 6–11. Types of butt joints.

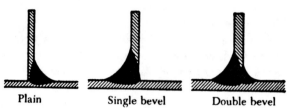

FIGURE 6–12. Types of tee joints.

bevels may be either single- or double-bevel type or single- or double-V type. A filler rod is used to add strength and reinforcement to the weld.

Cracks

Repair of cracks by welding may be considered just another type of butt joint. A stop drill hole is made at either end of the crack, then the two edges are brought together. The use of a filler rod is necessary.

Tee Joints

A tee joint is formed when the edge or end of one piece is welded to the surface of another, as shown in figure 6–12. These joints are quite common in aircraft work, particularly in tubular structures. The plain tee joint is suitable for most aircraft metal thicknesses, but heavier thicknesses require the vertical member to be either single or double beveled to permit the heat to penetrate deeply enough. The dark areas in figure 6–12 show the depth of heat penetration and fusion required.

Edge Joints

An edge joint may be used when two pieces of sheet metal must be fastened together and load stresses are not important. Edge joints are usually made by bending the edges of one or both parts upward, placing the two bent ends parallel to each other or placing one bent end parallel to the upright unbent end, and welding along the outside of the seam formed by the two joined edges. Figure 6–13 shows two types of edge joints. The type shown in fig. 6–13A requires no filler rod, since the edges can be melted down to fill theseam. The type shown in fig 6–13B, being thicker material, must be beveled for heat penetration; filler rod is added for reinforcement.

Corner Joints

A corner joint is made when two pieces of metal are brought together so that their edges form a corner of a box or enclosure as shown in figure 6–14. The corner joint shown in fig. 6–14A requires little or no filler rod, since the edges fuse to make the weld. It is used where load stress is unimportant. The joint shown in fig 6–14B is used on heavier metals, and filler rod is added for roundness and strength. If much stress is to be placed on

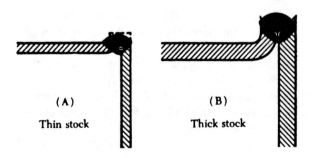

FIGURE 6–13. Edge joints.

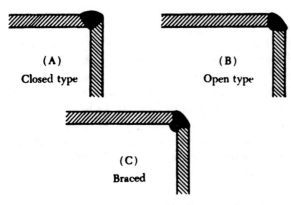

(A)
Closed type

(B)
Open type

(C)
Braced

FIGURE 6–14. Corner joints.

the corner, the inside is reinforced as shown in fig. 6–14C.

Lap Joints

The lap joint is seldom used in aircraft structures when welding with oxyacetylene, but is commonly used when spot welding. The single lap joint (figure 6–15) has very little resistance to bending, and will not withstand the shearing stress to which the weld may be subjected under tension or compression loads. The double lap joint (figure 6–15) offers more strength, but requires twice the amount of welding required on the simpler, more efficient butt weld.

Single lap   Double lap

FIGURE 6–15. Single and double lap joints.

EXPANSION AND CONTRACTION OF METALS

Heat causes metals to expand; cooling causes them to contract. Uneven heating, therefore, will cause uneven expansion, or uneven cooling will cause uneven contraction. Under such conditions, stresses are set up within the metal. These forces must be relieved, and unless precautions are taken, warping or buckling of the metal will take place. Likewise, on cooling, if nothing is done to take up the stress set up by the contraction forces, further warping may result; or if the metal is too heavy to permit this change in shape, the stresses remain within the metal itself.

The coefficient of linear expansion of a metal is the amount in inches that a 1 in. piece of metal will expand when its temperature is raised 1° F. The amount that a piece of metal will expand when heat

is applied is found by multiplying the coefficient of linear expansion by the temperature rise, and that product by the length of the metal in inches. For example, if a 10 ft. aluminum rod is to be raised to a temperature of 1,200° F. from a room temperature of 60° F., the rod will expand 1.75 in.— 0.00001280 (aluminum's coefficient of linear expansion) X 120 (length in inches) X 1140 (temperature rise).

Expansion and contraction have a tendency to buckle and warp thin sheet metal ⅛ in. or thinner. This is the result of having a large surface area that spreads heat rapidly and dissipates it soon after the source of heat is removed. The most effective method of alleviating this situation is to remove the heat from the metal near the weld, and thus prevent it from spreading across the whole surface area. This can be done by placing heavy pieces of metal, known as "chill bars," on either side of the weld; they absorb the heat and prevent it from spreading. Copper is most ofen used for chill bars because of its ability to absorb heat readily. Welding jigs sometimes use this same principle to remove heat from the base metal.

Expansion can also be controlled by tack welding at intervals along the joint.

The effect of welding a long seam (over 10 or 12 in.) is to draw the seam together as the weld progresses. If the edges of the seam are placed in contact with each other throughout their length before welding starts, the far ends of the seam will actually overlap before the weld is completed. This tendency can be overcome by setting the pieces to be welded with the seam spaced correctly at one end and increasing the space at the opposite end as shown in figure 6–16. The amount of space depends on the type of material, the thickness of the material, the welding process being used, and the shape and size of the pieces to be welded.

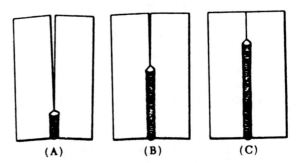

(A)   (B)   (C)

FIGURE 6–16. Allowance for a straight butt
weld when welding steel sheets.

The weld is started at the correctly spaced end and proceeds toward the end that has the increased gap. As the seam is welded, the space will close and should provide the correct gap at the point of welding. Sheet metal under $\frac{1}{16}$ in. can be handled by flanging the edges, tack welding at intervals, and then by welding between the tacks.

There is less tendency for plate stock over $\frac{1}{8}$ in. to warp and buckle when welded because the greater thickness limits the heat to a narrow area and dissipates it before it travels far on the plate.

Preheating the metal before welding is another method of controlling expansion and contraction. Preheating is especially important when welding tubular structures and castings. Great stress can be set up in tubular welds by contraction. When two members of a tee joint are welded, one tube tends to draw up because of the uneven contraction. If the metal is preheated before the welding operation begins, contraction still takes place in the weld, but the accompanying contraction in the rest of the structure is at almost the same rate, and internal stress is lessened.

CORRECT FORMING OF A WELD

The form of the weld metal has considerable bearing upon the strength and fatigue resistance of a joint. The strength of an improperly made weld is usually less than the strength for which the joint was designed. Low-strength welds are generally the result of insufficient penetration; undercutting of the base metal at the toe of the weld; poor fusion of the weld metal with the base metal; trapped oxides, slag, or gas pockets in the weld; overlap of the weld metal on the base metal; too much or too little reinforcement; and overheating of the weld.

Characteristics of a Good Weld

A completed weld should have the following characteristics:

(1) The seam should be smooth, the bead ripples evenly spaced, and of a uniform thickness.

(2) The weld should be built up, thus providing extra thickness at the joint.

(3) The weld should taper off smoothly into the base metal.

(4) No oxide should be formed on the base metal close to the weld.

(5) The weld should show no signs of blowholes, porosity, or projecting globules.

(6) The base metal should show no signs of burns, pits, cracks, or distortion.

Although a clean, smooth weld is desirable, this characteristic does not necessarily mean that the weld is a good one; it may be dangerously weak inside. However, when a weld is rough, uneven, and pitted, it is almost always unsatisfactory inside. Welds should never be filed to give them a better appearance, since filing deprives the weld of part of its strength. Welds should never be filled with solder, brazing material, or filler of any sort. Additional information about weld characteristics is contained in Chapter 10, of the Airframe and Powerplant Mechanics General Handbook, AC 65-9A.

When it is necessary to re-weld a joint, all old weld material must be removed before the operation is begun. It must be remembered, though, that reheating the area may cause the base metal to lose some of its strength and become brittle.

OXYACETYLENE WELDING OF FERROUS METALS
Steel

Low-carbon steel, low-alloy steel, cast steel, and wrought iron are easily welded with the oxyacetylene flame. Plain, low-carbon steel is the ferrous material that will be gas welded most frequently. As the carbon content of steel increases, it may be repaired by welding only under certain conditions. Factors involved are carbon content and hardenability. For corrosion- and heat-resistant nickel chromium steels, the allowed weldability depends upon their stability, carbon content, or re-heat treatment.

In order to make a good weld, the carbon content of the steel must not be altered, nor can other chemical constituents be added to or subtracted from the base metal without seriously altering the properties of the metal. Molten steel has a great affinity for carbon, and oxygen and nitrogen combine with the molten puddle to form oxides and nitrates, both of which lower the strength of steel. When welding with an oxyacetylene flame, the inclusion of impurities can be minimized by observing the following precautions:

(1) Maintain an exact neutral flame for most steels, and a slight excess of acetylene when welding alloys with a high nickel or chromium content, such as stainless steel.

(2) Maintain a soft flame and control the puddle.

(3) Maintain a flame sufficient to penetrate the metal and manipulate it so that the molten metal is protected from the air by the outer envelope of flame.

(4) Keep the hot end of the welding rod in

the weld pool or within the flame envelope.

Proper preparation for welding is an important factor in every welding operation. The edges of the parts must be prepared in accordance with the joint design chosen. The method chosen (bevel, groove, etc.) should allow for complete penetration of the base metal by the flame. The edges must be clean. Arrangements must be made for preheating, if this is required.

When preparing an aircraft part for welding, remove all dirt, grease or oil, and any protective coating such as cadmium plating, enamel, paint, or varnish. Such coatings not only hamper welding, but also mingle with the weld and prevent good fusion.

Cadmium plating can be chemically removed by dipping the edges to be welded in a mixture of 1 lb. of ammonium nitrate and 1 gal. of water.

Enamel, paint, or varnish may be removed from steel parts by a number of methods, such as a steel wire brush or emery cloth, by gritblasting, by using paint or varnish remover, or by treating the pieces with a hot, 10% caustic soda solution followed by a thorough washing with hot water to remove the solvent and residue. Gritblasting is the most effective method for removing rust or scale from steel parts. Grease or oil may be removed with a suitable grease solvent.

Enamel, paint, varnish, or heavy films of oxide on aluminum alloys can be removed using a hot 10% solution of either caustic soda or tri-sodium phosphate. After treatment, the parts should be immersed in a 10% nitric acid solution, followed with a hot water rinse to remove all traces of the chemicals. Paint and varnish can also be removed using paint and varnish remover.

The tip of the filler rod should be dipped below the surface of the weld puddle with a motion exactly opposite the motion of the torch. If the filler rod is held above the surface of the puddle, it will melt and fall into the puddle a drop at a time, ruining the weld.

Filler metal should be added until the surface of the joint is built up slightly above the edges of the parts being joined. The puddle of molten metal should be gradually advanced along the seam until the end of the material is reached.

As the end of the seam is approached, the torch should be raised slightly, chilling the molten steel to prevent it from spilling over the edge or melting through the work.

Chrome Molybdenum

The welding technique for chrome molybdenum is practically the same as that for carbon steels, except that the surrounding area must be preheated to a temperature between 300° and 400° F. before beginning to weld. If this is not done, the sudden application of heat causes cracks to form in the heated area.

A soft neutral flame should be used for welding; an oxidizing flame may cause the weld to crack when it cools, and a carburizing flame will make the metal brittle. The volume of the flame must be sufficient to melt the base metal, but not so hot as to weaken the grain structure of the surrounding area and set up strains in the metal. The filler rod should be the same as the base metal. If the weld requires high strength, a special chrome molybdenum rod is used and the piece is heat treated after welding.

Chrome molybdenum thicker than 0.093 in. is usually electric-arc welded, since for this thickness of metal, electric arc provides a narrow heat zone, fewer strains are developed, and a better weld is obtained, particularly when the part cannot be heat treated after welding.

Stainless Steel

The procedure for welding stainless steel is basically the same as that for carbon steels. There are, however, some special precautions that must be taken to obtain the best results.

Only stainless steel used for nonstructural members of aircraft can be welded satisfactorily; the stainless steel used for structural components is cold worked or cold rolled and, if heated, loses some of its strength. Nonstructural stainless steel is obtained in sheet and tubing form and is often used for exhaust collectors, stacks or manifolds. Oxygen combines very readily with this metal in the molten state, and extreme care must be taken to prevent this from occurring.

A slightly carburizing flame is recommended for welding stainless steel. The flame should be adjusted so that a feather of excess acetylene, about $1/16$ in. long, forms around the inner cone. Too much acetylene, however, will add carbon to the metal and cause it to lose its resistance to corrosion. The torch tip size should be one or two sizes smaller than that prescribed for a similar gage of low carbon steel. The smaller tip lessens the chances of overheating and subsequent loss of the corrosion-resistant qualities of the metal.

To prevent the formation of chromium oxide, a

flux should be spread on the underside of the joint and on the filler rod. Since oxidation is to be avoided as much as possible, sufficient flux should be used. Another method used to keep oxygen from reaching the metal is to surround the weld with a blanket of hydrogen gas. This method is discussed later. The filler rod used should be of the same composition as the base metal.

Since the coefficient of expansion of stainless steel is high, thin sheets which are to be butt-welded should be tacked at intervals of 1-¼ to 1-½ inches, as shown in figure 6–17. This is one means of lessening warping and distortion during the welding process.

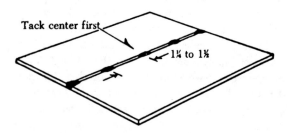

FIGURE 6–17. Tacking method for stainless steel welding.

When welding, hold the filler rod within the envelope of the torch flame so that the rod is melted in place or melted at the same time as the base metal. Add the filler rod by allowing it to flow into the molten pool. If the weld pool is stirred, air will enter the weld and increase oxidation. Avoid re-welding any portion or welding on the reverse side of the weld. Such practices result in warping and overheating of the metal.

WELDING NONFERROUS METALS USING OXY-ACETYLENE

Nonferrous metals are those that contain no iron. Examples of nonferrous metals are lead, copper, silver, magnesium, and most important in aircraft construction, aluminum. Some of these metals are lighter than the ferrous metals, but in most cases they are not as strong. Aluminum manufacturers have compensated for the lack of strength of pure aluminum by alloying it with other metals or by cold working it. For still greater strength, some aluminum alloys are also heat treated.

Aluminum Welding

The weldable aluminum alloys used in aircraft construction are 1100, 3003, 4043, and 5052. Alloy numbers 6053, 6061, and 6151 can also be welded, but since these alloys are in the heat-treated condition, welding should not be done unless the parts can be re-heat treated.

The equipment and technique used for aluminum welding differ only slightly from those of methods discussed earlier. As in all welding, the first step is to clean the surface to be welded—steel wool or a wire brush may be used, or a solvent in the case of paint or grease. The welder should be careful not to scratch the surface of the metal beyond the area to be welded; these scratches provide entry points for corrosion. The piece should then be preheated to lessen the strains caused by the large coefficient of expansion of aluminum.

Never preheat aluminum alloys to a temperature higher than 800° F. because the heat may melt some of the alloys and burn the metal. For thin sheet aluminum, merely passing the flame back and forth across the sheet three or four times should be sufficient.

Either of two types of filler rod can be used in welding aluminum alloys. Choosing the proper filler rod is important.

Aluminum and its alloys combine with air and form oxides very rapidly; oxides form doubly fast if the metal is hot. For this reason it is important to use a flux that will minimize or prevent oxides from forming.

Using the proper flux in welding aluminum is extremely important. Aluminum welding flux is designed to remove the aluminum oxide by chemically combining with it. Aluminum fluxes dissolve below the surface of the puddle and float the oxides to the top of the weld where they can be skimmed off. The flux can be painted directly on the top and bottom of the joint if no filler rod is required; if filler rod is used, it can be coated, and if the pieces to be welded are thick, both the metal and the rod should be coated with flux.

After welding is finished, it is important that all traces of flux be removed by using a brush and hot water. If aluminum flux is left on the weld, it will corrode the metal. A diluted solution of 10% sulfuric acid may be used if hot water is not available. The acid solution should be washed off with cold water.

Thickness of the aluminum alloy material determines the method of edge preparation. On material up to 0.062 in., the edges are usually formed to a 90° flange about the same height as the thickness of the material (figure 6–18A). The flanges should be straight and square. No filler rod is necessary when the edges are flanged in this manner.

Unbeveled butt welds are usually made on aluminum alloy from 0.062 to 0.188 in. thick. It may also be necessary to notch the edges with a saw or cold chisel in a manner similar to that shown in figure 6–18B. Edge notching is recommended in aluminum welding because it aids in getting full penetration and also prevents local distortion. All butt welds in material over 0.125 in. thick are generally notched in some manner.

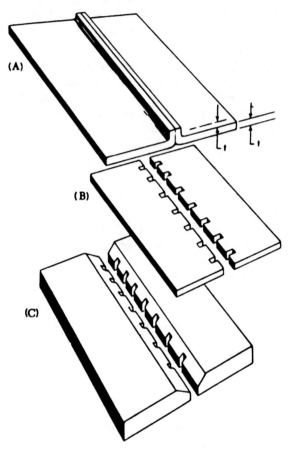

FIGURE 6–18. Edge preparation for welding aluminum.

In welding aluminum over 0.188 in. thick, the edges are usually beveled and notched as shown in figure 6–18C. The included angle of bevel may be from 90° to 120°.

A neutral flame should generally be used to weld aluminum alloys. In some cases a slightly carburizing flame can be used. However, the excess of acetylene should not be too great, as it will be absorbed into the molten metal, resulting in a weakened joint.

The torch must be adjusted to give the mildest flame that can be obtained without popping. The use of a strong, harsh flame makes it difficult to control the melting metal, and holes are often burned through the metal.

When starting to weld, the two joint edges should begin to melt before the filler rod is added. The work must be watched carefully for signs of melting. The melting point of aluminum is low and heat is conducted rapidly through the material. There is very little physical or color change to indicate that the metal is reaching the melting point. When the melting point is reached, the metal suddenly collapses and runs, leaving a hole in the aluminum.

A filler rod can be used to test the metal's condition. Aluminum begins to feel soft and plastic just before it reaches the melting point. Any tendency of the metal to collapse can be rectified by rapidly lifting the flame clear of the metal. With practice it is possible to develop enough skill to melt the metal surface without forming a hole.

The flame should be neutral and slanted at an approximate 45° angle to the metal. The inner cone should be about ⅛ in. from the metal. A constant and uniform movement of the torch is necessary to prevent burning a hole through the metal.

The correct integration of torch and rod action is important when welding aluminum. After heating the metal and when melting has begun, the filler rod is dipped into the pool and allowed to melt. The filler rod is lifted and the torch movement continues as the weld progresses. The rod is never lifted out of the outer envelope of flame, but is held there until almost melted and then added to the pool.

Magnesium Welding

Many aircraft parts are constructed of magnesium because of its light weight, strength, and excellent machinability. This metal is only two-thirds as heavy as aluminum and, like aluminum, is very soft in its pure state. For this reason, it is generally alloyed with zinc, manganese, tin, aluminum, or combinations of these metals. Repair of magnesium by welding is limited by two factors:

(1) If the magnesium is used as a structural member, it is usually heat treated and, like heat-treated aluminum, the welded section can never have the strength of the original metal. (As a rule, failures do not occur in the welded area, but in the areas adjacent to the weld, because the heat applied to

the metal weakens the grain structure in those areas.)

(2) It is necessary to use flux in making all magnesium welds, and to remove all the flux from the metal after welding or severe corrosion will take place.

The type of joint is limited to those that provide no possibility of trapping the flux—therefore, only butt welds can be made. Magnesium cannot be welded to other metals, and magnesium alloy castings are not considered suitable for stressed welds. If varying thicknesses of magnesium are to be welded, the thicker part must be preheated. The filler rod should be of the same composition as the base metal and one prepared by the manufacturer to fuse with his alloy. The filler rod comes with a protective plating that must be cleaned off before using.

The method of preparing the butt joint depends on the thickness of the metal. Sheet magnesium alloy up to 0.040 in. thick should be flanged by about $\frac{3}{32}$ in. to the angle as indicated in figure 6–19. Butt joints on metal from 0.040 to 0.125 in. thick are neither flanged nor beveled, but a $\frac{1}{16}$-in. space should be allowed between the edges of the joint. For butt joints in metal thicker than 0.125 in., each edge should be beveled down 45° to make a 90° included angle for the "V." A $\frac{1}{16}$-in. space should be allowed between the edges of the joint for metal 0.125 to 0.250 in. thick and a $\frac{1}{8}$-in. space for metal 0.250 in. and up (figure 6–19).

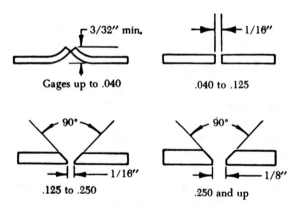

Gages up to .040   .040 to .125

.125 to .250   .250 and up

FIGURE 6–19. Preparation of edges for welding magnesium sheet.

Remove oil or grease with a suitable solvent, and then use a wire brush or abrasive cloth to clean and brighten the metal for a distance of $\frac{3}{4}$ in. back from the weld area. Select a filler rod of the same

material as the base metal. The filler rod and both sides of the seam should be covered with flux. Use a neutral or slightly carburizing flame, and hold it at a flat angle to the work to avoid burning through.

Two rod techniques are recommended for the welding of magnesium. One method requires that the filler rod be kept in the puddle at all times; the other method is the same as that used in welding aluminum.

It is preferable to make the weld on one uninterrupted pass, but if oxidation occurs, the weld should be stopped and scraped out before continuing. The joint edges should be tack-welded at the ends at intervals of $\frac{1}{2}$ to 3 in., depending upon the shape and thickness of the metal.

Welding should be accomplished as quickly and with as little heat as possible. Buckling and warping can be straightened while the metal is still hot by hammering with a soft-faced mallet. The metal should be allowed to cool slowly. When the weld is cool enough to handle, the accessible portions should be scrubbed lightly, using a bristle brush and hot water, to remove excess flux. The part should then be soaked in hot water (160° to 200° F.) to float off the flux adhering to any portions not reached by the scrub brush. When soaking is completed, the part should be immersed in a 1% solution of citric acid for approximately 10 min.

After the citric acid bath the part should be drained thoroughly and then rinsed clean in fresh water. The part must be dried quickly and completely to prevent oxidation.

TITANIUM

Titanium welding does not have the wide application which steel has, therefore this handbook will not treat titanium as extensively.

Titanium Welding

Titanium can be fusion welded with weld efficiencies of 100% by arc welding techniques which, in many respects, are quite similar to those used for other metals.

Certain characteristics must be understood in order to successfully weld titanium.

1. Titanium and its alloys are subject to severe embrittlement by relatively small amounts of certain impurities. Oxygen and nitrogen even in quantities as low as 0.5% will embrittle a weld so much that it is useless.

Titanium, as it is heated toward its melting point will absorb oxygen and nitrogen from the atmosphere. In order to successfully weld titanium, the weld zone must be shielded with an inert gas such as argon or helium.

2. Cleanliness is necessary as titanium is very reactive with most materials. The metal and the welding area must be clean and free of dust, grease, and other contaminants. Contact with ceramic blocks and other foreign materials must be avoided during welding. Coated arc-welding electrodes, and other fluxing compounds will cause contamination and embrittlement.

3. Titanium, when alloyed excessively with other structural metals, looses ductility and impact strength due to the formation of brittle intermetallic compounds and excessive solid solution hardening.

4. Any fusion welding cycle results in a weld zone containing "as-cast material". Additionally, the high heat will have reduced the ductilities of certain highly heat-treatable titanium alloys to an unacceptable condition.

Equipment

Either non-consumable electrode or consumable electrode arc welding equipment may be used for fusion welding of titanium. Whatever type is used the weld must be shielded with an inert gas such as argon or helium.

Titanium may be spot welded with any machine which has accurate control over the four basic spot welding parameters: welding current amperage, duration of welding current (cycles at 60 cycles per second), force applied to the welding electrodes (pounds per square inch), and electrode geometry.

The complexity of titanium welding processes and its limited application outside of specialized titanium fabrication shops does not justify detailed treatment in this handbook.

The foregoing discussion on titanium welding was extracted from Titanium Welding Techniques, published by Titanium Metals Corporation of America.

CUTTING METAL USING OXYACETYLENE

Cutting metals by the oxyacetylene process is fundamentally the rapid burning or oxidizing of the metal in a localized area. The metal is heated to a bright red (1,400° to 1,600° F.), which is the kindling or ignition temperature, and a jet of high-pressure oxygen is directed against it. This oxygen blast combines with the hot metal and forms an intensely hot oxide. The molten oxide is blown down the sides of the cut, heating the metal in its path to a kindling temperature. The metal thus heated also burns to an oxide which is blown away on the underside of the piece. The action is precisely that which the torch accomplishes when the mixing head is replaced with a cutting attachment or when a special cutting torch is used.

Figure 6–20 shows an example of a cutting torch. It has the conventional oxygen and acetylene needle valves, which control the flow of the two gases. Many cutting torches have two oxygen needle valves so that a finer adjustment of the neutral flame can be obtained.

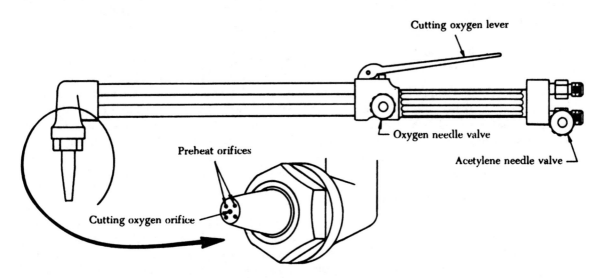

FIGURE 6–20. Cutting torch.

BRAZING METHODS

Brazing refers to a group of metal-joining processes in which the bonding material is a nonferrous metal or alloy with a melting point higher than 800° F., but is lower than that of the metals being joined. Brazing includes silver soldering, also called hard soldering, copper brazing, and aluminum brazing.

Brazing requires less heat than welding and can be used to join metals that are damaged by high heat. However, because the strength of brazed joints is not so great as welded joints, brazing is not used for structural repairs on aircraft. In deciding whether brazing of a joint is justified, it should be remembered that a metal which will be subjected to a sustained high temperature in use should not be brazed.

As the definition of brazing implies, the base metal parts are not melted. The brazing metal adheres to the base metal by molecular attraction and intergranular penetration; it does not fuse and amalgamate with them.

In brazing, the edges of the pieces to be joined are usually beveled as in welding steel. The surrounding surfaces must be cleaned of dirt and rust. Parts to be brazed must be securely fastened together to prevent any relative movement. The strongest brazed joint is one in which the molten filler metal is drawn in by capillary action, thus a close fit must be obtained.

A brazing flux is necessary to obtain a good union between the base metal and the filler metal. A good flux for brazing steel is a mixture containing two parts borax and one part boric acid. Application of the flux may be made in the powder form or dissolved in hot water to a highly saturated solution. A neutral torch flame should be used, moved with a slight semicircular motion.

The base metal should be preheated slowly with a mild flame. When it reaches a dull red heat (in the case of steel), the rod should be heated to a dark or purple color and dipped into the flux. Since enough flux adheres to the rod, it is not necessary to spread it over the surface of the metal.

A neutral flame is used for most brazing applications. However, a slightly oxidizing flame should be used when copper/zinc, copper/zinc/silicon, or copper/zinc/nickel/silicon filler alloys are used. When brazing aluminum and its alloys a neutral flame is preferred, but if difficulties are encountered, a slightly reducing flame is preferred to an oxidizing flame.

The filler rod can now be brought near the tip of the torch, causing the molten bronze to flow over a small area of the seam. The base metal must be at the flowing temperature of the filler metal before it will flow into the joint. The brazing metal melts when applied to the steel and runs into the joint by capillary attraction. The rod should continue to be added as the brazing progresses, with a rhythmic dipping action so that the bead will be built to a uniform width and height. The job should be completed rapidly and with as few passes as possible of the rod and torch.

When the job is finished, the weld should be allowed to cool slowly. After cooling, remove the flux from the parts by immersing them for 30 minutes in a lye solution.

Silver Solder

The principal use of silver solder in aircraft work is in the fabrication of high-pressure oxygen lines and other parts which must withstand vibration and high temperatures. Silver solder is used extensively to join copper and its alloys, nickel and silver, as well as various combinations of these metals, and thin steel parts. Silver soldering produces joints of higher strength than those produced by other brazing processes.

It is necessary to use flux in all silver soldering operations because of the necessity for having the base metal chemically clean without the slightest film of oxide to prevent the silver solder from coming into intimate contact with the base metal.

The joint must be physically clean, which means it must be free of all dirt, grease, oil, and/or paint, and also chemically clean. After removing the dirt, grease, and/or paint, any oxide should be renoved by grinding or filing the piece until bright metal can be seen. During the soldering operation, the flux continues the process of keeping oxide away from the metal, and aids the flow of the solder.

In figure 6–21, three types of joints for silver soldering are shown. Flanged, lap, and edge joints, in which the metal may be formed to furnish a seam wider than the base metal thickness, furnish the type of joint which will bear up under all kinds of loads. If a lap joint is used, the amount of lap should be determined according to the strength needed in the joint. For strength equal to that of the base metal in the heated zone, the amount of lap should be four to six times the metal thickness for sheet metal and small-diameter tubing.

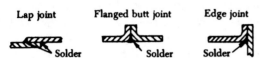

FIGURE 6–21. Silver soldering joints.

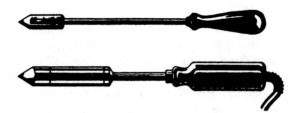

FIGURE 6–22. Soldering copper and soldering iron.

The oxyacetylene flame for silver soldering should be neutral, but may have a slight excess of acetylene. It must be soft, not harsh. During both preheating and application of the solder, the tip of the inner cone of the flame should be held about ½ in. from the work. The flame should be kept moving so that the metal will not become overheated.

When both parts of the base metal are at the right temperature (indicated by the flow of flux), solder can be applied to the surface of the under or inner part at the edge of the seam. It is necessary to simultaneously direct the flame over the seam and keep moving it so that the base metal remains at an even temperature.

SOFT SOLDERING

Soft soldering is used chiefly for copper, brass, and coated iron in combination with mechanical seams; that is, seams that are rivited, bolted, or folded. It is also used where a leakproof joint is desired, and sometimes for fitting joints to promote rigidity and prevent corrosion. Soft soldering is generally performed only in very minor repair jobs. This process is also used to join electrical connections. It forms a strong union with low electrical resistance.

Soft solder yields gradually under a steadily applied load and should not be used unless the transmitted loads are very low. It should never be used as a means of joining structural members.

A soldering copper (called a soldering iron if it is electrically heated) is the tool used in soldering. Its purpose is to act as a source of heat for the soldering operation. The bit, or working face, is made from copper, since this metal will readily take on heat and transmit it to the work. Figure 6–22 shows a correctly shaped bit.

To tin the copper, it is first heated to a bright red, then the point is cleaned by filing until it is smooth and bright. No dirt or pits should remain

on its surface. After the copper has been mechanically cleaned, it should be re-heated sufficiently to melt solder, and chemically cleaned by rubbing it lightly on a block of sal ammoniac. (If sal ammoniac is not available, powdered resin may be used.) Then solder is applied to the point and wiped with a clean cloth.

The last two operations may be combined by melting a few drops of solder on a block of sal ammoniac (cleaning compound) and then rubbing the soldering copper over the block until the tip is well coated with solder. A properly tinned copper has a thin unbroken film of solder over the entire surface of its point.

Soft solders are chiefly alloys of tin and lead. The percentages of tin and lead vary considerably in various solders, with a corresponding change in their melting points, ranging from 293° to 592° F. "half-and-half" (50–50) solder is a general purpose solder and is most frequently used. It contains equal proportions of tin and lead and melts at approximately 360° F.

The application of the melted solder requires somewhat more care than is apparent. The parts should be locked together or held mechanically or manually while tacking. To tack the seam, the hot copper is touched to a bar of solder, then the drops of solder adhering to the copper are used to tack the seam at a number of points. The film of solder between the surfaces of a joint must be kept thin to make the strongest joint.

A hot, well-tinned soldering copper should be held so that its point lies flat on the metal at the seam, while the back of the copper extends over the seam proper at a 45° angle, and a bar of solder is touched to the point. As the solder melts, the copper is drawn slowly along the seam. As much solder as necessary is added without raising the soldering copper from the job. The melted solder should run between the surfaces of the two sheets and cover the full width of the seam. Work should progress along the seam only as fast as the solder will flow into the joint.

265

ELECTRIC ARC WELDING

Electric arc welding is a fusion process based on the principle of generating heat with an electric arc jumping an airgap to complete an electrical circuit. This process develops considerably more heat than an oxyacetylene flame. In some applications, it reaches a temperature of approximately 10,000° F. Variations of the process are metallic arc welding, inert-gas (helium) welding, and multi-arc welding. The metallic arc and helium processes have the widest application in aircraft maintenance.

The welding circuit (figure 6–23) consists of a welding machine, two leads, an electrode holder, an electrode, and the work to be welded. The electrode, which is held in electrode holder, is connected to one lead, and the work to be welded is connected to the other lead. When the electrode is touched to the metal to be welded, the electrical circuit is completed and the current flows. When the electrode is withdrawn from the metal, an airgap is formed between the metal and the electrode. If this gap is of the proper length, the electric current will bridge this gap to form a sustained electric spark, called the electric arc.

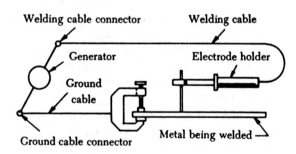

FIGURE 6–23. Typical arc-welding circuit.

Metallic Arc Welding

Metallic arc welding is used mainly for welding low-carbon and low-alloy steels. However, many nonferrous materials, such as aluminum and nickel alloys, can be welded using this method.

To form an arc between the electrode and the work, the electrode is applied to the work and immediately withdrawn. This initiates an arc of intense heat. To maintain the arc between the electrode and the work, the metal electrode must be fed at a uniform rate or maintained at a constant distance from the work as it melts.

Metallic arc welding is a nonpressure fusion welding process which develops welding heat through an arc produced between a metal electrode and the work to be welded. Under the intense heat developed by the arc, a small part of the base metal or work to be welded is brought to the melting point instantaneously. At the same time, the end of the metal electrode is also melted, and tiny globules or drops of molten metal pass through the arc to the base metal. The force of the arc carries the molten metal globules directly into the puddle formed on the base metal, and thus filler metal is added to the part being welded. By moving the metal electrode along the joint and down to the work, a controlled amount of filler metal can be deposited on the base metal to form a weld bead.

The instant the arc is formed, the temperature of the work at the point of welding and the welding electrode increases to approximately 6,500° F. This tremendous heat is concentrated at the point of welding and in the end of the electrode, and simultaneously melts the end of the electrode and a small part of the work to form a small pool of molten metal, commonly called the crater.

The heat developed is concentrated and causes less buckling and warping of work than gas welding. This localization of the heat is advantageous when welding cracks in heat-treated parts and when welding in close places.

Gas Shielded Arc-Welding

A good weld has the same qualities as the base metal. Such a weld has been made without the molten puddle being contaminated by atmospheric oxygen and/or nitrogen. In gas-shielded arc welding, a gas is used as a covering shield around the arc to prevent the atmosphere from contaminating the weld.

The original purpose of gas shielded arc welding was to weld corrosion resistant and other difficult-to-weld materials. Today the various gas shielded arc processes are being applied to all types of metal. See fig. 6–24 for typical applications.

The ease of operation, increased welding speed and the superiority of the weld, will lead to shielded arc-welding and oxy-acetelyne welding being replaced with gas shielded arc welding.

Advantages of Gas-Shielded Arc Welding

The shielding gas excludes the atmosphere from the molten puddle. The resulting weld is stronger, more ductile and more corrosion resistant. Also welding of nonferrous metal does not require the use of flux. This eliminates flux removal, and the possibilities of gas pockets or slag inclusions.

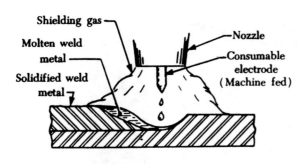

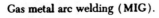

Gas metal arc welding (MIG).

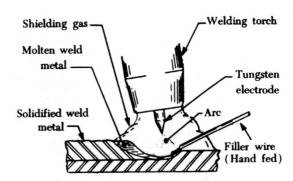

Gas tungsten – arc welding (TIG).

FIGURE 6-24A. Gas-shielded arc welding process.

Figures 6–24, 6–25, 6–26, and 6–27 courtesy Hobart Bros.

Another advantage of the gas shielded arc is that a neater and sounder weld can be made because there is very little smoke fumes or sparks to control. The weld may be observed at all times. Weld splatter is held to a minimum, therefore there is little or no metal finishing required. A gas-shielded weld does not distort the base metal near the weld. The completed weld is clean and free of the complications often encountered in other forms of metallic-arc or gas welding.

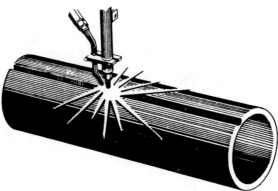

FIGURE 6-24(C). One of the many types of automatic welders.

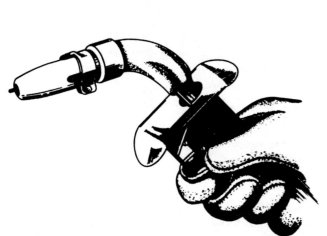

FIGURE 6-24(B). A semi-automatic welder.

Tungsten Inert Gas Welding (TIG)

In Tungsten Inert Gas (TIG) welding a virtually non-consumable tungsten electrode is used (figure 6–25) to provide the arc for welding. During the welding cycle a shield of inert gas expels the air from the welding area and prevents oxidation of the electrode, weld puddle and surrounding heat-affected zone. In TIG welding the electrode is used only to create the arc. If additional metal is needed, a filler rod is used in the same manner as in oxy-acetylene welding.

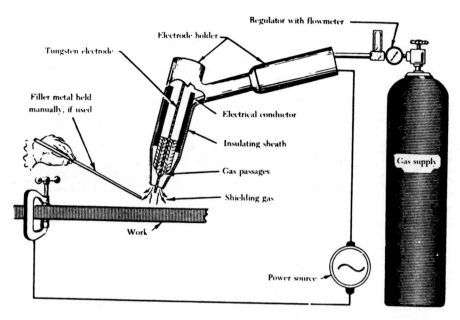

FIGURE 6-25. Typical TIG welding equipment.

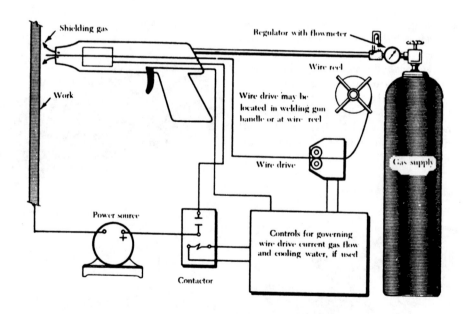

FIGURE 6-26. Typical MIG welding equipment.

The type gas used in TIG welding depends upon the metal being welded. Argon, helium or a mixture of the two gases is used. Argon is used more extensively than helium because it is cheaper. Argon is preferred for several reasons besides the cost. It is heavier therefore provides better cover. It provides better cleaning action when welding aluminum and magnesium. The welding arc is quieter and smoother. Vertical and overhead welding arcs are easier to control. Welding arcs are easier to start and for a given welding the weld produced is narrower with a smaller heat-affected zone.

Helium is used primarily in TIG machine welding or when welding heavy material having high heat conductivity. The arc voltage is higher when using helium, therefore a lower current flow is possible to get the same arc power.

Metal Inert Gas Arc Welding (MIG)

With the substitution of a continuous feed consumable wire electrode for the non-consumable tungsten electrode used in TIG, the welding process becomes Metal Inert Gas Arc Welding (figure 6–26). The wire electrode is fed continuously through the center of the torch at pre-set controlled speeds, shielding gas is fed through the torch, completely covering the weld puddle with a shield of gas. This tends to complete automation of the welding process. Power, gas flow, wire feed and travel over the work piece are preset when using a welding machine. In semi-automatic welding, the operator controls the travel over the work.

Argon is the commonly used gas. Some metals use small amounts of helium or oxygen. Low carbon steel uses carbon dioxide or argon plus 2% O_2.

Plasma Arc Welding

Plasma welding is a process which utilizes a central core of extreme temperature, surrounded by a sheath of coal gas. The required heat for fusion is generated by an electric arc which has been highly intensified by the injection of a gas into the arc stream.

The super-heated columnar arc is concentrated into a narrow stream and when directed on metal makes possible butt welds up to one-half inch in thickness or more in a single pass without filler rods or edge preparation.

In many respects plasma welding may be considered as an extension of the conventional TIG welding. In plasma arc welding the arc column is constricted and it is this constriction that produces the much higher heat transfer rate.

The arc plasma actually becomes a jet of high current density. The arc gas upon striking the metal cuts through the piece producing a small hole which is carried along the weld seam. During this cutting action, the melted metal in front of the arc flows around the arc column, then is drawn together immediately behind the hole by surface tension forces and reforms in a weld bead.

Plasma is often considered the fourth state of matter. The other three are gas, liquid, and solid. Plasma results when a gas is heated to a high temperature, and changes into positive ions, neutral atoms and negative electrons. When matter passes from one state to another latent heat is generated.

In a plasma torch the electrode is located within the nozzle. The nozzle has a relatively small orifice which constricts the arc. The high-pressure gas flows through the arc where it is heated to the plasma temperature range. Since the gas cannot expand due to the constriction of the nozzle, it is forced through the opening, and emerges in the form of a supersonic jet. This heat melts any known metal and its velocity blasts the molten metal through the kerf (figure 6–27).

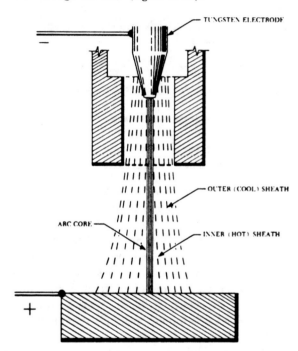

FIGURE 6-27. Plasma welding uses a central core of extreme temperature surrounded by a sheath of cool gas.

WELDING PROCEDURES AND TECHNIQUES

The first step in preparing to arc weld is to make certain that the necessary equipment is available and that the welding machine is properly connected and in good working order. Particular attention should be given to the ground connection, since a poor connection will result in a fluctuating arc, which is difficult to control.

The electrode should be clamped to its holder at right angles to the jaws. Shielded electrodes have one end of the electrode free of coating to provide

269

good electrical contact. The electrode holder should be handled with care to prevent accidental contact with the bench or work, since such contact may weld it fast.

Before starting to weld, the following typical list of items should be checked:

(1) Is the machine in good working order?

(2) Have all connections been properly made? Will the ground connection make good contact?

(3) Has the proper type and size electrode been selected for the job?

(4) Is the electrode properly secured in the holder?

(5) Has sufficient protective clothing been provided, and is it in good condition?

(6) Is the work metal clean?

(7) Does the polarity of the machine coincide with that of the electrode?

(8) Is the machine adjusted to provide the necessary current for striking the arc?

The welding arc is established by touching the plate with the electrode and immediately withdrawing it a short distance. At the instant the electrode touches the plate, a rush of current flows through the point of contact. As the electrode is withdrawn, an electric arc is formed, melting a spot on the base metal and the end of the electrode.

The main difficulty confronting a beginner in striking the arc is freezing; that is, sticking or welding the electrode to the work. If the electrode is not withdrawn promptly upon contact with the plate, the high amperage will flow through the electrode and practically short circuit the welding machine. The heavy current melts the electrode which sticks to the plate before it can be withdrawn.

There are two essentially similar methods of striking the arc. The first is a touch method, illustrated in figure 6–28, and the second is a scratch method, shown in figure 6–29.

When using the touch method, the electrode should be held in a vertical position, and lowered until it is an inch or so above the point where the arc is to be struck. Then the electrode is touched

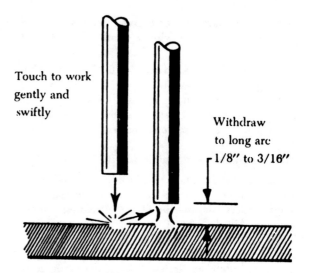

FIGURE 6–28. Touch method of starting the arc.

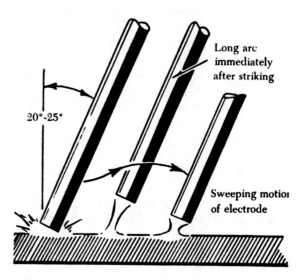

FIGURE 6–29. Scratch method of starting the arc.

very gently and swiftly to the work, using a downward motion of the wrist, followed immediately by withdrawing the electrode to form a long arc (approximately ⅛ to 3/16 in. long).

To strike the arc by the scratch method, the electrode is moved downward until it is just above the plate and at an angle of 20° to 25°, as illustrated in figure 6–29. The arc should be struck gently, with a swiftly sweeping motion, scratching the electrode on the work with a wrist motion. The electrode is then immediately withdrawn to form a long arc. The purpose of holding an excessively long arc immediately after striking is to prevent the large drops of metal, passing across the arc at this

270

time, from shorting out the arc and thus causing freezing.

To form a uniform bead, the electrode must be moved along the plate at a constant speed in addition to the downward feed of the electrode. The rate of advance, if too slow, will form a wide bead resulting in overlapping, with no fusion at the edges. If the rate of advance is too fast, the bead will be too narrow and have little or no fusion at the plate. When proper advance is made, no overlapping occurs, and good fusion is assured.

In advancing the electrode, it should be held at an angle of about 20° to 25° in the direction of travel, as illustrated in figure 6–30.

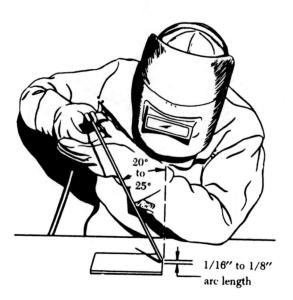

FIGURE 6–30. Angle of electrode.

If the arc is broken during the welding of a bead, a crater will be formed at the point where the arc ends. The arc may be broken by feeding the electrode too slowly or too fast, or when the electrode should be replaced. The arc should not be re-started in the crater of the interrupted bead, but just ahead of the crater on the work metal. Then, the electrode should be returned to the back edge of the crater. From this point, the weld may be continued by welding right through the crater and down the line of weld, as originally planned. Figure 6–31 illustrates the procedure for re-starting the arc.

Every particle of slag must be removed from the vicinity of the crater before re-starting the arc. This prevents the slag from becoming trapped in the weld.

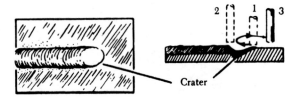

FIGURE 6–31. Re-starting the arc.

Multiple-Pass Welding

Groove and fillet welds in heavy metals often require the deposit of a number of beads to complete a weld. It is important that the beads be deposited in a predetermined sequence to produce the soundest welds with the best proportions. The number of beads is, of course, determined by the thickness of the metal being welded.

The sequence of the bead deposits is determined by the kind of joint and the position of the metal. All slag must be removed from each bead before another bead is deposited. Typical multiple-pass groove welding of butt joints is shown in figure 6–32.

Techniques of Position Welding

Each time the position of a welding joint or the type of joint is changed, it may be necessary to change any one or a combination of the following: (1) Current value, (2) electrode, (3) polarity, (4) arc length, or (5) welding technique.

Current values are determined by the electrode size as well as the welding position. Electrode size is governed by the thickness of the metal and the joint preparation, and the electrode type by the welding position. Manufacturers specify the polarity to be used with each electrode. Arc length is controlled by a combination of the electrode size, welding position, and welding current.

Since it is impractical to cite every possible variation occasioned by different welding conditions, only the information necessary for the commonly used positions and welds is discussed here.

Flat Position Welding

There are four types of welds commonly used in flat position welding. They are the bead, groove, fillet, and lap joint welds. Each type is discussed separately in the following paragraphs.

Bead Welds

Welding a square butt joint by means of stringer beads involves the same techniques as depositing

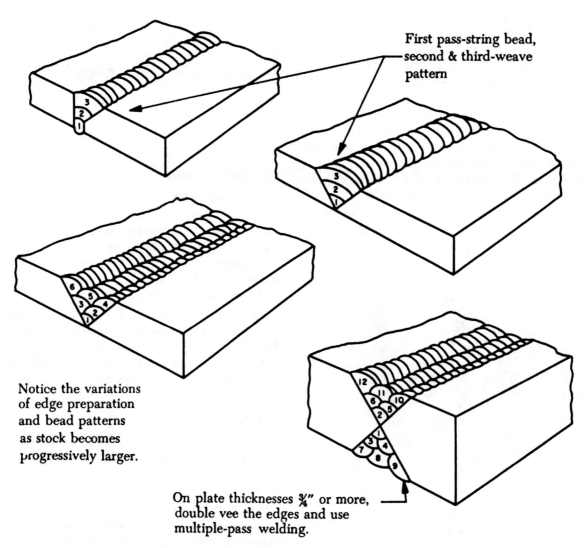

First pass-string bead, second & third-weave pattern

Notice the variations of edge preparation and bead patterns as stock becomes progressively larger.

On plate thicknesses ¾″ or more, double vee the edges and use multiple-pass welding.

FIGURE 6–32. Multiple-pass groove welding of butt joints.

stringer beads on a flat metal surface. Square butt joints may be welded in one, two, or three pases. If the joint is welded with the deposition of one stringer bead, complete fusion is obtained by welding from one side. If the thickness of metal is such that complete fusion cannot be obtained by welding from one side, the joint must be welded from both sides.

When the metals to be welded are butted squarely together, two passes are necessary. If the metals must be spaced, three passes are required to complete the weld. In the latter case, the third pass is made directly over the first and completely envelops it.

It must be constantly kept in mind that beads, either the stringer or weave type, are used to weld all types of joints. Even though the bead may not be deposited on the same type of surface, its action

in the different welding positions and joints is basically the same as its action on the surface of flat metal. The same fundamental rules apply regarding electrode size and manipulation, current values, polarity, and arc lengths.

Bead welds can be made by holding a short arc and welding in a straight line at a constant speed, with the electrode inclined 5° to 15° in the direction of welding. The proper arc can best be judged by recognizing a sharp cracking sound heard all during the time the electrode is being moved to and above the surface of the plate. Some of the characteristics of good bead welds are as follows:

(1) They should leave very little spatter on the surface of the plate.
(2) The arc crater, or depression, in the bead when the arc has been broken should be approximately $\frac{1}{16}$ in. deep.

272

(3) The depth of the crater at the end of the bead can be used as a measure of penetration into the base metal.

(4) The bead weld should be built up slightly, without any weld metal overlap at the top surface, which would indicate poor fusion.

Figure 6–33 illustrates a properly made bead weld.

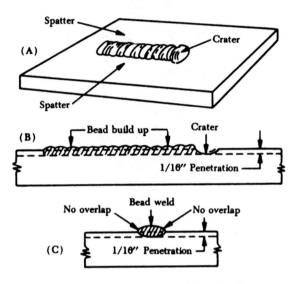

FIGURE 6–33.   Properly made bead weld.

Groove Welds (Butt Joint)

Groove welding may be executed in either a butt joint or an outside corner joint. A outside corner joint corresponds to a single vee butt joint, and the same welding technique is used for both. For this reason, these two types of joints are classified under the heading of grooved welding. There are certain fundamentals which are applicable to groove welds, regardless of the position of the joint.

Groove welds are made on butt joints where the metal to be welded is ¼ in. or more in thickness. Butt joints with a metal thickness of less than ¼ in. require no special edge preparation and can be joined with a bead weld on one or both sides.

Groove welds can be classified as either single groove or double groove. This is true whether the shape of the groove is a V, U, J, or any other form. Regardless of the position in which a single-groove weld is made, it can be welded with or without a backing strip. If a backing strip is used, the joint may be welded from only one side. When a single-groove weld is made without a backing strip, the weld may be made from one side, if necessary, although welding from both sides assures better fusion.

The first pass of the weld deposit may be from either side of the groove. The first bead should be deposited to set the space between the two plates and to weld the root of the joint. This bead, or layer off weld metal, should be thoroughly cleaned to remove all slag before the second layer of metal is deposited. After the first layer is cleaned, each additional layer should be applied with a weaving motion, and each layer should be cleaned before the next one is applied.

The number of passes required to complete a weld is determined by the thickness of the metal being welded and the electrode size being used. As in bead welding, the tip of the electrode should be inclined between 5° and 15° in the direction of welding.

Double-groove welds are welded from both sides. This type of weld is used primarily on heavy metals to minimize distortion. This is best accomplished by alternately welding from each side; i.e., depositing a bead from one side and then from the other. However, this necessitates turning the plates over several times (six times for ¾-in. plate.)

Distortion may be effectively controlled if the plates are turned over twice, as follows: (1) Weld half the passes on the first side; (2) turn the plate over and weld all the passes on the second side; and (3) turn the plates over and complete the passes on the first side.

The root of a double-groove weld should be made with a narrow bead, making sure that the bead is uniformly fused into each root face. When a few passes have been made on one side, the root on the opposite side should be chipped to sound metal to make the groove and then welded with a single-bead weld.

Any groove weld made in more than one pass must have the slag, spatter, and oxide carefully removed from all previous weld deposits before welding over them.   Figure 6–34 shows some of the common types of groove welds performed on butt joints in the flat position.

Fillet Welds

Fillet welds are used to make tee and lap joints. In welding tee joints in the flat position, the two plates are placed to form an angle of 90° between their surfaces, as shown in figure 6–35.   The electrode should be held at an angle of 45° to the plate surface. The top of the electrode should be tilted at an angle of about 15° in the direction of welding. Light plates should be welded with little or no

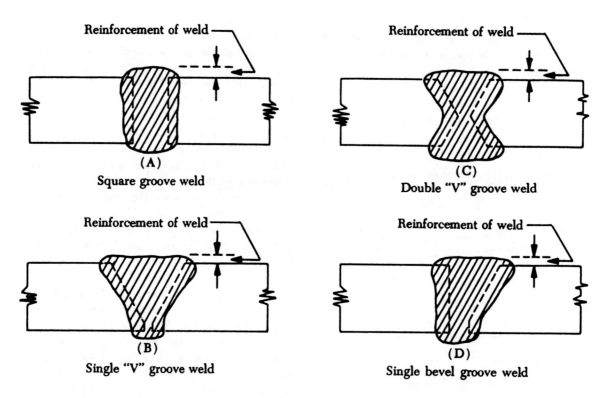

FIGURE 6–34. Groove welds on butt joints in the flat position.

weaving motion of the electrode, and the weld is made in one pass. Fillet welding of heavier plates may require two or more passes. In that case, the second pass or layer is made with a semicircular weaving motion. In making the weave bead, there should be a slight pause at the end of each weaving motion to obtain good fusion to the edges of the two plates without undercutting them.

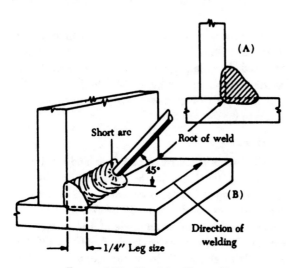

FIGURE 6–35. Tee joint fillet weld.

The procedure for making the lap joint fillet weld is similar to that used for making the fillet weld in a tee joint. The electrode should be held at an angle of 30° to the vertical. The top of the electrode should be tilted to an angle of 15° in the direction of welding. Figure 6–36 illustrates a typical lap joint. The weaving motion is the same as that used for tee joints, except that the hesitation at the edge of the top plate is prolonged to obtain good fusion with no undercut. When welding plates of different thickness, the electrode is held at an angle of 20° to the vertical. Care must be taken not to overheat and undercut the thinner plate edge. The arc must be controlled to wash up the molten metal to the edge of this plate.

Overhead Position Welding

The overhead position is one of the most difficult in welding, since a very short arc must be maintained constantly to retain complete control of the molten metal.

The force of gravity tends to cause the molten metal to drop down or sag on the plate. If a long arc is held, the difficulty in transferring metal from the electrode to the base metal is increased, and

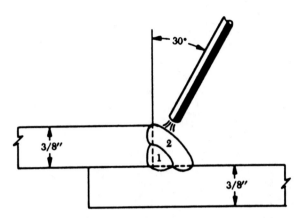

FIGURE 6-36. Typical lap joint fillet weld.

large globules of molten metal will drop from the electrode and the base metal. The transfer of metal is aided by first shortening and then lengthening the arc. However, care should be taken not to carry too large a pool of molten metal in the weld. The procedures for making bead, groove, and fillet welds in the overhead position are discussed in the following paragraphs.

Bead Welds

For bead welds, the electrode should be held at an angle of 90° to the base metal. In some cases, however, where it is desirable to observe the arc and the crater of the weld, the electrode may be held at an angle of 15° in the direction of welding. Weave beads can be made by using the weaving motion. A rather rapid motion is necessary at the end of each semicircular weave to control the molten metal deposit. Care should be taken to avoid excessive weaving. This will cause overheating of the weld deposit and form a large pool of metal, which is hard to control.

Groove Welds (Butt Joints)

Improved overhead groove welds can be made by using a backup strip. The plates should be prepared in a manner similar to preparing plates for welding butt joints in the flat position. If no backup strip is used and the plates are beveled with a featheredge, the weld will burn through repeatedly, unless the operator is extremely careful.

Fillet Welds

When making fillet welds on overhead tee or lap joints, a short arc should be held, and there should be no weaving of the electrode, The electrode should be held at an angle of about 30° to the

vertical plate, and moved uniformly in the direction of welding.

The arc motion should be controlled to secure good penetration to the root of the weld and good fusion with the sidewalls of the vertical and horizontal plates. If the molten metal becomes too fluid and tends to sag, the electrode should be whipped away quickly from the crater ahead of the weld to lengthen the arc and allow the metal to solidify. The electrode should then be returned immediately to the crater of the weld and the welding continued.

Welding on heavy plates requires several passes to make the joint. The first pass is a string bead with no weaving motion of the electrode. The second, third, and fourth passes are made with a slight circular motion of the end of the electrode, while the top of the electrode is held tilted at an angle of about 15°.

Vertical Position Welding

The vertical position, like the overhead position just discussed, is also more difficult than welding in the flat position. Because of the force of gravity, the molten metal will always have a tendency to run down. To control the flow of molten metal, a short arc is necessary, as well as careful arc voltage and welding current adjustments.

In metallic arc welding, current settings for welds made in the vertical position should be less than those used for the same electrode size and type on welds made in the flat position. The currents used for welding upward on vertical plate are slightly lower than those used for welding downward on vertical plate. The procedure for making bead, groove, and fillet welds in the vertical position are discussed in the following paragraphs.

Bead Welds

When making vertical bead welds, it is necessary to maintain the proper angle between the electrode and the base metal to deposit a good bead. In welding upward, the electrode should be held at an angle of 90° to the vertical. When weaving is necessary, the electrode should be oscillated with a "whipping up" motion. In welding downward, bead welds should be made by holding the top end of the electrode at an angle of about 15° below the horizontal to the plate with the arc pointed upward toward the oncoming molten metal. When a weave bead is necessary, in welding downward, a slight semicircular movement of the electrode is necessary.

In depositing a bead weld in the horizontal plane on a vertical plate, the electrode should be held at right angles to the vertical. The top of the electrode should be tilted at an angle at about 15° toward the direction of welding to obtain a better view of the arc and crater. The welding currents used should be slightly less than those required for the same type and size of electrode in flat position welding.

Groove Welds (Butt Joints)

Butt joints in the vertical position are "groove welded" in a manner similar to the welding of butt joints in the flat position. To obtain good fusion with no undercutting, a short arc should be held, and the motion of the electrode should be carefully controlled.

Butt joints on beveled plates ¼ in. in thickness can be groove welded by using a triangular weaving motion. In groove welding butt joints in the horizontal position on identical plates, a short arc is necessary at all times. The first pass is made from left to right or right to left, with the electrode held at an angle of 90° to the vertical plate. The second, third, and, if required, any additional passes are made in alternate steps, with the electrode held approximately parallel to the beveled edge opposite to the one being welded.

Fillet Welds

When making fillet welds in either tee or lap joints in the vertical position the electrode should be held at an angle of 90° to the plates or at an angle of up to 15° below the horizontal, for better control of the molten puddle. The arc should also be held short to obtain good penetration, fusion, and molten metal control.

In welding tee joints in the vertical position, the electrode should be moved in a triangular weaving motion. The joint should be started at the bottom and welded upwards. A slight hesitation in the weave, as shown in figure 6–37, will improve sidewall penetration and allow good fusion at the root of the joint. If the weld metal overheats, the electrode should be lifted away quickly at short rapid intervals without breaking the arc. This will allow the molten metal to solidify without running down. The electrode should be returned immediately to the crater of the weld to maintain the desired size of the weld.

When more than one layer of metal is needed to make a vertical tee weld, different weaving motions may be used. A slight hesitation at the end of the weave will result in good fusion without undercutting the plate at the edges of the weld. When welding lap welds in the vertical position, the same pro-

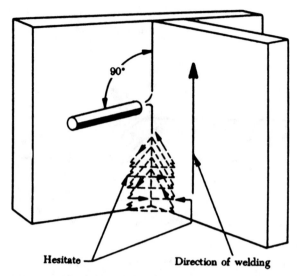

For 1/4", 3/8", and 1/2" size fillets

FIGURE 6–37. Vertical tee joint fillet weld.

cedure is followed as that outlined for welding vertical tee joints, except that the electrode is directed more toward the one vertical plate. Care should be taken not to undercut either plate, or to allow the molten metal to overlap the edges of the weave. On heavy plate, lap joints require more than one layer of metal.

WELDING OF AIRCRAFT STEEL STRUCTURES

Oxyacetylene or electric arc welding may be utilized for repair of some aircraft structures, since most aircraft structures are fabricated from one of the weldable alloys; however, careful consideration should be given to the alloy being welded since all alloys are not readily weldable. Also, certain structural parts may be heat treated and therefore could require special handling. In general, the more responsive an alloy steel is to heat treatment, the less suitable it is for welding because of its tendency to become brittle and lose its ductility in the welded area. The following steels are readily weldable: (1) Plain carbon of the 1000 series, (2) nickel steel of the SAE 2300 series, (3) chrome/nickel alloys of the SAE 3100 series, (4) chrome/molybdenum steels of the SAE 4100 series, and (5) low-chrome/molybdenum steel of the SAE 8600 series.

Aircraft Steel Parts Not To Be Welded

Welding repairs should not be performed on aircraft parts whose proper function depends on strength properties developed by cold working, such as streamlined wires and cables.

276

Brazed or soldered parts should never be repaired by welding, since the brazing mixture or solder can penetrate the hot steel and weaken it.

Aircraft parts such as turnbuckle ends and aircraft bolts which have been heat treated to improve their mechanical properties should not be welded.

Repair of Tubular Members

Welded steel tubing can usually be spliced or repaired at any joint along the length of the tube, but particular attention should be given to the proper fit and alignment to avoid distortion. Some of the many acceptable practices are outlined in the following paragraphs.

Dents at a steel tube cluster-joint can be repaired by welding a specially formed steel patch plate over the dented area and surrounding tubes, as shown in figure 6–38.

To prepare the patch plate, a section of steel sheet is cut from the same material and thickness as the heaviest tube damaged. The reinforcement plate is trimmed so that the fingers extend over the tubes a minimum of one and one-half times the respective tube diameters (figure 6–38). The reinforcement plate may be formed before any welding is attempted, or it may be cut and tack welded to one or more of the tubes in the cluster-joint, then heated and formed around the joint to produce a smooth contour. Sufficient heat should be applied to the plate during the forming process so that no gap exists. If a gap exists it should not exceed $\frac{1}{16}$ in. from the contour of the joint to the plate. After the plate is formed and tack welded to the cluster-joint, all the reinforcement plate edges are welded to the cluster-joint.

Repair by Welded Sleeve

This type of repair of a dented or bent tube is illustrated in figure 6–39. The repair material selected should be a length of steel tube sleeving having an inside diameter approximately equal to the outside diameter of the damaged tube and of the same material and wall thickness. This sleeve reinforcement should be cut at a 30° angle on both ends so that the minimum distance of the sleeve from the edge of the crack or dent is not less than one and one-half times the diameter of the damaged tube.

After the angle cuts have been made to the ends, the entire length of the reinforcement sleeve should be cut, separating the sleeve into half-sections (figure 6–39). The two sleeve sections are then clamped to the proper positions on the affected areas of the original tube. The sleeve is welded along the length of the two sides, and both ends are welded to the damaged tube, as shown in figure 6–39.

Repair by Bolted Sleeve

Bolted sleeve repairs on welded steel tubular structure are not recommended unless specifically authorized by the manufacturer or the Federal Aviation Administration. The material removed by drilling the boltholes in this type of repair may prove to weaken the tubular structure critically.

Welded-Patch Repair

Dents or holes in tubing can be safely repaired by a welded patch of the same material but one gage thicker, as illustrated in figure 6–40, with the following exceptions:

(1) Do not use a welded patch to repair dents deeper than one-tenth of the tube diameter, dents that involve more than one-fourth of the tube circumference, or those longer than the tube diameter.

(2) Use welded-patch repairs only if dents are free from cracks, abrasions, and sharp corners.

(3) Use welded-patch repairs only when the dented tubing can be substantially reformed without cracking before application of the patch.

(4) In the case of punctured tubing, use welded-patch repairs if the holes are not longer than the tube diameter and involve not more than one-fourth of the tube circumference.

Splicing Tubing by Inner Sleeve Method

If the damage to a structural tube is such that a partial replacement of the tube is necessary, the inner sleeve splice shown in figure 6–41 is recommended, especially where a smooth tube surface is desired. A diagonal cut is made to remove the damaged portion of the tube, and the burrs are removed from the edges of the cut by filing or similar means. A replacement steel tube of the same material and diameter, and at least the same wall thickness is then cut to match the length of the removed portion of the damaged tube. At each end of the replacement tube a $\frac{1}{8}$-in. gap should be allowed from the diagonal cuts to the stubs of the original tube.

A length of steel tubing should next be selected of at least the same wall thickness and of an outside

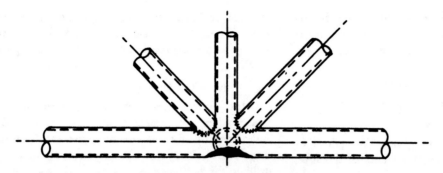

Longeron dented at a station.

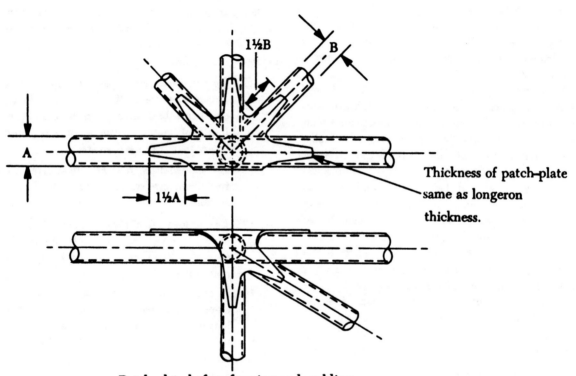

1½B

B

A

1½A

Thickness of patch-plate same as longeron thickness.

Patch plate before forming and welding.

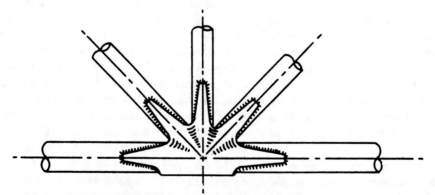

Patch plate formed and welded to tubes.

FIGURE 6–38. Repair of members dented at a cluster.

Dented or bent tube.

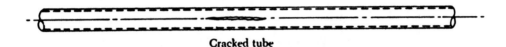

Cracked tube

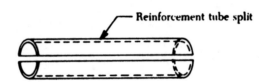

Reinforcement tube split

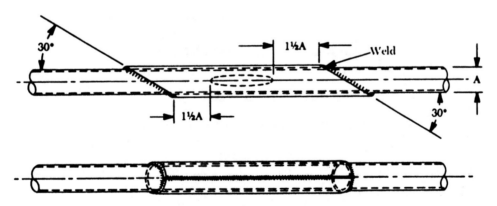

FIGURE 6–39. Repair by welded sleeve.

diameter equal to the inside diameter of the damaged tube. This inner tube material should be fitted snugly within the original tube. Cut two sections of tubing from this inner-sleeve tube material, each of such a length that the ends of the inner sleeve will be a minimum distance of one and one-half tube diameters from the nearest end of the diagonal cut.

If the inner sleeve fits very tightly in the replacement tube, the sleeve can be chilled with dry ice or in cold water. If this procedure is inadequate, the diameter of the sleeve can be polished down with emery cloth. The inner sleeve can be welded to the tube stubs through the ⅛-in. gap, forming a weld bead over the gap.

Engine Mount Repairs

All welding on an engine mount should be of the highest quality, since vibration tends to accentuate any minor defect. Engine-mount members should preferably be repaired by using a larger diameter replacement tube telescoped over the stub of the original member, using fishmouth and rosette welds. However, 30° scarf welds in place of the fishmouth welds are usually considered acceptable for engine mount repair work.

Repaired engine mounts must be checked for accurate alignment. When tubes are used to replace bent or damaged ones, the original alignment of the

279

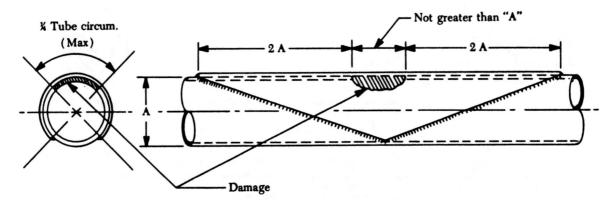

FIGURE 6–40. Welded-patch repair.

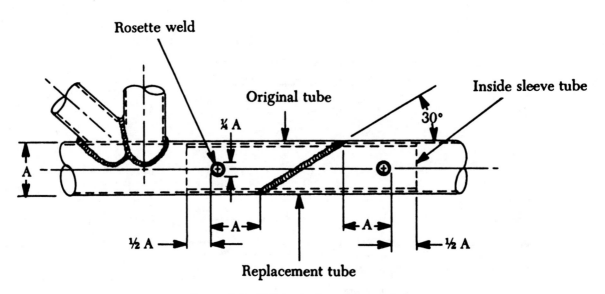

FIGURE 6–41. Splicing by inner-sleeve method.

structure must be maintained. This can be done by measuring the distance between points of corresponding members that have not been distorted, and by reference to the manufacturer's drawings.

If all members are out of alignment, the engine mount should be replaced by one supplied by the manufacturer or one built to conform to the manufacturer's drawings. The method of checking the alignment of the fuselage or nacelle points can be requested from the manufacturer.

Minor damage, such as a crack adjacent to an engine attachment lug, can be repaired by re-welding the ring and extending a gusset or a mounting lug past the damaged area. Engine mount rings which are extensively damaged must not be repaired, unless the method of repair is specifically approved by an authorized representative of the Federal Aviation Administration, or is accomplished using instructions furnished by the aircraft manufacturer.

Repair at Built-In Fuselage Fittings

An example of a recommended repair at built-in fuselage fittings is illustrated in figure 6–42. There are several acceptable methods for effecting this type of repair. The method illustrated in figure 6–42 utilizes a tube (sleeve) of larger diameter than the original. This necessitates reaming the fitting holes in the longeron to a larger diameter. The forward splice is a 30° scarf splice. The rear longeron is cut approximately 4 in. from the center line of the joint, and a spacer 1 in. long is fitted over the longeron. The spacer and longeron are edge-welded. A tapered "V" cut approximately 2 in. long is made in the aft end of the outer sleeve, and the end of the outer sleeve is swaged to fit the longeron and then welded.

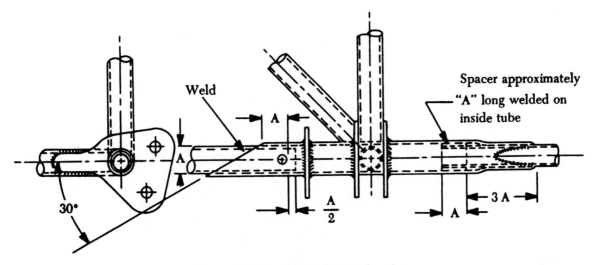

FIGURE 6-42. Repair at built-in fuselage fitting.

Landing Gear Repair

Landing gear made of round tubing is generally repaired using repairs and splices illustrated in figures 6-39 and 6-42. One method of repairing landing gear made of streamlined tubing is shown in figure 6-43.

Representative types of repairable and nonrepairable landing gear axle assemblies are shown in figure 6-44. The types shown in A, B, and C of this figure are formed from steel tubing and may be repaired by any of the methods described in this section. However, it will always be necessary to

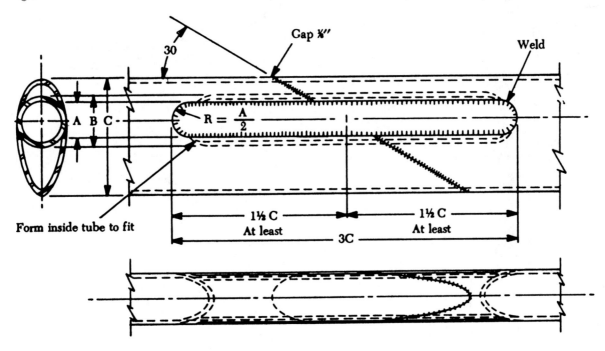

A — Slot width (original tube)

B — Outside diameter (insert tube)

C — Streamline tube length of major axis

FIGURE 6-43. Streamlined tube splice on landing gear using round tube.

281

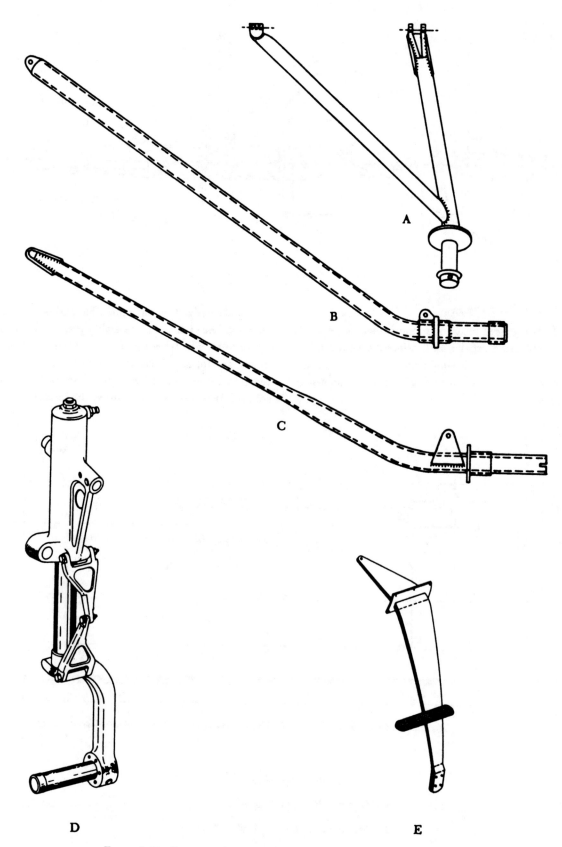

FIGURE 6–44. Representative types of repairable and nonrepairable assemblies.

ascertain whether or not the members are heat treated. Assemblies originally heat treated must be re-heat treated after welding.

The axle assembly shown in D of figure 6–44 is, in general, of a nonrepairable type for the following reasons:

(1) The axle stub is usually made from a highly heat treated nickel alloy steel and carefully machined to close tolerances. These stubs are usually replaced when damaged.

(2) The oleo portion of the structure is generally heat treated after welding and is perfectly machined to assure proper functioning of the shock absorber. These parts would be distorted by welding after the machining process.

A spring-steel leaf, shown in E of figure 6–44, supports each main landing gear wheel assembly on many light aircraft. These springs are, in general, nonrepairable and should be replaced when they become excessively sprung or are otherwise damaged.

Built-Up Tubular Wing or Tail Surface Spar Repair

Built-up tubular wing or tail surface spars can be repaired by using any of the splices and methods outlined in the discussion on welding of aircraft steel structures, provided the spars are not heat treated. In the case of heat treated spars, the entire spar assembly must be re-heat treated to the manufacturer's specifications after completion of the repair.

Wing and Tail Brace Struts

In general, it is advantageous to replace damaged wing-brace struts made either from rounded or streamlined tubing with new members purchased from the aircraft manufacturer. However, there is usually no objection from an airworthiness point of view to repairing such members properly. Members made of round tubes using a standard splice can be repaired as shown in figures 6–39 or 6–41.

Steel brace struts may be spliced at any point along the length of the strut, provided the splice does not overlap part of an end fitting. The jury strut attachment is not considered an end fitting; therefore, a splice may be made at this point. The repair procedure and workmanship should be such as to minimize distortion due to welding and the necessity for subsequent straightening operations. The repaired strut should be observed carefully during initial flights to ascertain that the vibration characteristics of the strut and attaching components are not adversely affected by the repair. A wide range of speed and engine power combinations must be covered during this check.

GENERAL

Rain, snow, and ice are transportation's ancient enemies. Flying has added a new dimension, particularly with respect to ice. Under certain atmospheric conditions, ice can build rapidly on airfoils and air inlets.

The two types of ice encountered during flight are rime and glaze. Rime ice forms a rough surface on the aircraft leading edges. It is rough because the temperature of the air is very low and freezes the water before it has time to spread. Glaze ice forms a smooth, thick coating over the leading edges of the aircraft. When the temperature is just slightly below freezing, the water has more time to flow before it freezes.

Ice may be expected to form whenever there is visible moisture in the air and the temperature is near or below freezing. An exception is carburetor icing which can occur during warm weather with no visible moisture present. If ice is allowed to accumulate on the wings and empennage leading edges, it destroys the lift characteristics of the airfoil. Ice or rain accumulations on the windshield interfere with vision.

Icing Effects

Ice on an aircraft affects its performance and efficiency in many ways. Ice buildup increases drag and reduces lift. It causes destructive vibration, and hampers true instrument readings. Control surfaces become unbalanced or frozen. Fixed slots are filled and movable slots jammed. Radio reception is hampered and engine performance is affected (figure 7–1).

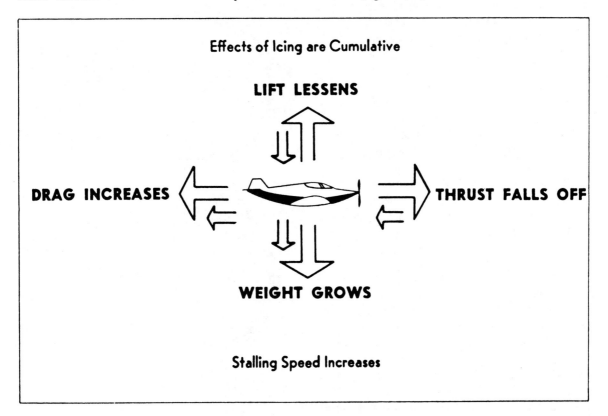

FIGURE 7–1. Effects of structural icing.

The methods used to prevent icing (anti-icing) or to eliminate ice that has formed (deicing) vary with the aircraft make and model. In this Chapter ice prevention and ice elimination using pneumatic pressure, application of heat, and the application of fluid will be discussed.

Ice Prevention

Several means to prevent or control ice formation are used in aircraft today: (1) Heating surfaces using hot air, (2) heating by electrical elements, (3) breaking up ice formations, usually by inflatable boots, and (4) alcohol spray. A surface may be anti-iced either by keeping it dry by heating to a temperature that evaporates water upon impingement; or by heating the surface just enough to prevent freezing, maintaining it running wet; or the surface may be deiced by allowing ice to form and then removing it.

Ice prevention or elimination systems ensure safety of flight when icing conditions exist. Ice may be controlled on aircraft structure by the following methods.

Location of Ice	Method of Control
1. Leading edge of the wing	Pneumatic, Thermal
2. Leading edges of vertical and horizontal stabilizers	Pneumatic, Thermal
3. Windshields, windows, and radomes	Electrical, Alcohol
4. Heater and Engine air inlets	Electrical
5. Stall warning transmitters	Electrical
6. Pitot tubes	Electrical
7. Flight controls	Pneumatic, Thermal
8. Propeller blade leading edges	Electrical, Alcohol
9. Carburetors	Thermal, Alcohol
10. Lavatory drains	Electrical

PNEUMATIC DEICING SYSTEMS

Pneumatic deicing systems use rubber deicers, called boots or shoes, attached to the leading edge of the wing and stabilizers. The deicers are composed of a series of inflatable tubes. During operation, the tubes are inflated with pressurized air, and deflated in an alternating cycle as shown in figure 7–2. This inflation and deflation causes the ice to crack and break off. The ice is then carried away by the airstream.

Deicer tubes are inflated by an engine-driven air pump (vacuum pump), or by air bled from gas turbine engine compressors. The inflation sequence is controlled by either a centrally located distributor valve or by solenoid operated valves located adjacent to the deicer air inlets.

Deicers are installed in sections along the wing with the different sections operating alternately and symmetrically about the fuselage. This is done so that any disturbance to airflow caused by an inflated tube will be kept to a minimum by inflating only short sections on each wing at a time.

Deicing system not operating. Cells lie close to airfoil section. Ice is permitted to form.

Flexible hose

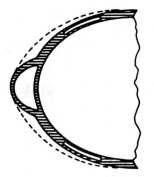

After deicer system has been put into operation, center cell inflates, cracking ice.

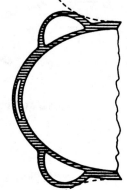

When center cell deflates, outer cells inflate. This raises cracked ice causing it to be blown off by air stream.

Figure 7–2. Deicer boot inflation cycle.

286

DEICER BOOT CONSTRUCTION

Deicer boots are made of soft, pliable rubber or rubberized fabric and contain tubular air cells. The outer ply of the deicer is of conductive neoprene to provide resistance to deterioration by the elements and many chemicals. The neoprene also provides a conductive surface to dissipate static electricity charges. These charges, if allowed to accumulate, would eventually discharge through the boot to the metal skin beneath, causing static interference with the radio equipment.

Deicer boots are attached to the leading edge of wing and tail surfaces with cement or fairing strips and screws, or a combination of both.

Deicer boots which are secured to the surface with fairing strips and screws or a combination of fairing strips, screws, and cement have a bead and bead wire on each lengthwise edge. On this type installation, screws pass through a fairing strip and the deicer boot just ahead of the bead wire and fit into Rivnuts located permanently in the skin of the aircraft.

The new type deicer boots (figure 7–3) are completely bonded to the surface with cement. The trailing edges of this type boot are tapered to provide a smooth airfoil. By eliminating the fairing strips and screws, this type installation cuts down on the weight of the deicer system.

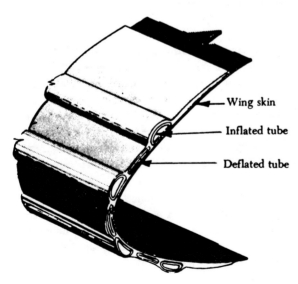

FIGURE 7–3. Deicer boot cross section.

Wing skin

Inflated tube

Deflated tube

The deicer boot air cells are connected to system pressure and vacuum lines by nonkinking flexible hose.

In addition to the deicer boots, the major components of a representative pneumatic deicing system are a source of pressurized air, an oil separator, air pressure and suction relief valves, a pressure regulator and shutoff valve, an inflation timer, and a distributor valve or a control valve. A schematic of a typical system is shown in figure 7–4.

In this system, air pressure for system operation is supplied by air bled from the engine compressor. The bleed air from the compressor is ducted to a pressure regulator. The regulator reduces the pressure of the turbine bleed air to the deicer system pressure. An ejector, located downstream of the regulator, provides the vacuum necessary to keep the boots deflated.

The air pressure and suction relief valves and regulators maintain the pneumatic system pressure and suction at the desired settings. The timer is essentially a series of switch circuits actuated successively by a solenoid-operated rotating step switch. The timer is energized when the deicing switch is placed in the "on" position.

When the system is operated, the deicer port in the distributor valve is closed to vacuum and system operating pressure is applied to the deicers connected to that port. At the end of the inflation period the deicer pressure port is shut off, and air in the deicer flows overboard through the exhaust port. When the air flowing from the deicers reaches a low pressure (approximately 1 p.s.i.), the exhaust port is closed. Vacuum is re-applied to exhaust the remaining air from the deicer.

This cycle is repeated as long as the system is operating. If the system is turned off, the system timer automatically returns to its starting position.

A pneumatic deicing system that uses an engine-driven air pump is shown in figure 7–5. The right hand side of the system is illustrated, however, the left hand side is identical. Notice that inflatable deicers are provided for the wing leading edges and the horizontal stabilizer leading edges. Included in the system are two engine-driven air (vacuum) pumps, two primary oil separators, two combination units, six distributor valves, an electronic timer, and the control switches on the deicing control panel. To indicate system pressure, a suction indicator and a pressure indicator are included in the system.

Pneumatic System Operation

As shown in figure 7–5, the deicer boots are arranged in sections. The right-hand wing boots include two sections: (1) An inboard (inner boot

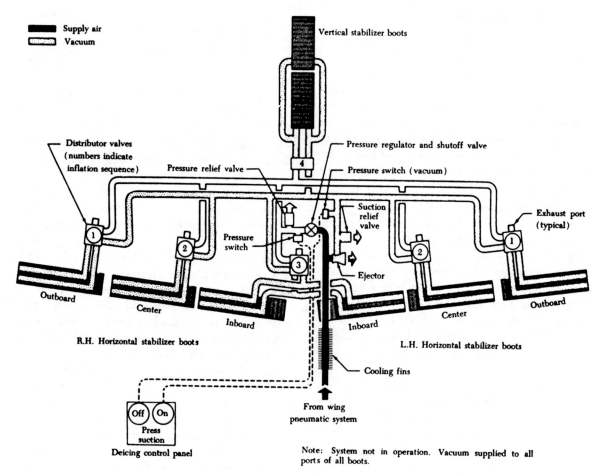

Supply air
Vacuum

Vertical stabilizer boots

Distributor valves
(numbers indicate
inflation sequence)

Pressure regulator and shutoff valve

Pressure relief valve

Pressure switch (vacuum)

Suction
relief
valve

Pressure
switch

Ejector

Exhaust port
(typical)

Outboard

Center

Inboard

Inboard

Center

Outboard

R.H. Horizontal stabilizer boots

L.H. Horizontal stabilizer boots

Cooling fins

From wing
pneumatic system

Off   On

Press
suction

Deicing control panel

Note: System not in operation. Vacuum supplied to all
ports of all boots.

FIGURE 7–4. Schematic of a pneumatic deicing system.

A1 and outer boot B2) section and (2) an out-board (inner boot A3 and outer boot B4) section. The right-hand horizontal stabilizer has two boot sections (inner boot A5 and outer boot B6). A distributor valve serves each wing boot section and another distributor valve serves both horizontal stabilizer boot sections. Notice that each distributor valve has a pressure inlet port, a suction outlet port, a dump port, and two additional ports (A and B). Distributor valve A and B ports are connected to related boot A and B ports. Pressure and suction can be alternated through A and B ports by the movement of a distributor valve solenoid servo valve. Note also that each distributor valve is connected to a common pressure manifold and a common suction manifold. When the pneumatic deicing system is on, pressure or suction is applied by either or both engine-driven air (vacuum) pumps. The suction side of each pump is connected to the suction manifold. The pressure side of each

pump is connected through a pressure relief valve to the pressure manifold. The pressure relief valve maintains the pressure in the pressure manifold at 17 p.s.i. The pressurized air then passes to the primary oil separator. The oil separator removes any oil from the air. Oil-free air is then delivered to the combination unit. The combination unit directs, regulates to 15 p.s.i., and filters the air supply to the distributor valves.

When the pneumatic deicing system is off, air pump suction, regulated at 4 in. Hg by an adjustable suction relief valve, holds the deicing boots deflated. Air pump pressure is then directed overboard by the combination unit.

DEICING SYSTEM COMPONENTS

Engine-Driven Air Pump

The engine-driven air pump is of the rotary, four vane, positive displacement type and is mounted on

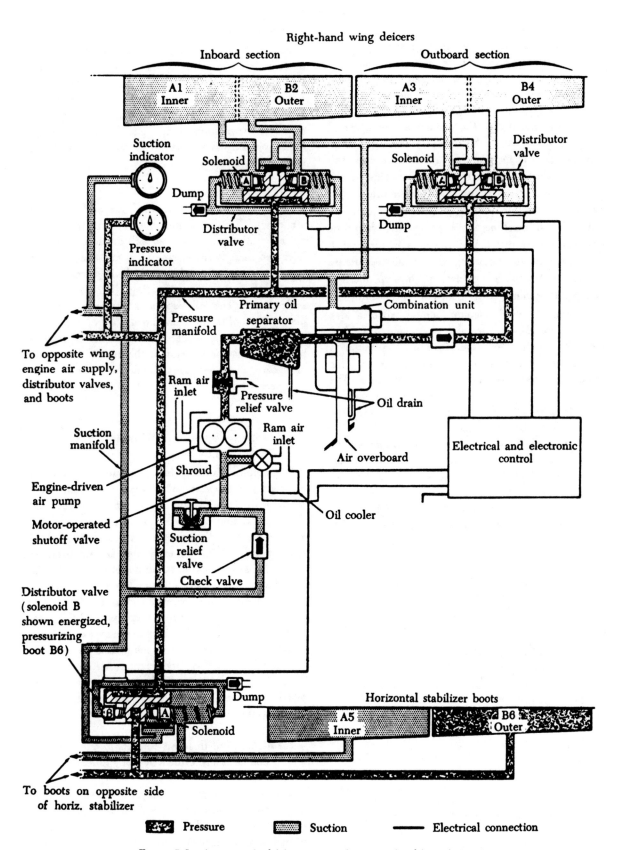

FIGURE 7-5. A pneumatic deicing system using an engine-driven air pump.

the accessory drive gear box of the engine. The compression side of each pump supplies air pressure to inflate the wing and tail deicer boots. Suction is supplied from the inlet side of each pump to hold down the boots, when not being inflated, while in flight. One type of pump uses engine oil for lubrication and is mounted so that the drive gear is mated with the drive gear in the accessory drive gear box. The oil taken in the pump for lubrication and sealing is discharged through the pressure side, to the oil separator. At this point, most of the oil is separated from the air and fed back to the engine oil sump. When installing a new pump, care should be taken to ensure that the oil passages in the gasket, pump, and engine mounting pad are aligned (figure 7–6). If this oil passage does not line up, the new pump will be damaged from lack of lubrication.

Another type of vacuum pump, called a dry pump, depends on specially compounded carbon parts to provide pump lubrication. The pump is constructed with carbon vanes for the rotor. This material is also used for the rotor bearings. The carbon vane material wears at a controlled rate to provide adequate lubrication. This eliminates the need for external lubricants.

When using the dry type of pump, oil, grease, or degreasing fluids must be prevented from entering the system. This is important at installation and during subsequent maintenance.

Maintenance of the engine-driven pump is limited to inspection for loose connections and security of mounting.

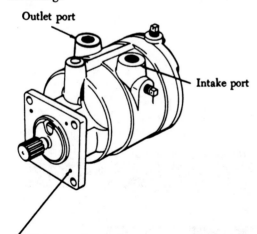

Outlet port

Intake port

Holes in adapter flange for engine pad lubrication. If this type pump is used, be sure the holes are open and not covered by the flange gasket at installation.

FIGURE 7–6. Lubrication of wet type vacuum pump.

Safety Valves

An air pressure safety valve is installed on the pressure side of some types of engine-driven air pumps. Schematically, this valve is placed on the air pressure side of the pump between the primary oil separators and the pump. The safety valve exhausts excessive air at high pump r.p.m. when a predetermined pressure is reached. The valve is preset and is not adjustable.

Oil Separator

An oil separator is provided for each wet-type air pump. Each separator has an air inlet port, an air outlet port, and an oil drain line which is routed back to the engine oil sump. Since the air pump is internally lubricated, it is necessary to provide this means of separating oil from the pressurized air. The oil separator removes approximately 75% of the oil from the air.

The only maintenance required on the oil separator is flushing the interior of the unit with a suitable cleaning solvent. This should be done at intervals prescribed by the applicable maintenance manual.

Combination Regulator, Unloading Valve, and Oil Separator

The combination regulator, unloading valve, and oil separator consists of a diaphragm-controlled, spring-loaded unloading valve, an oil filter and drain; a diaphragm type air pressure regulator valve with an adjustment screw; and a solenoid selector valve. The assembly has an air pressure inlet port, an exhaust port, an outlet to the solenoid distributor valves, an outlet to the suction side of the engine-driven air pumps, and an oil drain. The combination unit has three functions: (1) To remove all residual oil left in the air by the primary oil separator before it enters the pressure manifold; (2) to control, direct, and regulate air pressure in the system; and (3) to discharge air to the atmosphere when the deicer system is not in use, thereby allowing the air pump to operate at no pressure load.

Maintenance of this unit consists of changing the filter element as prescribed by the applicable maintenance manual.

The pressure regulator may be adjusted, if the deicer system pressure gage does not register the specified pressure. The adjusting screw should be turned counterclockwise to increase the pressure and clockwise to decrease the system pressure.

Suction Regulating Valve

An adjustable suction regulating valve is installed in each engine nacelle. One side of each valve is piped to the inlet (suction) side of the engine-driven air pump and the other side to the main suction manifold line. The purpose of the suction valve is to maintain the deicer system suction automatically.

Maintenance of this valve consists of removing the air intake filter screen and cleaning it with a cleaning solvent as prescribed by the applicable maintenance manual.

This valve may be adjusted to obtain the desired deicing system suction. The deicer system suction is increased by turning the adjusting screw counterclockwise and decreased by turning it clockwise.

Solenoid Distributor Valve

The solenoid distributor valve is normally located near the group of deicer boots which it serves. Each distributor valve incorporates a pressure inlet port, suction outlet port, two ports ("A" and "B") to the boots, and a port piped overboard to a low pressure area. Each distributor also has two solenoids, A and B. The pressure inlet port is integral with the manifold pressure line, thereby making approximately 15 p.s.i. pressure available at all times when the deicer system is operating. The suction port is connected to the main suction line. This allows approximately 4 in. Hg suction available at all times in the distributor valve. Ports "A" and "B" connect suction and pressure to the boots, as controlled by the distributor valve. The port piped to the low pressure area allows the air under pressure in the boots to be dumped overboard as controlled by the distributor valve servo.

The distributor valve normally allows suction to be supplied to the boots for holddown in flight. However, when the solenoid in the distributor valve is energized by the electronic timer cycle control, it moves a servo valve, changing the inlet to that section of the boot from suction to pressure. This allows the boot to inflate fully for a predetermined time. This interval is controlled by the electronic timer. When the solenoid is de-energized, the airflow through the valve is cut off. The air then discharges out of the boot through an integral check valve until the pressure reaches approximately 1 in. Hg, the boot is ported to the suction manifold and the remaining air is evacuated, thus again holding the boot down by suction.

Electronic Timer

An electronic timer is used to control the operating sequence and the time intervals of the deicing system. When the deicing system is turned on, the electronic timer energizes a solenoid in the unloading valve. The solenoid closes a servo valve, thereby directing air pressure to the unloading valve and closing it until the regulator valve of the combination unit takes over. The regulator valve then tends to keep the entire manifold system at approximately 15 p.s.i. pressure and unloads any surplus air at the separator by dumping it overboard. The pressure manifold line is then routed to the distributor valves. The electronic timer then controls the operating sequence of the distributor valves.

PNEUMATIC DEICING SYSTEM MAINTENANCE

Maintenance on pneumatic deicing systems varies with each aircraft model. The instructions of the airframe or system components manufacturer should be followed in all cases. Depending on the aircraft, maintenance usually consists of operational checks, adjustments, troubleshooting, and inspection.

Operational Checks

An operational check of the system can be made by operating the aircraft engines, or by using an external source of air. Most systems are designed with a test plug to permit ground checking the system without operating the engines. When using an external air source, make certain that the air pressure does not exceed the test pressure established for the system.

Before turning the deicing system on, observe the vacuum-operated instruments. If any of the gages begin to operate, it is an indication that one or more check valves have failed to close and that reverse flow through the instruments is occurring. Correct the difficulty before continuing the test. If no movement of the instrument pointers occurs, turn on the deicing system.

With the deicer system controls in their proper positions, check the suction and pressure gages for proper indications. The pressure gage will fluctuate as the deicer tubes inflate and deflate. A relatively steady reading should be maintained on the vacuum gage. It should be noted that not all systems use a vacuum gage. If the operating pressure and vacuum are satisfactory, observe the deicers for actuation.

With an observer stationed outside the aircraft, check the inflation sequence to be certain that it agrees with the sequence indicated in the aircraft maintenance manual. Check the timing of the system through several complete cycles. If the cycle time varies more than is allowable, determine the difficulty and correct it. Inflation of the deicers must be rapid to provide efficient deicing. Deflation of the boot being observed should be completed before the next inflation cycle.

Adjustments

Examples of adjustments that may be required include adjusting the deicing system control cable linkages, adjusting system pressure relief valves and deicing system vacuum (suction) relief valves.

A pressure relief valve acts as a safety device to relieve excess pressure in the event of regulator valve failure. To adjust this valve, operate the aircraft engines and adjust a screw on the valve until the deicing pressure gage indicates the specified pressure at which the valve should relieve.

Vacuum relief valves are installed in a system that uses a vacuum pump to maintain constant suction during varying vacuum pump speeds. To adjust a vacuum relief valve, operate the engines. While watching the vacuum (suction) gage, an assistant should adjust the suction relief valve adjusting screw to obtain the correct suction specified for the system.

Troubleshooting

Not all troubles that occur in a deicer system can be corrected by adjusting system components. Some troubles must be corrected by repair or replacement of system components or by tightening loose connections. Several troubles common to pneumatic deicing systems are shown in the left-hand column of the chart in figure 7–7. Note the probable causes and the remedy of each trouble listed in the chart. In addition to using troubleshooting charts, operational checks are sometimes necessary to determine the possible cause of trouble.

Inspection

During each preflight and scheduled inspection, check the deicer boots for cuts, tears, deterioration, punctures, and security; and during periodic inspections go a little further and check deicer components and lines for cracks. If weather cracking of rubber is noted, apply a coating of conductive cement. The cement in addition to sealing the boots against weather, dissipates static electricity so that it will not puncture the boots by arcing to the metal surfaces.

Deicer Boot Maintenance

The life of the deicers can be greatly extended by storing them when they are not needed and by observing these rules when they are in service:

(1) Do not drag gasoline hoses over the deicers.

(2) Keep deicers free of gasoline, oil, grease, dirt and other deteriorating substances.

(3) Do not lay tools on or lean maintenance equipment against the deicers.

(4) Promptly repair or re-surface the deicers when abrasion or deterioration is noted.

(5) Wrap the deicer in paper or canvas when storing it.

Thus far preventive maintenance has been discussed. The actual work on the deicers consists of cleaning, re-surfacing, and repairing. Cleaning should ordinarily be done at the same time the aircraft is washed, using a mild soap and water solution. Grease and oil can be removed with a cleaning agent, such as naptha, followed by soap and water scrubbing.

Whenever the degree of wear is such that it indicates that the electrical conductivity of the deicer surface has been destroyed, it may be necessary to re-surface the deicer. The re-surfacing substance is a black, conductive neoprene cement. Prior to applying the re-surfacing material, the deicer must be cleaned thoroughly and the surface roughened.

Cold patch repairs can be made on a damaged deicer. The deicer must be relieved of its installed tension before applying the patch. The area to be patched must be clean and buffed to roughen the surface slightly.

One disadvantage of a pneumatic deicer system is the disturbance of airflow over the wing and tail caused by the inflated tubes. This unwanted feature

Trouble	Probable Cause	Remedy
Pressure gage oscillates.	Faulty lines or connections.	Repair or replace lines. Tighten loose connections.
	Deicing boots torn or punctured.	Repair faulty boots.
	Faulty gage.	Replace gage.
	Faulty air relief valve.	Adjust or replace relief valve.
	Faulty air regulator.	Adjust or replace regulator.
Pressure gage oscillates; peaks at a specified pressure while instrument vacuum gage shows no reading.	Vacuum check valves installed improperly.	Re-install correctly.
	Vacuum relief valve improperly adjusted or faulty.	Adjust or replace valve as necessary.
	Faulty lines between pump and gage.	Tighten, repair, or replace faulty lines or connections.
Pressure gage shows no pressure while vacuum gage shows normal reading.	Faulty pressure gage line.	Repair or replace line.
	Faulty pressure gage.	Replace gage.
	Pressure relief valve faulty.	Adjust or replace as necessary.
	Pressure regulator faulty.	Adjust or replace as necessary.
Cycling period irregular.	Loose or faulty tubing and connection.	Tighten, repair, or replace as necessary.
	Boots torn or punctured.	Repair faulty boots.
	Faulty electronic timer.	Replace timer.

FIGURE 7–7. Troubleshooting pneumatic deicing systems.

of the deicer boot system has led to the development of other methods of ice control, one of which is the thermal anti-icing system.

THERMAL ANTI-ICING SYSTEMS

Thermal systems used for the purpose of preventing the formation of ice or for deicing airfoil leading edges, usually use heated air ducted spanwise along the inside of the leading edge of the airfoil and distributed around its inner surface. However, electrically heated elements are also used for anti-icing and deicing airfoil leading edges.

There are several methods used to provide heated air. These include bleeding hot air from the turbine compressor, engine exhaust heat exchangers, and ram air heated by a combustion heater.

In installations where protection is provided by preventing the formation of ice, heated air is supplied continuously to the leading edges as long as the anti-icing system is "on." When a system is designed to deice the leading edges, much hotter air is supplied for shorter periods on a cyclic system.

The systems incorporated in some aircraft include an automatic temperature control. The temperature is maintained within a predetermined range by mixing heated air with cold air.

A system of valves is provided in some installations to enable certain parts of the anti-icing system to be shut off. In the event of an engine failure these valves also permit supplying the entire anti-icing system with heated air from one or more of the remaining engines. In other installations the valves are arranged so that when a critical portion of the wing has been deiced, the heated air can be diverted to a less critical area to clear it of ice. Also, should icing conditions of unusual severity be encountered, the entire flow of air can be directed to the most critical areas.

The portions of the airfoils which must be protected from ice formation are usually provided with a closely spaced double skin (see figure 7–8). The heated air carried through the ducting is passed into the gap. This provides sufficient heat to the outer skin to melt the layer of ice next to the skin or to prevent its formation. The air is then exhausted to the atmosphere at the wing tip or at points where ice formation could be critical; for example, at the leading edge of control surfaces.

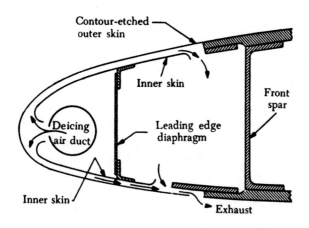

FIGURE 7–8. A typical heated leading edge.

When the air is heated by combustion heaters, usually one or more heaters are provided for the wings. Another heater is located in the tail area to provide hot air for the leading edges of the vertical and horizontal stabilizers.

When the engine is the source of heat, the air is routed to the empennage through ducting which is usually located under the floor.

Anti-Icing Using Combustion Heaters

Anti-icing systems using combustion heaters usually have a separate system for each wing and the empennage. A typical system of this type has the required number of combustion heaters located in each wing and in the empennage. A system of ducting and valves controls the airflow.

The anti-icing system is automatically controlled by overheat switches, thermal cycling switches, a balance control, and a duct pressure safety switch. The overheat and cycling switches allow the heaters to operate at periodic intervals, and they also stop heater operation completely if overheating occurs. A complete description of combustion heaters and their operation is discussed in Chapter 14, Cabin

Atmosphere Control Systems, of this handbook. The balance control is used to maintain equal heating in both wings. The duct pressure safety switch interrupts the heater ignition circuits if ram air pressure falls below a specified amount. This protects the heaters from overheating when not enough ram air is passing through.

An airflow diagram of a typical wing and empennage anti-icing system using combustion heaters is shown in figure 7–9.

Anti-Icing Using Exhaust Heaters

Anti-icing of the wing and tail leading edges is accomplished by a controlled flow of heated air from heat muffs around a reciprocating engine's tail pipe. In some installations this assembly is called an augmentor. As illustrated in figure 7–10, an adjustable vane in each augmentor aft section can be controlled through a range of positions from closed to open. Partially closing each vane restricts the flow of cooling air and exhaust gases. This causes the temperature to rise in the heat muff forward of the vane. This provides a source of heat for the anti-icing system.

Normally, heated air from either engine supplies the wing leading edge anti-icing system in the same wing section. During single engine operation, a crossover duct system interconnects the left and right wing leading edge ducts. This duct supplies heated air to the wing section normally supplied by the inoperative engine. Check valves in the crossover duct prevent the reverse flow of heated air and also prevent cold air from entering the anti-icing system from the inoperative engine.

Figure 7–11 is a schematic of a typical anti-icing system that uses exhaust heaters. Note that, normally, the wing and tail anti-icing system is controlled electrically by operating the heat anti-ice button. When the button is in the "off" position, the outboard heat source valves and the tail anti-ice valve are closed. While the anti-ice system is off, the inboard heat source valves are controlled by the cabin temperature control system. Furthermore the augmentor vanes can be controlled by the augmentor vane switch.

Pushing the heat anti-ice button to the "on" position opens the heat source valves and the tail anti-icing valve. A holding coil keeps the button in the "on" position. In addition, the augmentor vane control circuits are automatically armed. The vanes can be closed by positioning the augmentor vane switch

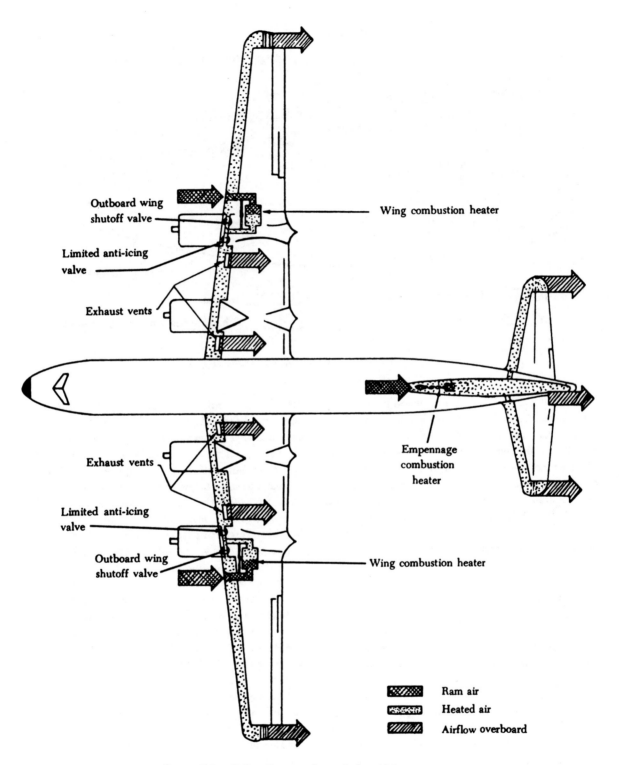

FIGURE 7–9. Airflow diagram of a typical anti-icing system.

to "close." This provides for maximum heat from the system. A safety circuit, controlled by thermostatic limit switches (not shown) in the anti-icing system ducts, releases the anti-ice button to the

"off" position whenever a duct becomes overheated. When overheating occurs, the heat source valves and the tail anti-icing shutoff valve close and the augmentor vanes go to the trail (open) position.

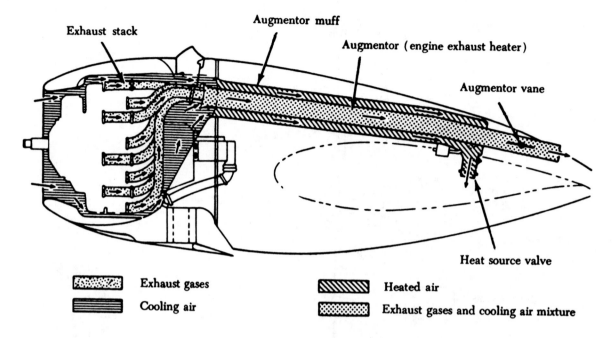

Exhaust stack

Augmentor muff

Augmentor (engine exhaust heater)

Augmentor vane

Heat source valve

▨ Exhaust gases

▨ Cooling air

▨ Heated air

▨ Exhaust gases and cooling air mixture

FIGURE 7–10. Heat source for thermal anti-icing system.

The heat source valves can be closed manually by the manual heat anti-ice shutoff handle. Manual operation may be necessary if the electrical control circuits for the valves fail. In this system, the handle is connected to the valves by a cable system and clutch mechanism. Once the heat source valves have been operated manually, they cannot be operated electrically until the manual override system is re-set. Most systems provide for re-setting ·of the manual override system in flight.

Anti-Icing Using Engine Bleed Air

Heated air for anti-icing is obtained by bleeding air from the engine compressor. The reason for the use of such a system is that relatively large amounts of very hot air can be tapped off the compressor, providing a satisfactory source of anti-icing and deicing heat.

A typical system of this type is shown in figure 7–12. This system is divided into six sections. Each section includes (1) a shutoff valve, (2) a temperature indicator, and (3) an overheat warning light.

The shutoff valve for each anti-icing section is a pressure regulating type. The valve controls the flow of air from the bleed air system to the ejectors, where it is ejected through small nozzles into mixing chambers. The hot bleed air is mixed with ambient air. The resultant mixed air at approximately

350° F. flows through passages next to the leading edge skin. Each of the shutoff valves is pneumatically actuated and electrically controlled. Each shutoff valve acts to stop anti-icing and to control airflow when anti-icing is required. A thermal switch connected to the control solenoid of the shutoff valve causes the valve to close and shut off the flow of bleed air when the temperature in the leading edge reaches approximately 185° F. When the temperature drops, the valve opens, and hot bleed air enters the leading edge.

The temperature indicator for each anti-icing section is located on the anti-icing control panel. Each indicator is connected to a resistance-type temperature bulb located in the leading edge area. The temperature bulb is placed so that it senses the temperature of the air in the area aft of the leading edge skin, not the hot air passed next to the skin.

Overheat warning systems are provided to protect the aircraft structure from damage due to excessive heat. If the normal cyclic system fails, temperature sensors operate to open the circuit controlling the anti-ice shutoff valves. The valves close pneumatically to shut off the flow of hot air.

PNEUMATIC SYSTEM DUCTING

The ducting usually consists of aluminum alloy, titanium, stainless steel, or molded fiber glass tubes.

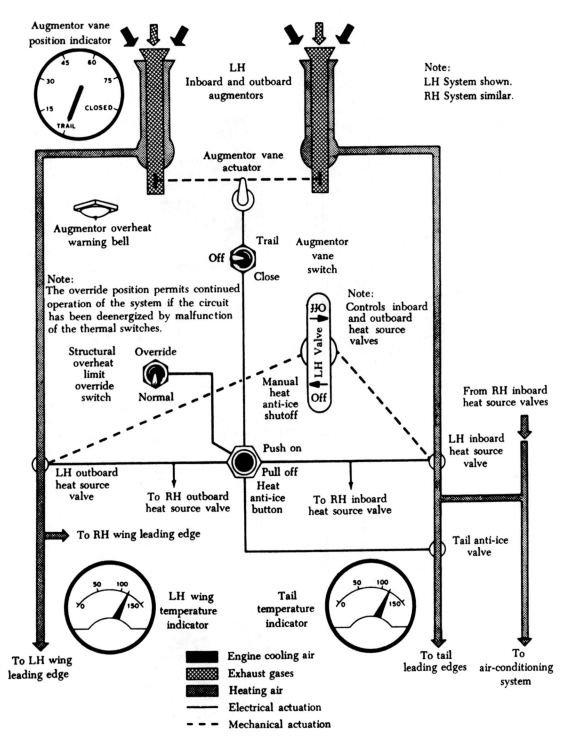

Augmentor vane position indicator

45 60
30 75
15 CLOSED
TRAIL

LH Inboard and outboard augmentors

Note:
LH System shown.
RH System similar.

Augmentor vane actuator

Augmentor overheat warning bell

Off Trail
Close

Augmentor vane switch

Note:
The override position permits continued operation of the system if the circuit has been deenergized by malfunction of the thermal switches.

Structural overheat limit override switch

Override
Normal

Note:
Controls inboard and outboard heat source valves

H₂O
LH Valve
Off

Manual heat anti-ice shutoff

From RH inboard heat source valves

LH inboard heat source valve

Push on
Pull off
Heat anti-ice button

LH outboard heat source valve

To RH outboard heat source valve

To RH inboard heat source valve

To RH wing leading edge

Tail anti-ice valve

50 100
70 150

LH wing temperature indicator

Tail temperature indicator

50 100
70 150

To LH wing leading edge

Engine cooling air
Exhaust gases
Heating air
——— Electrical actuation
- - - Mechanical actuation

To tail leading edges

To air-conditioning system

FIGURE 7-11. Wing and tail anti-icing system schematic.

The tube, or duct, sections are attached to each other by bolted end flanges or by band-type vee-clamps. The ducting is lagged with a fire-resistant, heat insulating material, such as fiber glass.

In some installations the ducting is interposed with thin stainless steel expansion bellows. These

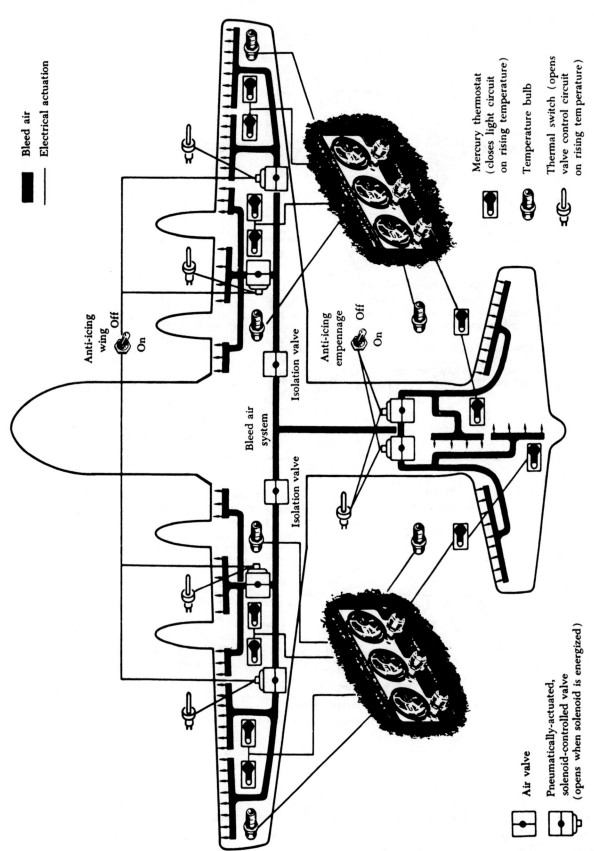

Bleed air
Electrical actuation

Mercury thermostat (closes light circuit on rising temperature)

Temperature bulb

Thermal switch (opens valve control circuit on rising temperature)

Anti-icing wing Off On

Anti-icing empennage Off On

Isolation valve

Isolation valve

Bleed air system

Air valve

Pneumatically-actuated, solenoid-controlled valve (opens when solenoid is energized)

FIGURE 7–12. Schematic of a typical thermal anti-icing system.

298

bellows are located at strategic positions to absorb any distortion or expansion of the ducting which may occur due to temperature variations.

The joined sections of ducting are hermetically sealed by sealing rings. These seals are fitted into annular recesses in the duct joint faces. When installing a section of duct, make certain that the seal bears evenly against, and is compressed by, the adjacent joint's flange.

When specified, the ducts should be pressure tested at the pressure recommended by the manufacturer of the aircraft concerned. Pressure testing is particularly important with pressurized aircraft, since a leak in the ducting may result in the inability to maintain cabin pressure. However, pressure tests are more often made to detect defects in the duct which would permit the escape of heated air. The rate of leakage at a given pressure should not exceed that recommended in the aircraft maintenance or service manual.

Air leaks can often be detected audibly, and are sometimes revealed by holes in the lagging or thermal insulation material. However, if difficulty arises in locating leaks, a soap and water solution may be used.

All ducting should be inspected for security, general condition, or distortion. Lagging or insulating blankets must be checked for security and must be free of flammable fluids such as oil or hydraulic fluid.

GROUND DEICING OF AIRCRAFT

The presence of ice on an aircraft may be the result of direct precipitation, formation of frost on integral fuel tanks after prolonged flight at high altitude, or accumulations on the landing gear following taxiing through snow or slush.

Any deposits of ice, snow, or frost on the external surfaces of an aircraft may drastically affect its performance. This may be due to reduced aerodynamic lift and increased aerodynamic drag resulting from the disturbed airflow over the airfoil surfaces, or it may be due to the weight of the deposit over the whole aircraft. The operation of an aircraft may also be seriously affected by the freezing of moisture in controls, hinges, valves, microswitches, or by the ingestion of ice into the engine.

When aircraft are hangared to melt snow or frost, any melted snow or ice may freeze again if the aircraft is subsequently moved into subzero temperatures. Any measures taken to remove frozen deposits while the aircraft is on the ground must also prevent the possible re-freezing of the liquid.

Frost Removal

Frost deposits can be removed by placing the aircraft in a warm hangar or by using a frost remover or deicing fluid. These fluids normally contain ethylene glycol and isopropyl alcohol and can be applied either by spray or by hand. It should be applied within 2 hrs. of flight.

Deicing fluids may adversely affect windows or the exterior finish of the aircraft. Therefore, only the type of fluid recommended by the aircraft manufacturer should be used.

Removing Ice and Snow Deposits

Probably the most difficult deposit to deal with is deep, wet snow when ambient temperatures are slightly above the freezing point. This type of deposit should be removed with a brush or squeegee. Use care to avoid damage to antennas, vents, stall warning devices, vortex generators, etc., which may be concealed by the snow.

Light, dry snow in subzero temperatures should be blown off whenever possible; the use of hot air is not recommended, since this would melt the snow, which would then freeze and require further treatment.

Moderate or heavy ice and residual snow deposits should be removed with a deicing fluid. No attempt should be made to remove ice deposits or break an ice bond by force.

After completion of deicing operations, inspect the aircraft to ensure that its condition is satisfactory for flight. All external surfaces should be examined for signs of residual snow or ice, particularly in the vicinity of control gaps and hinges. Check the drain and pressure sensing ports for obstructions. When it becomes necessary to physically remove a layer of snow, all protrusions and vents should be examined for signs of damage.

Control surfaces should be moved to ascertain that they have full and free movement. The landing gear mechanism, doors, and bay, and wheel brakes should be inspected for snow or ice deposits and the operation of uplocks and microswitches checked.

Snow or ice can enter turbine engine intakes and

freeze in the compressor. If the compressor cannot be turned by hand for this reason, hot air should be blown through the engine until the rotating parts are free.

WINDSHIELD ICING CONTROL SYSTEMS

In order to keep window areas free of ice, frost, etc., window anti-icing, deicing, defogging, and demisting systems are used. The systems vary according to the type of aircraft and its manufacturer. Some windshields are built with double panels having a space between, which will allow the circulation of heated air between the surfaces to control icing and fogging. Others use windshield wipers and anti-icing fluid which is sprayed on.

One of the more common methods for controlling ice formation and fog on modern aircraft windows is the use of an electrical heating element built into the window. When this method is used with pressurized aircraft, a layer of tempered glass gives strength to withstand pressurization. A layer of transparent conductive material (stannic oxide) is the heating element and a layer of transparent vinyl plastic adds a nonshattering quality to the window. The vinyl and glass plies (figure 7–13) are bonded by the application of pressure and heat. The bond is achieved without the use of a cement as vinyl has a natural affinity for glass. The conductive coating dissipates static electricity from the windshield in addition to providing the heating element.

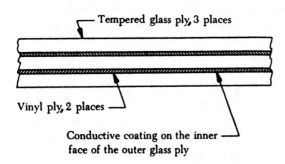

FIGURE 7–13. Section through a laminated windshield.

On some aircraft, thermal electric switches automatically turn the system on when the air temperature is low enough for icing or frosting to occur. The system may stay on all the time during such temperatures, or on some aircraft it may operate with a pulsating on-and-off pattern. Thermal overheat switches automatically turn the systems off in case of an overheating condition which could damage the transparent area.

An electrically heated windshield system includes:

(1) Windshield autotransformers and heat control relays.
(2) Heat control toggle switches.
(3) Indicating lights.
(4) Windshield control units.
(5) Temperature-sensing elements (thermistors) laminated in the panel.

A typical system is shown in figure 7–14. The system receives power from the 115 v. a.c. buses through the windshield heat control circuit breakers. When the windshield heat control switch is set to "high," 115 v., 400 Hz a.c. is supplied to the left and right amplifiers in the windshield control unit.

The windshield heat control relay is energized, thereby applying 200 v., 400 Hz a.c. to the windshield heat autotransformers. These transformers provide 218 v. a.c. power to the windshield heating current bus bars through the windshield control unit relays. The sensing element in each windshield has a positive temperature coefficient of resistance and forms one leg of a bridge circuit. When the windshield temperature is above calibrated value, the sensing element will have a higher resistance value than that needed to balance the bridge. This decreases the flow of current through the amplifiers and the relays of the control unit are de-energized. As the temperature of the windshield drops, the resistance value of the sensing elements also drops and the current through the amplifiers will again reach sufficient magnitude to operate the relays in the control unit, thus energizing the windshield heaters.

When the windshield heat control switch is set to "low," 115 v., 400 Hz a.c. is supplied to the left and right amplifiers in the windshield control unit and to the windshield heat autotransformers. In this condition, the transformers provide 121 v. a.c. power to the windshield heating current bus bars through the windshield control unit relays. The sensing elements in the windshield operate in the same manner as described for high-heat operation to maintain proper windshield temperature control.

The temperature control unit contains two hermetically sealed relays and two three-stage electronic amplifiers. The unit is calibrated to maintain a windshield temperature of 40° – 49° C. (105° – 120° F.). The sensing element in each windshield panel has a positive temperature coefficient of resistance and forms one leg of a bridge which controls the flow of current in its associated ampli-

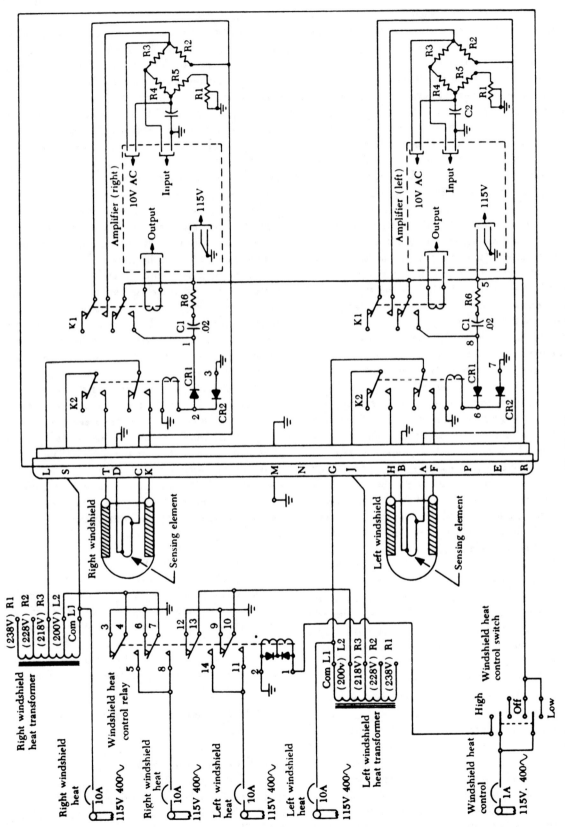

FIGURE 7-14. Windshield temperature control circuit.

301

fier. The final stage of the amplifier controls the hermetically sealed relay which provides a.c. power to the windshield heating current bus bars. When the windshield temperature is above calibrated value, the sensing elements will have a higher resistance value than that needed to balance the bridge. This decreases the flow of current through the amplifiers and the relays of the control unit are de-energized. As the temperature of the windshield drops, the resistance value of the sensing elements also drops and the current through the amplifiers will again reach sufficient magnitude to operate the relays in the control unit, thus energizing the circuit.

There are several problems associated with electrically heated windshields. These include delamination, scratches, arcing, and discoloration.

Delamination (separation of the plies) although undesirable, is not harmful structurally provided it is within the limits established by the aircraft manufacturer, and is not in an area where it affects the optical qualities of the panel.

Arcing in a windshield panel usually indicates that there is a breakdown in the conductive coating. Where chips or minute surface cracks are formed in the glass plies, simultaneous release of surface compression and internal tension stresses in the highly tempered glass can result in the edges of the crack and the conductive coating parting slightly. Arcing is produced where the current jumps this gap, particularly where these cracks are parallel to the window bus bars. Where arcing exists, there is invariably a certain amount of local overheating which, depending upon its severity and location, can cause further damage to the panel. Arcing in the vicinity of a temperature-sensing element is a particular problem since it could upset the heat control system.

Electrically heated windshields are transparent to directly transmitted light but they have a distinctive color when viewed by reflected light. This color will vary from light blue, yellow tints, and light pink depending upon the manufacturer of the window panel. Normally, discoloration is not a problem unless it affects the optical qualities.

Windshield scratches are more prevalent on the outer glass ply where the windshield wipers are indirectly the cause of this problem. Any grit trapped by a wiper blade can convert it into an extremely effective glass cutter when the wiper is set in motion. The best solution to scratches on the windshield is a preventive one; clean the windshield wiper blades as frequently as possible. Incidentally, windshield wipers should never be operated on a dry panel, since this increases the chances of damaging the surface.

Assuming that visibility is not adversely affected, scratches or nicks in the glass plies are allowed within the limitations set forth in the appropriate service or maintenance manuals. Attempting to improve visibility by polishing out nicks and scratches is not recommended. This is because of the unpredictable nature of the residual stress concentrations set up during the manufacture of tempered glass. Tempered glass is stronger than ordinary annealed glass due to the compression stresses in the glass surface which have to be overcome before failure can occur from tension stresses in the core. Polishing away any appreciable surface layer can destroy this balance of internal stresses and can even result in immediate failure of the glass.

Determining the depth of scratches has always presented some difficulty. An optical micrometer can be used for this purpose. It is essentially a microscope supported on small legs rather than the more familiar solid base mounting. When focused on any particular spot, the focal length of the lens (distance from the lens to the object) can be read from a micrometer scale on the barrel of the instrument. The depth of a scratch or fissure in a windshield panel, for example, can thus be determined by obtaining focal length readings to the surface of the glass and to the bottom of the scratch. The difference between these two readings gives the depth of the scratch. The optical micrometer can be used on flat, convex, and concave surfaces of panels whether they are installed on the aircraft or not.

Window Defrost System

The window defrost system directs heated air from the cabin heating system (or from an auxiliary heater, depending on the aircraft) to the pilot's and copilot's windshield and side windows by means of a series of ducts and outlets. In warm weather when heated air is not needed for defrosting, the system can be used to defog the windows. This is done by blowing ambient air on the windows using the blowers.

Windshield and Carburetor Alcohol Deicing Systems

An alcohol deicing system is provided on some aircraft to remove ice from the windshield and the carburetor. Figure 7–15 illustrates a typical two-engine system in which three deicing pumps (one for each carburetor and one for the windshield) are used. Fluid from the alcohol supply tank is controlled by a solenoid valve which is energized when any of the alcohol pumps are on. Alcohol flow from the solenoid valve is filtered and directed to the alcohol pumps and distributed through a system of plumbing lines to the carburetors and windshield.

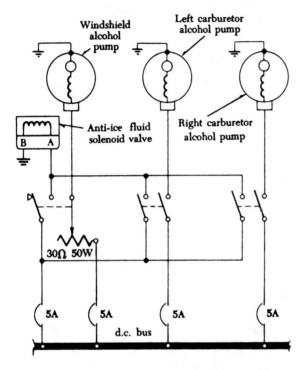

FIGURE 7–15. Carburetor and windshield deicing system.

Toggle switches control the operation of the carburetor alcohol pumps. When the switches are placed in the "on" position, the alcohol pumps are turned on and the solenoid-operated alcohol shutoff valve is opened. Operation of the windshield deicer pump and the solenoid-operated alcohol shutoff valve is controlled by a rheostat-type switch, located in the pilot's station. When the rheostat is moved away from the "off" position, the shutoff valve is opened and the alcohol pump will pump fluid to the windshield at the rate selected by the rheostat. When the rheostat is returned to the "off" position, the shutoff valve closes and the pump stops operating.

Pitot Tube Anti-Icing

To prevent the formation of ice over the opening in the pitot tube, a built-in electric heating element is provided. A switch, located in the cockpit, controls power to the heater. Use caution when ground checking the pitot tube since the heater must not be operated for long periods unless the aircraft is in flight. Additional information concerning pitot

tubes is found in Chapter 2 of this handbook, Aircraft Instrument Systems.

Heating elements should be checked for functioning by ensuring that the pitot head begins to warm up when power is applied. If an ammeter or loadmeter is installed in the circuit, the heater operation can be verified by noting the current consumption when the heater is turned on.

WATER AND TOILET DRAIN HEATERS

Heaters are provided for toilet drain lines, water lines, drain masts, and waste water drains when they are located in an area that is subjected to freezing temperatures in flight. The types of heaters used are integrally heated hoses, ribbon, blanket, or patch heaters that wrap around the lines, and gasket heaters (see figure 7–16). Thermostats are provided in heater circuits where excessive heating is undesirable or to reduce power consumption. The heaters have a low voltage output and continuous operation will not cause overheating.

RAIN ELIMINATING SYSTEMS

When rain forms on a windshield during flight, it becomes a hazard and must be eliminated. To provide a clear windshield, rain is eliminated by wiping it off or blowing it off. A third method of rain removal involves chemical rain repellants. Rain is blown from the windshield of some aircraft by air from jet nozzles located beneath the windshield. On other aircraft, windshield wipers are used to eliminate the rain. The windshield wipers of an aircraft accomplish the same function as those of an automobile. In each instance, rubber blades wipe across the windshield to remove rain and slushy ice.

Electrical Windshield Wiper Systems

In an electrical windshield wiper system, the wiper blades are driven by an electric motor(s) which receive(s) power from the aircraft's electrical system. On some aircraft the pilot's and copilot's windshield wipers are operated by separate systems to ensure that clear vision is maintained through one of the windows should one system fail.

Figure 7–17 shows a typical electrical windshield wiper installation. An electrically operated wiper is installed on each windshield panel. Each wiper is driven by a motor-converter assembly. The converters change the rotary motion of the motor to reciprocating motion to operate the wiper arms. A shaft protruding from the assembly provides an attachment for the wiper arm.

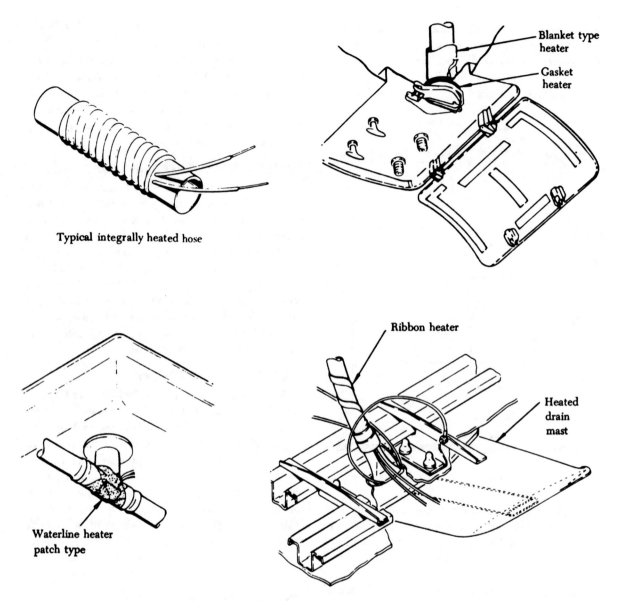

Typical integrally heated hose

Blanket type heater

Gasket heater

Ribbon heater

Heated drain mast

Waterline heater patch type

FIGURE 7-16. Typical water and drain line heaters.

The windshield wiper is controlled by setting the wiper control switch to the desired wiper speed. When the "high" position is selected (figure 7-18), relays 1 and 2 are energized. With both relays energized, field 1 and field 2 are energized in parallel. The circuit is completed and the motors operate at an approximate speed of 250 strokes per minute. When the "low" position is selected, relay 1 is energized. This causes field 1 and field 2 to be energized in series. The motor then operates at approximately 160 strokes per minute. Setting the switch to the "off" position, allows the relay contacts to return to their normal positions. However, the wiper motor will continue to run until the wiper arm reaches the "park" position. When both relays are open and the park switch is closed, the excitation to the motor is reversed. This causes the wiper to move off the lower edge of the windshield, opening the cam-operated park switch. This de-energizes the motor and releases the brake solenoid applying the brake. This ensures that the motor will not coast and re-close the park switch.

Hydraulic Windshield Wiper Systems

Hydraulic windshield wipers are driven by pressure from the aircraft's main hydraulic system. Figure 7-19 shows the components making up a representative hydraulic windshield wiper system.

304

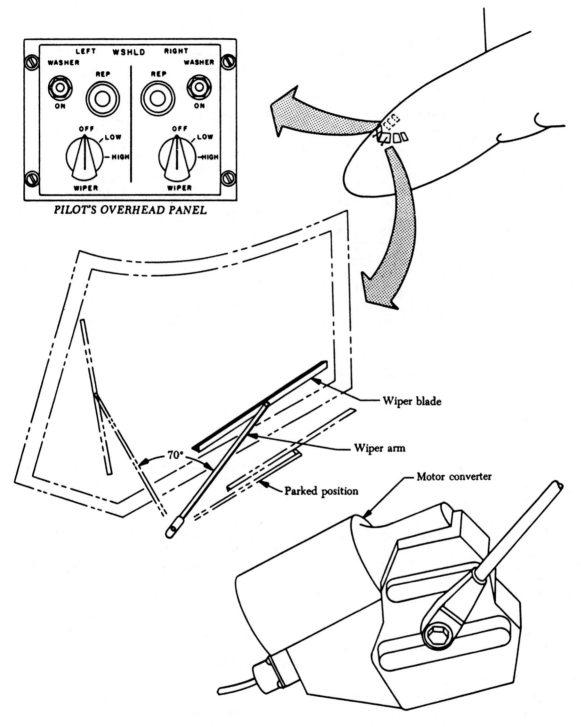

PILOT'S OVERHEAD PANEL

70°

Wiper blade

Wiper arm

Parked position

Motor converter

FIGURE 7-17. Electrical windshield wiper system.

The speed control valve is used to start, stop, and control the operating speed of the windshield wipers. The speed control valve is a type of variable restrictor. Turning the handle of this valve counterclockwise increases the size of the fluid opening, the flow of fluid to the control unit, and therefore the speed of the windshield wipers.

The control unit directs the flow of hydraulic fluid to the wiper actuators and returns fluid discharged from the actuators to the main hydraulic

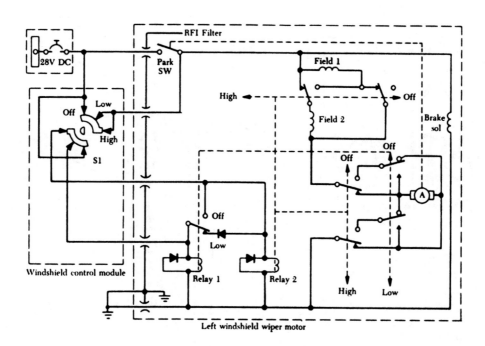

FIGURE 7–18. Windshield wiper circuit diagram.

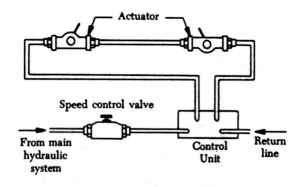

FIGURE 7–19. Hydraulic windshield wiper schematic.

system. The control unit also alternates the direction of hydraulic fluid flow to each of the two wiper actuators. The wiper actuators convert hydraulic energy into reciprocating motion to drive the wiper arms back and forth.

Figure 7–20 shows the construction and the plumbing of the actuators. Notice that each actuator consists of a two-port housing, a piston rack, and a pinion gear. The teeth of the pinion gear mesh with those on the piston rack. Thus, whenever pressurized fluid enters the actuator and moves the piston

rack, the pinion gear is rotated a fraction of a turn. Since the pinion gear connects through a shaft to the wiper blade, the blade rotates through an arc. Notice that one line from the control unit connects to port No. 1 of actuator A, while the other line connects to port No. 4 of actuator B. Notice, too that a line connects ports No. 2 and No. 3 of the actuators.

Turning on the speed control valve allows fluid to flow from the main hydraulic system into the control unit, which directs pressure first into one line and then into the other. When line No. 1 is placed under pressure, fluid flows into port No. 1 and into the chamber at the left of actuator A. This drives the piston rack to the right, causing the pinion and wiper blade to rotate through a counterclockwise arc. As the piston rack moves to the right, it forces fluid in the right chamber of actuator A to move out port No. 2, through the connecting line to port No. 3 and into actuator B. This causes the piston rack in actuator B to move to the right, causing its pinion and wiper blade to rotate counterclockwise. As the piston rack moves to the right, it forces fluid in the right chamber of actuator B to move out of

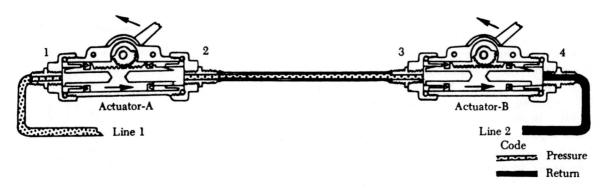

FIGURE 7–20. Windshield wiper actuators.

port No. 4, into line No. 2, through the control unit, and into the hydraulic system return line. When line No. 2 is pressurized by fluid from the control unit, the flow of fluid and the action of the actuators are reversed.

Pneumatic Rain Removal Systems

Windshield wipers characteristically have two basic problem areas. One is the tendency of the slipstream aerodynamic forces to reduce the wiper blade loading pressure on the window, causing ineffective wiping or streaking. The other is in achieving fast enough wiper oscillation to keep up with high rain impingement rates during heavy rain falls. As a result, most aircraft wiper systems fail to provide satisfactory vision in heavy rain.

With the advent of turbine-powered aircraft, the pneumatic rain removal system became feasible. This method uses high pressure, high temperature engine compressor bleed air which is blown across the windshields (figure 7–21). The air blast forms a barrier that prevents raindrops from striking the windshield surface.

Windshield Rain Repellant

When water is poured onto clean glass, it spreads out evenly. Even when the glass is held at a steep angle or subjected to air velocity, the glass remains wetted by a thin film of water.

However, when glass is treated with certain chemicals, a transparent film is formed which causes the water to behave very much like mercury on glass. The water draws up into beads which cover only a portion of the glass and the area between beads is dry. The water is readily removed from the glass.

This principle lends itself quite naturally to removing rain from aircraft windshields. The high velocity slipstream continually removes the water beads, leaving a large part of the window dry.

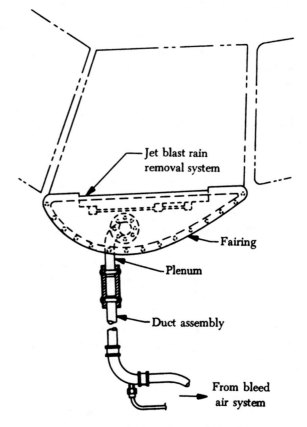

FIGURE 7–21. Typical pneumatic rain removal system.

A rain repellant system permits application of the chemical repellant by a switch or push button in the cockpit. The proper amount of repellant is applied regardless of how long the switch is held. The repellant is marketed in pressurized disposable cans which screw into the aircraft system and provide the propelling force for application. Actuating the control switch opens an electrically-operated solenoid valve which allows repellant to flow to the discharge nozzles. The liquid repellant is squirted onto the exterior of the windshield and uses the rain itself as the carrying agent to distribute the chemicals over the windshield surface.

The rain repellant system should not be operated on dry windows because heavy undiluted repellant will restrict window visibility. Should the system be operated inadvertently, do not operate the windshield wipers or rain clearing system as this tends to increase smearing. Also the rain repellant residues caused by application in dry weather or very light rain can cause staining or minor corrosion of the aircraft skin. To prevent this, any concentrated repellant or residue should be removed by a thorough fresh-water rinse at the earliest opportunity.

After application, the repellant film slowly deteriorates with continuing rain impingement. This makes periodic re-application necessary. The length of time between applications depends upon rain intensity, the type of repellant used, and whether windshield wipers are in use.

MAINTENANCE OF RAIN ELIMINATING SYSTEMS

Windshield Wiper Systems

Maintenance performed on windshield wiper systems consists of operational checks, adjustments, and troubleshooting.

An operational check should be performed whenever a system component is replaced or whenever the system is suspected of not working properly. During the check, make sure that the windshield area covered by the wipers is free of foreign matter and is kept wet with water.

Adjustment of a windshield wiper system consists of adjusting the wiper blade tension, the angle at which the blade sweeps across the windshield, and proper parking of the wiper blades. Figure 7–22 illustrates the adjustment points on a typical wiper blade installation.

One adjustment is that of the tie rod length. The tie rod length adjustment nut is shown in figure 7–22. The tie rod connects the wiper blade holder to a pivot bolt next to the drive shaft. With the drive arm and the tie rod connected to the wiper blade holder, a parallelogram linkage is formed between the wiper blade holder and the wiper converter. This linkage permits the wiper blade to remain parallel to the windshield posts during its travel from one side of the windshield to the other.

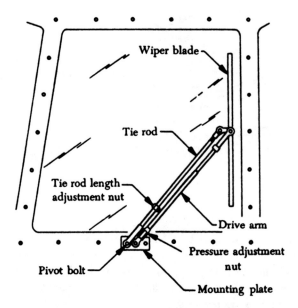

FIGURE 7–22. Adjustment of windshield wiper components.

The length of the tie rod may be adjusted to vary the angle at which the wiper blade sweeps across the windshield.

Another adjustment is that which is required for proper parking of the windshield wiper blades. When they are not operating, the wiper blades should move to a position where they will not interfere with vision. If the wipers do not park as they should, the cam which actuates the microswitch on the converter can be adjusted.

The other adjustment to be made is that of the windshield wiper spring tension. To make the adjustment, place a lightweight spring scale under the drive arm at its point of attachment to the wiper blade, and lift the scale up at a 90° angle to the drive, to a point at which the blade is just ready to leave the glass. (If tension is properly adjusted, the spring scale should indicate between 5 and 6 lbs.) If the scale reading does not fall within this limit, adjust the pressure adjustment nut shown in figure 7–22 until the proper tension is indicated on the scale.

Pneumatic (Jet Blast) Systems

Maintenance of a jet blast system includes the replacement of defective components, the checking (by hand) of duct-to-valve connections for leakage, and an operational checkout.

CHAPTER 8

HYDRAULIC AND PNEUMATIC POWER SYSTEMS

AIRCRAFT HYDRAULIC SYSTEMS

The word hydraulics is based on the Greek word for water, and originally meant the study of the physical behavior of water at rest and in motion. Today the meaning has been expanded to include the physical behavior of all liquids, including hydraulic fluid.

Hydraulic systems are not new to aviation. Early aircraft had hydraulic brake systems. As aircraft became more sophisticated newer systems with hydraulic power were developed.

Although some aircraft manufacturers make greater use of hydraulic systems than others, the hydraulic system of the average modern aircraft performs many functions. Among the units commonly operated by hydraulic systems are landing gear, wing flaps, speed and wheel brakes, and flight control surfaces.

Hydraulic systems have many advantages as a power source for operating various aircraft units. Hydraulic systems combine the advantages of light weight, ease of installation, simplification of inspection, and minimum maintenance requirements. Hydraulic operations are also almost 100% efficient, with only a negligible loss due to fluid friction.

All hydraulic systems are essentially the same, regardless of their function. Regardless of application, each hydraulic system has a minimum number of components, and some type of hydraulic fluid.

HYDRAULIC FLUID

Hydraulic system liquids are used primarily to transmit and distribute forces to various units to be actuated. Liquids are able to do this because they are almost incompressible. Pascal's Law states that pressure applied to any part of a confined liquid is transmitted with undiminished intensity to every other part. Thus, if a number of passages exist in a system, pressure can be distributed through all of them by means of the liquid.

Manufacturers of hydraulic devices usually specify the type of liquid best suited for use with their equipment, in view of the working conditions, the

service required, temperatures expected inside and outside the systems, pressures the liquid must withstand, the possibilities of corrosion, and other conditions that must be considered.

If incompressibility and fluidity were the only qualities required, any liquid not too thick might be used in a hydraulic system. But a satisfactory liquid for a particular installation must possess a number of other properties. Some of the properties and characteristics that must be considered when selecting a satisfactory liquid for a particular system are discussed in the following paragraphs.

Viscosity

One of the most important properties of any hydraulic fluid is its viscosity. Viscosity is internal resistance to flow. A liquid such as gasoline flows easily (has a low viscosity) while a liquid such as tar flows slowly (has a high viscosity). Viscosity increases with temperature decreases.

A satisfactory liquid for a given hydraulic system must have enough body to give a good seal at pumps, valves, and pistons; but it must not be so thick that it offers resistance to flow, leading to power loss and higher operating temperatures. These factors will add to the load and to excessive wear of parts. A fluid that is too thin will also lead to rapid wear of moving parts, or of parts which have heavy loads.

The viscosity of a liquid is measured with a viscosimeter or viscometer. There are several types, but the instrument most often used by engineers in the U.S. is the Saybolt universal viscosimeter (figure 8–1). This instrument measures the number of seconds it takes for a fixed quantity of liquid (60 cc. (cubic centimeters)) to flow through a small orifice of standard length and diameter at a specific temperature. This time of flow is taken in seconds, and the viscosity reading is expressed as SSU (seconds, Saybolt universal). For example, a certain liquid might have a viscosity of 80 SSU at 130° F.

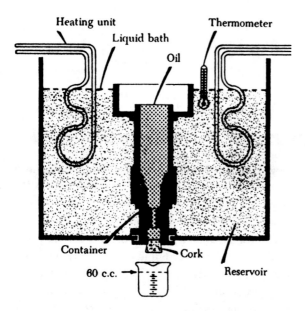

Heating unit
Liquid bath
Oil
Thermometer

Container
Cork
Reservoir

60 c.c. →

FIGURE 8–1. Saybolt viscosimeter.

Chemical Stability

Chemical stability is another property which is exceedingly important in selecting a hydraulic liquid. It is the liquid's ability to resist oxidation and deterioration for long periods. All liquids tend to undergo unfavorable chemical changes under severe operating conditions. This is the case, for example, when a system operates for a considerable period of time at high temperatures.

Excessive temperatures have a great effect on the life of a liquid. It should be noted that the temperature of the liquid in the reservoir of an operating hydraulic system does not always represent a true state of operating conditions. Localized hot spots occur on bearings, gear teeth, or at the point where liquid under pressure is forced through a small orifice. Continuous passage of a liquid through these points may produce local temperatures high enough to carbonize or sludge the liquid, yet the liquid in the reservoir may not indicate an excessively high temperature. Liquids with a high viscosity have a greater resistance to heat than light or low viscosity liquids which have been derived from the same source. The average hydraulic liquid has a low viscosity. Fortunately, there is a wide choice of liquids available for use within the viscosity range required of hydraulic liquids.

Liquids may break down if exposed to air, water, salt, or other impurities, especially if they are in constant motion or subject to heat. Some metals, such as zinc, lead, brass, and copper, have an undesirable chemical reaction on certain liquids.

These chemical processes result in the formation of sludge, gums, and carbon or other deposits which clog openings, cause valves and pistons to stick or leak, and give poor lubrication to moving parts. As soon as small amounts of sludge or other deposits are formed, the rate of formation generally increases more rapidly. As they are formed, certain changes in the physical and chemical properties of the liquid take place. The liquid usually becomes darker in color, higher in viscosity, and acids are formed.

Flash Point

Flash point is the temperature at which a liquid gives off vapor in sufficient quantity to ignite momentarily or flash when a flame is applied. A high flash point is desirable for hydraulic liquids because it indicates good resistance to combustion and a low degree of evaporation at normal temperatures.

Fire Point

Fire point is the temperature at which a substance gives off vapor in sufficient quantity to ignite and continue to burn when exposed to a spark or flame. Like flash point, a high fire point is required of desirable hydraulic liquids.

TYPES OF HYDRAULIC FLUIDS

To assure proper system operation and to avoid damage to non-metallic components of the hydraulic system, the correct fluid must be used.

When adding fluid to a system, use the type specified in the aircraft manufacturer's maintenance manual or on the instruction plate affixed to the reservoir or unit being serviced.

There are three types of hydraulic fluids currently being used in civil aircraft.

Vegetable Base Hydraulic Fluid

Vegetable base hydraulic fluid (MIL–H–7644) is composed essentially of caster oil and alcohol. It has a pungent alcoholic odor and is generally dyed blue. Although it has a similar composition to automotive type hydraulic fluid, it is not interchangeable. This fluid is used primarily in older type aircraft. Natural rubber seals are used with vegetable base hydraulic fluid. If it is contaminated with petroleum base or phosphate ester base fluids, the seals will swell, break down and block the system. This type fluid is flammable.

310

Mineral Base Hydraulic Fluid

Mineral base hydraulic fluid (MIL-H-5606) is processed from petroleum. It has an odor similar to penetrating oil and is dyed red. Synthetic rubber seals are used with petroleum base fluids. Do not mix with vegetable base or phosphate ester base hydraulic fluids. This type fluid is flammable

PHOSPHATE ESTER BASE FLUIDS

Non-petroleum base hydraulic fluids were introduced in 1948 to provide a fire-resistant hydraulic fluid for use in high performance piston engines and turboprop aircraft.

These fluids were fire-resistance tested by being sprayed through a welding torch flame (6000°). There was no burning, but only occasional flashes of fire. These and other tests proved non-petroleum base fluids (Skydrol ®) would not support combustion. Even though they might flash at exceedingly high temperatures, Skydrol ® fluids could not spread a fire because burning was localized at the source of heat. Once the heat source was removed or the fluid flowed away from the source, no further flashing or burning occurred.

Several types of phosphate ester base (Skydrol ®) hydraulic fluids have been discontinued. Currently used in aircraft are Skydrol ® 500B—a clear purple liquid having good low temperature operating characteristics and low corrosive side effects; and, Skydrol ® LD—a clear purple low weight fluid formulated for use in large and jumbo jet transport aircraft where weight is a prime factor.

Intermixing of Fluids

Due to the difference in composition, vegetable base, petroleum base and phosphate ester fluids *will not mix.* Neither are the seals for any one fluid useable with or tolerant of any of the other fluids. Should an aircraft hydraulic system be serviced with the wrong type fluid, immediately drain and flush the system and maintain the seals according to the manufacturer's specifications.

Compatibility With Aircraft Materials

Aircraft hydraulic systems designed around Skydrol ® fluids should be virtually trouble-free if properly serviced. Skydrol ® does not appreciably affect common aircraft metals—aluminum, silver, zinc, magnesium, cadmium, iron, stainless steel, bronze, chromium, and others—as long as the fluids are kept free of contamination.

Due to the phosphate ester base of Skydrol ® fluids, thermoplastic resins, including vinyl compositions, nitrocellulose lacquers, oil base paints, linoleum and asphalt may be softened chemically by Skydrol ® fluids. However, this chemical action usually requires longer than just momentary exposure; and spills that are wiped up with soap and water do not harm most of these materials.

Paints which are Skydrol ® resistant include epoxies and polyurethanes. Today polyurethanes are the standard of the aircraft industry because of their ability to keep a bright, shiny finish for long periods of time and for the ease with which they can be removed.

Skydrol ® is a registered trademark of Monsanto Company. Skydrol ® fluid is compatible with natural fibers and with a number of synthetics, including nylon and polyester, which are used extensively in most aircraft.

Petroleum oil hydraulic system seals of neoprene or Buna-N are not compatible with Skydrol ® and must be replaced with seals of butyl rubber or ethylene-propylene elastomers. These seals are readily available from any suppliers.

Health and Handling

Skydrol ® fluid does not present any particular health hazard in its recommended use. Skydrol ® fluid has a very low order of toxicity when taken orally or applied to the skin in liquid form. It causes pain on contact with eye tissue, but animal studies and human experience indicate Skydrol ® fluid causes no permanent damage. First aid treatment for eye contact includes flushing the eyes immediately with large volumes of water and the application of any anesthetic eye solution. If pain persists, the individual should be referred to a physician.

In mist or fog form, Skydrol ® is quite irritating to nasal or respiratory passages and generally produces coughing and sneezing. Such irritation does not persist following cessation of exposure.

Silicone ointments, rubber gloves, and careful washing procedures should be utilized to avoid excessive repeated contact with Skydrol ® in order to avoid solvent effect on skin.

Hydraulic Fluid Contamination

Experience has shown that trouble in a hydraulic system is inevitable whenever the liquid is allowed to become contaminated. The nature of the trouble, whether a simple malfunction or the complete destruction of a component, depends to some extent on the type of contaminant.

Two general contaminants are:

(1) Abrasives, including such particles as core sand, weld spatter, machining chips, and rust.

(2) Nonabrasives, including those resulting from oil oxidation, and soft particles worn or shredded from seals and other organic components.

Contamination Check

Whenever it is suspected that a hydraulic system has become contaminated, or the system has been operated at temperatures in excess of the specified maximum, a check of the system should be made. The filters in most hydraulic systems are designed to remove most foreign particles that are visible to the naked eye. Hydraulic liquid which appears clean to the naked eye may be contaminated to the point that it is unfit for use.

Thus, visual inspection of the hydraulic liquid does not determine the total amount of contamination in the system. Large particles of impurities in the hydraulic system are indications that one or more components in the system are being subjected to excessive wear. Isolating the defective component requires a systematic process of elimination. Fluid returned to the reservoir may contain impurities from any part of the system. To determine which component is defective, liquid samples should be taken from the reservoir and various other locations in the system.

Samples should be taken in accordance with the applicable manufacturer's instructions for a particular hydraulic system. Some hydraulic systems are equipped with permanently installed bleed valves for taking liquid samples, whereas on other systems, lines must be disconnected to provide a place to take a sample. In either case, while the fluid is being taken, a small amount of pressure should be applied to the system. This ensures that the liquid will flow out of the sampling point and thus prevent

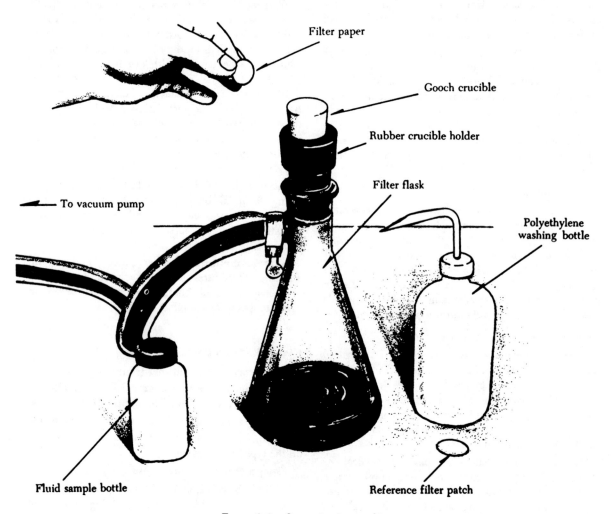

FIGURE 8–2. Contamination test kit.

dirt from entering the hydraulic system. Some contamination test kits have hypodermic syringes for taking samples.

Various test procedures are used to determine the contamination level in hydraulic liquids. The filter patch test provides a reasonable idea of the condition of the fluid. This test consists basically of filtration of a sample of hydraulic system liquid through a special filter paper. This filter paper darkens in degree in relation to the amount of contamination present in the sample, and is compared to a series of standardized filter disks which, by degree of darkening, indicate the various contamination levels. The equipment provided with one type of contamination test kit is illustrated in figure 8–2.

When using this type of contamination test kit, the liquid samples should be poured through the filter paper, and the test filter paper should be compared with the test patches supplied with the test kit. More expensive test kits have a microscope for making this comparison.

To check liquid for decomposition, pour new hydraulic liquid into a sample bottle of the same size and color as the bottle containing the liquid to be checked. Visually compare the color of the two bottles. Liquid which is decomposed will be darker in color.

At the same time the contamination check is made, it may be necessary to make a chemical test. This test consists of a viscosity check, a moisture check, and a flash point check. However, since special equipment is required for these checks, the liquid samples must be sent to a laboratory, where a technician will perform the test.

Contamination Control

Filters provide adequate control of the contamination problem during all normal hydraulic system operations. Control of the size and amount of contamination entering the system from any other source is the responsibility of the people who service and maintain the equipment. Therefore, precautions should be taken to minimize contamination during maintenance, repair, and service operations. Should the system become contaminated, the filter element should be removed and cleaned or replaced.

As an aid in controlling contamination, the following maintenance and servicing procedures should be followed at all times:

(1) Maintain all tools and the work area (workbenches and test equipment) in a clean, dirt-free condition.

(2) A suitable container should always be provided to receive the hydraulic liquid that is spilled during component removal or disassembly procedures.

(3) Before disconnecting hydraulic lines or fittings, clean the affected area with dry cleaning solvent.

(4) All hydraulic lines and fittings should be capped or plugged immediately after disconnecting.

(5) Before assembly of any hydraulic components, wash all parts in an approved dry cleaning solvent.

(6) After cleaning the parts in the dry cleaning solution, dry the parts thoroughly and lubricate them with the recommended preservative or hydraulic liquid before assembly. Use only clean, lint-free cloths to wipe or dry the component parts.

(7) All seals and gaskets should be replaced during the re-assembly procedure. Use only those seals and gaskets recommended by the manufacturer.

(8) All parts should be connected with care to avoid stripping metal slivers from threaded areas. All fittings and lines should be installed and torqued in accordance with applicable technical instructions.

(9) All hydraulic servicing equipment should be kept clean and in good operating condition.

FILTERS

A filter is a screening or straining device used to clean the hydraulic fluid, thus preventing foreign particles and contaminating substances from remaining in the system. If such objectionable material is not removed, it may cause the entire hydraulic system of the aircraft to fail through the breakdown or malfunctioning of a single unit of the system.

The hydraulic fluid holds in suspension tiny particles of metal that are deposited during the normal wear of selector valves, pumps, and other system components. Such minute particles of metal may injure the units and parts through which they pass if they are not removed by a filter. Since tolerances within the hydraulic system components are quite small, it is apparent that the reliability and efficiency of the entire system depends upon adequate filtering.

Filters may be located within the reservoir, in the pressure line, in the return line, or in any other location where the designer of the system decides that they are needed to safeguard the hydraulic system against impurities.

There are many models and styles of filters. Their position in the aircraft and design requirements determine their shape and size.

Most filters used in modern aircraft are of the inline type. The inline filter assembly is comprised of three basic units: head assembly, bowl, and element. The head assembly is that part which is secured to the aircraft structure and connecting lines. Within the head there is a bypass valve which routes the hydraulic fluid directly from the inlet to the outlet port if the filter element becomes clogged with foreign matter. The bowl is the housing which holds the element to the filter head and is that part which is removed when element removal is required.

The element may be either a micronic, porous metal, or magnetic type. The micronic element is made of a specially treated paper and is normally thrown away when removed. The porous metal and magnetic filter elements are designed to be cleaned by various methods and replaced in the system.

Micronic Type Filters

A typical micronic type filter is shown in figure 8–3. This filter utilizes an element made of specially treated paper which is formed in vertical convolutions (wrinkles). An internal spring holds the elements in shape.

The micronic element is designed to prevent the passage of solids greater than 10 microns (0.000394 inch) in size (figure 8–4). In the event that the filter element becomes clogged, the spring loaded relief valve in the filter head will bypass the fluid after a differential pressure of 50 p.s.i. has been built up.

Hydraulic fluid enters the filter through the inlet port in the filter body and flows around the element inside the bowl. Filtering takes place as the fluid passes through the element into the hollow core, leaving the foreign material on the outside of the element.

Maintenance of Filters

Maintenance of filters is relatively easy. It mainly involves cleaning the filter and element or cleaning the filter and replacing the element.

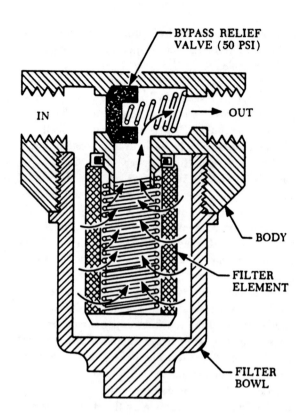

FIGURE 8–3. Hydraulic filter, micronic type.

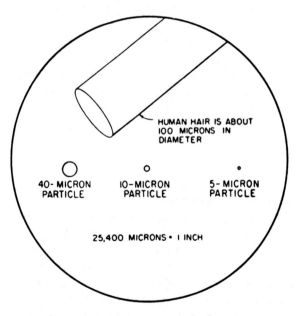

FIGURE 8–4. Enlargement of small particles.

Filters using the micronic-type element should have the element replaced periodically according to applicable instructions. Since reservoir filters are of the micronic type, they must also be periodically changed or cleaned. Filters using other than

314

the micronic-type element, cleaning the filter and element is usually all that is necessary. However, the element should be inspected very closely to insure that it is completely undamaged. The methods and materials used in cleaning all filters are too numerous to mention. Consult the manufacturer's instructions for this information.

Some hydraulic filters have been equipped with an indicator pin that will visually indicate a clogged element. When this pin protrudes from the filter housing, the element should be removed and cleaned; also, the fluid downstream of the filter should be checked for contamination and flushed if required. All remaining filters should be checked for contamination and cleaned (if required) to determine the cause of contamination.

BASIC HYDRAULIC SYSTEM

Regardless of its function and design, every hydraulic system has a minimum number of basic components in addition to a means through which the fluid is transmitted.

Hand Pump System

Figure 8–5 shows a basic hydraulic system. The first of the basic components, the reservoir, stores the supply of hydraulic fluid for operation of the system. It replenishes the system fluid when needed, provides room for thermal expansion, and in some systems provides a means for bleeding air from the system.

A pump is necessary to create a flow of fluid. The pump shown in figure 8–5 is hand operated; however, aircraft systems are, in most instances equipped with engine-driven or electric motor-driven pumps.

The selector valve is used to direct the flow of fluid. These valves are normally actuated by solenoids or manually operated, either directly or indirectly through use of mechanical linkage. An actuating cylinder converts fluid pressure into useful work by linear or reciprocating mechanical motion, whereas a motor converts fluid pressure into useful work by rotary mechanical motion.

The flow of hydraulic fluid can be traced from the reservoir through the pump to the selector valve in figure 8–5. With the selector valve in the position shown, the hydraulic fluid flows through the selector valve to the right-hand end of the actuating cylinder. Fluid pressure then forces the piston to the left, and at the same time the fluid which is on the left side of the piston (figure 8–5) is forced out, up through the selector valve, and back to the reservoir through the return line.

When the selector valve is moved to the opposite position, the fluid from the pump flows to the left

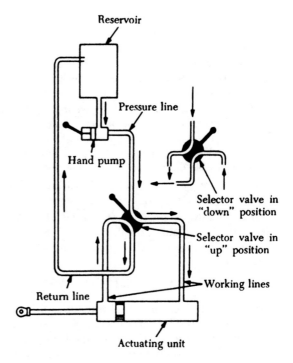

FIGURE 8–5. Basic hydraulic system with hand pump.

side of the actuating cylinder, thus reversing the process. Movement of the piston can be stopped at any time by moving the selector valve to neutral. In this position, all four ports are closed and pressure is trapped in both working lines.

Power Driven Pump System

Figure 8–6 shows a basic system with the addition of a power-driven pump and filter, pressure regulator, accumulator, pressure gage, relief valve, and two check valves. The function of each of these components is described in the following paragraphs.

The filter removes foreign particles from the hydraulic fluid, preventing dust, grit, or other undesirable matter from entering the system.

The pressure regulator unloads or relieves the power-driven pump when the desired pressure in the system is reached. Thus, it is often referred to as an unloading valve. When one of the actuating units is being operated and pressure in the line between the pump and selector valve builds up to the desired point, a valve in the pressure regulator automatically opens and fluid is bypassed back to the reservoir. This bypass line is shown in figure 8–6 leading from the pressure regulator to the return line.

Many hydraulic systems do not use a pressure regulator, but have other means of unloading the pump and maintaining the desired pressure in the system. These methods are described in this chapter.

The accumulator (figure 8–6) serves a twofold purpose: (1) It acts as a cushion or shock absorber by maintaining an even pressure in the system, and (2) It stores enough fluid under pressure to provide for emergency operation of certain actuating units. Accumulators are designed with a compressed air chamber which is separated from the fluid by a flexible diaphragm or movable piston.

The pressure gage (figure 8–6) indicates the amount of hydraulic pressure in the system.

The relief valve is a safety valve installed in the system to bypass fluid through the valve back to the reservoir in case excessive pressure is built up in the system.

The check valves allow the flow of fluid in one direction only. Check valves are installed at various points in the lines of all aircraft hydraulic systems. In figure 8–4, one check valve prevents power-pump pressure from entering the hand-pump line; the other prevents hand-pump pressure from being directed to the accumulator.

The units of a typical hydraulic system used most commonly are discussed in detail in the following paragraphs. Not all models or types are included, but examples of typical components are used in all cases.

RESERVOIRS

There is a tendency to envision a reservoir as an individual component; however, this is not always true. There are two types of reservoirs and they are:

(1) In-Line—this type has its own housing, is complete within itself, and is connected with other components in a system by tubing or hose.

(2) Integral—this type has no housing of its own but is merely a space set aside within some major component to hold a supply of operational fluid. A familiar example of this type is the reserve fluid space found within most automobile brake master cylinders.

In an in-line reservoir (figure 8–7), a space is provided in the reservoir, above the normal level of the fluid, for fluid expansion and the escape of entrapped air. Reservoirs are never intentionally filled to the top with fluid. Most reservoirs are designed so the rim of the filler neck is somewhat below the top of the reservoir to prevent over filling during servicing. Most reservoirs are equipped with

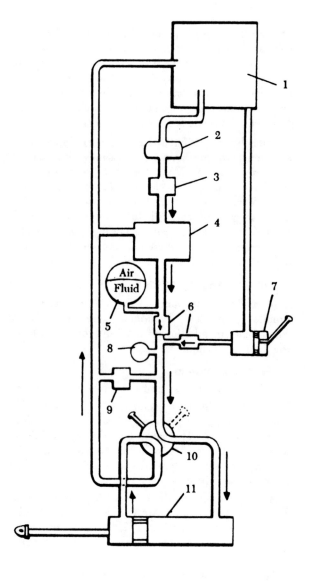

1. Reservoir	7. Hand pump
2. Power pump	8. Pressure gage
3. Filter	9. Relief valve
4. Pressure regulator	10. Selector valve
5. Accumulator	11. Actuating unit
6. Check valves	

FIGURE 8–6. Basic hydraulic system with power pump and other hydraulic components.

a dipstick or a glass sight gage by which fluid level can be conveniently and accurately checked.

Reservoirs are either vented to the atmosphere or closed to the atmosphere and pressurized. In vented reservoirs, atmospheric pressure and gravity are the forces which cause fluid to flow from the reservoir

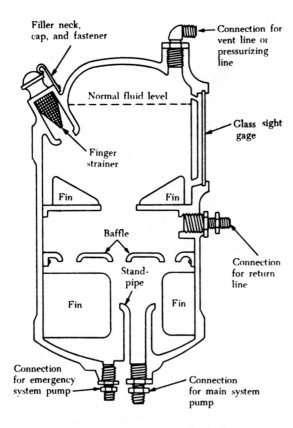

FIGURE 8-7. Reservoir, "in-line".

of hydraulic system, by using an air pressure regulator.

Reservoirs that are pressurized with hyrdaulic fluid (figure 8–8) are constructed somewhat differently from reservoirs pressurized with air. A flexible, coated-fabric bag, called a "bellowfram" or diaphragm, is attached to the reservoir head. The bag hangs inside a metal barrel to form a fluid container. The bottom of the diaphragm rests on a large piston. Attached to the large piston is an indicator rod. The other end of the indicator rod is machined to form a small piston which is exposed to fluid pressure from the hydraulic pump. This pressure forces the small piston upward, causing the larger piston to move upward, thus producing a reservoir pressure of approximately 30 to 32 p.s.i. in normal operation. If the internal pressure should exceed 46 p.s.i., the reservoir relief valve will open allowing fluid to escape through the drilled head of the valve retainer. This type of reservoir must be completely filled with hydraulic fluid and have all the air bled from it.

Reservoir Components

Baffles and/or fins are incorporated in most reservoirs to keep the fluid within the reservoir from

into the pump intake. On many aircraft, atmospheric pressure is the principal force causing fluid to flow to the pump intake. However, for some aircraft, atmospheric pressure becomes too low to supply the pump with adequate fluid, and the reservoirs must be pressurized.

There are several methods of pressurizing a reservoir. Some systems use air pressure directly from the aircraft cabin pressurization system; or from the engine compressor in the case of turbine-powered aircraft. Another method used is an aspirator or venturi-tee. In other systems an additional hydraulic pump is installed in the supply line at the reservoir outlet to supply fluid under pressure to the main hydraulic pump.

Pressurizing with air is accomplished by forcing air into the reservoir above the level of the fluid. In most cases, the initial source of the air pressure is the aircraft engine from which it is bled. Usually, air coming directly from the engine is at a pressure of approximately 100 p.s.i. This pressure is reduced to between 5 and 15 p.s.i., depending upon the type

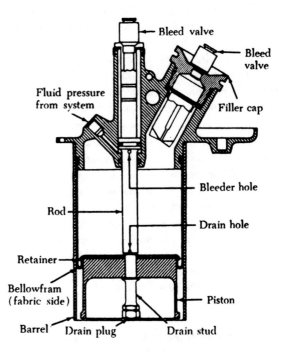

FIGURE 8-8. Hydraulic reservoir pressurized with hydraulic fluid.

having random movement such as vortexing (swirling) and surging. These conditions can cause fluid to foam and air to enter the pump along with the fluid.

Many reservoirs incorporate strainers in the filler neck to prevent the entry of foreign matter during servicing. These strainers are made of fine mesh screening and are usually referred to as finger strainers because of their shape. Finger strainers should never be removed or punctured as a means of speeding up the pouring of fluid into the reservoir.

Some reservoirs incorporate filter elements. They may be used either to filter air before it enters the reservoir or to filter fluid before it leaves the reservoir. A vent filter element, when used, is located in the upper part of the reservoir, above the fluid level. A fluid filter element, when used, is located at or near the bottom of the reservoir. Fluid, as it returns to the reservoir, surrounds the filter element and flows through the wall of the element. This leaves any fluid contaminant on the outside of the filter element.

Reservoirs with filter elements incorporate a by-pass valve normally held closed by a spring. The bypass valve ensures that the pump will not be starved of fluid if the filter element becomes clogged. A clogged filter causes a partial vacuum to develop and the spring-loaded bypass valve opens. The filter element most commonly used in reservoirs is the micronic type. These filter elements are made of treated cellulose formed into accordion-like pleats. The pleats expose the fluid to the maximum amount of filter surface within a given amount of space. These micronic elements are capable of re-moving small particles of contamination.

Some aircraft have emergency hydraulic systems that take over if main systems fail. In many such systems, the pumps of both systems obtain fluid from a single reservoir. Under such circumstances a supply of fluid for the emergency pump is ensured by drawing the hydraulic fluid from the bottom of the reservoir. The main system draws its fluid through a standpipe located at a higher level. With this arrangement, adequate fluid is left for operation of the emergency system should the main system's fluid supply become depleted.

Double Action Hand Pumps

The double-action hydraulic hand pump is used in some older aircraft and in a few newer systems

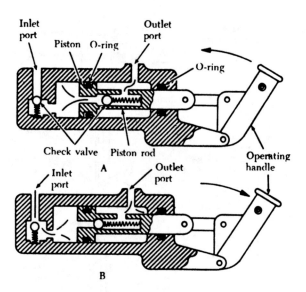

FIGURE 8-9. Double-action hand pump.

as a backup unit. Double-action hand pumps produce fluid flow and pressure on each stroke of the handle.

The double-action hand pump (figure 8-9) consists essentially of a housing which has a cylinder bore and two ports, a piston, two spring-loaded check valves, and an operating handle. An O-ring on the piston seals against leakage between the two chambers of the piston cylinder bore. An O-ring in a groove in the end of the pump housing seals against leakage between the piston rod and housing.

Power-Driven Pumps

Many of the power-driven hydraulic pumps of current aircraft are of variable-delivery, compensator-controlled type. There are some constant-delivery pumps in use. Principles of operation are the same for both types of pumps. Because of its relative simplicity and ease of understanding, the constant-delivery pump is used to describe the principles of operation of power-driven pumps.

Constant-Delivery Pump

A constant-delivery pump, regardless of pump r.p.m., forces a fixed or unvarying quantity of fluid through the outlet port during each revolution of the pump. Constant-delivery pumps are sometimes called constant-volume or fixed-delivery pumps. They deliver a fixed quantity of fluid per revolution, regardless of the pressure demands. Since the constant-delivery pump provides a fixed quantity of fluid during each revolution of the pump, the quantity of fluid delivered per minute will depend upon pump r.p.m. When a constant-delivery pump is used in a hydraulic system in which the pressure must be kept at a constant value, a pressure regulator is required.

Variable-Delivery Pump

A variable-delivery pump has a fluid output that is varied to meet the pressure demands of the system by varying its fluid output. The pump output is changed automatically by a pump compensator within the pump.

Pumping Mechanisms

Various types of pumping mechanisms are used in hydraulic pumps, such as gears, gerotors, vanes, and pistons. The piston-type mechanism is commonly used in power-driven pumps because of its durability and capability to develop high pressure. In 3,000 p.s.i. hydraulic systems, piston-type pumps are nearly always used.

Gear Type Pump

A gear-type power pump (figure 8-10) consists of two meshed gears that revolve in a housing. The driving gear is driven by the aircraft engine or some other power unit. The driven gear meshes with, and is driven by, the driving gear. Clearance between the teeth as they mesh, and between the teeth and the housing, is very small. The inlet port of the pump is connected to the reservoir, and the outlet port is connected to the pressure line. When the driving gear turns in a counterclockwise direction, as shown in figure 8-10, it turns the driven gear in a clockwise direction. As the gear teeth pass the inlet port, fluid is trapped between the gear teeth and the housing, and is then carried around the housing to the outlet port.

Gerotor Type Pump

A gerotor-type power pump (figure 8-11) consists essentially of a housing containing an eccentric-shaped stationary liner, an internal gear rotor having five wide teeth of short height, a spur driving gear having four narrow teeth, and a pump cover which contains two crescent-shaped openings. One opening extends into an inlet port, and the other extends into an outlet port. The pump cover as shown in figure 8-11 has its mating face turned up to clearly show the crescent-shaped openings. When the cover is turned over and properly installed on the pump housing, it will have its inlet port on the left and the outlet port on the right.

During the operation of the pump, the gears turn

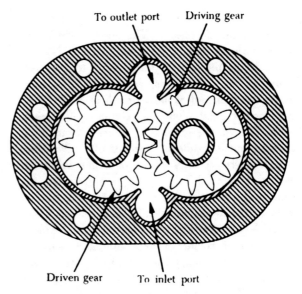

FIGURE 8-10. Gear-type power pump.

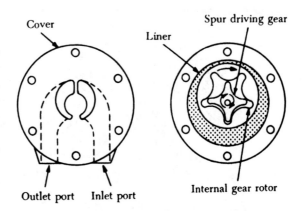

FIGURE 8-11. Gerotor-type power pump.

clockwise. As the pockets on the left side of the pump move from a lowermost position toward a topmost position, the pockets increase in size (figure 8-11) resulting in the production of a partial vacuum within these pockets. As the pockets open at the inlet port, fluid is drawn into them. As these same pockets (now full of fluid) rotate over to the right side of the pump, moving from the topmost position toward the lowermost position, they decrease in size. This results in the fluid being expelled from the pockets through the outlet port.

Vane Type Pump

The vane-type power pump (figure 8-12) consists of a housing containing four vanes (blades), a hollow steel rotor with slots for the vanes, and a coupling to turn the rotor. The rotor is positioned off center within the sleeve. The vanes, which are mounted in the slots in the rotor, together with the rotor, divide the bore of the sleeve into four sections. As the rotor turns, each section, in turn, passes one point where its volume is at a minimum, and another point where its volume is at a maximum. The volume gradually increases from minimum to maximum during one-half of a revolution, and gradually decreases from maximum to minimum during the second half of the revolution. As the volume of a given section is increasing, that section is connected to the pump inlet port through a slot in the sleeve. Since a partial vacuum is produced by the increase in volume of the section, fluid is drawn into the section through the pump inlet port and the slot in the sleeve. As the rotor turns through the second half of the revolution, and the volume of the given section is decreasing, fluid is displaced out of the section, through the slot in the sleeve, through the outlet port, and out of the pump.

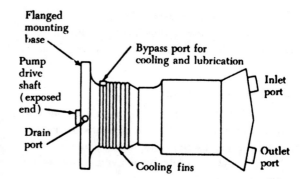

FIGURE 8–13.   Typical piston-type hydraulic pump.

Torque from the driving unit is transmitted to the pump drive shaft by a drive coupling (figure 8–14). The drive coupling is a short shaft with a set of male splines on both ends. The splines on one end engage with female splines in a driving gear; the splines on the other end engage with female

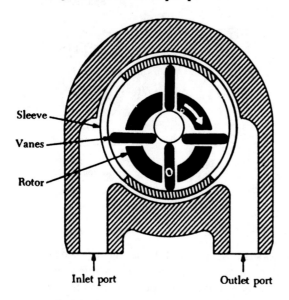

FIGURE 8–12.   Vane-type power pump.

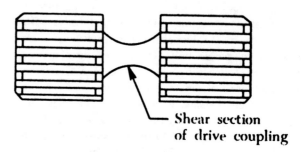

FIGURE 8–14.   Pump drive coupling.

Piston Type Pump

The common features of design and operation that are applicable to all piston-type hydraulic pumps are described in the following paragraphs.

Piston-type power-driven pumps have flanged mounting bases for the purpose of mounting the pumps on the accessory drive cases of aircraft engines and transmissions. A pump drive shaft, which turns the mechanism, extends through the pump housing slightly beyond the mounting base (figure 8–13).

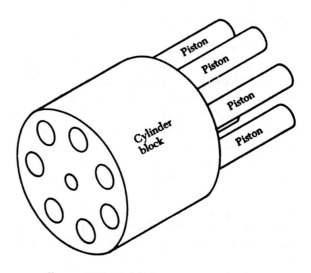

FIGURE 8–15.   Axial-piston pump mechanism.

320

splines in the pump drive shaft. Pump drive couplings are designed to serve as safety devices. The shear section of the drive coupling, located midway between the two sets of splines, is smaller in diameter than the splines. If the pump becomes unusually hard to turn or becomes jammed, this section will shear, preventing damage to the pump or driving unit.

The basic pumping mechanism of piston-type pumps (figure 8–15) consists of a multiple-bore cylinder block, a piston for each bore, and a valving arrangement for each bore. The purpose of the valving arrangement is to let fluid into and out of the bores as the pump operates. The cylinder bores lie parallel to and symmetrically around the pump axis. The term "axial-piston pump" is often used in

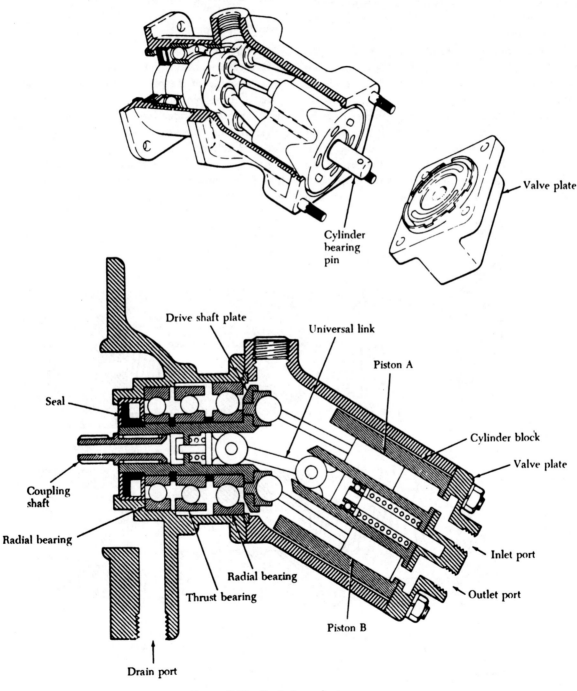

FIGURE 8–16. Typical angular-type pump.

321

referring to pumps of this arrangement. All aircraft axial-piston pumps have an odd number of pistons (5, 7, 9, 11, etc.).

Angular Type Piston Pump

A typical angular-type pump is shown in figure 8–16. The angular housing of the pump causes a corresponding angle to exist between the cylinder block and the drive shaft plate to which the pistons are attached. It is this angular configuration of the pump that causes the pistons to stroke as the pump shaft is turned.

When the pump is operated, all parts within the pump (except the outer races of the bearings which support the drive shaft, the cylinder bearing pin on which the cylinder block turns, and the oil seal) turn together as a rotating group. Because of the angle between the drive shaft and the cylinder block, at one point of rotation of the rotating group a minimum distance exists between the top of the cylinder block and the upper face of the drive shaft plate. At a point of rotation 180° away, the dis-

tance between the top of the cylinder block and the upper face of the drive shaft plate is at a maximum.

At any given moment of operation, three of the pistons will be moving away from the top face of the cylinder block, producing a partial vacuum in the bores in which these pistons operate. Fluid will be drawn into these bores at this time.

Movement of the pistons when drawing in and expelling fluid is overlapping in nature, and results in a practically non-pulsating discharge of fluid from the pump.

Cam Type Pump

Cam-type pumps (figure 8–17) utilize a cam to cause stroking of the pistons. There are two variations of cam-type pumps; one in which the cam turns and the cylinder block is stationary, and the other in which the cam is stationary and the cylinder block rotates.

As an example of the manner in which the pistons of a cam-type pump are caused to stroke, the operation of a rotating cam-type pump is described

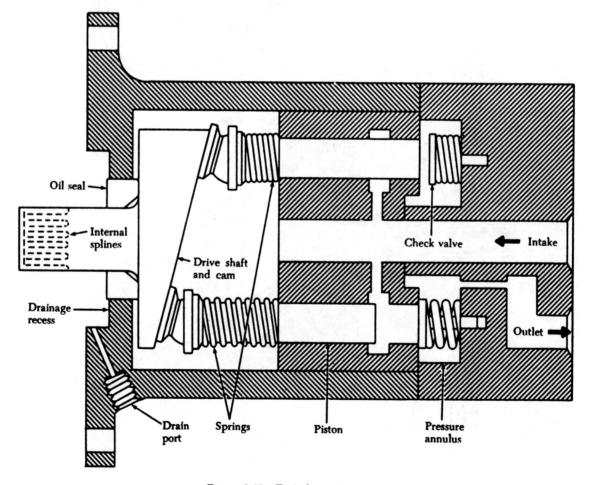

FIGURE 8–17. Typical cam-type pump.

322

as follows: As the cam turns, its high and low points pass alternately and, in turn, under each piston. When the rising ramp of the cam passes under a piston, it pushes the piston further into its bore, causing fluid to be expelled from the bore. When the falling ramp of the cam comes under a piston, the piston's return spring extends the piston outward of its bore. This causes fluid to be drawn into the bore. Because the movement of the pistons when drawing in and expelling fluid is overlapping in nature, the discharge of fluid from the cam-type pump is practically non-pulsating.

Each bore has a check valve that opens to allow fluid to be expelled from the bore by movement of the piston. These valves close during the inlet strokes of the pistons. Because of this, inlet fluid can be drawn into the bores only through the central inlet passage.

PRESSURE REGULATION

Hydraulic pressure must be regulated in order to use it to perform the desired tasks. Pressure regulating systems will always use three elemental devices; a pressure relief valve, a pressure regulator and a pressure gage.

Pressure Relief Valves

A pressure relief valve is used to limit the amount of pressure being exerted on a confined liquid. This is necessary to prevent failure of components or rupture of hydraulic lines under excessive pressures. The pressure relief valve is, in effect, a system safety valve.

The design of pressure relief valves incorporates adjustable spring-loaded valves. They are installed in such a manner as to discharge fluid from the pressure line into a reservoir return line when the pressure exceeds the predetermined maximum for which the valve is adjusted. Various makes and designs of pressure relief valves are in use, but, in general, they all employ a spring-loaded valving device operated by hydraulic pressure and spring tension. Pressure relief valves are adjusted by increasing or decreasing the tension on the spring to determine the pressure required to open the valve. Two general forms of pressure relief valves, the two-port and the four-port, are illustrated in figure 8–18.

Pressure relief valves may be classified as to their type of construction or uses in the system. However,

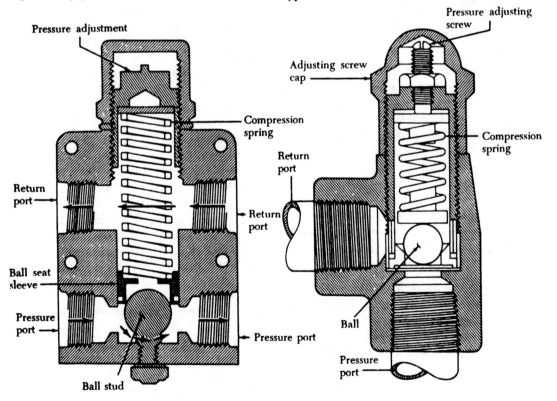

Four-port pressure relief valve   Two-port pressure relief valve

FIGURE 8–18. Pressure relief valves.

the general purpose and operation of all pressure relief valves are the same. The basic difference in construction of pressure relief valves is in the design of the valving. The most common types of valve are:

(1) *Ball type.* In pressure relief valves with a ball-type valving device, the ball rests on a contoured seat. Pressure acting on the bottom of the ball pushes it off its seat, allowing the fluid to bypass.

(2) *Sleeve type.* In pressure relief valves with a sleeve-type valving device, the ball remains stationary and a sleeve-type seat is moved up by the fluid pressure. This allows the fluid to bypass between the ball and the sliding sleeve-type seat.

(3) *Poppet type.* In pressure relief valves with a poppet-type valving device, a cone-shaped poppet may have any of several design configurations; however, it is basically a cone and seat machined at matched angles to prevent leakage. As the pressure rises to its predetermined setting, the poppet is lifted off its seat, as in the ball-type device. This allows the fluid to pass through the opening created and out the return port.

Pressure relief valves cannot be used as pressure regulators in large hydraulic systems that depend upon engine-driven pumps for the primary source of pressure because the pump is constantly under load, and the energy expended in holding the pressure relief valve off its seat is changed into heat. This heat is transferred to the fluid and in turn to the packing rings causing them to deteriorate rapidly. Pressure relief valves, however, may be used as pressure regulators in small, low-pressure systems or when the pump is electrically driven and is used intermittently. Pressure relief valves may be used as:

(1) *System relief valve.* The most common use of the pressure relief valve is as a safety device against the possible failure of a pump compensator or other pressure regulating device. All hydraulic systems which have hydraulic pumps incorporate pressure relief valves as safety devices.

(2) *Thermal relief valve.* The pressure relief valve is used to relieve excessive pressures that may exist due to thermal expansion of the fluid.

Pressure Regulators

The term "pressure regulator" is applied to a device used in hydraulic systems that are pressurized by constant-delivery type pumps. One purpose of the pressure regulator is to manage the output of the pump to maintain system operating pressure within a predetermined range. The other purpose is to permit the pump to turn without resistance (termed unloading the pump) at times when pressure in the system is within normal operating range. The pressure regulator is so located in the system that pump output can get into the system pressure circuit only by passing through the regulator. The combination of a constant-delivery type pump and the pressure regulator is virtually the equivalent of a compensator-controlled, variable-delivery type pump.

Pressure Gage

The purpose of this gage is to measure the pressure, in the hydraulic system, used to operate hydraulic units on the aircraft. The gage uses a Bourdon tube and a mechanical arrangement to transmit the tube expansion to the indicator on the face of the gage. A vent in the bottom of the case maintains atmospheric pressure around the Bourdon tube. It also provides a drain for any accumulated moisture. There are several ranges of pressure used in hydraulic systems and gages are calibrated to match the system they accommodate.

Accumulator

The accumulator is a steel sphere divided into two chambers by a synthetic rubber diaphragm. The upper chamber contains fluid at system pressure, while the lower chamber is charged with air.

The function of an accumulator is to:

a. Dampen pressure surges in the hydraulic system caused by actuation of a unit and the effort of the pump to maintain pressure at a preset level.

b. Aid or supplement the power pump when several units are operating at once by supplying extra power from its "accumulated" or stored power.

c. Store power for the limited operation of a hydraulic unit when the pump is not operating.

d. Supply fluid under pressure to compensate for small internal or external (not desired) leaks which would cause the system to cycle continuously by action of the pressure switches continually "kicking in."

Diaphragm Accumulator

Diaphragm type accumulators consist of two hollow half-ball metal sections fastened together at the centerline. One of these halves has a fitting for attaching the unit to the system; the other half is equipped with an air valve for charging the unit with compressed air. Mounted between the two halves is a synthetic rubber diaphragm which divides the tank into two compartments. A screen covers the outlet on the fluid side of the accumulator. This prevents a part of the diaphragm from being pushed up into the system pressure port and being damaged. This could happen whenever there is an air charge in the unit and no balancing fluid pressure. In some units, a metal disc attached to the center of the diaphragm is used in place of the screen. (See figure 8–19).

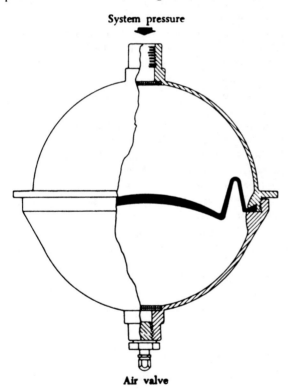

System pressure

Air valve

FIGURE 8–19. Diaphragm-type accumulator.

Bladder-Type Accumulators

The bladder-type accumulator operates on the same principle as the diaphragm type. It serves the same purpose, but varies in construction. This unit consists of a one-piece metal sphere with a fluid pressure inlet at the top. There is an opening at the bottom for inserting the bladder. A large screw-type plug at the bottom of the accumulator retains the bladder and also seals the unit. The high-pressure air valve is also mounted in the re-

tainer plug. A round metal disc attached to the top of the bladder prevents air pressure from forcing the bladder out through the pressure port. As fluid pressure rises, it forces the bladder downward against the air charge, filling the upper chamber with fluid pressure. The broken lines in figure 8–20 show the approximate shape of the bladder when the accumulator is charged.

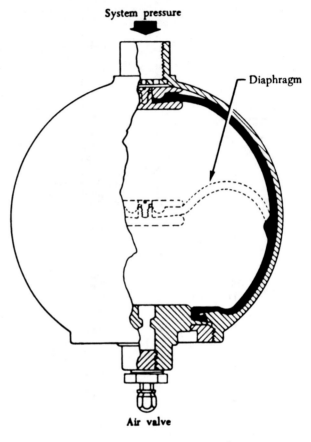

System pressure

Diaphragm

Air valve

FIGURE 8–20. Bladder-type accumulator.

Piston-Type Accumulators

The piston-type accumulator also serves the same purpose and operates much like the diaphragm and bladder accumulators. As shown in figure 8–21 this unit is a cylinder (B) and piston assembly (E) with openings on each end. System fluid pressure enters the top port (A), and forces the piston down against the air charge in the bottom chamber (D). A high-pressure air valve (C) is located at the bottom of the cylinder for servicing the unit. There are two rubber seals (represented by the black dots) which prevent leakage between the two chambers (D and G). A passage (F) is drilled from the fluid side of the piston to the space between the seals. This provides lubrication between the cylinder walls and the piston.

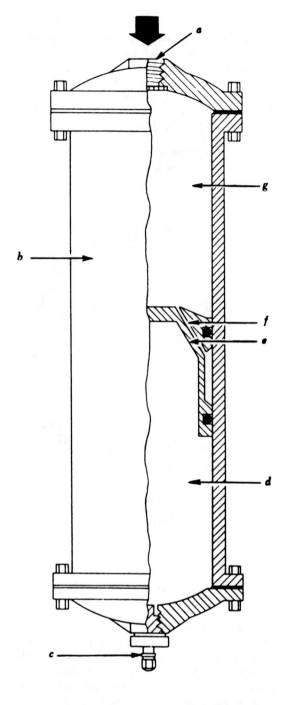

a. Fluid port
b. Cylinder
c. High-pressure air
   vent
d. Air chamber
e. Piston assembly
f. Drilled passage
g. Fluid chamber

FIGURE 8–21. Piston-type accumulator.

Maintenance of Accumulators

Maintenance consists of inspections, minor repairs, replacement of component parts, and testing. There is an element of danger in maintaining accumulators. Therefore, proper precautions must be strictly observed to prevent injury and damage.

BEFORE DISASSEMBLING ANY ACCUMULATOR, MAKE SURE THAT ALL PRELOAD AIR (OR NITROGEN) PRESSURE HAS BEEN DISCHARGED. Failure to release the air could result in serious injury to the mechanic. (Before making this check, however, be certain you know the type of high-pressure air valve used.) When you know that all air pressure has been removed, go ahead and take the unit apart. Be sure, though, that you follow manufacturer's instructions for the specific unit you have.

Check Valves

For hydraulic components and systems to operate as intended, the flow of fluid must be rigidly controlled. Fluid must be made to flow according to definite plans. Many kinds of valve units are used for exercising such control. One of the simplest and most commonly used is the check valve which allows free flow of fluid in one direction, but no flow or a restricted flow in the opposite direction.

Check valves are made in two general designs to serve two different needs. In one, the check valve is complete within itself. It is inter-connected with other components, with which it operates, by means of tubing or hose. Check valves of this design are commonly called in-line check valves. There are two types of in-line check valves, the simple-type in-line check valve and the orifice-type in-line valve. (See figure 8–22.)

In the other design, the check valve is not complete within itself because it does not have a housing exclusively its own. Check valves of this design are commonly called integral check valves. This valve is actually an integral part of some major component and, as such, shares the housing of that component.

In-Line Check Valve

The simple-type in-line check valve (often called check valve) is used when a full flow of fluid is desired in only one direction (figure 8–22A). Fluid enters the inlet port of the check valve forcing the valve off its seat against the restraint of the spring. This permits fluid to flow through the passageway thus opened. The instant fluid stops moving in this direction, the spring returns the valve to its seat. This blocks the opening in the valve seat, and therefore blocks reverse flow of fluid through the valve.

Orifice-Type Check Valve

The orifice-type in-line check valve (figure 8–22B) is used to allow normal operating speed

326

Simple-type in-line check valve (ball-type)

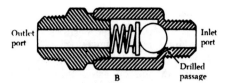

Orifice-type in-line check valve (ball-type)

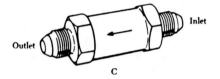

Flow direction marking on simple-
type in-line check valve

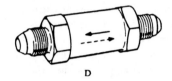

Flow direction marking on orifice-
type in-line check valve

FIGURE 8-22. Typical in-line check valves.

of a mechanism by providing free flow of fluid in one direction, while allowing limited operating speed through restricted flow of fluid in the opposite direction. The operation of the orifice-type in-line check valve is the same as the simple-type in-line check valve, except for the restricted flow allowed when closed. This is accomplished by a second opening in the valve seat that is never closed, so that some reverse flow can take place through the valve. The second opening is much smaller than the opening in the valve seat. As a rule, this opening is of a specific size thus maintaining close control over the rate at which fluid can flow back through the valve. This type of valve is sometimes called a damping valve.

The direction of fluid flow through in-line check valves is normally indicated by stamped arrow markings on the housings (figure 8-22C and D). On the simple-type in-line check valve, a single arrow points in the direction which fluid can flow. The orifice-type in-line check valve is usually marked with two arrows. One arrow is more pronounced than the other, and indicates the direction of unrestricted flow. The other arrow is either of smaller size than the first or of broken-line construction, and points in the direction of restricted reverse fluid flow.

In addition to the ball-type in-line check valves shown in figure 8-22, other types of valves, such as disks, needles, and poppets are used.

The operating principles of integral check valves are the same as the operating principles of in-line check valves.

Line-Disconnect or Quick-Disconnect Valves

These valves are installed in hydraulic lines to prevent loss of fluid when units are removed. Such valves are installed in the pressure and suction lines of the system just in front of and immediately behind the power pump. These valves can also be used in other ways than just for unit replacement. A power pump can be disconnected from the system and a hydraulic test stand connected in its place. These valve units consist of two interconnecting sections coupled together by a nut when installed in the system. Each valve section has a piston and poppet assembly. These are spring loaded to the CLOSED position when the unit is disconnected.

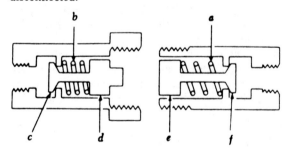

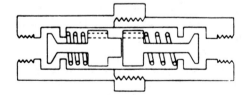

a. Spring   d. Piston
b. Spring   e. Piston
c. Poppet   f. Poppet

FIGURE 8-23. Line disconnect valve.

327

The top illustration of figure 8-23 shows the valve in the LINE-DISCONNECTED position. The two springs (a and b) hold both poppets (c and f) in the CLOSED position as shown. This prevents loss of fluid through the disconnected line. The bottom illustration of figure 8-23 shows the valve in the LINE-CONNECTED position. When the valve is being connected, the coupling nut draws the two sections together. The extension (d or e) on one of the pistons forces the opposite piston back against its spring. This action moves the poppet off its seat and permits the fluid to flow through that section of the valve. As the nut is drawn up tighter, one piston hits a stop; now the other piston moves back against its spring and, in turn, allows fluid to flow. Thus, fluid is allowed to continue through the valve and on through the system.

Bear in mind that the above disconnect valve is only one of the many types presently used. Although all line-disconnect valves operate on the same principle, the details will vary. Each manufacturer has his own design features.

A very important factor in the use of the line-disconnect valve is its proper connection. Hydraulic pumps can be seriously damaged if the line disconnects are not properly connected. If you are in doubt about the line disconnect's operation, consult the aircraft maintenance manual.

The extent of maintenance to be performed on a quick disconnect valve is very limited. The internal parts of this type valve are precision built and factory assembled. They are made to very close tolerances, therefore, no attempt should be made to disassemble or replace internal parts in either coupling half. However, the coupling halves, lock-springs, union nuts, and dust caps may be replaced. When replacing the assembly or any of the parts, follow the instructions in the applicable maintenance manual.

ACTUATING CYLINDERS

An actuating cylinder transforms energy in the form of fluid pressure into mechanical force, or action, to perform work. It is used to impart powered linear motion to some movable object or mechanism.

A typical actuating cylinder consists fundamentally of a cylinder housing, one or more pistons and piston rods, and some seals. The cylinder housing contains a polished bore in which the piston operates, and one or more ports through which fluid enters and leaves the bore. The piston and rod form an assembly. The piston moves forward and backward within the cylinder bore and an attached piston rod moves into and out of the cylinder housing through an opening in one end

of the cylinder housing. Seals are used to prevent leakage between the piston and the cylinder bore, and between the piston rod and the end of the cylinder. Both the cylinder housing and the piston rod have provisions for mounting and for attachment to an object or mechanism which is to be moved by the actuating cylinder.

Actuating cylinders are of two major types: (1) Single-action and (2) Double-action. The single-action (single port) actuating cylinder is capable of producing powered movement in one direction only. The double-action (two port) actuating cylinder is capable of producing powered movement in two directions.

Single-Action Actuating Cylinder

A single-action actuating cylinder is illustrated in figure 8-24. Fluid under pressure enters the port at the left and pushes against the face of the piston, forcing the piston to the right. As the piston moves, air is forced out of the spring chamber through the vent hole, compressing the spring. When pressure on the fluid is released to the point that it exerts less force than is present in the compressed spring, the spring pushes the piston toward the left. As the piston moves to the left, fluid is forced out of the fluid port. At the same time, the moving piston pulls air into the spring chamber through the venthole. A three-way control valve is normally used for controlling the operation of a single-action actuating cylinder

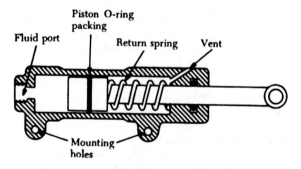

FIGURE 8-24. Single-action actuating cylinder.

Double-Action Actuating Cylinder

A double-action (two-port) actuating cylinder is illustrated in figure 8-25. The operation of a double-action actuating cylinder is usually controlled by a four-way selector valve. Figure 8-26 shows an actuating cylinder interconnected with a selector valve. Operation of the selector valve and actuating cylinder is discussed below.

Placing the selector valve in the "on" position (figure 8-26A) admits fluid pressure to the left-hand chamber of the actuating cylinder. This results in the piston being forced toward the right.

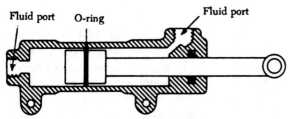

FIGURE 8-25. Double-action actuating cylinder.

As the piston moves toward the right, it pushes return fluid out of the right-hand chamber and through the selector valve to the reservoir.

When the selector valve is placed in its other "on" position, as illustrated in figure 8-26B, fluid pressure enters the right-hand chamber, forcing the piston toward the left. As the piston moves toward the left, it pushes return fluid out of the left-hand chamber and through the selector valve to the reservoir. Besides having the ability to move a load into position, a double-acting cylinder also has the ability to hold a load in position. This capability exists because when the selector valve used to control operation of the actuating cylinder is placed in the "off" position, fluid is trapped in the chambers on both sides of the actuating cylinder piston.

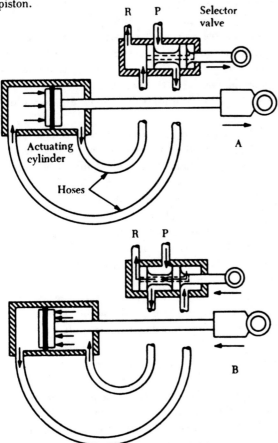

FIGURE 8-26. Control of actuating cylinder movement.

In addition to the two general design types of actuating cylinders discussed (single-action and double-action), other types are available. Figure 8-27 shows three additional types.

SELECTOR VALVES

Selector valves are used to control the direction of movement of an actuating unit. A selector valve provides a pathway for the simultaneous flow of hydraulic fluid into and out of a connected actuating unit. A selector valve also provides a means of immediately and conveniently switching the directions in which the fluid flows through the actuator, reversing the direction of movement.

One port of the typical selector valve is connected with a system pressure line for the input of fluid pressure. A second port of the valve is connected to a system return line for the return of fluid to the reservoir. The ports of an actuating unit through which fluid enters and leaves the actuating unit are connected by lines to other ports of the selector valve.

Selector valves have various numbers of ports. The number of ports is determined by the particular requirements of the system in which the valve is

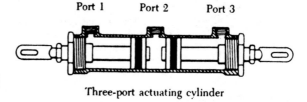

Three-port actuating cylinder

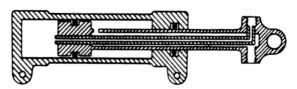

Actuating cylinder having ports in piston rod

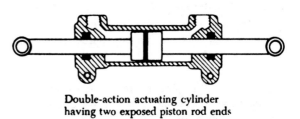

Double-action actuating cylinder having two exposed piston rod ends

FIGURE 8-27. Types of actuating cylinders.

used. Selector valves having four ports are the most commonly used. The term four-way is often used instead of four-port in referring to selector valves.

The ports of selector valves (figure 8–28) are individually marked to provide ready identification. The most commonly used markings are: PRESSURE (or PRESS, or P), RETURN (or RET, or R), CYLINDER 1 (or CYL 1), and CYLINDER 2 (or CYL 2). The use of the word "cylinder" in the designation of selector valve ports does not indicate, as it may suggest, that only hydraulic cylinders are to be connected to the ports so marked. In fact, any type of hydraulic actuating unit may be connected to the ports. The numbers 1 and 2 are a convenient means of differentiating between two ports of the selector valve.

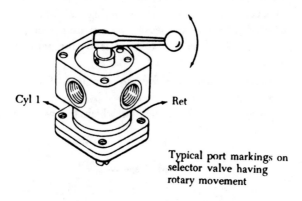

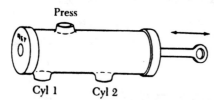

Typical port markings on selector valve having rotary movement

Typical port markings on selector valve having slide movement

FIGURE 8–28. Typical port markings on selector valves.

Four-Way Closed-Center Selector Valve

Because the four-way, closed-center selector valve is one of the most commonly used selector valves in an aircraft hydraulic system, it is discussed in detail in the following paragraphs. Valving devices of various kinds, such as balls, poppets, rotors, or spools, are used in the four-way, closed-center selector valves.

Figure 8–29A illustrates a four-way, closed-center selector valve in the "off" position. All of the valve ports are blocked, and fluid can not flow into or out of the valve.

In figure 8–29B, the selector valve is placed in one of its "on" positions. The PRESS port and CYL 1 port become interconnected within the valve. As a result, fluid flows from the pump into the selector valve PRESS port, out of the selector valve CYL 1 port, and into port A of the motor. This flow of fluid causes the motor to turn in a clockwise direction. Simultaneously, return fluid is forced out of port B of the motor and enters the selector valve CYL 2 port. Fluid then proceeds through the passage in the valve rotor and leaves the valve through the RET port.

In figure 8–29C, the selector valve is placed in the other "on" position. The PRESS port and CYL 2 port become interconnected. This causes fluid pressure to be delivered to port B of the motor, which results in the motor turning counterclockwise. Return fluid leaves port A of the motor, enters the selector valve CYL 1 port, and leaves through the selector valve RET port.

Spool-Type Selector Valve

The valving device of the spool-type selector valve is spool-shaped (figure 8–30). The spool is a one-piece, leak-tight, free-sliding fit in the selector valve housing and can be moved lengthwise in the housing by means of the extended end which projects through the housing. A drilled passage in the spool interconnects the two end chambers of the selector valve. Selector valve spools are sometimes called pilot valves.

When the spool is moved to the selector valve "off" position, the two cylinder ports are directly blocked by the lands (flanges) of the spool (figure 8–30A). This indirectly blocks the PRESS and RET ports and fluid can not flow into or out of the valve.

Moving the spool toward the right moves the spool lands away from the CYL 1 and CYL 2 ports (figure 8–23B). The PRESS port and CYL 2 port then become interconnected. This permits fluid pressure to pass on to the actuating unit. The RET port and CYL 1 port also become interconnected. This provides an open route for the return of fluid from the actuating unit to the system reservoir.

Moving the spool toward the left moves the spool lands away from the CYL 1 and CYL 2 ports (figure 8–30C). The PRESS port and CYL 1 port then become interconnected. This permits fluid pressure to flow to the actuating unit. The RET port and CYL 2 port also become interconnected, providing a route for the return of fluid from the actuating unit to the reservoir.

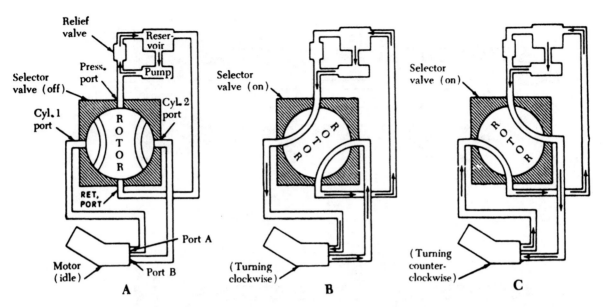

FIGURE 8–29. Typical rotor-type, closed-center selector valve operation.

AIRCRAFT PNEUMATIC SYSTEMS

Some aircraft manufacturers equip their aircraft with a pneumatic system. Such systems operate a great deal like hydraulic systems, except they employ air instead of a liquid for transmitting power. Pneumatic systems are sometimes used for:

(1) Brakes.
(2) Opening and closing doors.
(3) Driving hydraulic pumps, alternators, starters, water injection pumps, etc.
(4) Operating emergency devices.

Both pneumatic and hydraulic systems are similar units and use confined fluids. The word "confined" means trapped or completely enclosed. The word "fluid" implies such liquids as water, oil, or anything that flows. Since both liquids and gases will flow, they are considered as fluids; however, there is a great deal of difference in the characteristics of the two. Liquids are practically incompressible; a quart of water still occupies about a quart of space regardless of how hard it is compressed. But gases are highly compressible; a quart of air can be compressed into a thimbleful of space. In spite of this difference, gases and liquids are both fluids and can be confined and made to transmit power.

The type of unit used to provide pressurized air for pneumatic systems is determined by the system's air pressure requirements.

High Pressure System

For high-pressure systems, air is usually stored in metal bottles (figure 8–31) at pressures ranging from 1,000 to 3,000 p.s.i., depending on the particular system. This type of air bottle has two valves, one of which is a charging valve. A ground-operated compressor can be connected to this valve to add air to the bottle. The other valve is a control valve. It acts as a shutoff valve, keeping air trapped inside the bottle until the system is operated.

Although the high-pressure storage cylinder is light in weight, it has a definite disadvantage. Since the system cannot be re-charged during flight, operation is limited by the small supply of bottled air. Such an arrangement can not be used for the continuous operation of a system. Instead, the supply of bottled air is reserved for emergency operation of such systems as the landing gear or brakes. The usefulness of this type of system is increased, however, if other air-pressurizing units are added to the aircraft.

On some aircraft, permanently installed air compressors have been added to re-charge air bottles whenever pressure is used for operating a unit. Several types of compressors are used for this purpose. Some have two stages of compression, while others have three. Figure 8–32 shows a simplified schematic of a two-stage compressor; the pressure of the incoming air is boosted first by cylinder No. 1 and again by cylinder No. 2.

The compressor in figure 8–32 has three check valves. Like the check valves in a hydraulic hand pump, these units allow fluid to flow in only one direction. Some source of power, such as an electric

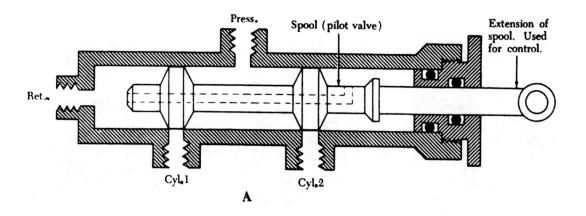

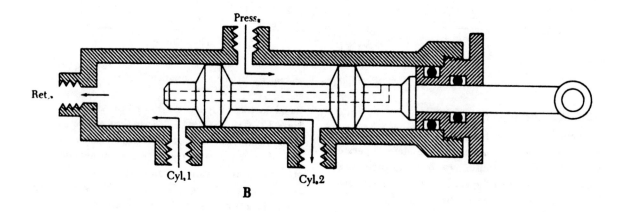

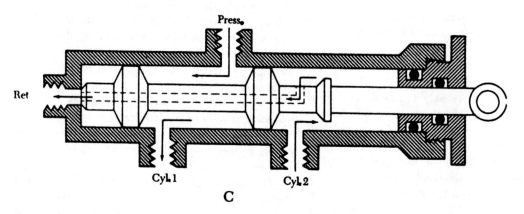

FIGURE 8–30. Typical spool-type, closed-center selector valve.

motor or aircraft engine, operates a drive shaft. As the shaft turns, it drives the pistons in and out of their cylinders. When piston No. 1 moves to the right, the chamber in cylinder No. 1 becomes larger, and outside air flows through the filter and check valve into the cylinder. As the drive shaft continues to turn, it reverses the direction of piston movement. Piston No. 1 now moves deeper into its cylinder, forcing air through the pressure line and into cylinder No. 2. Meanwhile piston No. 2 is moving out of cylinder No. 2 so that cylinder No. 2 can receive the incoming air. But cylinder No. 2 is smaller than cylinder No. 1; thus, the air must be highly compressed to fit into cylinder No. 2.

FIGURE 8–31. Steel cylinder for high-pressure air storage.

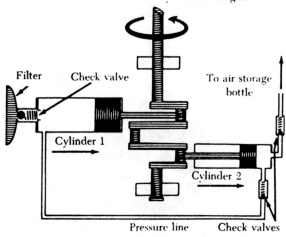

FIGURE 8–32. Schematic of two-stage air compressor.

Because of the difference in cylinder size, piston No. 1 gives the air its first stage of compression. The second stage occurs as piston No. 2 moves deeper into its cylinder, forcing high-pressure air to flow through the pressure line and into the air storage bottle.

Medium Pressure System

A medium-pressure pneumatic system (100 — 150 p.s.i.) usually does not include an air bottle. Instead, it generally draws air from a jet engine compressor section. In this case, air leaves the engine through a takeoff and flows into tubing, carrying air first to the pressure-controlling units and then to the operating units. Figure 8–33 shows a jet engine compressor with a pneumatic system takeoff.

Low Pressure System

Many aircraft equipped with reciprocating engines obtain a supply of low-pressure air from vane-type pumps. These pumps are driven by electric motors or by the aircraft engine. Figure 8–34 shows a schematic view of one of these pumps, which consists of a housing with two ports, a drive shaft, and two vanes. The drive shaft and the vanes contain slots so the vanes can slide back and forth through the drive shaft. The shaft is eccentrically mounted in the housing, causing the vanes to form four different sizes of chambers (A, B, C, and D). In the position shown, B is the largest chamber and is connected to the supply port. As depicted in the illustration, outside air can enter chamber B of the pump.

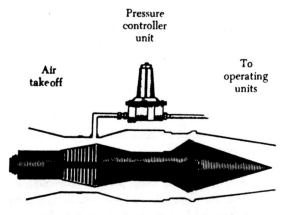

FIGURE 8–33. Jet engine compressor with pneumatic system takeoff.

333

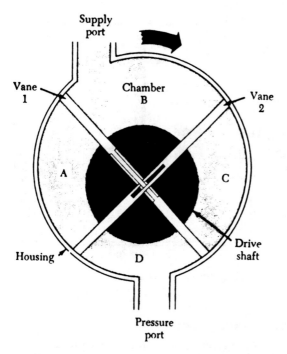

Supply
port

Vane
1

Chamber
B

Vane
2

A

C

Housing

D

Drive
shaft

Pressure
port

FIGURE 8–34. Schematic of vane-type air pump.

When the pump begins to operate, the drive shaft rotates and changes positions of the vanes and sizes of the chambers. Vane No. 1 then moves to the right (figure 8–34), separating chamber B from the supply port. Chamber B now contains trapped air. As the shaft continues to turn, chamber B moves downward and becomes increasingly smaller, gradually compressing its air. Near the bottom of the pump, chamber B connects to the pressure port and sends compressed air into the pressure line. Then chamber B moves upward again becoming increasingly larger in area. At the supply port it receives another supply of air. There are four such chambers in this pump, and each goes through this same cycle of operation. Thus, the pump delivers to the pneumatic system a continuous supply of compressed air at from 1 to 10 p.s.i.

PNEUMATIC SYSTEM COMPONENTS

Pneumatic systems are often compared to hydraulic systems, but such comparisons can only hold true in general terms. Pneumatic systems do not utilize reservoirs, hand pumps, accumulators, regulators, or engine-driven or electrically-driven power pumps for building normal pressure. But similarities do exist in some components.

Relief Valves

Relief valves are used in pneumatic systems to prevent damage. They act as pressure-limiting units and prevent excessive pressures from bursting lines and blowing out seals. Figure 8–35 illustrates a cutaway view of a pneumatic system relief valve.

At normal pressures, a spring holds the valve closed (figure 8–35), and air remains in the pressure line. If pressure grows too high, the force it creates on the disk overcomes spring tension and opens the relief valve. Then, excess air flows through the valve and is exhausted as surplus air into the atmosphere. The valve remains open until the pressure drops to normal.

Control Valves

Control valves are also a necessary part of a typical pneumatic system. Figure 8–36 illustrates how a valve is used to control emergency air brakes. The control valve consists of a three-port housing, two poppet valves, and a control lever with two lobes.

In figure 8–36A, the control valve is shown in the "off" position. A spring holds the left poppet closed so that compressed air entering the pressure port cannot flow to the brakes. In figure 8–36B, the control valve has been placed in the "on" position. One lobe of the lever holds the left poppet open,

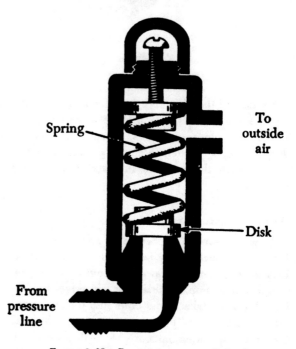

Spring

To
outside
air

Disk

From
pressure
line

FIGURE 8–35. Pneumatic system relief valve.

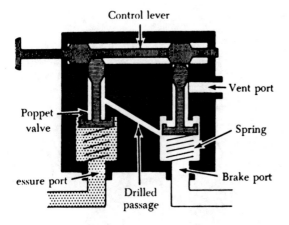

A. Control valve "off"

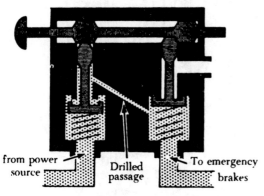

B. Control valve "on"

▓▓▓ Compressed air

☐ Atmospheric pressure

FIGURE 8–36. Flow diagram of a pneumatic control valve.

and a spring closes the right poppet. Compressed air now flows around the opened left poppet, through a drilled passage, and into a chamber below the right poppet. Since the right poppet is closed, the high-pressure air flows out of the brake port and into the brake line to apply the brakes.

To release the brakes, the control valve is returned to the "off" position (figure 8–36A). The left poppet now closes, stopping the flow of high-pressure air to the brakes. At the same time, the right poppet is opened, allowing compressed air in the brake line to exhaust through the vent port and into the atmosphere.

Check Valves

Check valves are used in both hydraulic and pneumatic systems. Figure 8–37 illustrates a flap-type pneumatic check valve. Air enters the left port of the check valve, compresses a light spring, forcing the check valve open and allowing air to flow out the right port. But if air enters from the right, air pressure closes the valve, preventing a flow of air out the left port. Thus, a pneumatic check valve is a one-direction flow control valve.

Restrictors

Restrictors are a type of control valve used in pneumatic systems. Figure 8–38 illustrates an orifice type restrictor with a large inlet port and a small outlet port. The small outlet port reduces the rate of airflow and the speed of operation of an actuating unit.

Variable Restrictor

Another type of speed-regulating unit is the variable restrictor shown in figure 8–39. It contains an adjustable needle valve, which has threads around the top and a point on the lower end. Depending on the direction turned, the needle valve moves the sharp point either into or out of a small opening to decrease or increase the size of the opening. Since air entering the inlet port must pass through this opening before reaching the outlet port, this adjustment also determines the rate of airflow through the restrictor.

Filters

Pneumatic systems are protected against dirt by means of various types of filters. A micronic filter (figure 8–40) consists of a housing with two ports, a replaceable cartridge, and a relief valve. Normally, air enters the inlet, circulates around the cellulose cartridge, then flows to the center of the cartridge and out the outlet port. If the cartridge becomes clogged with dirt, pressure forces the relief valve open and allows unfiltered air to flow out the outlet port.

A screen-type filter (figure 8–41) is similar to the micronic filter but contains a permanent wire screen instead of a replaceable cartridge. In the screen filter a handle extends through the top of the housing and can be used to clean the screen by rotating it against metal scrapers.

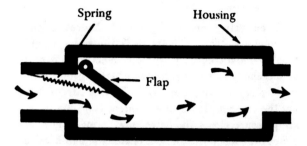

FIGURE 8-37. Pneumatic system check valve.

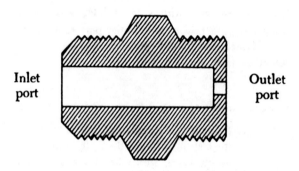

FIGURE 8-38. Orifice restrictor.

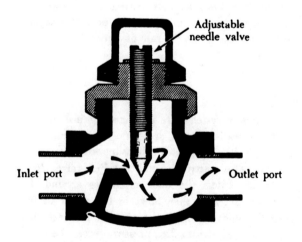

FIGURE 8-39. Variable pneumatic restrictor.

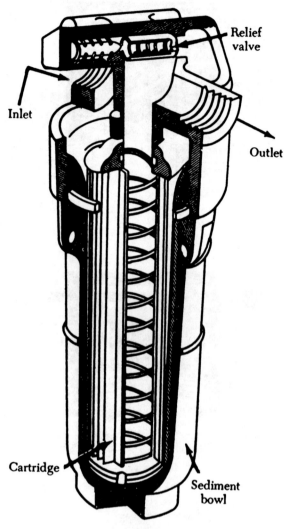

FIGURE 8-40. Micronic filter.

are compressed air systems. Figure 8-42 illustrates one type of system which uses compressed air.

Air Bottle

The air bottle usually stores enough compressed air for several applications of the brakes. A high-pressure air line connects the bottle to an air valve which controls operation of the emergency brakes.

If the normal brake system fails, place the control handle for the air valve in the "on" position. The valve then directs high-pressure air into lines leading to the brake assemblies. But before air enters the brake assemblies, it must first flow through a shuttle valve.

If the main hydraulic braking system fails, power brakes are usually equipped with some type of emergency pressurizing system for stopping the aircraft. In many instances, these emergency systems

336

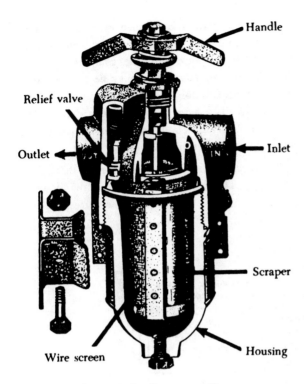

FIGURE 8–41. Screen-type filter.

Brake Shuttle Valve

The circled inset at the upper right of figure 8–42 shows one type of shuttle valve. The valve consists of a shuttle enclosed by a four-port hous-

ing. The shuttle is a sort of floating piston that can move up or down in the hollow housing. Normally, the shuttle is down, and in this position it seals off the lower air port and directs hydraulic fluid from the upper port into the two side ports, each of which leads to a brake assembly. But when the emergency pneumatic brakes are applied, high-pressure air raises the shuttle, seals off the hydraulic line, and connects air pressure to the side ports of the shuttle valve. This action sends high-pressure air into the brake cylinder to apply the brakes.

After application and when the emergency brakes are released, the air valve closes, trapping pressure in the air bottle. At the same time, the air valve vents the pneumatic brake line to outside air pressure. Then as air pressure in the brake line drops, the shuttle valve moves to the lower end of the housing, again connecting the brake cylinders to the hydraulic line. Air pressure remaining in the brake cylinders then flows out the upper port of the shuttle valve and into the hydraulic return line.

Lines and Tubing

Lines for pneumatic systems consist of rigid metal tubing and flexible rubber hose. Fluid lines and fittings are covered in detail in Chapter 5 of the Airframe and Powerplant Mechanics General Handbook, AC 65–9A.

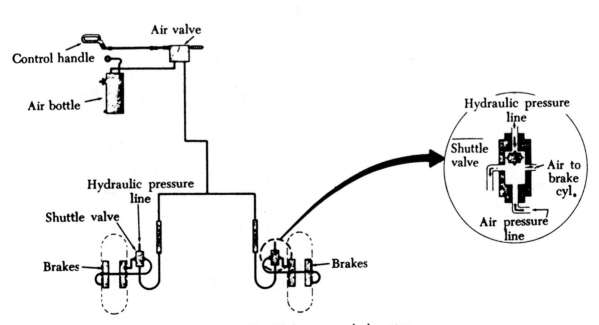

FIGURE 8–42. Simplified emergency brake system.

TYPICAL PNEUMATIC POWER SYSTEM

A typical turbine-engine pneumatic power system supplies compressed air for various normal and emergency actuating systems. The compressed air is stored in storage cylinders in the actuating systems until required by actuation of the system. These cylinders and the power system manifold are initially charged with compressed air or nitrogen from an external source through a single air-charge valve. In flight, the air compressor replaces the air pressure and volume lost through leakage, thermal contraction, and actuating system operation.

The air compressor is supplied with supercharged air from the engine bleed air system. This ensures an adequate air supply to the compressor at all altitudes. The air compressor may be driven either by an electric motor or a hydraulic motor. The system described here is hydraulically driven. The following description is illustrated by the pneumatic power system shown in figure 8–43.

The compressor inlet air is filtered through a high-temperature, 10-micron filter and the air pressure is regulated by an absolute pressure regulator

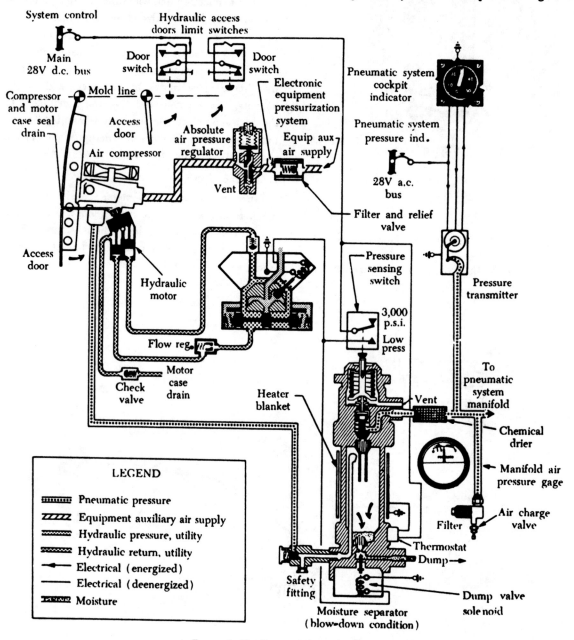

FIGURE 8–43. Pneumatic power system.

338

to provide a stabilized source of air for the compressor. (See figure 8–43.)

The aircraft utility hydraulic system provides power to operate the hydraulic-motor-driven air compressor. The air compressor hydraulic actuating system consists of a solenoid-operated selector valve, flow regulator, hydraulic motor, and motor bypass (case drain) line check valve. When energized, the selector valve allows the system to be pressurized to run the hydraulic motor; when de-energized the valve blocks off utility system pressure, stopping the motor. The flow regulator, compensating for the varying hydraulic system flow and pressures, meters the flow of fluid to the hydraulic motor to prevent excessive speed variation and/or overspeeding of the compressor. A check valve in the motor bypass line prevents system return line pressures from entering the motor and stalling it.

The air compressor is the pneumatic system's pressurizing air source. The compressor is activated or deactivated by the manifold pressure-sensing switch, which is an integral part of the moisture separator assembly.

The moisture separator assembly is the pneumatic system's pressure-sensing regulator and relief valve. The manifold (system) pressure switch governs the operation of the air compressor. When the manifold pressure drops below 2,750 p.s.i., the pressure-sensing switch closes, energizing the separator's moisture dump valve and the hydraulic selector valve which activates the air compressor. When the manifold pressure builds up to 3,150 p.s.i., the pressure-sensing switch opens, de-energizing the hydraulic selector valve to deactivate the air compressor and dump valve, thus venting overboard any moisture accumulated in the separator.

The safety fitting, installed at the inlet port of the moisture separator, protects the separator from internal explosions caused by hot carbon particles or flames that may be emitted from the air compressor.

A chemical drier further reduces the moisture content of the air emerging from the moisture separator.

A pressure transmitter senses and electrically transmits a signal to the pneumatic pressure indicator located in the cockpit. The indicating system is an "autosyn" type that functions exactly like the hydraulic indicating systems.

An air-charge valve provides the entire pneumatic system with a single external ground servicing point. An air pressure gage, located near the air-charge valve, is used in servicing the pneumatic system. This gage indicates the manifold pressure.

An air filter (10-micron element) in the ground air-charge line prevents the entry of particle impurities into the system from the ground servicing source.

The high-pressure air exiting from the fourth stage of the air compressor is directed through a bleed valve (controlled by an oil pressure tap on the pressure side of the oil pump) to the high-pressure air outlet. The oil pressure applied to the piston of the bleed valve maintains the valve piston in the "closed" position. When the oil pressure drops (due either to restriction of oil flow or to stopping of the compressor), the spring within the bleed valve re-positions the bleed valve piston, thereby connecting the inlet port and the drain port of the valve. This action unloads the pressure from the compressor and purges the line of moisture.

The air filter through which the ground-charge air passes is located immediately upstream of the air-charge valve. Its purpose is to prevent the entry of particle impurities into the system from the ground servicing source. The filter assembly is made up off three basic components—body, element, and bowl.

The pneumatic system air compressor inlet air is filtered through a high-temperature filter. Its purpose is to prevent particles of foreign matter from entering the compressor's absolute pressure regulator, thereby causing it to malfunction. The filter is an in-line, full-flow type (with integral relief valve) housed in a cylindrical body.

The moisture separator is the pneumatic power system's pressure-sensing regulator and relief valve, and is capable of removing up to 95% of the moisture from the air compressor discharge line. The automatically operated, condensation dump valve purges the separator's oil/moisture chamber by means of a blast of air (3,000 p.s.i.) each time the compressor shuts down. The separator assembly is made up of several basic components, each of which performs a specific function.

Components

The pressure switch controls system pressurization by sensing the system pressure between the check valve and the relief valve. It electrically energizes the air compressor solenoid-operated selector valve when the system pressure drops below 2,750 p.s.i., and de-energizes the selector valve when the system pressure reaches 3,100 p.s.i.

The condensation dump valve solenoid is energized and de-energized by the pressure switch.

339

When energized, it prevents the air compressor from dumping air overboard; when de-energized, it completely purges the separator's reservoir and lines up to the air compressor.

The filter protects the dump valve port from becoming clogged and thus ensures proper sealing of the passage between the reservoir and the dump port.

The check valve protects the system against pressure loss during the dumping cycle and prevents backflow through the separator to the air compressor during the relief condition.

The relief valve protects the system against over-pressurization (thermal expansion). The relief valve opens when the system pressure reaches 3,750 p.s.i. and re-seats at 3,250 p.s.i.

The thermostatically controlled wraparound-blanket type heating element prevents freezing of the moisture within the reservoir due to low-temperature atmospheric conditions. The thermostat closes at 40° F. and opens at 60° F.

Pneumatic Power System Maintenance

Maintenance of the pneumatic power system consists of servicing, troubleshooting, removal and installation of components, and operational testing.

The air compressor's lubricating oil level should be checked daily in accordance with the applicable manufacturer's instructions. The oil level is indicated by means of a sight gage or dipstick. When re-filling the compressor oil tank, the oil (type specified in the applicable instructions manual) is added until the specified level. After the oil is added, ensure that the filler plug is torqued and safety wire is properly installed.

The pneumatic system should be purged periodically to remove the contamination, moisture, or oil from the components and lines. Purging the system is accomplished by pressurizing it and removing the plumbing from various components throughout the system. Removal of the pressurized lines will cause a high rate of airflow through the system causing foreign matter to be exhausted from the system. If an excessive amount of foreign matter, particularly oil, is exhausted from any one system, the lines and components should be removed and cleaned or replaced.

Upon completion of pneumatic system purging and after re-connecting all the system components, the system air bottles should be drained to exhaust any moisture or impurities which may have accumulated there.

After draining the air bottles, service the system with nitrogen or clean, dry compressed air. The system should then be given a thorough operational check and an inspection for leaks and security.

GENERAL

The landing gear of a fixed-wing aircraft consists of main and auxiliary units, either of which may or may not be retractable. The main landing gear forms the principal support of the aircraft on land or water and may include any combination of wheels, floats, skis, shock-absorbing equipment, brakes, retracting mechanism with controls and warning devices, cowling, fairing, and structural members necessary to attach any of the foregoing to the primary structure. The auxiliary landing gear consists of tail or nose wheel installations, outboard pontoons, skids, etc., with necessary cowling and reinforcement.

Landing Gear Arrangement

Many aircraft are equipped with a tricycle gear arrangement. This is almost universally true of large aircraft, the few exceptions being older model aircraft. Component parts of a tricycle gear arrangement are the nose gear and the main gear. Nose gear equipped aircraft are protected at the fuselage tail section with a tail skid or bumper. The nose gear arrangement has at least three advantages:

(1) It allows more forceful application of the brakes for higher landing speeds without nosing over.

(2) It permits better visibility for the pilot during landing and taxiing.

(3) It tends to prevent aircraft ground-looping by moving the aircraft c.g. forward of the main wheels. Forces acting on the c.g. tend to keep the aircraft moving forward on a straight line rather than ground-looping.

The number and location of wheels on the main gear vary. Some main gears have two wheels as shown in figure 9–1. Multiple wheels spread the aircraft's weight over a larger area in addition to providing a safety margin if one tire should fail.

Heavier aircraft may use four or more wheels. When more than two wheels are attached to one strut. the attaching mechanism is referred to as a "bogie," as shown in figure 9–2. The number of wheels that are included in the bogie is determined by the gross design weight of the aircraft and the surfaces on which the loaded aircraft may be required to land.

The tricycle arrangement of the landing gear is made up of many assemblies and parts. These consist of air/oil shock struts, main gear alignment units, support units, retraction and safety devices, auxiliary gear protective devices, nose wheel steering system, aircraft wheels, tires, tubes, and aircraft brake systems. The airframe mechanic should know all about each of these assemblies, their inspection procedures, and their relationships to the total operation of the landing gear.

Shock Struts

Shock struts are self-contained hydraulic units that support an aircraft on the ground and protect the aircraft structure by absorbing and dissipating

FIGURE 9–1. Dual main landing gear wheel arrangement.

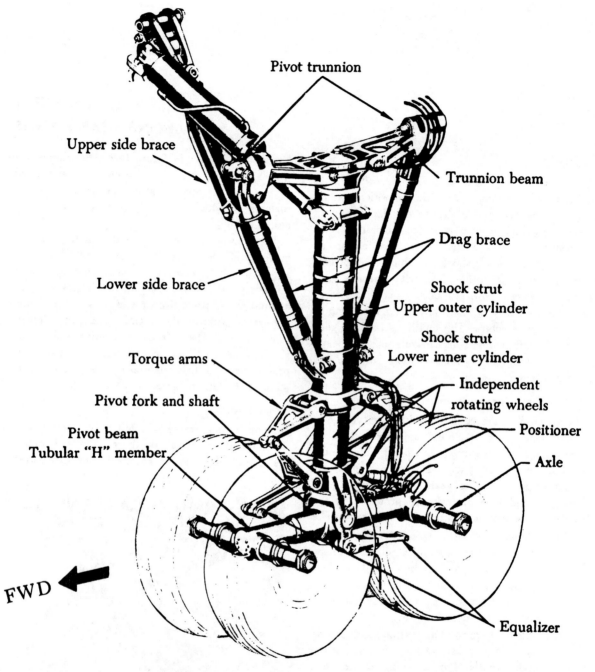

Upper side brace

Pivot trunnion

Trunnion beam

Lower side brace

Drag brace

Shock strut
Upper outer cylinder

Shock strut
Lower inner cylinder

Torque arms

Pivot fork and shaft

Pivot beam
Tubular "H" member

Independent
rotating wheels

Positioner

Axle

FWD

Equalizer

FIGURE 9–2. "Bogie" truck main landing gear assembly.

the tremendous shock loads of landing. Shock struts must be inspected and serviced regularly to function efficiently.

Since there are many different designs of shock struts, only information of a general nature is included in this section. For specific information about a particular installation, consult the applicable manufacturer's instructions.

A typical pneumatic/hydraulic shock strut

(figure 9–3) uses compressed air combined with hydraulic fluid to absorb and dissipate shock loads, and is often referred to as the air/oil or oleo strut.

A shock strut is made up essentially of two telescoping cylinders or tubes, with externally closed ends (figure 9–3). The two cylinders, known as cylinder and piston, when assembled, form an upper and lower chamber for movement of the fluid. The lower chamber is always filled with fluid, while the

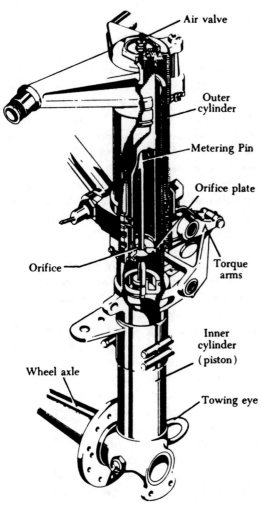

FIGURE 9–3. Landing gear shock strut of the metering pin type.

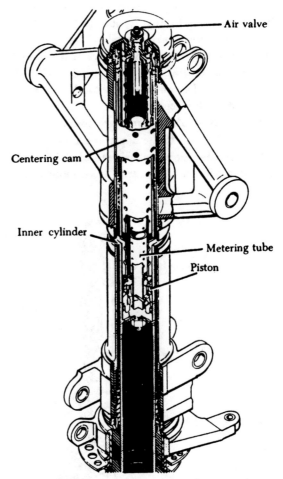

FIGURE 9–4. Landing gear shock strut of the metering tube type.

upper chamber contains compressed air. An orifice is placed between the two chambers and provides a passage for the fluid into the upper chamber during compression and return during extension of the strut.

Most shock struts employ a metering pin similar to that shown in figure 9–3 for controlling the rate of fluid flow from the lower chamber into the upper chamber. During the compression stroke, the rate of fluid flow is not constant, but is controlled automatically by the variable shape of the metering pin as it passes through the orifice.

On some types of shock struts, a metering tube replaces the metering pin, but shock strut operation is the same (figure 9–4).

Some shock struts are equipped with a damping or snubbing device consisting of a recoil valve on the piston or recoil tube to reduce the rebound during the extension stroke and to prevent too rapid extension of the shock strut. This could result in a sharp impact at the end of the stroke and possible damage to the aircraft and the landing gear.

The majority of shock struts are equipped with an axle attached to the lower cylinder to provide for installation of the wheels. Shock struts not equipped with axles have provisions on the end of the lower cylinder for easy installation of the axle assembly. Suitable connections are also provided on all shock struts to permit attachment to the aircraft.

A fitting consisting of a fluid filler inlet and air valve assembly is located near the upper end of each shock strut to provide a means of filling the strut with hydraulic fluid and inflating it with air.

A packing gland designed to seal the sliding joint between the upper and lower telescoping cylinders is installed in the open end of the outer cylinder. A packing gland wiper ring is also installed in a groove in the lower bearing or gland nut on most shock struts to keep the sliding surface of the piston or inner cylinder free from dirt, mud, ice, and snow. Entry of foreign matter into the packing gland would result in leaks.

The majority of shock struts are equipped with torque arms attached to the upper and lower cylinders to maintain correct alignment of the wheel. Shock struts without torque arms have splined pis-

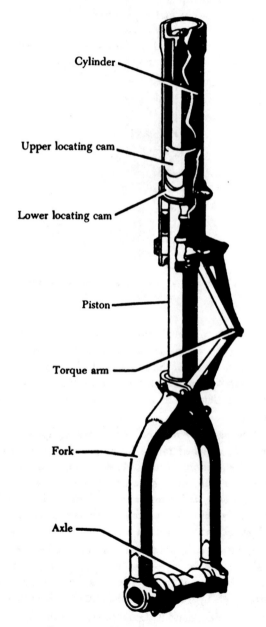

FIGURE 9-5. Nose gear shock strut.

ton heads and cylinders which maintain correct wheel alignment.

Nose gear shock struts are provided with an upper locating cam attached to the upper cylinder and a mating lower locating cam attached to the lower cylinder (figure 9-5). These cams line up the wheel and axle assembly in the straight-ahead position when the shock strut is fully extended. This prevents the nosewheel from being cocked to one side when the nose gear is retracted, thus preventing possible structural damage to the aircraft. The mating cams also keep the nosewheel in a straight-ahead position prior to landing when the strut is fully extended. Some nose gear shock struts have attachments for installing an external shimmy damper.

Generally, nose gear struts are equipped with a locking (or disconnect) pin to enable quick turning of the aircraft when it is standing idle on the ground or in the hangar. Disengagement of this pin will allow the wheel fork spindle to rotate 360°, thus enabling the aircraft to be turned in a very small space, as in a crowded hangar.

Nose and main gear shock struts are usually provided with jacking points and towing lugs. Jacks should always be placed under the prescribed points; and when towing lugs are provided, the towing bar should be attached only to these lugs.

All shock struts are provided with an instruction plate which gives brief instructions for filling the strut with fluid and for inflating the strut. The instruction plate attached near the filler inlet and air valve assembly also specifies the correct type of hydraulic fluid to use in the strut. It is of utmost importance to become familiar with these instruction prior to filling a shock strut with hydraulic fluid or inflating it with air.

Figure 9-6 shows the inner construction of a shock strut and illustrates the movement of the fluid during compression and extension of the strut.

The compression stroke of the shock strut begins as the aircraft wheels touch the ground; the center of mass of the aircraft continues to move downward, compressing the strut and sliding the inner cylinder into the outer cylinder. The metering pin is forced through the orifice and, by its variable shape, controls the rate of fluid flow at all points of the compression stroke. In this manner the greatest possible amount of heat is dissipated through the walls of the shock strut. At the end of the downward stroke, the compressed air is further compressed, limiting the compression stroke of the strut.

344

The extension stroke occurs at the end of the compression stroke as the energy stored in the compressed air causes the aircraft to start moving upward in relation to the ground and wheels. At this instant, the compressed air acts as a spring to return the strut to normal. It is at this point that a snubbing or damping effect is produced by forcing the fluid to return through the restrictions of the snubbing device. If this extension were not snubbed, the aircraft would rebound rapidly and tend to oscillate up and down, due to the action of the compressed air. A sleeve, spacer, or bumper ring incorporated in the strut limits the extension stroke.

For efficient operation of shock struts, the proper fluid level and air pressure must be maintained. To check the fluid level, the shock strut must be deflated and in the fully compressed position. Deflating a shock strut can be a dangerous operation unless servicing personnel are thoroughly familiar with high-pressure air valves. Observe all the necessary safety precautions. Refer to manufacturer's instructions for proper deflating technique.

Two of the various types of high-pressure air valves currently in use on shock struts are illustrated in figure 9–7. Although the two air valves

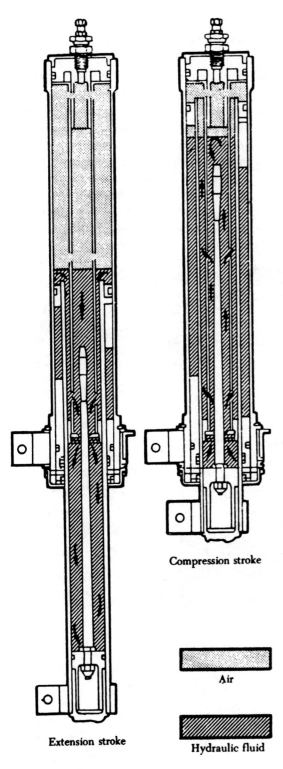

Compression stroke

Air

Hydraulic fluid

Extension stroke

FIGURE 9–6. Shock strut operation.

If there was an insufficient amount of fluid and/or air in the strut, the compression stroke would not be limited and the strut would "bottom."

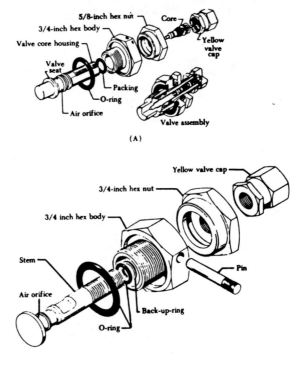

FIGURE 9–7. High-pressure air valves.

345

are interchangeable, there are some important differences in their construction. One valve (figure 9–7A) contains a valve core and has a 5/8-in. swivel hex nut. The other air valve (figure 9–7B) has no valve core and has a 3/4-in. swivel hex nut.

Servicing Shock Struts

The following procedures are typical of those used in deflating a shock strut, servicing with hydraulic fluid, and re-inflating (figure 9–8):

(1) Position the aircraft so the shock struts are in the normal ground operating position. Make certain that personnel, workstands, and other obstacles are clear of the aircraft. (Some aircraft must be placed on jacks to service the shock struts.)

(2) Remove the cap from the air valve (figure 9–8A).

(3) Check the swivel hex nut for tightness with a wrench (figure 9–8B).

(4) If the air valve is equipped with a valve core, release any air pressure that may be trapped between the valve core and the valve seat by depressing the valve core (figure 9–8C). Always stand to one side of the valve, since high-pressure air can cause serious injury, e.g., loss of eyesight.

(5) Remove the valve core (figure 9–8D).

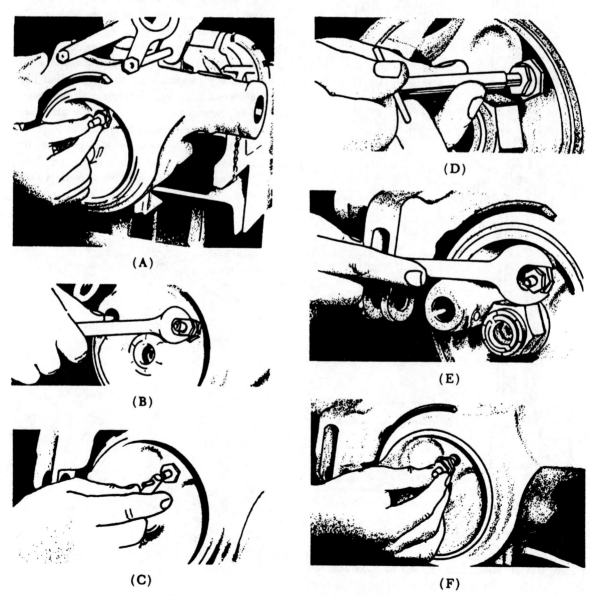

(A)

(B)

(C)

(D)

(E)

(F)

FIGURE 9–8. Steps in servicing shock struts.

(6) Release the air pressure in the strut by slowly turning the swivel nut counterclockwise (figure 9–8E).

(7) Ensure that the shock strut compresses as the air pressure is released. In some cases, it may be necessary to rock the aircraft after deflating to ensure compression of the strut.

(8) When the strut is fully compressed, the air valve assembly may be removed (figure 9–8F).

(9) Fill the strut to the level of the air valve opening with an approved type of hydraulic fluid.

(10) Re-install the air valve assembly, using a new O-ring packing. Torque the air valve assembly to the values recommended in the applicable manufacturer's instructions.

(11) Install the air valve core.

(12) Inflate the strut, using a high-pressure source of dry air or nitrogen. Bottled gas should not be used to inflate shock struts. On some shock struts the correct amount of inflation is determined by using a high-pressure air gage. On others it is determined by measuring the amount of extension (in inches) between two given points on the strut. The proper procedure can usually be found on the instruction plate attached to the shock strut. Shock struts should always be inflated slowly to avoid excessive heating and over inflation.

(13) Tighten the swivel hex nut, using the torque values specified in the applicable manufacturer's instructions.

(14) Remove the high-pressure air line chuck and install the valve cap. Tighten the valve cap fingertight.

Bleeding Shock Struts

If the fluid level of a shock strut has become extremely low, or if for any other reason air is trapped in the strut cylinder, it may be necessary to bleed the strut during the servicing operation. Bleeding is usually performed with the aircraft placed on jacks. In this position the shock struts can be extended and compressed during the filling operation, thus expelling all the entrapped air. The following is a typical bleeding procedure:

(1) Construct a bleed hose containing a fitting suitable for making an airtight connection to the shock strut filler opening. The base should be long enough to reach from the shock strut filler opening to the ground when the aircraft is on jacks.

(2) Jack the aircraft until all shock struts are fully extended.

(3) Release the air pressure in the strut to be bled.

(4) Remove the air valve assembly.

(5) Fill the strut to the level of the filler port with an approved type hydraulic fluid.

(6) Attach the bleed hose to the filler port and insert the free end of the hose into a container of clean hydraulic fluid, making sure that this end of the hose is below the surface of the hydraulic fluid.

(7) Place an exerciser jack (figure 9–9) or other suitable single-base jack under the shock strut jacking point. Compress and extend the strut fully by raising and lowering the jack until the flow of air bubbles from the strut has completely stopped. Compress the strut slowly and allow it to extend by its own weight.

(8) Remove the exerciser jack, and then lower and remove all other jacks.

(9) Remove the bleed hose from the shock strut.

(10) Install the air valve and inflate the strut.

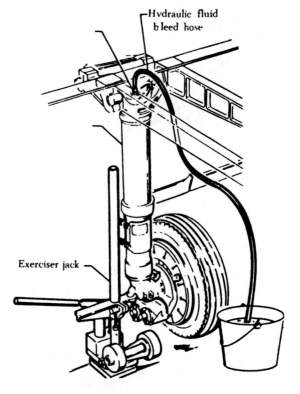

FIGURE 9–9. Bleeding a shock strut using an exerciser jack.

Shock struts should be inspected regularly for leakage of fluid and for proper extension. Exposed portions of the strut pistons should be wiped clean daily and inspected closely for scoring or corrosion.

MAIN LANDING GEAR ALIGNMENT, SUPPORT, RETRACTION

The main landing gear consists of several components that enable it to function. Typical of these are the torque links, trunnion and bracket arrangements, drag strut linkages, electrical and hydraulic gear-retraction devices, and gear indicators.

Alignment

Torque links (figure 9–10) keep the landing gear pointed in a straight-ahead direction; one torque link connects to the shock strut cylinder, while the other connects to the piston. The links are hinged at the center so that the piston can move up or down in the strut.

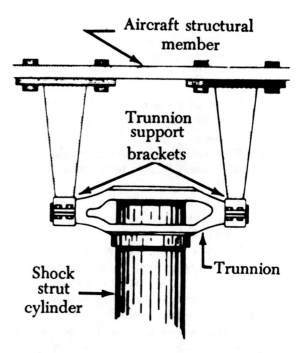

FIGURE 9–11.   Trunnion and bracket arrangement.

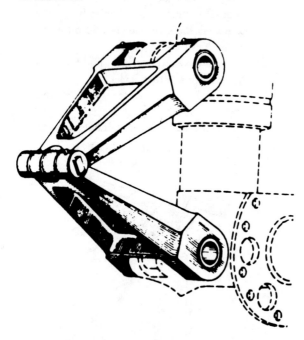

FIGURE 9–10.   Torque links.

Support

To anchor the main gear to the aircraft structure, a trunnion and bracket arrangement (figure 9–11) is usually employed. This arrangement is constructed to enable the strut to pivot or swing forward or backward as necessary when the aircraft is being steered or the gear is being retracted. To restrain this action during ground movement of the aircraft, various types of linkages are used, one being the drag strut.

The upper end of the drag strut (figure 9–12) connects to the aircraft structure, while the lower end connects to the shock strut. The drag strut is hinged so that the landing gear can be retracted.

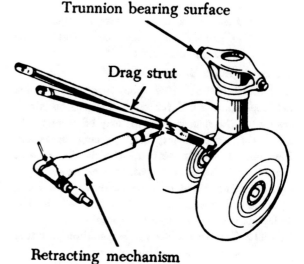

FIGURE 9–12.   Drag strut linkage.

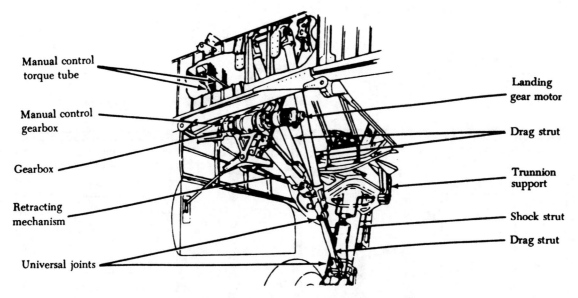

Manual control torque tube

Manual control gearbox

Gearbox

Retracting mechanism

Universal joints

Landing gear motor

Drag strut

Trunnion support

Shock strut

Drag strut

FIGURE 9–13. Electrical retraction system.

Electrical Landing Gear Retraction System

An electrical landing gear retraction system, such as that shown in figure 9–13, has the following features:

(1) A motor for converting electrical energy into rotary motion.

(2) A gear reduction system for decreasing the speed and increasing the force of rotation.

(3) Other gears for changing rotary motion (at a reduced speed) into push-pull movement.

(4) Linkage for connecting the push-pull movement to the landing gear shock struts.

Basically, the system is an electrically driven jack for raising or lowering the gear. When a switch in the cockpit is moved to the "up" position, the electric motor operates. Through a system of shafts, gears, adapters, an actuator screw, and a torque tube, a force is transmitted to the drag strut linkages. Thus, the gear retracts and locks. If the switch is moved to the "down" position, the motor reverses and the gear moves down and locks. The sequence of operation of doors and gears is similar to that of a hydraulically operated landing gear system.

Hydraulic Landing Gear Retraction Systems

Devices used in a typical hydraulically operated landing gear retraction system include actuating cylinders, selector valves, uplocks, downlocks, sequence valves, tubing, and other conventional hydraulic components. These units are so interconnected that they permit properly sequenced retraction and extension of the landing gear and the landing gear doors.

The operation of a hydraulic landing gear retraction system is of such importance that is must be covered in some detail. First, consider what happens when the landing gear is retracted. As the selector valve (figure 9–14) is moved to the "up" position, pressurized fluid is directed into the gear up line. The fluid flows to each of eight units; to sequence valves C and D, to the three gear downlocks, to the nose gear cylinder, and to the two main actuating cylinders.

Notice what happens to the fluid flowing to sequence valves C and D in figure 9–14. Since the sequence valves are closed, pressurized fluid cannot flow to the door cylinders at this time. Thus, the doors cannot close. But the fluid entering the three downlock cylinders is not delayed; therefore, the gear is unlocked. At the same time, fluid also enters the up side of each gear-actuating cylinder and the gears begin to retract. The nose gear completes

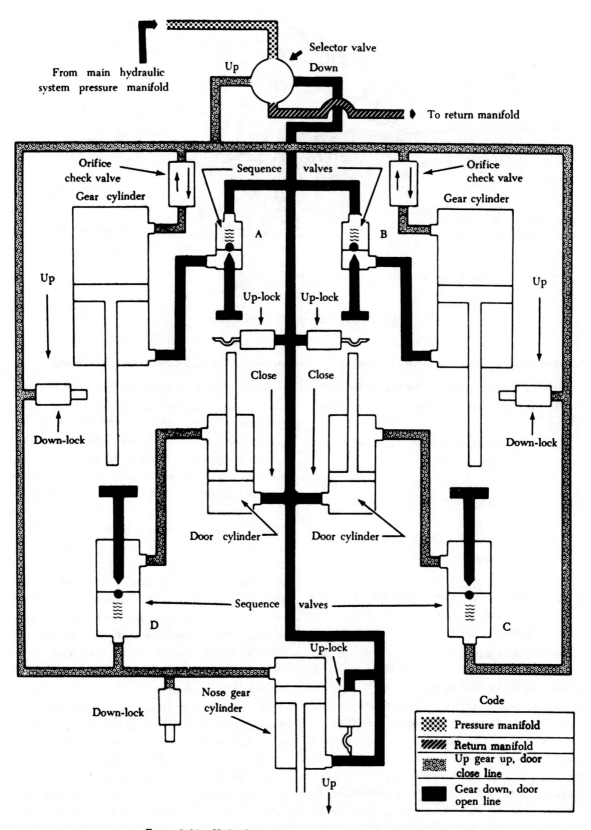

Figure 9-14. Hydraulic landing gear retraction system schematic.

350

retraction and engages its uplock first, because of the small size of its actuating cylinder. Also, since the nose gear door is operated solely by linkage from the nose gear, this door closes. Meanwhile, the main landing gear is still retracting, forcing fluid to leave the downside of each main gear cylinder. This fluid flows unrestricted through an orifice check valve, opens the sequence check valve A or B, and flows through the landing-gear selector valve into the hydraulic system return line. Then, as the main gear reaches the fully retracted position and engages the spring-loaded uplocks, gear linkage strikes the plungers of sequence valves C and D. This opens the sequence check valves and allows pressurized fluid to flow into the door cylinders, closing the landing gear doors.

Wing Landing Gear Operation

A typical wing landing gear operating sequence is illustrated in figure 9–15. The wing landing gear retracts or extends when hydraulic pressure is applied to the up or down side of the gear actuator. The gear actuator applies the force required to raise and lower the gear. The actuator works in conjunction with a walking beam to apply force to the wing gear shock strut, swinging it inboard and forward into the wheel well. Both the actuator and the walking beam are connected to lugs on the landing gear trunnion. The outboard ends of the actuator and walking beam pivot on a beam hanger which is attached to the aircraft structure. A wing landing gear locking mechanism located on the outboard side of the wheel well locks the gear in the "up" position. Locking of the gear in the "down" position is accomplished by a downlock bungee which positions an upper and lower jury strut so that the upper and lower side struts will not fold.

EMERGENCY EXTENSION SYSTEMS

The emergency extension system lowers the landing gears if the main power system fails. Some aircraft have an emergency release handle in the cockpit, which is connected through a mechanical linkage to the gear uplocks. When the handle is operated, it releases the uplocks and allows the gears to free-fall, or extend, under their own weight. On other aircraft, release of the uplock is accomplished using compressed air which is directed to uplock release cylinders.

In some aircraft, design configurations make emergency extension of the landing gear by gravity and airloads alone impossible or impractical. In such aircraft, provisions are included for forceful gear extension in an emergency. Some installations are designed so that either hydraulic fluid, or compressed air provides the necessary pressure; while others use a manual system for extending the landing gears under emergency conditions.

Hydraulic pressure for emergency operation of the landing gear may be provided by an auxiliary hand pump, an accumulator, or an electrically powered hydraulic pump, depending upon the design of the aircraft.

LANDING GEAR SAFETY DEVICES

Accidental retraction of a landing gear may be prevented by such safety devices as mechanical downlocks, safety switches, and ground locks. Mechanical downlocks are built-in parts of a gear-retraction system and are operated automatically by the gear-retraction system. To prevent accidental operation of the downlocks, electrically operated safety switches are installed.

Safety Switch

A landing gear safety switch (figure 9–16) in the landing gear safety circuit is usually mounted in a bracket on one of the main gear shock struts. This switch is actuated by a linkage through the landing gear torque links. The torque links spread apart or move together as the shock strut piston extends or retracts in its cylinder. When the strut is compressed (aircraft on the ground), the torque links are close together, causing the adjusting links to open the safety switch. During takeoff, as the weight of the aircraft leaves the struts, the struts and torque links extend, causing the adjusting links to close the safety switch. As shown in figure 9–16, a ground is completed when the safety switch closes. The solenoid then energizes and unlocks the selector valve so that the gear handle can be positioned to raise the gear.

Ground Locks

In addition to this safety device, most aircraft are equipped with additional safety devices to prevent collapse of the gear when the aircraft is on the ground. These devices are called ground locks. One

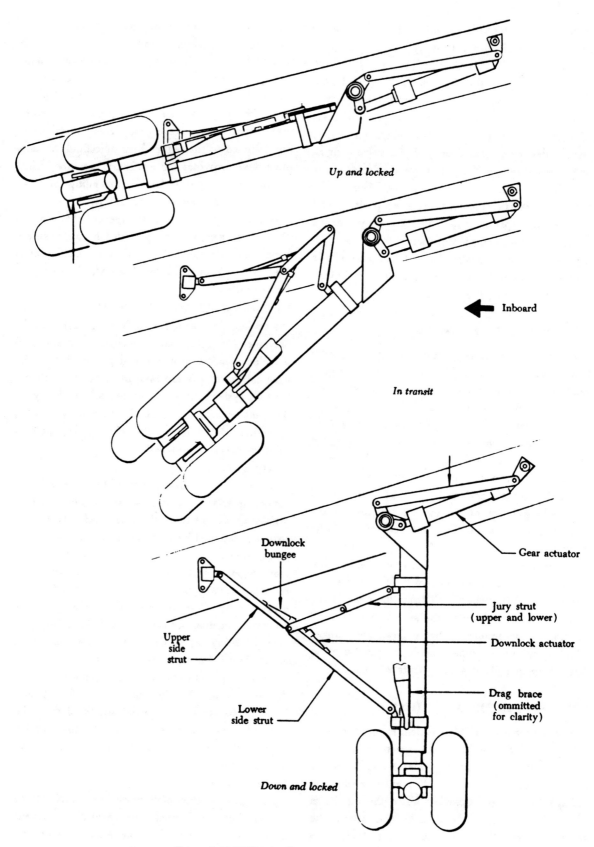

Up and locked

Inboard

In transit

Downlock
bungee

Gear actuator

Jury strut
(upper and lower)

Upper
side
strut

Downlock actuator

Drag brace
(ommitted
for clarity)

Lower
side strut

Down and locked

FIGURE 9–15.   Wing landing gear operating sequence.

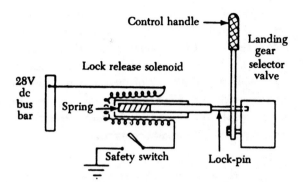

FIGURE 9–16. Typical landing gear safety circuit.

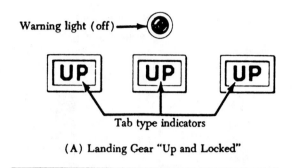

(A) Landing Gear "Up and Locked"

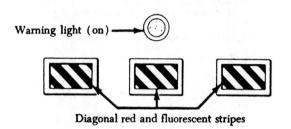

(B) Gear unlocked and in an intermediate position

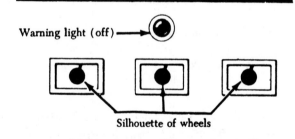

(C) Landing gear "Down and Locked"

FIGURE 9–17. A typical gear position indicator
and warning light.

common type is a pin installed in aligned holes drilled in two or more units of the landing gear support structure. Another type is a spring-loaded clip designed to fit around and hold two or more units of the support structure together. All types of ground locks usually have red streamers permanently attached to them to readily indicate whether or not they are installed.

Gear Indicators

To provide a visual indication of landing gear position, indicators are installed in the cockpit or flight compartment.

Gear warning devices are incorporated on all retractable gear aircraft and usually consist of a horn or some other aural device and a red warning light. The horn blows and the light comes on when one or more throttles are retarded and the landing gear is in any position other than down and locked.

Several designs of gear position indicators are available. One type displays movable miniature landing gears which are electrically positioned by movement of the aircraft gear. Another type consists of two or three green lights which burn when the aircraft gear is down and locked. A third type (figure 9–17) consists of tab-type indicators with markings "up" to indicate that the gear is up and locked, a display of red and white diagonal stripes to show when the gear is unlocked, or a silhouette of each gear to indicate when it locks in the "down" position.

Nosewheel Centering

Centering devices include such units as internal centering cams (figure 9–18) to center the nose wheel as it retracts into the wheel well. If a centering unit were not included in the system, the fuselage wheel well and nearby units could be damaged.

During retraction of the nose gear, the weight of the aircraft is not supported by the strut. The strut is extended by means of gravity and air pressure within the strut. As the strut extends, the raised area of the piston strut contacts the slopping area of the fixed centering cam and slides along it. In so doing, it aligns itself with the centering cam and rotates the nose gear piston into a straight-ahead direction.

The internal centering cam is a feature common to most large aircraft. However, other centering

353

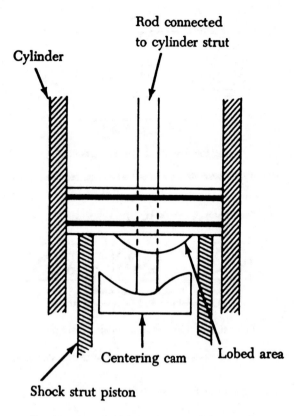

Cylinder

Rod connected to cylinder strut

Centering cam

Lobed area

Shock strut piston

FIGURE 9–18.   Cutaway view of a nose gear internal centering cam.

devices are commonly found on small aircraft. Small aircraft characteristically incorporate an external roller or guide pin on the strut. As the strut is folded into the wheel well on retraction, the roller or guide pin will engage a ramp or track mounted to the wheel well structure. The ramp/track guides the roller or pin in such a manner that the nose wheel is straightened as it enters its well.

In either the internal cam or external track arrangement, once the gear is extended and the weight of the aircraft is on the strut, the nose wheel may be turned for steering.

NOSEWHEEL STEERING SYSTEM
Light Aircraft

Light aircraft are commonly provided nosewheel steering capabilities through a simple system of mechanical linkage hooked to the rudder pedals. Most common applications utilize push-pull rods to connect the pedals to horns located on the pivotal portion of the nosewheel strut.

Heavy Aircraft

Large aircraft, with their larger mass and a need for positive control, utilize a power source for nosewheel steering. Even though large aircraft nosewheel steering system units differ in their construction features, basically all of these systems work in approximately the same manner and require the same sort of units. For example, each steering system (figure 9–19) usually contains:

(1) A cockpit control, such as a wheel, handle, lever, or switch to allow starting, stopping, and to control the action of the system.

(2) Mechanical, electrical, or hydraulic connections for transmitting cockpit control movements to a steering control unit.

(3) A control unit, which is usually a metering or control valve.

(4) A source of power, which is, in most instances, the aircraft hydraulic system.

(5) Tubing for carrying fluid to and from various parts of the system.

(6) One or more steering cylinders, together with the required linkages, for using pressurized fluid to turn the nose gear.

(7) A pressurizing assembly to keep fluid in each steering cylinder always under pressure, thereby preventing shimmy.

(8) A followup mechanism, consisting of gears, cables, rods, drums, and/or bellcranks, for returning the steering control unit to NEUTRAL and thus holding the nose gear at the correct angle of turn.

(9) Safety valves to allow the wheels to trail or swivel, in the event of hydraulic failure.

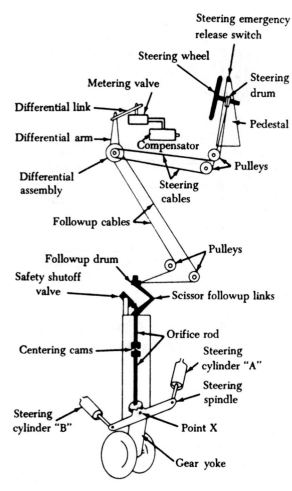

FIGURE 9–19. Nosewheel system mechanical
and hydraulic units.

Nosewheel Steering Operation

The nosewheel steering wheel connects through a
shaft to a steering drum located inside the cockpit
control pedestal. The rotation of this drum transmits the steering signal by means of cables and
pulleys to the control drum of the differential assembly. Movement of the differential assembly is
transmitted by the differential link to the metering
valve assembly, where it moves the selector valve to
the selected position. Then hydraulic pressure
provides the power for turning the nose gear.

As shown in figure 9–20, pressure from the aircraft hydraulic system is directed through the
opened safety shutoff valve and into a line leading
to the metering valve. This metering valve then
routes pressurized fluid out of port A, through the
right-turn alternating line, and into steering cylinder A. This is a one-port cylinder, and pressure

forces the piston to start extending. Since the rod of
this piston connects to the nose steering spindle,
which pivots at point X, the extension of the piston
turns the steering spindle gradually toward the
right. This action turns the nose gear slowly toward
the right, because the spindle is connected to the
nose gear shock strut. As the nose gear turns right,
fluid is forced out of cylinder B, through the left-
turn alternating line, and into port B of the metering valve. The metering valve sends this return fluid
into a compensator, which routes the fluid into the
aircraft system return manifold.

Thus, hydraulic pressure starts the nose gear
turning. However, the gear should not be turned
too far. The nose gear steering system contains
devices to stop the gear at the selected angle of turn
and hold it there.

Followup Linkage

As already pointed out, the nose gear is turned
by the steering spindle as the piston of cylinder A
(figure 9–20) extends. But the rear of the spindle
has gear teeth which mesh with a gear on the bottom of the orifice rod. Thus, as the nose gear and
its spindle turn, the orifice rod also turns, although
in the opposite direction. This rotation is transmitted by the two sections of the orifice rod to the
scissor followup links (figure 9–19), located at the
top of the nose gear strut. As the followup links
return, they rotate the connected followup drum,
which transmits the movement by cables and pulleys
to the differential assembly. Operation of the differential assembly causes the differential arm and links
to move the metering valve back toward the neutral
position.

The compensator unit (figure 9–21) which is a
part of the nosewheel system keeps fluid in the
steering cylinders pressurized at all times. This hydraulic unit consists of a three-port housing, which
encloses a spring-loaded piston and poppet. The left
port is an air vent, which prevents trapped air at
the rear of the piston from interfering with movements of the piston. The second port, located at the
top of the compensator, connects through a line to
the metering valve return port. The third port is
located at the right side of the compensator. This
port, which is connected to the hydraulic return

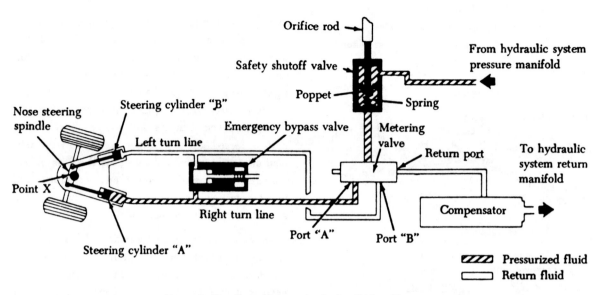

FIGURE 9–20. Nosewheel steering hydraulic flow diagram.

manifold, routes the steering system return fluid into the manifold when the poppet valve is open.

The compensator poppet opens when pressure acting on the piston becomes high enough to compress the spring. This requires 100 p.s.i.; therefore, fluid in the metering valve return line contains fluid trapped under that pressure. Now, since pressure in a confined fluid is transmitted equally and undiminished in all directions (Pascal's law), 100 p.s.i. also exists in metering valve passage H and in chambers

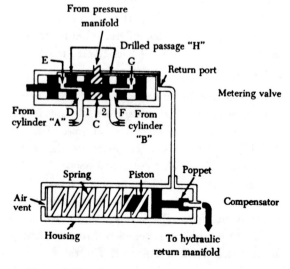

FIGURE 9–21. Cutaway view of metering valve and compensator.

E, D, G, and F (figure 9–21). This same pressure is also applied in the right- and left-turn alternating lines as well as in the steering cylinders themselves.

SHIMMY DAMPERS

A shimmy damper controls vibration, or shimmy, through hydraulic damping. The damper is either attached to or built integrally with the nose gear and prevents shimmy of the nosewheel during taxiing, landing, or takeoff. There are three types of shimmy dampers commonly used on aircraft: (1) The piston type, (2) vane type, and (3) features incorporated in the nosewheel power steering system of some aircraft.

Piston-Type Shimmy Damper

The piston-type shimmy damper shown in figure 9–22 consists of two major components: (1) The cam assembly and (2) the damper assembly. The shimmy damper is mounted on a bracket at the lower end of the nose gear shock strut outer cylinder.

The cam assembly is attached to the inner cylinder of the shock strut and rotates with the nosewheel. Actually, the cam consists of two cams that are mirror images of each other. Lobes on the cams are so placed that the damping effect offers the greatest resistance to rotation when the wheel is centered. The cam follower crank is a U-shaped casting which incorporates the rollers that follow the cam lobes to restrict rotation. The arm of the crank connects to the operating piston shaft.

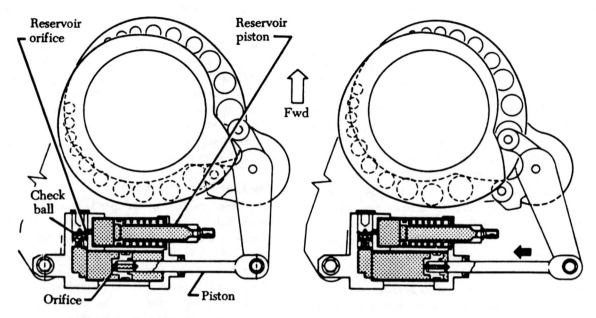

Nose wheel straight ahead   Nose wheel castering

Shimmy damper operational schematic

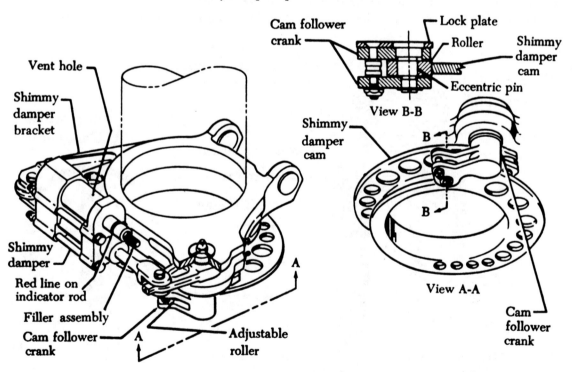

FIGURE 9–22. Typical piston-type shimmy damper.

The damper assembly consists of a spring-loaded reservoir piston to maintain the confined fluid under constant pressure, and an operating cylinder and piston. A ball check permits the flow of fluid from the reservoir to the operating cylinder to make up for any fluid loss in the operating cylinder. Because of the rod on the operating piston, its stroke away from the filler end of the piston displaces more fluid than its stroke toward the filler end. This difference is taken care of by the reservoir orifice, which permits a small flow both ways between the reservoir and operating cylinder.

A red mark (figure 9–22) on the reservoir indicator rod indicates the fluid level in the reservoir. When the piston goes into the reservoir so far that the mark is not visible, the reservoir must be serviced. The operating cylinder houses the operating piston. A small orifice in the piston head allows fluid to flow from one side of the piston to the other. The piston shaft is connected to the arm of the cam follower crank.

As the nosewheel fork rotates in either direction (figure 9–22), the shimmy damper cam displaces the cam follower rollers, causing the operating piston to move in its chamber. This movement forces fluid through the orifice in the piston. Since the orifice is very small, rapid movements of the piston, which commonly occur during landing and takeoff, are restricted, and nosewheel shimmy is eliminated. Gradual rotation of the nosewheel fork is not resisted by the damper. This enables the aircraft to be steered at slow speeds. If the fork rotates in either direction until the rollers are over the high spots on

their portions of the cam, further movement of the nosewheel is practically unrestricted.

The piston-type shimmy damper generally requires a minimum of servicing and maintenance; however, it should be checked periodically for evidence of hydraulic leaks around the damper assembly, and the reservoir fluid level must be properly maintained at all times. The cam assembly should be checked for evidence of binding and for worn, loose, or broken parts.

Vane-Type Shimmy Damper

The vane-type shimmy damper is located on the nosewheel shock strut just above the nosewheel fork and may be mounted either internally or externally. If mounted internally, the housing of the shimmy damper is fitted and secured inside the shock strut, and the shaft is splined to the nosewheel fork. If mounted externally, the housing of the shimmy damper is bolted to the side of the shock strut, and the shaft is connected by mechanical linkage to the nosewheel fork.

The housing (figure 9–23) is divided into three

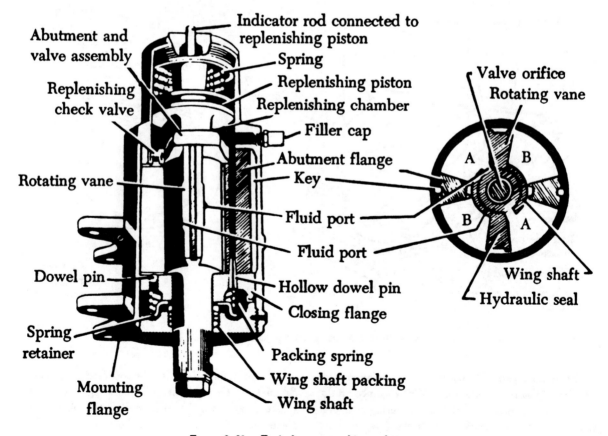

FIGURE 9–23.   Typical vane-type shimmy damper.

358

main parts: (1) The replenishing chamber, (2) the working chamber, and (3) the lower shaft packing chamber.

The replenishing chamber is in the top part of the housing and stores a supply of fluid under pressure. Pressure is applied to the fluid by the spring-loaded replenishing piston and piston shaft which extends through the upper housing and serves as a fluid level indicator. The area above the piston contains the piston spring and is open to atmosphere to prevent a hydraulic lock. Fluid is prevented from leaking past the piston by O-ring packings. A grease fitting provides the means for filling the replenishing chamber with fluid.

The working chamber is separated from the replenishing chamber by the abutment and valve assembly. The working chamber contains two one-way ball check valves, which will allow fluid to flow from the replenishing chamber to the working chamber only. This chamber is divided into four sections by two stationary vanes called abutment flanges, which are keyed to the inner wall of the housing, and two rotating vanes, which are an integral part of the wing shaft. The shaft contains the valve orifice through which the fluid must pass in going from one chamber to another.

Turning the nosewheel in either direction causes the rotating vanes to move in the housing. This results in two sections of the working chamber growing smaller, while the opposite two chambers grow larger. The rotating vane can move only as fast as the fluid can be displaced from one chamber to the other. All of the fluid being displaced must pass through the valve orifice in the shaft. Resistance to the flow of fluid through the orifice is proportional to the velocity of flow. This means that the shimmy damper offers little resistance to slow motion, such as that encountered during normal steering of the nose gear or ground handling, but offers high resistance to shimmy on landing, takeoff, and high-speed taxiing. An automatic orifice adjustment compensates for temperature changes. A bimetallic thermostat in the shaft opens and closes the orifice as the temperature and viscosity change. This results in a constant resistance over a wide temperature range. In case an exceptionally high pressure is suddenly built up in the working chamber by a severe twisting force on the nosewheel, the closing flange moves down, compressing the lower shaft packing spring, allowing fluid to pass around the lower ends of the vanes, preventing structure damage.

Maintaining the proper fluid level is necessary to the continued functioning of a vane-type shimmy damper. If a shimmy damper is not operating properly, the fluid level is the first item which should be checked by measuring the protrusion of the indicator rod from the center of the reservoir cover. Inspection of a shimmy damper should include a check for evidence of leakage and a complete examination of all fittings and connections between the moving parts of the shock strut and the damper shaft for loose connections.

Fluid should be added only when the indicator rod protrudes less than the required amount. The distance the rod should protrude varies with different models. A shimmy damper should not be overfilled. If the indicator rod is above the height specified on the nameplate, fluid should be bled out of the damper.

Steer Damper

A steer damper is hydraulically operated and accomplishes the two separate functions of steering and/or eliminating shimmying. The type discussed here is designed for installation on nose gear struts and is connected into the aircraft hydraulic system. A typical steer damper is shown in figure 9–24.

FIGURE 9–24. Typical steer damper.

Basically, a steer damper consists of a closed cylinder containing rotary vane-type working chambers (similar to the vane-type shimmy damper) and a valving system.

Steer dampers may contain any even number of

working chambers. A steer damper with one vane on the wing shaft and one abutment leg on the abutment flange would have two chambers. Similarly, a unit with two vanes on the wingshaft and two abutment legs on the abutment flange would have four chambers. The single- or double-vaned units are the ones most commonly used.

A mechanical linkage is connected from the protruding splined portion of the wing shaft to the wheel fork and is used as a means for transmitting force. The linkage on the steer damper may be connected to a heavy coil spring on the outside of the reservoir for automatic wheel centering.

The steer damper accomplishes two separate functions: one is steering the nosewheel and the other is shimmy damping. Only the damping function of the steer damper is discussed in this section.

The steer damper automatically reverts to damping when, for any reason, the flow of high-pressure fluid is removed from the inlet of the steer damper. This high pressure, which activates the steer damper valving system, is removed from the control passages by one of two methods, depending on the installation. When the inlet line is supplied with a three-way solenoid valve and the high-pressure supply is shut off, the fluid bleeds out of the unit through the outlet port of the valve to the discharge line. When a two-way solenoid valve is furnished, high-pressure fluid leaves the control passages through an orifice specially provided for this type of installation which is located in the center of the return line plunger.

Effective damping is assured by maintaining unaerated hydraulic fluid in the working chambers. This is accomplished by allowing air and a very small amount of hydraulic fluid to leave the working chambers through strategically located vent grooves while unaerated fluid is allowed to enter through replenishing valves from the hydraulic return line. Excessive pressure in the unit due to temperature changes is prevented by the thermal relief valve in the inlet flange.

Daily inspection of a steer damper should include a check for leakage and a complete inspection of all hydraulic connections, steer damper mounting bolts for tightness, and all fittings and connections between the moving parts of the shock strut and the steer damper wing shaft.

BRAKE SYSTEMS

Proper functioning of the brakes is of utmost importance on aircraft. The brakes are used for slowing, stopping, holding, or steering the aircraft.

They must develop sufficient force to stop the aircraft in a reasonable distance; brakes must hold the aircraft for normal engine turnup; and brakes must permit steering of the aircraft on the ground. Brakes are installed in each main landing wheel and they may be actuated independently of each other. The right-hand landing wheel is controlled by applying toe pressure to the right rudder pedal and the left-hand wheel is controlled by the left rudder pedal.

For the brakes to function efficiently, each component in the brake system must operate properly, and each brake assembly on the aircraft must operate with equal effectiveness. It is therefore important that the entire brake system be inspected frequently and an ample supply of hydraulic fluid be maintained in the system. Each brake assembly must be adjusted properly and friction surfaces kept free of grease and oil.

Three types of brake systems are in general use: (1) Independent systems, (2) power control systems, and (3) power boost systems. In addition, there are several different types of brake assemblies in widespread use.

Independent Brake Systems

In general, the independent brake system is used on small aircraft. This type of brake system is termed "independent" because it has its own reservoir and is entirely independent of the aircraft's main hydraulic system.

Independent brake systems are powered by master cylinders similar to those used in the conventional automobile brake system. The system is composed of a reservoir, one or two master cylinders,

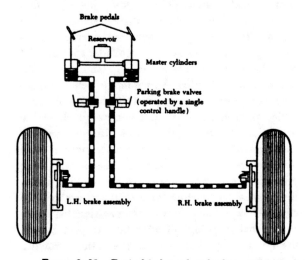

FIGURE 9–25. Typical independent brake system.

360

mechanical linkage which connects each master cylinder with its corresponding brake pedal, connecting fluid lines, and a brake assembly in each main landing gear wheel (figure 9–25).

Each master cylinder is actuated by toe pressure on its related pedal. The master cylinder builds up pressure by the movement of a piston inside a sealed, fluid-filled cylinder. The resulting hydraulic pressure is transmitted to the fluid line connected to the brake assembly in the wheel. This results in the friction necessary to stop the wheel.

When the brake pedal is released, the master cylinder piston is returned to the "off" position by a return spring. Fluid that was moved into the brake assembly is then pushed back to the master cylinder by a piston in the brake assembly. The brake assembly piston is returned to the "off" position by a return spring in the brake. Some light aircraft are equipped with a single master cylinder which is hand-lever operated and applies brake action to both main wheels simultaneously. Steering on this system is accomplished by nosewheel linkage.

The typical master cylinder has a compensating port or valve that permits fluid to flow from the brake chamber back to the reservoir when excessive pressure is developed in the brake line due to temperature changes. This ensures that the master cylinder won't lock or cause the brakes to drag.

Various manufacturers have designed master cylinders for use on aircraft. All are similar in operation, differing only in minor details and construction. Two well-known master cylinders—the Goodyear and the Warner—are described and illustrated in this section.

In the Goodyear master cylinder (figure 9–26) fluid is fed from an external reservoir by gravity to the master cylinder. The fluid enters through the cylinder inlet port and compensating port and fills the master cylinder casting ahead of the piston and the fluid line leading to the brake actuating cylinder.

Application of the brake pedal, which is linked to the master cylinder piston rod, causes the piston rod to push the piston forward inside the master cylinder casting. A slight forward movement blocks the compensating port, and the buildup of pressure begins. This pressure is transmitted to the brake assembly.

When the brake pedal is released and returns to the "off" position, the piston return spring pushes the front piston seal and the piston back to full "off" position against the piston return stop. This again clears the compensating port. Fluid that was moved into the brake assembly and brake connecting line is then pushed back to the master cylinder by the brake piston which is returned to the "off" position by the pressure of the brake piston return springs. Any pressure or excess volume of fluid is relieved through the compensating port and passes back to the fluid reservoir. This prevents the master cylinder from locking or causing the brakes to drag.

If any fluid is lost back of the front piston seal due to leakage, it is automatically replaced from the fluid reservoir by gravity. Any fluid lost in front of the piston from leaks in the line or at the brake assembly is automatically replaced through the piston head ports, and around the lip of the front piston seal when the piston makes the return stroke to the full "off" position. The front piston seal functions as a seal only during the forward stroke.

These automatic fluid replacement arrangements always keep the master cylinder, brake connecting line, and brake assembly fully supplied with fluid as long as there is fluid in the reservoir.

The rear piston seal seals the rear end of the cylinder at all times to prevent leakage of fluid, and the flexible rubber boot serves only as a dust cover.

The brakes may be applied for parking by a ratchet-type lock built into the mechanical linkage between the master cylinder and the foot pedal. Any change in the volume of fluid due to expansion while the parking brake is on is taken care of by a spring incorporated in the linkage. The brakes are unlocked by application of sufficient pressure on the brake pedals to unload the ratchet.

Brake systems employing the Goodyear master cylinder must be bled from the top down. In no case should bleeding be attempted from the bottom up, because it is impossible to remove the air behind the piston seal.

The Warner master cylinder (figure 9–27) incorporates a reservoir, pressure chamber, and compen-

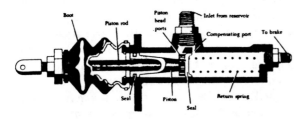

FIGURE 9–26. Goodyear master brake cylinder.

sating device in a single housing. The reservoir is vented to the atmosphere through the filler plug, which also contains a check valve. A fluid level tube is located in the side of the reservoir housing.

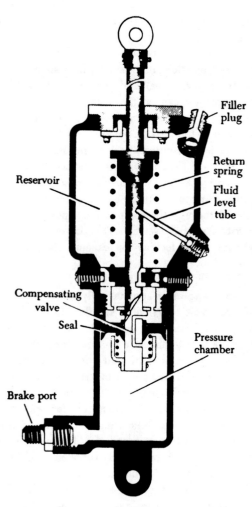

FIGURE 9–27. Warner master brake cylinder.

Toe pressure on the brake pedal is transferred to the cylinder piston by mechanical linkage. As the piston moves downward, the compensating valve closes and pressure is trapped in the pressure chamber. Further movement of the piston forces fluid into the brake assembly, creating the braking action. When toe pressure is removed from the brake pedal, the piston return spring returns the piston to the "off" position. The compensating device allows fluid to flow to and from the reservoir and pressure chamber when the brakes are in the "off" position and the entire system is under atmospheric pressure.

Certain models of the Warner master cylinder

have a parking feature which consists of a ratchet and spring arrangement. The ratchet locks the unit in the "on" position, and the spring compensates for expansion and contraction of fluid.

Power Brake Control Systems

Power brake control valve systems (figure 9–28) are used on aircraft requiring a large volume of fluid to operate the brakes. As a general rule, this applies to many large aircraft. Because of their weight and size, large wheels and brakes are required. Larger brakes mean greater fluid displacement and higher pressures, and for this reason independent master cylinder systems are not practical on heavy aircraft.

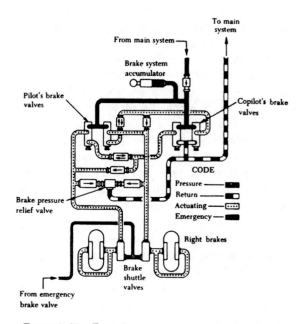

FIGURE 9–28. Typical power brake control valve system.

In this system a line is tapped off from the main hydraulic system pressure line. The first unit in this line is a check valve which prevents loss of brake system pressure in case of main system failure.

The next unit is the accumulator which stores a reserve supply of fluid under pressure. When the brakes are applied and pressure drops in the accumulator, more fluid enters from the main system and is trapped by the check valve. The accumulator also acts as a surge chamber for excessive loads imposed upon the brake hydraulic system.

Following the accumulator are the pilot's and copilot's control valves. The control valves regulate and control the volume and pressure of the fluid which actuates the brakes.

362

Four check valves and two orifice check valves are installed in the pilot's and copilot's brake actuating lines. The check valves allow the flow of fluid in one direction only. The orifice check valves allow unrestricted flow of fluid in one direction from the pilot's brake control valve; flow in the opposite direction is restricted by an orifice in the poppet. Orifice check valves help prevent chatter.

The next unit in the brake actuating lines is the pressure relief valve. In this particular system, the pressure relief valve is preset to open at 825 p.s.i. to discharge fluid into the return line, and closes at 760 p.s.i. minimum.

Each brake actuating line incorporates a shuttle valve for the purpose of isolating the emergency brake system from the normal brake system. When brake actuating pressure enters the shuttle valve, the shuttle is automatically moved to the opposite end of the valve. This closes off the hydraulic brake system actuating line. Fluid returning from the brakes travels back into the system to which the shuttle was last open.

Pressure Ball-Check Brake Control Valve

A pressure ball-check power brake control valve (figure 9–29) releases and regulates main system pressure to the brakes and relieves thermal expansion when the brakes are not being used. The main

parts of the valve are the housing, piston assembly, and tuning fork. The housing contains three chambers and ports: pressure inlet, brake, and return.

When toe pressure is applied to the brake pedal, the motion is transmitted through linkage to the tuning fork. The tuning fork swivels, moving the piston upward in the cylinder. The first movement upward causes the piston head to contact a flange on the pilot pin, closing the return fluid passage. Further movement upward unseats the ball-check valve allowing main system pressure to flow into the brake line. As the pressure increases in the brake actuating cylinder and line, the pressure also increases on the top side of the piston. When the total force on top of the piston is greater than the force applied at the brake pedal, the piston is forced downward against the bar spring tension. This allows the ball-check valve to seat, closing off system pressure. In this position, the pressure and return ports are both closed and the power brake valve is balanced. This balancing action cuts system pressure down to brake pressure by closing off the pressure from the main system when the desired brake pressure is attained. As long as the valve is balanced, fluid under pressure is trapped in the brake assembly and line.

Power Brake Control Valve (Sliding Spool Type)

A sliding spool power brake control valve (figure 9–30) basically consists of a sleeve and spool installed in a housing. The spool moves inside the sleeve, opening or closing either the pressure or return port of the brake line. Two springs are provided. The large spring, referred to in figure 9–30 as the plunger spring, provides "feel" to the brake pedal. The small spring returns the spool to the "off" position.

When the plunger is depressed the large spring moves the spool closing the return port and opening the pressure port to the brake line. When the pressure enters the valve, fluid flows to the opposite end of the spool through a hole when the pressure pushes the spool back far enough toward the large spring to close the pressure port, but does not open the return port. The valve is then in the static condition. This movement partially compresses the large spring, giving "feel" to the brake pedal. When the brake pedal is released the small spring moves the spool back and opens the return port. This allows fluid pressure in the brake line to flow out through the return port.

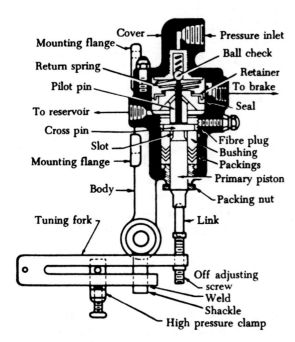

FIGURE 9–29. Pressure ball-check power brake control valve.

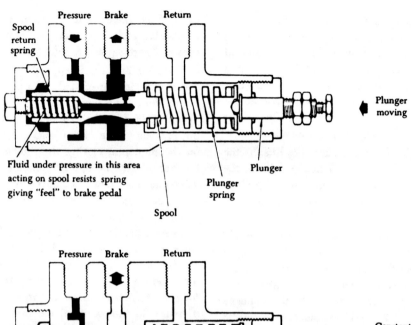

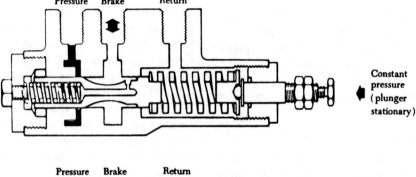

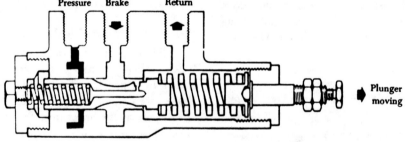

FIGURE 9–30. Sliding spool power brake control valve.

Brake Debooster Cylinders

In some power brake control valve systems, debooster cylinders are used in conjunction with the power brake control valves. Debooster units are generally used on aircraft equipped with a high-pressure hydraulic system and low-pressure brakes. Brake debooster cylinders reduce the pressure to the brake and increase the volume of fluid flow. Figure 9–31 shows a typical debooster cylinder installation, mounted on the landing gear shock strut in the line between the control valve and the brake.

As shown in the schematic diagram of the unit, the cylinder housing contains a small chamber and a large chamber, a piston with a small head and a large head, a spring-loaded ball-check valve, and a piston return spring.

In the "off" position, the piston assembly is held at the inlet (or small) end of the debooster by the piston return spring. The ball-check valve is held on its seat in the small piston head by a light spring. Fluid displaced by thermal expansion in the brake unit can easily push the ball-check valve off its seat to escape through the debooster back to the power control valve.

When the brakes are applied, fluid under pressure enters the inlet port to act on the small end of the piston. The ball check prevents the fluid from passing through the shaft. Force is transmitted through the small end of the piston to the large end of the piston. As the piston moves downward in the housing a new flow of fluid is created from the

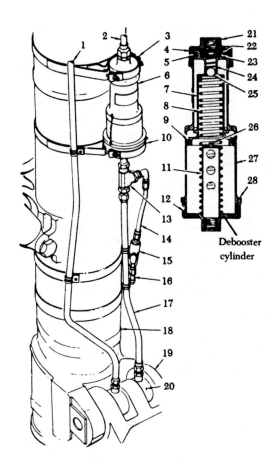

1. Emergency system pressure line
2. Main brake pressure line
3. Upper support clamp
4. Packing
5. Packing
6. Debooster cylinder assembly
7. Piston
8. Piston spring
9. Packing
10. Lower support clamp
11. Riser tube
12. Packing
13. Tee fitting
14. Brake line (to pressure relief valve)
15. Brake pressure relief valve
16. Overflow line
17. Brake line (debooster to shuttle valve)
18. Shock strut
19. Torque link
20. Brake shuttle valve
21. Upper end cap
22. Snapring
23. Spring retainer
24. Valve spring
25. Ball
26. Ball pedestal
27. Barrel
28. Lower end cap

FIGURE 9–31. Brake debooster cylinder.

large end of the housing through the outlet port to the brakes. Because the force from the small piston head is distributed over the greater area of the

large piston head, pressure at the outlet port is reduced. At the same time, a greater volume of fluid is displaced by the large piston head than that used to move the small piston head.

Normally, the brakes will be fully applied before the piston has reached the lower end of its travel. However, if the piston fails to meet sufficient resistance to stop it (due to a loss of fluid from the brake unit or connecting lines), the piston will continue to move downward until the riser unseats the ball-check valve in the hollow shaft. With the ball-check valve unseated, fluid from the power control valve will pass through the piston shaft to replace the lost fluid. Since the fluid passing through the piston shaft acts on the large piston head, the piston will move up, allowing the ball-check valve to seat when pressure in the brake assembly becomes normal.

When the brake pedals are released pressure is removed from the inlet port, and the piston return spring moves the piston rapidly back to the top of the debooster. The rapid movement causes a suction in the line to the brake assembly, resulting in faster release of the brakes.

Power Boost Brake Systems

As a general rule, power boost brake systems are used on aircraft that land too fast to employ the independent brake system, but are too light in weight to require power brake control valves.

In this type of system, a line is tapped off the main hydraulic system pressure line, but main hydraulic system pressure does not enter the brakes. Main system pressure is used only to assist the pedals through the use of power boost master cylinders.

A typical power boost brake system (figure 9–32) consists of a reservoir, two power boost master cylinders, two shuttle valves, and the brake assembly in each main landing wheel. A compressed air bottle with a gage and release valve is installed for emergency operation of the brakes. Main hydraulic system pressure is routed from the pressure manifold to the power master cylinders. When the brake pedals are depressed, fluid for actuating the brakes is routed from the power boost master cylinders through the shuttle valves to the brakes.

When the brake pedals are released, the main system pressure port in the master cylinder is closed. Fluid that was moved into the brake assem-

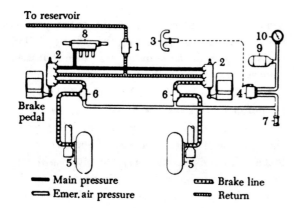

— Main pressure   ⊞⊞⊞ Brake line
⊟ Emer. air pressure   ⊞⊞⊞ Return

1. Brake reservoir
2. Power boost
   master cylinder
3. Emergency
   brake control
4. Air release valve
5. Wheel brake

6. Shuttle valve
7. Air vent
8. Main system
   pressure manifold
9. Emergency air
   bottle
10. Emergency air gage

FIGURE 9–32. Power boost master cylinder brake system.

bly is forced out the return port by a piston in the brake assembly, through the return line to the brake reservoir. The brake reservoir is connected to the main hydraulic system reservoir to assure an adequate supply of fluid to operate the brakes.

Nose Wheel Brakes

Many transport type aircraft such as the B–727 have brakes installed on the nose wheel. Movement of either right or left brake pedal will actuate the corresponding right or left main gear brake metering valve.

Movement of both brake pedals together will apply both main gear brakes and the nose gear brakes after approximately half the pedal stroke. Actuation of one brake pedal for directional control braking will not actuate the nose gear brakes until nearly the end of the pedal stroke. Nose gear brake application is controlled through the brake differential linkage.

When the brake pedals are depressed the differential directs the force through linkage to the main gear metering valve first. After this valve is opened, continued movement of the brake pedals is directed to the nose gear metering valve, opening it and activating the brakes. Nose wheel braking is available above 15 mph from straight ahead to approximately 6° of steering. At this point the nose wheel steering brake cutoff switch activates the anti skirt valve and shuts off nose wheel braking. There is no nose wheel braking below 15 mph.

BRAKE ASSEMBLIES

Brake assemblies commonly used on aircraft are the single-disk, dual-disk, multiple-disk, segmented rotor, or expander tube types. The single- and dual-disk types are more commonly used on small aircraft; the multiple-disk type is normally used on medium-sized aircraft; and the segmented rotor and expander tube types are commonly found on heavier aircraft.

Single-Disk Brakes

With the single-disk brake, braking is accomplished by applying friction to both sides of a rotating disk which is keyed to the landing gear wheel. There are several variations of the single-disk brake; however, all operate on the same principle and differ mainly in the number of cylinders and the type of brake housing. Brake housings may be either the one-piece or divided type. Figure 9–33 shows a single-disk brake installed on an aircraft, with the wheel removed. The brake housing is attached to the landing gear axle flange by mounting bolts.

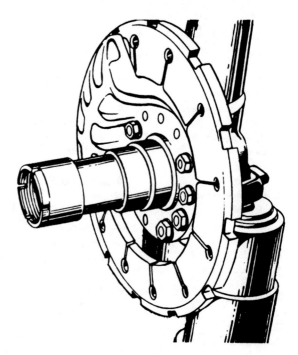

FIGURE 9–33. Typical single-disk brake installation.

Figure 9–34 shows an exploded view of a typical single-disk brake assembly. This brake assembly has a three-cylinder, one-piece housing. Each cylinder

366

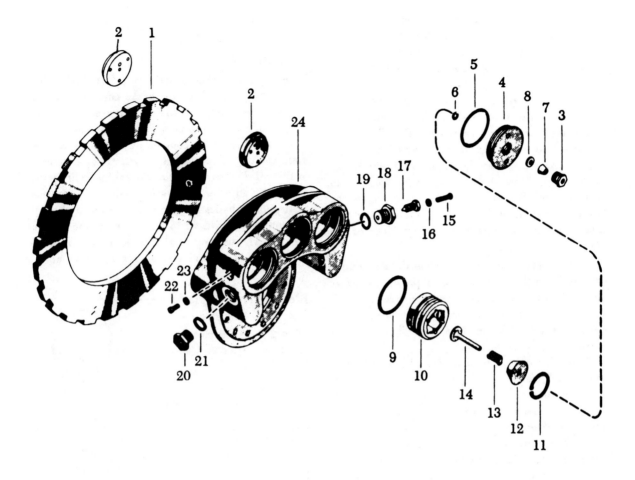

1. Brake disk
2. Lining puck
3. Adjusting pin nut
4. Cylinder head
5. O-ring gasket
6. O-ring packing
7. Adjusting pin grip
8. Washer
9. O-ring packing
10. Piston
11. Internal retainer ring
12. Spring guide
13. Brake return spring
14. Adjusting pin
15. Bleeder screw
16. Washer
17. Bleeder valve
18. Bleeder adapter
19. Gasket
20. Fluid inlet bushing
21. Gasket
22. Screw
23. Washer
24. Brake housing

FIGURE 9–34. Exploded view of single-disk brake assembly.

in the housing contains a piston, a return spring, and an automatic adjusting pin.

There are six brake linings, three on the inboard side of the rotating disk and three on the outboard side of the rotating disk. These brake linings are often referred to as "pucks." The outboard lining pucks are attached to the three pistons and move in and out of the three cylinders when the brakes are operated. The inboard lining pucks are mounted in recesses in the brake housing and are therefore stationary.

Hydraulic pressure from the brake control unit enters the brake cylinders and forces the pistons and their pucks against the rotating disk. The rotating disk is keyed to the landing gear wheel so that it is free to move laterally within the brake cavity of the wheel. Thus, the rotating disk is forced into contact with the inboard pucks mounted in the housing. The lateral movement of the rotating disk ensures equal braking action on both sides of the disk.

When brake pressure is released, the return springs force the pistons back to provide a preset clearance between the pucks and the disk. The self-adjusting feature of the brake will maintain the desired puck-to-disk clearance, regardless of lining wear.

When the brakes are applied, hydraulic pressure moves each piston and its puck against the disk. At the same time, the piston pushes against the adjusting pin (through the spring guide) and moves the pin inboard against the friction of the adjusting pin grip. When pressure is relieved, the force of the return spring is sufficient to move the piston away from the brake disk but not enough to move the adjusting pin, which is held by the friction of the pin grip. The piston moves away from the disk until it stops against the head of the adjusting pin. Thus, regardless of the amount of wear, the same travel of the piston will be required to apply the brake.

Maintenance of the single-disk brake may include bleeding, performing operational checks, checking lining wear, checking disk wear, and replacing worn linings and disks.

A bleeder valve is provided on the brake housing for bleeding the single-disk brake. Always bleed brakes in accordance with the applicable manufacturer's instructions.

Operational checks are made during taxiing. Braking action for each main landing gear wheel should be equal, with equal application of pedal pressure and without any evidence of soft or spongy action. When pedal pressure is released, the brakes should release without any evidence of drag.

Dual-Disk Brakes

Dual-disk brakes are used on aircraft when more braking friction is desired. The dual-disk brake is very similar to the single-disk type, except that two rotating disks instead of one are used.

Multiple-Disk Brakes

Multiple-disk brakes are heavy-duty brakes, designed for use with power brake control valves or power boost master cylinders. Figure 9–35 is an exploded view of the complete multiple-disk brake assembly. The brake consists of a bearing carrier, four rotating disks called rotors, three stationary disks called stators, a circular actuating cylinder, an automatic adjuster, and various minor components.

Regulated hydraulic pressure is applied through the automatic adjuster to a chamber in the bearing carrier. The bearing carrier is bolted to the shock strut axle flange and serves as a housing for the

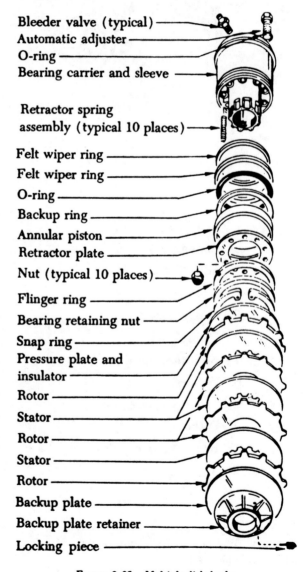

Bleeder valve (typical)
Automatic adjuster
O-ring
Bearing carrier and sleeve

Retractor spring assembly (typical 10 places)

Felt wiper ring
Felt wiper ring
O-ring
Backup ring
Annular piston
Retractor plate
Nut (typical 10 places)
Flinger ring
Bearing retaining nut
Snap ring
Pressure plate and insulator
Rotor
Stator
Rotor
Stator
Rotor
Backup plate
Backup plate retainer
Locking piece

FIGURE 9–35. Multiple-disk brake.

annular actuating piston. Hydraulic pressure forces the annular piston to move outward, compressing the rotating disks, which are keyed to the landing wheel and compressing the stationary disks, which are keyed to the bearing carrier. The resulting friction causes a braking action on the wheel and tire assembly.

When the hydraulic pressure is relieved, the retracting springs force the actuating piston to retract into the housing chamber in the bearing carrier. The hydraulic fluid in the chamber is forced out by the returning of the annular actuating piston, and is bled through the automatic adjuster to the return line. The automatic adjuster traps a predetermined amount of fluid in the brake, an amount just sufficient to give correct clearances between the rotating disks and stationary disks.

Maintenance of the multiple-disk brake may include bleeding, checking disks for wear, replacing disks, and performing operational checks.

Bleeder valves are provided, making it possible to bleed the brake in any position. Bleeding should be accomplished according to the instructions for the specific aircraft.

The disks are checked for wear using a gage equipped with a movable slide, a stop pin, and an anvil.

Segmented Rotor Brakes

Segmented rotor brakes are heavy-duty brakes, especially adapted for use with high-pressure hydraulic systems. These brakes may be used with either power brake control valves or power boost master cylinders. Braking is accomplished by means of several sets of stationary, high-friction type brake linings making contact with rotating (rotor) segments. A cutaway view of the brake is shown in figure 9–36.

The segmented rotor brake is very similar to the multiple-disk type described previously. The brake

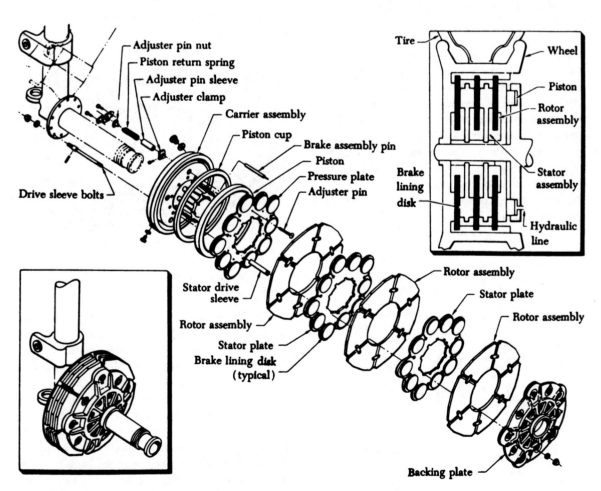

FIGURE 9–36. Segmented rotor brake assembly units.

369

assembly consists of a carrier, two pistons and piston cup seals, a pressure plate, an auxiliary stator plate, roto segments, stator plates, a compensating shim, automatic adjusters, and a backing plate.

The carrier assembly is the basic unit of the brake. It is the part which is attached to the landing gear shock strut flange upon which the other components are assembled. Two grooves, or cylinders, are machined in the carrier to receive the piston cups and pistons. Hydraulic fluid is admitted to these cylinders through a line which is attached to a threaded boss on the outside of the carrier.

The automatic adjusters are threaded into equally spaced holes (figure 9–36) located in the face of the carrier. The adjusters compensate for lining wear by maintaining a fixed running clearance with the brake in the "off" position. Each automatic adjuster is composed of an adjuster pin, adjuster clamp, return spring, sleeve, nut, and clamp hold-down assembly.

The pressure plate is a flat, circular, nonrotating plate, notched on the inside diameter to fit over the stator drive sleeves.

Following the pressure plate is the auxiliary stator plate. This is also a nonrotating plate, notched on the inside diameter. Brake lining is riveted to one side of the auxiliary stator plate.

The next unit in the assembly is the first of several rotor segments. Each rotor plate is notched on the outside circumference. This enables it to be keyed to the landing gear wheel and rotate with it. This particular model of the segmented rotor brake has four sets of these rotor segments.

Mounted between each of the rotor segments is a stator plate (figure 9–37). The stator plates are non-rotating plates and have brake linings riveted

to both sides. The linings are in the form of multiple blocks, separated to aid in the dissipation of heat.

Following the last rotor segment is the compensating shim. The compensating shim is provided so that all the brake lining available for wear can be used. Without the shim, only about one-half of the lining could be used due to the limited travel of the pistons. After approximately one-half of the brake lining has been used, the shim is removed. The adjuster clamp is then re-positioned on the automatic adjuster pin, restoring piston travel so that the remaining lining can be used.

The backing plate (figure 9–38) is the final unit in the assembly and is a non-rotating plate with brake linings riveted to one side. The backing plate receives the ultimate hydraulic force resulting from brake application.

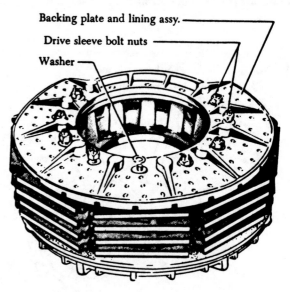

Backing plate and lining assy.
Drive sleeve bolt nuts
Washer

FIGURE 9–38. Backing plate installed.

Hydraulic pressure released from the brake control unit enters the brake cylinders and acts on the piston cups and pistons forcing them outward from the carrier. The pistons apply force against the pressure plate, which in turn pushes against the auxiliary stator. The auxiliary stator contacts the first rotor segment, which in turn engages the first stator plate. Lateral movement continues until all the braking surfaces are in contact. The auxiliary stator plate, the stator plates, and the backing plate are prevented from rotating by the stator drive sleeves. Thus, the non-rotating linings are all forced

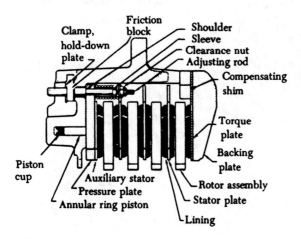

Clamp, hold-down plate
Friction block
Shoulder
Sleeve
Clearance nut
Adjusting rod
Compensating shim
Torque plate
Backing plate
Rotor assembly
Stator plate
Lining
Piston cup
Auxiliary stator
Pressure plate
Annular ring piston

FIGURE 9–37. Cross section of a segmented rotor brake.

370

in contact with the rotors, creating enough friction to stop the wheel to which the rotors are keyed.

The function of the automatic adjuster is dependent upon the correct friction between the adjuster pin and the adjuster clamp. Adjustment of brake running clearance is governed by the distance obtained between the adjuster washer and the end of the adjuster nut when the brake is assembled.

During brake application, the pressure plate moves toward the rotors. The washer moves with the pressure plate, causing the spring to compress. As piston travel increases and as the pressure plate moves farther, the lining then comes in contact with the rotor segments. As the linings wear, the pressure plate continues to move and eventually comes into direct contact with the adjuster sleeve through the adjuster washer. Thus, no further force is applied to the spring. Additional travel of the pressure plate caused by lining wear will force the adjuster pin to slide through the adjuster clamp.

When hydraulic pressure in the brake is released, the return spring forces the pressure plate to return until it bottoms on the shoulder of the adjuster pin. As this cycle is repeated during brake application and release, the adjuster pin will advance through the adjuster clamp due to lining wear, but the running clearance will remain constant.

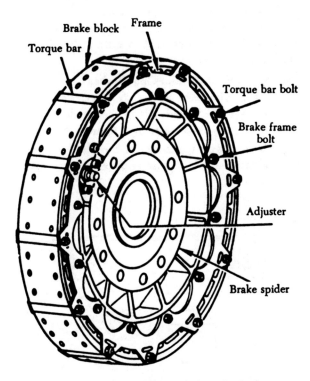

FIGURE 9-39. Assembled expander tube brake.

Expander Tube Brakes

The expander tube brake (figure 9–39) is a low-pressure brake with 360° of braking surface. It is light in weight, has few moving parts, and may be used on large aircraft as well as on small aircraft.

An exploded view of the expander tube brake is shown in figure 9–40. The main parts of the brake are the frame, expander tube, brake blocks, return springs, and clearance adjuster.

The brake frame is the basic unit around which the expander tube brake is built. The main part of the frame is a casting which is bolted to the torque flange of the landing gear shock strut. Detachable metal sides form a groove around the outer circumference into which moving parts of the brake are fitted.

The expander tube is made of neoprene reinforced with fabric, and has a metal nozzle through which fluid enters and leaves the tube.

The brake blocks are made of special brake lining material, the actual braking surface being strengthened by a backing plate of metal. The brake blocks are held in place around the frame and are prevented from circumferential movement by the torque bars.

The brake return springs are semi-elliptical or half-moon in shape. One is fitted between each separation in the brake blocks. The ends of the return springs push outward against the torque bars, while the bowed center section pushes inward, retracting the brake blocks when the brakes are released.

When hydraulic fluid under pressure enters the expander tube the tube expands. Since the frame prevents the tube from expanding inward and to the sides, all movement is outward. This forces the brake blocks against the brake drum, creating friction. The tube shields prevent the expander tube from extruding out between the blocks, and the torque bars prevent the blocks from rotating with the drum. Friction created by the brake is directly proportional to brake line pressure.

The clearance adjuster (figure 9–40) consists of a spring-loaded piston acting behind a neoprene diaphragm. It closes off the fluid passage in the inlet manifold when the spring tension is greater than the fluid pressure in the passage. Tension on the spring may be increased or decreased by turning the adjustment screw. Some of the older models of the expander tube brake are not equipped with clearance adjusters.

For brakes equipped with adjusters, clearance between the brake blocks and drum is usually set to a

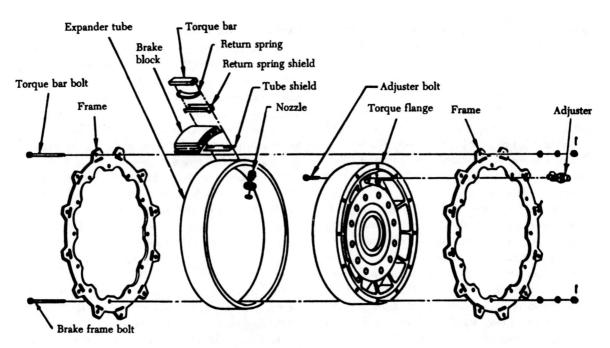

FIGURE 9–40. Exploded view of expander tube brakes.

minimum of 0.002 to 0.015 in., the exact setting depending on the particular aircraft concerned. All brakes on the same aircraft should be set to the same clearance.

To decrease clearance, turn the adjuster knob clockwise; to increase clearance, turn the adjuster knob counterclockwise. It should be kept in mind, however, that turning the adjuster knob alone does not change the clearance. The brakes must be applied and released after each setting of the adjuster knob to change the pressure in the brake and thereby change the brake clearance.

INSPECTION AND MAINTENANCE OF BRAKE SYSTEMS

Proper functioning of the brake system is of the utmost importance. Hence, conduct inspections at frequent intervals, and perform needed maintenance promptly and carefully.

When checking for leaks, make sure the system is under operating pressure. However, tighten any loose fittings with the pressure off. Check all flexible hoses carefully for swelling, cracking, or soft spots, and replace if evidence of deterioration is noted.

Maintain the proper fluid level at all times to prevent brake failure or the introduction of air into the system. Air in the system is indicated by a spongy action of the brake pedals. If air is present

in the system, remove it by bleeding the system.

There are two general methods of bleeding brake systems—bleeding from the top downward (gravity method) and bleeding from the bottom upward (pressure method). The method used generally depends on the type and design of the brake system to be bled. In some instances it may depend on the bleeding equipment available. A general description of each method follows.

Gravity Method of Bleeding Brakes

In the gravity method, the air is expelled from the brake system through one of the bleeder valves provided on the brake assembly (figure 9–41).

A bleeder hose is attached to the bleeder valve, and the free end of the hose is placed in a recepta-

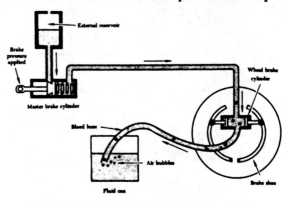

FIGURE 9–41. Gravity method of bleeding brakes.

372

cle containing enough hydraulic fluid to cover the end of the hose. The air-laden fluid is then forced from the system by operating the brake. If the brake system is a part of the main hydraulic system, a portable hydraulic test stand may be used to supply the pressure. If the system is an independent master cylinder system, the master cylinder will supply the necessary pressure. In either case, each time the brake pedal is released the bleeder valve must either be closed or the bleeder hose pinched off; otherwise, more air will be drawn back into the system. Bleeding should continue until no more air bubbles come through the bleeder hose into the container.

Pressure Method of Bleeding Brakes

In the pressure method, the air is expelled through the brake system reservoir or other specially provided location. Some aircraft have a bleeder valve located in the upper brake line. In using this method of bleeding, pressure is applied using a bleed tank (figure 9–42). A bleed tank is a portable tank containing hydraulic fluid under pressure. The bleeder tank is equipped with an air valve, air gage, and a connector hose. The connector hose attaches to the bleeder valve on the brake assembly and is provided with a shutoff valve.

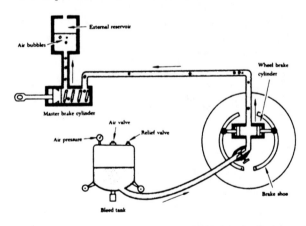

FIGURE 9–42. Pressure method of bleeding brakes.

Perform this method of bleeding strictly in accordance with the specific manufacturer's instructions for the aircraft concerned.

Although the bleeding of individual systems presents individual problems, observe the following precautions in all bleeding operations:

   (1) Be certain that the bleeding equipment to be used is absolutely clean and is filled with the proper type of hydraulic fluid.

   (2) Maintain an adequate supply of fluid dur-

ing the entire operation. A low fluid supply will allow air to be drawn into the system.

   (3) Bleeding should continue until no more air bubbles are expelled from the system, and a firm brake pedal is obtained.

   (4) After the bleeding operation is completed, check the reservoir fluid level. With brake pressure on, check the entire system for leaks.

Brakes which have become overheated from excessive braking are dangerous and should be treated accordingly. Excessive brake heating weakens the tire and wheel structure and increases tire pressure.

AIRCRAFT LANDING WHEELS

Aircraft wheels provide the mounting for tires which absorb shock on landing. support the aircraft on the ground and assist in ground control during taxi, takeoff and landing. Wheels are usually made from either aluminum or magnesium. Either of these materials provides a strong, light weight wheel requiring very little maintenance.

  1. Split wheel—the most popular type. (Figures 9–43 and 9–44 Heavy aircraft wheel. Figures 9–45 and 9–46 Light aircraft wheel.)

  2. Removable flange type Figure 9–47.

  3. Drop center fixed flange type Figure 9–48.

The split wheel is used on most aircraft today. Illustrations of wheels used on light civilian type and heavy transport type aircraft are shown to illustrate the similarities and differences.

Split Wheels

Figures 9–43 and 9–44 and the description that follows were extracted from the B.F. Goodrich maintenance manual on wheels. The wheel illustrated in figure 9–43 is used on the Boeing B–727 transport aircraft.

Description—The numbers in parenthesis refer to figures 9–43 and 9–44.

  A. The main landing gear wheel is a tubeless, split-type assembly made of forged aluminum.

  B. The inner and outer wheel half assemblies are fastened together by 18 equally spaced tie bolts (11), secured with nuts (9). A tubeless tire valve assembly installed in the web of the inner wheel half (48), with the valve stem (7) protruding through a vent hole in the outer wheel half (30), is used to inflate the tubeless tire used with

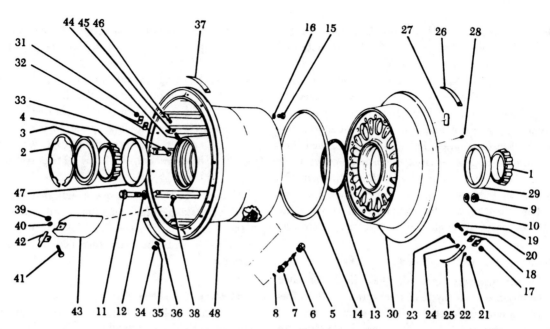

FIGURE 9-43. Split wheel—heavy aircraft.

Index No.	Description		
	WHEEL, LANDING GEAR, 49 x 17, TUBELESS, MAIN	25	IDENTIFICATION PLATE
1	CONE BEARING	26	INSTRUCTION PLATE
2	RING, RETAINING	27	PLATE, IDENTIFICATION
3	SEAL	28	INSERT, HELI-COIL
4	CONE, BEARING	29	CUP, BEARING
	VALVE ASSY, TUBELESS TIRE	30	WHEEL HALF, OUTER
5	CAP, VALVE		WHEEL HALF, ASSY, INNER
6	VALVE, INSIDE	31	NUT
7	STEM, VALVE	32	WEIGHT, WHEEL BALANCE, 1/4 oz.
8	GROMMET, RUBBER (TIRE AND RIM ASSOC.)	33	BOLT, MACHINE
9	NUT	34	NUT
10	WASHER	35	WASHER, FLAT
11	BOLT	36	IDENTIFICATION PLATE
12	WASHER	37	INSTRUCTION PLATE
13	PACKING, PREFORMED	38	BOLT, MACHINE
14	PACKING, PREFORMED	39	NUT
15	PLUG, MACHINE THD, THERMAL PRESSURE RELIEF, ASSY OF	40	WASHER, FLAT
		41	BOLT, MACHINE
16	PACKING, PREFORMED	42	BRACKET
	WHEEL HALF ASSY, OUTER	43	SHIELD, HEAT
17	NUT	44	SCREW
18	WEIGHT, WHEEL BALANCE, 1/4 oz.	45	INSERT
19	BOLT, MACHINE	46	INSERT, HELI-COIL
20	WASHER	47	CUP, BEARING
21	NUT	48	WHEEL HALF, INNER
22	WASHER, FLAT		
23	BOLT, MACHINE		
24	WASHER, FLAT		

FIGURE 9-44. Parts list—split wheel, heavy aircraft.

this wheel. Leakage of air from the tubeless tire through the wheel half mating surfaces is prevented by a rubber packing (14) mounted on the register surface of the inner wheel half. Another packing (13) mounted on the inner register surface of the inner wheel half seals the hub area of the wheel against dirt and moisture.

C. A retaining ring (2) installed in the hub of the inner wheel half holds the seal (3)

and bearing cone (4) in place when wheel is removed from axle. The seal retains the bearing lubricant and keeps out dirt and moisture. Tapered roller bearings (1, 4, 29, 47) in the wheel half hubs support the wheel on the axle.

D. Inserts (45) installed over bosses in the inner wheel half (48) engage the drive slots in the brake disks, rotating the disks as the wheel turns. A heat shield (43), mounted underneath and between the inserts, keeps excessive heat, generated by the brake, from the wheel and the tire. Two alignment brackets (42), 160° apart, are attached with the heat shield to the wheel half. The brackets prevent brake disk misalignment during wheel installation.

E. Three thermal relief plugs (15), equally spaced and mounted in the web of the inner wheel half directly under the mating surfaces, protect against excessive brake heat expanding the air pressure in the tire and causing a blowout. The inner core of the thermal relief plug is made of fusible metal that melts at a predetermined temperature, releasing the air in the tire. A packing (16) is installed underneath the head of each thermal relief plug to prevent leakage of air from the tires.

Figures 9–45 and 9–46 were extracted from the B.F. Goodrich maintenance manual on wheels. The wheel illustrated is typical of split wheels used on light aircraft.

Description—Numbers in parentheses refer to figures 9–45 and 9–46.

A. This main wheel is a tubeless split-type assembly made of forged aluminum.

B. The inner (24) and outer (16) wheel half assemblies are fastened together by 8 equally spaced tie bolts (11), secured with nuts (9). A tubeless tire valve assembly

installed in the outer wheel half (16) is used to inflate the 6.50–8 tubeless tire used with this wheel. Leakage of air from the tubeless tire through the wheel half mating surfaces is prevented by a rubber packing (12) mounted in the mating surface of the outer wheel half.

C. A seal (1) retains grease in the bearing (2) which is installed into bearing cup (23) inner wheel half, and (15) outer wheel half. Tapered bearings (2) installed in the bearing cups in the wheel halves support the wheel on the axle.

D. Torque keys (19) installed in cutouts in the inner rim of the wheel engage the drive tabs in the brake disks, rotating the disks as the wheel turns.

Removable Flange Wheels

The drop-center and flat base removable flange wheels (figure 9–47) have a one-piece flange held in place by a retainer snap ring. Wheels of the removable-flange type are used with low-pressure casings and may have either the drop center or a flat base. A flat-base rim may be removed quickly from the tire by removing the retaining lock ring that holds the one-piece removable flange in place, and lifting the flange from the seat. When a brake drum of the conventional type is installed on each side of the wheel, this provides a dual-brake assembly.

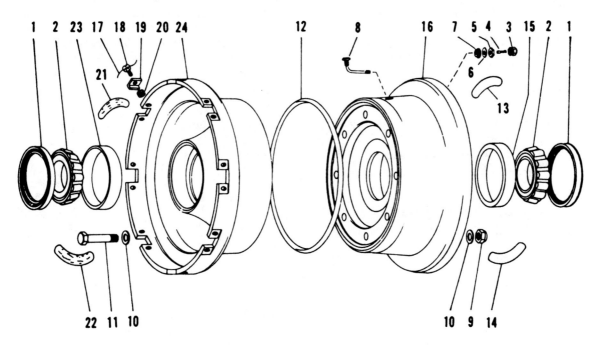

FIGURE 9–45. Split wheel—light aircraft.

375

Index No.	Description
1	WHEEL ASSEMBLY (With Valve Assy)
2	SEAL ASSY
	CONE, Bearing
	VALVE ASSY, Tubeless tire
3	CAP, Valve
4	CODE, Valve
5	NUT
6	SPACER
7	GROMMET
8	STEM, Valve
9	NUT
10	WASHER
11	BOLT
12	PACKING
13	WHEEL-HALF ASSY, Outer
14	PLATE, Identification
15	PLATE, Instruction
16	CUP, Bearing
	WHEEL-HALF ASSY, Inner
17	WIRE, Lock
18	SCREW
19	KEY, Torque
20	INSERT, Heli-coil
21	PLATE, Identification
22	PLATE, Instruction
23	CUP, Bearing
24	WHEEL-HALF, Inner

FIGURE 9–46. Parts list—split wheel, light aircraft.

A brake-drum liner may be held in place by means of steel bolts projecting through the casting with lock nuts on the inner side. These can be tightened easily through spokes in the wheel.

The bearing races are usually shrink-fitted into the hub of the wheel casting and provide the surfaces on which the bearings ride. The bearings are the tapered roller type. Each bearing is made up of a cone and rollers. Bearings should be cleaned and repacked with grease periodically in accordance with applicable manufacturer's instructions.

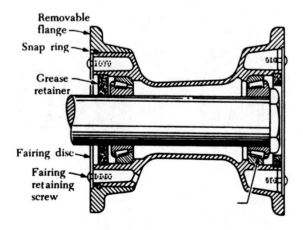

A. Drop center.

FIGURE 9–47. Removable flange wheels.

Fixed Flange Wheels

Drop center fixed flange aircraft wheels (figure 9–48) are special use wheels such as military, for high pressure tires. Some may be found installed on older type aircraft.

Outboard radial ribs are provided generally to give added support to the rim at the outboard bead seat. The principal difference between wheels used for streamline tires and those used for smooth contour tires is that the latter are wider between the flanges.

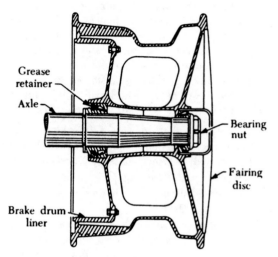

FIGURE 9–48. Fixed flange, drop center wheel.

Wheel Bearings

The bearings of an airplane wheel are of the tapered roller type and consist of a bearing cone, rollers with a retaining cage, and a bearing cup, or outer race. Each wheel has the bearing cup, or race, pressed into place and is often supplied with a hub cap to keep dirt out of the outside bearing.

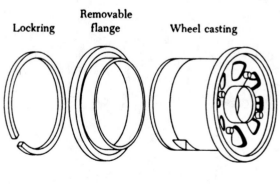

B. Flat base.

Suitable retainers are supplied inboard of the inner bearing to prevent grease from reaching the brake lining. Felt seals are provided to prevent dirt from fouling multiple-disk brakes. Seals are also supplied on amphibian airplanes to keep out water.

AIRCRAFT TIRES

Aircraft tires, tubeless or tube type, provide a cushion of air that helps absorb the shocks and roughness of landings and takeoffs: they support the weight of the aircraft while on the ground and provide the necessary traction for braking and stopping aircraft on landing. Thus, aircraft tires must be carefully maintained to meet the rigorous demands of their basic job . . . to accept a variety of static and dynamic stresses dependably—in a wide range of operating conditions.

Aircraft Tire Construction

Dissect an aircraft tire and you'll find that it's one of the strongest and toughest pneumatic tires made. It must withstand high speeds and very heavy static and dynamic loads. For example, the main gear tires of a four-engine jet transport are required to withstand landing speeds up to 250 mph, as well as static and dynamic loads as high as 22 and 33 tons respectively. Typical construction is shown in figure 9–49.

Tread

Made of rubber compound for toughness and durability, the tread is patterned in accordance with aircraft operational requirements. The circumferential ribbed pattern is widely used today because it provides good traction under widely varying runway conditions.

Tread Reinforcement

One or more layers of reinforced nylon cord fabric strengthens the tread for high speed operation. Used mainly for high speed tires.

Breakers

Not always used, these extra layers of reinforcing nylon cord fabric are placed under the tread rubber to protect casing plies and strengthen tread area. They are considered an integral part of the carcass construction.

Casing Plies/Cord Body

Diagonal layers of rubber-coated nylon cord fabric (running at opposite angles to one another) provide the strength of a tire. Completely encompassing the tire body, the carcass plies are folded around the wire beads and back against the tire sidewalls (the "ply turnups").

Beads

Made of steel wires embedded in rubber and wrapped in fabric, the beads anchor the carcass plies and provide firm mounting surfaces on the wheel.

Flippers

These layers of fabric and rubber insulate the carcass from the bead wires and improve the durability of the tire.

Chafers

Layers of fabric and rubber that protect the carcass from damage during mounting and demounting. They insulate the carcass from brake heat and provide a good seal against movement during dynamic operations.

Bead Toe

The inner bead edge closest to the tire center line.

Bead Heel

The outer bead edge which fits against the wheel flange.

Innerliner

On tubeless tires, this inner layer of less permeable rubber acts as a built-in tube, it prevents air from seeping through casing plies. For tube type tires, a thinner rubber liner is used to prevent tube chafing against the inside ply.

Tread Reinforcing Ply

Rubber compound cushion between tread and casing plies, provides toughness and durability. It adds protection against cutting and bruising throughout the life of the tread.

Sidewall

Sidewalls are primarily covers over the sides of the cord body to protect the cords from injury and exposure. Little strength is imparted to the cord body by the sidewalls. A special sidewall construction, the "chine tire," is a nose wheel tire designed with built-in deflector to divert runway water to the side, thus reducing water spray in the area of rear mounted jet engines.

Apex Strip

The apex strip is additional rubber formed around the bead to give a conture for anchoring the ply turnups.

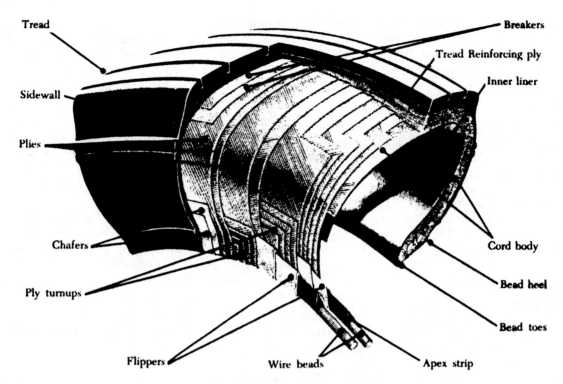

FIGURE 9–49. Aircraft tire construction.

Aircraft Tire Care

Tires are as vital to the operation of aircraft as they are to the operation of an automobile. During ground operation tires can be considered as ground control surfaces. The same rules of safe driving and careful inspection apply on the runway as on the highway.

They include control of speed, braking, and cornering, and inspection for proper inflation, cuts, bruises, and signs of tread wear. Contrary to what most people think—including many beginning pilots—the toughest demand on aircraft tires is rapid heat buildup during lengthy ground operations, not the impact of hard landings.

Aircraft tires are designed to flex more than automotive tires—over twice as much. This flexing causes internal stress and friction as tires roll on the runway. High temperatures are generated, damaging the body of the tire.

The best safeguards against heat buildup in aircraft tires are short ground rolls, slow taxi speeds, minimum braking, and proper tire inflation.

Excessive braking increases tread abrasion. Likewise, fast cornering accelerates tread wear. Proper inflation assures the correct amount of flexing and keeps heat buildup to a minimum, increasing tire life and preventing excessive tread wear.

Inflation pressure should always be maintained as specified in the aircraft maintenance manual or according to information available from a tire dealer.

Even though using a tire gage is the only accurate way to spot-check inflation, a quick visual inspection of the tread can reveal if air pressure has been consistently high or low. Excessive wear in the shoulder area of the tire is an indication of under inflation. Excessive wear in the center of the tire suggests over inflation.

Tires should also be carefully inspected for cuts or bruises. The best way to avoid aircraft tire cuts and bruises is to slow down when unsure of runway or taxiing surface conditions.

Since airplane tires have to grip the runway in the same way car tires grip the road, tread depth is also important. Tread grooves must be deep enough to permit water to pass under the tires, minimizing the danger of skidding or hydroplaning on wet runways. Tire treads should be inspected visually or with an approved depth gage according to manufacturers' specifications.

Another inspection goal is detection and removal of any traves of gasoline or oil on the tires. Such mineral fluids damage rubber, reducing tire life. Likewise, tires should be inspected for ozone or weather checking. Electricity changes oxygen in the air to ozone, which also prematurely ages rubber.

MATCHING DUAL TIRES

Matching tires on dual wheels, or dual wheels on a multi-wheel gear configuration, is necessary so that each tire will have the same contact area with the ground and thereby carry an equal share of the load. Only those tires having inflated diameters within the tolerances listed below should be paired together on dual wheels.

Tires should not be measured until they have been mounted and kept fully inflated for at least 12 hours, at normal room temperatures.

O. D. Range of Tires	Maximum Tolerance Permissable
Up to 24"	1/4"
25" to 32"	5/16"
33" to 40"	3/8"
41" to 48"	7/16"
49" to 55"	1/2"
56" to 65"	9/16"
66" and up	5/8"

FIGURE 9–50. Matching tires on dual wheel installations.

Aircraft tires should be stored in a cool dry place away from electric motors. The manufacturers specifications should be followed at all times when performing tire maintenance.

Tires on dual wheel aircraft will have a longer operational life if matched as suggested in figure 9–50.

AIRCRAFT TIRE MAINTENANCE

All aircraft tire manufacturers publish maintenance manuals and instruction manuals.

The following discussion on aircraft tires is excerpted from the B.F. Goodrich publication Care and Maintenance of Aircraft Tires, Fourth Edition, and is published with their permission.

Proper Inflation For Satisfactory Service

Proper inflation is undoubtedly the most necessary maintenance function for safe, long service from aircraft tires.

Tire pressures should be checked with an accurate gage at least once a week or oftener, and it is recommended that they be checked before each flight. Otherwise, if a slow leak should develop, it could cause severe loss of air within two or three days, with resulting damage to the tire and tube. Air pressures should only be checked when tires are cool. Wait at least two hours after a flight before checking pressures (three hours in hot weather).

For New Mountings

A newly mounted tire and/or tube should be checked at least daily for several days, after which the regular inflation control schedule may be followed. This is necessary because air is usually trapped between the tire and tube at the time of mounting, giving a false pressure reading. As this trapped air seeps out under the beads of the tire and around the valve hole in the wheel, the tire may become severely under inflated within a day or two.

Allow For Nylon Stretch

All aircraft tires are now made with nylon cord, and the initial 24-hour stretch of a newly mounted nylon tire may result in a 5 percent to 10 percent drop in air pressure. Thus, such a tire should not be placed in service until it has been left to stand at least 12 hours after being mounted and inflated to regular operating pressure. The air pressure should then be adjusted to compensate for the decrease in pressure caused by the stretching of the cord body.

Tubeless Air Diffusion Loss

Maximum allowable diffusion is 5 percent for any 24-hour period. However, no accurate tests can be made until after the tire has been mounted and inflated for at least 12 hours, and air added to compensate for pressure drop due to normal nylon cord body expansion, and any changes in tire temperature. A pressure drop of over 10 percent during this initial period should be sufficient reason to not place the tire and wheel assembly into service.

For Duals: Equalize Pressures

Differences of air pressure in tires mounted as duals, whether main or nose, should be cause for concern, as it ordinarily means that one tire is carrying more of the load than the other. If there is a difference of more than 5 lbs., it should be noted in the log. The log-book should then be referred to on each subsequent inflation check. Impending tire or tube failure can often be detected by this method. Should a pressure difference be found, check the valve core by spreading a little water over the end of the valve. If no bubble appears, it can be assumed that the valve core is holding pressure satisfactorily.

Sources of Pressure Data

Inflation of nose wheel tires should follow the recommendations of the aircraft manufacturers, because they take into consideration both the extra load transferred to the nose wheel by the braking effect and by the static load. Air pressure in the nose wheel tire, based on the static load only, would result in under inflation for the load carried when the brakes are applied. Tail wheel tires, however, should be inflated in accordance with the axle static load.

When tires are inflated under load, the recommended pressure should be increased by 4 percent. The reason is that the deflected portion of the tire causes the volume of the air chamber to be reduced, and increases the inflation pressure reading, which must then be offset in accordance with the above rule.

Effects of Under Inflation

Under inflation results in harmful and potentially dangerous effects. Aircraft tires which are under inflated are much more likely to creep or slip on the wheels on landing or when brakes are applied. Tube valves can shear off, and the complete tire, tube and wheel assembly can be destroyed under such conditions. Too-low pressures can also cause rapid or uneven wear at or near the edge of the tread.

Under inflation provides more opportunity for the sidewalls or the shoulders of the tire to be crushed by the wheel's rim flanges on landing or upon striking the edge of a runway while maneuvering the aircraft. Tires may flex over the wheel flange, with greater possibility of damage to the bead and lower sidewall areas. A bruise break or rupture of the tire's cord body can result.

Severe under inflation may result in cords being loosened and the tire destroyed because of extreme heat and strain produced by the excessive flexing action. This same condition could cause inner tube chafing and a resultant blowout.

Observe Load Recommendations

Since the beginning of air transportation, aircraft tires have been doing their required job with increasing efficiency and safety. But there is a limit to the load that any aircraft tire can safely and efficiently carry.

Loading aircraft tires above the limit can result in these undesirable effects:

1. Undue strain is put on the cord body and beads of the tire, reducing the factor of safety and reducing service life.
2. There is greater chance of bruising upon striking an obstruction or upon landing (bruise breaks, impact breaks and flex breaks in the sidewall or shoulder).
3. Possibility of damaging wheels. Under the severe strain of an extra load, a wheel may fail before the tire does.

Note: While additional air pressure (inflation) to offset increased loads can reduce excessive tire deflection, it puts an added strain on the cord body, and increases its susceptibility to cutting, bruises and impact breaks.

Nylon Flat-Spotting

Nylon aircraft tires will develop temporary flat spots under static load. The degree of this flat-spotting will vary according to the drop in the internal pressure in the tire and the amount of weight being sustained by the tire. Naturally, flat-spotting can be more severe during cold weather and is more difficult to work out of a tire at lower temperatures.

Under normal conditions, a flat spot will disappear by the end of the taxi run. If it doesn't, the tire can generally be reshaped by overinflating it 25 percent or 50 percent and moving the aircraft until the low spot is on the upper side. This pres-

---

PREVENTIVE MAINTENANCE SUMMARY

- Check tire pressures with an accurate gage at least once a week, and before each flight. Tires should be at ambient temperature.
- Check newly mounted tires or tubes daily for several days.
- Newly mounted tires should not be placed in service until cord body stretch has been compensated with re-inflation.
- Check for abnormal diffusion loss.
- Follow inflation recommendations carefully — guard against underinflation.
- Observe load recommendations.
- Move aircraft regularly or block up when out of service for long periods.

FIGURE 9–51. Preventive maintenance summary.

sure should be left in the tire for one hour. It may even be necessary to taxi or tow the aircraft until the reshaping is accomplished. Needless to say, any flat spot can cause severe vibration and other unpleasant sensations to both pilot and passengers.

Aircraft that is to remain idle for a period longer than three days should either be moved every 48 hours or blocked up so that no weight is on the tires. Aircraft in storage (out of service for more than 14 days) should be blocked up so that there is no weight on the tires.

Figure 9–51 gives a brief, summary checklist of tire preventive maintenance.

TIRE INSPECTION—MOUNTED ON WHEEL

Leaks Or Damage At Valve

To check valves for leaks, put a small amount of water on the end of the valve stem and watch for air bubbles. If bubbles appear, replace the valve core and repeat this check.

Always inspect the valve to be sure the threads are not damaged; otherwise, the valve core and valve cap will not fit properly. If threads are damaged, the valve can usually be rethreaded, inside or outside, by use of a valve repair tool, without demounting the tire from the wheel.

Make certain that every valve has a valve cap on it—screwed on firmly with the fingers. The cap prevents dirt, oil and moisture from getting inside the valve and damaging the core. It also seals in air and serves as protection in case a leak develops in the valve core.

Check the valve to be sure it is not rubbing against the wheel. If it is bent, cracked, or severely worn, demount the tire and replace the tube or valve at once.

Tread Injuries

Carefully inspect the tread area for cuts or other injuries. Be sure to remove any glass, stones, metal or other foreign objects that might be embedded in the tread or that have penetrated the cord body.

Use a blunt awl for this purpose, although a medium size screwdriver can be used if a blunt awl is not available. When probing an injury for foreign material, be careful not to enlarge the injury or drive the point of the awl or screwdriver into the cord body beyond the depth of the injury. When prying out foreign material that might be embedded, the other hand should be held over the injury in such a way that the object will not fly out and strike the person conducting the inspection in the face.

Tires with cuts or other injuries which expose or penetrate the cord body, should be removed and repaired, recapped or scrapped. Where a cut does not expose the carcass cord body, taking the tire out of service is not required.

Remove any tires that show signs of a bulge in the tread or sidewall. This may be the result of an injury to the cord body, or may indicate tread or ply separation. Always mark such a bulged area with tire crayon before deflating the tire; otherwise, it may be very difficult, if not impossible, to locate the area after the tire is deflated.

Sidewall Injuries

Inspect both sidewalls for evidence of weather or ozone checking and cracking, radial cracks, cuts, snags, gouges, etc. If cords are exposed or damaged, the tires should be removed from service.

When To Remove For Recapping

Check tires for possible need of recapping. They should be taken out of service when:

(a) They have one or more flat spots. Generally, a single flat spot or skid burn does not expose the carcass cord body and the tire may remain in service, unless severe unbalance is reported by the aircraft crew.

(b) They show 80 percent or more tread wear.

(c) There are numerous cuts that would require repair. In other words, if the cost of repairing the cuts would amount to 50 percent or more of the recapping cost, it would be considered more economical to have the tire recapped.

Uneven Wear

Check tires for evidence of wheel misalignment. Tires showing such wear should be demounted, turned around and remounted, in order to even up the wear. Also, check for spotty, uneven wear due to faulty brakes so mechanical corrections can be made as soon as possible.

Wheel Damage

Inspect the entire wheel for damage. Wheels which are cracked or injured should be taken out of service and laid aside for further checking, repair or replacement.

When inspecting a mounted tire on the wheel of a plane, always be sure that nothing is caught between the landing gear and the tire; and that no parts of the landing gear are rubbing against the tire.

At this time, also check the nacelle into which the tire fits, when the landing gear is retracted. Clearances are sometimes close and any foreign

A. A valve cap for every valve.   B. Depth gage shows wear.   C. Mark and remove foreign objects.

FIGURE 9–52. Basic tire maintenance

material or loose or broken parts in the nacelle can cause severe damage to the tire and even cause it to fail upon landing.

Figure 9–52 shows checks to be accomplished while tire is mounted.

TIRE INSPECTION—TIRE DEMOUNTED

Periodic Demounting

A definite schedule can be set up to provide for regular inspections of tires and tubes, after a certain number of hours or landings, and that each individual tire and tube be removed from the wheel for that inspection. However, if an airplane has made an emergency or particularly rough landing, the tire and tube should be demounted and inspected as soon as possible to determine whether any hidden damage exists. The wheel should also be examined at the same time.

Probe Injuries

Probe all cuts, punctures and other tread injuries with a blunt awl and remove any foreign material. When prying out the foreign material imbedded in the tread, the hand should be held over the injury in such a position that the object will not fly out and strike the person conducting the inspection in the face.

Squeezing the sidewalls together will also assist, as this will open up the cuts and injuries. Size and depth of cuts and injuries can then be determined by probing with the awl. Do not push the awl point into the cord body beyond the depth of the injury.

Repairable Injuries

Injuries into, or through, the carcass cord body, measuring no more than $\frac{1}{4}$" on the outside and $\frac{1}{8}$" on the inside, would be considered punctures and are easily repaired without the need for fabric reinforcement inside the tire. Tires qualified to speeds over 160 mph can be repaired if they meet the following qualifications: injuries through the tread must not penetrate more than 40 percent of the actual body plies; they should measure no more than $1\frac{1}{2}$" long and $\frac{1}{4}$" wide before tread is removed; and after the tread has been removed, injuries should not be over 1" long and $\frac{1}{8}$" wide on the surface.

Tires qualified to speeds of less than 160 mph can be repaired if the injury penetrates through more than 40 percent of the actual plies, and is no larger than 1" in length. Naturally, there must be a limit as to the number of such injuries that there can be in the tire. The decision to recondition or not recondition such a tire should be left up to the tire manufacturer.

Sidewall Conditions

Inspect both sidewalls for evidence of weather or ozone checking and cracking, radial cracks, cuts or snags.

(a) Scrap any tires with radial cracks extending to the cords.

(b) Scrap any tires with weather-checking, ozone-checking, or cracking, which extends to the cords. Weather-checking is a normal condition affecting all tires and, until cords are exposed, will in no way affect the serviceability or safety of the tire.

(c) Tires with cuts or snags in the sidewall area which have damaged the outer ply should be scrapped.

Bead Damage

Check the entire bead and the area just above the heel of the bead on the outside of the tire for

382

A. Probe with blunt awl, shield work.　　B. Are sidewall cords exposed?　　C. Finishing & trip damage, repairable.

FIGURE 9–53. Basic tire maintenance

chafing from the wheel flange or damage from tire tools. Any blistering or separation of the chafer strips from the first ply beneath requires repairing or replacement of the chafer strip. If cords in the first ply under the chafer strip are damaged, the tire should be scrapped.

If protruding bead wire, bead wire separation, or a badly kinked bead is found, the tire should be scrapped.

Loose or blistered finishing strips can generally be replaced during the retreading process, but tires should not be continued in service in that condition.

Figure 9–53 shows checks to be made with tire demounted.

Bulges—Broken Cords

Check any tires which were marked for bulges when the tire was mounted and inflated. If no break is found on the inside of the tire, probe with an awl to determine whether separation exists. If separation is found, the tire should be discarded, unless there is only a small localized separation between the tread or sidewall rubber and the cord body. In this case, spot repair or retreading may be satisfactory.

Any tire found with loose, frayed or broken cords inside should be scrapped.

IMPORTANT: Do not use an awl or any pointed tool on the inside of a tubeless tire for probing or inspection purposes.

Tubeless Tires—Bead Area

A tubeless tire fits tighter on the wheel than a tube type tire, in order to properly retain air pressure. Therefore, the face of the bead (the flat surface between the toe and heel of the bead) must not be demaged as this may cause the tire to leak. The primary sealing surface of the tubeless tire is this area, so examine it carefully for evidence of damage by tire tools, slippage while in service, or damage that would permit air to escape from inside the tire. Bare cords on the face of the bead normally will cause no trouble, and such a tire should be fit for continued service either as is, or after retreading.

Liner Blisters

Tubeless tires with liner blisters or liner separation areas, larger than 4″ x 8″, should be scrapped. Generally, small blisters (not over two inches in diameter) will cause no trouble and do not need to be repaired. However, **do not pierce the blisters or cut them,** as this will destroy the air-retaining ability of the tire.

Thermal Fuze

Some aircraft wheels have fusible plugs that are designed to melt at specific elevated temperatures and relieve air pressure to prevent the tire's blowing out or breaking of the wheel. Should air be lost due to the melting of one of these plugs, it is recommended that the tire involved be scrapped. However, an effort should be made to determine whether the plug melted at a much lower temperature than it should have, or whether air could have been lost around the plug because of improper installation.

If a tire has been subjected to a temperature high enough to melt one of these fuze plugs, it should be carefully inspected for evidence of reversion of the rubber coating in the rim contact area.

Figure 9–54 shows typical marking for bulged area, rubber reversion around rim, and outer sidewall damage.

383

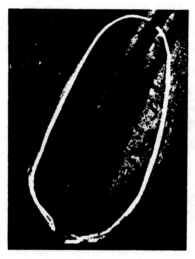  

A. Mark bulged areas.   B. Outer sidewall ply damaged, scrap tire.   C. Reversion of rubber from high wheel heat.

FIGURE 9–54. Basic tire maintenance

TUBE INSPECTION

Proper Size

In tube type tires, failure of an inner tube can easily cause irreparable damage to the tire in which it is mounted, as well as to the wheel and the aircraft itself. It is very important that tubes be of the proper size and equipped with the correct valves.

When inspecting tubes **do not** inflate with more air than that which is required to simply round out the inner circumference of the tube. Too much air places a strain on splices and areas around valve stems. In addition, excessive air will damage fabric base tubes by causing the fabric base to pull away from the outside of the tube.

Inspect the tube carefully for leaks while under pressure, preferably by inflating and submerging in water. If the tube is too large to be submerged in an available water tank, spread water over the surface as you inspect the tube, and look for bubbles.

Valve Stems

Examine the tube carefully around the valve stem for leaks, signs of valve pad separation, and bent or damaged valve stems.

Wrinkles

Tubes with severe wrinkles should be removed from service and scrapped. These wrinkles are evidence of improper fitting of the tube within the tire, and wherever a wrinkle occurs, chafing takes place. A blowout could result.

Chafing

Inspect tubes for evidence of chafing by the toes of the tire beads. If considerable evidence of this chafing is present, remove the tube from service and scrap.

Thinning

Where the heat is greatest, the tube has a tendency to be stretched over the rounded edge of the bead seat of the wheel. This is one of the reasons why, when mounting, tubes should always be inflated until the tire beads are in position, and then completely deflated and reinflated to the final pressure. The stretch on the tube is then equalized throughout its inner and outer periphery.

Also check tubes for possible thinning out due to brake drum heat in the area where they contact the wheel and bead toes.

In figure 9–55 it can be seen that the "set" or shape of the tube can assist in determining when it should be removed from service because of thinning in the bead area. In addition, feeling the tube with the fingers in that area will tell, after a little practice, when the life has gone out of the tube and it should be scrapped.

On wheels with only one brake drum this heat-set condition will normally show up on only one side of the tube. In those cases where the brake drum is a considerable distance from the rim, it is unlikely that this condition will ever be experienced.

A. Natural contour.   B. Taking a "set".   C. Thinned out at edges.

FIGURE 9–55. Inner tube inspection.

Fabric Base Tubes

In cases where brake drum heat is a recognizable factor, careful checking of the tube, as well as the tire beads, should be made to prevent a failure which might be disastrous. In such instances, fabric base tubes should always be used. These have a layer of nylon cord directly imbedded in the inner circumference of the tubes to protect them from thinning out under brake drum heat. Additional protection is also provided against chafing action of the tire bead toes and from damage during mounting and demounting.

MOUNTING AND DEMOUNTING

The object of these instructions is to show how to do the job as easily, and safely as possible, using proper tools, without damaging tires, tubes or wheels.

Almost every experienced aircraft tire service man has developed methods which are more or less his own, and undoubtedly some of these methods are as practical as the suggestions given here.

These instructions are intended to be simple, so that they can be carried out with tools which are commonly available, in contrast to the specialized equipment that is usually available only at larger airports or military installations.

Inspection, Tube Installation

Before mounting any tire, examine the wheel carefully to make sure there are no cracked or injured parts. Naturally, the tire and tube should be carefully inspected, as described in the pages dealing with tire and tube inspection. A quick check should always be made to be certain that no foreign material is inside the tire or clinging to the inner tube.

Dust the inside of the tire and the inner tube outer surface with tire talc or soapstone before the tube is installed in the tire. This will prevent its sticking to the inside of the tire or to the tire beads.

Dusting also helps the tube assume its normal shape inside the tire during inflation, and lessens the chances of wrinkling or thinning out.

It is good practice to always mount the tube in the tire with the valve projecting on the serial side of the tire.

Lubrication

Tubeless tires fit tighter on the wheel than the tube type. It is therefore desirable to lubricate the toes of the beads with an approved 10 percent vegetable oil soap solution, or plain water. This will facilitate mounting and accomplish proper seating of the tire beads against the wheel flanges so there will be no air loss. Care must be used, however, to make certain that none of the solution gets on the area of the bead making contact with the wheel flange.

On tube type tires, lubrication of tire beads may or may not be necessary, depending upon the type of wheel being used. An approved mounting solution, such as the 10 percent vegetable oil soap solution, or water mentioned above, can be used on the bead toes, and even on the rim side of the inner tube, to facilitate mounting.

Balance

Balance in an aircraft wheel assembly is very important. From a wear standpoint, when the wheels are in landing position a heavy spot in a wheel assembly will have a tendency to remain at the bottom and thus will always strike the ground or runway first. This results in severe wear at one area of the tire tread and can necessitate early replacement. In addition, unbalanced tires can cause severe vibration which may affect the operation of the aircraft. In fact, pilots have reported that sometimes instruments could not be read because of such vibration.

A. Valve projecting on serial side of tire.

B. Tube balance mark aligned with tire balance mark.

FIGURE 9–56. Basic tire and tube assembly.

Balance marks appear on certain aircraft tubes to indicate the heavy portion of the tube. These marks are approximately ½″ wide by 2″ long. When the tube is inserted in the tire, the balance mark on the tube should be located at the balance mark on the tire (figure 9–56). If the tube has no balance mark, place the valve at the balance mark on the tire.

When mounting tubeless aircraft tires, the "red dot" balance mark on the tire must always be placed at the valve that is mounted in the wheel.

Inflation Safety

After the tire and tube are mounted on the wheel the assembly should be placed in a safety cage for inflation. The cage should be placed against an outside wall, constructed so as to withstand, if necessary, the effects of an explosion of either the tire, tube or wheel (figure 9–57).

The air line from the compressor or other air source should be run to a point at least 20 to 30 ft. away from the safety cage and a valve and pressure gage installed at that point. The line should

A. Clip-on chuck permits inflation at safe distance from tire safety cage.

B. Recommended for all aircraft tire shops.

FIGURE 9–57. Inflation precautions.

then be continued and fastened to the safety cage with a rubber hose extending from that connection. A clip-on chuck is then fitted on the end of the hose for actual inflation purposes. This arrangement makes it unnecessary to reach into the cage to check air pressures or to be anywhere near the safety cage while the tire is being inflated.

Seating Tube Type Tires

To seat the tire beads properly on the wheel, first inflate the tire to the pressure recommended for that particular size and for the aircraft on which it is to be mounted. Then the tire should be completely deflated and finally reinflated to the correct pressure (do not fasten valve to rim until this has been done). Use the valve extension for inflation purposes, if necessary.

This procedure accomplishes the following: it helps to remove any wrinkles in the tube and to prevent pinching the tube under the toe of the bead; it eliminates the possibility of one section of the tube stretching more than the rest and thinning out in that area; and it assists in the removal of air that might be trapped between the inner tube and the tire.

NOTE: With tubeless tires, it is not necessary to go through this inflation-deflation-reinflation procedure.

Let Stand—Then Recheck

It is recommended that a newly-mounted assembly be stored away from work areas for at least 12 hours and preferably for 24 hours. This is for the purpose of determining if there is any structural weakness in either the tire, tube or wheel. This also permits rechecking of the tire, after the 12- or 24-hour period, to determine any drop in pressure and to judge whether this drop in pressure is in accordance with normal tire growth.

By such a test, when the assembly is mounted on the aircraft, it can be done with the assurance that each part of the assembly is satisfactory for service.

Demounting Safety

Always be sure to deflate tires completely before demounting. There have been many serious accidents caused by failure to follow this important step. For even a better practice, it is recommended that tires be deflated before wheels are removed from the aircraft.

NOTE: Use caution when unscrewing valve cores, as the air pressure within the tube or tire can cause a valve core to be ejected like a bullet.

Handle Beads and Wheels With Care

With any type of wheel, the tire beads must be loosened from the wheel flange and bead seat before any further steps are taken in demounting. Be very careful not to injure the beads of the tire or the relatively soft metal of the wheel. Even with approved tools, extreme care must be taken.

A. Tubeless—Split Wheels

In the tubeless design, the tire and wheel are used to retain air pressure. Inflation is accomplished through a tubeless tire inflation valve installed in the wheel. The wheel valve hole, in which the tubeless tire inflation valve is mounted, is sealed against loss of air by a packing ring or by an "O" ring. (See figure 9-58.)

Split wheels are sealed against loss of air by an "O" ring mounted in a groove in the mating surface of one of the wheel halves.

Demountable flange wheels are similarly sealed against loss of air by an "O" ring installed in a groove on the wheel base under the area covered by the demountable flange.

The air pressure contained in the tubeless tire seals the tire bead against the wheel bead seat to prevent loss of air.

Wheels used with disc brakes have thermal fuze (relief) plugs installed in the rotor drive area of the wheel as a protective measure against the tire blowouts due to excessive heat. The plugs have a fusible metal core that melts at a predetermined temperature to relieve the high pressure build-up.

Mounting

Check tubeless tire inflation valve and thermal relief plugs for proper installation, and absence of damage. Refer to wheel manufacturer's manual for installation procedure.

Inspect "O" ring used to seal wheel for damage and replace if necessary.

Lubricate "O" ring (as specified by the wheel manufacturer) and install in wheel groove. Make sure the "O" ring is free of kinks or twists and is seated properly.

Mounting a demountable flange on a wheel base, be careful not to dislocate or damage the "O" ring previously installed in the wheel base.

Mount tubeless tires in the same manner as tube type tires. Make sure that the wheel bead seats are clean and dry to insure proper sealing of the tubeless tire bead.

Assemble wheel halves of split type wheels with the light sides (impression stamped "L" on the

A. "O"-ring seal is vital — handle and install carefully.

B. Inspect valve seals for signs of damage or deterioration.

FIGURE 9–58. Split wheel seal inspection.

flanges) 180° apart from each other to insure minimum out-of-balance condition.

Be sure that nuts, washers, and bolts used to assemble split type wheels are in proper order and that the bearing surfaces of these parts are properly lubricated. Tighten to recommended torque values. See wheel manufacturer's manual for recommended procedures.

Demounting

The procedure for demounting tubeless tires is generally the same as for tires with tubes. However, care must be taken to avoid damaging: (1) the wheel's "O" ring groove and mating register surfaces; (2) the flange area that seats the tire bead; and (3) the tubeless tire inflation valve hole sealing area. These areas of the wheel are critical and if damaged will result in failure of the wheel and tire unit to maintain required air pressure.

B. Tube Type Tires

Mounting

With the tube entirely deflated, insert it in the tire (folding makes this easier, particularly in small diameter tires), and inflate until the tube is just rounded out. The valve core should be in the valve during this operation.

Apply with brush or swab a 10 percent solution of vegetable oil soap to the rim of the tube extending well up into the tire. Use care not to lubricate any part of the bead which comes in contact with the rim flange.

Insert the valve hole section of the wheel into the tire and push the valve through the valve hole in the wheel.

Insert the other side of the wheel while holding the valve in position. Be careful during this operation not to pinch the tube between the wheel sections.

Inflate, deflate and reinflate to the recommended pressure.

Install the locking nut or nuts and tighten securely. Put on valve cap and tighten with fingers.

Demounting

Remove valve core and fully deflate.

Do not use a pry bar, tire irons or any other sharp tool to loosen tire bead as the wheel may be damaged. Break bead before loosening tie bolts to prevent damage to register surfaces.

Use a bead breaker only to loosen tire bead from both wheel half flanges by applying pressure around the entire circumference of each sidewall.

Remove the tie bolts and the bolt nuts from the wheel and pull out both parts of the wheel from the tire.

C. Removable Side Flange Drop Center Wheels

Mounting

Fully deflate tube and line up tube balance mark with tire balance mark. Start tire over flange on an angle, being careful not to damage valve.

Be sure to remove valve extensions or valve fishing tools before wheel is installed.

Demounting

Be sure to deflate fully.

Make full use of the wheel in pulling bead over flange on removable side.

Work wheel up and down to ease it out of tire.

For one-man demounting, tire can be leaned against a wall or bench, valve side out.

D. Smooth Contour Tail Wheel Tires

Note: Smooth contour tires are usually harder to handle because of stiffer beads, small clearances and small diameters.

Mounting

Inflate tube sufficiently to round out and be sure tube is worked in all around to avoid pinching.

Work bead opposite valve over edge of wheel first. Deflate tube.

Keep second bead on edge of wheel to allow for inserting valve in valve hole.

Inflate, deflate, reinflate.

Demounting

Be sure to use tools with good leverage.

Be careful not to damage soft metal of rim flange.

Keep tube inflated after lock ring is removed.

Work tube out carefully, using water as a lubricant before completing demounting.

E. One Piece Drop Center Wheels

Mounting

Insert wheel in tire, reversing the customary procedure. (Valve hole side goes in first.)

Pry bead over flange with small bites—use a fairly thin tool.

When first bead is on wheel, insert tube. Be sure no part of tube is caught under bead.

Inflate, deflate, reinflate.

Demounting

After loosening bead, lay tire flat with wooden block, 3 to 4 inches high under sidewall. Work bead off in small bites.

F. Flat Base Wheels Removable Flange Locking Ring

Mounting

Examine wheel and flanges carefully for burrs or gouges.

Line up tube and tire balance marks.

Dust tube with talc.

Lean tire against bench or wall—flange outward. Pry flange loose evenly to prevent binding.

Demounting

Loosen bead carefully.

Use lead hammer or rubber mallet to loosen side ring.

Pry up side ring carefully and evenly.

Place wheel and tire on a wooden block about 14" high and large enough to fit over the wheel hub, so that wheel can be removed easily.

CAUSES OF AIR PRESSURE LOSS IN TUBELESS AIRCRAFT TIRES

There can be numerous causes for loss of air pressure within an aircraft wheel and tire assembly, therefore, it is economical and wise to follow a systematic check list. Without such a procedure, trial and error substitution of parts can needlessly increase tire maintenance costs.

For example, complaints of air loss in tubeless aircraft tire assemblies, while more common during cold weather, have no seasonal limitations. Factors which may seem distantly related to the problem—changes in tire maintenance personnel, inaccurate gages, air temperature fluctuations—are often the underlying causes of unsatisfactory tire service, further emphasizing the need for simple check procedures.

For guidance in setting up uniform inspection methods, there are general areas of the tire and wheel assembly which can be involved in air pressure loss. (See figure 9–59.)

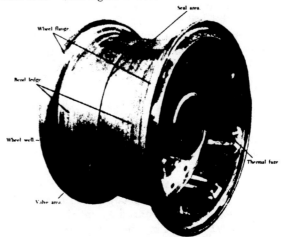

FIGURE 9–59. Air loss in split wheels.

Damaged Beads—Check for exposure of the carcass cord body in bead toe area or under face of the bead.

Improperly Seated Beads—Condition can be caused by: (a) insufficient air pressure; (b) beads not lubricated; (c) kinked or distorted beads.

Cut or Puncture—Check for cut or puncture entirely through the carcass cord body and liner.

Air Temperature

Was tire inflated in heated room and then stored outside? Air pressure will drop approximately 1 psi for every 4° drop in temperature. Tires should be checked and pressure adjusted for specific requirements when tires have reached the outside ambient temperature.

Venting of Tubeless Tires

Tubeless aircraft tires are vented in the sidewall area to permit any air that has diffused through the liner and cord body to escape, thus preventing pressure build up within the carcass cord body and possible tread or ply separation. Rate of diffusion will vary by manufacturer and the maximum permissible is no more than 5 percent in any 24-hour period.

Vent holes penetrate the sidewall rubber to, or into, the carcass cord body and may vary in size, depth and angle. Therefore, the amount of air diffused through these holes will vary. Thus, when water or a soapy solution is brushed over the outside of an inflated tubeless tire, air bubbles form. Some vent holes may emit a continuous stream of bubbles, where others may produce intermittent bubbles. This is normal and does not mean that there is anything wrong with the tire. In fact, as long as a tubeless tire is in an inflated condition, air will be coming out of these vent holes. Where the rate of loss exceeds 5 percent 24 hours, recheck for possible injuries. Vent holes may be covered or closed by spilled solvent or by the tire paint. They may also be covered during the retreading process. Check for evidence that tubeless tires have been revented after being retreaded.

Initial Stretch Period

All aircraft tires are of nylon construction and a certain amount of stretch occurs after the tire has been inflated. This, in itself, will reduce air pressure within the tire. It is absolutely necessary that the tire be inflated to its regular air pressure and let stand at least 12 hours in order to permit this expansion of the cord body. This may result in as much as 10 percent drop in air pressure. Compensate by reinflating the tire to original pressure. Only after this initial stretch period can it be determined if there is any true air loss within the tire.

THE WHEEL

Any of the following wheel conditions can contribute to air loss in the bead area of the tire:

Cracks Or Scratches In The Bead Ledge Or Flange Area

Cracks can usually be traced to fatigue failure while scratches and gouges are the result of handling damage or the improper use of tire irons.

Exceptionally Smooth Enamel Surface On Bead Seat Ledges

Corrosion Or Wear In Bead Ledge Area

Usually occurs at the toe area of the tire bead.

Poor Seating In Bead Area

May be caused by accumulation of rubber from the tire or dirt.

Knurls

Wheels converted from tube-type use should have knurls removed.

Porous Wheel Assemblies

Can be protected either by proper paint procedure and/or an impregnation process.

Holes for attachment of components of wheel assembly. In case of through bolts used to attach such items as drive lugs for brake assembly, etc., the mounting screws or bolts must be properly sealed. The recommendations of the wheel manufacturer should be followed.

Cracks in the wheel well area, in most cases, cannot be repaired.

Sealing Surfaces

Look for damage or improper machining of sealing surfaces. Care should be taken to see that there is no handling damage. Any irregularities should be corrected before remounting wheel and tire. (See figure 9–60.)

Foreign material or paint can impair the sealing surface. Thus, all foreign material should be cleaned from the sealing surface before assembly of the wheel. A light, even coat of primer is permissible. However, surface must be free of runs or dirt inclusions.

FIGURE 9-60. Wheel inspection.

Improper Installation of "O" Rings

Twisting or failure to provide lubricant when specified, may cause loss of air. The wrong size or type of "O" ring, or an "O" ring of the wrong compound for special low temperature service, may also cause leakage.

Inspect used "O" rings carefully. Be sure they are not thinned out, deformed, chipped, damaged or otherwise deteriorated.

Wheel Tie Bolts

Proper torque and torquing procedure, as specified by the wheel manufacturer, should be followed to assure adequate compression of sealing "O" ring under all temperature conditions. Low torques, low temperatures, and shrinkage of the wheel halves may cause a significant lessening of compression on the "O" ring seal.

Tubeless Wheel Valve Holes

Tubeless wheel valve holes and surrounding area must be free of scratches, gouges and foreign material.

The proper rubber grommet or "O" ring must be used as specified by the wheel manufacturer. Seals other than those specified may not function properly under the compression loads and low temperatures required for good sealing. Tightening of tubeless valve should follow specific instructions of wheel manufacturer.

Valve core should be checked and replaced when found leaking.

Valve caps should be used and tightened finger tight.

Thermal Fuze Installation

A faulty thermal fuze may cause leakage and require replacement. Usually this is the result of a poor bond between thermal melting material and bolt body.

Sealing surface for thermal fuze gasket must be clean and free from scratches and dirt. In some cases, surface can be repaired in accordance with manufacturer's instruction.

Be sure that sealing gasket is the one specified by the wheel manufacturer, sized and compounded for its specific job. Gasket should be free of distortion, cuts, etc.

To guard against air pressure losses before assembly the best insurance is a careful and complete inspection. After assembly, if air loss occurs, the use of a soap solution (or, if possible, complete immersion of wheel-tire assembly) may pinpoint the exact source of leakage.

GOOD PRESSURE GAGE PRACTICE

Quite often it is found that the differences in reported air pressures are entirely due to the difference in accuracy in different gages, rather than in any change in air pressure.

It is not unusual to find an inaccurate tire gage in constant use with a tag that states that the gage reads a certain number of pounds too high, or too low. Unfortunately, this error will change as different pressures are checked. A tire gage reading 10 lbs. high at 80 lbs. pressure may very well read 25 lbs. too high at 150 lbs. pressure. Therefore, incorrect tire gages should either be repaired or replaced. They should not be continued in service.

Cold temperatures may also affect tire gages and cause pressure readings lower than they actually are. Occasionally, too, a gage has been mistakenly treated with oil or some other lubricant in expectation of making it work better. This, of course, will actually cause incorrect readings and probably render the gage unfit for further service.

It is good practice to have gages recalibrated periodically and to use the same gage for performing an inflation cycle—for the original 12- or 24-hour stretch period. Dial type gages, of good quality, are highly recommended for all tire maintenance installations—regardless of size!

Storing Aircraft Tires and Tubes

The ideal location for tire and tube storage is a cool, dry and reasonably dark location, free from air currents and dirt. While low temperatures (not below 32° F.) are not objectionable, high room temperatures (80° F.) are detrimental and should be avoided.

Avoid Moisture, Ozone

Wet or moist conditions have a rotting effect and may be even more damaging when the moisture contains foreign elements that are further detrimental to rubber and cord fabric. Strong air currents should be avoided, since they increase the supply of oxygen and quite often carry ozone, both of which cause rapid aging of rubber. Also, particular care should be taken to store tires away from electric motors, battery chargers, electric welding equipment, electric generators and similar equipment as they all create ozone.

Fuel and Solvent Hazards

Care should be taken that tires do not come in contact with oil, gasoline, jet fuel, hydraulic fluids, or any type of rubber solvent, since all of these are natural enemies of rubber and cause it to disintegrate rapidly. Be especially careful not to stand or lay tires on floors that are covered with oil or grease. When working on engines or landing gear, tires should be covered so that oil does not drip on them.

Store In Dark

The storage room should be dark, or at least free from direct sunlight. Windows should either be given a coat of blue paint or covered with black plastic. Either of these will provide some diffused lighting during the daytime. Black plastic is preferred as it will lower the temperature in the room during the warm months and permit tires to be stored closer to the windows.

Tire Racks Preferred

Whenever possible, tires should be stored in regular tire racks which hold them up vertically. The surface of the tire rack against which the weight of the tire rests should be flat and, if possible, 3 to 4 inches wide in order that no permanent distortion to the tire is caused. If tires are piled on top of one another, they should not be stacked too high, as this will cause distortion of the tire and might result in trouble when the tires are put into service. This is particularly true of tubeless tires, as those on the bottom of the stack may have the beads pressed so closely together that a bead seating tool will have to be used to force the tire beads onto the wheel far enough to retain air pressure for inflation (figure 9–61).

Safe Tube Storage

Tubes should always be stored in their original cartons, so that they are protected from light and air currents. They should never be stored in bins or on shelves without being wrapped, preferably in several layers of heavy paper.

Tubes can also be stored by inflating slightly and inserting them in the same size of tire. This of course, should only be done as a more or less temporary measure. However, before using such an assembly, the tube should be removed from the tire and the inside of the tire carefully examined, since quite often foreign material will get between the two and if not removed, could cause irreparable damage to both tire and tube.

Under no circumstances should tubes be hung over nails or pegs, or over any other object that might form a crease in the tube. Such a crease will eventually produce a crack in the rubber.

REPAIRING

Many aircraft tires and tubes which become injured in service can be successfully repaired. Likewise, aircraft tires which have become worn out in service, or flat spotted and removed prematurely,

Size	Without Tube	With Tube Inserted	Size	Without Tube	With Tube Inserted
			56" SC and larger	3	4
26x6	5	6	12.50-16	4	5
33"	4	5	15.00-16	3	4
36"	4	5	17.00-16	3	4
44"	4	5	15.50-20	3	4
47"	3	4	17.00-20	3	4

Smaller sizes of tubess tires may be stacked five high. This would include sizes through 39 x 13. Larger sizes of tubless tires should not be stacked more than four high.

FIGURE 9–61. Permissible tire stacking.

can be recapped so that the new tread will give service comparable to the original tread. The recapping and repairing of aircraft tires has been practiced for many years and has saved aircraft operators considerable sums of money. Tires which might otherwise have been discarded have been safely reconditioned (many repeatedly) for continued service.

Recapping Aircraft Tires

Recapping is a general term meaning reconditioning of a tire by renewing the tread, or renewing the tread plus one or both sidewalls. (See figure 9-62.) There are actually four different types of recapping for aircraft tires.

Top Capping—For tires worn to the bottom of the tread design, with no more than slight flat spotting and/or shoulder wear, the old tread is roughened and a new tread is applied.

Full Capping—For tires worn all around, those flat spotted to the cords, or those with numerous cuts in the tread area, the new tread material is wider than that used on a top cap, and comes down over the shoulder of the tire for several inches.

Three-Quarter Retread—For tires needing a new tread, plus renewing of the sidewall rubber on one side, due to damage or weather checking, a full cap is applied, and in addition, approximately 1/16″ of the thickness of the old sidewall rubber is buffed off one side. New sidewall rubber is then applied from the bead to the edge of the new tread, on the buffed side only.

Bead-to-Bead Retread—A new tread and both new sidewalls are applied by this method.

Tires That May Be Recapped

Tires with sound cord bodies and beads, or which meet injury limitations described under "Repairable Aircraft Tires."

Tires which are worn to 80 percent or more of their total tread depth.

Tires with one or more flat spots severe enough to cause an out-of-balance condition, regardless of the percentage of wear. Tires having so many tread cuts that repairing the tread rubber would be uneconomical.

Nonrecappable Tires

Tires having injuries which would make them nonrepairable.

Tires with six full plies or more having any spot worn through more than one body ply. (It is generally not considered economical to retread 4 and 6 ply aircraft tires.)

Tires with weather checking or ozone cracking of tread or sidewall that exposes the cords.

Repairable Aircraft Tires

When considering a tire for repairing only, the amount of service remaining in the tire is important. Any tire with at least 30 percent of tread life remaining, normally would be considered as having enough service left in it to warrant repair only.

Nonrepairable Aircraft Tires

The following conditions disqualify a tire for repair:

Any injury to the beads, or in the bead area (except injuries limited to the bead cover or finishing strip as previously mentioned under repairable aircraft tires).

A. Sidewall cords damaged beyond repair.

B. Sidewall cords OK for retread.

C. Worn through breaker only. OK for retread.

FIGURE 9-62. Operational damage

Any tire with protruding bead wire or badly kinked bead.

Any tire which shows evidence of ply or tread separation.

Any tire with loose, damaged, or broken cords on the inside.

Tires with broken or cut cords in the outside of the sidewall or shoulder area.

Tires that have gone flat, or partially flat, due to the melting or failure of fuze plugs in the wheels should normally be scrapped, even though there may be no visible evidence of damage to the inside or outside of the tire. The only exceptions would be where it was known that the fuze plug leaked air because of its being defective.

Spot Repairs

When considered economical, spot repairs can be made to take care of tread injuries such as cuts, snags, etc., which are through no more than 25 percent of the actual body plies of the tire, and not over 2″ in length at the surface. Vulcanized spot repairs are also made at times to fill in tread gouges that do not go deeper than the tread rubber and do not penetrate the cord body.

Low speed tires (under 160 mph) with tread injuries that penetrate no more than 25 percent of the actual body (breaker strips not included), and have a maximum surface length of 2″, may be repaired. If the injury penetrates beyond 25 percent of the actual body plies, it may still be repairable but the surface length of the injury should be no more than 1″.

High speed tires (over 160 mph) with tread injuries that penetrate no more than 40 percent of the actual body plies (breaker strips and fabric reinforcement strips in the tread not included) and have a maximum surface length of 1½″ with maximum width no more than ¼″, may be repaired.

Injuries through the cord body in the tread area, measuring ⅛″ or less at the largest point, are considered punctures and are easily repaired.

Shallow cuts in the sidewall or shoulder rubber only are repairable if cords are exposed but not damaged.

Tires having minor injuries through the finishing strip, or slight injuries caused by tire tools in the general bead area, are repairable if the injury does not extend into the plies of the tire, and provided there is no sign of separation in the bead area. If the finishing strips are loose or blistered, they can only be replaced by bead-to-bead retreading.

Liner blisters smaller than 4″ x 8″ may be repaired if there are no more than two in any one quarter section of the tire, and no more than five in the complete tire. Normally, however, it would be more economical to do this repair at the time the tire is recapped.

It is generally considered uneconomical to repair 4- and 6-ply aircraft tires.

OPERATING AND HANDLING TIPS

Taxiing

Needless tire damage or excessive wear can be prevented by proper handling of the aircraft during taxiing.

Most of the gross weight of any aircraft is on the main landing gear wheels—on two, four, eight or more tires. The tires are designed and inflated to absorb shock of landing and will deflect (bulge at the sidewall) about two and one-half times more than a passenger car or truck tire. The greater deflection causes more working of the tread, produces a scuffing action along the outer edges of the tread and results in more rapid wear.

Also, if an aircraft tire strikes a chuck hole, a stone, or some foreign objects lying on the runway, taxi strip, or ramp, there is more possibility of its being cut, snagged or bruised because of the percentage of deflection. Or, one of the main landing gear wheels, when making a turn, may drop off the edge of the paved surface causing severe sidewall or shoulder damage. The same type of damage may also occur when the wheel rolls back over the edge of the paved surface.

With dual main landing gear wheels, one tire might be forced to take a damaging impact (which two could withstand without damage) simply because all the weight on one side of the plane is concentrated on one tire instead of being divided between two.

As airports grow in size and taxi runs become longer, chances for tire damage and wear increase. Taxi runs should be no longer than absolutely necessary and should be made at speeds no greater than 25 mph, particularly for aircraft not equipped with nose wheel steering.

For less damage in taxiing, all personnel should see that ramps, parking areas, taxi strips, runways and other paved areas are regularly cleaned and cleared of all objects that might cause tire damage.

Braking and Pivoting

Increasing airport traffic, longer taxi runs and longer runs on takeoff and landing are subjecting tires to more abrasion resulting from braking, turning and pivoting.

Severe use of brakes can wear flat spots on tires and cause them to be out of balance, making premature recapping or replacement necessary.

Severe or prolonged application of the brakes can be avoided when ground speed is reduced.

Careful pivoting of aircraft also helps to prolong tire tread life. If an aircraft turned as an automobile or truck does—in a rather wide radius—the wear on the tire tread would be materially reduced. However, when an aircraft is turned by locking one wheel (or wheels), the tire on the locked wheel is twisted with great force against the pavement. A small piece of rock or stone that would ordinarily cause no damage, can, in such a case, be literally screwed into the tire. This scuffing or grinding action takes off tread rubber and places a very severe strain on the sidewalls and beads of the tire at the same time.

To keep this action at a minimum, it is recommended that whenever a turn is made, the inside wheel (wheels) be allowed to roll on a radius of 20 to 25 feet and up to 40 feet for aircraft with bogies.

Takeoffs and Landings

Aircraft tire assemblies are always under severe strain on takeoff or landing. But under normal conditions, with proper control and maintenance of tires, they are able to withstand many such stresses without damage.

Tire damage on takeoff, up to the point of being airborne, is generally the result of running over some foreign object. Flat spots or cuts incurred in pivoting can also be a cause of damage during takeoff or landing.

Tire damage at the time of landing can be traced to errors in judgment or unforeseen circumstances. Smooth landings result in longer tread wear and eliminate much of the excessive strain on tires at the moment of impact.

Landings with brakes locked, while almost a thing of the past, can result in flat spotting. Removal of the tire for recapping or replacement is almost invariably indicated. Brakes-on landings also cause very severe heat at the point of contact on the tire tread and may even melt the tread rubber (skid burn). Heat has a tendency to weaken the cord body and places severe strain on the beads. In addition, heat buildup in the brakes may literally devulcanize the tire in the bead area. Under these circumstances, blowouts are not uncommon because air under compression must expand when heated.

On aircraft equipped with tail wheels, a two point landing is usually somewhat smoother than a three point landing, but is ordinarily made at considerably higher speed. As a result, more braking may be required to bring the aircraft to a stop. If tires are skidded on the runway at high speed, the action is similar to tires being ground against a fast turning emery wheel.

Sometimes an aircraft will be brought in so fast that full advantage will not be taken of runway length and brakes are applied so severely that flat spots are produced on the tires. Or, if brakes are applied when the plane is still traveling at high speed and still has considerable lift, tires may skid on the runway and become damaged beyond further use or reconditioning.

The same thing may occur during a rough landing if brakes are applied after the first bounce. For maximum tire service, delay brake application until the plane is definitely settled into its final roll.

More tires fail on takeoff than on landing and such failures on takeoff can be extremely dangerous. For that reason, emphasis must be placed on proper preflight inspection of tires and wheels.

Condition of Landing Field

Regardless of the preventive maintenance and extreme care taken by the pilot and ground crew in handling the aircraft, tire damage is almost sure to result if runways, taxi strips, ramps and other paved field areas are in a bad condition or poorly maintained.

Chuck holes, pavement cracks or step-offs from these areas to the ground, all can cause tire damage. In cold climates, especially during the winter, all pavement breaks should be repaired immediately.

Another hazardous condition often overlooked is accumulated loose material on paved areas and hangar floors. Stones and other foreign materials should also be swept off all the paved areas. In addition, tools, bolts, rivets and other repair materials are sometimes left lying on the aircraft and when the aircraft is moved, these materials drop off. These objects picked up by the tires of another aircraft can cause punctures, cuts, or complete failure of the tire, tube and even the wheel. With jet aircraft, it is even more important that foreign material be kept off areas used by aircraft.

Hydroplaning

This is a condition whereby on wet runways, a wave of water can build up in front of spinning tires and when over-run, tires will no longer make contact with the runway. This results in the complete loss of steering capability and braking action. Hydroplaning can also be caused by a thin film of water on the runway mixing with the contaminants present. (See figure 9–63.)

Cross cutting of runways has been completed at some of the major airports and reportedly has greatly reduced the danger of hydroplaning. However, the ridges of concrete created by this cross-cutting can cause a chevron type of cutting of tread ribs, particularly with the higher pressure tires used on jet aircraft. These cuts are at right angles to the ribs and rarely penetrate to the fabric tread reinforcing strip. Such damage would not be considered cause for removal unless fabric was exposed due to a piece of tread rib being torn out.

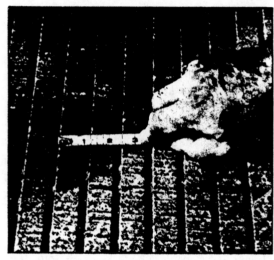

FIGURE 9–63. Cross cutting runways reduces danger of hydroplaning.

TUBE REPAIR

Most tube repairs are necessary because of valves being broken off or otherwise damaged. However, occasionally, a tube might be cut, punctured, or damaged by the tire tools during mounting or de-mounting. Injuries larger than one inch can be repaired by using a reinforcement patch inside the tube. This reinforcement would be the same material that is used for repairing the tube on the outside. An injury smaller than one inch should not need a reinforcement piece.

Chaffing damage caused by the toe of the tire bead, or thinning out of the tube from brake heat should be cause for immediate scrapping of the tube.

There are three general types of valves used in aircraft inner tubes:

1. The rubber valve, which has a rubber stem and a rubber base cured to the outside surface of the inner tube. This valve is similar to those used on inner tubes in passenger car service.

Replacement of this valve can be made by most any gasoline service station or garage, providing they have the proper valve for replacement.

2. Metal valve with rubber base. These are easily recognized since the rubber base is similar to the one used on the all-rubber valve just described. Generally, replacement is made by the same method but it is absolutely necessary that the replacement valve have the same dimensions as the original.

3. Metal valve with a fabric-reinforced rubber base. The base may be cured on top of the tube, or it may be cured into the tube. The metal valve may have to be bent to the proper angle, or angles. Repairing of this type of tube is more difficult because it is necessary to replace the valve pad. Experienced personnel and special equipment are necessary to effect the proper cure of the replacement valve pad to the inner tube.

Repair type valves are also available for valves mentioned in 2 and 3 above. These are applied by cutting off the original valve and screwing on a replacement repair valve to the spud of the original valve. Follow instructions provided by the manufacturer of these repair type valves.

SIDEWALL-INFLATED AIRCRAFT TIRES

Some tires for small aircraft are manufactured with a valve in the sidewall, thus eliminating the need for machining the wheel to take a conventional valve (see figure 9–64).

FIGURE 9–64. Sidewall-inflated tire.

Inflation as well as checking air pressure, is accomplished by inserting a needle through the rubber sidewall valve, similar to the way footballs and other sporting goods equipment are inflated. Care should be given these needles. If damaged, they may injure the valve, resulting in air loss, particularly when the tire is under load.

Replacing these valves is easy. The only equipment needed is a knife or scissors to cut off the old valve inside the tire, and a piece of string for inserting the replacement valve. It is even possible to replace these valves without completely removing the tire from the wheel.

TIRE INSPECTION SUMMARY

Tires in service should be inspected regularly for excessive wear or other conditions which may render the tire unsafe. This will reduce tire costs noticeably and may prevent a serious accident. Figure 9-65 shows the most common types of tire wear and damage.

A

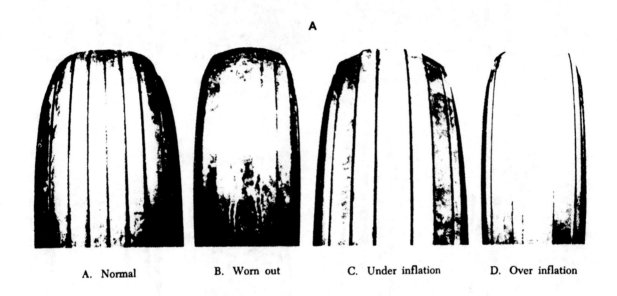

A. Normal   B. Worn out   C. Under inflation   D. Over inflation

B

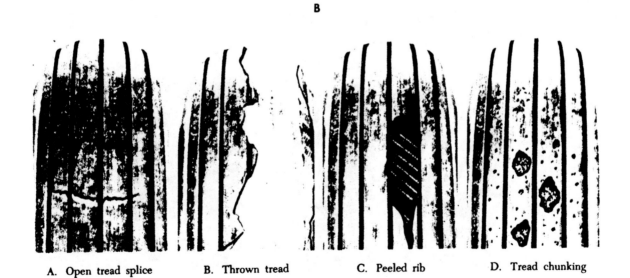

A. Open tread splice   B. Thrown tread   C. Peeled rib   D. Tread chunking

FIGURE 9-65. Common types of tire damage: A, Tread wear; B, Tread; C, Sidewall;
D, Carcass; and E, Bead.

A. Cut   B. Blister and tread   C. Groove cracking   D. Flaking and chipping
                   separation   and rib undercutting

A. Skid   B. Tread rubber reversion   C. Chevron cutting   D. Fabric fraying

C

A. Circumferential   B. Radial cracks   C. Weather checking.
    cracks

Figure 65—Cont.

D

A. Impact break   B. Liner breakdown   C. Contamination

E

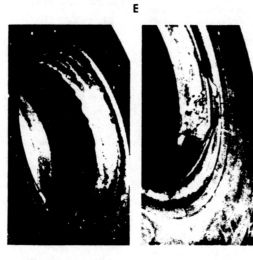

A. Brake heat damage   B. Chaff damage

FIGURE 65—Cont.

ANTISKID SYSTEM

The purpose of a wheel brake is to bring a rapidly moving aircraft to a stop during ground roll. It does this by changing the energy of movement into heat energy through the friction developed in the brakes. A feature found in high performance aircraft braking systems is skid control or antiskid protection. This is an important system because if a wheel goes into a skid, its braking value is greatly reduced.

The skid control system performs four functions: (1) normal skid control, (2) locked wheel skid control, (3) touchdown protection, and (4) fail-safe protection. The main components of the system consist of two skid control generators, a skid control box, two skid control valves, a skid control switch, a warning lamp, and an electrical control harness with a connection to the squat switch.

Normal Skid Control

Normal skid control comes into play when wheel rotation slows down but has not come to a stop. When this slowing down happens, the wheel sliding action has just begun but has not yet reached a full scale slide. In this situation the skid control valve removes some of the hydraulic pressure to the wheel. This permits the wheel to rotate a little faster and stop its sliding. The more intense the skid is, the more braking pressure is removed. The skid detection and control of each wheel is completely independent of the others. The wheel skid intensity is measured by the amount of wheel slow-down.

Skid Control Generator

The skid control generator is the unit that measures the wheel rotational speed. It also senses any changes in the speed. It is a small electrical generator, one for each wheel, mounted in the wheel axle. The generator armature is coupled to, and driven by, the main wheel through the drive cap in the wheel. As it rotates, the generator develops a voltage and current signal. The signal strength indicates the wheel rotational speed. This signal is fed to the skid control box through the harness.

Skid Control Box

The box reads the signal from the generator and senses change in signal strength. It can interpret these as developing skids, locked wheels, brake applications, and brake releases. It analyses all it reads, then sends appropriate signals to solenoids in the skid control valves.

Skid Control Valves

The two skid control valves mounted on the brake control valve are solenoid operated. Electric signals from the skid control box actuate the solenoids. If there is no signal (because there is no wheel skidding), the skid control valve will have no effect on brake operation. But, if a skid develops, either slight or serious, a signal is sent to the skid control valve solenoid. This solenoids' action lowers the metered pressure in the line between the metering valve and the brake cylinders. It does so by dumping fluid into the reservoir return line whenever the solenoid is energized. Naturally, this immediately relaxes the brake application. The pressure flow into the brake lines from the metering valves continues as long as the pilot depresses the brake pedals. But the flow and pressure is rerouted to the reservoir instead of to the wheel brakes.

The utility system pressure enters the brake control valve where it is metered to the wheel brakes in proportion to the force applied on the pilot's foot pedal. However, before it can go to the brakes, it must pass through a skid control valve. There, if the solenoid is actuated, a port is opened in the line between the brake control valve and the brake. This port vents the brake application pressure to the utility system return line. This reduces the brake application, and the wheel rotates faster again. The system is designed to apply enough force to operate just below the skid point. This gives the most effective braking.

Pilot Control

The pilot can turn off the operation of the anti-skid system by a switch in the cockpit. A warning lamp lights when the system is turned off or if there is a system failure.

Locked Wheel Skid Control

The locked wheel skid control causes the brake to be fully released when its wheel locks. A locked wheel easily occurs on a patch of ice due to lack of tire friction with the surface. It will occur if the normal skid control does not prevent the wheel from reaching a full skid. To relieve a locked wheel skid, the pressure is bled off longer than in normal skid function. This is to give the wheel time to regain speed. The locked wheel skid control is out of action during aircraft speeds of less than 15–20 mph.

Touchdown Protection

The touchdown protection circuit prevents the brakes from being applied during the landing approach even if the brake pedals are depressed. This prevents the wheels from being locked when they contact the runway. The wheels have a chance to begin rotating before they carry the full weight of the aircraft. Two conditions must exist before the skid control valves permit brake application. Without them the skid control box will not send the proper signal to the valve solenoids. The first is that the squat switch must signal that the weight of the aircraft is on the wheels. The second is that the wheel generators sense a wheel speed of over 15–20 mph.

Fail-Safe Protection

The fail-safe protection circuit monitors operation of the skid control system. It automatically returns the brake system to full manual in case of system failure. It also turns on a warning light.

LANDING GEAR SYSTEM MAINTENANCE

Because of the stresses and pressures acting on the landing gear, inspection, servicing, and other maintenance becomes a continuous process. The most important job in the maintenance of the aircraft landing gear system is thorough, accurate inspections. To properly perform the inspections, all surfaces should be cleaned to ensure that no trouble spots go undetected.

Periodically, it will be necessary to inspect shock struts, shimmy dampers, wheels, wheel bearings, tires, and brakes. During this inspection, check for the presence of installed ground safety locks. Check landing gear position indicators, lights, and warning horns for operation. Check emergency control handles and systems for proper position and condition. Inspect landing gear wheels for cleanliness, corrosion, and cracks. Check wheel tie bolts for looseness. Examine anti-skid wiring for deterioration. Check tires for wear, cuts, deterioration, presence of grease or oil, alignment of slippage marks, and proper inflation. Inspect the landing gear mechanism for condition, operation, and proper adjustment. Lubricate the landing gear, including the nose wheel steering. Check steering system cables for wear, broken strands, alignment, and safetying. Inspect landing gear shock struts for such conditions as cracks, corrosion, breaks, and security. Where applicable, check the brake clearances.

Various types of lubricants are required to lubricate points of friction and wear on the landing gear. These lubricants are applied by hand, an oil can, or a pressure-type grease gun. Before using the pressure-type grease gun, wipe the lubrication fittings clean of old grease and dust accumulations,

because dust and sand mixed with a lubricant produce a very destructive abrasive compound. As each fitting is lubricated, the excess lubricant on the fitting and any that is squeezed out of the assembly should be wiped off. Wipe the piston rods of all exposed actuating cylinders; clean them frequently, particularly prior to operation, to prevent damage to seals and polished surfaces.

Periodically, wheel bearings must be removed, cleaned, inspected, and lubricated. When cleaning a wheel bearing, use a suitable cleaning solvent. (*Leaded gasoline should not be used.*) Dry the bearing by directing a blast of dry air between the rollers. Do not direct the air so that it will spin the bearing, as it may fly apart and injure nearby persons. When inspecting the bearing, check for defects that would render it unserviceable, such as flaked, cracked, or broken bearing surfaces; roughness due to impact pressure or surface wear; corrosion or pitting of the bearing surfaces; discoloration from excessive heat; cracked or broken cages;

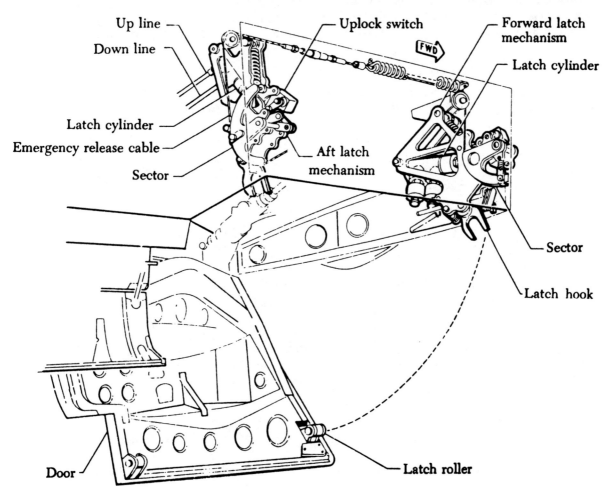

FIGURE 9-66. Main gear door latch mechanisms.

401

or scored or loose bearing cups or cones which would affect proper seating on the axle or wheel. If any of these defects exist replace the bearing with a servicable one. To prevent rust or corrosion, lubricate the bearing immediately after cleaning and inspecting it.

To apply lubricant to a tapered roller bearing, place a small amount of the proper lubricant on the palm of the hand. Grasp the cone of the bearing assembly with the thumb and first two fingers of the other hand, keeping the larger diameter of the bearing next to the palm. Move the bearing assembly across the hand toward the thumb, forcing the lubricant into the space between the cone and rollers. Turn the assembly after each stroke until all openings between the rollers are filled with lubricant. Remove the excess lubricant from the cone and the outside of the cage.

Landing Gear Rigging and Adjustment

Occasionally it becomes necessary to adjust the landing gear switches, doors, linkages, latches, and locks to assure proper operation of the landing gears and doors. When landing gear actuating cylinders are replaced and when length adjustments are made, overtravel must be checked. Overtravel is that action of the cylinder piston beyond the movement necessary for landing gear extension and retraction. The additional action operates the landing gear latch mechanism.

Because of the wide variety of aircraft types and designs, procedures for rigging and adjusting landing gear will vary. Uplock and downlock clearances, linkage adjustments, limit switch adjustments, and other landing gear adjustments vary widely with landing gear design. For this reason, always consult the applicable manufacturers maintenance or service manual before performing any phase of landing gear rigging or installation.

Adjusting Landing Gear Latches

The adjustment of latches are of prime concern to the airframe mechanic. A latch is used in landing gear systems to hold a unit in a certain position after the unit has traveled through a part of, or all of, its cycle. For example, on some aircraft, when the landing gear is retracted, each gear is held in the up position by a latch. The same holds true

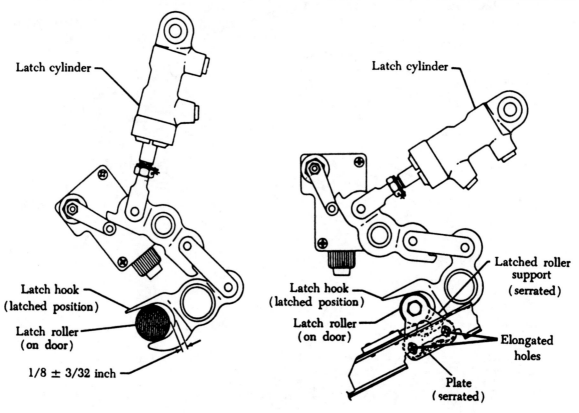

(A) Latch roller clearance

(B) Latch roller support adjustment

FIGURE 9-67. Landing gear door latch installations.

402

when the landing gear is extended. Latches are also used to hold the landing gear doors in the open and closed positions.

There are many variations in latch design. However, all latches are designed to accomplish the same thing. They must operate automatically, at the proper time, and hold the unit in the desired position. A typical landing gear door latch is described in the following paragraphs.

On this particular aircraft, the landing gear door is held closed by two door latches. As shown in figure 9-66, one is installed near the rear of the door. To have the door locked securely, both locks must grip and hold the door tightly against the aircraft structure. The principal components of each latch mechanism shown in figure 9-66 are a hydraulic latch cylinder, a latch hook, a spring loaded crank-and-lever linkage, and a sector. The latch cylinder is hydraulically connected with the landing gear control system and mechanically connected, through linkage, with the latch hook. When hydraulic pressure is applied, the cylinder operates the linkage to engage (or disengage) the hook with (or from) the latch roller on the door. In the gear-down sequence the hook is disengaged by the spring load on the linkage. In the gear-up sequence, spring action is reversed when the closing door is in contact with the latch hook and the cylinder operates the linkage to engage the hook with the latch roller.

Cables on the landing gear emergency extension system are connected to the sector to permit emergency release of the latch rollers. An uplock switch is installed on, and actuated by, each latch to provide a gear-up indication in the cockpit.

With the gear up and the door latched, inspect the latch roller for proper clearance as shown in figure 9-67, view A. On this installation the required clearance is 1/8 ±3/32 in. If the roller is not within tolerance it may be adjusted by loosening its mounting bolts and raising or lowering the latch roller support. This may be done due to the elongated holes and serrated locking surfaces of the latch roller support and serrated plate (view B).

Landing Gear Door Clearances

Landing gear doors have specific allowable clearances which must be maintained between doors and the aircraft structure or other landing gear doors. These required clearances can be maintained by adjusting the door hinges and connecting links and

trimming excess material from the door if necessary.

On some installations, door hinges are adjusted by placing the serrated hinge and serrated washers in the proper position and torquing the mounting bolts. Figure 9-68 illustrates this type of mounting, which allows linear adjustments. The amount of linear adjustment is controlled by the length of the elongated bolthole in the door hinge.

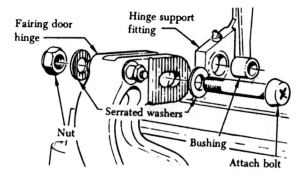

FIGURE 9-68.   Adjustable door hinge installation.

The distance the landing gear doors open or close depends upon the length of door linkage and the adjustment of the door stops. The manufacturer's maintenance manuals specify the length of door linkages and adjustment of stops or other procedures whereby correct adjustments may be made.

Landing Gear Drag and Side Brace Adjustment

The landing gear side brace illustrated in figure 9-69 consists of an upper and lower link, hinged at the center to permit the brace to jackknife during retraction of the landing gear. The upper end pivots on a trunnion attached to the wheel well overhead. The lower end is connected to the shock strut.

On the side brace illustrated, a locking link is incorporated between the upper end of the shock strut and the lower drag link. Usually in this type installation, the locking mechanism is adjusted so that it is positioned slightly overcenter. This provides positive locking of the side brace and the locking mechanism, and as an added safety feature, prevents inadvertent gear collapse caused by the side brace folding.

To adjust the overcenter position of the side brace locking link illustrated in figure 9-69, place the landing gear in the down position and adjust the lock link end fitting so that the side brace lock link is held firmly overcenter. Manually break the

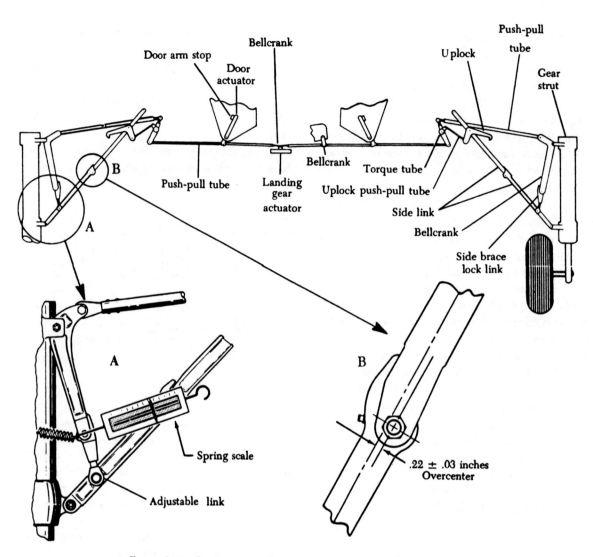

FIGURE 9–69. Landing gear schematic showing overcenter adjustments.

lock link and move the landing gear to a position 5 or 6 in. inboard from the down locked position, then release the gear. The landing gear must free-fall and lock down when released from this position.

In addition to adjusting overcenter travel, the down lock spring tension must be checked using a spring scale. The tension should be between 40 to 60 lbs. for the lock link illustrated. The specific tension and procedure for checking will vary on other aircraft.

Landing Gear Retraction Check

There are several occasions when a retraction check should be performed. First, a retraction check should be performed during an annual inspection of the landing gear system. Second, when performing maintenance that might affect the landing gear link-

age or adjustment, such as changing an actuator, make a retraction check to see whether everything is connected and adjusted properly. Third, it may be necessary to make a retraction check after a hard or overweight landing has been made which may have damaged the landing gear. Closely inspect the gear for obvious damage and then make the retraction check. And finally, one method of locating malfunctions in the landing gear system is to perform a gear retraction check.

There are a number of specific inspections to perform when making a retraction check of landing gear. Included are:

(1) Landing gears for proper retraction and extension.
(2) Switches, lights, and warning horn or buzzer for proper operation.

404

(3) Landing gear doors for clearance and freedom from binding.

(4) Landing gear linkage for proper operation, adjustment, and general condition.

(5) Latches and locks for proper operation and adjustment.

(6) Alternate extension or retraction systems for proper operation.

(7) Any unusual sounds such as those caused by rubbing, binding, chafing, or vibration.

The procedures and information presented herein were intended to provide familiarization with some of the details involved in landing gear rigging, adjustments, and retraction checks and do not have general application. For exact information regarding a specific aircraft landing gear system, consult the applicable manufacturer's instructions.

GENERAL

Because fire is one of the most dangerous threats to an aircraft, the potential fire zones of modern multi-engine aircraft are protected by a fixed fire protection system. A "fire zone" is an area or region of an aircraft designed by the manufacturer to require fire detection and/or fire extinguishing equipment and a high degree of inherent fire resistance. The term "fixed" describes a permanently installed system in contrast to any type of portable fire extinguishing equipment, such as a hand-held CO_2 fire extinguisher.

A complete fire protection system on modern aircraft and on many older model aircraft includes both a fire 'detection and a fire extinguishing system.

To detect fires or overheat conditions, detectors are placed in the various zones to be monitored. Fires are detected in reciprocating engine aircraft, using one or more of the following:

(1) Overheat detectors.

(2) Rate-of-temperature-rise detectors.

(3) Flame detectors.

(4) Observation by crewmembers.

In addition to these methods, other types of detectors are used in aircraft fire protection systems, but are seldom used to detect engine fires; for example, smoke detectors are better suited to monitor areas such as baggage compartments, where materials burn slowly, or smolder. Other types of detectors in this category include carbon monoxide detectors and chemical sampling equipment capable of detecting combustible mixtures that can lead to accumulations of explosive gases.

Detection Methods

The following list of detection methods includes those most commonly used in turbine engine aircraft fire protection systems. The complete aircraft fire protection system of most large turbine engine aircraft will incorporate several of these different detection methods.

(1) Rate-of-temperature-rise detectors.

(2) Radiation sensing detectors.

(3) Smoke detectors.

(4) Overheat detectors.

(5) Carbon monoxide detectors.

(6) Combustible mixture detectors.

(7) Fiber-optic detectors.

(8) Observation of crew or passengers.

The three types of detectors most commonly used for fast detection of fires are the rate-of-rise, radiation sensing, and overheat detectors.

Detection System Requirements

Fire protection systems on modern aircraft do not rely on observation by crewmembers as a primary method of fire detection. An ideal fire detection system will include as many as possible of the following features:

(1) A system which will not cause false warnings under any flight or ground operating conditions.

(2) Rapid indication of a fire and accurate location of the fire.

(3) Accurate indication that a fire is out.

(4) Indication that a fire has re-ignited.

(5) Continuous indication for duration of a fire.

(6) Means for electrically testing the detector system from the aircraft cockpit.

(7) Detectors which resist exposure to oil, water, vibration, extreme temperatures, maintenance handling.

(8) Detectors which are light in weight and easily adaptable to any mounting position.

(9) Detector circuitry which operates directly from the aircraft power system without inverters.

(10) Minimum electrical current requirements when not indicating a fire.

(11) Each detection system should actuate a cockpit light indicating the location of the fire and an audible alarm system.

(12) A separate detection system for each engine.

There are a number of detectors or sensing devices available. Many older model aircraft still operating are equipped with some type of thermal switch system or thermocouple system.

FIRE DETECTION SYSTEMS

A fire detection system should signal the presence of a fire. Units of the system are installed in locations where there are greater possibilities of a fire. Three detector systems in common use are the thermal switch system, thermocouple system, and the continuous-loop detector system.

Thermal Switch System

A thermal switch system consists of one or more lights energized by the aircraft power system and thermal switches that control operation of the light(s). These thermal switches are heat-sensitive units that complete electrical circuits at a certain temperature. They are connected in parallel with each other but in series with the indicator lights (figure 10–1). If the temperature rises above a set value in any one section of the circuit, the thermal switch will close, completing the light circuit to indicate the presence of a fire or overheat condition.

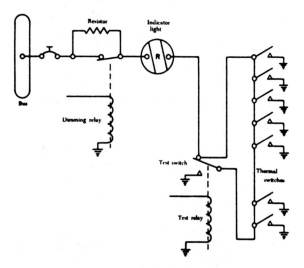

FIGURE 10–1. Thermal switch fire circuit.

No set number of thermal switches is required; the exact number is usually determined by the aircraft manufacturer. On some installations several thermal detectors are connected to one light; on others there may be only one thermal switch for an indicator light.

Some warning lights are the "push-to-test" type. The bulb is tested by pushing it in to complete an auxiliary test circuit. The circuit in figure 10–1 includes a test relay. With the relay contact in the position shown, there are two possible paths for current flow from the switches to the light. This is an additional safety feature. Energizing the test relay completes a series circuit and checks all the wiring and the light bulb.

Also included in the circuit shown in figure 10–1 is a dimming relay. By energizing the dimming relay, the circuit is altered to include a resistor in series with the light. In some installations several circuits are wired through the dimming relay, and all the warning lights may be dimmed at the same time.

The thermal switch system uses a bimetallic thermostat switch or spot detector similar to that shown in figure 10–2. Each detector unit consists of a bimetallic thermoswitch. Most spot detectors are dual-terminal thermoswitches.

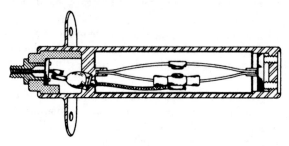

FIGURE 10–2. Fenwal spot detector.

Fenwal Spot Detector

Fenwal spot detectors are wired in parallel between two complete loops of wiring, as illustrated in figure 10–3. Thus, the system can withstand one fault, either an electrical open circuit or a short to ground, without sounding a false fire warning. A double fault must exist before a false fire warning can occur. In case of a fire or overheat condition, the spot-detector switch closes and completes a circuit to sound an alarm.

The Fenwal spot-detector system operates without a control unit. When an overheat condition or a fire causes the switch in a detector to close, the alarm bell sounds and a warning light for the affected area is lighted.

Thermocouple Systems

The thermocouple fire warning system operates on an entirely different principle than the thermal switch system. A thermocouple depends upon the rate of temperature rise and will not give a warning when an engine slowly overheats or a short circuit

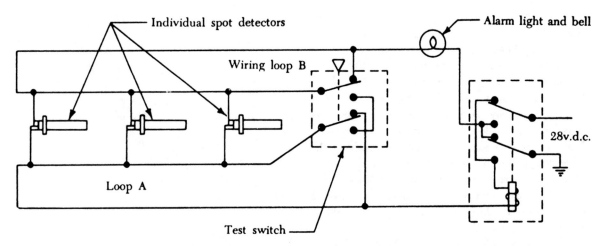

Figure 10-3. Fenwal spot-detector circuit.

develops. The system consists of a relay box, warning lights, and thermocouples. The wiring system of these units may be divided into the following circuits (figure 10-4): (1) The detector circuit, (2) the alarm circuit, and (3) the test circuit.

The relay box contains two relays, the sensitive relay and the slave relay, and the thermal test unit. Such a box may contain from one to eight identical circuits, depending on the number of potential fire zones. The relays control the warning lights. In turn, the thermocouples control the operation of the relays. The circuit consists of several thermocouples in series with each other and with the sensitive relay.

The thermocouple is constructed of two dissimilar metals such as chromel and constantan. The point where these metals are joined and will be exposed

to the heat of a fire is called a hot junction. There is also a reference junction enclosed in a dead air space between two insulation blocks. A metal cage surrounds the thermocouple to give mechanical protection without hindering the free movement of air to the hot junction.

If the temperature rises rapidly, the thermocouple produces a voltage because of the temperature difference between the reference junction and the hot junction. If both junctions are heated at the same rate, no voltage will result and no warning signal is given.

If there is a fire, however, the hot junction will heat more rapidly than the reference junction. The ensuing voltage causes a current to flow within the detector circuit. Any time the current is greater than 4 milliamperes (0.004 ampere), the sensitive relay will close. This will complete a circuit from

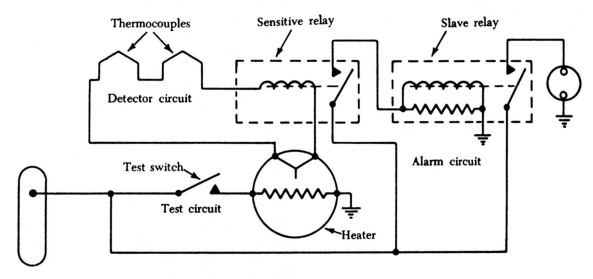

Figure 10-4. Thermocouple fire warning circuit.

the aircraft power system to the coil of the slave relay which closes and completes the circuit to the fire-warning light.

The total number of thermocouples used in individual detector circuits depends on the size of the fire zone and the total circuit resistance. The total resistance usually does not exceed 5 ohms. As shown in figure 10–4, the circuit has two resistors. The resistor connected across the terminals of the slave relay absorbs the coil's self-induced voltage. This is to prevent arcing across the points of the sensitive relay, since the contacts of the sensitive relay are so fragile they would burn or weld if arcing were permitted.

When the sensitive relay opens, the circuit to the slave relay is interrupted and the magnetic field around its coil collapses. When this happens, the coil gets a voltage through self-induction, but with the resistor across the coil terminals, there is a path for any current flow as a result of this voltage. Thus, arcing at the sensitive relay contacts is eliminated.

Continuous-Loop Detector Systems

A continuous-loop detector or sensing system permits more complete coverage of a fire hazard area than any type of spot-type temperature detectors. Continuous-loop systems are versions of the thermal switch system. They are overheat systems, heat-sensitive units that complete electrical circuits at a certain temperature. There is no rate-of-heat-rise sensitivity in a continuous-loop system. Two widely used types of continuous-loop systems are the Kidde and the Fenwal systems.

In the Kidde continuous-loop system (figure 10–5), two wires are imbedded in a special ceramic core within an Inconel tube.

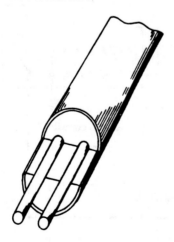

FIGURE 10–5. Kidde sensing element.

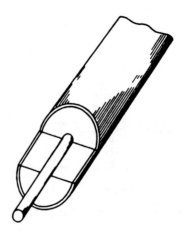

FIGURE 10–6. Fenwal sensing element.

One of the two wires in the Kidde sensing system is welded to the case at each end and acts as an internal ground. The second wire is a hot lead (above ground potential) that provides a current signal when the ceramic core material changes its resistance with a change in temperature.

Another continuous-loop system, the Fenwal system (figure 10–6), uses a single wire surrounded by a continuous string of ceramic beads in an Inconel tube.

The beads in the Fenwal detector are wetted with a eutectic salt which possesses the characteristic of suddenly lowering its electrical resistance as the sensing element reaches its alarm temperature. In both the Kidde and the Fenwal systems, the resistance of the ceramic or eutectic salt core material prevents electrical current from flowing at normal temperatures. In case of a fire or overheat condition, the core resistance drops and current flows between the signal wire and ground, energizing the alarm system.

The Kidde sensing elements are connected to a relay control unit. This unit constantly measures the total resistance of the full sensing loop. The system senses the average temperature, as well as any single hot spot.

The Fenwal system uses a magnetic amplifier control unit. This system is non-averaging but will sound an alarm when any portion of its sensing element reaches the alarm temperature.

Both systems continuously monitor temperatures in the affected compartments, and both will automatically reset following a fire or overheat alarm after the overheat condition is removed or the fire extinguished.

410

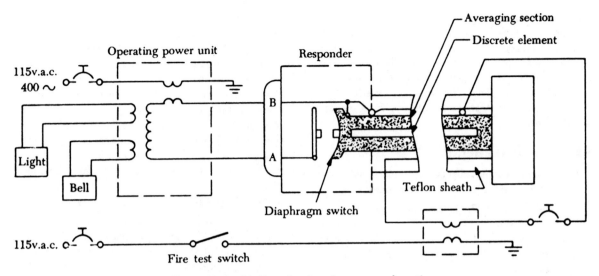

FIGURE 10–7. Lindberg fire detection system schematic.

Continuous Element System

The Lindberg fire detection system (figure 10–7) is a continuous-element type detector consisting of a stainless steel tube containing a discrete element. This element has been processed to absorb gas in proportion to the operating temperature set point. When the temperature rises (due to a fire or overheat condition) to the operating temperature set point, the heat generated causes the gas to be released from the element. Release of the gas causes the pressure in the stainless steel tube to increase. This pressure rise mechanically actuates the diaphragm switch in the responder unit, activating the warning lights and an alarm bell.

A fire test switch is used to heat the sensors, expanding the trapped gas. The pressure generated closes the diaphragm switch, activating the warning system.

Overheat Warning Systems

Overheat warning systems are used on some aircraft to indicate high area temperatures that may lead to a fire. The number of overheat warning systems varies with the aircraft. On some aircraft, they are provided for each engine turbine and each nacelle, on others they are provided for wheel well areas and for the pneumatic manifold.

When an overheat condition occurs in the detector area, the system causes a light on the fire control panels to flash.

In most systems the detector is a type of thermal switch. Each detector is operated when the heat rises to a specified temperature. This temperature depends upon the system and the type and model of the aircraft. The switch contacts of the detector are on spring struts, which close whenever the meter case is expanded by heat. One contact of each detector is grounded through the detector mounting bracket. The other contacts of all detectors connect in parallel to the closed loop of the warning light circuit. Thus, the closed contacts of any one detector can cause the warning lights to burn.

When the detector contacts close, a ground is provided for the warning light circuit. Current then flows from an electrical bus through the warning lights and a flasher or keyer to ground. Because of the flasher in the circuit, the lights flash on and off to indicate an overheat condition.

TYPES OF FIRES

The National Fire Protection Association has classified fires in three basic types:

a. Class A fires, defined as fires in ordinary combustible materials such as wood, cloth, paper, upholstery materials, etc.

b. Class B fires, defined as fires in flammable petroleum products or other flammable or combustible liquids, greases, solvents, paints, etc.

c. Class C fires, defined as fires involving energized electrical equipment where the electrical non-conductivity of the extinguishing media is of importance. In most cases where electrical equipment is de-energized, extinguishers suitable for use on Class A or B fires may be employed effectively.

Aircraft fires, in flight or on the ground, may encompass either or all of these type fires. Therefore, detection systems, extinguishing systems and extinguishing agents as applied to each type fire must be considered. Each type fire has characteristics that require special handling. Agents usable on Class A fires are not acceptable on Class B or C fires. Agents effective on Class B or C fires will have some effect on Class A fires but are not the most efficient.

FIRE ZONE CLASSIFICATION

Powerplant compartments are classified into zones based on the air flow through them.

a. *Class A Zone.* Zones having large quantities of air flowing past regular arrangements of similarly shaped obstructions. The power section of a reciprocating engine is usually of this type.

b. *Class B Zone.* Zones having large quantities of air flowing past aerodynamically clean obstructions. Heat exchanger ducts and exhaust manifold shrouds are usually of this type. Also, zones where the inside of the enclosing cowling or other closure is smooth, free of pockets, and adequately drained so leaking flammables cannot puddle are of this type. Turbine engine compartments may be considered in this class if engine surfaces are aerodynamically clean and all airframe structural formers are covered by a fireproof liner to produce an aerodynamically clean enclosure surface.

c. *Class C Zone.* Zones having relatively small air flow. An engine accessory compartment separated from the power section is an example of this type of zone.

d. *Class D Zone.* Zones having very little or no air flow. These include wing compartments and wheel wells where little ventilation is provided.

e. *Class X Zone.* Zones having large quantities of air flowing through them and are of unusual construction making uniform distribution of the extinguishing agent very difficult. Zones containing deeply recessed spaces and pockets between large structural formers are of this type. Tests indicate agent requirements to be double those for Class A zones.

EXTINGUISHING AGENT CHARACTERISTICS

Aircraft fire extinguishing agents have some common characteristics which make them compatible to aircraft fire extinguishing systems. All agents may be stored for long time periods without adversely affecting the system components or agent quality. Agents in current use will not freeze at normally expected atmospheric temperatures. The nature of the devices inside a powerplant compartment require agents that are not only useful against flammable fluid fires but also effective on electrically caused fires. The various agents' characteristics are narratively described and then summarized in tabular form in Figures 10–8, 10–9, and 10–10. Agents are classified into two general categories based on the mechanics of extinguishing action: the halogenated hydrocarbon agents and the inert cold gas agents.

a. **The Halogenated Hydrocarbon Agents.**

(1) The most effective agents are the compounds formed by replacement of one or more of the hydrogen atoms in the simple hydrocarbons methane and ethane by halogen atoms.

(a) The halogens used to form extinguishing compounds are fluorine, chlorine, and bromine. Iodine may be used but is more expensive with no compensating advantage. The extinguishing compounds are made up of the element carbon in all cases, along with different combinations of hydrogen, fluorine, chlorine, and bromine. Completely halogenated agents contain no hydrogen atoms in the compound, are thus more stable in the heat associated with fire, and are considered safer. Incompletely halogenated compounds, those with one or more hydrogen atoms, are classed as fire extinguishing agents but under certain conditions may become flammable.

(b) The probable extinguishing mechanism of halogenated agents is a "chemical interference" in the combustion process between fuel and oxidizer. Experimental evidence indicates that the most likely method of transferring energy in the combustion process is by "molecule fragments" resulting from the chemical reaction of the constituents. If these fragments are blocked from transferring their energy to the unburned fuel molecules the combustion process may be slowed or stopped completely (extinguished). It is believed that the halogenated agents react with the molecular fragments, thus preventing the energy transfer. This may be termed "chemical cooling" or "energy transfer blocking." This extinguishing mechanism is much more effective than oxygen dilution and cooling.

AGENT	ADVANTAGES	DISADVANTAGES
Bromotrifluoromethane CBr_3F "BT" Halon 1301	Excellent extinguishant, about 4x as effective as "CB" Nontoxic at normal temperatures Noncorrosive Compatible with conventional system, excellent with HRD	Moderately high cost Heavier storage containers required
Bromochlorodi-fluoromethane $CBrClF_2$ "BCF" Halon 1211	Very effective extinguishant Lightweight storage containers	Low relative toxicity Requires N_2 pressure for expellant
Bromochloromethane CH_2BrCl "CB" Halon 1011	Very effective extinguishant when used in conventional systems Noncorrosive to steel and brass Lightweight storage containers	Relatively toxic at normal temperatures Very toxic when pyrolized Requires mechanical vaporization during discharge Corrosive to aluminum and magnesium
Methyl bromide CH_3Br "MB" Halon 1001	More effective than CO_2 Lightweight storage containers Readily available Low cost Compatible with conventional and HRD systems	Relatively toxic Rapidly corrodes aluminum, zinc, and magnesium
Carbon tetrachloride CCl_4 "CTC" Halon 104	Liquid at normal temperatures Readily available Low cost	Relatively toxic Severely toxic when pyrolized Corrosive to iron and other metals Requires propellant charge
Dibromodifluoromethane CBr_2F_2 Halon 1202	Very effective extinguishant Noncorrosive to aluminum, steel and brass Lightweight storage containers Conventional or HRD system	Relatively toxic at normal temperatures Very toxic when pyrolized High cost.
Carbon dioxide CO_2	Conventional or HRD system Relatively nontoxic Noncorrosive Readily available Low cost Under normal temperatures it provides its own propellant	Can cause suffocation of personnel from lengthy exposure Requires heavy storage containers Requires N_2 booster in cold climates
Nitrogen N_2	Very effective extinguishant Noncorrosive Basically nontoxic System may provide large quantities of extinguishant N_2 provides greatest O_2 dilution	Can cause suffocation from lengthly exposure Requires dewar to maintain liquid

FIGURE 10-8. Extinguishing Agent Comparison.

(c) Since halogenated agents react with the molecular fragments, new compounds are formed which in some cases present hazards much greater than the agents themselves. Carbon tetrachloride, for example, may form phosgene, used in warfare as a poison gas. However, most agents generate relatively harmless halogen acids. This chemical reaction caused by heat (pyrolysis) makes some of these agents quite toxic in extinguishing use while being essentially nontoxic under normal room conditions. To evaluate the relative toxic hazard for

413

GROUP	DEFINITION	EXAMPLES
1 (Highest)	Gases or vapors which in concentration of the order of 1/2 to 1 per cent by volume for duration of exposure of the order of 5 minutes are lethal or produce serious injury.	Sulfur dioxide
2	Gases or vapors which in concentrations of the order of 1/2 to 1 per cent by volume for durations of exposure of the order of 1/2 hour are lethal or produce serious injury.	Ammonia Methyl bromide
3	Gases or vapors which in concentrations of the order of 2 to 2½ per cent by volume for durations of exposure of the order of 1 hour are lethal or produce serious injury.	Carbon tetrachloride Chloroform
4	Gases or vapors which in concentrations of the order of 2 to 2½ per cent by volume for durations of exposure of the order of 2 hours are lethal or produce serious injury.	Methyl chloride Ethyl bromide
5	Gases or vapors less toxic than Group 4 but more toxic than Group 6.	Methylene chloride Carbon dioxide Ethane, Propane, Butane
6 (Lowest)	Gases or vapors which in concentrations up to at least about 20 per cent by volume for durations of exposure of the order of 2 hours do not produce injury.	Bromotrifluoromethane

FIGURE 10-9. Comparative life hazard of various refrigerants and other vaporizing liquids and gases. (Classified by the Underwriters' Laboratories, Inc.).

each agent, some consideration must be given to the effectiveness of the individual agent. The more effective the agent, the less quantity of agent required and the quicker will be the extinguishment, with less generation of decomposition products.

(d) These agents are classified through a system of "halon numbers" which describes the several chemical compounds making up this family of agents. The first digit represents the number of carbon atoms in the compound molecule; the second digit, the number of fluorine atoms; the third digit, the number of chlorine atoms; the fourth digit, the number of bromine atoms; and the fifth digit, the number of iodine atoms, if any. Terminal zeroes are not expressed. For example, bromotrifluoromethane ($CBrF_3$), is referred to as Halon 1301.

(e) At ordinary room temperatures some agents are liquids that will vaporize readily though not instantaneously, and are referred to as "vaporizing liquid" extinguishing agents. Other agents are gaseous at normal room temperature but may be liquefied by compression and cooling, enabling them to be stored under pressure as liquids; these are called "liquefied gas" extinguishing agents. Both types of agents may be expelled from extinguishing system storage vessels by using nitrogen gas as a propellant.

(2) Characteristics of some halogenated agents follow:

(a) Bromotrifluoromethane, $CBrF_3$, was developed by the research laboratories of of E. I. DuPont de Nemours & Co. in a program sponsored by the U.S. Armed Forces for the development of an aircraft fire extinguishing agent. It is very effective as an extinguishant, is relatively non-toxic and requires no pressurizing agent. This recently developed agent is gaining in usage because of its obvious advantages.

(b) Bromochlorodifluoromethane, $CBrClF_2$, another very effective agent that has been extensively tested by the U.S. Air Force. It has relatively low toxicity but it requires pressurization by nitrogen to expel it from storage at a satisfactory rate for extinguishment.

(c) Chlorobromomethane, CH_2BrCl, was originally developed in Germany in World War II for military aircraft. It is a more effective extinguishing agent than carbon tetrachloride and is somewhat less toxic although it is classed in the same hazard group.

(d) Methyl bromide, CH_3Br, has been used in the extinguishing systems in British-built aircraft engine installations for many years. Its natural vapor is more toxic than carbon tetrachloride and this characteristic hinders its use. Methyl bromide, as an incompletely halogenated compound with three hydrogen atoms per molecule, is a "borderline" material which may be flammable in itself at elevated temperatures. Tests indicate, however, that it is quite effective in its flame quenching power. Under the conditions found in an aircraft engine nacelle, the explosion suppression characteristic is dominant.

(e) Dibromodifluoromethane, CBr_2F_2 is generally considered more effective than methyl bromide and at least twice as effective as carbon tetrachloride as a flame suppressant. However, its relative toxicity limits its use where it may enter inhabited compartments.

(f) Carbon tetrachloride, CCl_4, is described in this manual primarily because of its historical interest and to provide a comparison with the other agents. CCl_4 is seldom used in aircraft extinguishing systems. It was the first generally accepted agent of the halogenated family and has been used commercially during the past 60 years, particularly for electrical hazards. In recent years, however, use of CCl_4 has declined due principally to the development of more effective agents and in part to the growing concern about the toxic nature of the CCl_4 vapors, especially when decomposed by heat.

b. **Inert Cold Gas Agents.**

Both carbon dioxide (CO_2) and nitrogen (N_2) are effective extinguishing agents. Both are readily available in gaseous and liquid forms; their main difference is in the temperatures and pressures required to store them in their compact liquid phase.

(1) Carbon dioxide, CO_2, has been used for many years to extinguish flammable fluid fires and fires involving electrical equipment. It is noncombustible and does not react with most substances. It provides its own pressure for discharge from the storage vessel except in extremely cold climates where a booster charge of nitrogen may be added to "winterize" the system. Normally, CO_2 is a gas, but it is easily liquefied by compression and cooling. After liquefaction, CO_2 will remain in a closed container as both liquid and gas. When CO_2 is then discharged to the atmosphere, most of the liquid expands to gas. Heat absorbed by the gas during vaporization cools the remaining liquid to $-110°$ F. and it becomes a finely divided white solid, dry ice "snow."

CO_2 is about 1½ times as heavy as air which gives it the ability to replace air above burning surfaces and maintain a smothering atmosphere. CO_2 is effective as an extinguishant primarily because it dilutes the air and reduces the oxygen content so that the air will no longer support combustion. Under certain conditions some cooling effect is also realized. CO_2 is considered only mildly toxic, but it can cause unconsciousness and death by suffocation if the victim is allowed to breath CO_2 in fire extinguishing concentrations for 20 to 30 minutes.

CO_2 is not effective as an extinguishant on fires involving chemicals containing their own oxygen supply, such as cellulose nitrate (some aircraft paints). Also fires involving magnesium and titanium (used in aircraft structures and assemblies) cannot be extinguished by CO_2.

(2) Nitrogen, N_2, is an even more effective extinguishing agent. Like CO_2, N_2 is an inert gas of low toxicity. N_2 extinguishes by oxygen dilution and smothering. It is hazardous to humans in the same way as CO_2. But more cooling is provided by N_2 and pound for pound, N_2 provides almost twice the volume of inert gas to the fire as CO_2 resulting in greater dilution of oxygen.

The main disadvantage of N_2 is that it must be stored as a cyrogenic liquid which requires a dewar and associated plumbing to maintain the $-320°$ F. temperature of liquid nitrogen (LN_2). Some large Air Force aircraft already in service use LN_2 in several ways. LN_2 systems are primarily utilized to inert the

AGENT	SYMBOL	CHEMICAL FORMULA	TYPE OF AGENT	HALON NUMBER	UL TOXICITY GROUPING (3)	SPECIFIC GRAVITY OF LIQUID @ 68°F	SPECIFIC WEIGHT lb./in.³	BOILING POINT, °F	FREEZING POINT, °F	HEAT OF VAPORIZATION (BTU/lb.)	APPROX. LETHAL CON ppm (4)	Vapor Pressure, psi @ 70°F	Vapor Pressure, psi @ 160°F
Carbon dioxide	CO₂	CO_2	Gas / Liquid	—	5	1.529 (1)	0.1234 (2)	−110	−110	112.5	658,000 / 658,000	750	
Carbon tetrachloride	CTC	CCl_4	Liquid	104	3	1.60	0.059	170	−8	83.5	28,000 / 300	1.9	12.5
Methyl bromide	MB	CH_3Br	Liquid	1001	2	1.73	0.0625	39	−139	108.2	5,900 / 9,600	27	120
Bromochloromethane	CB BCM, CMB	CH_2BrCl	Liquid	1011	3	1.94	0.069/ 0.070	149	−124	99.8	65,000 / 4,000	2.7	17.0
Bromochlorodifluoromethane	BCF	$CBrClF_2$	Liquefied Gas	1211	5	1.83	0.0663	25	−257	57.6	324,000 / 7,650	35	135
Dibromodifluoromethane		CBr_2F_2	Liquid	1202	4	2.28	0.0822	76	−112	52.4	54,000 / 1,850	13	58
Bromotrifluoromethane (Bromotri)	BT	$CBrF_3$	Liquefied Gas	1301	6	1.57	0.057	−72	−270.4	47.7	800,000 / 20,000	212	550
Nitrogen	N₂	N_2	—	—	5	0.97 (1)	0.078 (2)	−320		85			

(1) Dry gas compared to dry air at same temperature and pressure.
(2) Specific weight, lb/ft³ @ 1 atmosphere pressure and 0°C.
(3) See Figure 10-9 for definitions.
(4) 1st value represents unheated agent.
    2nd value represents agent heated to 1475°F.

FIGURE 10-10. Characteristics of extinguishing agents.

atmosphere in the fuel tank ullage by replacing most of the air with dry gaseous nitrogen, thereby diluting the oxygen content. With the large quantities of LN_2 thus available, N_2 is also being used for aircraft fire control and is feasible as a practical powerplant fire extinguishing agent.

A long-duration LN_2 system discharge can provide greater safety than conventional short-duration system by cooling potential reignition sources and reducing the vaporization rate of any flammable fluids remaining after extinguishment. Liquid nitrogen systems are expected to see commercial usage in the near future.

FIRE EXTINGUISHING SYSTEMS

a. *High-Rate-of-Discharge Systems.* This term, abbreviated HRD, is applied to the highly effective systems most currently in use. Such HRD systems provide high discharge rates through high pressurization, short feed lines, large discharge valves and outlets. The extinguishing agent is usually one of the halogenated hydrocarbons (halons) sometimes boosted by high-pressure dry nitrogen (N_2). Because the agent and pressurizing gas of an HRD system are released into the zone in one second or less, the zone is temporarily pressurized, and interrupts the ventilating air flow. The few, large sized outlets are carefully located to produce high velocity swirl effects for best distribution.

b. *Conventional Systems.* This term is applied to those fire extinguishing installations first used in aircraft. Still used in some older aircraft, the systems are satisfactory for their intended use but are not as efficient as newer designs. Typically these systems utilize the perforated ring and the so-called distributor-nozzle discharge arrangement. One application is that of a perforated ring in the accessory section of a reciprocating engine where the air flow is low and distribution requirements are not severe. The distributor-nozzle arrangements are used in the power section of reciprocating engine installations with nozzles placed behind each cylinder and in other areas necessary to provide adequate distribution. This system usually uses carbon dioxide (CO_2) for extinguishant but may use any other adequate agent.

RECIPROCATING ENGINE CONVENTIONAL CO_2 SYSTEM

CO_2 is one of the earliest types of fire extinguisher systems for transport aircraft and is still used on many older aircraft.

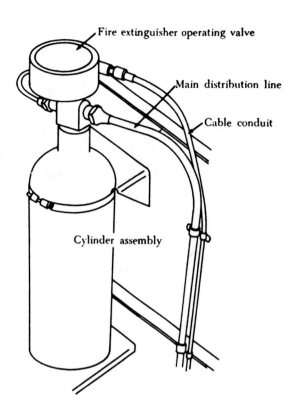

FIGURE 10–11. Carbon dioxide (CO_2) cylinder installation.

This fire extinguisher system is designed around a cylinder (figure 10–11) that stores the flame-smothering CO_2 under pressure and a remote control valve assembly in the cockpit to distribute the extinguishing agent to the engines. The gas is distributed through tubing from the CO_2 cylinder valve to the control valve assembly in the cockpit, and then to the engines via tubing installed in the fuselage and wing tunnels. The tubing terminates in perforated loops which encircle the engines (figure 10–12).

To operate this type of engine fire extinguisher system, the selector valve must be set for the engine which is on fire. An upward pull on the T-shaped control handle located adjacent to the engine selector valve actuates the release lever in the CO_2 cylinder valve. The compressed liquid in the CO_2 cylinder flows in one rapid burst to the outlets in the distribution line (figure 10–12) of the affected engine. Contact with the air converts the liquid into gas and "snow" which smothers the flame.

A more sophisticated type of CO_2 fire protection system is used on many four-engine aircraft. This system is capable of delivering CO_2 twice to any one of the four engines. Fire warning systems are installed at all fire hazardous locations of the aircraft to provide an alarm in case of fire.

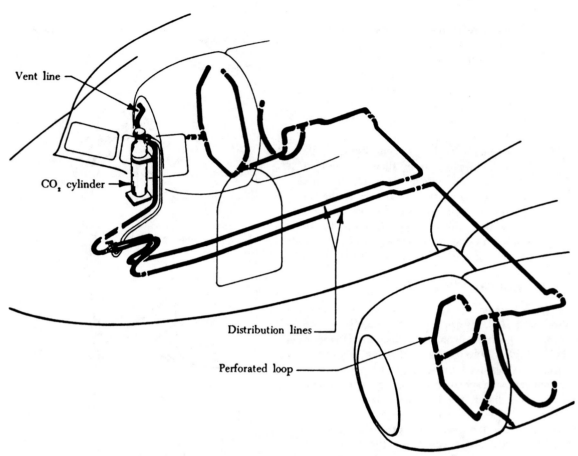

FIGURE 10–12. CO₂ fire extinguisher system on a twin-engine transport aircraft.

The various warning systems operate fire warning lights on the cockpit fire control panels and also energize a cockpit warning bell.

In one type of aircraft the CO_2 system consists of six cylinders, mounted three to a row in each side of the nose wheel well with flood valves installed on each CO_2 bottle. The flood valves of each row are connected with gas interconnect lines. The valves on the two aft bottles in each bank are designed to be opened mechanically by means of a cable connected to the discharge control handles on the main fire control panel in the cockpit. In case of discharge by mechanical means, the forward bottle flood valve in each bank is operated by the released CO_2 pressure from the two aft bottles through the interconnect lines. The flood valve on the forward bottle of each bank contains a solenoid. It is designed to be operated electrically by energizing the solenoid by depressing a button on the control panel. In case of a discharge by electrical means, the valves on the two aft bottles of each bank are operated by the released CO_2 pressure from the forward bottle through the interconnector lines.

Each bank of CO_2 bottles is equipped with a red thermosafety-discharge indicator disk set to rupture at or above a pressure of 2,650 p.s.i. Discharge overboard will occur at temperatures above 74° C. Each bank of bottles is also equipped with a yellow system-discharge indicator disk. Mounted adjacent to the red disk, the yellow disk indicates which bank of bottles has been emptied by a normal discharge.

This type of CO_2 fire protection system includes a fire warning system. It is a continuous-loop, low-impedance, automatic re-setting type for the engine and nacelle areas.

A single fire detector circuit is provided for each engine and nacelle area. Each complete circuit consists of a control unit, sensing elements, a test relay, a fire warning signal light, and a fire warning signal circuit relay. Associated equipment such as flexible connector assemblies, wire, grommets, mounting brackets, and mounting clamps are used in various quantities, depending upon individual installation requirements. On a four-engine aircraft, four warning light assemblies, one for each engine and nacelle area, give corresponding warning indications when an alarm is initiated by a respective engine fire warning circuit. Warning light assemblies in the CO_2 manual release handles are connected into all four engine fire detector circuits, along with a fire warning bell with its guarded cutoff switch and indicating light.

The insulated wire of the detector circuit runs from the control unit in the radio compartment, through the fuselage and wing, to the test relay. The wire is then routed through the nacelle and engine sections and back to the test relay, where it is joined to itself to form a loop.

The control units are normally located on a radio compartment rack. Each unit contains tubes or transistors, transformers, resistors, capacitors, and a potentiometer. It also contains an integrated circuit which introduces a time delay that desensitizes the warning system to short-duration transient signals that would otherwise cause momentary false alarms. When a fire or overheat condition exists in an engine or nacelle area, the resistance of the sensing loop decreases below a preset value determined by the setting of the control unit potentiometer which is in the bias circuit of the control unit detector and amplifier circuit. The output of this circuit is used to energize the fire warning bell and fire warning light.

TURBOJET FIRE PROTECTION SYSTEM

A fire protection system for a large multi-engine turbojet aircraft is described in detail in the following paragraphs. This system is typical of most turbojet transport aircraft and includes components and systems typically encountered on all such aircraft.

The fire protection system of most large turbine engine aircraft consists of two subsystems: (1) A fire detection system and (2) a fire extinguishing system. These two subsystems provide fire protection not only to the engine and nacelle areas but also to such areas as the baggage compartments and wheel wells.

Each turbine engine installed in a pod and pylon configuration contains an automatic heat-sensing fire detection circuit. This circuit consists of a heat-sensing unit, a control unit, a relay, and warning devices. The warning devices normally include a warning light in the cockpit for each circuit and a common alarm bell used with all such circuits.

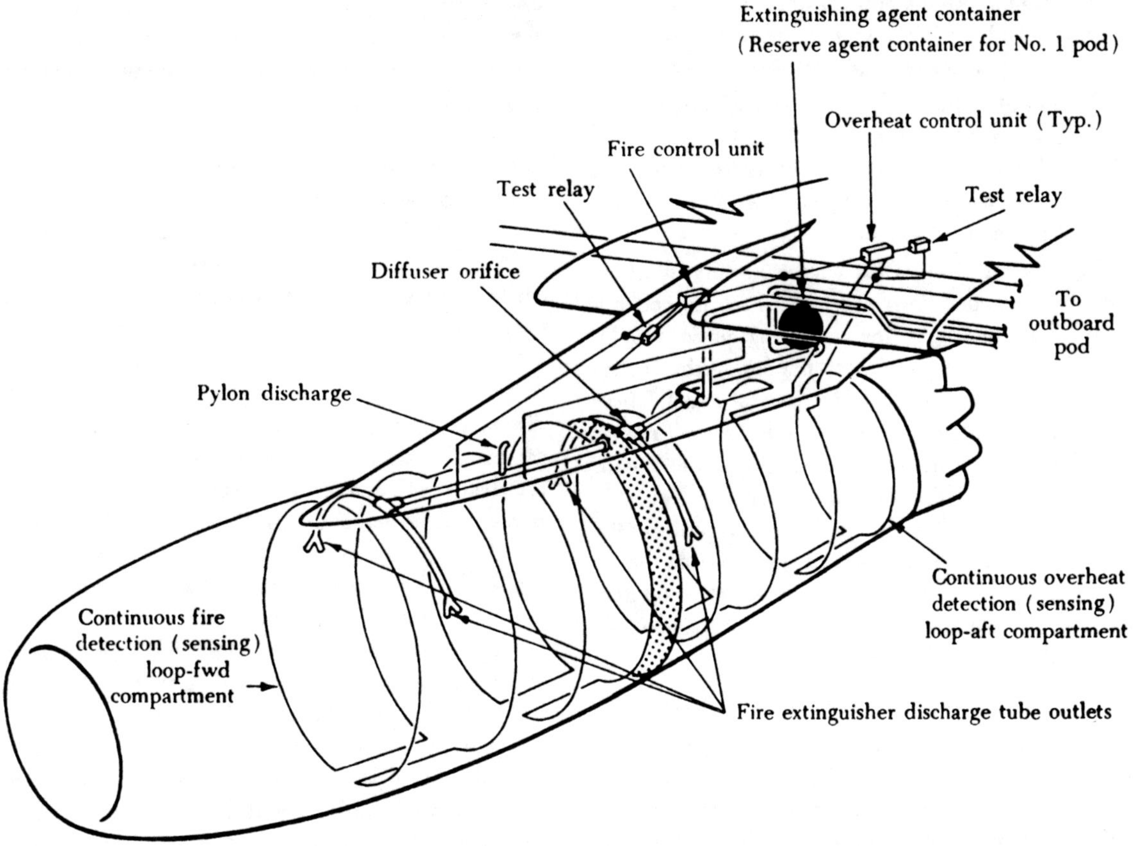

FIGURE 10–13. Typical pod and pylon fire protection installation.

419

The heat-sensing unit of each circuit is a continuous loop routed around the areas to be protected. These areas are the burner and tailpipe areas. Also included in most turbine engine aircraft are the compressor and accessory areas, which in some installations may be protected by a separate fire protection circuit. Figure 10–13 illustrates the typical routing of a continuous-loop fire detection circuit. A typical continuous loop is made up of sensing elements joined to each other by moistureproof connectors, which are attached to the aircraft structure. In most installations, the loop is supported by attachments or clamps every 10 to 12 in. of its length. Too great a distance between supports may permit vibration or chafing of the unsupported section and become a source of false alarms.

In a typical turbine engine fire detection system, a separate control unit is provided for each sensing circuit. The control unit contains an amplifier, usually a transistorized or magnetic amplifier, which produces an output when a predetermined input current flow is detected from the sensing loop. Each control unit also contains a test relay, which is used to simulate a fire or overheat condition to test the circuit. All the control units are mounted in a relay shield or junction box located in a radio compartment or in a special area of the cockpit.

The output of the control unit amplifier is used to energize a warning relay, often called a fire relay. Usually located near the control units, these fire relays, when energized, complete the circuit to appropriate warning devices.

The warning devices for engine and nacelle fires and overheat conditions are located in the cockpit. A fire warning light for each engine is usually located in a special fire switch handle on the instrument panel, light shield, or fire control panel. These fire switches are sometimes referred to as fire-pull T-handles. As illustrated in figure 10–14, the T-handle contains the fire detection warning light. In some models of this fire-pull switch, pulling the T-handle exposes a previously inaccessible extinguishing agent switch and also actuates microswitches which energize the emergency fuel shutoff valve and other pertinent shutoff valves.

TURBINE ENGINE FIRE EXTINGUISHING SYSTEM

The typical fire extinguishing portion of a complete fire protection system includes a cylinder or container of extinguishing agent for each engine and nacelle area. One type of installation provides for a container in each of four pylons on a multi-

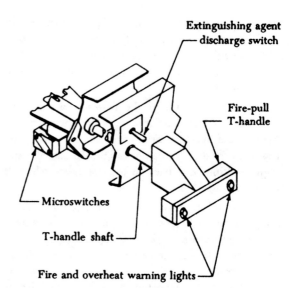

FIGURE 10–14. Fire-pull T-handle switch.

engine aircraft. This type of system uses an extinguishing agent container similar to the type shown in figure 10–15. This type of container is equipped with two discharge valves which are operated by electrically discharged cartridges. These two valves are the main and the reserve controls which release and route the agent to the pod and pylon in which the container is located or to the other engine on the same wing. This type of two-shot, crossfeed configuration permits the release of a second charge of fire extinguishing agent to the same engine if another fire breaks out, without providing two containers for each engine area.

Another type of four-engine installation uses two independent fire extinguisher systems. The two engines on one side of the aircraft are equipped with two fire extinguisher containers, but they are located together in the inboard pylon (figure 10–16). A pressure gage, a discharge plug, and a safety discharge connection are provided for each container. The discharge plug is sealed with a breakable disk combined with an explosive charge which is electrically detonated to discharge the contents of the bottle. The safety discharge connection is capped at the inboard side of the strut with a red indicating disk. If the temperature rises beyond a predetermined safe value, the disk will rupture, dumping the agent overboard, and the discharge will be indicated in the cockpit.

The manifold connecting the two containers of the dual installation (figure 10–16) includes a double check valve and a tee-fitting from which tubing connects to the discharge indicator. This indicator is capped at the inboard side of the strut with a

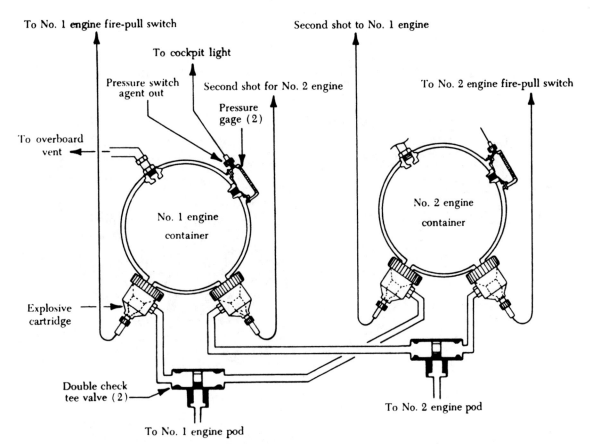

To No. 1 engine fire-pull switch

To cockpit light

Pressure switch agent out

Second shot for No. 2 engine

Pressure gage (2)

Second shot to No. 1 engine

To No. 2 engine fire-pull switch

To overboard vent

No. 1 engine container

No. 2 engine container

Explosive cartridge

Double check tee valve (2)

To No. 2 engine pod

To No. 1 engine pod

FIGURE 10–15. Fire extinguisher system for a multi-engine aircraft.

yellow disk, which is blown out when the manifold is pressurized from either container. The discharge line has two branches (figure 10–16), a short line to the inboard engine and a long one extending along the wing leading edge to the outboard engine. Both of the branches terminate in a tee-fitting near the forward engine mount.

Discharge tube configuration may vary with the type and size of turbine engine installations. In figure 10–17, a semicircular discharge tube with Y-outlet terminations encircles the top forward area of both the forward and aft engine compartments. Diffuser orifices are spaced along the diffuser tubes. A pylon discharge tube is incorporated in the inlet line to discharge the fire extinguishing agent into the pylon area.

Another type of fire extinguisher discharge configuration is shown in figure 10–18. The inlet discharge line terminates in a discharge nozzle, which is a tee-fitting near the forward engine mount. The tee-fitting contains diffuser holes which allow the fire extinguishing agent to be released along the top of the engine and travel downward along both sides of the engine.

When any section of the continuous-loop circuit

is exposed to an overheat condition or fire, the detector warning lights in the cockpit illuminate and the fire warning bell sounds. The warning light may be located in the fire-pull T-handle, or in some installations the fire switch may incorporate the associated fire warning light for a particular engine under a translucent plastic cover, as shown in figure 10–19. In this system, a transfer switch is provided for the left and right fire extinguisher system. Each transfer switch has two positions: "TRANS" and "NORMAL." If a fire occurs in the No. 4 engine, the warning light in the No. 4 fire switch will illuminate; and with the transfer switch in the "NORMAL" position, the No. 4 fire switch is pulled and the No. 4 push button discharge switch located directly under the fire switch will be accessible. Activating the discharge switch will discharge a container of fire extinguishing agent into the No. 4 engine area.

If more than one shot of the agent is required, the transfer switch is placed in the "TRANS" position so that the second container can be discharged into the same engine.

An alarm bell control permits any one of the engine fire detection circuits to energize the

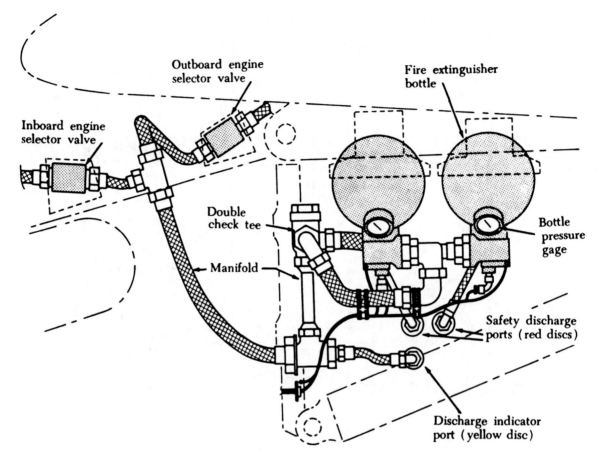

FIGURE 10–16. Dual container installation and fittings.

common alarm bell. After the alarm bell sounds, it can be silenced by activating the bell cutout switch (figure 10–19). The bell can still respond to a fire signal from any of the other circuits.

Most fire protection systems for turbine engine aircraft also include a test switch and circuitry which permit the entire detection system to be tested at one time. The test switch is located in the center of the panel in figure 10–19.

TURBINE ENGINE GROUND FIRE PROTECTION

The problem of ground fires has increased in seriousness with the increased size of turbine engine aircraft. For this reason, means are usually

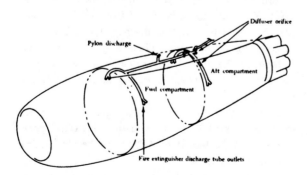

FIGURE 10–17. Fire extinguisher discharge tubes.

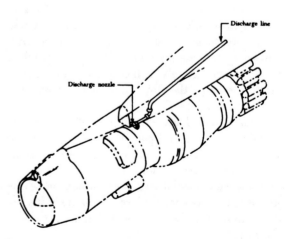

FIGURE 10–18. Fire extinguisher discharge nozzle location.

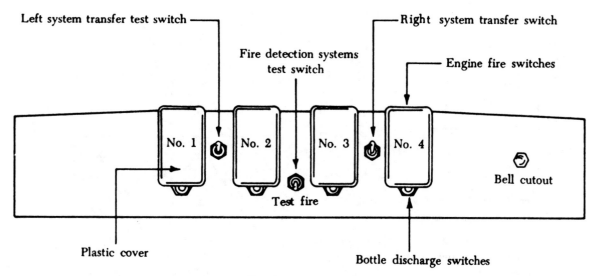

Left system transfer test switch ——

—— Right system transfer switch

Fire detection systems test switch

—— Engine fire switches

No. 1   No. 2   No. 3   No. 4

Bell cutout

Test fire

Plastic cover

Bottle discharge switches

FIGURE 10–19. Fire detection system and fire switches.

provided for rapid access to the compressor, tailpipe, and/or burner compartments. Thus, many aircraft systems are equipped with spring-loaded access doors in the skin of the various compartments. Such doors are usually located in accessible areas, but not in a region where opening a door might spill burning liquids on the fire fighter.

Internal engine tailpipe fires that take place during engine shutdown or false starts can be blown out by motoring the engine with the starter. If the engine is running, it can be accelerated to a higher r.p.m. to achieve the same result. If such a fire persists, a fire extinguishing agent can be directed into the tailpipe. It should be remembered that excessive use of CO_2 or other agents which have a cooling effect can shrink the turbine housing onto the turbine and may damage the engine.

FIRE DETECTION SYSTEM MAINTENANCE PRACTICES

Fire detector sensing elements are located in many high-activity areas around aircraft engines. Their location, together with their small size, increases the chances of damage to the sensing elements during maintenance. The installation of the sensing elements inside the aircraft cowl panels provides some measure of protection not afforded elements attached directly to the engine. On the other hand, the removal and re-installation of cowl panels can easily cause abrasion or structural defects to the elements. A well-rounded inspection and maintenance program for all types of continuous-loop systems should include the following visual checks. These procedures are provided as examples and

should not be used to replace approved local maintenance directives or the applicable manufacturer's instructions.

Sensing elements should be inspected for:

(1) Cracked or broken sections caused by crushing or squeezing between inspection plates, cowl panels, or engine components.

(2) Abrasion caused by rubbing of element on cowling, accessories, or structural members.

(3) Pieces of safety wire or other metal particles which may short the spot detector terminals.

(4) Condition of rubber grommets in mounting clamps, which may be softened from exposure to oils, or hardened from excessive heat.

(5) Dents and kinks in sensing element sections. Limits on the element diameter, acceptable dents or kinks, and degree of smoothness of tubing contour are specified by manufacturers. No attempt should be made to straighten any acceptable dent or kink, since stresses may be set up that could cause tubing failure. (See illustration of kinked tubing in figure 10–20.)

(6) Loose nuts or broken safety wire at the end of the sensing elements (figure 10–21). Loose nuts should be re-torqued to the value specified in the manufacturer's instructions. Some types of sensing element connections require the use of

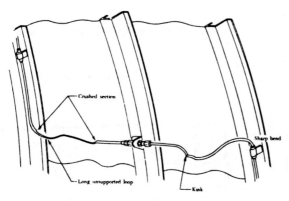

Figure 10–20. Sensing element defects.

copper crush gaskets. These gaskets should be replaced any time a connection is separated.

(7) Broken or frayed flexible leads, if used. The flexible lead is made up of many fine metal strands woven into a protective covering surrounding the inner insulated wire. Continuous bending of the cable or rough treatment can break these fine wires, especially those near the connectors. Broken strands can also protrude into the insulated gasket and short the center electrode.

(8) Proper sensing element routing and clamping (figure 10–22). Long unsupported sections may permit excessive vibration which can cause breakage. The distance between clamps on straight runs is usually about 8 to 10 in., and is specified by each manufacturer. At end connectors, the first support clamp is usually located about 4 to 6 in. from the end

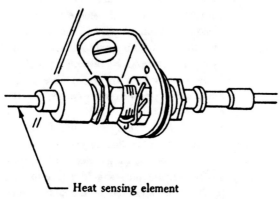

Figure 10–21. Connector joint fitting attached to structure.

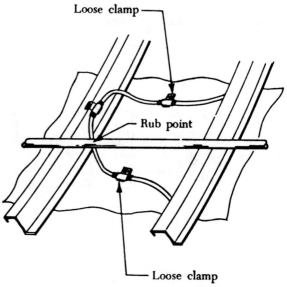

Figure 10–22. Rubbing interference.

connector fittings. In most cases, a straight run of 1 in. is maintained from all connectors before a bend is started, and an optimum bend radius of 3 in. is normally adhered to.

(9) Rubbing between a cowl brace and a sensing element (figure 10–22). This interference, in combination with loose rivets holding the clamps to skin, may cause wear and short the sensing element.

(10) Correct grommet installation. The grommets are installed on the sensing element to prevent the element from chafing on the clamp. The slit end of the grommet should face the outside of the nearest bend. Clamps and grommets (figure 10–23) should fit the element snugly.

(11) Thermocouple detector mounting brackets should be repaired or replaced when cracked, corroded, or damaged. When replacing a thermocouple detector, note which wire is connected to the identified plus terminal of the defective unit and connect the replacement in the same way.

(12) Test the fire detection system for proper operation by turning on the power supply and placing the fire detection test switch in the "TEST" position. The red warning light should flash on within the time period established for the system. On some aircraft an audible alarm will also sound.

In addition, the fire detection circuits are checked for specified resistance and for an open or grounded condition. Tests required after repair of replacement of units in a fire detection system or when the system is inoperative include: (1) Checking the polarity, ground, resistance and continuity of systems that use thermocouple detector units, and (2) resistance and continuity tests performed on systems with sensing elements or cable detector units. In all situations follow the recommended practices and procedures of the manufacturer of the type system with which you are working.

FIRE DETECTION SYSTEM TROUBLESHOOTING

The following troubleshooting procedures represent the most common difficulties encountered in engine fire detection systems.

(1) Intermittent alarms are most often caused by an intermittent short in the detector system wiring. Such shorts may be caused by a loose wire which occasionally touches a nearby terminal, a frayed wire brushing against a structure, or a sensing element rubbing long enough against a structural member to wear through the insulation. Intermittent faults can often best be located by moving wires to recreate the short.

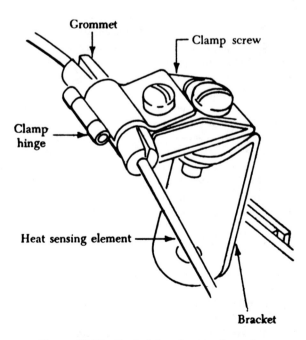

FIGURE 10–23. Typical fire detector loop clamp.

(2) Fire alarms and warning lights can occur when no engine fire or overheat condition exists. Such false alarms can most easily be located by disconnecting the engine sensing loop from the aircraft wiring. If the false alarm continues, a short must exist between the loop connections and the control unit. If, however, the false alarm ceases when the engine sensing loop is disconnected, the fault is in the disconnected sensing loop, which should be examined for areas which have been bent into contact with hot parts of the engine. If no bent element can be found, the shorted section can be located by isolating and disconnecting elements consecutively around the entire loop.

(3) Kinks and sharp bends in the sensing element can cause an internal wire to short intermittently to the outer tubing. The fault can be located by checking the sensing element with a megger while tapping the element in the suspected areas to produce the short.

(4) Moisture in the detection system seldom causes a false fire alarm. If, however, moisture does cause an alarm, the warning will persist until the contamination is removed or boils away and the resistance of the loop returns to its normal value.

(5) Failure to obtain an alarm signal when the test switch is actuated may be caused by a defective test switch or control unit, the lack of electrical power, inoperative indicator light, or an opening in the sensing element or connecting wiring. When the test switch fails to provide an alarm, the continuity of a two-wire sensing loop can be determined by opening the loop and measuring the resistance. In a single-wire, continuous-loop system, the center conductor should be grounded.

FIRE EXTINGUISHER SYSTEM MAINTENANCE PRACTICES

Regular maintenance of fire extinguisher systems typically includes such items as the inspection and servicing of fire extinguisher bottles (containers), removal and re-installation of cartridge and discharge valves, testing of discharge tubing for leakage, and electrical wiring continuity tests. The following paragraphs contain details of some of the most typical maintenance procedures, and are in-

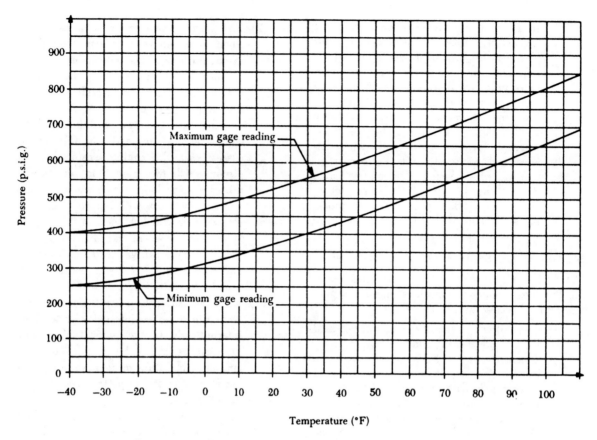

FIGURE 10-24. Fire extinguisher container pressure/temperature curve.

cluded to provide an understanding of the operations involved.

Fire extinguisher system maintenance procedures vary widely according to the design and construction of the particular unit being serviced. The detailed procedures outlined by the airframe or system manufacturer should always be followed when performing maintenance.

Container Pressure Check

A pressure check of fire extinguisher containers is made periodically to determine that the pressure is between the minimum and maximum limits prescribed by the manufacturer. Changes of pressure with ambient pressure must also fall within · prescribed limits. The graph shown in figure 10-24 is typical of the pressure/temperature curve graphs that provide maximum and minimum gage readings. If the pressure does not fall within the graph limits, the extinguisher container should be replaced.

Freon Discharge Cartridges

The service life of fire extinguisher discharge cartridges is calculated from the manufacturer's date stamp, which is usually placed on the face of the cartridge. The manufacturer's service life is usually recommended in terms of hours below a predetermined temperature limit. Many cartridges are available with a service life of approximately 5,000 hours. To determine the unexpired service life of a discharge cartridge, it is necessary to remove the electrical leads and discharge hose from the plug body, which can then be removed from the extinguisher container.

Care must be taken in the replacement of cartridge and discharge valves. Most new extinguisher containers are supplied with their cartridge and discharge valve disassembled. Before installation on the aircraft, the cartridge must be properly assembled into the discharge valve and the valve connected to the container, usually by means of a swivel nut that tightens against a packing ring gasket.

If a cartridge is removed from a discharge valve for any reason, it should not be used in another discharge valve assembly, since the distance the contact point protrudes may vary with each unit. Thus, continuity might not exist if a used plug which had been indented with a long contact point were installed in a discharge valve with a shorter contact point.

When actually performing maintenance, always refer to the applicable maintenance manuals and other related publications pertaining to a particular aircraft.

Freon Containers

Bromochloromethane and freon extinguishing agents are stored in steel spherical containers. There are four sizes in common use today ranging from 224 cu. in. (small) to 945 cu. in. (large). The large containers weigh about 33 lbs. The small spheres have two openings, one for the bonnet assembly (sometimes called an operating head), and the other for the fusible safety plug (figure 10–25). The larger containers are usually equipped with two firing bonnets and a two-way check valve as shown in figure 10–26.

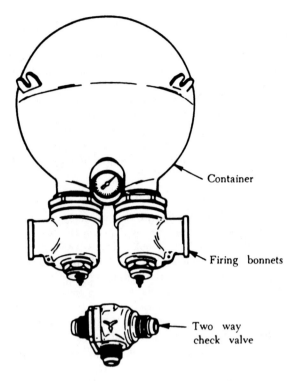

FIGURE 10–26. Typical double bonnet extinguisher assembly.

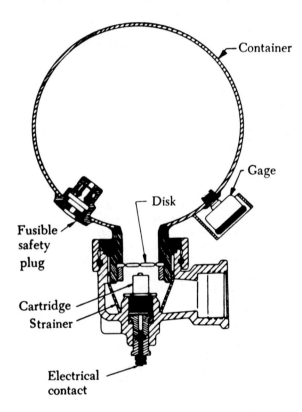

FIGURE 10–25. Single bonnet sphere assembly.

The containers are charged with dry nitrogen in addition to a specified weight of the extinguishing agent. The nitrogen charge provides sufficient pressure for complete discharge of the agent. The bonnet assembly contains an electrically ignited power cartridge which breaks the disk, allowing the extinguishing agent to be forced out of the sphere by the nitrogen charge.

A single bonnet sphere assembly is illustrated in figure 10–25. The function of the parts shown, other than those described in the preceding paragraph, are as follows: (1) The strainer prevents pieces of the broken disk from entering the system, (2) the fusible safety plug melts and releases the liquid when the temperature is between 208° and 220° F., and (3) the gage shows the pressure in the container. In this type of design, there is no need for siphon tubes.

In some installations the safety plug is connected to a discharge indicator mounted in the fuselage skin, while others simply discharge the fluid into the fire extinguisher container storage compartment.

The gage on the container should be checked for an indication of the specified pressure as given in the applicable aircraft maintenance manual. In addition make certain that the indicator glass is unbroken and that the bottle is securely mounted.

Some types of extinguishing agents rapidly corrode aluminum alloy and other metals, especially under humid conditions. When a system that uses a corrosive agent has been discharged, the system must be purged thoroughly with clean, dry, compressed air as soon as possible.

Almost all types of fire extinguisher containers require re-weighing at frequent intervals to determine the state of charge. In addition to the weight check, the containers must be hydrostatically tested, usually at 5-year intervals.

The circuit wiring of all electrically discharged containers should be inspected visually for condition. The continuity of the entire circuit should be checked following the procedures in the applicable maintenance manual. In general this consists of checking the wiring and the cartridge, by using a resistor in the test circuit that limits the circuit current to less than 35 milliamperes to prevent detonating the cartridge.

Carbon Dioxide Cylinders

These cylinders come in various sizes, are made of stainless steel, and are wrapped with steel wire to make them shatterproof. The normal storage pressure of the gas ranges from 700 to 1,000 p.s.i. However, the state of the cylinder charge is determined by the weight of the CO_2. In the container, about two-thirds to three-fourths of the CO_2 is liquefied. When the CO_2 is released, it expands about 500 times as it converts to gas.

The cylinder does not have to be protected against cold weather, for the freezing point of carbon dioxide is minus 110° F. However, it can discharge prematurely in hot climates. To prevent this, manufacturers put in a charge of dry nitrogen, at about 200 p.s.i., before they fill the cylinder with carbon dioxide. When treated in this manner, most CO_2 cylinders are protected against premature discharge up to 160° F. With a temperature increase, the pressure of the nitrogen does not rise as much as that of the CO_2 because of its stability with regard to temperature changes. The nitrogen also provides additional pressure during normal release of the CO_2 at low temperature during cold weather.

Carbon dioxide cylinders are equipped internally with one of three types of siphon tubes, as shown in figures 10–27 and 10–28. Aircraft fire extinguishers have either the straight rigid or the short flexible siphon tube installed. The tube is used to make certain that the CO_2 is transmitted to the discharge nozzle in the liquid state.

Cylinders containing either the straight rigid or short flexible types of siphoning tubes should be mounted as shown in figure 10–28. Notice that the straight rigid siphon tube is allowed a 60° tolerance, while the tolerance for the short flexible tube is only 30°.

CO_2 cylinders are equipped with metal safety disks designed to rupture at 2,200 to 2,800 p.s.i. This disk is attached to the cylinder release valve body by a threaded plug. A line leads from the fitting to a discharge indicator installed in the fuselage skin. Rupture of the red disk means that

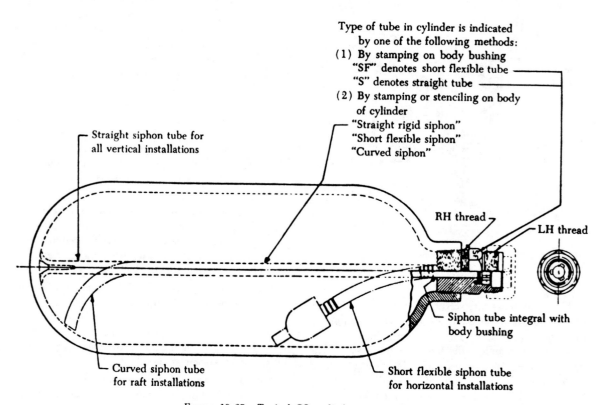

FIGURE 10–27. Typical CO_2 cylinder construction.

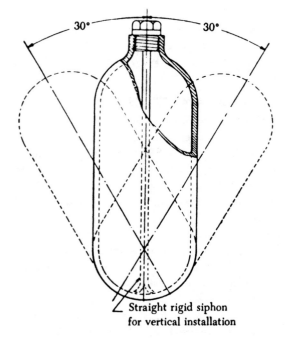

Straight rigid siphon
for vertical installation

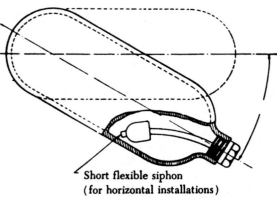

Short flexible siphon
(for horizontal installations)

FIGURE 10-28. Mounting positions of CO₂ cylinders.

the container safety plug has ruptured because of an overheat condition. A yellow disk is also installed in the fuselage skin. Rupture of this disk indicates that the system has been discharged normally.

FIRE PREVENTION AND PROTECTION

Leaking fuel and hydraulic, de-icing, or lubricating fluids, can be sources of fire in an aircraft. This condition should be noted, and corrective action taken, when inspecting aircraft systems. Minute pressure leaks of these fluids are particularly dangerous for they quickly produce an explosive atmospheric condition.

Carefully inspect fuel tank installations for signs of external leaks. With integral fuel tanks the external evidence may occur at some distance from where the fuel is actually escaping.

Many hydraulic fluids are flammable and should not be permitted to accumulate in the structure. Sound-proofing and lagging materials may become highly flammable if soaked with oil of any kind.

Any leakage or spillage of flammable fluid in the vicinity of combustion heaters is a serious fire risk, particularly if any vapor is drawn into the heater and passes over the hot combustion chamber.

Oxygen system equipment must be kept absolutely free from traces of oil or grease, since these substances will spontaneously ignite in contact with oxygen under pressure. Oxygen servicing cylinders should be clearly marked so that they cannot be mistaken for cylinders containing air or nitrogen, as explosions have resulted from this error during maintenance operations.

Fire prevention is much more rewarding than fire extinguishing.

COCKPIT AND CABIN INTERIORS

All wool, cotton, and synthetic fabrics used in interior trim are treated to render them flame resistant. Tests conducted have shown foam and sponge rubber to be highly flammable. However, if they are covered with a flame-resistant fabric which will not support combustion, there is little danger from fire as a result of ignition produced by accidental contact with a lighted cigarette or burning paper.

Fire protection for the aircraft interior is usually provided by hand-held extinguishers. Four types of fire extinguishers are available for extinguishing interior fires: (1) water, (2) carbon dioxide, and (3) dry chemical, and (4) halogenated hydrocarbons.

Extinguisher Types

(1) Water extinguishers are for use primarily on nonelectrical fires such as smoldering fabric, cigarettes, or trash containers. Water extinguishers should not be used on electrical fires because of the danger of electrocution. Turning the handle of a water extinguisher clockwise punctures the seal of a CO₂ cartridge which pressurizes the container. The water spray from the nozzle is controlled by a trigger on top of the handle.

(2) Carbon dioxide fire extinguishers are provided to extinguish electrical fires. A long, hinged tube with a non-metallic megaphone-shaped nozzle permits discharge of the CO₂ gas close to the fire source to smother the fire. A trigger type release is normally lockwired and the lockwire can be broken by a pull on the trigger.

(3) A dry chemical fire extinguisher can be used to extinguish any type of fire. However, the

dry chemical fire extinguisher should not be used in the cockpit due to possible interference with visibility and the collection of nonconductive powder on electrical contacts of surrounding equipment. The extinguisher is equipped with a fixed nozzle which is directed toward the fire source to smother the fire. The trigger is also lockwired but can be broken by a sharp squeeze of the trigger.

(4) The development of halogenated hydrocarbons (freons) as fire extinguishing agents with low toxicity for airborne fire extinguishing protection systems logically directed attention to its use in hand type fire extinguishers.

Bromotrifluoromethane (Halon 1301) having a rating of 6 on the toxicity scale is the logical successor to CO_2 as a hand type fire extinguisher agent. It is effective on fires in lower concentrations. Halon 1301 can extinguish a fire with a concentration of 2% by volume. This compares with about 40% by volume concentration required for CO_2 to extinguish the same fire.

This quality allows Halon 1301 to be used in occupied personnel compartments without depriving people of the oxygen they require. Another advantage is that no residue or deposit remains after use. Halogen 1301 is the ideal agent to use in airborne hand held fire extinguishers because: (1) its low concentration is very effective, (2) it may be used in occupied personnel compartments, (3) it is effective on all 3 type fires, and (4) no residue remaining after its use.

Extinguishers Unsuitable as Cabin or Cockpit Equipment

The common aerosol can type extinguishers are definitely not acceptable as airborne hand type extinguishers. In one instance, an aerosol type foam extinguisher located in the pilot's seat back pocket exploded and tore the upholstery from the seat. The interior of the aircraft was damaged by the foam. This occurred when the aircraft was on the ground and the outside air temperature was 90° F. In addition to the danger from explosion, the size is inadequate to combat even the smallest fire.

A dry chemical extinguisher was mounted near a heater vent on the floor. For an unknown reason, the position of the unit was reversed. This placed the extinguisher directly in front of the heater vent. During flight, with the heater in operation, the extinguisher became overheated and exploded filling the compartment with dry chemical powder. The proximity of heater vents should be considered when selecting a location for a hand fire extinguisher.

Additional information relative to airborne hand fire extinguishers may be obtained from the local FAA District Office and from the National Fire Protection Association, 470 Atlantic Ave., Boston, MA 02210.

SMOKE DETECTION SYSTEMS

A smoke detection system monitors the cargo and baggage compartments for the presence of smoke, which is indicative of a fire condition. Smoke detection instruments, which collect air for sampling, are mounted in the compartments in strategic locations. A smoke detection system is used where the type of fire anticipated is expected to generate a substantial amount of smoke before temperature changes are sufficient to actuate a heat detection system.

Smoke detection instruments are classified by method of detection as follows: Type I – Measurement of carbon monoxide gas (CO detectors), Type II – Measurement of light transmissibility in air (photoelectric devices), and Type III – Visual detection of the presence of smoke by directly viewing air samples (visual devices).

To be reliable, smoke detectors must be maintained so that smoke in a compartment will be indicated as soon as it begins to accumulate. Smoke detector louvers, vents, and ducts must not be obstructed.

Carbon Monoxide Detectors

The CO detectors, which detect concentrations of carbon monoxide gas, are rarely used to monitor cargo and baggage compartments. However, they have gained widespread use in conducting tests for the presence of carbon monoxide gas in aircraft cabins and cockpits.

Carbon monoxide is a colorless, odorless, tasteless, non-irritating gas. It is the byproduct of incomplete combustion, and is found in varying degrees in all smoke and fumes from burning carbonaceous substances. Exceedingly small amounts of the gas are dangerous. A concentration of .02% (2 parts in 10,000) may produce headache, mental dullness, and physical loginess within a few hours.

There are several types of portable testers (sniffers) in use. One type has a replaceable indicator tube which contains a yellow silica gel, impregnated with a complex silico-molybdate compound and is catalyzed using palladium sulfate.

In use, a sample of air is drawn through the detector tube. When the air sample contains carbon monoxide, the yellow silica gel turns to a shade of green. The intensity of the green color is proportional to the concentration of carbon monoxide in the air sample at the time and location of the tests.

Another type indicator may be worn as a badge or installed on the instrument panel or cockpit wall. It is a button using a tablet which changes from a normal tan color to progressively darker shades of gray to black. The transition time required is relative to the concentration of CO. At a concentration of 50 ppm CO (0.005%), the indication will be apparent within 15 to 30 minutes. A concentration of 100 ppm CO (0.01%) will change color of the tablet from tan to gray in 2–5 minutes, from tan to dark gray in 15 to 20 minutes.

Photoelectric Smoke Detectors

This type of detector consists of a photoelectric cell, a beacon lamp, a test lamp, and a light trap, all mounted on a labyrinth. An accumulation of 10% smoke in the air causes the photoelectric cell to conduct electric current. Figure 10–29 shows the details of the smoke detector, and indicates how the smoke particles refract the light to the photoelectric cell. When activated by smoke, the detector supplies a signal to the smoke detector amplifier. The amplifier signal activates a warning light and bell.

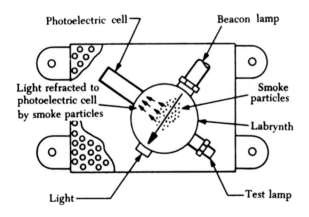

FIGURE 10–29. Photoelectric smoke detector.

A test switch (figure 10–30) permits checking the operation of the smoke detector. Closing the switch connects 28 v. d.c. to the test relay. When the test relay energizes, voltage is applied through the beacon lamp and test lamp in series to ground. A fire indication will be observed only if the beacon and test lamp, the photoelectric cell, the smoke detector amplifier, and associated circuits are operable.

A functional check of the detector should be made after installation, and at frequent intervals thereafter.

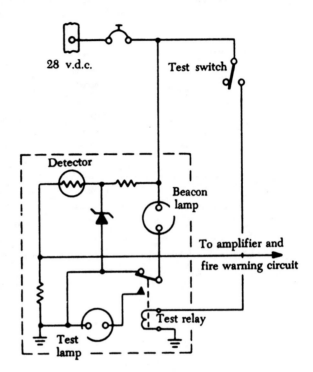

FIGURE 10–30. Smoke detector test circuit.

Visual Smoke Detectors

On a few aircraft visual smoke detectors provide the only means of smoke detection. Indication is provided by drawing smoke through a line into the indicator, using either a suitable suction device or cabin pressurization.

When smoke is present a lamp within the indicator is illuminated automatically by the smoke detector. The light is scattered so that the smoke is rendered visible in the appropriate window of the indicator. If no smoke is present the lamp will not be illuminated. A switch is provided to illuminate the lamp for test purposes. A device is also provided in the indicator to show that the necessary airflow is passing through the indicator.

The efficiency of any detection system depends on the positioning and serviceability of all the components of the system. The foregoing information is intended to provide familiarization with the various systems. For details of a particular installation, refer to the relevant manuals for the aircraft concerned.

The maximum allowable concentration, under Federal Law, for continuing exposure is 50 ppm (parts per million) which is equal to 0.005% of carbon monoxide. (See figure 10–31.)

The maximum allowable concentration under Federal Law for continuing exposure is 50 ppm (parts per million) which is equal to 0.005% of carbon monoxide.

Parts Per Million	Percentage	Reaction
50	0.005%	Maximum allowable concentration under Federal Law.
100	0.01 %	Tiredness, mild dizziness.
200	0.02 %	Headaches, tiredness, dizziness, nausea after 2 or 3 hours.
800	0.08 %	Unconsciousness in 1 hour and death in 2 to 3 hours.
2,000	0.20 %	Death after 1 hour.
3,000	0.30 %	Death within 30 minutes.
10,000	1.00 %	Instantaneous **death**.

FIGURE 10–31. Human reactions to carbon monoxide poisoning.

GENERAL

The satisfactory performance of any modern aircraft depends to a very great degree on the continuing reliability of electrical systems and subsystems. Improperly or carelessly installed wiring or improperly or carelessly maintained wiring can be a source of both immediate and potential danger. The continued proper performance of electrical systems depends on the knowledge and techniques of the mechanic who installs, inspects, and maintains the electrical system wires and cables.

Procedures and practices outlined in this section are general recommendations and are not intended to replace the manufacturer's instructions and approved practices.

For the purpose of this discussion, a wire is described as a single, solid conductor, or as a stranded conductor covered with an insulating material. Figure 11–1 illustrates these two definitions of a wire.

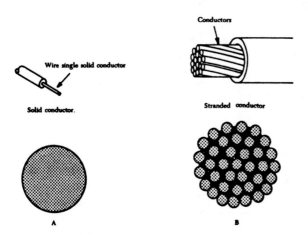

FIGURE 11–1. Two types of aircraft wire.

The term cable, as used in aircraft electrical installations, includes:

(1) Two or more separately insulated conductors in the same jacket (multi-conductor cable).

(2) Two or more separately insulated conductors twisted together (twisted pair).

(3) One or more insulated conductors, covered with a metallic braided shield (shielded cable).

(4) A single insulated center conductor with a metallic braided outer conductor (radio frequency cable). The concentricity of the center conductor and the outer conductor is carefully controlled during manufacture to ensure that they are coaxial.

Wire Size

Wire is manufactured in sizes according to a standard known as the AWG (American wire gage). As shown in figure 11–2, the wire diameters become smaller as the gage numbers become larger. The largest wire size shown in figure 11–2 is number 0000, and the smallest is number 40. Larger and smaller sizes are manufactured but are not commonly used.

A wire gage is shown in figure 11–3. This type of gage will measure wires ranging in size from number zero to number 36. The wire to be measured is inserted in the smallest slot that will just accommodate the bare wire. The gage number corresponding to that slot indicates the wire size. The slot has parallel sides and should not be confused with the semicircular opening at the end of the slot. The opening simply permits the free movement of the wire all the way through the slot.

Gage numbers are useful in comparing the diameter of wires, but not all types of wire or cable can be accurately measured with a gage. Large wires are usually stranded to increase their flexibility. In such cases, the total area can be determined by multiplying the area of one strand (usually computed in circular mils when diameter or gage number is known) by the number of strands in the wire or cable.

Factors Affecting the Selection of Wire Size

Several factors must be considered in selecting the size of wire for transmitting and distributing electric power.

Gage number	Diameter (mils)	Cross section		Ohms per 1,000 ft.	
		Circular mils	Square inches	25°C. (=77°F.)	65°C. (=149°F.)
0000	460.0	212,000.0	0.166	0.0500	0.0577
000	410.0	168,000.0	.132	.0630	.0727
00	365.0	133,000.0	.105	.0795	.0917
0	325.0	106,000.0	.0829	.100	.116
1	289.0	83,700.0	.0657	.126	.146.
2	258.0	66,400.0	.0521	.159	.184
3	229.0	52,600.0	.0413	.201	.232
4	204.0	41,700.0	.0328	.253	.292
5	182.0	33,100.0	.0260	.319	.369
6	162.0	26,300.0	.0206	.403	.465
7	144.0	20,800.0	.0164	.508	.586
8	128.0	16,500.0	.0130	.641	.739
9	114.0	13,100.0	.0103	.808	.932
10	102.0	10,400.0	.00815	1.02	1.18
11	91.0	8,230.0	.00647	1.28	1.48
12	81.0	6,530.0	.00513	1.62	1.87
13	72.0	5,180.0	.00407	2.04	2.36
14	64.0	4,110.0	.00323	2.58	2.97
15	57.0	3,260.0	.00256	3.25	3.75
16	51.0	2,580.0	.00203	4.09	4.73
17	45.0	2,050.0	.00161	5.16	5.96
18	40.0	1,620.0	.00128	6.51	7.51
19	36.0	1,290.0	.00101	8.21	9.48
20	32.0	1,020.0	.000802	10.4	11.9
21	28.5	810.0	.000636	13.1	15.1
22	25.3	642.0	.000505	16.5	19.0
23	22.6	509.0	.000400	20.8	24.0
24	20.1	404.0	.000317	26.2	30.2
25	17.9	320.0	.000252	33.0	38.1
26	15.9	254.0	.000200	41.6	48.0
27	14.2	202.0	.000158	52.5	60.6
28	12.6	160.0	.000126	66.2	76.4
29	11.3	127.0	.0000995	83.4	96.3
30	10.0	101.0	.0000789	105.0	121.0
31	8.9	79.7	.0000626	133.0	153.0
32	8.0	63.2	.0000496	167.0	193.0
33	7.1	50.1	.0000394	211.0	243.0
34	6.3	39.8	.0000312	266.0	307.0
35	5.6	31.5	.0000248	335.0	387.0
36	5.0	25.0	.0000196	423.0	488.0
37	4.5	19.8	.0000156	533.0	616.0
38	4.0	15.7	.0000123	673.0	776.0
39	3.5	12.5	.0000098	848.0	979.0
40	3.1	9.9	.0000078	1,070.0	1,230.0

FIGURE 11–2. American wire gage for standard annealed solid copper wire.

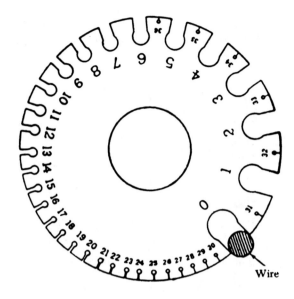

FIGURE 11–3. A wire gage.

One factor is the allowable power loss (I^2R loss) in the line. This loss represents electrical energy converted into heat. The use of large conductors will reduce the resistance and therefore the I^2R loss. However, large conductors are more expensive initially than small ones; they are heavier and require more substantial supports.

A second factor is the permissible voltage drop (IR drop) in the line. If the source maintains a constant voltage at the input to the lines, any variation in the load on the line will cause a variation in line current and a consequent variation in the IR drop in the line. A wide variation in the IR drop in the line causes poor voltage regulation at the load. The obvious remedy is to reduce either current or resistance. A reduction in load current lowers the amount off power being transmitted, whereas a reduction in line resistance increases the size and weight of conductors required. A compromise is generally reached whereby the voltage variation at the load is within tolerable limits and the weight of line conductors is not excessive.

A third factor is the current-carrying ability of the conductor. When current is drawn through the conductor, heat is generated. The temperature of the wire will rise until the heat radiated, or otherwise dissipated, is equal to the heat generated by the passage of current through the line. If the conductor is insulated, the heat generated in the conductor is not so readily removed as it would be if the conductor were not insulated. Thus, to protect the insulation from too much heat, the current

through the conductor must be maintained below a certain value.

When electrical conductors are installed in locations where the ambient temperature is relatively high, the heat generated by external sources constitutes an appreciable part of the total conductor heating. Allowance must be made for the influence of external heating on the allowable conductor current, and each case has its own specific limitations. The maximum allowable operating temperature of insulated conductors varies with the type of conductor insulation being used.

Tables are available that list the safe current ratings for various sizes and types of conductors covered with various types of insulation. Figure 11–5 shows the current-carrying capacity, in amperes, of single copper conductors at an ambient temperature of below 30°C. This example provides measurements for only a limited range of wire sizes.

Factors Affecting Selection of Conductor Material

Although silver is the best conductor, its cost limits its use to special circuits where a substance with high conductivity is needed.

The two most generally used conductors are copper and aluminum. Each has characteristics that make its use advantageous under certain circumstances. Also, each has certain disadvantages.

Copper has a higher conductivity; it is more ductile (can be drawn), has relatively high tensile strength, and can be easily soldered. It is more expensive and heavier than aluminum.

Although aluminum has only about 60% of the conductivity of copper, it is used extensively. Its lightness makes possible long spans, and its relatively large diameter for a given conductivity reduces corona (the discharge of electricity from the wire when it has a high potential). The discharge is greater when small diameter wire is used than when large diameter wire is used. Some bus bars are made of aluminum instead of copper where there is a greater radiating surface for the same conductance. The characteristics of copper and aluminum are compared in figure 11–4.

FIGURE 11–4. Characteristics of copper and aluminum.

Characteristic	Copper	Aluminum
Tensile strength (lb./in.²)	55,000	25,000
Tensile strength for same conductivity (lb.)	55,000	40,000
Weight for same conductivity (lb.)	100	48
Cross section for same conductivity (C. M.)	100	160
Specific resistance (Ω/mil ft.)	10.6	17

Size	Rubber or thermoplastic	Thermoplastic asbestos, var-cam, or asbestos var-cam	Impregnated asbestos	Asbestos	Slow-burning or weather-proof
0000	300	385	475	510	370
000	260	330	410	430	320
00	225	285	355	370	275
0	195	245	305	325	235
1	165	210	265	280	205
2	140	180	225	240	175
3	120	155	195	210	150
4	105	135	170	180	130
6	80	100	125	135	100
8	55	70	90	100	70
10	40	55	70	75	55
12	25	40	50	55	40
14	20	30	40	45	30

FIGURE 11-5. Current-carrying capacity of wire.

Voltage Drop in Aircraft Wire and Cable

It is recommended that the voltage drop in the main power cables from the aircraft generation source or the battery to the bus should not exceed 2% of the regulated voltage when the generator is carrying rated current or the battery is being discharged at a 5-min. rate. The tabulation in figure 11-6 shows the recommended maximum voltage drop in the load circuits between the bus and the utilization equipment.

FIGURE 11-6. Recommended maximum voltage drop in load circuits.

Nominal system voltage	Allowable voltage drop	
	Continuous operation	Intermittent operation
14	0.5	1
28	1	2
115	4	8
200	7	14

The resistance of the current return path through the aircraft structure is always considered negligible. However, this is based on the assumption that adequate bonding of the structure or a special electric current return path has been provided and is capable of carrying the required electric current with a negligible voltage drop. A resistance measurement of 0.005 ohm from the ground point of the generator or battery to the ground terminal of any electrical device is considered satisfactory. Another satisfactory method of determining circuit resistance is to check the voltage drop across the circuit. If the voltage drop does not exceed the limit established by the aircraft or product manufacturer, the resistance value for the circuit is considered satisfactory. When using the voltage drop method of checking a circuit, the input voltage must be maintained at a constant value.

Instructions For Use of Electric Wire Chart

The charts in figures 11-7 and 11-8 apply to copper conductors carrying direct current. Curves 1, 2, and 3 are plotted to show the maximum ampere rating for the specified conductor under the specified conditions shown. To select the correct size of conductor, two major requirements must be met. First, the size must be sufficient to prevent an excessive voltage drop while carrying the required current over the required distance. Secondly, the size must be sufficient to prevent overheating of the cable while carrying the required current. The charts in figures 11-7 and 11-8 can simplify these determinations. To use these charts to select the proper size of conductor, the following must be known:

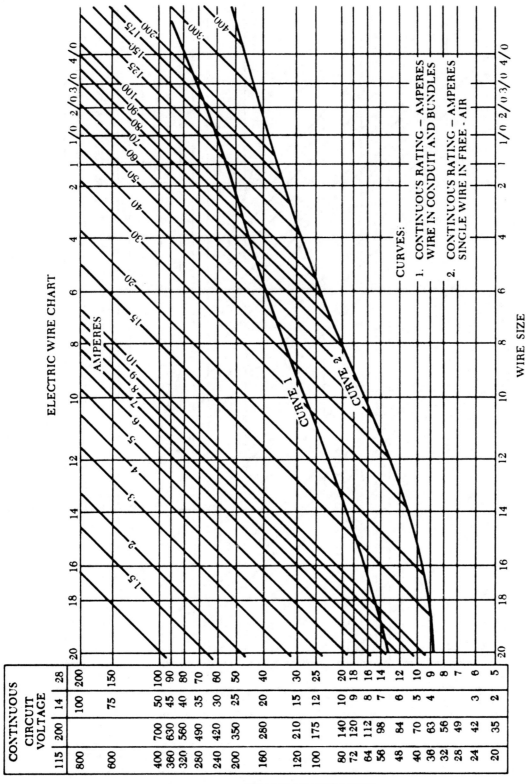

FIGURE 11-7. Conductor chart, continuous flow. (Applicable to copper conductors.)

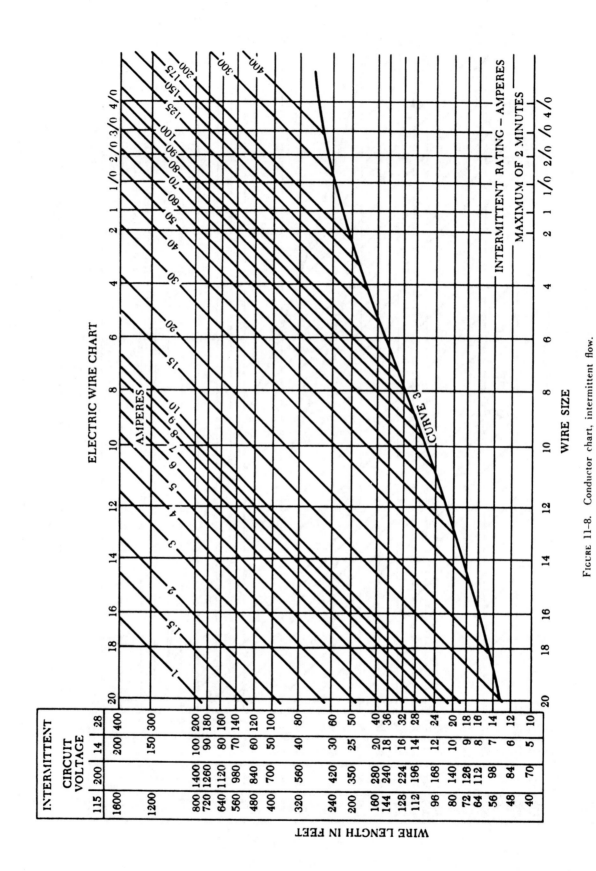

FIGURE 11–8. Conductor chart, intermittent flow.

438

(1) The conductor length in feet.

(2) The number of amperes of current to be carried.

(3) The amount of voltage drop permitted.

(4) Whether the current to be carried will be intermittent or continuous, and if continuous, whether it is a single conductor in free air, in a conduit, or in a bundle.

Assume that it is desired to install a 50-ft. conductor from the aircraft bus to the equipment in a 28-volt system. For this length, a 1-volt drop is permissible for continuous operation. By referring to the chart in figure 11–7, the maximum number of feet a conductor may be run carrying a specified current with a 1–volt drop can be determined. In this example the number 50 is selected.

Assuming the current required by the equipment is 20 amperes, the line indicating the value of 20 amperes should be selected from the diagonal lines. Follow this diagonal line downward until it intersects the horizontal line number 50. From this point, drop straight downward to the bottom of the chart to find that a conductor between size No. 8 and No. 10 is required to prevent a greater drop than 1 volt. Since the indicated value is between two numbers, the larger size, No. 8, should be selected. This is the smallest size conductor which should be used to avoid an excessive voltage drop.

To determine that the conductor size is sufficient to preclude overheating, disregard both the numbers along the left side of the chart and the horizontal lines. Assume that the conductor is to be a single wire in free air carrying continuous current. Place a pointer at the top of the chart on the diagonal line numbered 20 amperes. Follow this line until the pointer intersects the diagonal line marked "curve 2." Drop the pointer straight downward to the bottom of the chart. This point is between numbers 16 and 18. The larger size, No. 16, should be selected. This is the smallest size conductor acceptable for carrying 20-ampere current in a single wire in free air without overheating.

If the installation is for equipment having only an intermittent (max. 2 min.) requirement for power, the chart in figure 11–8 is used in the same manner.

Conductor Insulation

Two fundamental properties of insulation materials (for example, rubber, glass, asbestos, or plastic) are insulation resistance and dielectric strength. These are entirely different and distinct properties.

Insulation resistance is the resistance to current leakage through and over the surface of insulation materials. Insulation resistance can be measured with a megger without damaging the insulation, and data so obtained serves as a useful guide in determining the general condition of the insulation. However, the data obtained in this manner may not give a true picture of the condition of the insulation. Clean, dry insulation having cracks or other faults might show a high value of insulation resistance but would not be suitable for use.

Dielectric strength is the ability of the insulator to withstand potential difference and is usually expressed in terms of the voltage at which the insulation fails because of the electrostatic stress. Maximum dielectric strength values can be measured by raising the voltage of a test sample until the insulation breaks down.

Because of the expense of insulation and its stiffening effect, together with the great variety of physical and electrical conditions under which the conductors are operated, only the necessary minimum insulation is applied for any particular type of cable designed to do a specific job.

The type of conductor insulation material varies with the type of installation. Such types of insulation as rubber, silk, and paper are no longer used extensively in aircraft systems. More common today are such materials as vinyl, cotton, nylon, Teflon, and Rockbestos.

Identifying Wire and Cable

Aircraft electrical system wiring and cable may be marked with a combination of letters and numbers to identify the wire, the circuit it belongs to, the gage number, and other information necessary to relate the wire or cable to a wiring diagram. Such markings are called the identification code.

There is no standard procedure for marking and identifying wiring; each manufacturer normally develops his own identification code. One identification system (figure 11–9) shows the usual spacing in marking a wire. The number 22 in the code refers to the system in which the wire is installed, *e.g.*, the VHF system. The next set of numbers, .013, is the wire number, and the 18 indicates the wire size.

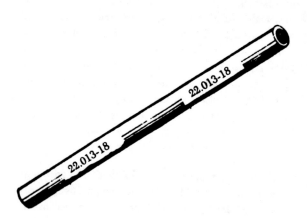

FIGURE 11-9. Wire identification code.

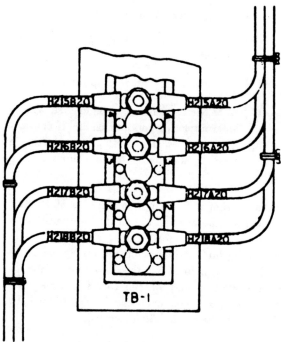

FIGURE 11-10. Wire identification at a terminal block.

Some system components, especially plugs and jacks, are identified by a letter or group of letters and numbers added to the basic identification number. These letters and numbers may indicate the location of the component in the system. Interconnected cables are also marked in some systems to indicate location, proper termination, and use.

In any system, the marking should be legible, and the stamping color should contrast with the color of the wire insulation. For example, black stamping should be used with light-colored backgrounds, or white stamping on dark-colored backgrounds.

Wires are usually marked at intervals of not more than 15 in. lengthwise and within 3 in. of each junction or terminating point. Figure 11-10 shows wire identification at a terminal block.

Coaxial cable and wires at terminal blocks and junction boxes are often identified by marking or stamping a wiring sleeve rather than the wire itself. For general purpose wiring, a flexible vinyl sleeving, either clear or white opaque, is commonly used. For high-temperature applications, silicone rubber or silicone fiber glass sleeving is recommended. Where resistance to synthetic hydraulic fluids or

other solvents is necessary, either clear or white opaque nylon sleeving can be used.

While the preferred method is to stamp the identification marking directly on the wire or on the sleeving, other methods are often employed. Figure 11-11 shows two alternate methods: one method uses

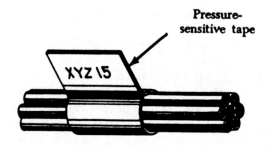

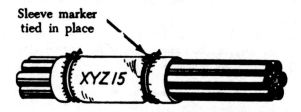

FIGURE 11-11. Alternate methods of identifying wire bundles.

a marked sleeve tied in place; the other uses a pressure-sensitive tape.

Electrical Wiring Installation

The following recommended procedures for installing aircraft electrical wiring are typical of those used on most aircraft. For purposes of this discussion, the following definitions are applicable:

(1) Open wiring—any wire, wire group, or wire bundle not enclosed in conduit.

(2) Wire group—two or more wires going to the same location tied together to retain identity of the group.

(3) Wire bundle—two or more wire groups tied together because they are going in the same direction at the point where the tie is located.

(4) Electrically protected wiring—wires which include (in the circuit) protection against overloading, such as fuses, circuit breakers, or other limiting devices.

(5) Electrically unprotected wiring—wires (generally from generators to main bus distribution points) which do not have protection, such as fuses, circuit breakers, or other current-limiting devices.

Wire Groups and Bundles

Grouping or bundling certain wires, such as electrically unprotected power wiring and wiring going to duplicate vital equipment, should be avoided.

Wire bundles should generally be less than 75 wires, or 1-1/2 to 2 in. in diameter where practicable. When several wires are grouped at junction boxes, terminal blocks, panels, etc., identity of the group within a bundle (figure 11–12) can be retained.

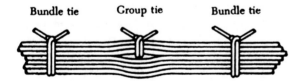

Bundle tie   Group tie   Bundle tie

Figure 11–12. Group and bundle ties.

Twisting Wires

When specified on the engineering drawing, or when accomplished as a local practice, parallel wires must sometimes be twisted. The following are the most common examples:

(1) Wiring in the vicinity of magnetic compass or flux valve.

(2) Three-phase distribution wiring.

(3) Certain other wires (usually radio wiring) as specified on engineering drawings.

Twist the wires so that they will lie snugly against each other, making approximately the number of twists per foot as shown in figure 11–13. Always check wire insulation for damage after twisting. If the insulation is torn or frayed, replace the wire.

Figure 11–13. Recommended number of twists per foot.

	Wire Size									
	#22	#20	#18	#16	#14	#12	#10	#8	#6	#4
2 Wires	10	10	9	8	7½	7	6½	6	5	4
3 Wires	10	10	8½	7	6½	6	5½	5	4	3

Spliced Connections in Wire Bundles

Spliced connections in wire groups or bundles should be located so that they can be easily inspected. Splices should also be staggered (figure 11–14) so that the bundle does not become excessively enlarged. All noninsulated splices should be covered with plastic, securely tied at both ends.

Slack in Wiring Bundles

Single wires or wire bundles should not be installed with excessive slack. Slack between supports should normally not exceed a maximum of 1/2 in. deflection with normal hand force (figure 11–15). However, this may be exceeded if the wire bundle is thin and the clamps are far apart. Slack should never be so great that the wire bundle could abrade against any surface. A sufficient amount of slack should be allowed near each end of a bundle to:

(1) Permit easy maintenance.

(2) Allow replacement of terminals.

(3) Prevent mechanical strain on the wires, wire junctions, and supports.

(4) Permit free movement of shock and vibration-mounted equipment.

(5) Permit shifting of equipment for purposes of maintenance.

Bend Radii

Bends in wire groups or bundles should be not less than 10 times the outside diameter of the wire group or bundle. However, at terminal strips, where wire is suitably supported at each end of the bend, a minimum radius of three times the outside diameter of the wire, or wire bundle, is normally acceptable. There are, of course, exceptions to these guidelines in the case of certain types of cable; for

FIGURE 11-14. Staggered splices in a wire bundle.

example, coaxial cable should never be bent to a smaller radius than ten times the outside diameter.

Routing and Installations

All wiring should be installed so that it is mechanically and electrically sound and neat in appearance. Whenever practicable, wires and bundles should be routed parallel with, or at right angles to, the stringers or ribs of the area involved. An exception to this general rule is coaxial cable, which is routed as directly as possible.

The wiring must be adequately supported throughout its length. A sufficient number of supports must be provided to prevent undue vibration of the unsupported lengths. All wires and wire groups should be routed and installed to protect them from:

(1) Chafing or abrasion.
(2) High temperature.
(3) Being used as handholds, or as support for personal belongings and equipment.
(4) Damage by personnel moving within the aircraft.
(5) Damage from cargo stowage or shifting.
(6) Damage from battery acid fumes, spray, or spillage.
(7) Damage from solvents and fluids.

Protection Against Chafing

Wires and wire groups should be protected against chafing or abrasion in those locations where contact with sharp surfaces or other wires would damage the insulation. Damage to the insulation can cause short circuits, malfunction, or inadvertent operation of equipment. Cable clamps should be used to support wire bundles at each hole through a bulkhead (figure 11-16). If wires come closer than 1/4 in. to the edge of the hole, a suitable grommet is used in the hole as shown in figure 11-17.

Sometimes it is necessary to cut nylon or rubber grommets to facilitate installation. In these instances, after insertion, the grommet can be secured in place with general-purpose cement. The cut should be at the top of the hole, and made at an angle of 45° to the axis of the wire bundle hole.

Protection against High Temperature

To prevent insulation deterioration, wires should be kept separate from high-temperature equipment, such as resistors, exhaust stacks, or heating ducts. The amount of separation is normally specified by engineering drawings. Some wires must invariably be run through hot areas. These wires must be insulated with high-temperature material such as asbestos, fiber glass, or Teflon. Additional protection is also often requred in the form of conduits. A low-temperature insulation wire should never be used to replace a high-temperature insulation wire.

Many coaxial cables have soft plastic insulation, such as polyethylene, which is especially subject to deformation and deterioration at elevated temperatures. All high-temperature areas should be avoided when installing these cables insulated with plastic or polyethylene.

Additional abrasion protection should be given to asbestos wires enclosed in conduit. Either conduit with a high-temperature rubber liner should be used, or asbestos wires can be enclosed individually in high-temperature plastic tubes before being installed in the conduit.

Protection Against Solvents and Fluids

Wires should not be installed in areas where they will be subjected to damage from fluids or in the lowest 4 in. of an aircraft fuselage, except those that must terminate in that area. If there is a possibility that wire may be soaked with fluids, plastic tubing should be used to protect the wire. This tubing should extend past the exposure area in both directions and should be tied at each end. If the wire has a low point between the tubing ends, provide a 1/8-in. drain hole, as shown in figure 11-18. This hole should be punched into the tubing after the installation is complete and the low point definitely established by using a hole punch to cut a half circle. Care should be taken not to damage any wires inside the tubing when using the punch.

Wire should never be routed below an aircraft

442

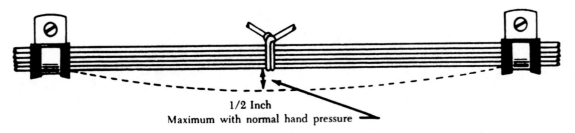

FIGURE 11-15. Slack in wire bundle between supports.

1/2 Inch
Maximum with normal hand pressure →

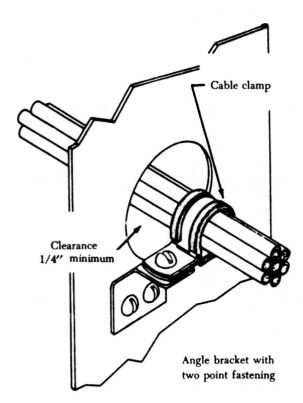

Cable clamp

Clearance
1/4" minimum

Angle bracket with
two point fastening

FIGURE 11-16. Cable clamp at bulkhead hole.

battery. All wires in the vicinity of an aircraft battery should be inspected frequently and wires discolored by battery fumes should be replaced.

Protection of Wires in Wheel Well Area

Wires located in wheel wells are subject to many additional hazards, such as exposure to fluids, pinching, and severe flexing in service. All wire bundles should be protected by sleeves of flexible tubing securely held at each end, and there should be no relative movement at points where flexible tubing is secured. These wires and the insulating tubing should be inspected carefully at frequent intervals, and wires or tubing should be replaced at the first sign of wear. There should be no strain on attachments when parts are fully extended, but slack should not be excessive.

Routing Precautions

When wiring must be routed parallel to combustible fluid or oxygen lines for short distances, as much fixed separation as possible should be maintained. The wires should be on a level with, or above, the plumbing lines. Clamps should be spaced so that if a wire is broken at a clamp it will not contact the line. Where a 6-in. separation is not possible, both the wire bundle and the plumbing line can be clamped to the same structure to prevent any relative motion. If the separation is less than 2 in. but more than 1/2 in., a polyethylene sleeve may be used over the wire bundle to give further protection. Also two cable clamps back-to-back, as shown in figure 11-19, can be used to maintain a rigid separation only, and not for support of the bundle. No wire should be routed so that it is located nearer than 1/2 in. to a plumbing line. Neither should a wire or wire bundle be supported from a

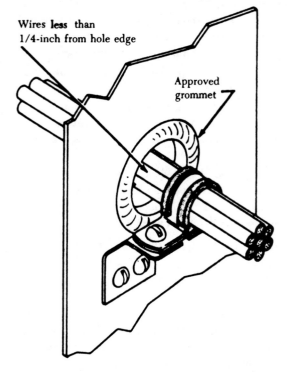

Wires less than
1/4-inch from hole edge

Approved
grommet

FIGURE 11-17. Cable clamp and grommet at bulkhead hole.

443

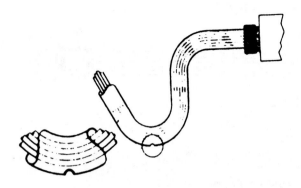

Drainage hole 1/8-inch diameter at
lowest point in tubing. Make the
hole after installation is complete
and lowest point is firmly established

FIGURE 11–18. Drain hole in low point of tubing.

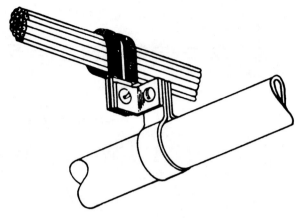

FIGURE 11–19. Separation of wires from plumbing lines.

plumbing line that carries flammable fluids or oxygen.

Wiring should be routed to maintain a minimum clearance of at least 3 in. from control cables. If this cannot be accomplished, mechanical guards should be installed to prevent contact between wiring and control cables.

Installation of Cable Clamps

Cable clamps should be installed with regard to the proper angle, as shown in figure 11–20. The mounting screw should be above the wire bundle. It is also desirable that the back of the cable clamp rest against a structural member where practicable.

Figure 11–21 shows some typical mounting hardware used in installing cable clamps.

Care should be taken that wires are not pinched in cable clamps. Where possible, mount the cables directly to structural members, as shown in figure 11–22.

Clamps can be used with rubber cushions to secure wire bundles to tubular structures as shown in figure 11–23. Such clamps must fit tightly, but should not be deformed when locked in place.

LACING AND TYING WIRE BUNDLES

Wire groups and bundles are laced or tied with cord to provide ease of installation, maintenance, and inspection. This section describes and illustrates recommended procedures for lacing and tying wires with knots which will hold tightly under all conditions. For the purposes of this discussion, the following terms are defined:

(1) Tying is the securing together of a group

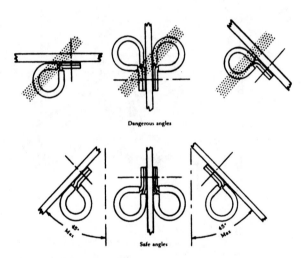

FIGURE 11–20. Proper mounting angles for cable clamps.

or bundle of wires by individual pieces of cord tied around the group or bundle at regular intervals.

(2) Lacing is the securing together of a group or bundle of wires by a continuous piece of cord forming loops at regular intervals around the group or bundle.

(3) A wire group is two or more wires tied or laced together to give identity to an individual system.

(4) A wire bundle is two or more wires or groups tied or laced together to facilitate maintenance.

The material used for lacing and tying is either cotton or nylon cord. Nylon cord is moisture- and fungus-resistant, but cotton cord must be waxed before using to give it these necessary protective characteristics.

Single-Cord Lacing

Figure 11–24 shows the step in lacing a wire bundle with a single cord. The lacing procedure is started at the thick end of the wire group or bundle with a knot consisting of a clove hitch with an extra loop. The lacing is then continued at regular intervals with half hitches along the wire group or bundle and at each point where a wire or wire group branches off. The half hitches should be spaced so that the bundle is neat and secure. The lacing is ended by tying a knot consisting of a clove hitch with an extra loop. After the knot is tied, the free

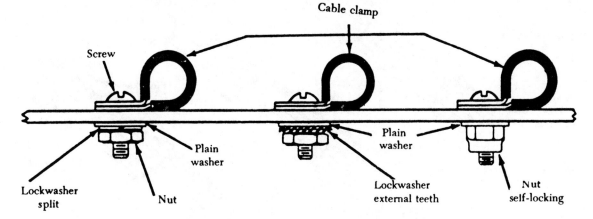

FIGURE 11–21. Typical mounting hardware for cable clamps.

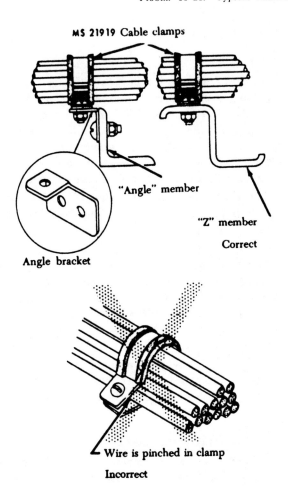

FIGURE 11–22. Mounting cable clamp to structure.

ends of the lacing cord should be trimmed to approximately 3/8 in.

Double-Cord Lacing

Figure 11–25 illustrates the procedure for dou-

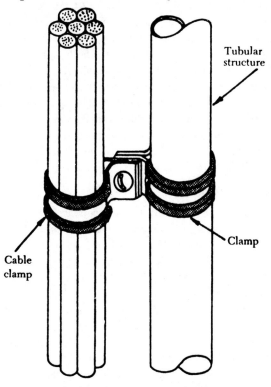

FIGURE 11–23. Installing cable clamp to tubular structure.

445

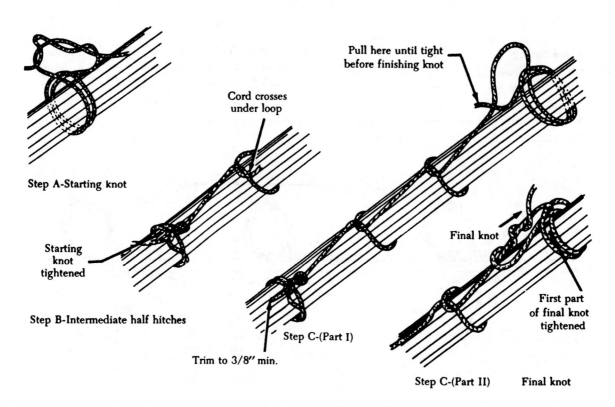

Step A-Starting knot

Starting knot tightened

Step B-Intermediate half hitches

Cord crosses under loop

Trim to 3/8″ min.

Step C-(Part I)

Pull here until tight before finishing knot

Final knot

First part of final knot tightened

Step C-(Part II)   Final knot

FIGURE 11-24. Single-cord lacing.

ble-cord lacing. The lacing is started at the thick end of the wire group or bundle with a bowline-on-a-bight knot (A of figure 11-25). At regular intervals along the wire group or bundle, and at each point where a wire branches off, the lacing is continued using half hitches, with both cords held firmly together. The half hitches should be spaced so that the group or bundle is neat and secure. The lacing is ended with a knot consisting of a half hitch, continuing one of the cords clockwise and the other counterclockwise and then tying the cord ends with a square knot. The free ends of the lacing cord should be trimmed to approximately 3/8 in.

Lacing Branch-Offs

Figure 11-26 illustrates a recommended procedure for lacing a wire group that branches off the main wire bundle. The branch-off lacing is started with a knot located on the main bundle just past the branch-off point. Continue the lacing along the branched-off wire group, using regularly spaced half hitches. If a double cord is used, both cords should be held snugly together. The half hitches should be spaced to lace the bundle neatly and securely. The lacing is ended with the regular terminal knot used in single- or double-cord lacing. The free ends of the lacing cord should be neatly trimmed.

Tying

All wire groups or bundles should be tied where supports are more than 12 in. apart. Figure 11-28 illustrates a recommended procedure for tying a wire group or bundle. The tie is started by wrapping the cord around the wire group to tie a clove-hitch knot. Then a square knot with an extra loop is tied, and the free ends of the cord are trimmed.

Temporary ties are sometimes used in making up and installing wire groups and bundles. Colored cord is normally used to make temporary ties, since they are removed when the installation is complete.

Whether laced or tied, bundles should be secured to prevent slipping, but not so tightly that the cord cuts into or deforms the insulation. This applies especially to coaxial cable, which has a soft dielectric insulation between the inner and outer conductor.

The part of a wire group or bundle located inside a conduit is not tied or laced, but wire groups or bundles inside enclosures, such as junction boxes, should be laced only.

CUTTING WIRE AND CABLE

To make installation, maintenance, and repair easier, wire and cable runs in aircraft are broken at specified locations by junctions, such as connectors,

446

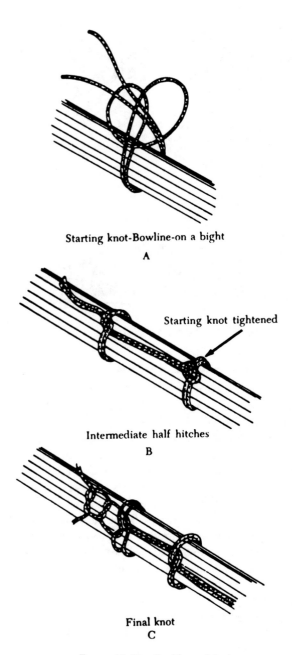

Starting knot-Bowline-on a bight

A

Starting knot tightened

Intermediate half hitches

B

Final knot

C

FIGURE 11-25. Double-cord lacing.

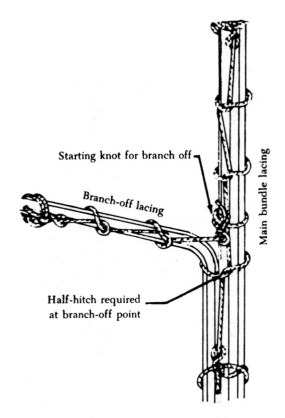

Starting knot for branch off

Branch-off lacing

Half-hitch required
at branch-off point

Main bundle lacing

FIGURE 11-26. Lacing a branch-off.

Stripping Wire and Cable

Before wire can be assembled to connectors, terminals, splices, etc., the insulation must be stripped from connecting ends to expose the bare conductor.

Copper wire can be stripped in a number of ways depending on the size and insulation. Figure 11-27 lists some types of stripping tools recommended for various wire sizes and types of insulation.

FIGURE 11-27. Wire strippers for copper wire.

Stripper	Wire Size	Insulations
Hot-blade	#26—#4	All except asbestos
Rotary, electric	#26—#4	All
Bench	#20—#6	All
Hand pliers	#26—#8	All
Knife	#2 —#0000	All

Aluminum wire must be stripped very carefully, using extreme care, since individual strands will break very easily after being nicked.

The following general precautions are recommended when stripping any type of wire:

(1) When using any type of wire stripper, hold the wire so that it is perpendicular to cutting blades.

(2) Adjust automatic stripping tools care-

terminal blocks, or buses. Before assembly to these junctions, wires and cables must be cut to length.

All wires and cables should be cut to the lengths specified on drawings and wiring diagrams. The cut should be made clean and square, and the wire or cable should not be deformed. If necessary, large-diameter wire should be re-shaped after cutting. Good cuts can be made only if the blades of cutting tools are sharp and free from nicks. A dull blade will deform and extrude wire ends.

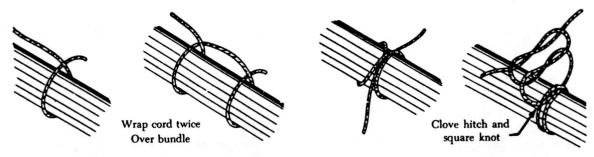

Wrap cord twice
Over bundle

Clove hitch and
square knot

FIGURE 11-28. Tying a wire group or bundle.

fully; follow the manufacturer's instructions to avoid nicking, cutting, or otherwise damaging strands. This is especially important for aluminum wires and for copper wires smaller than No. 10. Examine stripped wires for damage. Cut off and re-strip (if length is sufficient), or reject and replace any wires having more than the allowable number of nicked or broken strands listed in the manufacturer's instructions.

(3) Make sure insulation is clean-cut with no frayed or ragged edges. Trim if necessary.

(4) Make sure all insulation is removed from stripped area. Some types of wires are supplied with a transparent layer of insulation between the conductor and the primary insulation. If this is present, remove it.

(5) When using hand-plier strippers to remove lengths of insulation longer than 3/4 in., it is easier to accomplish in two or more operations.

(6) Re-twist copper strands by hand or with pliers, if necessary, to restore natural lay and tightness of strands.

A pair of hand wire strippers is shown in figure 11-29. This tool is commonly used to strip most types of wire.

The following general procedures describe the steps for stripping wire with a hand stripper. (Refer to figure 11-30.)

(1) Insert wire into exact center of correct cutting slot for wire size to be stripped. Each slot is marked with wire size.

(2) Close handles together as far as they will go.

(3) Release handles, allowing wire holder to return to the "open" position.

(4) Remove stripped wire.

FIGURE 11-29. Light-duty hand wire strippers.

Solderless Terminals and Splices

Splicing of electrical cable should be kept to a minimum and avoided entirely in locations subject to extreme vibrations. Individual wires in a group or bundle can usually be spliced, provided the completed splice is located so that it can be inspected periodically. Splices should be staggered so that the bundle does not become excessively enlarged. Many types of aircraft splice connectors are available for splicing individual wires. Self-insulated splice connectors are usually preferred; however, a noninsulated splice connector can be used if the splice is covered with plastic sleeving secured at both ends. Solder splices may be used, but they are particularly brittle and not recommended.

Electric wires are terminated with solderless terminal lugs to permit easy and efficient connection to and disconnection from terminal blocks, bus bars, or other electrical equipment. Solderless splices join electric wires to form permanent continuous runs. Solderless terminal lugs and splices are made of copper or aluminum and are preinsulated or uninsulated, depending on the desired application.

Terminal lugs are generally available in three types for use in different space conditions. These are the flag, straight, and right-angle lugs. Terminal lugs are "crimped" (sometimes called "staked" or "swaged") to the wires by means of hand or power crimping tools.

The following discussion describes recommended

Select correct
hole to match
wire gauge

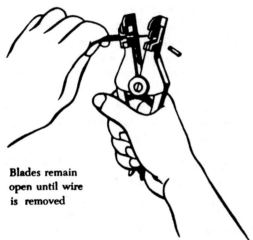

Blades remain
open until wire
is removed

FIGURE 11–30. Stripping wire with hand stripper.

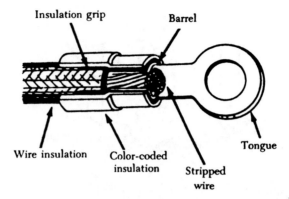

FIGURE 11–31. Preinsulated terminal lug.

methods for terminating copper and aluminum wires using solderless terminal lugs. It also describes the method for splicing copper wires using solderless splices.

Copper Wire Terminals

Copper wires are terminated with solderless, preinsulated straight copper terminal lugs. The insulation is part of the terminal lug and extends beyond its barrel so that it will cover a portion of the wire insulation, making the use of an insulation sleeve unnecessary (figure 11–31).

In addition, preinsulated terminal lugs contain an insulation grip (a metal reinforcing sleeve) beneath the insulation for extra gripping strength on the wire insulation. Preinsulated terminals accommodate more than one size of wire; the insulation is usually color-coded to identify the wire sizes that can be terminated with each of the terminal lug sizes.

Crimping Tools

Hand, portable power, and stationary power tools are available for crimping terminal lugs. These tools crimp the barrel of the terminal lug to the conductor and simultaneously crimp the insulation grip to the wire insulation.

Hand crimping tools all have a self-locking ratchet that prevents opening the tool until the crimp is complete. Some hand crimping tools are equipped with a nest of various size inserts to fit different size terminal lugs. Others are used on one terminal lug size only. All types of hand crimping tools are checked by gages for proper adjustment of crimping jaws.

Figure 11–32 shows a terminal lug inserted into a hand tool. The following general guidelines outline the crimping procedure.

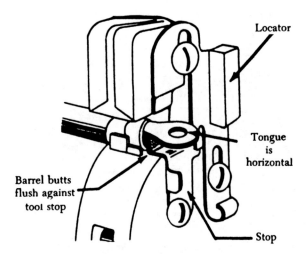

FIGURE 11-32. Inserting terminal lug into hand tool.

(1) Strip the wire insulation to proper length.

(2) Insert the terminal lug, tongue first, into hand tool barrel crimping jaws until the terminal lug barrel butts flush against the tool stop.

(3) Insert the stripped wire into the terminal lug barrel until the wire insulation butts flush against the end of the barrel.

(4) Squeeze the tool handles until the ratchet releases.

(5) Remove the completed assembly and examine it for proper crimp.

Some types of uninsulated terminal lugs are insulated after assembly to a wire by means of pieces of transparent flexible tubing called "sleeves." The sleeve provides electrical and mechanical protection at the connection. When the size of the sleeving used is such that it will fit tightly over the terminal lug, the sleeving need not be tied; otherwise, it

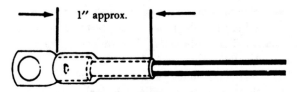

Tight or shrunk sleeve

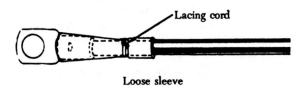

Loose sleeve

FIGURE 11-33. Insulating sleeve.

should be tied with lacing cord as illustrated in figure 11-33.

Aluminum Wire Terminals

The use of aluminum wire in aircraft systems is increasing because of its weight advantage over copper. However, bending aluminum will cause "work hardening" of the metal, making it brittle. This results in failure or breakage of strands much sooner than in a similar case with copper wire. Aluminum also forms a high-resistant oxide film immediately upon exposure to air. To compensate for these disadvantages, it is important to use the most reliable installation procedures.

Only aluminum terminal lugs are used to terminate aluminum wires. They are generally available in three types: (1) Straight, (2) right-angle, and (3) flag. All aluminum terminals incorporate an inspection hole (figure 11-34) which permits checking the depth of wire insertion. The barrel of aluminum terminal lugs is filled with a petrolatum-zinc dust compound. This compound removes the oxide film from the aluminum by a grinding process during the crimping operation. The compound will also minimize later oxidation of the completed connection by excluding moisture and air. The compound is retained inside the terminal lug barrel by a plastic or foil seal at the end of the barrel.

Splicing Copper Wires Using Preinsulated Splices

Preinsulated permanent copper splices join small wires of sizes 22 through 10. Each splice size can be used for more than one wire size. Splices are usually color-coded in the same manner as preinsulated small copper terminal lugs. Some splices are insulated with white plastic. Splices are also used to reduce wire sizes (figure 11-35).

Crimping tools are used to accomplish this type of splice. The crimping procedures are the same as those used for terminal lugs, except that the crimping operation must be done twice, once for each end of the splice.

EMERGENCY SPLICING REPAIRS

Broken wires can be repaired by means of crimped splices, by using terminal lugs from which the tongue has been cut off, or by soldering together and potting broken strands. These repairs are applicable to copper wire. Damaged aluminum wire must not be temporarily spliced. These repairs are for temporary emergency use only and should be replaced as soon as possible with permanent repairs. Since some manufacturers prohibit splicing,

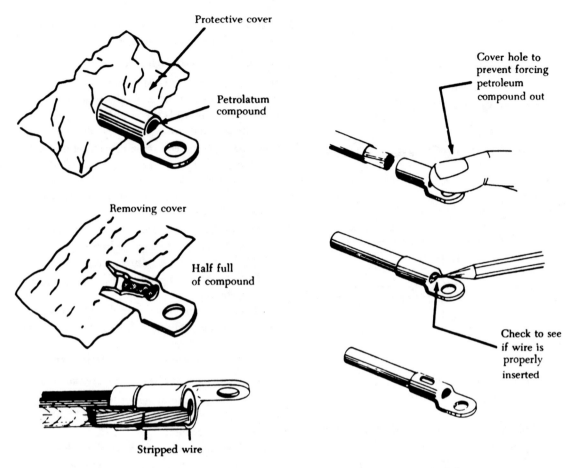

FIGURE 11-34. Inserting aluminum wire into aluminum terminal lugs.

the applicable manufacturer's instructions should always be consulted.

Splicing with Solder and Potting Compound

When neither a permanent splice nor a terminal lug is available, a broken wire can be repaired as follows (figure 11–36):

(1) Install a piece of plastic sleeving about 3 in. long, and of the proper diameter to fit loosely over the insulation, on one piece of the broken wire.

Thinner wire doubled over   Heavy wire

Cover with vinyl tube
tied at both ends

FIGURE 11–35. Reducing wire size with a permanent splice.

(2) Strip approximately 1–1/2 in. from each broken end of the wire.

(3) Lay the stripped ends side by side and twist one wire around the other with approximately four turns.

(4) Twist the free end of the second wire around the first wire with approximately four turns. Solder wire turns together, using 60/40 tin-lead resin-core solder.

(5) When solder is cool, draw the sleeve over the soldered wires and tie at one end. If potting compound is available, fill the sleeve with potting material and tie securely.

(6) Allow the potting compound to set without touching for 4 hrs. Full cure and electrical characteristics are achieved in 24 hrs.

CONNECTING TERMINAL LUGS TO TERMINAL BLOCKS

Terminal lugs should be installed on terminal blocks so that they are locked against movement in the direction of loosening (figure 11–37).

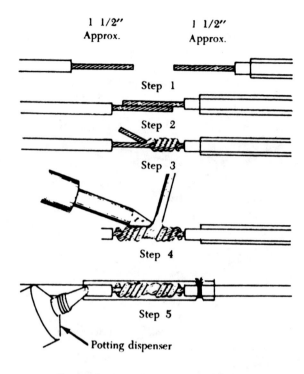

FIGURE 11–36. Repairing broken wire by
soldering and potting.

Terminal blocks are normally supplied with studs
secured in place by a plain washer, an external
tooth lockwasher, and a nut. In connecting terminals, a recommended practice is to place copper
terminal lugs directly on top of the nut, followed
with a plain washer and elastic stop nut, or with a
plain washer, split steel lockwasher, and plain nut.

Aluminum terminal lugs should be placed over a
plated brass plain washer, followed with another
plated brass plain washer, split steel lockwasher,
and plain nut or elastic stop nut. The plated brass
washer should have a diameter equal to the tongue
width of the aluminum terminal lug. Consult the
manufacturer's instructions for recommended dimensions of these plated brass washers. Do not
place any washer in the current path between two
aluminum terminal lugs or between two copper terminal lugs. Also, do not place a lockwasher directly
against the tongue or pad of the aluminum terminal.

To join a copper terminal lug to an aluminum
terminal lug, place a plated brass plain washer over
the nut which holds the stud in place; follow with
the aluminum terminal lug, a plated brass plain
washer, the copper terminal lug, plain washer, split
steel lockwasher, and plain nut or self-locking, all-
metal nut. As a general rule use a torque wrench to
tighten nuts to ensure sufficient contact pressure.

Manufacturer's instructions provide installation
torques for all types of terminals.

BONDING AND GROUNDING

Bonding is the electrical connecting of two or
more conducting objects not otherwise adequately
connected. Grounding is the electrical connecting of
a conducting object to the primary structure for a
return path for current. Primary structure is the
main frame, fuselage, or wing structure of the aircraft, commonly referred to as ground. Bonding
and grounding connections are made in aircraft
electrical systems to:

(1) Protect aircraft and personnel against hazards from lightning discharge.
(2) Provide current return paths.
(3) Prevent development of radio-frequency
potentials.
(4) Protect personnel from shock hazards.
(5) Provide stability of radio transmission
and reception.
(6) Prevent accumulation of static charge.

General Bonding and Grounding Procedures

The following general procedures and precautions are recommended when making bonding or
grounding connections:

(1) Bond or gronud parts to the primary aircraft structure where practicable.
(2) Make bonding or grounding connections
so that no part of the aircraft structure is
weakened.
(3) Bond parts individually if possible.
(4) Install bonding or grounding connections
against smooth, clean surfaces.
(5) Install bonding or grounding connections
so that vibration, expansion or contraction, or relative movement in normal service will not break or loosen the connection.

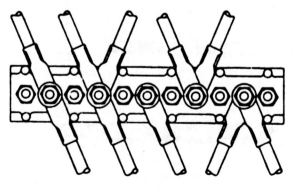

FIGURE 11–37. Connecting terminals to terminal block.

452

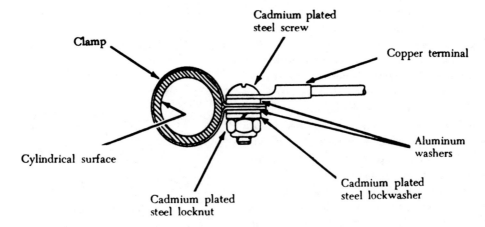

A. Copper jumper connection to tubular structure.

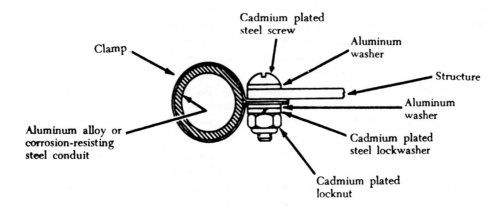

B. Bonding conduit to structure.

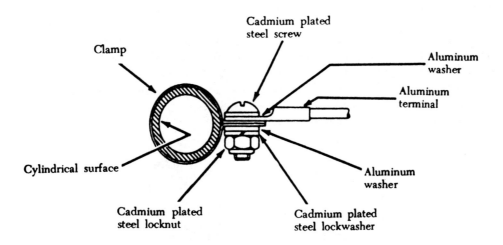

C. Aluminum jumper connection to tubular structure.

FIGURE 11-38. Hardware combinations used in making bonding connections.

(6) Install bonding and grounding connections in protected areas whenever possible.

Bonding jumpers should be kept as short as practicable. The jumper should not interfere with the operation of movable aircraft elements, such as surface controls; normal movement of these elements should not result in damage to the bonding jumper.

Electrolytic action can rapidly corrode a bonding connection if suitable precautions are not observed. Aluminum alloy jumpers are recommended for most cases; however, copper jumpers can be used to bond together parts made of stainless steel, cadimum-plated steel, copper, brass, or bronze. Where contact between dissimilar metals cannot be avoided, the choice of jumper and hardware should be such that corrosion is minimized, and the part most likely to corrode will be the jumper or associated hardware. Figure 11-38 illustrates some proper hardware combinations for making bonding connections. At locations where finishes are removed, a protective finish should be applied to the completed connection to prevent corrosion.

The use of solder to attach bonding jumpers should be avoided. Tubular members should be bonded by means of clamps to which the jumper is attached. The proper choice of clamp material minimizes the probability of corrosion. When bonding jumpers carry a substantial amount of ground return current, the current rating of the jumper should be adequate, and it should be determined that a negligible voltage drop is produced.

Bonding and grounding connections are normally made to flat surfaces by means of through-bolts or screws where there is easy access for installation. Other general types of bolted connections are as follows:

(1) In making a stud connection (figure 11-39), a bolt or screw is locked securely to the structure, thus becoming a stud. Grounding or bonding jumpers can be removed or added to the shank of the stud without removing the stud from the structure.

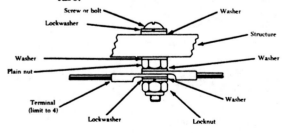

FIGURE 11-39. Stud bonding or grounding to a flat surface.

(2) Nut plates are used where access to the nut for repairs is difficult. Nut plates are riveted or welded to a clean area of the structure (figure 11-40).

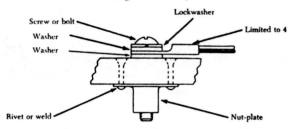

FIGURE 11-40. Nut plate bonding or grounding to a flat surface.

Bonding and grounding connections are also made to a tab riveted to a structure. In such cases it is important to clean the bonding or grounding surface and make the connection as through the connection were being made to the structure. If it is necessary to remove the tab for any reason, the rivets should be replaced with rivets one size larger, and the mating surfaces of the structure and the tab should be clean and free of anodic film.

Bonding or grounding connections can be made to aluminum alloy, magnesium, or corrosion-resistant steel tubular structure as shown in figure 11-41, which shows the arrangement of hardware for bonding with an aluminum jumper. Because of the ease with which aluminum is deformed, it is necessary to distribute the screw and nut pressure by means of plain washers.

Hardware used to make bonding or grounding connections should be selected on the basis of mechanical strength, current to be carried, and ease of installation. If connection is made by aluminum or copper jumpers to the structure of a dissimilar material, a washer of suitable material should be installed between the dissimilar metals so that any corrosion will occur on the washer, which is expendable.

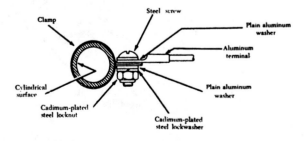

FIGURE 11-37. Bonding or grounding connections to a cylindrical surface.

454

Hardware material and finish should be selected on the basis of the material of the structure to which attachment is made and on the material of the jumper and terminal specified for the bonding or grounding connection. Either a screw or bolt of the proper size for the specified jumper terminal should be used. When repairing or replacing existing bonding or grounding connections, the same type of hardware used in the original connection should always be used.

Testing Grounds and Bonds

The resistance of all bond and ground connections should be tested after connections are made before re-finishing. The resistance of each connection should normally not exceed 0.003 ohm. Resistance measurements need to be of limited nature only for verification of the existence of a bond, but should not be considered as the sole proof of satisfactory bonding. The length of jumpers, methods, and materials used, and the possibility of loosening the connections in service should also be considered.

CONNECTORS

Connectors (plugs and receptacles) facilitate maintenance when frequent disconnection is required. Since the cable is soldered to the connector inserts, the joints should be individually installed and the cable bundle firmly supported to avoid damage by vibration. Connectors have been particularly vulnerable to corrosion in the past, due to condensation within the shell. Special connectors with waterproof features have been developed which may replace non-waterproof plugs in areas where mositure causes a problem. A connector of the same basic type and design should be used when replacing a connector. Connectors susceptible to corrosion difficulties may be treated with a chemically inert waterproof jelly. When replacing connector assemblies, the socket-type insert should be used on the half which is "live" or "hot" after the connector is disconnected, to prevent unintentional grounding.

Types of Connectors

Connectors are identified by AN numbers and are divided into classes with the manufacturer's variations in each class. The manufacturer's variations are differences in appearance and in the method of meeting a specification. Some commonly used connectors are shown in figure 11–42. There are five basic classes of AN connectors used in most aircraft. Each class of connector has slightly different construction characteristics. Classes A, B, C, and D are made of aluminum, and class K is made of steel.

(1) Class A—Solid, one-piece back shell, general-purpose connector.

(2) Class B—Connector back shell separates into two parts lengthwise. Used primarily where it is important that the soldered connectors be readily accessible. The back shell is held together by a threaded ring or by screws.

(3) Class C—A pressurized connector with inserts that are not removable. Similar to a class A connector in appearance, but the inside sealing arrangement is sometimes different. It is used on walls of bulkheads of pressurized equipment.

(4) Class D—Moisture- and vibration-resistant connector which has a sealing grommet in the back shell. Wires are threaded through tight-fitting holes in the grommet, thus sealing against moisture.

(5) Class K—A fireproof connector used in areas where it is vital that the electric current is not interrupted, even though the connector may be exposed to continuous open flame. Wires are crimped to the pin or socket contacts and the shells are made of steel. This class of connector is normally longer than other classes of connectors.

Connector Identification

Code letters and numbers are marked on the coupling ring or shell to identify a connector. This code (figure 11–43) provides all the information necessary to obtain the correct replacement for a defective or damaged part.

Many special-purpose connectors have been designed for use in aircraft applications. These include subminiature and rectangular shell connectors, and connectors with short body shells or split-shell construction.

AN3100
wall receptacle

AN3101
cable receptacle

AN3102
box receptacle

AN3107
MCK disconnect
plug

AN3106
straight plug

AN3106
straight plug

AN3108
angle plug

AN3106
angle plug

FIGURE 11–42.   AN connectors.

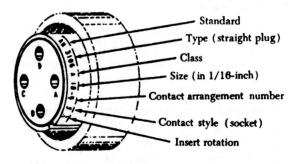

Standard

Type (straight plug)

Class

Size (in 1/16-inch)

Contact arrangement number

Contact style (socket)

Insert rotation

FIGURE 11–43.   AN connector marking.

Installation of Connectors

The following procedures outline one recommended method of assembling connectors to receptacles.

(1) Locate the proper position of the plug in relation to the receptable by aligning the key of one part with the groove or keyway of the other part.

(2) Start the plug into the receptacle with a light forward pressure and engage the

threads of the coupling ring and receptacle.

(3) Alternately push in the plug and tighten the coupling ring until the plug is completely seated.

(4) Use connector pliers to tighten coupling rings one sixteenth to one eighth turn beyond fingertight if space around the connector is too small to obtain a good finger grip.

(5) Never use force to mate connectors to receptacles. Do not hammer a plug into its receptacle; and never use a torque wrench or pliers to lock coupling rings.

A connector is generally disassembled from a receptacle in the following manner:

(1) Use connector pliers to loosen coupling rings which are too tight to be loosened by hand.

(2) Alternately pull on the plug body and unscrew the coupling ring until the connector is separated.

(3) Protect disconnected plugs and receptacles with caps or plastic bags to keep debris from entering and causing faults.

(4) Do not use excessive force, and do not pull on attached wires.

CONDUIT

Conduit is used in aircraft installations for the mechanical protection of wires and cables. It is available in metallic and nonmetallic materials in both rigid and flexible form.

When selecting conduit size for a specific cable bundle application, it is common practice to allow for ease in maintenance and possible future circuit expansion by specifying the conduit inner diameter about 25% larger than the maximum diameter of the conductor bundle. The nominal diameter of a rigid metallic conduit is the outside diameter. Therefore, to obtain the inside diameter, subtract twice the tube wall thickness.

From the abrasion standpoint, the conductor is vulnerable at the conduit ends. Suitable fittings are affixed to the conduit ends in such a manner that a smooth surface comes in contact with the conductor within the conduit. When fittings are not used, the conduit end should be flared to prevent wire insulation damage. The conduit is supported by clamps along the conduit run.

Many of the common conduit installation problems can be avoided by proper attention to the following details:

(1) Do not locate conduit where it can be used as a handhold or footstep.

(2) Provide drain holes at the lowest point in a conduit run. Drilling burrs should be carefully removed from the drain holes.

(3) Support the conduit to prevent chafing against the structure and to avoid stressing its end fittings.

Damaged conduit sections should be repaired to prevent damage to the wires or wire bundle. The minimum acceptable tube bend radii for rigid conduit as prescribed by the manufacturer's instructions should be followed carefully. Kinked or wrinkled bends in a rigid conduit are normally not acceptable.

Flexible aluminum conduit is widely available in two types: (1) Bare flexible and (2) rubber-covered conduit. Flexible brass conduit is normally used instead of flexible aluminum conduit, where necessary to minimize radio interference. Flexible conduit may be used where it is impractical to use rigid conduit, such as areas that have motion between conduit ends or where complex bends are necessary. Transparent adhesive tape is recommended when cutting flexible tubing with a hacksaw to minimize fraying of the braid.

ELECTRICAL EQUIPMENT INSTALLATION

This section provides general procedures and safety precautions for installation of commonly used aircraft electrical equipment and components. Electrical load limits, acceptable means of controlling or monitoring electrical loads, and circuit protection devices are subjects with which mechanics must be familiar to properly install and maintain aircraft electrical systems.

Electrical Load Limits

When installing additional electrical equipment that consumes electrical power in an aircraft, the total electrical load must be safely controlled or managed within the rated limits of the affected components of the aircraft's power-supply system.

Before any aircraft electrical load is increased, the associated wires, cables, and circuit protection devices (fuses or circuit breakers) should be checked to determine that the new electrical load (previous maximum load plus added load) does not exceed the rated limits of the existing wires, cables, or protection devices.

The generator or alternator output ratings prescribed by the manufacturer should be compared with the electrical loads which can be imposed on the affected generator or alternator by installed equipment. When the comparison shows that the probable total connected electrical load can exceed the output load limits of the generator(s) or alternator(s), the load should be reduced so that an overload cannot occur. When a storage battery is part of the electrical power system, ensure that the battery is continuously charged in flight, except when short, intermittent loads are connected such as a radio transmitter, a landing-gear motor, or other similar devices which may place short-time demand loads on the battery.

Controlling or Monitoring the Electrical Load

Placards are recommended to inform crewmembers of an aircraft about the combination of electrical loads that can safely be connected to the power source.

In installations where the ammeter is in the battery lead, and the regulator system limits the maximum current that the generator or alternator can deliver, a voltmeter can be installed on the system bus. As long as the ammeter does not read "discharge" (except for short, intermittent loads such as operating the gear and flaps) and the voltmeter remains at "system voltage," the generator or alternator will not be overloaded.

In installations where the ammeter is in the generator or alternator lead, and the regulator system does not limit the maximum current that the generator or alternator can deliver, the ammeter can be redlined at 100% of the generator or alternator rating. If the ammeter reading is never allowed to exceed the red line, except for short, inermittent loads, the generator or alternator will not be overloaded.

Where the use of placards or monitoring devices is not practicable or desired, and where assurance is needed that the battery in a typical small aircraft generator/battery power source will be charged in flight, the total continuous connected electrical load may be held to approximately 80% of the total rated generator output capacity. (When more than one generator is used in parallel, the total rated output is the combined output of the installed generators.)

When two or more generators are operated in parallel and the total connected system load can exceed the rated output of one generator, means must be provided for quickly coping with the sudden overloads which can be caused by generator or engine failure. A quick load reduction system, or a specified procedure whereby the total load can be reduced to a quantity which is within the rated capacity of the remaining operable generator(s), can be employed.

Electrical loads should be connected to inverters, alternators, or similar aircraft electrical power sources in such a manner that the rated limits of the power source are not exceeded, unless some type of effective monitoring means is provided to keep the load within prescribed limits.

Circuit Protection Devices

Conductors should be protected with circuit breakers or fuses located as close as possible to the electrical power source bus. Normally, the manufacturer of the electrical equipment specifies the fuse or circuit breaker to be used when installing equipment.

The circuit breaker or fuse should open the circuit before the conductor emits smoke. To accomplish this, the time current characteristic of the protection device must fall below that of the associated conductor. Circuit protector characteristics should be matched to obtain the maximum utilization of the connected equipment.

Figure 11–44 shows an example of the chart used in selecting the circuit breaker and fuse protection for copper conductors. This limited chart is applicable to a specific set of ambient temperatures and wire bundle sizes, and is presented as a typical example only. It is important to consult such guides before selecting a conductor for a specific purpose. For example, a wire run individually in the open air may be protected by the circuit breaker of the next higher rating to that shown on the chart.

Wire AN gage copper	Circuit breaker amperage	Fuse amp.
22	5	5
20	7.5	5
18	10	10
16	15	10
14	20	15
12	30	20
10	40	30
8	50	50
6	80	70
4	100	70
2	125	100
1		150
0		150

FIGURE 11–44. Wire and circuit protector chart.

All re-settable circuit breakers should open the circuit in which they are installed regardless of the position of the operating control when an overload or circuit fault exists. Such circuit breakers are referred to as "trip-free." Automatic re-set circuit breakers automatically re-set themselves. They should not be used as circuit protection devices in aircraft.

Switches

A specifically designed switch should be used in all circuits where a switch malfunction would be hazardous. Such switches are of rugged construction and have sufficient contact capacity to break, make, and carry continuously the connected load current. Snap-action design is generally preferred to obtain rapid opening and closing of contacts regardless of the speed of the operating toggle or plunger, thereby minimizing contact arcing.

The nominal current rating of the conventional aircraft switch is usually stamped on the switch housing. This rating represents the continuous current rating with the contacts closed. Switches should be derated from their nominal current rating for the following types of circuits:

(1) High rush-in circuits—Circuits containing incandescent lamps can draw an initial current which is 15 times greater than the continuous current. Contact burning or welding may occur when the switch is closed.

(2) Inductive circuits—Magnetic energy stored in solenoid coils or relays is released and appears as an arc when the control switch is opened.

(3) Motors—Direct-current motors will draw several times their rated current during starting, and magnetic energy stored in their armature and field coils is released when the control switch is opened.

The chart in figure 11–45 is typical of those available for selecting the proper nominal switch rating when the continuous load current is known. This selection is essentially a derating to obtain reasonable switch efficiency and service life.

Hazardous errors in switch operation can be avoided by logical and consistent installation. Two-position "on-off" switches should be mounted so that the "on" position is reached by an upward or forward movement of the toggle. When the switch controls movable aircraft elements, such as landing gear or flaps, the toggle should move in the same

Nominal system voltage	Type of load	Derating factor
24 v. d.c.	Lamp	8
24 v. d.c.	Inductive (Relay-Solenoid)	4
24 v. d.c.	Resistive (Heater)	2
24 v. d.c.	Motor	3
12 v. d.c.	Lamp	5
12 v. d.c.	Inductive (Relay-Solenoid)	2
12 v. d.c.	Resistive (Heater)	1
12 v. d.c.	Motor	2

FIGURE 11–45. Switch derating factors.

direction as the desired motion. Inadvertent operation of a switch can be prevented by mounting a suitable guard over the switch.

Relays

Relays are used as switching devices where a weight reduction can be achieved or electrical controls can be simplified. A relay is an electrically operated switch and is therefore subject to dropout under low system voltage conditions. The foregoing discussion of switch ratings is generally applicable to relay contact ratings.

AIRCRAFT LIGHTING SYSTEMS

Aircraft lighting systems provide illumination for both exterior and interior use. Lights on the exterior provide illumination for such operations as landing at night, inspection of icing conditions, and safety from midair collision. Interior lighting provides illumination for instruments, cockpits, cabins, and other sections occupied by crewmembers and passengers. Certain special lights, such as indicator and warning lights, indicate the operational status of equipment.

Exterior Lights

Position, anti-collision, landing, and taxi lights are common examples of aircraft exterior lights. Some lights, such as position lights and anti-collision lights, are required for night operations. Other types of exterior lights, such as wing inspection lights, are of great benefit for specialized flying operations.

Position Lights

Aircraft operating at night must be equipped with position lights that meet the minimum requirements specified by the Federal Aviation Regulations. A set of position lights consist of one red, one

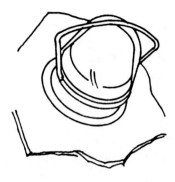

A. Tail position light unit.

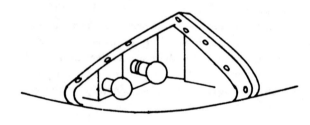

B. Wingtip position light unit.

Figure 11–46. Position lights.

green, and one white light. Position lights are sometimes referred to as "navigation" lights. On many aircraft each light unit contains a single lamp mounted on the surface of the aircraft (A of figure 11–46). Other types of position light units contain two lamps (B of figure 11–46), and are often streamlined into the surface of the aircraft structure

The green light unit is always mounted at the extreme tip of the right wing The red unit is mounted in a similar position on the left wing The white unit is usually located on the vertical stabilizer in a position where it is clearly visible through a wide angle from the rear of the aircraft

The wingtip lamps and the tail lamps are controlled by a double-pole, single-throw switch in the pilot's compartment. On "dim", the switch connects a resistor in series with the lamps Since the resistor decreases current flow, the light intensity is reduced. On "bright", the resistor is shorted out of the circuit, and the lamps glow at full brilliance

On some types of installations a switch in the pilot's compartment provides for steady or flashing operation of the position lights. For flashing operation, a flasher mechanism is usually installed in the position light circuit. It consists essentially of a motor-driven camshaft on which two cams are mounted and a switching mechanism made up of two breaker arms and two contact screws. One breaker arm supplies d.c. current to the wingtip light circuit through one contact screw, and the other breaker arm supplies the tail light circuit through the other contact screw. When the motor rotates, it turns the camshaft through a set of reduction gears and causes the cams to operate the

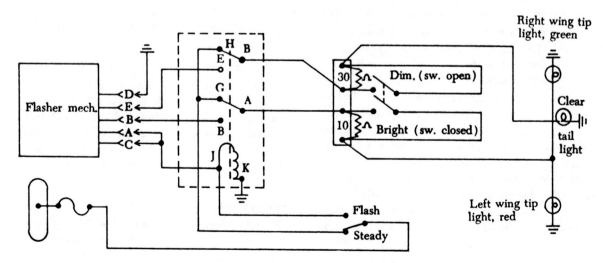

Figure 11–47. Position light circuitry.

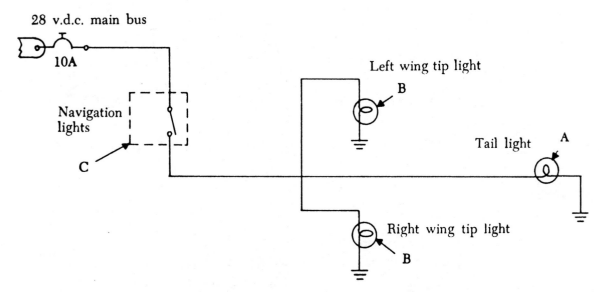

28 v.d.c. main bus

10A

Navigation lights

C

Left wing tip light

B

Tail light

A

Right wing tip light

B

FIGURE 11-48. Single-circuit position light circuitry without flasher.

breaker which opens and closes the wing and tail light circuits alternately. Figure 11–47 is a simplified schematic diagram of a navigation light circuit which illustrates one type of position light circuitry.

The schematic diagram of another type of position light circuitry is shown in figure 11–48. Control of the position lights by a single on-off toggle switch provides only a steady illumination. There is no flasher and no dimming rheostat.

There are, of course, many variations in the position light circuits used on different aircraft. All circuits are protected by fuses or circuit breakers, and many circuits include flashing and dimming equipment. Still others are wired to energize a special warning light dimming relay, which causes all the cockpit warning lights to dim perceptibly when the position lights are illuminated.

Small aircraft are usually equipped with a simplified control switch and circuitry. In some cases, one control knob or switch is used to turn on several sets of lights; for example, one type utilizes a control knob, the first movement of which turns on the position lights and the instrument panel lights. Further rotation of the control knob increases the intensity of only the panel lights. A flasher unit is seldom included in the position light circuitry of very light aircraft, but is used in small twin-engine aircraft.

Anti-collision Lights

An anti-collision light system may consist of one or more lights. They are rotating beam lights which are usually installed on top of the fuselage or tail in

such a location that the light will not affect the vision of the crewmember or detract from the conspicuousness of the position lights. In some cases one of the lights is mounted on the underside of the fuselage.

The simplest means of installing an anti-collision light is to secure it to a reinforced fuselage skin panel, as shown in figure 11–49.

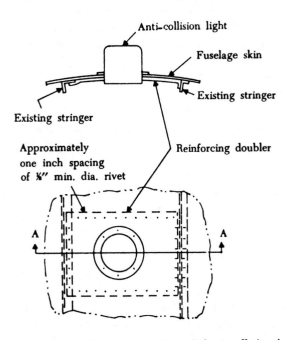

FIGURE 11–49. Typical anti-collision light installation in an unpressurized skin panel.

461

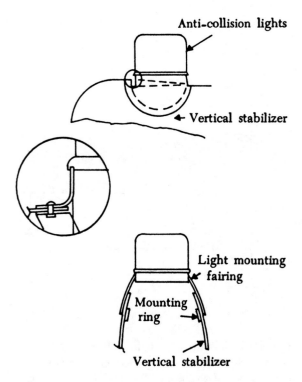

FIGURE 11-50. Typical anti-collision light installation in a vertical stabilizer.

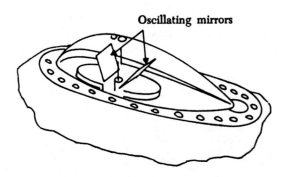

FIGURE 11-51. Anti-collision light.

An anti-collision light is often installed on top of the vertical stabilizer if the cross section of the stabilizer is large enough to accommodate the installation, and if aircraft flutter and vibration characteristics are not adversely affected. Such installations should be located near a spar, and formers should be added as required to stiffen the structure near the light. Figure 11-50 shows a typical anti-collision light installation in a vertical stabilizer.

An anti-collision light unit usually consists of one or two rotating lights operated by an electric motor. The light may be fixed, but mounted under rotating mirrors inside a protruding red glass housing. The mirrors rotate in an arc, and the resulting flash rate is between 40 and 100 cycles per minute. (See figure 11-51.) The anti-collision light is a safety light to warn other aircraft, especially in congested areas.

Landing Lights

Landing lights are installed in aircraft to illuminate runways during night landings. These lights are very powerful and are directed by a parabolic reflector at an angle providing a maximum range of illumination. Landing lights are usually located midway in the leading edge of each wing or stream-

lined into the aircraft surface. Each light may be controlled by a relay, or it may be connected directly into the electric circuit.

Since icing of the lamp lenses reduces the illumination quality of a lamp, some installations use retractable landing lamps (figure 11-52). When the lamps are not in use, a motor retracts them into receptacles in the wing where the lenses are not exposed to the weather.

As shown in figure 11-53, one type of retractable landing light motor has a split-field winding. Two of the field winding terminals connect to the two outer terminals of the motor control switch through the points of contacts C and D, while the center terminal connects to one of two motor brushes. The brushes connect the motor and magnetic brake sole-

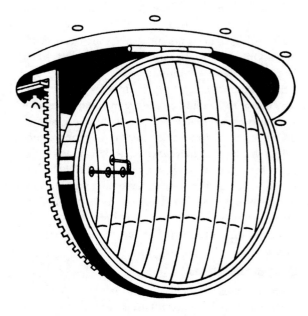

FIGURE 11-52. Retractable landing light.

462

noid into the electric circuit. The points of contact C are held open by the geared quadrant of the landing lamp mechanism. The points of contact D are held closed by the tension of the spring to the right of the contacts. This is a typical arrangement of a landing lamp circuit when the landing lamp is retracted and the control switch is in the "off" position. No current flows in the circuit, and neither the motor nor the lamp can be energized.

When the control switch is placed in the upper, or "extend," position (figure 11–53), current from the battery flows through the closed contacts of the switch, the closed contacts of contact D, the center terminal of the field winding, and the motor itself. Current through the motor circuit energizes the brake solenoid, which withdraws the brake shoe from against the motor shaft, allowing the motor to turn and lower the lamp mechanism. After the lamp mechanism moves about 10°, contact A touches and rides along the copper bar B. In the meantime, relay F is energized, and its contacts close. This permits current to flow through the copper bar B, contact A, and the lamp. When the lamp mechanism

is completely lowered, the projection at the top of the gear quadrant pushes the D contacts apart, opens the circuit to the motor, and causes the de-energized brake solenoid to release the brake. The brake is pushed against the motor shaft by the spring, stopping the motor and completing the lowering operation.

To retract the landing lamp, the control switch is placed in the "retract" position (figure 11–53). The motor and brake circuits are completed through the points of contact C, since these contacts are closed when the gear quadrant is lowered. This action completes the circuit, the brake releases, the motor turns (this time in the opposite direction) and the landing light mechanism is retracted. Since switching to "retract" breaks the circuit to relay F, the relay contacts open, disconnecting the copper bar and causing the landing lamp to go out. When the mechanism is completely retracted, contact points C open, and the circuit to the motor is again broken, the brake applied, and the motor stopped.

Retractable landing lights that can be extended to any position of their extension are employed on some aircraft. Landing lights used on high-speed aircraft are usually equipped with an airspeed pressure switch which prevents extension of landing lights at excessive airspeeds. Such switches also cause retraction of landing lights if the aircraft exceeds a predetermined speed.

Many large aircraft are equipped with four landing lights, two of which are fixed and two retractable. Fixed lights are usually located in either the wing root areas or just outboard of the fuselage in the leading edge of each wing. The two retractable lights are usually located in the lower outboard surface of each wing, and are normally controlled by separate switches. On some aircraft, the fixed

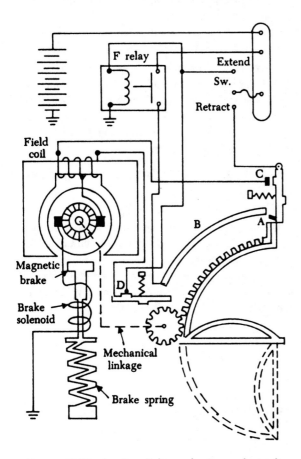

FIGURE 11–53. Landing light mechanism and circuit.

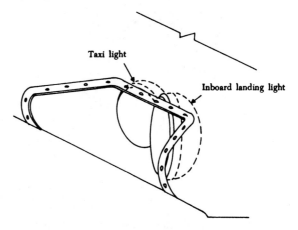

FIGURE 11–54. Fixed landing light and taxi light.

landing light is mounted in an area with a taxi light, as shown in figure 11–54.

Taxi Lights

Taxi lights are designed to provide illumination on the ground while taxiing or towing the aircraft to or from a runway, taxi strip, or in the hangar area.

Taxi lights are not designed to provide the degree of illumination necessary for landing lights; 150- to 250-watt taxi lights are typical on many medium and heavy aircraft.

On aircraft with tricycle landing gear, either single or dual taxi lights are often mounted on the non-steerable part of the nose landing gear. As illustrated in figure 11–55, they are positioned at an oblique angle to the center line of the aircraft to provide illumination directly in front of the aircraft and also some illumination to the right and left of the aircraft's path. On some aircraft the dual taxi lights are supplemented by wingtip clearance lights controlled by the same circuitry.

Taxi lights are also mounted in the recessed areas of the wing leading edge, often in the same area with a fixed landing light.

Many small aircraft are not equipped with any type of taxi light, but rely on the intermittent use of a landing light to illuminate taxiing operations. Still other aircraft utilize a dimming resistor in the landing light circuit to provide reduced illumination for taxing. A typical circuit for dual taxi lights is shown in figure 11–56.

Some large aircraft are equipped with alternate taxi lights located on the lower surface of the aircraft, aft of the nose radome. These lights, operated by a separate switch from the main taxi lights, illuminate the area immediately in front of and below the aircraft nose.

Wing Inspection Lights

Some aircraft are equipped with wing inspection lights to illuminate the leading edge of the wings to permit observation of icing and general condition of these areas in flight. On some aircraft, the wing inspection light system (also called wing ice lights) consists of a 100-watt light mounted flush on the outboard side of each nacelle forward of the wing. These lights permit visual detection of ice formation on wing leading edges while flying at night. They are also often used as floodlights during ground servicing. They are usually controlled through a relay by an "on-off" toggle switch in the cockpit.

Some wing inspection light systems may include or be supplemented by additional lights, sometimes called nacelle lights, that illuminate adjacent areas such as cowl flaps or the landing gear. These are normally the same type of lights and can be controlled by the same circuits.

MAINTENANCE AND INSPECTION OF LIGHTING SYSTEMS

Inspection of an aircraft's lighting systems normally includes checking the condition and security of all visible wiring, connections, terminals, fuses, and switches. A continuity light or meter can be used in making these checks, since the cause of many troubles can often be located by systematically testing each circuit for continuity.

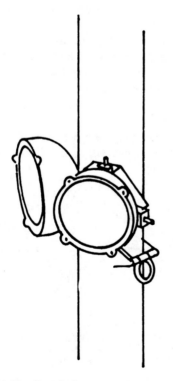

FIGURE 11–55. Taxi lights mounted on non-steerable portion of nose landing gear.

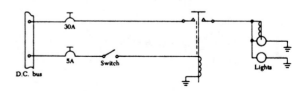

FIGURE 11–56. Typical taxi light circuit.

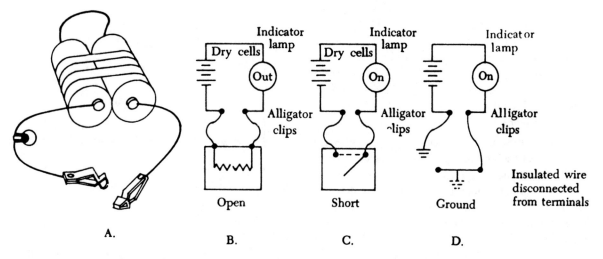

FIGURE 11-57. Continuity testing with a continuity tester.

All light covers and reflectors should be kept clean and polished. Cloudy reflectors are often caused by an air leak around the lens.

The condition of the sealing compound around position light frames should be inspected regularly. Leaks or cracks should be filled with an approved sealing compound.

Care should be exercised in installing a new bulb in a light assembly, since many bulbs fit into a socket in only one position and excessive force can cause an incomplete or open circuit in the socket area.

Circuit testing, commonly known as troubleshooting is a means of systematically locating faults in an electrical system. These faults are usually of three kinds:

(1) Open circuits in which leads or wires are broken.

(2) Shorted circuits in which grounded leads cause current to be returned by shortcuts to the source of power.

(3) Low power in circuits causing lights to burn dimly and relays to chatter. Electrical troubles may develop in the unit or in the wiring. If troubles such as these are carefully analyzed and systematic steps are taken to locate them, much time and energy not only can be saved, but damage to expensive testing equipent often can be avoided. (For a more extensive treatment of circuit testing than the summary provided here, refer to Chapter 8, Airframe and Powerplant Mechanics General Handbook. AC 65-9A.)

The equipment generally used in testing lighting circuits in an aircraft consists of a voltmeter, test light, continuity meter, and ohmmeter.

Although any standard d.c. voltmeter with flexible leads and test prods is satisfactory for testing circuits, portable voltmeters especially designed for test purposes are usually used.

The test lamp consists of a low wattage aircraft light. Two leads are used with this light.

Continuity testers vary somewhat. One type consists of a small lamp connected in series with two small batteries (flashlight batteries are very suitable) and two leads. (See A of figure 11-57.) Another type of continuity tester contains two batteries connected in series with a d.c. voltmeter and two test leads. A completed circuit will be registered by the voltmeter.

Whenever generator or battery voltage is available, the voltmeter and the test light can be used in circuit testing, since these sources of power will activate the test light and the voltmeter.

If no electrical power is available (the circuit is dead), then the continuity tester is used. The self-contained batteries of the continuity tester force current through the circuit, causing the continuity meter to indicate when the circuit being tested is completed. When using the continuity meter, the circuit being tested should always be isolated from all other circuits by removing the fuse, by opening the switch, or by disconnecting the wires.

Figure 11-57 illustrates techniques which may be

465

used in checking circuits. The continuity tester contains a light to serve as an indicator. When the test leads are touched together, a complete circuit is established and the indicator light illuminates. When the leads are brought into contact with a resistor or other circuit element, as shown in B of figure 11–57, and the light does not illuminate, then the circuit being tested is open.

For the open test to be conclusive, be sure the resistance of the unit tested is low enough to permit the lamp to light. In a test in which the resistance is too high, usually more than 10 ohms, connect a voltmeter in the circuit in place of the lamp. If the voltmeter pointer fails to deflect, an open circuit is indicated.

The test for shorts (C of figure 11–57) shows the continuity tester connected across the terminals of a switch in the "open" position. If the tester lamp lights, there is a short circuit in the switch.

To determine whether a length of wire is grounded at some point between its terminals, disconnect the wire at each end and hook one test clip to the wire at one end and ground the other test clip (D of figure 11–57). If the wire is grounded, the lamp will light. To locate the ground, check back at intervals toward the other end. The lighting of the lamp will indicate the section of the wire that is grounded.

The ohmmeter, although primarily designed to measure resistance, is useful for checking continuity. With an ohmmeter, the resistance of a lighting circuit can be determined directly by scale. Since an open circuit has infinite resistance, a zero reading on the ohmmeter indicates circuit continuity.

As illustrated in figure 11–58, the ohmmeter uses a battery as the source of voltage. There are fixed resistors, which are of such value that when the test prods are shorted together, the meter will read full scale. The variable resistor, in parallel with the meter, and the fixed resistors compensate for changes in voltage of the battery. The variable resistor provides for zero adjustment on the meter control panel.

On the meter there may be several scales, which are made possible by various values of resistance and battery voltage. The desired scale is selected by a selector switch on the face of the ohmmeter. Each scale reads low resistances at the upper end. Greater resistance in a circuit is indicated by less deflection of the indicator on the scale.

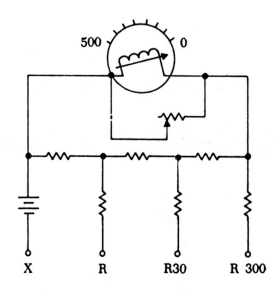

FIGURE 11–58. Typical ohmmeter internal circuitry.

When using an ohmmeter to check continuity, connect the leads across the circuit. A zero ohm reading indicates circuit continuity. For checking resistance, a scale should be chosen which will contain the resistance of the element being measured. In general, a scale should be selected on which the reading will fall in the upper half of the scale. Short the leads together and set the meter to read zero ohm by the zero adjustment. If a change in scales is made anytime, remember to re-adjust the meter to zero ohm.

When making circuit tests with the ohmmeter, never attempt to check continuity or measure the resistance in a circuit while it is connected to a source of voltage. Disconnect one end of an element when checking resistance, so that the ohmmeter will not read the resistance of parallel paths.

The following summary of continuity testing of lighting circuits is recommended, using either an ohmmeter or any other type of continuity tester.

(1) Check the fuse or circuit breaker. Be sure it is the correct one for the circuit being tested.

(2) Check the electrical unit (light).

(3) If fuse or circuit breaker and light are in good condition, check at the most accessible point for an open or short in the circuit.

(4) Never guess. Always locate the trouble in the positive lead of a circuit, the operating

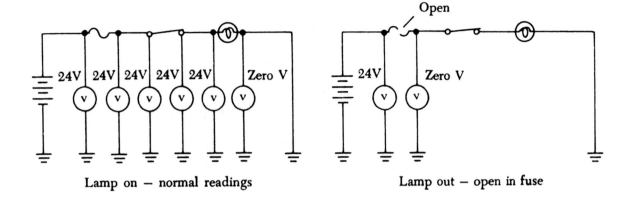

Lamp on — normal readings

Lamp out — open in fuse

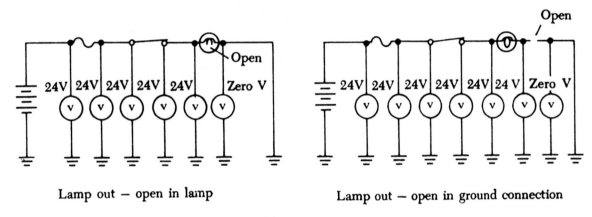

Lamp out — open in lamp

Lamp out — open in ground connection

FIGURE 11-59. Continuity testing with a voltmeter.

unit, or the negative lead before removing any equipment or wires.

A voltmeter with long flexible leads provides a satisfactory, though different, method of checking the continuity of lighting system wiring in an aircraft. The voltage to be checked by the voltmeter is furnished by the storage battery in the aircraft.

The following procedure indicates the steps for continuity checking by a voltmeter in a circuit which consists of a 24-volt battery, a fuse, a switch, and a landing lamp.

(1) Draw a simple wiring diagram of the circuitry to be tested, as shown in figure 11-59.

(2) Check the fuse by touching the positive voltmeter lead to the load end of the fuse and the negative lead to ground. If the fuse is good, there will be an indication on the voltmeter. If it is burned out, it must be replaced. If it burns out again,

the circuit is grounded. Check for the ground at the lamp by removing the connector and replacing the fuse; if it burns out, the short is in the line. However, if the fuse does not burn out this time, the short is within the lamp.

(3) If the fuse tests good, the circuit has an open. Then with the negative clip of the voltmeter connected to ground, move the positive clip from point to point along the circuit, following the diagram as a guide. Test each unit and length of wire. The first zero reading on the voltmeter indicates that there is an open circuit between the last point at which the voltage was normal and the point of the first zero reading. In the illustration of figure 11-59, open circuits are caused by an open fuse, an open lamp filament, and an open lamp-to-ground connection.

467

GENERAL

Safe, economical, and reliable operation of modern aircraft is dependent upon the use of instruments. The first aircraft instruments were fuel and oil pressure instruments to warn of engine trouble so that the aircraft could be landed before the engine failed. As aircraft that could fly over considerable distances were developed, weather became a problem. Instruments were developed that helped to fly through bad weather conditions.

Instrumentation is basically the science of measurement. Speed, distance, altitude, attitude, direction, temperature, pressure, and r.p.m. are measured and these measurements are displayed on dials in the cockpit.

There are two ways of grouping aircraft instruments. One is according to the job they perform. Within this grouping they can be classed as flight instruments, engine instruments, and navigation instruments. The other method of grouping aircraft instruments is according to the principle on which they work. Some operate in relation to changes in temperature or air pressure and some by fluid pressure. Others are activated by magnetism and electricity, and others depend on gyroscopic action.

The instruments that aid in controlling the in-flight attitude of the aircraft are known as flight instruments. Since these instruments must provide information instantaneously, they are located on the main instrument panel within ready visual reference of the pilot. Basic flight instruments in an aircraft are the airspeed indicator, altimeter and the magnetic direction indicator. In addition, some aircraft may have a rate-of-turn indicator, a bank indicator, and an artificial horizon indicator. Flight instruments are operated by atmospheric, impact, differential, or static pressure or by a gyroscope.

Engine instruments are designed to measure the quantity and pressure of liquids (fuel and oil) and gases (manifold pressure), r.p.m., and temperature. The engine instruments usually include a tachometer, fuel and oil pressure gages, oil temperature gage, and a fuel quantity gage. In addition some aircraft that are powered by reciprocating engines are equipped with manifold pressure gage(s), cylinder head temperature gage(s), and carburetor air-temperature gage(s). Gas turbine powered aircraft will have a turbine or tailpipe temperature gage(s), and may have an exhaust pressure ratio indicator(s).

Navigational instruments provide information that enables the pilot to guide the aircraft accurately along definite courses. This group of instruments includes a clock, compasses (magnetic compass and gyroscopic directional indicator), radios, and other instruments for presenting navigational information to the pilot.

INSTRUMENT CASES

A typical instrument can be compared to a clock, in that the instrument has a mechanism, or works; a dial, or face; pointers, or hands; and a cover glass. The instrument mechanism is protetced by a one; or two-piece case. Various materials, such as aluminum alloy, magnesium alloy, iron, steel, or plastic are used in the manufacture of instrument cases. Bakelite is the most commonly used plastic. Cases for electrically operated instruments are made of iron or steel; these materials provide a path for stray magnetic force fields that would otherwise interfere with radio and electronic devices.

Some instrument mechanisms are housed in air-tight cases, while other cases have a vent hole. The vent allows air pressure inside the instrument case to vary with the aircraft's change in altitude.

DIALS

Numerals, dial markings, and pointers of instruments are frequently coated with luminous paint. Some instruments are coated with luminous calcium sulphide, a substance that glows for several hours after exposure to light. Other instruments have a phosphor coating that glows only when excited by a small ultraviolet lamp in the cockpit. Some instruments are marked with a combination of radioactive

469

salts, zinc oxide, and shellac. In handling these instruments, care should be taken against radium poisoning. The effects of radium are cumulative and can appear after a long period of continued exposure to small amounts of radiation. Poisoning usually results from touching the mouth or nose after handling instrument dials or radioactive paint. After handling either, the hands should be kept away from the mouth and nose, and washed thoroughly with hot water and soap as soon as possible.

RANGE MARKINGS

Instrument range markings indicate, at a glance, whether a particular system or component is operating in a safe and desirable range of operation or in an unsafe range.

Instruments should be marked and graduated in accordance with the Aircraft Specifications or Type Certificate Data Sheets and the specific aircraft maintenance or flight manual. Instrument markings usually consist of colored decalcomanias or paint applied to the outer edges of the cover glass or over the calibrations on the dial face. The colors generally used as range markings are red, yellow, green, blue, or white. The markings are usually in the form of an arc or a radial line.

A red radial line may be used to indicate maximum and minimum ranges: operations beyond these markings are dangerous and should be avoided. A blue arc marking indicates that operation is permitted under certain conditions. A green arc indicates the normal operating range during continuous operation. Yellow is used to indicate caution.

A white index marker is placed near the bottom of all instruments that have range markings on the cover glass. The index marker is a line extending from the cover glass onto the instrument case. The marker shows if glass slippage has occurred. Glass slippage would cause the range markings to be in error.

INSTRUMENT PANELS

With a few exceptions, instruments are mounted on a panel in the cockpit so that the dials are plainly visible to the pilot or copilot. Instrument panels are usually made of sheet aluminum alloy strong enough to resist flexing. The panels are nonmagnetic and are painted with a nonglare paint to eliminate glare or reflection.

In aircraft equipped with only a few instruments, only one panel is necessary; in some aircraft, additional panels are required. In such cases the for-

ward instrument panel is usually referred to as the "main" instrument panel to distinguish it from additional panels on the cockpit overhead or along the side of the flight compartment. On some aircraft the main instrument panel is also referred to as the pilot's or copilot's panel, since many of the pilot's instruments on the left side of the panel are duplicated on the right side.

The method of mounting instruments on their respective panels depends on the design of the instrument case. In one design, the bezel is flanged in such a manner that the instrument can be flush-mounted in its cutout from the rear of the panel. Integral self-locking nuts are provided at the rear faces of the flange corners to receive mounting screws from the front of the panel. The flanged type case can also be mounted from the front of the panel.

The mounting of instruments that have flangeless cases is a simpler process. The flangeless case is mounted from the front of the panel. A special expanding type of clamp, shaped and dimensioned to fit the instrument case, is secured to the rear face of the panel. As actuating screw is connected to the clamp and is accessible from the front of the panel. The screw can be rotated to loosen the clamp, permitting the instrument to slide freely into the clamp. After the instrument is positioned, the screw is rotated to tighten the clamp around the instrument case.

Instrument panels are usually shock-mounted to absorb low-frequency, high-amplitude shocks. Shock mounts are used in sets of two, each secured to

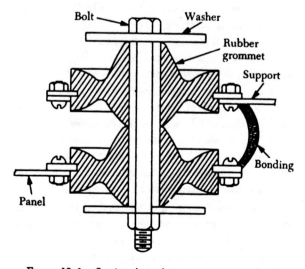

FIGURE 12–1. Section through instrument panel shock.

470

separate brackets. The two mounts absorb most of the vertical and horizontal vibration, but permit the instruments to operate under conditions of minor vibration. A cross sectional view of a typical shock mount is shown in figure 12–1.

The type and number of shock mounts to be used for instrument panels are determined by the weight of the unit. The weight of the complete unit is divided by the number of suspension points. For example, an instrument panel weighing 16 lbs. which is supported at four points would require eight shock absorbers, each capable of supporting 4 lbs. When the panel is mounted, the weight should deflect the shock absorbers approximately ⅛ in.

Shock-mounted instrument panels should be free to move in all directions and have sufficient clearance to avoid striking the supporting structure. When a panel does not have adequate clearance, inspect the shock mounts for looseness, cracks, or deterioration.

REPAIR OF AIRCRAFT INSTRUMENTS

The repair of aircraft instruments is highly specialized, requiring special tools and equipment. Instrument repairmen must have had specialized training or extensive on-the-job training in instrument repair. For these reasons, the repair of instruments must be performed by a properly certificated instrument repair facility. However, mechanics are responsible for the installation, connection, removal, servicing, and functional checking of the instruments.

AIRCRAFT PRESSURE GAGES

Pressure gages are used to indicate the pressure at which engine oil is forced through the bearings,

FIGURE 12–2. Engine gage unit.

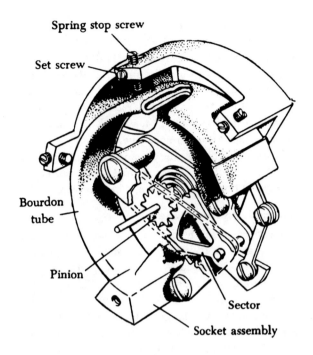

FIGURE 12–3. Bourdon tube pressure gage.

oil passages, and moving parts of the engine and the pressure at which fuel is delivered to the carburetor or fuel control. They are also used to measure the pressure of air in de-icer systems and gyroscope drives, of fuel/air mixtures in the intake manifold, and of liquid or gases in several other systems.

Engine Gage Unit

The engine gage unit is comprised of three separate instruments housed in a single case. A typical engine gage unit, containing gages for oil and fuel pressure and oil temperature, is shown in figure 12–2.

Two types of oil temperature gages are available for use in an engine gage unit. One type consists of an electrical resistance type oil thermometer, supplied electrical current by the aircraft d.c. power system. The other type, a capillary oil thermometer, is a vapor pressure type thermometer consisting of a bulb connected by a capillary tube to a Bourdon tube. A pointer, connected to the Bourdon tube through a multiplying mechanism, indicates on a dial the temperature of the oil.

The Bourdon tube is an aircraft instrument made of metal tubing, oval or somewhat flattened in cross section (figure 12–3). The metal tubing is closed at one end and mounted rigidly in the instrument case at its other end.

The fluid whose pressure is to be measured is introduced into the fixed end of the Bourdon tube by a small tube leading from the fluid system to the instrument. The greater the pressure of the fluid, the more the Bourdon tube tends to become straight. When the pressure is reduced or removed, the inherent springiness of the metal tube causes it to curve back to its normal shape.

If an indicator needle or pointer is attached to the free end of the Bourdon tube, its reactions to changes in the fluid pressure can be observed.

Hydraulic Pressure Gage

The mechanisms used in raising and lowering the landing gear or flaps in most aircraft are operated by a hydraulic system. A pressure gage to measure the differential pressure in the hydraulic system indicates how this system is functioning. Hydraulic pressure gages are designed to indicate either the pressure of the complete system or the pressure of an individual unit in the system.

A typical hydraulic gage is shown in figure 12–4. The case of this gage contains a Bourdon tube and a gear-and-pinion mechanism by which the Bourdon tube's motion is amplified and transferred to the pointer. The position of the pointer on the calibrated dial indicates the hydraulic pressure in p.s.i.

The pumps which supply pressure for the operation of an aircraft's hydraulic units are driven either by the aircraft's engine or by an electric motor, or both. Some installations use a pressure

accumulator to maintain a reserve of fluid under pressure at all times. In such cases the pressure gage registers continuously. With other installations, operating pressure is built up only when needed, and pressure registers on the gage only during these periods.

De-icing Pressure Gage

The rubber expansion boots, which de-ice the leading edges of wings and stabilizers on some aircraft, are operated by a compressed air system. The de-icing system pressure gage measures the difference between prevailing atmospheric pressure and the pressure inside the de-icing system, indicating whether there is sufficient pressure to operate the de-icer boots. The gage also provides a method of measurement when adjusting the relief-valve and the regulator of the de-icing system.

A typical de-icing pressure gage is shown in figure 12–5. The case is vented at the bottom to keep the interior at atmospheric pressure, as well as to provide a drain for any moisture which might accumulate.

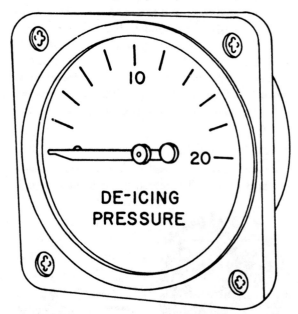

FIGURE 12–5. De-icing pressure gage.

The pressure-measuring mechanism of the de-icing pressure gage consists of a Bourdon tube and a sector gear, with a pinion for amplifying the motion of the tube and transferring it to the pointer. The de-icing system pressure enters the Bourdon tube through a connection at the back of the case.

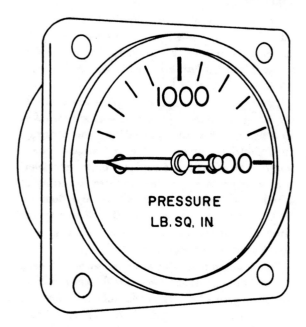

FIGURE 12–4. Hydraulic pressure gage

472

The range of the gage is typically from zero p.s.i. to 20 p.s.i., with the scale marked in 2-p.s.i. graduations as shown in figure 12–5.

When installed and connected into an aircraft's de-icing pressure system, the gage reading always remains at zero unless the de-icing system is operating. The gage pointer will fluctuate from zero p.s.i. to approximately 8 p.s.i. under normal conditions, because the de-icer boots are periodically inflated and deflated. This normal fluctuation should not be confused with oscillation.

Diaphragm-Type Pressure Gages

This type of pressure gage uses a diaphragm for measuring pressure. The pressure or suction to be measured is admitted to the pressure-sensitive diaphragm through an opening in the back of the instrument case (figure 12–6).

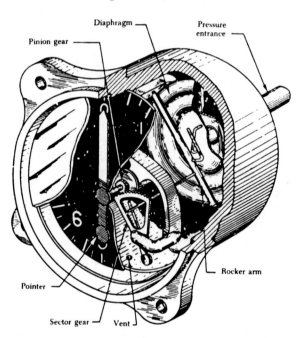

FIGURE 12–6. Diaphragm-type pressure gage.

An opposing pressure, such as that of the atmosphere, is admitted through a vent in the case (figure 12–6). Since the walls of the diaphragm are very thin, an increase of pressure will cause it to expand, and a decrease in pressure will cause it to contract. Any movement of the diaphragm is transferred to the pointer by means of the rocker shaft, sector, and pinion, which are connected to the front side of the diaphragm. This gage is also a differential-pressure measuring device since it indicates the difference between the pressure applied at the vent

of the case and the pressure or suction inside the diaphragm.

Suction Gages

Suction gages are used on aircraft to indicate the amount of suction that actuates the air-driven gyroscopic instruments. The spinning rotors of gyroscopic instruments are kept in motion by streams of air directed against the rotor vanes. These airstreams are produced by pumping air out of the instrument cases by the vacuum pump. Atmospheric pressure then forces air into the cases through filters, and it is this air that is directed against the rotor vanes to turn them.

The suction gage indicates whether the vacuum system is working properly. The suction gage case is vented to the atmosphere or to the line of the air filter, and contains a pressure-sensitive diaphragm plus the usual multiplying mechanism which amplifies the movement of the diaphragm and transfers it to the pointer. The reading of a suction gage indicates the difference between atmospheric pressure and the reduced pressure in the vacuum system.

Manifold Pressure Gage

The manifold pressure gage is an important instrument in an aircraft powered by a reciprocating engine. The gage is designed to measure absolute pressure. This pressure is the sum of the air pressure and the added pressure created by the supercharger. The dial of the instrument is calibrated in inches of mercury (Hg).

When the engine is not running, the manifold pressure gage records the existing atmospheric pressure. When the engine is running, the reading obtained on the manifold pressure gage depends on the engine's r.p.m. The manifold pressure gage indicates the manifold pressure immediately before the cylinder intake ports.

The schematic of one type of manifold pressure gage is shown in figure 12–7. The outer shell of the gage protects and contains the mechanism. An opening at the back of the case provides for the connection to the manifold of the engine.

The gage contains an aneroid diaphragm and a linkage for transmitting the motion of the diaphragm to the pointer. The linkage is completely external to the pressure chamber, and thus is not exposed to the corrosive vapors of the manifold. The pressure existing in the manifold enters the sealed chamber through a damping tube, which is a short length of capillary tubing at the rear of the

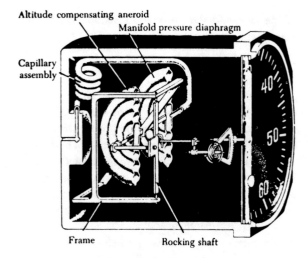

Figure 12-7. Manifold pressure gage.

case. This damping tube acts as a safety valve to prevent damage to the instrument by engine backfire. The sudden surge of pressure caused by backfire is considerably reduced by the restricted capillary tubing.

When installing a manifold pressure gage, care should be taken to ensure that the pointer is vertical when registering 30 in. Hg.

When an engine is not running, the manifold pressure gage reading should be the same as the local barometric pressure. It can be checked against a barometer known to be in proper operating condition. In most cases the altimeter in the aircraft can be used, since it is a barometric instrument. With the aircraft on the ground, the altimeter hands should be set to zero and the instrument panel tapped lightly a few times to remove any possible frictional errors. The barometer scale on the altimeter face will indicate local atmospheric pressure when the altimeter hands are at zero. The manifold pressure gage should agree with this pressure reading. If it does not, the gage should be replaced with a gage that is operating properly.

If the pointer fails to respond entirely, the mechanism is, in all probability, defective. The gage should be removed and replaced. If the pointer responds but indicates incorrectly, there may be moisture in the system, obstruction in the lines, a leak in the system, or a defective mechanism.

When doubt exists about which of these items is the cause of the malfunction, the engine should be operated at idle speed and the drain valve (usually located near the gage) opened for a few minutes. This will usually clear the system of moisture. To

clear an obstruction, the lines may be disconnected and blown clear with compressed air. The gage mechanism may be checked for leaks by disconnecting the line at the engine end and applying air pressure until the gage indicates 50 in. Hg. Then the line should be quickly closed. A leak is present if the gage pointer returns to atmospheric pressure. If a leak is evident but cannot be found, the gage should be replaced.

PITOT-STATIC SYSTEM

Three of the most important flight instruments are connected into a pitot-static system. These instruments are the airspeed indicator, the altimeter, and the rate-of-climb indicator. Figure 12-8 shows these three instruments connected to a pitot-static tube head.

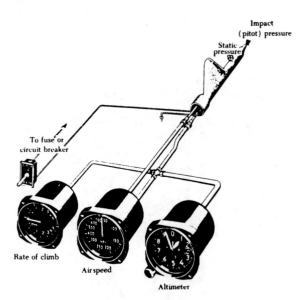

Figure 12-8. Pitot-static system.

The pitot-static system head, or pitot-static tube as it is sometimes called, consists of two sections. As shown in figure 12-9, the forward section is open at the front end to receive the full force of the impact air pressure. At the back of this section is a baffle plate to protect the pitot tube from moisture and dirt that might otherwise be blown into it. Moisture can escape through a small drain hole at the bottom of the forward section.

The pitot, or pressure, tube leads back to a chamber in the "shark-fin" projection near the rear of the assembly. A riser, or upright tube, leads the air from this chamber through tubing to the airspeed indicator.

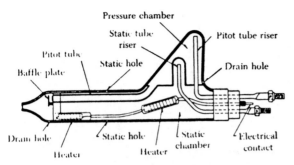

FIGURE 12-9. Pitot-static system head.

The rear, or static, section of the pitot-static tube head is pierced by small openings on the top and bottom surfaces. These openings are designed and located so that this part of the system will provide accurate measurements of atmospheric pressure in a static, or still, condition. The static section contains a riser tube which is connected to the airspeed indicator, the altimeter, and the rate-of-climb indicator.

Many pitot-static tubes are provided with heating elements to prevent icing during flight (figure 12-9). During ice-forming conditions, the electrical heating elements can be turned on by means of a switch in the cockpit. The electrical circuit for the heater element may be connected through the ignition switch. Thus, in case the heater switch is inadvertently left in the "on" position, there will be no drain on the battery when the engine is not operating.

The pitot-static tube head is mounted on the outside of the aircraft at a point where the air is least likely to be turbulent. It is pointed in a forward direction parallel to the aircraft's line of flight. One general type of tube head is designed for mounting on a streamlined mast extending below the nose of the aircraft fuselage. Another type is designed for installation on a boom extending forward of the leading edge of the wing. Both types are shown in figure 12-10. Although there is a slight difference in their construction, they operate identically.

Most pitot-static tubes are manufactured with a union connection in both lines from the head, near the point at which the tube head is attached to the mounting boom or mast (figure 12-10). These connections simplify removal and replacement, and are usually reached through an inspection door in the wing or fuselage. When a pitot-static tube head is to be removed, these connections should be disconnected before any mounting screws and lockwashers are removed.

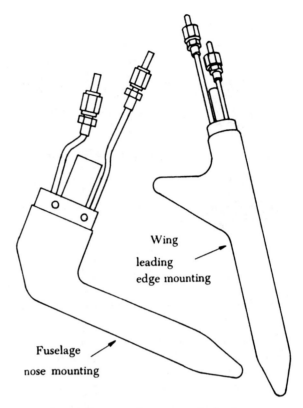

FIGURE 12-10. Pitot-static tube heads.

In many aircraft equipped with a pitot-static tube, an alternate source of static pressure is provided for emergency use. A schematic diagram of a typical system is shown in figure 12-11. As shown in the diagram, the alternate source of static pressure may be vented to the interior of the aircraft.

Another type of pitot-static system provides for the location of the pitot and static sources at sepa-

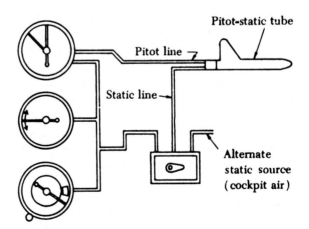

FIGURE 12-11. Pitot-static system with alternate source of static pressure.

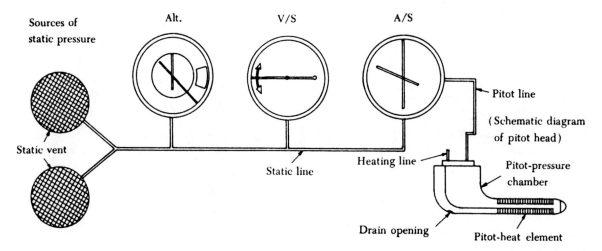

FIGURE 12–12. Pitot-static system with separate sources of pressure.

rate positions on the aircraft. This type of system is illustrated in figure 12–12.

Impact pressure is taken from the pitot-head (figure 12–12) which is mounted parallel to the longitudinal axis of the aircraft and generally in line with the relative wind. The leading edge of the wing, nose section, or vertical stabilizer are the usual mounting positions, since at those points there is usually a minimum disturbance of air due to motion of the aircraft.

Static pressure in this type of pitot-static system is taken from the static line attached to a vent or vents mounted flush with the fuselage or nose section. On aircraft using a flush-mounted static source, there may be two vents, one on each side of the aircraft. This compensates for any possible variation in static pressure on the vents due to erratic changes in aircraft attitude. The two vents are usually connected by a Y-type fitting. In this type of system, clogging of the pitot opening by ice or dirt (or failure to remove the pitot cover) affects the airspeed indicator only.

A pitot-static system used on a pressurized, multi-engine aircraft is shown in figure 12–13. Three additional units, the cabin pressure controller, the cabin differential pressure gage, and the autopilot system are integrated into the static system. Both heated and unheated flushmounted static ports are used.

Altimeters

There are many kinds of altimeters in general use today. However, they are all constructed on the same basic principle as an aneroid barometer. They all have pressure-responsive elements (aneroids) which expand or contract with the pressure change of different flight levels. The heart of an altimeter is its aneroid mechanism (figure 12–14). The expansion or contraction of the aneroid with pressure changes actuates the linkage, and the indicating hands show altitude. Around the aneroid mechanism of most altimeters is a device called the bi-metal yoke. As the name implies, this device is composed of two metals and performs the function of compensating for the effect that temperature has on the metals of the aneroid mechanism.

The presentation of altitude by altimeters in current use varies from the multi-pointer type to the drum and single pointer, and the digital counter and single pointer types.

The dial face of the typical altimeter is graduated with numerals from zero to 9 inclusive, as shown in figure 12–15. Movement of the aneroid element is transmitted through a gear train to the three hands on the instrument face. These hands sweep the calibrated dial to indicate the altitude of the aircraft. The shortest hand indicates altitude in tens of thousands of feet; the intermediate hand, in thousands of feet; and the longest hand, in hundreds of feet in 20-ft. increments. A barometric scale, located at the right of the instrument face, can be set by a knob located at the lower left of the instrument case. The barometric scale indicates barometric pressure in inches of mercury.

Since atmospheric pressure continually changes, the barometric scale must be re-set to the local station altimeter setting before the altimeter will indicate the correct altitude of the aircraft above

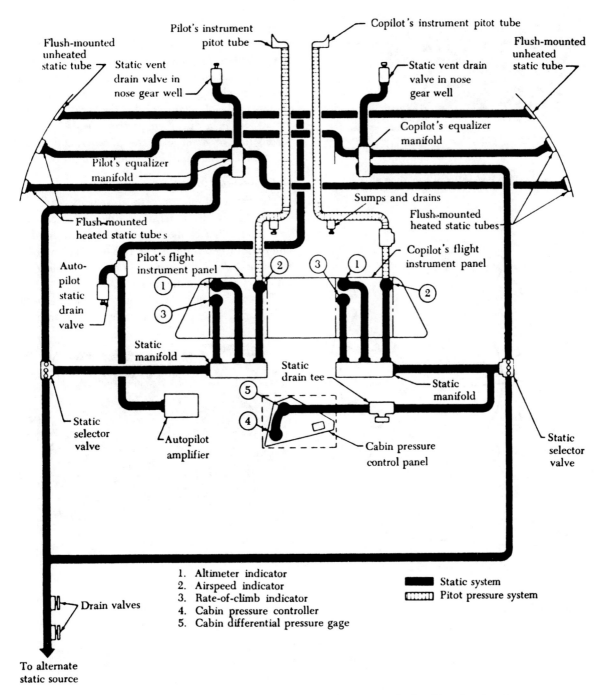

Flush-mounted unheated static tube

Static vent drain valve in nose gear well

Pilot's instrument pitot tube

Copilot's instrument pitot tube

Static vent drain valve in nose gear well

Flush-mounted unheated static tube

Copilot's equalizer manifold

Pilot's equalizer manifold

Sumps and drains

Flush-mounted heated static tubes

Flush-mounted heated static tubes

Auto-pilot static drain valve

Pilot's flight instrument panel

Copilot's flight instrument panel

Static manifold

Static drain tee

Static manifold

Static selector valve

Autopilot amplifier

Cabin pressure control panel

Static selector valve

Drain valves

To alternate static source

1. Altimeter indicator
2. Airspeed indicator
3. Rate-of-climb indicator
4. Cabin pressure controller
5. Cabin differential pressure gage

▮▮▮ Static system
▥▥▥ Pitot pressure system

FIGURE 12–13.   Schematic of typical pitot-static system on pressurized multi-engine aircraft.

sea level. When the setting knob is turned, the barometric scale, the hands, and the aneroid element move to align the instrument mechanism with the new altimeter setting.

Two setting marks, inner and outer, indicate barometric pressure in feet of altitude. They operate in conjunction with the barometric scale, and indications are read on the altimeter dial. The outer mark indicates hundreds of feet, and the inner mark thousands of feet. Since there is a limit to the graduations which can be placed on the barometric scale, the setting marks are used when the barometric pressure to be read is outside the limits of the scale.

477

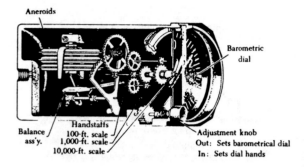

FIGURE 12–14. Mechanism of a sensitive altimeter.

FIGURE 12–15. Sensitive altimeter.

Altimeter Errors

Altimeters are subject to various mechanical errors. A common one is that the scale is not correctly oriented to standard pressure conditions. Altimeters should be checked periodically for scale errors in altitude chambers where standard conditions exist.

Another mechanical error is the hysteresis error. This error is induced by the aircraft maintaining a given altitude for an extended period of time, then suddenly making a large altitude change. The resulting lag or drift in the altimeter is caused by the elastic properties of the materials which comprise the instrument. This error will eliminate itself with slow climbs and descents or after maintaining a new altitude for a reasonable period of time.

In addition to the errors in the altimeter mechanism, another error called installation error affects the accuracy of indications. The error is caused by the change of alignment of the static pressure port with the relative wind. The change of alignment is caused by changes in the speed of the aircraft and in the angle of attack, or by the location of the static port in a disturbed pressure field. Improper installation or damage to the pitot-static tube will also result in improper indications of altitude.

Rate-of-Climb Indicators

The rate-of-climb, or vertical velocity, indicator (figure 12–16) is a sensitive differential pressure gage that indicates the rate at which an aircraft is climbing or descending. The rate-of-climb indicator is connected to the static system and senses the rate of change of static pressure.

The rate of altitude change, as shown on the indicator dial, is positive in a climb and negative

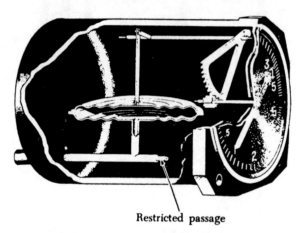

Restricted passage

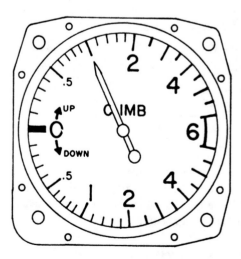

FIGURE 12–16. Typical rate-of-climb indicator.

when descending in altitude. The dial pointer moves in either direction from the zero point, depending on whether the aircraft is going up or down. In level flight the pointer remains at zero.

The operation of a climb indicator is illustrated in figure 12–16. The case of the instrument is airtight except for a small connection through a restricted passage to the static line of the pitot-static system.

Inside the sealed case of the rate-of-climb indicator is a diaphragm with connecting linkage and gearing to the indicator pointer. Both the diaphragm and the case receive air at atmospheric pressure from the static line. When the aircraft is on the ground or in level flight, the pressures inside the diaphragm and the instrument case remain the same and the pointer is at the zero indication. When the aircraft climbs the pressure inside the diaphragm decreases but, due to the metering action of the restricted passage, the case pressure will remain higher and cause the diaphragm to contract. The diaphragm movement actuates the mechanism, causing the pointer to indicate a rate of climb.

When the aircraft levels off, the pressure in the instrument case is equalized with the pressure in the diaphragm. The diaphragm returns to its neutral position and the pointer returns to zero.

In a decent, the pressure conditions are reversed. The diaphragm pressure immediately becomes greater than the pressure in the instrument case. The diaphragm expands and operates the pointer mechanism to indicate the rate of descent.

When the aircraft is climbing or descending at a constant rate, a definite ratio between the diaphragm pressure and the case pressure is maintained through the calibrated restricted passage, which requires approximately 6 to 9 sec. to equalize the two pressures, causing a lag in the proper reading. Any sudden or abrupt changes in the aircraft's attitude may cause erroneous indications due to the sudden change of airflow over the static ports.

The instantaneous rate-of-climb indicator is a more recent development which incorporates acceleration pumps to eliminate the limitations associated with the calibrated leak. For example, during an abrupt climb, vertical acceleration causes the pumps to supply extra air into the diaphragm to stabilize the pressure differential without the usual lag time. During level flight and steady-rate climbs and descents, the instrument operates on the same principles as the conventional rate-of-climb indicator.

A zero-setting system, controlled by a setscrew or an adjusting knob permits adjustment of the pointer to zero. The pointer of an indicator should indicate zero when the aircraft is on the ground or maintaining a constant pressure level in flight.

Airspeed Indicator

Airspeed indicators are sensitive pressure gages which measure the difference between the pitot and static pressures, and present such difference in terms of indicated airspeed. Airspeed indicators are made by various manufacturers and vary in their mechanical construction. However, the basic construction and operating principle is the same for all types.

The airspeed indicator (figure 12–17) is a sensitive, differential pressure gage which measures and indicates promptly the differential between the impact and the static air pressures surrounding an airplane at any moment of flight. The airspeed indicator consists primarily of a sensitive metallic diaphragm whose movements, resulting from the slightest difference in impact and static air pressures, are multiplied by means of a link, a rocking shaft, a sector with hairspring and pinion, and a tapered shaft to impart rotary motion to the pointer, which indicates the aircraft velocity on the dial face in terms of knots or m.p.h.

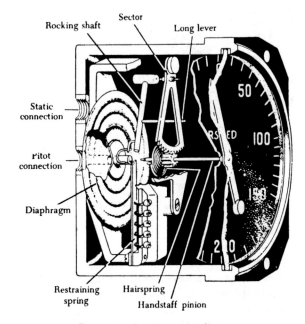

FIGURE 12–17. Airspeed indicator.

479

Most airspeed indicators are marked to show speed limitations at a glance. The never-exceed velocity is designated by a red radial line. A yellow arc designates the cautionary range, and a white arc is used to indicate the range of permissible limits of flap operation.

The dial numbers used on different airspeed indicators are indicative of the type of aircraft in which they are used; for example, an airspeed indicator with a range of zero to 160 knots is commonly used in many light aircraft. Other types, such as a 430-knot indicator, are used on larger and faster aircraft.

Another type of airspeed indicator in use is the maximum allowable airspeed indicator shown in figure 12–18. This indicator includes a maximum allowable needle, which shows a decrease in maximum allowable airspeed with an increase in altitude. It operates from an extra diaphragm in the airspeed indicator which senses changes in altitude and measures this change on the face of the instrument. Its purpose is to indicate maximum allowable indicated airspeed at any altitude.

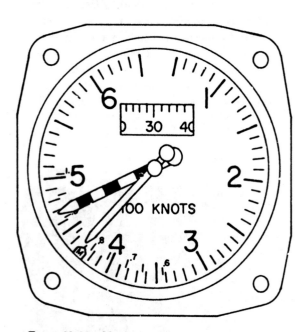

FIGURE 12–18. Maximum allowable airspeed indicator.

The type of airspeed indicator known as a true airspeed indicator is shown in figure 12–19. It uses an aneroid, a differential pressure diaphragm, and a bulb temperature diaphragm, which respond respectively to changes in barometric pressure, impact pressure, and free air temperature. The actions of the diaphragms are mechanically resolved to indicate true airspeed in knots. A typical true airspeed indicator is designed to indicate true airspeed from 1,000 ft. below sea level to 50,000 ft. above sea level under free air temperature conditions from $+40°$ to $-60°$ C.

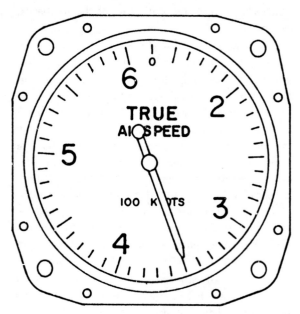

FIGURE 12–19. True airspeed indicator.

Mach Indicator

Machmeters indicate the ratio of aircraft speed to the speed of sound at the particular altitude and temperature existing at any time during flight.

Construction of a Mach indicator is much the same as that of an airspeed indicator. It will usually contain a differential pressure diaphragm which senses pitot-static pressure, and an aneroid diaphragm which senses static pressure. By mechanical means, changes in pressures are then displayed on the instrument face in terms of Mach numbers.

The Machmeter shown in figure 12–20A is designed to operate in the range of 0.3 to 1.0 Mach and at altitudes from zero to 50,000 feet. The Machmeter shown in figure 12–20B is designed to operate in the range of 0.5 to 1.5 at altitudes up to 50,000 feet.

Combined Airspeed/Mach Indicator

Combined airspeed/Mach indicators are provided for aircraft where instrument space is at a premium and it is desirable to present information on a combined indicator. These instruments show indicated

A

B

FIGURE 12–20. Machmeters.

airspeed, Mach, and limiting Mach by use of impact and static pressures and an altitude aneroid.

These combined units utilize a dual-pointed needle which shows airspeed on a fixed scale and Mach indication on a rotating scale. A knurled knob located on the lower portion of the instrument is provided to set a movable index marker to reference

a desired speed. A combined airspeed/Mach indicator is shown in figure 12–21.

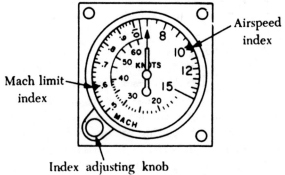

FIGURE 12–21. Combined airspeed/Mach indicator.

MAINTENANCE OF PITOT-STATIC SYSTEMS

The specific maintenance instructions for any pitot-static system are usually detailed in the applicable aircraft manufacturer's maintenance manual. However, there are certain inspections, procedures, and precautions to be observed that apply to all systems.

Pitot tubes and their supporting masts should be inspected for security of mounting and evidence of damage. Checks should also be made to ensure that electrical connections are secure. The pitot pressure entry hole, drain holes, and static holes or ports should be inspected to ensure that they are unobstructed. The size of the drain holes and static holes is aerodynamically critical. They must never be cleared of obstruction with tools likely to cause enlargement or burring.

Heating elements should be checked for functioning by ensuring that the pitot tube begins to warm up when the heater is switched "on." If an ammeter or loadmeter is installed in the circuit, a current reading should be taken.

The inspections to be carried out on the individual instruments are primarily concerned with security, visual defects, and proper functioning. The zero setting of pointers must also be checked. At the time of inspecting the altimeter, the barometric pressure scale should be set to read field barometric pressure. With this pressure set, the instrument should read zero within the tolerances specified for the type installed. No adjustment of any kind can be made, if the reading is not within limits, the instrument must be replaced.

Leak Testing Pitot-Static Systems

Aircraft pitot-static systems must be tested for leaks after the installation of any component parts, when system malfunction is suspected, and at the periods specified in the Federal Aviation Regulations.

The method of leak testing and the type of equipment to use depends on the type of aircraft and its pitot-static system. In all cases, pressure and suction must be applied and released slowly to avoid damage to the instruments. The method of testing consists basically of applying pressure and suction to pressure heads and static vents respectively, using a leak tester and coupling adapters. The rate of leakage should be within the permissible tolerances prescribed for the system. Leak tests also provide a means of checking that the instruments connected to a system are functioning properly. However, a leak test does not serve as a calibration test.

Upon completion of the leak test, be sure that the system is returned to the normal flight configuration. If it was necessary to blank off various portions of a system, check to be sure that all blanking plugs, adapters, or pieces of adhesive tape have been removed.

TURN-AND-BANK INDICATOR

The turn-and-bank indicator, figure 12–22, also referred to as the turn-and-slip or needle-and-ball indicator, shows the correct execution of a bank and turn and indicates the lateral attitude of the aircraft in level flight.

The turn needle is operated by a gyro, driven either by a vacuum, air pressure, or electricity. The turn needle indicates the rate, in number of degrees per second, at which an aircraft is turning about its vertical axis. It also provides information on the amount of bank. The gyro axis is horizontally mounted so that the gyro rotates up and away from the pilot. The gimbal around the gyro is pivoted fore and aft.

Gyroscopic precession causes the rotor to tilt when the aircraft is turned. Due to the direction of rotation, the gyro assembly tilts in the opposite direction from which the aircraft is turning. This prevents the rotor axis from becoming vertical to the earth's surface. The linkage between the gyro assembly and the turn needle, called the reversing mechanism, causes the needle to indicate the proper direction of turn.

Power for the electric gyro may be supplied from either an a.c. or d.c. source.

The principal value of the electric gyro in light aircraft is its safety factor. In single-engine aircraft equipped with vacuum-driven attitude and heading indicators, the turn needle is commonly operated by an electric gyro. In the event of vacuum system failure and loss of two gyro instruments, the pilot still has a reliable standby instrument for emergency operation. Operated on current directly from the battery, the electric turn indicator is reliable as long as current is available, regardless of generator or vacuum system malfunction. In the electric instrument, the gyro is a small electric motor and flywheel. Otherwise both electric and vacuum-driven turn-needles are de-

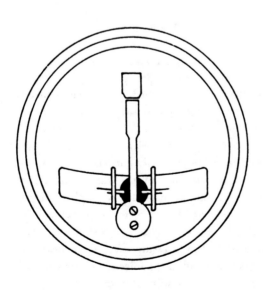

Two minute turn indicator

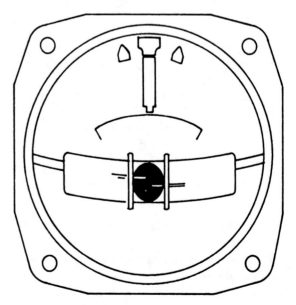

Four minute turn indicator

FIGURE 12–22. Two types of turn-and-bank indicators.

signed to use the same gyroscopic principle of precession.

Power for the suction-driven turn needle is regulated by a restrictor installed between the main suction line and the instrument to produce a desired suction and rotor speed. Since the needle measures the force of precession, excessivly high or low vacuum results in unreliable turn-needle operation. For a specific rate of turn, low vacuum produces less than normal rotor speed and, therefore, less needle deflection for this specific rate of turn. The reverse is true for the condition of high vacuum.

Of the two types of turn needles shown in figure 12–22, the 2-min. turn indicator is the older. If the instrument is accurately calibrated; a single needle-width deflection on the 2-min. indicator means that the aircraft is turning at 3° per sec., or standard (2 min. for a 360° turn). On the 4-min. indicator, a single needle-width deflection shows when the aircraft is turning at $1\frac{1}{2}$° per sec., or half standard rate (4 min. for a 360° turn). The 4-min. turn indicator was developed especially for high-speed aircraft.

The slip indicator (ball) part of the instrument is a simple inclinometer consisting of a sealed, curved glass tube containing kerosene and a black agate or a common steel ball bearing, which is free to move inside the tube. The fluid provides a damping action, ensuring smooth and easy movement of the ball. The tube is curved so that in a horizontal position the ball tends to seek the lowest point. A small projection on the left end of the tube contains a bubble of air which compensates for expansion of the fluid during changes in temperature. Two strands of wire wound around the glass tube fasten the tube to the instrument case and also serve as reference markers to indicate the correct position of the ball in the tube. During coordinated straight-and-level flight, the force of gravity causes the ball to rest in the lowest part of the tube, centered between the reference wires.

Maintenance Practices for Turn-and-Bank Indicators

Errors in turn needle indications are usually due to insufficient or excessive rotor speed or inaccurate adjustment of the calibrating spring. There is no practical operational test or checkout of this instrument, other than visually noting that the indicator pointer and the ball are centered.

SYNCHRO-TYPE REMOTE INDICATING INSTRUMENTS

A synchro system is an electrical system used for transmitting information from one point to another. Most position-indicating instruments are designed around a synchro system. The word "synchro" is a shortened form of synchronous, and refers to any one of a number of electrical devices capable of measuring and indicating angular deflection. Synchro systems are used as remote position indicators for landing gear and flap systems, in autopilot systems, in radar systems, and many other remote-indicating applications. There are different types of synchro systems. The three most common are: (1) Autosyn, (2) selsyn, and (3) magnesyn. These systems are similar in construction and all operate on identical electrical and mechanical principles.

D.C. Selsyn Systems

The d.c. selsyn system is a widely used electrical method of indicating a remote mechanical condition. Specifically, d.c. selsyn systems can be used to show the movement and position of retractable landing gear, wing flaps, cowl flaps, oil cooler doors, or similar movable parts of the aircraft.

A selsyn system consists of a transmitter, an indi-

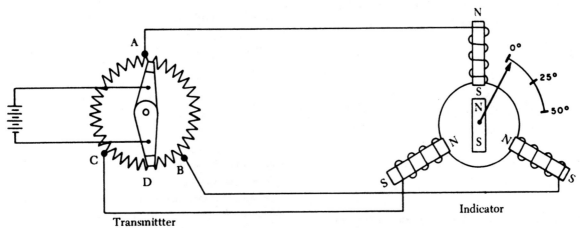

FIGURE 12–23. Schematic diagram of a d.c. selsyn system.

cator and connecting wires. The voltage required to operate the selsyn system is supplied from the aircraft's electrical system.

A selsyn system is shown schematically in figure 12–23. The transmitter consists of a circular resistance winding and a rotatable contact arm. The rotatable contact arm turns on a shaft in the center of the resistance winding. The two ends of the arm, or brushes, always touch the winding on opposite sides. The shaft to which the contact arm is fastened protrudes through the end of the transmitter housing and is attached to the unit (flaps, landing gear, etc.) whose position is to be transmitted. The transmitter is usually connected to the unit through a mechanical linkage. As the unit moves, it causes the transmitter shaft to turn. Thus, the arm can be turned so that voltage can be applied at any two points around the circumference of the winding.

As the voltage at the transmitter taps is varied, the distribution of currents in the indicator coils varies and the direction of the resultant magnetic field across the indicator is changed. The magnetic field across the indicating element corresponds in position to the moving arm in the transmitter. Whenever the magnetic field changes direction, the polarized motor turns and aligns itself with the new position of the field. The rotor thus indicates the position of the transmitter arm.

When the d.c. selsyn system is used to indicate the position of landing gear, an additional circuit is connected to the transmitter winding, which acts as a lock-switch circuit. The purpose of this circuit is to show when the landing gear is up and locked, or down and locked. Lock switches are shown connected into a three-wire system in figure 12–24.

A resistor is connected between one of the taps of the transmitter at one end and to the individual lock switches at the other end. When either lock switch is closed, the resistance is added into the transmit-

ter circuit to cause an unbalance in one section of the transmitter winding. This unbalance causes the current flowing through one of the indicator coils to change. The resultant movement in the indicator pointer shows that the lock switch has been closed. The lock switch is mechanically connected to the landing gear up- or down-locks, and when the landing gear locks either up or down, it closes the lock switch connected to the selsyn transmitter. This locking of the landing gear is repeated on the indicator.

Magnesyn System

The magnesyn system is an electrical self-synchronous device used to transmit the direction of a magnetic field from one coil to another. The magnesyn position system is essentially a method of measuring the extent of the movement of such elements as the wing and cowl flaps, trim tabs, landing gear, or other control surfaces. The two main units of the system are the transmitter and the indicator (figure 12–25).

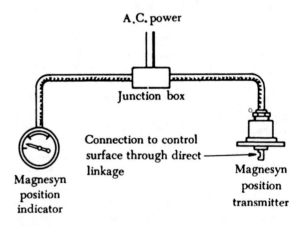

FIGURE 12–25.   Magnesyn position-indicating system.

In a magnesyn transmitter a soft iron ring is placed around a permanent magnet so that most of the magnet's lines of force pass within the ring. This circular core of magnetic material is provided with a single continuous electrical winding of fine wire. Figure 12–26 shows an electrical wiring schematic of a magnesyn system. The circular core of magnetic material and the winding are the essential components of the magnesyn stator. The rotor consists of the permanent magnet.

The movement of the control surface of the aircraft causes a proportional movement of the transmitter shaft. This in turn causes a rotary displacement of the magnet. Varying voltages are set up in

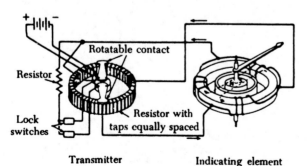

FIGURE 12–24.   A double-lock switch in a three-wire selsyn system.

484

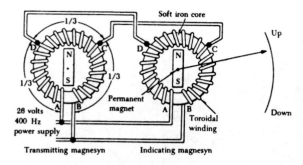

FIGURE 12–26. Magnesyn system.

the magnesyn stator, depending on the position of the magnet. The voltage is transmitted to a magnesyn indicator which indicates on a dial the values received from the transmitter. The indicator consists essentially of a magnesyn, a graduated dial, and a pointer. The pointer is attached to the shaft and the shaft is attached to the magnet; thus, movement of the magnet causes movement of the pointer.

REMOTE-INDICATING FUEL AND OIL PRESSURE GAGES

Fuel and oil pressure indications can be conveniently obtained through use of the various synchro systems. The type of synchro system used may be the same for either fuel or oil pressure measurement; however, an oil system transmitter is usually not interchangeable with a fuel system transmitter.

A typical oil pressure indicating system is shown in figure 12–27. A change in oil pressure introduced into the synchro transmitter causes an electrical signal to be transmitted through the interconnecting wiring to the synchro receiver. This signal causes the receiver rotor and the indicator pointer to move a distance proportional to the amount of pressure exerted by the oil.

Most oil pressure transmitters are composed of two main parts, a bellows mechanism for measuring pressure and a synchro assembly. The pressure of the oil causes linear displacement of the synchro rotor. The amount of displacement is proportional to the pressure, and varying voltages are set up in the synchro stator. These voltages are transmitted to the synchro indicator.

In some installations, dual indicators are used to obtain indications from two sources. On some aircraft, both oil and fuel pressure transmitters are joined through a junction and operate a synchro oil and fuel pressure indicator (dual side-by-side), thus combining both gages in one case.

CAPACITOR-TYPE FUEL QUANTITY SYSTEM

The capacitor-type fuel quantity system is an electronic fuel measuring device that accurately determines the weight of the fuel in the tanks of an aircraft. The basic components of the system are an indicator, a tank probe, a bridge unit, and an amplifier. In some systems the bridge unit and amplifier are one unit, mounted in the same box. More recent systems have been designed with the bridge unit and a transistorized amplifier built into the instrument case.

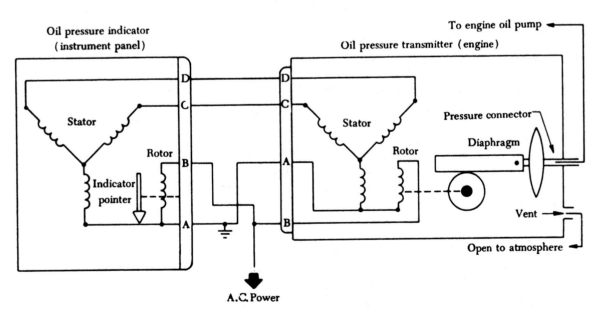

FIGURE 12–27. Oil pressure synchro system.

485

The fuel quantity indicator shown in figure 12–28 is a sealed, self balancing, motor-driven instrument containing a motor, pointer assembly, transistorized amplifier, bridge circuit, and adjustment potentiometers. A change in the fuel quantity of a tank causes a change in the capacitance of the tank unit. The tank unit is one arm of a capacitance bridge circuit. The voltage signal resulting from the unbalanced bridge is amplified by a phase-sensitive amplifier in the power unit. This signal energizes one winding of a two-phase induction motor in the indicator. The induction motor drives the wiper or a rebalancing potentiometer in the proper direction to balance the bridge, and at the same time positions an indicator pointer to show the quantity of fuel remaining in the tank.

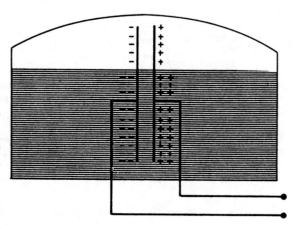

FIGURE 12–29. Simplified capacitance-tank circuit.

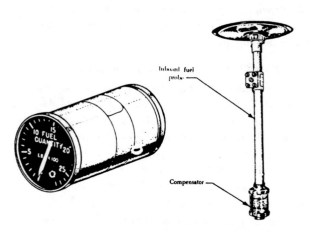

FIGURE 12–28. Indicator and probe of a capacitor type fuel quantity system.

A simplified capacitance bridge circuit is shown in figure 12–30. The fuel tank capacitor and a fixed reference capacitor are connected in series across a transformer secondary winding. A voltmeter is connected from the center of the transformer winding to a point between the two capacitors. If the two capacitances are equal, the voltage drop across them will be equal, and the voltage between the center tap and point P will be zero. As the fuel quantity increases, the capacitance of the tank unit increases, causing more current to flow in the tank unit leg of the bridge circuit. This will cause a voltage to exist across the voltmeter that is in phase with the voltage applied to the transformer. If the quantity of fuel in the tank decreases, there will be a smaller flow of current in the tank unit leg of the bridge.

A simplified version of a tank unit is shown in figure 12–29. The capacitance of a capacitor depends on three factors: (1) The area of the plates, (2) the distance between the plates, and (3) the dielectric constant of the material between the plates. The only variable factor in the tank unit is the dielectric of the material between the plates. When the tank is full, the dielectric material is all fuel. Its dielectric constant is about 2.07 at 0° C., compared to a dielectric constant of 1 for air. When the tank is half full, there is air between the upper half of the plates and fuel between the lower half. Thus, the capacitor has less capacitance than it had when the tank was full. When the tank is empty, there is only air between the plates; consequently, the capacitance is still less. Any change in fuel quantity between full and empty will produce a corresponding change in capacitance.

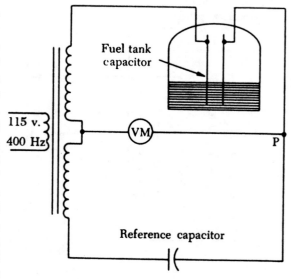

FIGURE 12–30. Simplified capacitance bridge circuit.

The voltage across the voltmeter will now be out of phase with the voltage applied to the transformer.

In an actual capacitor type fuel gage, the input to a two-stage amplifier is connected in place of the voltmeter. It amplifies the signal resulting from an unbalance in the bridge circuit. The output of the amplifier energizes a winding of the two-phase indicator motor. The other motor winding, called the line phase winding, is constantly energized by the same voltage that is applied to the transformer in the bridge circuit, but its phase is shifted 90° by a series capacitor. As a result, the indicator motor is phase sensitive; that is, it will operate in either direction, depending on whether the tank unit capacitance is increasing or decreasing.

As the tank unit capacitance increases or decreases because of a change in fuel quantity, it is necessary to readjust the bridge circuit to a balanced condition so the indicator motor will not continue to change the position of the indicating needle. This is accomplished by a balancing potentiometer connected across one-half of the transformer secondary, as shown in figure 12–31. The indicator motor drives this potentiometer wiper in the direction necessary to maintain continuous balance in the bridge.

The circuit shown in figure 12–31 is a self-balancing bridge circuit. An "empty" calibrating potentiometer and a "full" calibrating potentiometer are connected across portions of the transformer secondary winding at opposite ends of the winding. These potentiometers may be adjusted to balance the bridge voltages over the entire empty-to-full capacitance range of a specific system.

In some installations where the indicator shows the contents of only one tank and where the tank is fairly symmetrical, one unit is sufficient. However, for increased accuracy in peculiarly shaped fuel tanks, two or more units are connected in parallel to minimize the effects of changes in aircraft attitude and sloshing of fuel in the tanks.

ANGLE-OF-ATTACK INDICATOR

The angle-of-attack indicating system detects the local angle of attack of the aircraft from a point on the side of the fuselage and furnishes reference

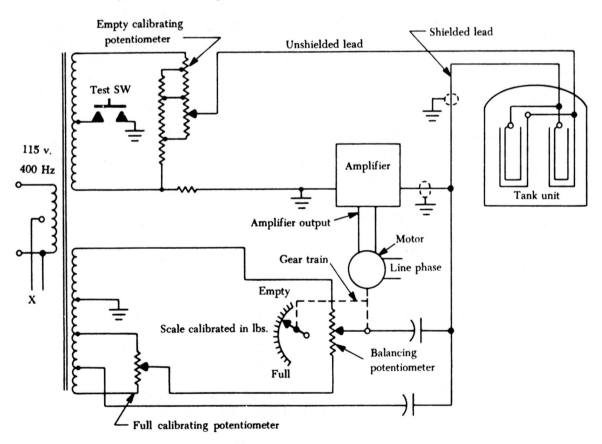

FIGURE 12–31. Self-balancing bridge circuit.

information for the control and actuation of other units and systems in the aircraft. Signals are provided to operate an angle-of-attack indicator (figure 12–32), located on the instrument panel, where a continuous visual indication of the local angel of attack is displayed.

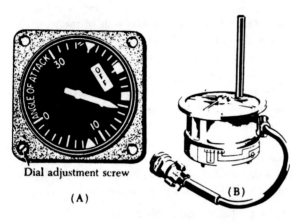

FIGURE 12–32. Angle-of-attack system. (A) Indicator (B) Transmitter.

A typical angle-of-attack system provides electrical signals for the operation of a rudder pedal shaker, which warns the operator of an impending stall when the aircraft is approaching the critical stall angle of attack. Electrical switches are actuated at the angle-of-attack indicator at various preset angles of attack.

The angle-of-attack indicating system consists of an airstream direction detector (transmitter) (figure 12–32B), and an indicator located on the instrument panel. The airstream direction detector contains the sensing element which measures local airflow direction relative to the true angle of attack by determining the angular difference between local airflow and the fuselage reference plane. The sensing element operates in conjunction with a servo-driven balanced bridge circuit which converts probe positions into electrical signals.

The operation of the angle-of-attack indicating system is based on detection of differential pressure at a point where the airstream is flowing in a direction that is not parallel to the true angle of attack of the aircraft. This differential pressure is caused by changes in airflow around the probe unit. The probe extends through the skin of the aircraft into the airstream.

The exposed end of the probe contains two parallel slots which detect the differential airflow pressure (figure 12–33). Air from the slots is transmitted through two separate air passages to separate compartments in a paddle chamber. Any differential pressure, caused by misalignment of the probe with respect to the direction of airflow, will cause the paddles to rotate. The moving paddles will rotate the probe, through mechanical linkage, until the pressure differential is zero. This occurs when the slots are symmetrical with the airstream direction.

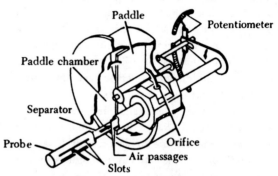

FIGURE 12–33. Airstream direction detector.

Two electrically separate potentiometer wipers, rotating with the probe, provide signals for remote indications. Probe position, or rotation, is converted into an electrical signal by one of the potentiometers which is the transmitter component of a self-balancing bridge circuit. When the angle of attack of the aircraft is changed and, subsequently, the position of the transmitter potentiometer is altered, an error voltage exists between the transmitter potentiometer and the receiver potentiometer the indicator. Current flows through a sensitive polarized relay to rotate a servomotor in the indicator. The servomotor drives a receiver/potentiometer in the direction required to reduce the voltage and restore the circuit to an electrically balanced condition. The indicating pointer is attached to, and moves with, the receiver/potentiometer wiper arm to indicate on the dial the relative angle of attack.

TACHOMETERS

The tachometer indicator is an instrument for indicating the speed of the crankshaft of a reciprocating engine and the speed of the main rotor assembly of a gas turbine engine.

The dials of tachometer indicators used with reciprocating engines are calibrated in r.p.m.; those used with turbine engines are calibrated in percentage of r.p.m. being used, based on the takeoff r.p.m. Figure 12–34 shows a typical dial for each of the indicators just described.

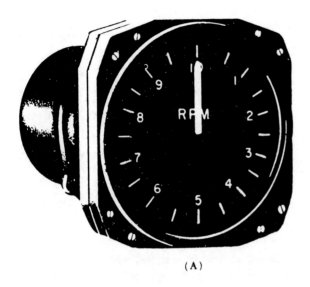

FIGURE 12–34. Tachometer. (A) Reciprocating engine type. (B) Turbine engine type.

There are two types of tachometer systems in wide use today: (1) The mechanical indicating system, and (2) the electrical indicating system.

Mechanical Indicating Systems

Mechanical indicating systems consist of an indicator connected to the engine by a flexible drive shaft. The indicator contains a flyweight assembly coupled to a gear mechanism that drives a pointer. As the drive shaft rotates, centrifugal force acts on the flyweights assembly and moves them to an angular position. This angular position varies with the r.p.m. of the engine. Movement of the flyweights is transmitted through the gear mechanism to the pointer. The pointer rotates to indicate the r.p.m. of the engine on the tachometer indicator.

Electric Indicating Systems

A number of different types and sizes of tachometer generators and indicators are used in aircraft electrical tachometer systems. Generally, the varous types of tachometer indicators and generators operate on the same basic principle. Thus, the system described will be representative of most electrical tachometer systems; the manufacturer's instructions should always be consulted for details of a specific tachometer system.

The typical tachometer system (figure 12–35) is a three-phase a.c. generator coupled to the aircraft engine, and connected electrically to an indicator mounted on the instrument panel. These two units are connected by a current-carrying cable. The generator transmits three-phase power to the synchron-

ous motor in the indicator. The frequency of the transmitted power is proportional to the engine speed. Through use of the magnetic drag principle, the indicator furnishes an accurate indication of engine speed.

Tachometer generators are small compact units, generally availble in three types: (1) The pad, (2) the swivel-nut, and (3) the screw type. These names are derived from the kind of mounting used in attaching the generator to the engine. The pad-type tachometer generator (figure 12–36A) is constructed with an end shield designed to permit attachment of the generator to a flat plate on the engine frame, or accessory reduction gearbox, with four bolts. The swivel-nut tachometer generator is constructed with a mounting nut which is free to turn in respect to the rest of the instrument. This type of generator can be held stationary while the mounting nut is screwed into place. The screw-type tachometer generator (figure 12–36B) is constructed with a mounting nut inserted in one of the generator end shields. The mounting nut is a rigid part of the instrument, and the whole generator must be turned to screw the nut onto its mating threads.

The dual tachometer consists of two tachometer indicator units housed in a single case. The indicator pointers show simultaneously on a single dial the r.p.m. of two engines. Some tachometer indicators are equipped with a flight-hour meter dial, usually located in the lower center area of the dial face, just below the pointer pivot.

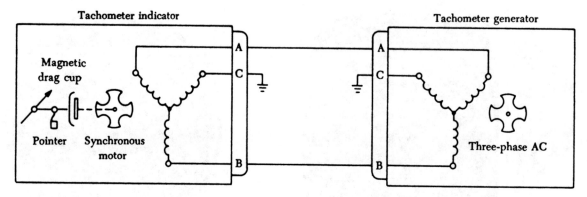

FIGURE 12-35. Schematic of a tachometer system.

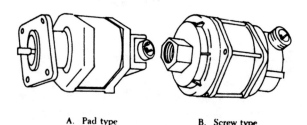

A. Pad type   B. Screw type

FIGURE 12-36. Tachometer generators.

Dual tachometers are also placed in the same case with a synchroscope for various purposes. One of these is the helicopter tachometer with synchroscope. This instrument shows simultaneously the speed of rotation of the engine crankshaft, the speed of rotation of the rotor shaft, and the slippage of the rotor due to malfunctioning of the clutch or excessive speed of the rotor when the clutch is disengaged in flight. The speed of both the rotor shaft and the engine shaft is indicated by a regular dual tachometer, and the slippage is indicated on a synchroscope (figure 12-37).

Tachometer Maintenance

Tachometer indicators should be checked for loose glass, chipped scale markings, or loose pointers. The difference in indications between readings taken before and after lightly tapping the instrument should not exceed approximately ±15 r.p.m. This value may vary, depending on the tolerance established by the indicator manufacturer. Both tachometer generator and indicator should be inspected for tightness of mechanical and electrical connections, security of mounting, and general condition. For detailed maintenance procedures, the manufacturer's instructions should always be consulted.

When an engine equipped with an electric tachometer is running at idle r.p.m. the tachometer indicator pointers may fluctuate and read low. This is an indication that the synchronous motor is not synchronized with the generator output. As the engine speed is increased the motor should synchronize and register the r.p.m. correctly. The r.p.m. at which synchronization occurs will vary with the design of the tachometer system.

If the instrument pointers oscillate at speeds above the synchronizing value, determine that the total oscillation does not exceed the allowable tolerance. If the oscillation exceeds the tolerance, determine if it is the instrument or one of the other components that is at fault.

Pointer oscillation can occur with a mechanical indicating system if the flexible drive is permitted to whip. The drive shaft should be secured at frequent intervals to prevent it from whipping.

When installing mechanical type indicators, be sure that the flexible drive has adequate clearance

FIGURE 12-37. Helicopter tachometer with synchroscope.

behind the panel. Any bends necessary to route the drive should not cause strain on the instrument when it is secured to the panel. Avoid sharp bends in the drive; an improperly installed drive can cause the indicator to fail to read, or to read incorrectly.

SYNCHROSCOPE

The synchroscope is an instrument that indicates whether two (or more) engines are synchronized; that is, whether they are operating at the same r.p.m. The instrument consists of a small electric motor which receives electrical current from the tachometer generators of both engines. The synchroscope is designed so that current from the faster-running engine controls the direction in which the synchroscope motor rotates.

If both engines are operating at exactly the same speed, the synchroscope motor does not operate. If, however, one engine is operating faster than the other, its generator signal will cause the synchroscope motor to turn in a given direction. If the speed of the other engine then becomes greater than that of the first engine, the signal from its generator will then cause the synchroscope motor to reverse itself and turn in the opposite direction.

The motor of the synchroscope is connected by means of a shaft to a double-ended pointer on the dial of the instrument (figure 12–38). It is necessary to designate one of the two engines a master engine if the synchroscope indicators are to be useful. The dial readings, with leftward rotation of the pointer indicating "slow" and rightward motion indicating "fast," then refer to the operation of the second engine in relation to the speed of the master engine.

For aircraft with more than two engines, additional synchroscopes are used. One engine is designated the master engine, and synchroscopes are connected between its tachometer and those of each of the other individual engines. On a complete installation of this kind, there will, of course, be one less instrument than there are engines, since the master engine is common to all the pairs.

One type of four-engine synchroscope is a special instrument that is actually three individual synchroscopes in one case (figure 12–39).

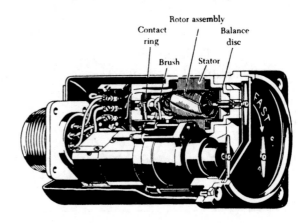

FIGURE 12–39. Four-engine synchroscope.

The rotor of each is electrically connected to the tachometer generator of the engine designated as the master, while each stator is connected to one of the other engine tachometers. There are three hands, each indicating the relative speed of the number two, three, or four engine, as shown in figure 12–40.

The separate hands revolve clockwise when their respective engine is running faster than the master and counterclockwise when it is running slower. Rotation of the hand begins as the speed difference reaches about 350 r.p.m., and as the engines approach synchronization the hand revolves at a ratio proportional to the speed difference.

TEMPERATURE INDICATORS

Various temperature indications must be known in order for an aircraft to be operated properly. It is important that the temperature of the engine oil, carburetor mixture, inlet air, free air, engine cylinders, heater ducts, and exhaust gas temperature of

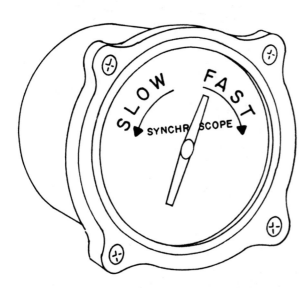

FIGURE 12–38. Synchroscope dial.

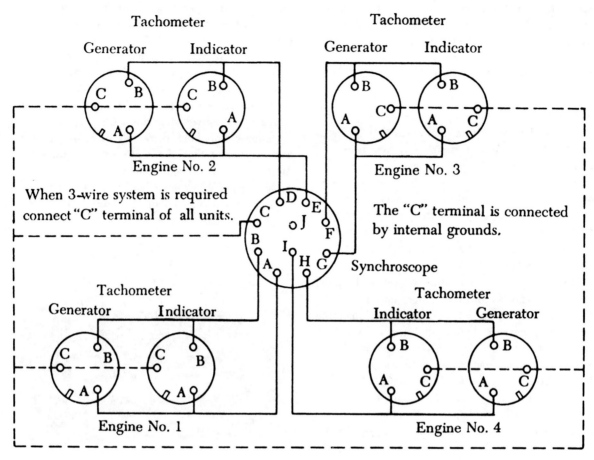

Tachometer

Generator   Indicator

Engine No. 2

Tachometer

Generator   Indicator

Engine No. 3

When 3-wire system is required connect "C" terminal of all units.

The "C" terminal is connected by internal grounds.

Synchroscope

Tachometer

Generator   Indicator

Engine No. 1

Tachometer

Indicator   Generator

Engine No. 4

FIGURE 12–40. Four-engine synchroscope schematic.

turbine engines be known. Many other temperatures must also be known, but these are some of the more important. Different types of thermometers are used to collect and present this information.

Electrical Resistance Thermometer

Electrical resistance thermometers are used widely in many types of aircraft to measure carburetor air, oil, and free air temperatures.

The principal parts of the electrical resistance thermometer are the indicating instrument, the temperature-sensitive element (or bulb), and the connecting wires and plug connectors.

Oil temperature thermometers of the electrical resistance type have typical ranges of from —10° to +120° C., or from —70° to +150° C. Carburetor air and mixture thermometers may have a range of from —50° to +50° C., as do many free air thermometers.

A typical electrical resistance thermometer is shown in figure 12–41. Indicators are also available

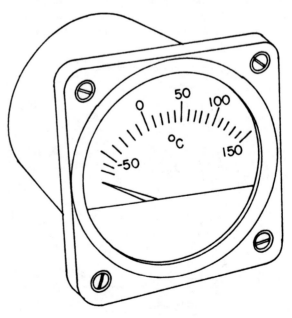

FIGURE 12–41. Typical electrical resistance temperature indicator.

492

in dual form for use in multi-engine aircraft. Most indicators are self-compensated for changes in cockpit temperature.

The electrical resistance thermometer operates on the principle of the change in the electrical resistance of most metals with changes in temperature. In most cases, the electrical resistance of a metal increases as the temperature rises. The resistance of some metals increases more than the resistance of others with a given rise in temperature. If a metallic resistor with a high temperature-resistant coefficient (a high rate of resistance rise for a given incresae in temperature) is subjected to a temperature to be measured, and a resistance indicator is connected to it, all the requirements for an electrical thermometer system are present.

The heat-sensitive resistor is the main element in the bulb. It is manufactured so that it has a definite resistance for each temperature value within its working range. The temperature-sensitive resistor element is a winding made of various alloys, such as nickel/manganese wire, in suitable insulating material. The resistor is protected by a closed-end metal tube attached to a threaded plug with a hexagon head (figure 12–42). The two ends of the winding are brazed or welded to an electrical receptacle designed to receive the prongs of the connector plug.

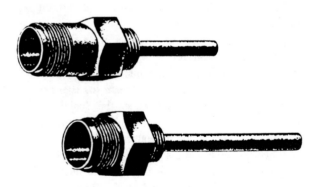

FIGURE 12–42. Two types of resistance thermometer bulb assemblies.

The electrical resistance indicator is a resistance-measuring instrument. Its dial is calibrated in degrees of temperature instead of ohms and measures temperature by using a modified form of the Wheatstone-bridge circuit.

The Wheatstone-bridge meter operates on the principle of balancing one unknown resistor against other known resistances. A simplified form of a Wheatstone-bridge circuit is shown in figure 12–43.

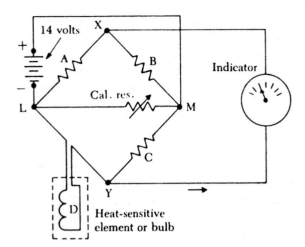

FIGURE 12–43. Wheatstone-bridge meter circuit.

Three equal values of resistances (A, B, and C, figure 12–43) are connected to a diamond-shaped bridge circuit with a resistance of unknown value (D).

The unknown resistance represents the resistance of the temperature bulb of the electrical resistance thermometer system. A galvanometer calibrated to read in degrees is attached across the circuit at point X and Y.

When the temperature causes the resistance of the bulb to equal that of the other resistances, no potential difference exists between points X and Y in the circuit, and no current flows in the galvanometer leg of the circuit. If the temperature of the bulb changes, its resistance will also change, and the bridge becomes unbalanced, causing current to flow through the galvanometer in one direction or the other.

The dial of the galvanometer is calibrated in degrees of temperature, converting it to a temperature-measuring instrument. Most indicators are provided with a zero adjustment screw on the face of the instrument to set the pointer at a balance point (the position of the pointer when the bridge is balanced and no current flows through the meter).

Thermocouple Thermometer Indicators

The cylinder temperature of most air-cooled reciprocating aircraft engines is measured by a thermometer which has its heat-sensitive element attached to some point on one of the cylinders (normally the hottest cylinder). In the case of turbojet engines, the exhaust temperature is measured by attaching thermocouples to the tailcone.

A thermocouple is a circuit or connection of two

493

unlike metals; such a circuit has two junctions. If one of the junctions is heated to a higher temperature than the other, an electromotive force is produced in the circuit. By including a galvanometer in the circuit, this force can be measured. The hotter the high-temperature junction (hot junction) becomes, the greater the electromotive force produced. By calibrating the galvanometer dial in degrees it becomes a thermometer.

A typical thermocouple thermometer system (figure 12–44) used to indicate engine temperature consists of a galvanometer indicator calibrated in degrees of centigrade, a thermocouple, and thermocouple leads.

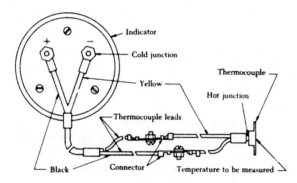

FIGURE 12–44. Reciprocating engine cylinder head temperature thermocouple system.

Thermocouple leads are commonly made from iron and constantan, but copper and constantan or chromel and alumel are other combinations of dissimilar metals in use. Iron/constantan is used mostly in radial engines, and chromel/alumel is used in jet engines.

Thermocouple leads are designed to provide a definite amount of resistance in the thermocouple circuit. Thus, their length or cross-sectional size cannot be altered unless some compensation is made for the change in total resistance.

The hot junction of the thermocouple varies in shape depending on its application. Two common types are shown in figure 12–45; they are the gasket type and the bayonet type. In the gasket type, two rings of dissimilar metals are pressed together to form a spark plug gasket. Each lead that makes a connection back to the galvanometer must be made of the same metal as the part of the thermocouple to which it is connected. For example, a copper wire is connected to the copper ring and a constantan wire is connected to the constantan ring. The bayonet type thermocouple fits into a hole or well in the

cylinder head. Here again, the same metal is used in the lead as in the part of the thermocouple to which it is connected.

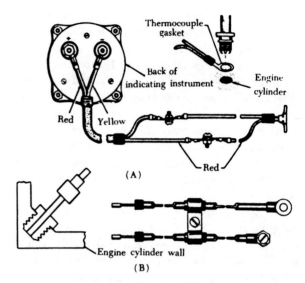

FIGURE 12–45. Thermocouples: (A) Gasket type, (B) Bayonet type.

The cylinder chosen for installing the thermocouple is the one which runs the hottest under most operating conditions. The location of this cylinder varies with different engines.

The cold junction of the thermocouple circuit is inside the instrument case.

Since the electromotive force set up in the circuit varies with the difference in temperature between the "hot" and "cold" junctions, it is necessary to compensate the indicator mechanism for changes in cockpit temperature which affect the "cold" junction. This is accomplished by using a bimetallic spring connected to the indicator mechanism.

When the leads are disconnected from the indicator, the temperature of the cockpit area around the instrument panel can be read on the indicator dial. This is because the bimetallic compensator spring continues to function as a thermometer.

Figure 12–46 shows the dials of two thermocouple temperature indicators.

Gas Temperature Indicating Systems

EGT (exhaust gas temperature) is a critical variable of turbine engine operation. The EGT indicating system provides a visual temperature indication in the cockpit of the turbine exhaust gases as they leave the turbine unit. In certain turbine engines the temperature of the exhaust gases is measured at

494

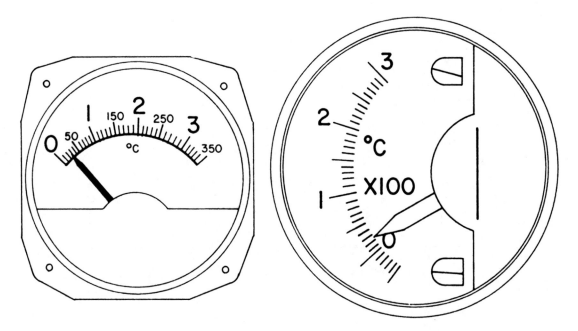

FIGURE 12–46. Two types of thermocouple temperature indicators.

the entrance to the turbine unit. This is usually referred to as a TIT (turbine inlet temperature) indicating system. The principal disadvantages of this method are that the number of thermocouples required become greater and the environmental temperatures in which they must operate are increased.

A gas temperature thermocouple is mounted in a ceramic insulator and encased in a metal sheath; the assembly forms a probe which projects into the exhaust gas stream. The thermocouple is made from chromel (a nickel/chromium alloy) and alumel (a nickel/aluminum alloy). The hot junction protrudes into a space inside the sheath. The sheath has transfer holes in the end of it which allow the exhaust gases to flow across the hot junction.

Several thermocouples are used and are spaced at intervals around the perimeter of the engine turbine casing or exhaust duct. The thermocouples measure engine EGT in millivolts, and this voltage is applied to the amplifier in the cockpit indicator, where it is amplified and used to energize the servomotor which drives the indicator pointer.

A typical EGT thermocouple system is shown in figure 12–47.

The EGT indicator shown is a hermetically sealed unit and has provisions for a mating electrical connector plug. The instrument's scale ranges from 0° C. to 1,200° C., with a vernier dial in the upper right-hand corner. A power "off" warning flag is located in the lower portion of the dial.

The TIT indicating system provides a visual indication at the instrument panel of the temperature of gases entering the turbine. On one type of turbine aircraft the temperature of each engine turbine inlet is measured by 18 dual-unit thermocouples installed in the turbine inlet casing. One set of these thermocouples is paralleled to transmit signals to the cockpit indicator. The other set of paralleled thermocouples provides temperature signals to the temperature datum control. Each circuit is electrically independent, providing dual system reliability.

The thermocouple assemblies are installed on pads around the turbine inlet case. Each thermocouple incorporates two electrically independent junctions within a sampling probe. The average voltage of the thermocouples at the thermocouple terminal blocks represents the TIT.

A schematic for the turbine inlet temperature system for one engine of a four-engine turbine aircraft is shown in figure 12–48. Circuits for the other three engines are identical to this system. The indicator contains a bridge circuit, a chopper circuit, a two-phase motor to drive the pointer, and a feedback potentiometer. Also included are a voltage reference circuit, an amplifier, a power "off" flag, a power supply, and an overtemperature warning light. Output of the amplifier energizes the variable field of the two-phase motor which positions the indicator main pointer and a digital indicator. The motor also drives the feedback potentiometer to pro-

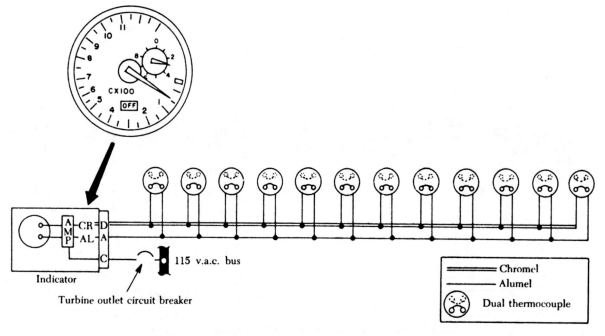

FIGURE 12–47. Typical exhaust gas temperature thermocouple system.

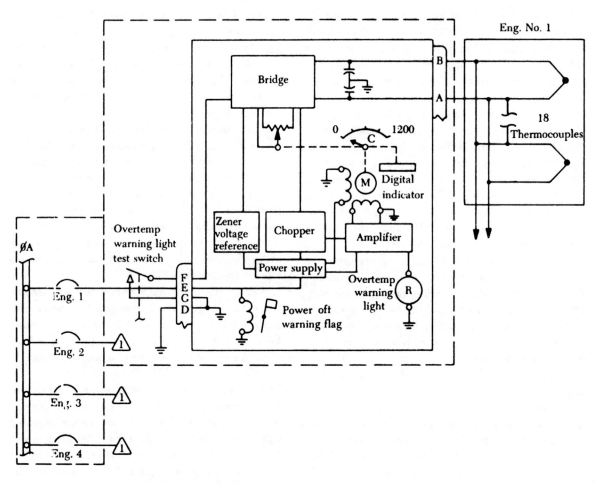

FIGURE 12–48. Turbine inlet temperature indicating system.

496

vide a huming signal to stop the drive motor when the correct pointer position, relative to the temperature signal, has been reached. The voltage reference circuit provides a closely regulated reference voltage in the bridge circuit to preclude error from input voltage variation to the indicator power supply.

The overtemperature warning light in the indicator illuminates when the TIT reaches a predetermined limit. An external test switch is usually installed so that overtemperature warning lights for all the engines can be tested at the same time. When the test switch is operated, an overtemperature signal is simulated in each indicator temperature control bridge circuit.

RATIOMETER ELECTRICAL RESISTANCE THERMOMETER

The basic Wheatstone-bridge, temperature-indicating system provides accurate indications when the pointer is at the balance point on the indicator dial. As the pointer moves away from the balance point, the Wheatstone-bridge indicator is increasingly affected by supply voltage variations. Greater accuracy can be obtained by inserting one of several types of automatic line voltage compensating circuits into the circuit. Some of these voltage regulators employ the filament resistance of lamps to achieve a more uniform supply voltage. The resistance of the lamp filaments helps regulate the voltage applied to the Wheatstone-bridge circuit since the filament resistance changes in step with supply voltage variation.

The ratiometer is a more sophisticated arrangement for obtaining greater accuracy in resistance-bulb indicators. The ratiometer measures the ratio of currents, using an adaptation of the basic Wheatstone-bridge with ratio circuitry for increased sensitivity.

A schematic of a ratiometer temperature circuit is shown in figure 12–49. The circuit contains two parallel branches, one with a fixed resistance in series with coil A, and the other a built-in resistance in series with coil B. The two coils are wound on a rotor pivoted in the center of the magnet air gap. The permanent magnet is arranged to provide a larger air gap between the magnet and the coils at the bottom than at the top. This produces a flux density that is progressively stronger from the bottom of the air gap to the top.

The direction of the current through each coil in respect to the polarity of the permanent magnet

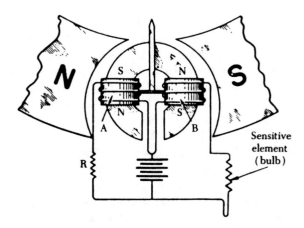

FIGURE 12–49. Ratiometer temperature-measuring system schematic.

causes the coil with the greater current flow to react in the weaker magnetic field. If the resistance of the temperature bulb is equal to the value of the fixed resistance, and equal values of current are flowing through the coils, the torque on the coils will be the same and the indicator points will be in the vertical (zero) position.

If the bulb temperature increases, its resistance will also increase, causing the current through the coil B circuit branch to decrease. Consequently, the torque on coil B decreases and coil A pushes downward into a weaker magnetic field; coil A, with its weaker current flow, moves into a stronger magnetic field. The torques on the coils still balance since the product of current times flux remains the same for both coils, but the pointer has moved to a new position on the calibrated scale. Just the opposite of this action would take place if the temperature of the heat-sensitive bulb should decrease.

Ratiometer temperature-measuring systems are used to measure engine oil, outside air, and carburetor air temperatures in many types of aircraft They are especially in demand to measure temperature conditions where accuracy is important or large variations of supply voltages are encountered.

FUEL FLOWMETER SYSTEMS

Fuel flowmeter systems are used to indicate fuel usage. They are most commonly installed on large multi-engine aircraft, but they may be found on any type of aircraft if fuel economy is an important factor.

A typical flowmeter system for a reciprocating engine consists of a flowmeter transmitter and an indicator. The transmitter is usually connected into the fuel line leading from the carburetor outlet to the fuel feed valve or discharge nozzle. The indicator is normally mounted in the instrument panel.

A cross sectional view of a typical transmitter fuel chamber is shown in figure 12–50. Fuel entering the inlet side of the fuel chamber is directed against the metering vane, causing the vane to swing on its shaft within the chamber. As the vane is moved from a closed position by the pressure of the fuel flow, the clearance between the vane and the fuel chamber wall becomes increasingly larger.

Figure 12–51 shows an exploded view of a fuel flowmeter system. Note that the metering vane moves against the opposing force of a hairspring. When the force created by a given fuel flow is balanced by spring tension, the vane becomes stationary. The vane is connected magnetically to the rotor of a transmitter, which generates electrical signals to position the cockpit indicator. The distance the metering vane moves is proportional to, and a measure of, the rate of fuel flow.

The damper vane of the transmitter cushions fluctuations caused by air bubbles. The relief valve bypasses fuel to the chamber outlet when the flow of fuel is greater than chamber capacity.

A simplified schematic of a vane-type flowmeter system (figure 12–52) shows the metering vane connected to the flowmeter transmitter and the rotor and stator of the indicator connected to a common power source with the transmitter.

The dial of a fuel-flow indicator is shown in figure 12–53. Some fuel-flow indicators are calibrated in gallons per hour, but most of them indicate the measurement of fuel flow in pounds.

The fuel flowmeter system used with turbine engine aircraft is usually a more complex system than that used in reciprocating engine aircraft.

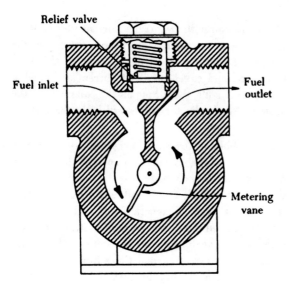

FIGURE 12–50. Flowmeter fuel chamber.

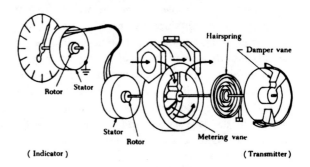

FIGURE 12–51. Fuel flowmeter system.

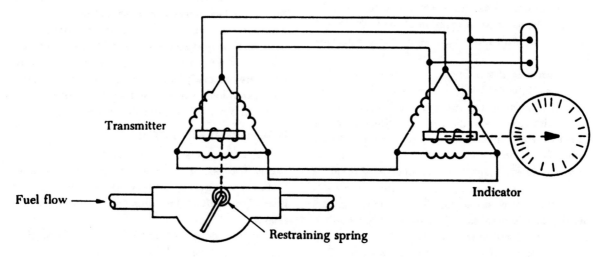

FIGURE 12–52. Schematic of vane-type flowmeter system.

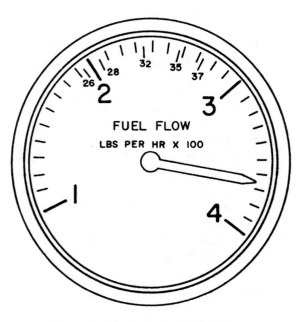

FIGURE 12-53. Typical fuel-flow indicator.

In the system shown schematically in figure 12-54, two cylinders, an impeller, and a turbine are mounted in the main fuel line leading to the engine. The impeller is driven at a constant speed by a special three-phase motor. The impeller imparts an angular momentum to the fuel, causing the turbine to rotate until the calibrated restraining spring force balances the force due to the angular momentum of the fuel. The deflection of the turbine positions the permanent magnet in the position transmitter to a position corresponding to the fuel flow in the line. This turbine position is transmitted electrically to the indicator in the cockpit.

GYROSCOPIC INSTRUMENTS

Three of the most common flight instruments, the attitude indicator, heading indicator, and the turn needle of the turn-and-bank indicator, are controlled by gyroscopes. To understand how these instruments operate requires a knowledge of gyroscopic principles, instrument power systems, and the operating principles of each instrument.

A gyroscope is a wheel or disk mounted to spin rapidly about an axis, and is also free to rotate about one or both of two axes perpendicular to

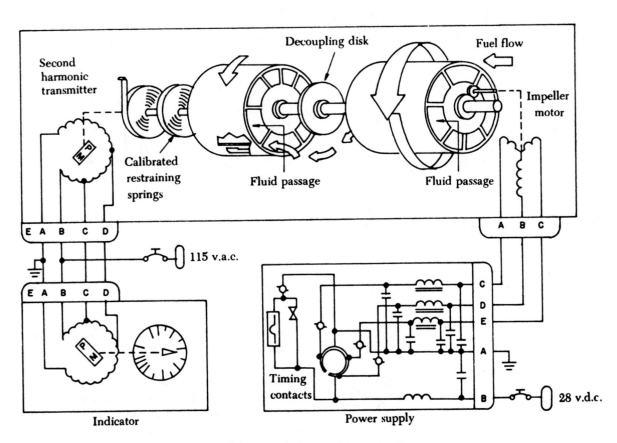

FIGURE 12-54. Schematic of a large turbine engine flowmeter system.

each other and to the axis of spin. A spinning gyroscope offers resistance to any force which tends to change the direction of the axis of spin.

A rotor and axle are the heart of a basic gyro (A of figure 12–55); a supporting ring with bearings on which the rotor and its axle can revolve are added to the basic unit (B of figure 12–55); and an outer ring with bearings at 90° to the rotor bearings has been added (C of figure 12–55). The inner ring with its rotor and axle can turn through 360° inside this outer ring.

A gyro at rest is shown in six different positions (figure 12–56) to demonstrate that unless the rotor is spinning a gyro has no unusual properties; it is simply a wheel universally mounted.

When the rotor is rotated at a high speed, the gyro exhibits one of its two gyroscopic characteristics. It acquires a high degree of rigidity, and its axle points in the same direction no matter how much its base is turned about (figure 12–57).

Gyrocsopic rigidity depends upon several design factors:

(1) **Weight.** For a given size, a heavy mass is more resistant to disturbing forces than a light mass.

(2) **Angular velocity.** The higher the rotational speed, the greater the rigidity or resistance to deflection.

(3) **Radius at which the weight is concentrated.** Maximum effect is obtained from a mass when its principal weight is concentrated near the rim rotating at high speed.

(4) **Bearing friction.** Any friction applies a deflecting force to a gyro. Minimum bearing friction keeps deflecting forces at a minimum.

A second gyroscopic characteristic, precession, is illustrated in figure 12–58A by applying a force or pressure to the gyro about the horizontal axis. The applied force is resisted, and the gyro, instead of turning about its horizontal axis, turns or "precesses" about its vertical axis in the direction indicated by the letter P. In a similar manner, if pressure is applied to the vertical axis, the gyro will precess about its horizontal axis in the direction shown by the arrow P in figure 12–58B.

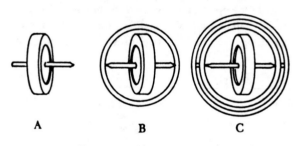

A   B   C

FIGURE 12–55. Basic gyroscope.

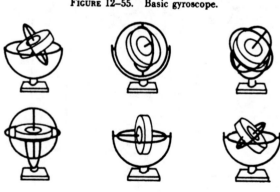

FIGURE 12–56. A gyro at rest.

FIGURE 12–57. Gyroscope inertia.

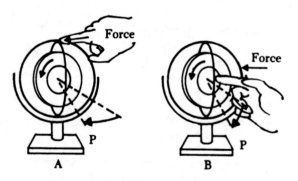

FIGURE 12–58. Gyrosocopic precession.

Two types of mountings are used, depending upon how the gyroscopic properties are to be used in the operation of an instrument. A freely or universally mounted gyro is set on three gimbals (rings), with the gyro free to rotate in any plane. Regardless of the position of the gyro base, the gyro tends to remain rigid in space. In the attitude indicator the horizon bar is gyro-controlled to remain parallel to the natural horizon, and changes in position of the aircraft are shown pictorially on the indicator.

The semirigid, or restricted, mounting employs two gimbals, limiting the rotor to two planes of rotation. In the turn-and-bank indicator, the semirigid mounting provides controlled precession of the rotor, and the precessing force exerted on the gyro by the turning aircraft causes the turn needle to indicate a turn.

SOURCES OF POWER FOR GYRO OPERATION

The gyroscopic instruments can be operated either by a vacuum system or an electrical power source. In some aircraft, all the gyros are either vacuum or electrically motivated; in others, vacuum (suction) systems provide the power for the attitude and heading indicators, while the electrical system drives the gyro for operation of the turn needle. Either alternating or direct current is used to power the gyroscopic instruments.

Vacuum System

The vacuum system spins the gyro by sucking a stream of air against the rotor vanes to turn the rotor at high speed, essentially as a water wheel or turbine operates. Air at atmospheric pressure drawn through a filter or filters drives the rotor vanes, and is sucked from the instrument case through a line to the vauuum source and vented to the atmosphere. Either a venturi or a vacuum pump can be used to provide the vacuum required to spin the rotors of the gyro instruments.

The vacuum value required for instrument operation is usually between 3½ in. to 4½ in. Hg and is usually adjusted by a vacuum relief valve located in the supply line. The turn-and-bank indicators used in some installations require a lower vacuum setting. This is obtained using an additional regulating valve in the individual instrument supply line.

Venturi-Tube Systems

The advantages of the venturi as a suction source are its relatively low cost and simplicity of installation and operation. A light, single-engine aircraft can be equipped with a 2-in. venturi (2 in. Hg vacuum capacity) to operate the turn needle. With an additional 8-in. venturi, power is available for the attitude and heading indicators. A venturi vacuum system is shown in figure 12–59.

The line from the gyro (figure 12–59) is connected to the throat of the venturi mounted on the exterior of the aircraft fuselage. Throughout the normal operating airspeed range the velocity of the

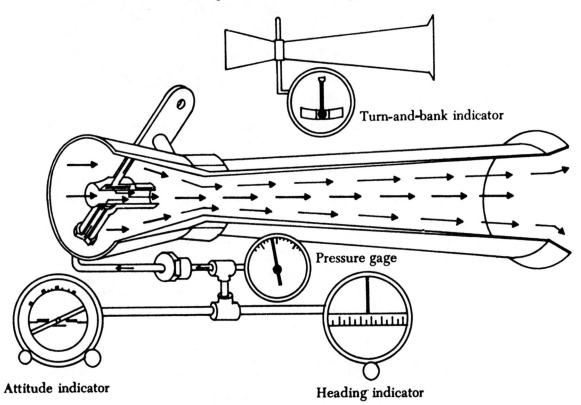

Turn-and-bank indicator

Pressure gage

Attitude indicator

Heading indicator

FIGURE 12–59. Venturi vacuum system.

air through the venturi creates sufficient suction to spin the gyro.

The limitations of the venturi system should be evident from the illustration in figure 12–59. The venturi is designed to produce the desired vacuum at approximately 100 m.p.h. under standard sea-level conditions. Wide variations in airspeed or air density, or restriction to airflow by ice accretion, will affect the pressure at the venturi throat and thus the vacuum driving the gyro rotor. And, since the rotor does not reach normal operating speed until after takeoff, preflight operational checks of venturi-powered gyro instruments cannot be made. For this reason the system is adequate only for light-aircraft instrument training and limited flying under instrument weather conditions. Aircraft flown throughout a wider range of speed, altitude, and weather conditions require a more effective source of power independent of airspeed and less susceptible to adverse atmospheric conditions.

Engine-Driven Vacuum Pump

The vane-type engine-driven pump is the most common source of vacuum for gyros installed in general aviation light aircraft. One type of engine-driven pump is mounted on the accessory drive shaft of the engine, and is connected to the engine lubrication system to seal, cool, and lubricate the pump.

Another commonly used source of vacuum is the dry vacuum pump, also engine-driven. The pump operates without lubrication, and the installation requires no lines to the engine oil supply, and no air-oil separator or gate check valve. In other respects, the dry pump system and oil lubricated system are the same.

The principal disadvantage of the pump-driven vacuum system relates to erratic operation in high-altitude flying. Apart from routine maintenance of the filters and plumbing, which are absent in the electric gyro, the engine-driven pump is as effective a source of power for light aircraft as the electrical system.

Typical Pump-Driven Vacuum System

Figure 12–61 shows the components of a vacuum system with a pump capacity of approximately 10″ Hg at engine speeds above 1000 rpm. Pump capacity and pump size vary in different aircraft, depending on the number of gyros to be operated.

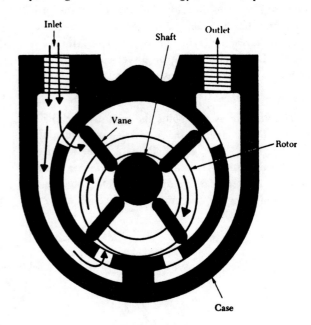

FIGURE 12–60. Cutaway view of a vane-type engine-driven vacuum pump.

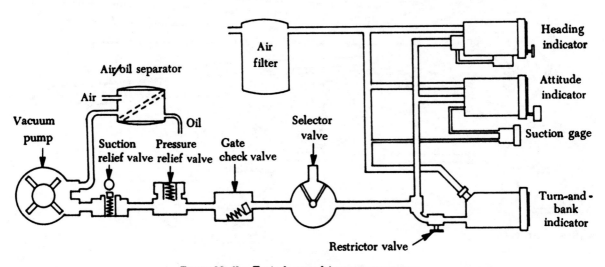

FIGURE 12–61. Typical pump-driven vacuum system.

Air-Oil Separator.—Oil and air in the vacuum pump are exhausted through the separator, which separates the oil from the air; the air is vented outboard, and the oil is returned to the engine sump.

Suction Relief Valve.—Since the system capacity is more than is needed for operation of the instruments, the adjustable suction relief valve is set for the vacuum desired for the instruments. Excess suction in the instrument lines is reduced when the spring-loaded valve opens to atmospheric pressure. (See fig. 12–62.)

Pressure Relief Valve.—Since a reverse flow of air from the pump would close both the gate check valve and the suction relief valve, the resulting pressure could rupture the lines. The pressure relief valve vents positive pressure into the atmosphere.

Gate Check Valve.—The gate check valve prevents possible damage to the instruments by engine back-fire, which would reverse the flow of air and oil from the pump. (See fig. 12–63.)

Selector Valve.—In twin-engine aircraft having vacuum pumps driven by both engines, the alternate pump can be selected to provide vacuum in the event of either engine or pump failure, with a check valve incorporated to seal off the failed pump.

Restrictor Valve.—Since the turn needle operates on less vacuum than that required for other gyro instruments, the vacuum in the main line must be reduced. This valve is either a needle valve adjusted to reduce the vacuum from the main line by approximately one-half, or a spring-loaded regulating valve that maintains a constant vacuum for the turn indicator, unless the main line vacuum falls below a minimum value.

Air Filter.—The master air filter screens foreign matter from the air flowing through all the gyro instruments, which are also provided with individual filters. Clogging of the master filter will reduce airflow and cause a lower reading on the suction gage. In aircraft having no master filter installed, each instrument has its own filter. With an individual filter system, clogging of a filter will not necessarily show on the suction gage.

Suction Gage.—The suction gage is a pressure gage, indicating the difference in inches of mercury, between the pressure inside the system and atmospheric or cockpit pressure. The desired vacuum, and the minimum and maximum limits, vary with gyro design. If the desired vacuum for the attitude and heading indicators is 5″ and the minimum in 4.6″, a reading below the latter value indicates that the airflow is not spinning the gyros fast enough for reliable operation. In many aircraft, the system provides a suction gage selector valve, permitting the pilot to check the vacuum at several points in the system.

Suction

Suction pressures discussed in conjunction with the operation of vacuum systems are actually minus or negative pressures (below sea level). For example, if sea level equals 17.5 p.s.i. then 1″ Hg (1 inch mercury) or 1 p.s.i. vacuum is equal to −1 p.s.i. negative pressure or 16.5 positive pressure. Likewise 3″ Hg = −3 p.s.i. negative pressure or +14.5 positive pressure.

Of course, for every action there is an equal and opposite reaction. Therefore when the vacuum pump develops a vacuum (negative pressure) it must also create pressure (positive). This pressure (compressed air) is sometimes utilized to operate pressure instruments, deicer boots and inflatable seals.

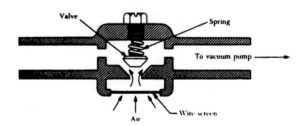

Figure 12–62. Vacuum regulator valve.

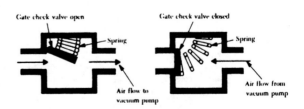

Figure 12–63. Gate check valve.

Typical System Operation

The schematic of a vacuum system for a twin-engine aircraft is shown in figure 12–64. This vacuum system consists of the following components; two engine-driven pumps, two vacuum relief valves, two flapper-type check valves, a vacuum manifold, a vacuum restrictor for each turn-and-bank indicator, an engine four-way selector valve, one vacuum gage, and a turn-and-bank selector valve.

The left and right engine-driven vacuum pumps and their associated lines and components are isolated from each other, and act as two independent vacuum systems. The vacuum lines are routed from each vacuum pump through a vacuum relief valve and through a check valve to the vacuum four-way selector valve.

From the engine four-way selector valve, which permits operation of the left or right engine vacuum system, the lines are routed to a vacuum manifold. From the manifold, flexible hose connects the vacuum-operated instruments into the system. From the instrument, lines routed to the vacuum gage pass through a turn-and-bank selector valve. This valve has three positions: main, left T&B, and right T&B. In the main position the vacuum gage indicates the vacuum in the lines of the artificial horizon and directional gyros. In the other positions, the lower value of vacuum for the turn-and-bank indicators can be read.

VACUUM-DRIVEN ATTITUDE GYROS

In a typical vacuum-driven attitude gyro system, air is sucked through the filter, then through pas-

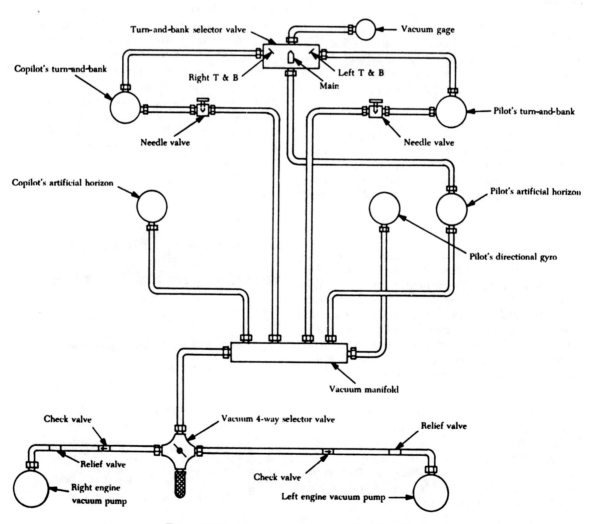

FIGURE 12–64. Vacuum system for a multi-engine aircraft.

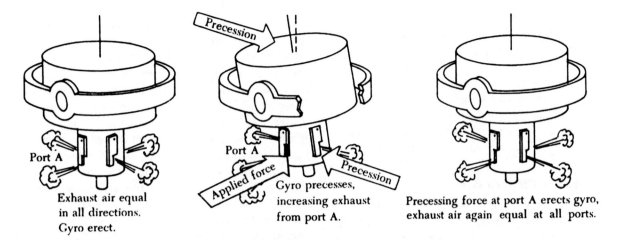

| Exhaust air equal in all directions. Gyro erect. | Gyro precesses, increasing exhaust from port A. | Precessing force at port A erects gyro, exhaust air again equal at all ports. |

FIGURE 12–65. Erecting mechanism of a vacuum-driven attitude indicator.

sages in the rear pivot and inner gimbal ring, then into the housing where it is directed against the rotor vanes through two openings on opposite sides of the rotor. The air then passes through four equally spaced ports in the lower part of the rotor housing and is sucked out into the vacuum pump or venturi (figure 12–65).

The chamber containing the ports is the erecting device that returns the spin axis to its vertical alignment whenever a precessing force, such as bearing friction, displaces the rotor from its horizontal plane. Four exhaust ports are each half-covered by a pendulous vane, which allows discharge of equal volumes of air through each port when the rotor is properly erected. Any tilting of the rotor disturbs the total balance of the pendulous vanes, tending to close one vane of an opposite pair while the opposite vane opens a corresponding amount. The increase in air volume through the opening port exerts a precessing force on the rotor housing to erect the gyro, and the pendulous vanes return to a balanced condition (figure 12–66).

The limits of the attitude indicator specified in the manufacturer's instructions refer to the maximum rotation of the gimbals beyond which the gyro will tumble. The bank limits of a typical vacuum-driven attitude indicator are from approximately 100° to 110°, and the pitch limits vary from approximately 60° to 70°, depending on the design of a specific unit. If, for example, the pitch limits are 60° with the gyro normally erected, the rotor will tumble when the aircraft climb or dive angle exceeds 60°. As the rotor gimbal hits the stops, the

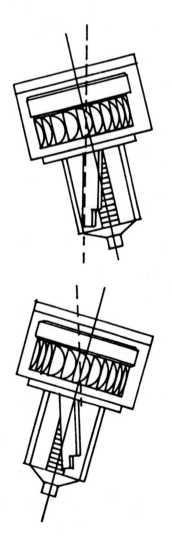

FIGURE 12–66. Action of pendulous vanes.

505

rotor precesses abruptly, causing excessive friction and wear on the gimbals. The rotor will normally precess back to the horizontal plane at a rate of approximately 8° per min.

Many gyros include a caging device, used to erect the rotor to its normal operating position prior to fight or after tumbling, and a flag to indicate that the gyro must be uncaged before use. Turning the caging knob prevents rotation of the gimbals and locks the rotor spin axis in its vertical position.

PRESSURE-OPERATED GYROS

The availability of pressure pumps on which no lubrication is necessary makes pressure-operated gyro systems feasible. In such installations, air is pushed under pressure through gyro instruments, rather than being sucked through the system. Positive-pressure pumps are more efficient than vacuum pumps, especially at higher altitudes.

VACUUM SYSTEM MAINTENANCE PRACTICES

Errors in the indications presented on the attitude indicator will result from any factor that prevents the vacuum system from operating within the design suction limits, or from any force that disturbs the free rotation of the gyro at design speed. These include poorly balanced components, clogged filters, improperly adjusted valves and pump malfunction. Such errors can be minimized by proper installation, inspection, and maintenance practices.

Other errors, inherent in the construction of the instrument, are caused by friction and worn parts. These errors, resulting in erratic precession and failure of the instrument to maintain accurate indications, increase with the service life of the instrument.

For the aviation mechanic the prevention or correction of vacuum system malfunctions usually consists of cleaning or replacing filters, checking

Possible Cause	Isolation Procedure	Correction
(1) No Vacuum Pressure or Insufficient Pressure		
Defective vacuum gage.	On multi-engine aircraft check opposite engine system on the gage.	Replace faulty vacuum gage.
Vacuum relief valve incorrectly adjusted.	Change valve adjustment.	Make final adjustment to proper setting of valve.
Vacuum relief valve installed backwards.	Visually inspect.	Install properly.
Broken line.	Visually inspect.	Replace line.
Lines crossed.	Visually inspect.	Install lines properly.
Obstruction in vacuum line.	Check for collapsed line.	Clean and test line. Replace defective part(s).
Vacuum pump failure.	Remove and inspect.	Replace faulty pump.
Vacuum regulator valve incorrectly adjusted.	Make valve adjustment and note pressure.	Adjust to proper pressure.
Vacuum relief valve dirty.	Clean and adjust relief valve.	Replace valve if adjustment fails.
(2) Excessive Vacuum		
Relief valve improperly adjusted.	-------------------------	Adjust relief valve to proper setting.
Inaccurate vacuum gage.	Check calibration of gage.	Replace faulty gage.
(3) Gyro Horizon Bar Fails to Respond		
Instrument caged.	Visually inspect.	Uncage instrument.
Instrument filter dirty.	Check filter.	Replace or clean as necessary.
Insufficient vacuum.	Check vacuum setting.	Adjust relief valve to proper setting.
Instrument assembly worn or dirty.	-------------------------	Replace instrument.
(4) Turn-and-Bank Indicator Fails to Respond		
No vacuum supplied to instrument.	Check lines and vacuum system.	Clean or replace lines and replace components of vacuum system as necessary.
Instrument filter clogged.	Visually inspect.	Replace filter.
Defective instrument.	Test with properly functioning instrument.	Replace faulty instrument.
(6) Turn-and-Bank Pointer Vibrates		
Defective instrument.	Test with properly functioning instrument	Replace defective instrument.

FIGURE 12–67. Vacuum system troubleshooting.

and correcting for insufficient vacuum, or removing and replacing the instruments. A list of the most common malfunctions, together with their correction, is included in figure 12–67.

ELECTRIC ATTITUDE INDICATOR

In the past, suction-driven gyros have been favored over the electric type for light aircraft because of their comparative simplicity and lower cost. However, the increasing importance of the attitude indicator has stimulated development of improved electric-driven gyros suited for light plane installation. Improvements relating to basic gyro design factors, easier readability, erection characteristics, reduction of induced errors, and instrument limitations are reflected in several available types. Depending upon the particular design improvements, the details for the instrument display and cockpit controls will vary among different instruments. All of them present, to a varying degree, the essential pitch-and-bank information for attitude reference.

The typical attitude indicator, or gyro horizon as it is sometimes referred to, has a vertical-seeking gyro, the axis of rotation tending to point toward the center of the earth. The gyro is linked with a horizon bar and stabilizes a kidney-shaped sphere having pitch attitude markings. The sphere, horizon bar, and bank index pointer move with changes of aircraft attitude. Combined readings of these presentations give a continuous pictorial presentation of the aircraft attitude in pitch and roll with respect to the earth's surface.

The gyroscope motor is driven by 115 v., 400 Hz alternating current. The gyro, turning at approximately 21,000 r.p.m., is supported by the yoke and pivot assembly (gimbals). Attached to the yoke and pivot assembly is the horizon bar, which moves up and down through an arc of approximately 27°. The kidney-shaped sphere provides a background for the horizon bar and has the words climb and dive and a bull's-eye painted on it. Climb and dive represent about 60° of pitch. Attached to the yoke and pivot assembly is the bank index pointer, which is free to rotate 360°. The dial face of the attitude indicator is marked with 0°, 10°, 20°, 30°, and 60° of bank, and is used with the bank index pointer to indicate the degree of bank left or right. The face of one type of gyro-horizon is shown in figure 12–68.

The function of the erection mechanism is to keep the gyro axis vertical to the surface of the earth. A magnet attached to the top of the gyro shaft spins at approximately 21,000 r.p.m. Around this magnet, but not attached, is a sleeve that is rotated by magnetic attraction at approximately 44 to 48 r.p.m. As illustrated in figure 12–69, the steel balls are free to move around the sleeve. If the pull of gravity is not aligned with the axis of the gyro, the balls will fall to the low side. The resulting precession re-aligns the axis of rotation vertically.

The gyro can be caged manually by a lever and cam mechanism to provide rapid erection. When the instrument is not getting sufficient power for normal operation, an "off" flag appears in the upper right-hand face of the instrument.

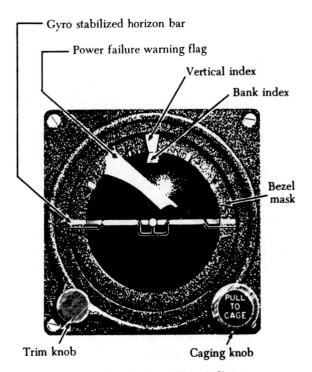

FIGURE 12–68. Gyro-horizon indicator.

Magnetic Compass

The magnetic compass is a simple, self-contained instrument which operates on the principle of magnetic attraction.

If a bar magnet is mounted on a pivot to be free to rotate in a horizontal plane, it will assume a position with one of its ends pointing toward the earth's north magnetic pole. This end of the magnet is called the north-seeking end of the magnet.

507

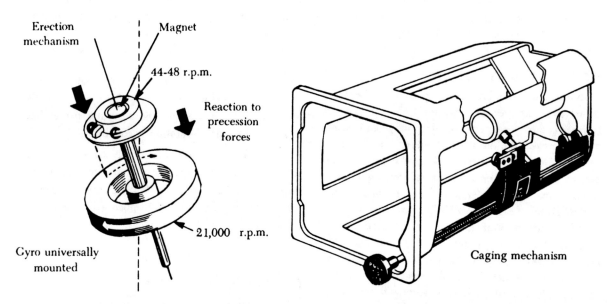

Erection mechanism

Magnet

44-48 r.p.m.

Reaction to precession forces

Gyro universally mounted

21,000 r.p.m.

Caging mechanism

FIGURE 12-69. Erecting and caging mechanisms of an electric attitude indicator.

The magnetic compass consists of a liquid-filled bowl containing a pivoted float element to which one or more bar magnets, called needles, are fastened. The liquid in the bowl dampens the oscillations of the float and decreases the friction of the pivot. A diaphragm and vent provide for expansion and contraction of the liquid as altitude and/or temperatures change.

If more than one magnet is used in a compass, the magnets are mounted parallel to each other, with like poles pointing in the same direction. The element on which the magnets are mounted is so suspended that the magnets are free to align themselves with the earth's north and south magnetic poles.

A compass card, usually graduated in 5° increments, is attached to the float element of the compass. A fixed reference marker, called a lubber line, is attached to the compass bowl. The lubber line and the graduations on the card are visible through a glass window. The magnetic heading of the aircraft is read by noting the graduation on which the lubber line falls. The two views of a magnetic compass in figure 12-70 show the face and the internal components of a magnetic compass.

A compensating device containing small permanent magnets is incorporated in the compass to correct for deviations of the compass which result from the magnetic influences of the aircraft structure and electrical system. Two screws on the face of the instrument are used to move the magnets and thus counterbalance the local magnetic influences acting on the main compass magnets. The two setscrews are labelled N–S and E–W.

Magnetic variation is the angular difference in degrees between the geographic north pole and the magnetic north pole. This variation is caused by the earth's magnetic field, which is constantly changing. Since variation differs according to geographic location, its effect on the compass cannot be removed by any type of compensation. Variation is called west variation when the earth's magnetic field draws the compass needle to the left of the geographic north pole and east variation when the needle is drawn to the right of the geographic north pole.

The compass needle is affected not only by the earth's magnetic field, but also by the magnetic fields generated when aircraft electrical equipment is operated, and by metal components in the aircraft. These magnetic disturbances within the aircraft, called deviation, deflect the compass needle from alignment with magnetic north.

To reduce this deviation, each compass in an aircraft is checked and compensated periodically by adjustment of the N–S and E–W magnets. The errors remaining after "swinging" the compass are recorded on a compass correction card mounted near the compass.

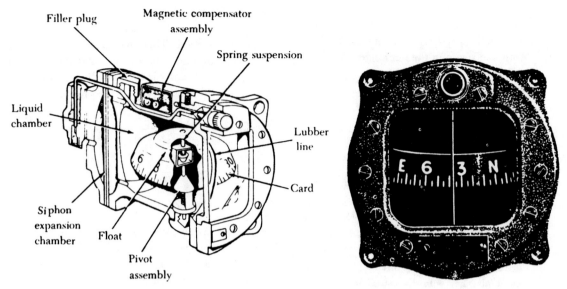

FIGURE 12-70. Magnetic compass.

The "swinging" (calibration) of a compass can be accomplished in flight or on the ground. Ground swinging of a compass is usually done with the aircraft at rest on a compass "rose." A compass rose (figure 12-71) is a circle laid out or painted on a level surface and graduated in degrees. The directions marked on the compass rose are magnetic directions, although true north is also marked on some compass roses.

Compass compensation procedures vary, depending on the type of aircraft. Requirements are often set up on a flight-hour and calendar basis. Most facilities perform compass checks anytime that equipment replacement, modification or relocation might cause compass deviation.

An example of compass compensation is outlined in the following paragraphs. These procedures are general in nature and do not have specific application.

(1) The compensator should be set either to zero or in a position where it has no effect on the main compass magnets.

(2) The aircraft is placed directly on a south magnetic heading on the compass rose. The tail of tailwheel aircraft should be raised to level-flying position.

(3) Note the compass reading and record it. The deviation is the algebraic difference between the magnetic heading and the compass reading.

EXAMPLE:

On the south (180°) heading, the compass reading is 175.5°. This would be recorded as a deviation of +4.5° (180° — 175.5° = 4.5°). If the compass reading is too low, the deviation is plus; if the reading is too high, the deviation is minus.

(4) Align the aircraft on a magnetic north heading. Record the compass reading and compute the deviation.

EXAMPLE:

On the north (000°) heading, the compass reads 006.5°. Since the deviation is 6.5° too high, it is recorded as a minus deviation (—6.5°).

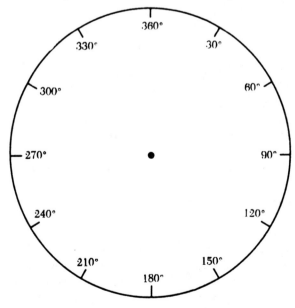

FIGURE 12-71. Typical compass rose.

509

(5) The coefficient of north-south deviation is determined by subtracting, algebraically, the south deviation from the north deviation and dividing the remainder by 2:

$$\text{Coefficient} = \frac{(-6.5°) - (4.5°)}{2}$$

$$= \frac{-11°}{2}$$

$$= -5.5°$$

The coefficient of north-south deviation, which is the average of the deviation on the two headings, is —5.5°. The north-south compensator is adjusted by this amount, and the reading on the north heading will now be 001°. This adjustment also corrects the south deviation by the same amount, so that on a south heading the compass will now read 181°.

(6) Align the aircraft on a magnetic west (270°) heading on the compass rose. Record the compass reading and compute the deviation. Suppose the compass reads 276°, a deviation of —6°.

(7) Align the aircraft on a magnetic east (090°) heading. Record the compass reading and compute the deviation. Suppose the compass reading is exactly 90° on the magnetic east heading, a deviation of 0°.

(8) Compute the coefficient of east-west deviation:

$$\text{Coefficient} = \frac{0° - (-6°)}{2}$$

$$= \frac{+6°}{2}$$

$$= +3°.$$

(9) While the aircraft is on the east heading, adjust the east-west compensator to add 3° to the compass reading. This reading then becomes 93° on the east heading and 273° on a west heading.

(10) Leaving the aircraft on an east magnetic heading, compute the coefficient of overall deviation. This coefficient is equal to the algebraic sum of the compass deviations on all four cardinal headings (north, east, south, and west), divided by 4:

$$\text{Coefficient} = \frac{(-6.5°) + 0° + 4.5° + (-6°)}{4}$$

$$= \frac{-8°}{4}$$

$$= -2°.$$

If the coefficient is greater than 1°, further compensation is usually accomplished. The compensation is not done with the magnetic compensation device. It is accomplished by re-aligning the compass, so that it is mounted parallel to the longitudinal axis of the aircraft.

(11) After the initial compensation is completed, the aircraft will be compensated again on headings of 30°, 60°, 120°, 150°, 210°, 240°, 300°, and 330°. The compass readings for each heading are recorded on a compass correction card. This card is then mounted as close as possible, to the instrument for ready reference. An example of a correction card is shown in figure 12–72.

The procedure described is a basic compensation procedure. Additional circuits around the compass rose should be made with the engine(s) and electrical and ratio equipment operating to verify the accuracy of the basic compensations.

Jacks, lifts, hoists, or any dolly needed to move and align the aircraft on the various headings of a compass rose should preferably be made of nonmagnetic material. When this is impossible, devices can be tested for their effect on the compass by moving them about the aircraft in a circle at the same distance that would separate them from the compass when they are being used. Equipment that causes a change in compass readings of more than one-quarter of a degree should not be used. Additionally, fuel trucks, tow tractors, or other aircraft containing magnetic metals should not be parked close enough to the compass rose to affect the compass of the aircraft being swung.

AIRCRAFT COMPASS

DATE..

		FOR	STEER
N		000°	000°
		030°	033°
		060°	060°
E		090°	095°
		120°	120°
		150°	149°
S		180°	175°
		210°	205°
		240°	334°
W		270°	265°
		300°	294°
		330°	326°

Calibrated by: ....

FIGURE 12–72. Compass correction card.

The magnetic compass is a simple instrument that does not require setting or a source of power. A minimum of maintenance is necessary, but the instrument is delicate and should be handled carefully during inspection. The following items are usually included in an inspection:

(1) The compass indicator should be checked for correct readings on various cardinal headings and re-compensated if necessary.

(2) Moving parts of the compass should work easily.

(3) The compass bowl should be correctly suspended on an anti-vibration device and should not touch any part of the metal container.

(4) The compass bowl should be filled with liquid. The liquid should not contain any bubbles nor have any discoloration.

(5) The scale should be readable and its illumination good.

AUTOPILOT SYSTEM

The automatic pilot is a system of automatic controls which holds the aircraft on any selected magnetic heading and returns the aircraft to that heading when it is displaced from it. The automatic pilot also keeps the aircraft stabilized around its horizontal and lateral axes.

The purpose of an automatic pilot system is primarily to reduce the work, strain, and fatigue of controlling the aircraft during long flights. To do this the automatic pilot system performs several functions. It allows the pilot to maneuver the aircraft with a minimum of manual operations. While under automatic control the aircraft can be made to climb, turn, and dive with small movements of the knobs on the autopilot controller.

Autopilot systems provide for one, two, or three axis control of the aircraft. Some autopilot systems control only the ailerons (one axis), others control ailerons and elevators or rudder (two axis). The three-axis system controls ailerons, elevators, and rudder.

All autopilot systems contain the same basic components: (1) Gyros, to sense what the airplane is doing; (2) servos, to move the control surfaces; and (3) an amplifier, to increase the strength of the gyro signals enough to operate the servos. A controller is also provided to allow manual control of the aircraft through the autopilot system.

Principle of Operation

The automatic pilot system flies the aircraft by using electrical signals developed in gyro-sensing units. These units are connected to flight instruments which indicate direction, rate-of-turn, bank, or pitch. If the flight attitude or magnetic heading is changed, electrical signals are developed in the gyros. These signals are used to control the operation of servo units which convert electrical energy into mechanical motion.

The servo is connected to the control surface and converts the electrical signals into mechanical force which moves the control surface in response to corrective signals or pilot commands. A basic autopilot system is shown in figure 12–73.

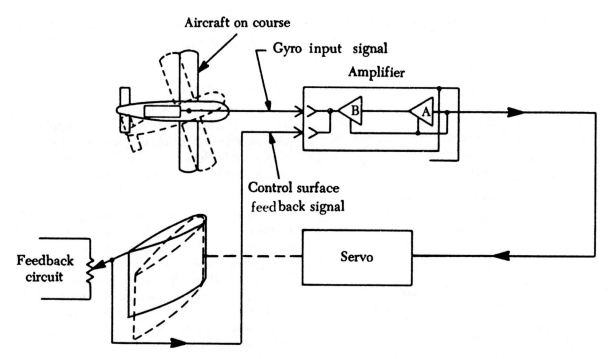

FIGURE 12–73. Basic autopilot system.

Most modern autopilots can be described in terms of their three major channels: (1) The rudder, (2) aileron, and (3) the elevator channels.

The rudder channel receives two signals that determine when and how much the rudder will move. The first signal is a course signal derived from a compass system. As long as the aircraft remains on the magnetic heading it was on when the autopilot was engaged, no signal will develop. But any deviation causes the compass system to send a signal to the rudder channel that is proportional to the angular displacement of the aircraft from the preset heading.

The second signal received by the rudder channel is the rate signal, which provides information anytime the aircraft is turning about the vertical axis. This information is provided by the turn-and-bank indicator gyro. When the aircraft attempts to turn off course, the rate gyro develops a signal proportional to the rate of turn, and the course gyro develops a signal proportional to the amount of displacement. The two signals are sent to the rudder channel of the amplifier, where they are combined and their strength is increased. The amplified signal is then sent to the rudder servo. The servo will turn the rudder in the proper direction to return the aircraft to the selected magnetic heading.

As the rudder surface moves, a followup signal is developed which opposes the input signal. When the two signals are equal in magnitude, the servo stops moving. As the aircraft arrives on course, the course signal will reach a zero value, and the rudder will be returned to the streamline position by the followup signal.

The aileron channel receives its input signal from a transmitter located in the gyro horizon indicator. Any movement of the aircraft about its longitudinal axis will cause the gyro-sensing unit to develop a signal to correct for the movement. This signal is amplified, phase-detected, and sent to the aileron servo which moves the aileron control surfaces to correct for the error.

As the aileron surfaces move, a followup signal builds up in opposition to the input signal. When the two signals are equal in magnitude, the servo stops moving. Since the ailerons are displaced from streamline, the aircraft will now start moving back toward level flight with the input signal becoming smaller and the followup signal driving the control surfaces back toward the streamline position. When the aircraft has returned to level flight in roll attitude, the input signal will again be zero. At the same time the control surfaces will be streamlined, and the followup signal will be zero.

The elevator channel circuits are similar to those of the aileron channel, with the exception that the

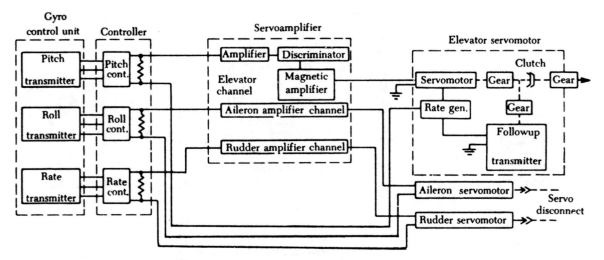

FIGURE 12-74. Autopilot block diagram.

elevator channel detects changes in pitch attitude of the aircraft. The circuitry of all three channels can be followed by referring to the block diagram in figure 12–74.

The foregoing autopilot system description was used to show the function of a simple autopilot. Most autopilots are far more sophisticated; however, many of the operating fundamentals are similar. Autopilot systems are capable of handling a variety of navigational inputs for automatic flight control.

BASIC AUTOPILOT COMPONENTS

The components of a typical autopilot system are illustrated in figure 12–75. Most systems consist of four basic types of units, plus various switches and auxiliary units. The four types of basic units are: (1) The sensing elements, (2) command elements, (3) output elements, and (4) the computing element.

Command Elements

The command unit (flight controller) is manually operated to generate signals which cause the aircraft to climb, dive, or perform coordinated turns. Additional command signals can be sent to the autopilot system by the aircraft's navigational equipment. The automatic pilot is engaged or disengaged electrically or mechanically, depending on system design.

While the automatic pilot system is engaged, the manual operation of the various knobs on the controller (figure 12–76) maneuvers the aircraft. By operating the pitch trim wheel, the aircraft can be made to climb or dive. By operating the turn knob, the aircraft can be banked in either direction. The engage switch is used to engage and disengage the autopilot. In addition, most systems have a disconnect switch located on the control wheel(s). This switch, operated by thumb pressure, can be used to disengage the autopilot system should a malfunction occur in the system.

One type of automatic pilot system has an engaging control that manually engages the clutch mechanism of the servomotor to the cable drum. A means of electrically disengaging the clutch is provided through a disconnect switch located on the control wheel(s).

Sensing Elements

The directional gyro, turn-and-bank gyro, attitude gyro, and altitude control are the sensing elements. These units sense the movements of the aircraft, and automatically generate signals to keep these movements under control.

Computer or Amplifier

The computing element consists of an amplifier or computer. The amplifier receives signals, determines what action the signals are calling for, and amplifies the signals received from the sensing elements. It passes these signals to the rudder, aileron, or elevator servos to drive the control surfaces to the position called for.

Output Elements

The output elements of an autopilot system are the servos which actuate the control surfaces. The

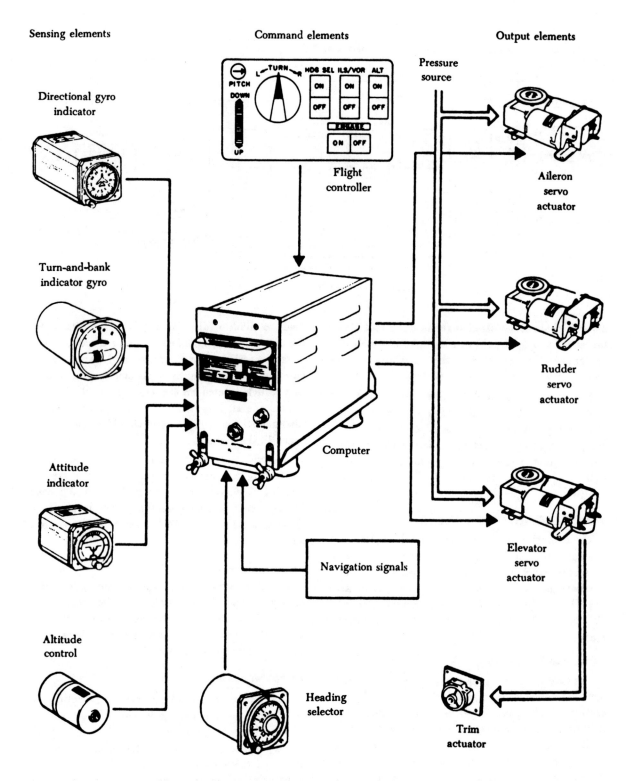

Sensing elements

Command elements

Output elements

Directional gyro
indicator

Turn-and-bank
indicator gyro

Attitude
indicator

Altitude
control

Flight
controller

Pressure
source

Aileron
servo
actuator

Rudder
servo
actuator

Computer

Navigation signals

Elevator
servo
actuator

Heading
selector

Trim
actuator

FIGURE 12–75. Typical autopilot system components.

majority of the servos in use today are either electric motors or electro/pneumatic servos.

An aircraft may have from one to three servos to operate the primary flight controls. One servo operates the ailerons, a second operates the rudder, and a third operates the elevators. Each servo drives its

FIGURE 12-76. Typical autopilot controller.

associated control surface to follow the directions of the particular automatic pilot channel to which the servo is connected.

Two types of electric motor-operated servos are in general use. In one, a motor is connected to the servo output shaft through reduction gears. The motor starts, stops, and reverses direction in reponse to the commands of the gyros or controller. The other type of electric servo uses a constantly running motor geared to the output shaft through two magnetic clutches. The clutches are arranged so that energizing one clutch transmits motor torque to turn the output shaft in one direction; energizing the other clutch turns the shaft in the opposite direction.

The electro/pneumatic servos are controlled by electrical signals from the autopilot amplifier and actuated by an appropriate air pressure source. The source may be a vacuum system pump or turbine engine bleed air. Each servo consists of an electro/magnetic valve assembly and an output linkage assembly.

FLIGHT DIRECTOR SYSTEMS

A flight director system is an instrument system consisting of electronic components that will compute and indicate the aircraft attitude required to attain and maintain a preselected flight condition. "Command" indicators on the instrument indicate how much and in what direction the attitude of the aircraft must be changed to achieve the desired result. The computed command indications relieve the operator of many of the mental calculations required for instrument flights, such as interception

angles, wind drift correction, and rates of climb and descent.

A flight director system has several components; the principal ones are the gyroscope, computer, and the cockpit presentaton. The gyro detects deviations from a preselected aircraft attitude. Any force exerted against the gyroscope is electrically transmitted to the computer, which in turn, sends a computed signal to the flight indicator, telling the operator what must be done with the controls. When using a flight director system, the operator is, in a sense, acting as a servo, following orders given by the command indicators.

The computers used in the various types of flight director systems are basically the same; however, the numbers and types of functions available will vary between systems because of the mission of a particular aircraft, the limited aircraft space available for installation, and the excessive cost of functions not absolutely required.

The instrument panel presentations and operating methods vary considerably between different systems. Command indications may be presented by several different symbols, such as bar-type command indicators with different types of movements, a phantom aircraft symbol, or two-element crossbar indicators.

Many flight director systems are equipped with an "altitude-hold" function, which permits selection of a desired altitude; the flight director computes the pitch attitude necessary to maintain this particular altitude.

A flight director greatly simplifies problems of aerial navigation. Selection of the VOR function electronically links the computer to the omnirange receiver. After selection of the desired omnicourse, the flight director will direct the bank attitude necessary to intercept and maintain this course.

Flight director systems are designed to offer the greatest assistance during the instrument approach phase of flight. ILS localizer and glide slope signals are transmitted through the receivers to the computer, and are presented as command indications. With the altitude-hold function, level flight can be maintained during the maneuvering and procedure turn phase of the approach. Once inbound on the localizer, the command signals of the flight director are maintained in a centered or zero condition.

Compensation for wind drift is automatic. Interception of the glide slope will cause a downward indication of the command pitch indicator. Any deviation from the proper glide slope path will cause a fly-up or fly-down indication on the flight director pitch command symbol. When altitude-hold is being used, it automatically disengages when the glide slope has been intercepted.

A flight director system not only shows the present situation, but also predicts the future consequences of this situation. For example, a momentary change in attitude is detected by the computer, and command symbol movement is made to correct this condition possibly before an altitude error can result. Thus, greater precision is achieved with less mental effort on the part of the aircraft operator.

AUTOPILOT SYSTEM MAINTENANCE

The information in this section does not apply to any particular autopilot system, but gives general information which relates to all autopilot systems. Maintenance of an autopilot system consists of visual inspection, replacement of components, cleaning, lubrication, and an operational checkout of the system.

With the autopilot disengaged, the flight controls should function smoothly. The resistance offered by the autopilot servos should not affect the control of the aircraft. The interconnecting mechanism between the autopilot system and the flight control system should be correctly aligned and smooth in operation. When applicable, the operating cables should be checked for tension.

An operational check is important to assure that every circuit is functioning properly. An autopilot operational check should be performed on new installations, after replacement of an autopilot component, or whenever a malfunction in the autopilot system is suspected.

After the aircraft's main power switch has been turned on, allow the gyros to come up to speed and the amplifier to warm up before engaging the autopilot. Some systems are designed with safeguards that prevent premature autopilot engagement.

While holding the control column in the normal flight position, engage the system using the engaging control (switch, handle).

After the system is engaged, perform the operational checks specified for the particular aircraft. In general, the checks consist of:

(1) Rotate the turn knob to the left; the left rudder pedal should move forward, and the control column wheel should move to and the control column wheel should move slightly aft.

(2) Rotate the turn knob to the right; the right rudder pedal should move forward, and the control column wheel should move to the right. The control column should move slightly aft. Return the turn knob to the center position; the flight controls should return to the level-flight position.

(3) Rotate the pitch-trim knob forward; the control column should move forward.

(4) Rotate the pitch-trim knob aft; the control column should move aft.

If the aircraft has a pitch-trim system installed, it should function to add downtrim as the control column moves forward, and add uptrim as the column moves aft. Many pitch-trim systems have an automatic and a manual mode of operation. The above action will occur only in the automatic mode.

Check to see if it is possible to manually override or overpower the autopilot system in all control positions. Center all the controls when the operational checks have been completed.

Disengage the autopilot system and check for freedom of the control surfaces by moving the control columns and rudder pedals. Then re-engage the system and check the emergency disconnect release circuit. The autopilot should disengae each time the release button is actuated.

When performing maintenance and operational checks on a specific autopilot system, always follow the procedure recommended by the aircraft or equipment manufacturer.

Annunciator System

Instruments are installed for two purposes, one to display current conditions, the other to notify of unsatisfactory conditions. Colored scales are used; usually green for satisfactory; yellow for caution or borderline conditions; red, for unsatisfactory conditions. As aircraft have become more complex with many systems to be monitored, the need for a centralized warning system became apparent. The necessity to coordinate engine and flight controls emphasized this need. What evolved is an annunciator or master warning system (figure 12–77).

Certain system failures are immediately indicated on an annunciator panel on the main instrument panel. A master caution light and a light indicating the faulting system flash on. The master light may be reset to "Off," but the indicating light will remain "On" until the fault is corrected or the equipment concerned is shut down. By resetting, the master caution light is ready to warn of a sub-

SYSTEM	ATA NUMBER	INDICATION
Aircraft Fuel	2800	Fuel Pressure Low
Engine Fuel	7300	Fuel Pressure Low
Electrical	2400	Inverter Out
Generator	2400	Generator Out
Generator	2400	Generator Overheated
Starting	8000	Starter Engaged
Engine Oil	7900	Oil Pressure Low
Landing Gear	3200	Brake Pressure Low
Landing Gear	3200	Not Locked Down
Landing Gear	3200	Anti-Skid Out
Air Conditioning	2100	Cabin Pressure High
Air Conditioning	2100	Cabin Pressure Low
Flight Control	2700	Speed Brake Extended
Stabilizer	5500	Not Set for Takeoff
Engine Exhaust	7800	Thrust Reversal Pressure Low
Aux Power	4900	APU Exhaust Door Not Open
Doors	5200	Cabin Door Unlocked
Doors	5200	Cargo Door Unlocked
Navigation	3400	Mach Trim Computer Out
Electrical	2400	Normal Bus Tie Open
Auto Flight	2200	Auto Pilot Off
Hydraulic	2900	Hydraulic Pressure Low
Firewarning	2600	AFT Compartment Overheated

FIGURE 12–77. Warning in annunciator system.

sequent fault even before correction of the initial fault. A press to test light is available for testing the circuits in this system.

One late model business jet has the sensing devices divided into groups, according to their method of operation. The fast group responds to heat and uses bimetallic strips set at predetermined temperatures. The second group responds to pressure changes and uses a flexible chamber that moves when pressurized. The third group consists of mechanically operated switches and/or contacts on a relay.

An annunciator system may include any or all of the following indications or others as applicable.

Aural Warning System

Aircraft with retractable landing gear use an aural warning system to alert the crew to an unsafe condition. A bell will sound if the throttle is retracted and the landing gear is not in a down and locked condition (figure 12–78).

Aural warning systems range in complexity from the simple one just described to that system necessary for safe operation of the most complex transport aircraft.

A typical transport aircraft has an aural warning system which will alert the pilot with audio signals to: An abnormal takeoff condition, landing condition, pressurization condition, mach-speed condition, an engine or wheel well fire, calls from the crew call system, and calls from the secal system. Shown in figure 12–78 are some of the problems which trigger warning signals in the aural warning system. For example: a continuous horn sounding during landing would indicate the landing gear is not down and locked when flaps are less than full up and the throttle is retarded. The corrective action would be to raise the flaps and advance the throttle.

(See figure 12–78 on next page)

STAGE OF OPERATION	WARNING SYSTEM	WARNING SIGNAL	CAUSE OF WARNING SIGNAL ACTIVATION	CORRECTIVE ACTION
Landing	Landing gear ATA 3200	Continuous horn	Landing gear is not down and locked when flaps are less than full up and throttle is retarded to idle.	Raise flaps Advance throttle
In flight	Mach warning ATA 3400	Clacker	Equivalent airspeed or mach number exceeds limits.	Decrease speed of aircraft
Takoff	Flight control ATA 2700 Aux power ATA 4900	Intermittent horn	Throttles are advanced and any of following conditions exist. 1. Speed brakes are not down 2. Flaps are not in takeoff range 3. Auxiliary power exhaust door is open 4. Stabilizer is not in the takeoff setting.	Correct the aircraft to proper takeoff conditions.
Inflight	Pressurization ATA 2100	Intermittent horn	If cabin pressure becomes equal to atmospheric pressure at the specific altitude (altitude at time of occurrence).	Correct the condition.
Any stage	Fire warning ATA 2600	Continuous bell	Any overheat condition or fire in any engine or nacelle, or main wheel or nose wheel well, APU engine or any compartment having firewarning system installed. Also whenever the firewarning system is tested.	1. Lower the heat in the area wherein the F/W was activated. 2. Signal may be silenced by pushing the F/W bell cutout switch or the APU cutout switch.
Any stage	Communications ATA 2300	High chime	Any time captain's call button is pressed at external power panel forward or rearward cabin attendant's panel	Release button or if button remains locked in, pull button out.
Any stage	Communications secal system* ATA 2300	Tow tone hi-low chime or single low chime.	Whenever a signal has been received by an HF or VHF communication system and decoded by the secal* decoder.	Press reset button on secal system control panel.

*NOTE: Secal system is the Selective Calling System: Each aircraft is assigned a particular four tone audio combination for identification purposes. A ground station will key the signal whenever contact with that particular aircraft is desired. The signal will be decoded by the airborne secal decoder and the crew alerted by the secal warning system.

FIGURE 12–78. Aural warning system.

518

CHAPTER 13
COMMUNICATIONS AND NAVIGATION SYSTEMS

GENERAL

Communications and navigation are the two major functions of airborne radio. Communication systems primarily involve voice transmission and reception between aircraft or aircraft and ground stations. Radios are used in aircraft as navigational aids in a number of applications. They range from a simple radio direction finder to navigational systems which use computers and other advanced electronic techniques to automatically solve the navigational problems for an entire flight. Marker beacon receivers, instrument landing systems (involving radio signals for glide slope and direction), distance measuring equipment, radar, area navigation systems, and omnidirectional radio receivers are but a few basic applications of airborne radio navigation systems available for installation and use in aircraft.

Safe aircraft operation is dependent to a large degree upon the satisfactory performance of the airborne communications and navigation systems. Reliability and performance of the radio and radar system are directly related to the skills of those who perform the maintenance.

Federal Aviation Regulations require an inspection of radio equipment installations at regular intervals. These inspections include a visual examination for security of attachment, condition of wiring, bonding, shock mounts, radio racks and supporting structure. In addition, a functional check is usually performed to determine that the equipment is operating properly and that its operation does not interfere with the operation of other systems.

Aircraft mechanics' responsibilities include the installation and inspection of radios, antennas, navigation equipment, and associated wiring. In addition, the FAA has certified radio repair facilities to perform maintenance on radios, antennas, navigation, and radar equipment. Transmitting equipment is calibrated by persons licensed by the FCC (Federal Communications Commission).

To be in a more favorable position to inspect system installations, a mechanic should possess some basic knowledge and understanding of the principles, purposes, and operation of the radio equipment used in aircraft.

Because of the many different makes and models of equipment and the various systems in use, it is not feasible to describe each in this handbook. The information presented in this chapter is general in nature and provides a broad introduction to radio, its principles, and application to aircraft from a mechanic's viewpoint.

BASIC RADIO PRINCIPLES

The principle of radio communication can be illustrated by using a simple transformer. As shown in figure 13–1, closing the switch in the primary circuit causes the lamp in the secondary circuit to be illuminated. Opening the switch extinguishes the light.

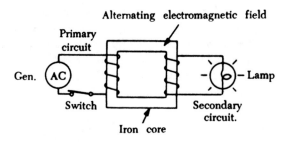

FIGURE 13–1. A simple transformer circuit.

There is no direct connection between the primary and secondary circuits. The energy that illuminates the light is transmitted by an alternating electromagnetic field in the core of the transformer. This is a simple form of wireless control of one circuit (the secondary) by another circuit (the primary).

The basic concept of radio communications involves the transmission and reception of electromagnetic (radio) energy waves through space. Alternating current passing through a conductor creates electromagnetic fields around the conductor. Energy is alternately stored in these fields and returned to the conductor. As the frequency of current alternation increases, less and less of the energy stored in the field returns to the conductor. Instead of returning, the energy is radiated into space in the form of electromagnetic waves. A conductor radiating in this manner is called the transmitting antenna.

For an antenna to radiate efficiently a transmitter must supply it with an alternating current of the selected frequency. The frequency of the radio wave radiated will be equal to the frequency of the applied current. When current flows through a transmitting antenna, radio waves are radiated in all directions in much the same way that waves travel on the surface of a pond into which a rock has been thrown. Radio waves travel at a speed of approximately 186,000 miles per second.

If a radiated electromagnetic field passes through a conductor, some of the energy in the field will set electrons in motion in the conductor. This electron flow constitutes a current that varies with changes in the electromagnetic field. Thus, a variation of the current in a radiating antenna causes a similar varying current in a conductor (receiving antenna) at a distant location. Any intelligence being produced as current in a transmitting antenna will be reproduced as current in a receiving antenna.

Frequency Bands

The radio frequency portion of the electromagnetic spectrum extends from approximately 30 kHz (kilohertz) to 30,000 MHz (Megahertz). As a matter of convenience, this part of the spectrum is divided into frequency bands. Each band or frequency range produces different effects in trans-

mission. The radio frequency bands proven most useful and presently in use are:

Frequency Range	Band
Low frequency (L/F)	30 to 300 kHz
Medium frequency (M/F)	300 to 3,000 kHz
High frequency (H/F)	3,000 kHz to 30 MHz
Very high frequency (VHF)	30 to 300 MHz
Ultra high frequency (UHF)	300 to 3,000 MHz
Superhigh frequency (SHF)	3,000 to 30,000 MHz

In practice, radio equipment usually covers only a portion of the designated band, e.g., civil VHF equipment normally operates at frequencies between 108.0 MHz and 135.95 MHz.

BASIC EQUIPMENT COMPONENTS

The basic components (figure 13–2) of a communication system are: microphone, transmitter, transmitting antenna, receiving antenna, receiver, and a headset or loudspeaker.

Transmitters

A transmitter may be considered as a generator which changes electrical power into radio waves. A transmitter must perform these functions: (1) Generate a RF (radio frequency) signal, (2) amplify the RF signal, and (3) provide a means of placing intelligence on the signal.

The transmitter contains an oscillator circuit to generate the RF signal (or a subharmonic of the transmitter frequency, if frequency doublers or multipliers are used) and amplifier circuits to increase the output of the oscillator to the power level required for proper operation.

The voice (audio) intelligence is added to the RF

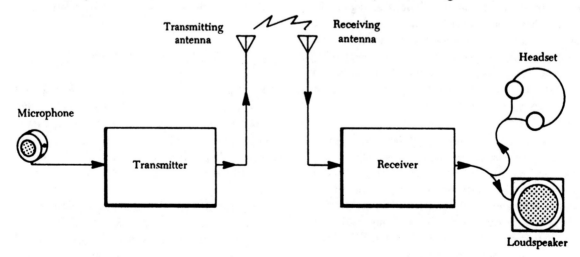

FIGURE 13–2. Basic communication equipment.

signal by a special circuit called the modulator. The modulator uses the audio signal to vary the amplitude or frequency of the RF signal. If the amplitude is varied, the process is called amplitude modulation or AM. If the frequency is varied, the process is known as frequency modulation or FM.

Transmitters take many forms, have varying degrees of complexity and develop various levels of power. The amount of power generated by a transmitter affects the strength of the electromagnetic field radiating from the antenna. Thus, it follows that the higher the power output from a transmitter, the greater the distance its signal may be received.

VHF transmitters used in single engine and light twin-engine aircraft vary in power output from 1 watt to 30 watts, depending on the particular model radio. However, radios having 3 to 5 watt ratings are used most frequently. Executive and large transport aircraft are usually equipped with VHF transmitters having a power output of 20 to 30 watts.

Aviation communication transmitters are crystal-controlled in order to meet the frequency tolerance requirements of the FCC. Most transmitters are selectable for more than one frequency. The frequency of the channel selected is determined by a crystal. Transmitters may have from one to 680 channels.

Receivers

The communications receiver must select radio frequency signals and convert the intelligence contained on these signals into a usable form; either audible signals for communication and audible or visual signal for navigation.

Radio waves of many frequencies are present in the air. A receiver must be able to select the desired frequency from all those present and amplify the small a.c. signal voltage.

The receiver contains a demodulator circuit to remove the intelligence. If the demodulator circuit is sensitive to amplitude changes, it is used in AM sets and called a detector. A demodulator circuit that is sensitive to frequency changes is used for FM reception and is known as a discriminator.

Amplifying circuits within the receiver increases the audio signal to a power level which will operate the headset or loudspeaker properly.

Antenna

An antenna is a special type of electrical circuit designed to radiate and receive electromagnetic energy. As mentioned previously, a transmitting antenna is a conductor which radiates electromagnetic waves when a radio frequency current is passed through it. Antennas vary in shape and design (figure 13–3) depending upon the frequency to be transmitted, and specific purposes they must serve. In general, communication transmitting stations radiate signals in all directions. However, special antennas are designed that radiate only in certain directions or certain beam patterns.

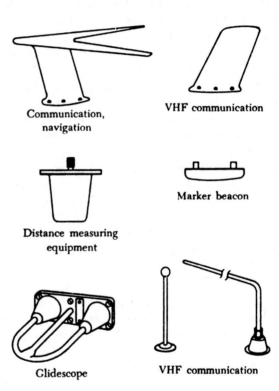

Communication, navigation

VHF communication

Distance measuring equipment

Marker beacon

Glidescope

VHF communication

FIGURE 13–3. Antennas.

The receiving antenna must intercept the electromagnetic waves that are present in the air. The shape and size of the receiving antenna will also vary according to the specific purpose for which it is intended. In airborne communications the same antenna is normally used for both transmission and reception of signals.

Microphones

A microphone is essentially an energy converter that changes acoustical (sound) energy into corresponding electrical energy. When spoken into a microphone, the audio pressure waves generated strike the diaphragm of the microphone causing it to move in and out in accordance with the instantaneous pressure delivered to it. The diaphragm is attached to a device that causes current to flow in proportion to the pressure applied.

For good quality sound, the electrical waves from a microphone must correspond closely in magnitude and frequency to the sound waves that cause them, so that no new frequencies are introduced. A desirable characteristic is the ability of the microphone to favor sounds coming from a nearby source over random sounds coming from a relatively greater distance. When talking into this type of microphone, the lips must be held as close as possible to the diaphragm.

Persons inexperienced in the use of the microphone are usually surprised at the quality of their own transmissions when they are taped and played back. Words quite clear when spoken to another person can be almost unintelligible over the radio. Readable radio transmissions depend on the following factors: (1) Voice amplitude, (2) rate of speech, and (3) pronunciation and phrasing. Clarity increases with amplitude up to a level just short of shouting. When using a microphone, speak loudly, without exerting extreme effort. Talk slowly enough so that each word is spoken distinctly. Avoid using unnecessary words.

POWER SUPPLY

The power supply is a component that furnishes the correct voltages and current needed to operate the communication equipment. The power supply can be a separate component or it may be contained within the equipment it supplies. Electromechanical devices used as electronic power supplies include dynamotors and inverters.

The dynamotor performs the dual functions of motor and generator, changing the relatively low voltage of the aircraft electrical system into a much higher value. The multi-vibrator is another type of voltage supply used to obtain a high a.c. or d.c. voltage from a comparatively low d.c. voltage.

In many aircraft, the primary source of electric power is direct current. An inverter is used to supply the required alternating current. Common aircraft inverters consist of a d.c. motor driving an a.c. generator. Static, or solid-state inverters are replacing the electromechanical inverters in many applications. Static inverters have no moving parts, but use semiconductor devices and circuits that periodically pulse d.c. current through the primary of a transformer to obtain an a.c. output from the secondary.

COMMUNICATION SYSTEMS

The most common communication system in use today is the VHF system. In addition to VHF equipment, large aircraft are usually equipped with HF communication systems.

Airborne communications systems vary considerably in size, weight, power requirements, quality of operation, and cost, depending upon the desired operation.

Many airborne VHF and HF communication systems use transceivers. A transceiver is a self-contained transmitter and receiver which share common circuits; i.e., power supply, antenna, and tuning. The transmitter and receiver both operate on the same frequency, and the microphone button determines when there is an output from the transmitter. In the absence of transmission, the receiver is sensitive to incoming signals. Since weight and space are of great importance in aircraft, the transceiver is widely used. Large aircraft may be equipped with transceivers or a communications system that uses separate transmitters and receivers.

The operation of radio equipment is essentially the same whether installed on large aircraft or small aircraft. In some radio installations the controls for frequency selection, volume, and the "on-off" switch are integral with the radio main chassis. In other installations, the controls are mounted on a panel located in the cockpit and the radio equipment is located in racks in another part of the aircraft.

Because of the many different types and models of radios in use, it is not possible to discuss the specific techniques for operating each in this manual. However, there are various practices of a nonspecific nature which apply to all radios. These general practices will be described.

VHF (Very High Frequency) Communications

VHF airborne communication sets operate in the frequency range from 108.0 MHz to 135.95 MHz. VHF receivers are manufactured that cover only the communications frequencies, or both communications and navigation frequencies. In general, the VHF radio waves follow approximately straight lines. Theoretically, the range of contact is the distance to the horizon and this distance is determined by the heights of the transmitting and receiving antennas. However, communication is sometimes possible many hundreds of miles beyond the assumed horizon range.

Many VHF radios have the transmitter, receiver, power supply, and operating controls built into a

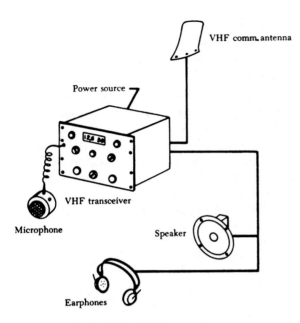

FIGURE 13-4. VHF system diagram.

communication system mounted on the instrument panel and the remainder remotely installed in a radio or baggage compartment.

To perform an operational check of a VHF communication system, a source of electric power must be available. After turning the radio control switch "on", allow sufficient time for the equipment to warm up before beginning the operational checks. Using the frequency selector, select the frequency of the ground station to be contacted. Adjust the volume control to the desired level.

With the microphone held close to the mouth, press the microphone button and speak directly into the microphone to transmit; when through talking, release the button. This action will return the communication receiver to operation. When the ground station acknowledges the initial transmission, request that an operational check be made on all frequencies or channels. Prior to transmitting, make certain that a station license is displayed in the aircraft. In addition, the person operating the transmitter must hold a current Restricted Radiotelephone Operator's permit. Both the station license and the operator's permit are issued by the FCC.

single unit. This unit is frequently installed in a cutout in the instrument panel. A system diagram of a typical panel-mounted VHF transceiver is shown in figure 13-4. Others have certain portions of the

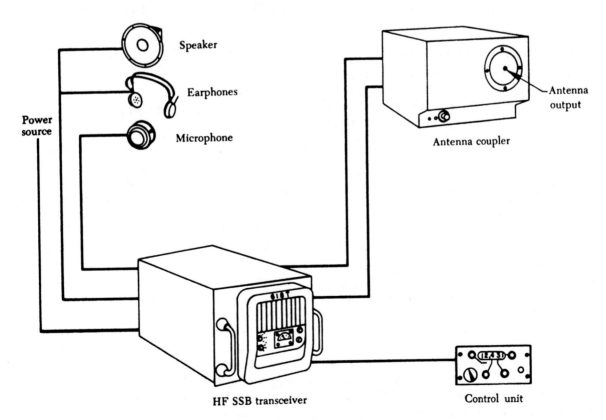

FIGURE 13-5. HF system diagram.

HF (High Frequency) Communications

A high frequency communication system (figure 13–5) is used for long-range communication. HF systems operate essentially the same as a VHF system, but operate in the frequency range from 3 MHz to 30 MHz. Communications over long distances are possible with HF radio because of the longer transmission range. HF transmitters have higher power outputs than VHF transmitters.

The design of antennas used with HF communication systems vary with the size and shape of the aircraft. Aircraft which cruise below 300 m.p.h. generally use a long wire antenna. Higher speed aircraft have specially designed antenna probes installed in the vertical stabilizer. Regardless of the type antenna, a tuner is used to match the impedance of the transceiver to the antenna.

An operational check of an HF radio consists of turning the control switch to "on," adjusting the RF gain and volume controls, selecting the desired channel and transmitting the appropriate message to the called station. Best adjustment of the gain control can be obtained with the volume control set at half range. The gain control is used to provide the strongest signal with the least amount of noise. The volume control is used to set sound level and affects only the loudness of the signal.

AIRBORNE NAVIGATION EQUIPMENT

"Airborne navigation equipment" is a phrase embracing many systems and instruments. These systems include VHF omnirange (VOR), instrument landing systems, distance-measuring equipment, automatic direction finders, doppler systems, and inertial navigation systems.

When applied to navigation, the radio receivers and transmitters handle signals which are used to determine bearing and in some cases distance, from geographical points or radio stations.

VHF OMNIRANGE SYSTEM

The VHF VOR (omnidirectional range) is an electronic navigation system. As the name implies, the omnidirectional or all-directional range station provides the pilot with courses from any point within its service range. It produces 360 usable radials or courses, any one of which is a radio path connected to the station. The radials can be considered as lines that extend from the transmitter antenna like spokes of a wheel. Operation is in the VHF portion of the radio spectrum (frequency range of 108.0 MHz – 117.95 MHz) with the result

that interference from atmospheric and precipitation static is negligible. The navigational information is visually displayed on an instrument in the cockpit.

The typical airborne VOR receiving system (figure 13–6) consists of a receiver, visual indicator, antennas, and a power supply. In addition, a unit frequency selector is required and in some cases located on the receiver unit front panel. Some manufacturers design a remote control frequency selector so the equipment may be installed in some other area of the aircraft. This frequency selector is used to tune the receiver to a selected VOR ground station.

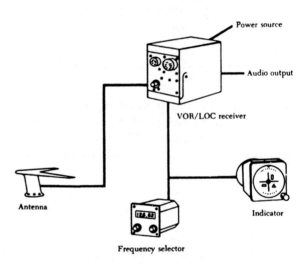

FIGURE 13–6. VOR system diagram.

The VOR receiver, in addition to course navigation, functions as a localizer receiver during ILS (instrument landing system) operation. Also, some VOR receivers include a glide slope receiver in a single case. Regardless of how individual manufacturers may design the VOR equipment, the intelligence from the VOR receiver is displayed on the CDI (course deviation indicator).

The CDI, figure 13–7, performs several functions. During VOR operation the vertical needle is used as the course indicator. The vertical needle also indicates when the aircraft deviates from the course and the direction the aircraft must be turned to attain the desired course. The "TO-FROM" indicator presents the direction to or from the station along the omniradial. The course deviation indicator also contains a "VOR-LOC" flag alarm. Normally this is a small arm which extends into view

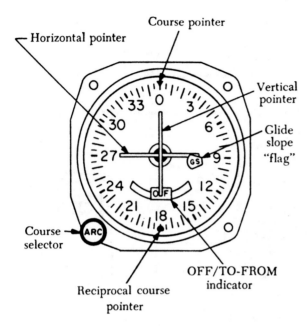

FIGURE 13-7. Course deviation indicator.

only in the case of a receiver malfunction or the loss of a transmitted signal.

When localizer signals are selected on the receiver, the indicator shows the position of the localizer beam relative to the aircraft and the direction the aircraft must be turned to intercept the localizer.

During VOR operation the VOR radial to be used is selected by rotating the OBS (omnibearing selector). The OBS is generally located on the CDI; however, in some installations it is a part of the navigation receiver. The OBS is graduated in degrees from zero to 360. Each degree is a VOR course to be flown in reference to a ground station.

The following steps are typical of those performed during a ground operational check. When checking a VOR system, follow the specific procedures recommended by the equipment manufacturer. The operational check can be performed using appropriate test equipment, VOT (Very High Frequency Omnirange Test) or the terminal VOR facility.

(1) Place the on/off switch in the "ON" position.

(2) Adjust the frequency selector to the desired station.

(3) Allow sufficient time for equipment to warm up.

(4) The VOR flag will disappear when the VOR station signal is received.

(5) Adjust the volume control to the desired level; assure that the selected VOR station identification is clear and correct.

(6) Check the CDI for vertical needle deflection.

(7) Center the vertical needle by rotating the OBS.

(8) Check "TO-FROM" indicator for "TO" indication.

(9) Rotate the OBS to read 10° higher than the setting at which the vertical needle was centered. The vertical needle should move left and cover the last dot which corresponds to 10° course displacement.

(10) Return the OBS to the original position. The vertical needle should return to the center position.

(11) Rotate the OBS to read 10° lower than the original setting. The pertical needle should move right and cover the last dot which corresponds to 10° course displacement.

(12) Assure that the vertical needle moves an equal distance in both directions. This total course width or course sensitivity should be 20°.

NOTE: When "TO-FROM" indicator reads "FROM", the vertical needle will deflect in the direction opposite that stated in the above procedures.

If the operational check is unsatisfactory, it will be necessary to remove the VOR receiver and associated instruments from the aircraft and have them calibrated.

INSTRUMENT LANDING SYSTEM

The ILS (instrument landing system), one of the facilities of the Federal airways, operates in the VHF portion of the electromagnetic spectrum. The ILS can be visualized as a slide made of radio signals on which the aircraft can be brought safely to the runway.

The entire system consists of a runway localizer, a glide slope signal, and marker beacons for position location. The localizer equipment produces a radio course aligned with the center of an airport runway. The on-course signals result from equal reception of two signals; one containing 90 Hz modulation and the other containing 150 Hz modulation. On one side of the runway center line the radio receiver develops an output in which the 150 Hz tone predominates. This area is called the blue sector. On the other side of the centerline the 90 Hz output is greater. This area is the yellow sector.

The localizer facility operates in the frequency range of 108.0 MHz to 112.0 MHz on the "odd tenths" of the megahertz steps. The VOR receiver also operates in this ferquency range on the "even tenths" of the megahertz steps. The airborne VOR receiver functions as the localizer receiver during ILS operation.

The glide slope is a radio beam which provides vertical guidance to the pilot, assisting him in making the correct angle of descent to the runway. Glide slope signals are radiated from two antennas located adjacent to the touchdown point of the runway. Each glide slope facility operates in the UHF frequency range from 329.3 MHz to 335.0 MHz.

The glide slope and VOR/localizer receivers may be separate receivers or combined in a single case. The glide slope receiver is paired to the localizer and one frequency selector is used to tune both receivers. A component diagram of an ILS is shown in figure 13–8.

The information from both localizer and glide slope receivers is presented on the CDI; the vertical needle displays localizer information and the horizontal needle displays glide slope information (figure 13–9). When both needles are centered, the aircraft is on course and descending at the proper

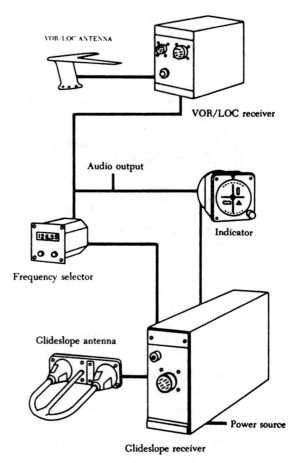

FIGURE 13–8.   Component diagram of an ILS.

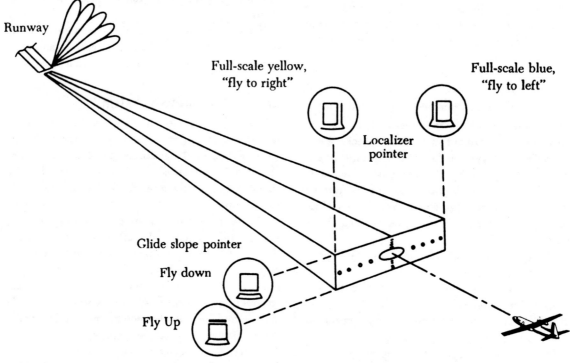

FIGURE 13–9.   Glide slope information.

rate. In addition the CDI contains a red warning flag for each system which comes into view when the receiver fails or the loss of a transmitted signal occurs.

Two antennas are usually required for ILS operation. One for the localizer receiver, also used for VOR navigation, and one for the glide slope. Some of the small aircraft use a single multi-element antenna for both glide slope and VOR/LOC operation. The VOR/localizer antenna is normally installed on the top of the aircraft fuselage or flush mounted in the vertical stabilizer. The glide slope antenna is, in most cases, installed on the nose of the aircraft. On aircraft equipped with radomes, the glide slope antenna is installed under the radome.

Marker Beacons

Marker beacons are used in connection with the instrument landing system. The markers are signals which indicate the position of the aircraft along the approach to the runway. Two markers are used in each installation. The location of each marker is identified by both an aural tone and a signal lamp. The marker beacon transmitters, operating on a fixed 75 MHz frequency, are placed at specific locations along the approach pattern of an ILS facility. The antenna radiation pattern is beamed straight up.

A marker receiver (figure 13–10) installed in the aircraft receives the antenna signals and converts them into power to illuminate a signal lamp and produce an audible tone in a headset. The outer marker marks the beginning of the approach path. The outer marker signal is modulated by a 400-Hz signal which produces a tone keyed in long dashes. In addition to providing aural identification, the signal lights a purple lamp in the cockpit. The middle marker is usually about 3,500 ft. from the end of the runway and is modulated at 1,300 Hz which produces a higher-pitched tone keyed with alternate dots and dashes. An amber lamp flashes to indicate that the aircraft is passing over the middle marker.

Marker beacon receivers vary in design from simple receivers that have no operating controls and no aural output to more sophisticated receivers that produce an aural tone and have an on/off switch and a volume control to adjust the sound level of the identification code.

Where three lights are used, a white light indicates the aircraft positions at various points along the airways. In addition to the light, a rapid series of tones (six dots per second) of 3,000 Hz is received in the headset. Distance-measuring equipment is rapidly replacing the "along-route" marker system. A 3,000-Hz tone and white light marker are also being used for inner markers (missed approach point) on some Category II, ILS-equipped runways.

The ILS system cannot be ground tested fully without using test equipment simulating localizer and glide slope signals.

If an aircraft is located at an airport which has an ILS-equipped runway, it may be possible to determine if the receiver is functioning by performing

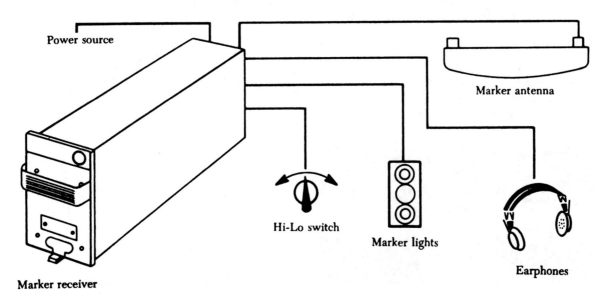

Power source

Marker antenna

Hi-Lo switch

Marker lights

Earphones

Marker receiver

FIGURE 13–10. Marker receiver system diagram.

the following. Place the on/off switch (if so equipped) in the "on" position and adjust the frequency selector to the proper ILS channel for the airport where the aircraft is located. Allow sufficient time for the equipment to warm up. In a strong signal area, both the localizer and glide slope warning flags will either start to move or go completely out of view. Observe that both cross pointers are deflected to their maximum displacement.

Some of the more sophisticated solid-state ILS equipment contains self-monitoring circuits. These circuits can be used for performing an operational test using the procedures in the aircraft or equipment manufacturer's service manuals.

DISTANCE-MEASURING EQUIPMENT

The purpose of DME (distance-measuring equipment) is to provide a constant visual indication of the distance the aircraft is from a ground station. A DME reading is not a true indication of point-to-point distance as measured over the ground. DME indicates the slant range between the aircraft and the ground station. Slant-range error increases as the aircraft approaches the station. At a distance of 30 to 60 nautical miles the slant range error is negligible.

DME operates in the UHF range of the radio frequency spectrum. The transmitting frequencies are in two groups, between 962 MHz to 1,024 MHz and 1,151 MHz to 1,212 MHz; the receiving frequencies are between 1,025 MHz to 1,149 MHz. Transmitting and receiving frequencies are given a channel number which is paired with a VOR channel. In some aircraft installations the DME channel selector is ganged with the VOR channel selector to simplify the radio operation. A typical DME control panel is shown in figure 13–11.

The aircraft is equipped with a DME transceiver which is tuned to a selected DME ground station. Usually DME ground stations are colocated with a

VOR facility (called VORTAC). The airborne transceiver transmits a pair of spaced pulses to the ground station. The pulse spacing serves to identify the signal as a valid DME interrogation. After reception of the challenging pulses, the ground station responds with a pulse transmission on a separate frequency to send a reply to the aircraft. Upon reception of the signal by the airborne transceiver, the elapsed time between the challenges and the reply is measured This time interval is a measure of the distance separating the aircraft and the ground station. This distance is indicated in nautical miles by a cockpit instrument similar to the one shown in figure 13–12.

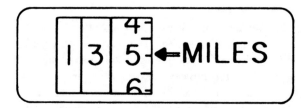

FIGURE 13–12.   DME digital indicator.

A typical DME antenna is shown in figure 13–13. Most DME antennas have a cover installed to protect them from damage. The DME antenna is usually a short, stub type mounted on the lower surface of the aircraft. To prevent an interruption in DME operation, the antenna must be located in a position that will not be blanked by the wing when the aircraft is banked.

FIGURE 13–13.   Typical DME antenna.

FIGURE 13–11.   Typical navigation DME control.

528

To determine if the DME operates, turn the on/off switch to the "on" position and select the appropriate channel. Allow sufficient time for the equipment to warm up. During this period, the distance indicator, both digital or pointer, will travel from minimum to maximum readings (sweep or search). When the DME has locked on a station, the indicator will stop searching and the red warning flag (if the indicator is equipped with one) will disappear. In most installations, no functional check can be made on the ground without a DME test set.

AUTOMATIC DIRECTION FINDERS

ADF (automatic direction finders) are radio receivers equipped with directional antennas which are used to determine the direction from which signals are received. Most ADF receivers provide controls for manual operation in addition to automatic direction finding. When an aircraft is within reception range of a radio station, the ADF equipment provides a means of fixing the position with reasonable accuracy. The ADF operates in the low and medium frequency spectrum from 190 kHz through 1,750 kHz. The direction to the station is displayed, on an indicator located in the cockpit, as a relative bearing to the station.

The airborne equipment (figure 13–14), consists of a receiver, loop antenna, sense or nondirectional antenna, indicator, and control unit. Most ADF receivers used in general aviation aircraft are panel mounted. Their operating controls appear on the front of the radio case.

In one type ADF system, the loop antenna (figure 13–15) rotates through 360° and receives maximum signal strength when in a parallel position with the direction of the transmitted signal. As the loop is rotated from this position, the signal becomes weaker and reaches a minimum when the plane of the loop is perpendicular to the direction of the transmitted signal. This position of the loop is called the null position. The null position of the loop is used for direction finding. When the loop is rotated to a null position the radio station is being received on a line perpendicular to the plane of the loop. However, the direction of the radio station from the aircraft may be either of two directions 180° apart. The inability of the loop antenna to determine from which of the two directions the transmitted signal is being received necessitates the installation of a sense antenna.

The loop and sense antenna are both connected to the ADF receiver. When the signal strength of the sense antenna is superimposed on the signal received from the loop antenna, it results in only

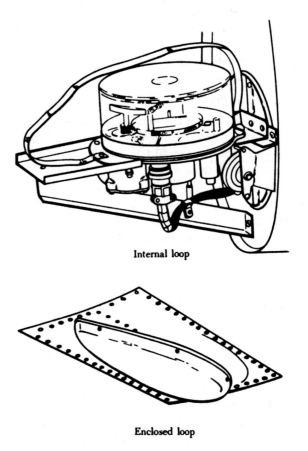

Internal loop

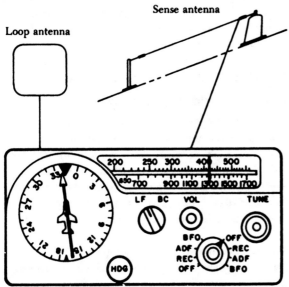

ADF receiver and control

FIGURE 13–14. Typical ADF installation.

Enclosed loop

FIGURE 13–15. Typical ADF antennas.

one null position of the loop. The one null position always indicates the direction to the transmitting facility.

Another type ADF system uses fixed, ferrite core loops in conjunction with a rotatable transformer called a resolver or goniometer. It operates essentially the same as the rotating loop, except that one of the windings of the goniometer rotates instead of the loop.

A general procedure for performing an operational check of the ADF system is as follows:

(1) Turn on/off switch to the "on" position and allow the radio to warm up. On installations that use the RMI (Radio Magnetic Indicator) pointer as an ADF indicator, assure that the switch has been positioned to present ADF information.

(2) Tune to the desired station.

(3) Adjust the volume control to an appropriate level.

(4) Rotate loop antenna and determine that only one null is received.

(5) Check that the ADF needle points towards the station. If the aircraft is situated among buildings or any other large reflecting surfaces, the ADF needle may indicate an error as a result of a reflected signal.

RADAR BEACON TRANSPONDER

The radar beacon transponder system is used in conjunction with a ground base surveillance radar to provide positive aircraft identification directly on the controller's radar scope.

The airborne equipment or transponder receives a ground radar interrogation for each sweep of the surveillance radar antenna and automatically dispatches a coded response. Civil transponders operate in two modes labeled "Mode A" and "Mode AC" which are switch controlled. The flight identification code, a four-digit number, is assigned during the flight planning procedure.

Some aircraft transponders are equipped with an altitude encoding feature. The aircraft's altitutde is transmitted to the ground station through the transponder. The mode selector switch is placed in the "AC" mode when it is necessary to transmit altitude information.

There are several different aircraft transponders in use. They all perform the same function and are basically the same electrically. The major differ-

ences are in construction; either a single unit or a control unit for remotely operating the transponder.

A typical transponder is shown in figure 13–16. The front panel of the illustrated transponder contains all the switches and dials needed for operation.

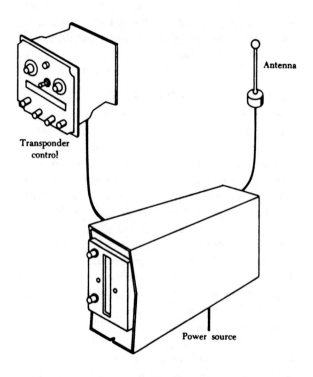

FIGURE 13–16. Typical transponder system.

A short stub or covered stub antenna is used for transponder operation and is usually mounted on the lower surface of the aircraft fuselage.

To ground check the radar beacon transponder the appropriate test equipment must be used.

DOPPLER NAVIGATION SYSTEMS

Doppler navigational radar automatically and continuously computes and displays ground speed and drift angle of an aircraft in flight without the aid of ground stations, wind estimates, or true airspeed data. The Doppler radar does not sense direction as search radar does. Instead, it is speed conscious and drift-conscious. It uses continuous carrier wave transmission energy and determines the forward and lateral velocity components of the aircraft by utilizing the principle known as Doppler effect.

The Doppler effect, or frequency change of a signal, can be explained in terms of an approaching and departing sound. As shown in figure 13–17, the sound emitter is a siren located on a moving ambulance and the receiver is the ear of a stationary person. Notice the spacing between the emitter when it is approaching and when it is departing from the stationary receiver. When the sound waves are closely spaced the listener hears a sound that is higher in pitch. The reverse is true when the emitter is moving away from the listener. Doppler radar uses the frequency change phenomenon just described, except in the radio frequency range.

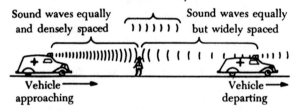

Median spacing if vehicle
were stationary

Sound waves equally
and densely spaced

Sound waves equally
but widely spaced

Vehicle →
approaching

Vehicle →
departing

FIGURE 13–17. Doppler effect with sound waves.

The Doppler radar emits narrow beams of energy at one frequency, and these waves of energy strike the earth's surface and are reflected. Energy waves returning from the earth are spaced differently than the waves striking the earth. The earth-returned energy is intercepted by the receiver and compared with the outgoing transmitter energy. The difference, due to Doppler effect, is used to develop ground speed and wind drift angle information.

A ground operational check of a Doppler system consists of setting a precise airspeed, and a deviation angle which will give a distance-off-course reading. Always refer to the equipment manufactur-er's instruction manual or the aircraft's operation manual for the proper test procedure.

INERTIAL NAVIGATION SYSTEM

The intertial navigation system is presently being used on large aircraft as a long-range navigation aid. It is a self-contained system and does not require signal inputs from ground navigational facilities. The system derives attitude, velocity, and heading information from measurement of the aircraft's accelerations. Two accelerometers are required, one referenced to north and the other to east. The accelerometers (figure 13–18) are mounted on a gyro-stabilized unit, called the stable platform, to avert the introduction of errors resulting from the acceleration due to gravity.

An inertial navigation system is a complex system containing four basic components. They are:

(1) A stable platform which is oriented to maintain accelerometers horizontal to the earth's surface and provide azimuth orientation.

(2) Accelerometers arranged on the platform to supply specific components of acceleration.

(3) Integrators which receive the output from the accelerometers and furnish velocity and distance.

(4) A computer which receives signals from the integrators and changes distance traveled to position in selected coordinates.

The diagram in figure 13–18 shows how these components are linked together to solve a navigation problem. Initial conditions are set into the system and the navigation process is begun. In inertial navigation, the term initialization is used to denote the process of bringing the system to a set of

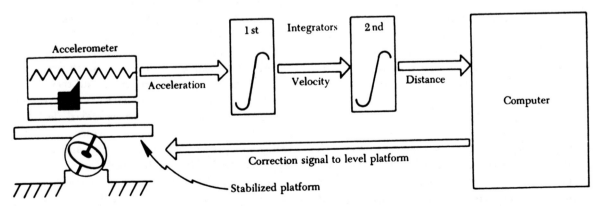

FIGURE 13–18. A basic inertial navigation system.

531

initial conditions from which it can proceed with the navigation process. These conditions include leveling the platform, aligning the azimuth reference, setting initial velocity and position, and making any computations required to start the navigation.

Although all inertial navigation systems must be initialized, the procedure varies according to the equipment and the type aircraft in which it is installed. The prescribed initialization procedures are detailed in the appropriate manufacturer's manuals.

From the diagram it can be seen that the accelerometers are maintained in a horizontal position to the earth's surface by a gyro-stabilized platform. As the aircraft accelerates, a signal from the accelerometer is sent to the integrators. The output from the integrators, or distance, is then fed into the computer, where two operations are performed. First, a position is determined in relation to the preset flight profile, and second, a signal is sent back to the platform to position the accelerometer horizontally to the earth's surface. The output from high-speed gyros and accelerometers, when connected to the flight controls of the aircraft, resists any changes in the flight profile.

AIRBORNE WEATHER RADAR SYSTEM

Radar (radio detection and ranging) is a device used to see certain objects in darkness, fog, or storms, as well as in clear weather. In addition to the appearance of these objects on the radar scope, their range and relative position are also indicated.

Radar is an electronic system using a pulse transmission of radio energy to receive a reflected signal from a target. The received signal is known as an echo; the time between the transmitted pulse and received echo is computed electronically and is displayed on the radar scope in terms of nautical miles.

A radar system (figure 13-19) consists of a transceiver and synchronizer, an antenna installed in the nose of the aircraft, a control unit installed in the cockpit, and an indicator or scope. A waveguide connects the receiver/transmitter to the antenna.

In the operation of a typical weather radar system, the transmitter feeds short pulses of radio-frequency energy through a waveguide to the dish antenna in the nose of the aircraft. In one typical installation the antenna radiates the energy in a beam 3.8° wide. Part of the transmitted energy is reflected from objects in the path of the beam and

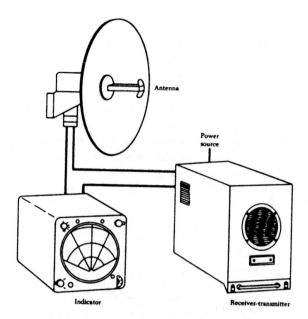

FIGURE 13-19. Weather radar system diagram.

is received by the dish antenna. Electronic switching simultaneously connects the antenna to the transmitter and disconnects the receiver during pulse transmission. Following the completion of pulse transmission, the antenna is switched from the transmitter to the receiver. The switching cycle is performed for each transmitted pulse.

The time required for radar waves to reach the target and reflect to the aircraft antenna is directly proportional to the distance of the target from the aircraft. The receiver measures the time interval between transmission of radar signals and reception of reflected energy and uses this interval to represent the distance, or range, of the target.

Rotation or sweep of the antenna and radar beam gives azimuth indications. The indicator sweep trace rotates in synchronization with the antenna. The indicator display shows the area and the relative size of targets, whose azimuthal position is shown relative to the line of flight.

The weather radar increases safety in flight by enabling the operator to detect storms in the flight path in order to chart a course around them. The terrain-mapping facilities of the radar show shorelines, islands, and other topographical features along the flight path. These indications are presented on the visual indicator in range and azimuth relative to the heading of the aircraft.

An operational check consists of the following:

(1) Tow or taxi the aircraft clear of all buildings and parked aircraft.

(2) Apply power to the equipment, and allow sufficient warmup time.

(3) Tilt the antenna to an upward position.

(4) Check the scan on the radar scope for an indication of targets.

RADIO ALTIMETER

Radio altimeters are used to measure the distance from the aircraft to the ground. This is accomplished by transmitting radio frequency energy to the ground and receiving the reflected energy at the aircraft. Most modern day altimeters are pulse type and the altitude is determined by measuring the time required for the transmitted pulse to hit the ground and return. The indicating instrument will indicate the true altitude of the aircraft, which is its height above water, mountains, buildings, or other objects on the surface of the earth.

The present day generation of radio altimeters are primarily used during landing and are a Category II requirement. The altimeter provides the pilot with the altitude of the aircraft during approach. Altimeter indications determine the decision point whether to continue to land, or execute a climb-out.

A radio altimeter system (figure 13–20) consists

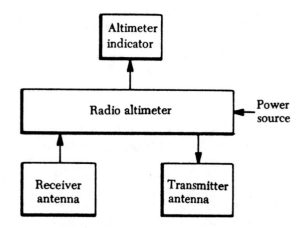

FIGURE 13–20. Typical radio altimeter system diagram.

of a transceiver, normally located in an equipment rack, an indicator installed in the instrument panel and two antennas located on the belly of the aircraft.

EMERGENCY LOCATOR TRANSMITTER (ELT)

Emergency locator transmitters are self-contained, self-powered radio transmitters designed to transmit a signal on the international distress bands of 121.5 MHz (civilian) and 243 MHz (military).

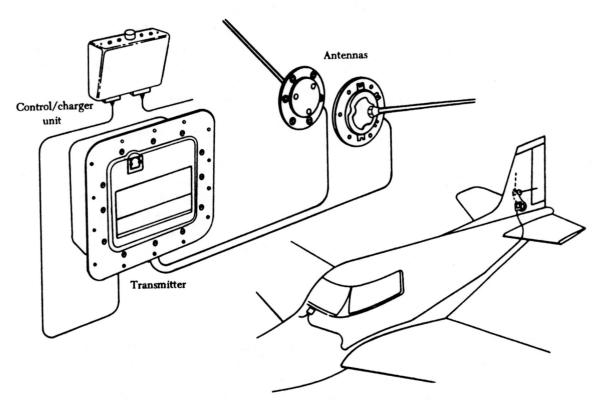

FIGURE 13–21. Emergency locator transmitter (ELT).

533

Operation is automatic on impact. The transmitter may also be activated by a remote switch in the cockpit or a switch integral with the unit. If the "G" force switch in the transmitter is activated from impact it can be turned off only with the switch on the case. (See figure 13–21.)

Transmitter

The transmitter may be located anywhere within the aircraft, but the ideal location is, as far aft as possible but just forward of the vertical fin. It must be accessible to permit monitoring the replacement date of the battery and for arming or disarming of the unit. A remote control arm/disarm switch may be installed in the cockpit.

The external antenna must be installed as far as practicable from other antennas to prevent interaction between avionics systems.

Batteries

Batteries are the power supply for emergency locator transmitters. When activated, the battery must be capable of furnishing power for signal transmission for at least 48 hours. The useful life of the battery is the length of time which the battery may be stored without losing its ability to continuously operate the ELT for 48 hours. This useful life is established by the battery manufacturer; batteries must be changed or recharged as required at 50% of the battery's useful life. This gives reasonable assurance that the ELT will operate if activated. The battery replacement date must be marked on the outside of the transmitter. This time is computed from the date of manufacture of the battery.

Batteries may be nickel-cadmium, lithium, magnesium dioxide, dry-cell batteries. Wet cell batteries have an unlimited shelf life until liquid is added. At that time their life in an ELT is regulated the same as dry cell batteries—change at 50% of shelf life. When replacing batteries use only those recommended by the manufacturer of the ELT. Do not use flashlight type batteries as their condition and useful life are unknown.

Testing

Testing of ELT's should be coordinated with the nearest FAA Tower or Flight Service Station and establish coordination for the test. Tests should be conducted only during the first five minutes of any hour and should be restricted to 3 audio sweeps. Any time maintenance is performed in the vicinity of the ELT, the VHF communication receiver should be tuned to 121.5 MHz and listen for ELT audio sweeps. If it is determined the ELT is operating, it must be turned off immediately.

False Alarms

False alarms have caused many of the problems with ELT's. Battery failures, with resulting corrosion of the unit results in either a complete failure or an unwanted transmission. Another type of unwanted transmission is the result of careless handling by the operators of the aircraft.

Test Equipment

Two monitors are available for identifying and/or locating unwanted ELT transmissions. A miniature scanning receiver may be mounted in the cockpit to warn the pilot if his ELT is transmitting. The other is a small portable ELT locator for use at general aviation airports to assist in finding an aircraft whose transmitter has accidentally become activated.

The operation of an ELT can be verified by tuning a communication receiver to the civil emergency frequency (121.5 MHz) and activating the ELT. Turn the ELT off immediately upon receiving a signal in the communication receiver.

In all maintenance and testing of ELT's the manufacturers instructions must be followed.

INSTALLATION OF COMMUNICATION AND NAVIGATION EQUIPMENT

There are many factors which the mechanic must consider prior to altering an aircraft by the addition of radio equipment. These factors include the space available, the size and weight of the equipment, and previously accomplished alterations. In addition, the power consumption of the added equipment must be calculated to determine the maximum continuous electrical load. Each installation should be planned to allow easy access for inspection, maintenance, and exchange of units.

The installation of radios is primarily mechanical, involving sheet metal work in mounting the radios, racks, antennas, and controls. Routing of the interconnecting wires, cables, antenna leads, etc., is also an important part of the installation process. When selecting a location for the equipment, first consider the areas designated by the airframe manufacturer. If such information is not available, or if the aircraft does not contain provisions for adding equipment, select an area that will carry the loads imposed by the weight of the equipment, and which is capable of withstanding the additional inertia forces.

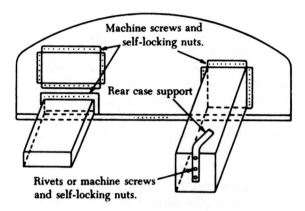

FIGURE 13–22. Typical radio installation in a stationary instrument panel.

If the radio is to be mounted in the instrument panel and no provisions have been made for such an installation, determine if the panel is primary structure prior to making any cutouts. To minimize the load on a stationary instrument panel, install a support bracket (figure 13–22) between the rear of the radio case or rack and a nearby structural member of the aircraft.

The radio equipment must be securely mounted to the aircraft. All mounting bolts must be secured by locking devices to prevent loosening from vibration.

Adequate clearance between the radio equipment and the adjacent structure must be provided to prevent mechanical damage to electric wiring or radio equipment from vibration, chafing, or shock landing.

Do not locate radio equipment and wiring near units containing combustible fluids. When separation is impractical, install baffles or shrouds to prevent contact of the combustible fluids with radio equipment in the event of a plumbing failure.

Cooling and Moisture

The performance and service life of most radio equipment is seriously limited by excessive ambient temperatures. The installation should be planned so that the radio equipment can dissipate its heat readily. In some installations it may be necessary to produce an airflow over the radio equipment, either with a blower or through the use of a venturi.

The presence of water in radio equipment promotes rapid deterioration of the exposed components. Some means must be provided to prevent the entry of water into the compartments housing the radio equipment.

Vibration Isolation

Vibration is a continued motion caused by an oscillating force. The amplitude and frequency of vibration of the aircraft structure will vary considerably with the type of aircraft.

Radio equipment is sensitive to mechanical shock and vibration and is normally "shock-mounted" to provide some protection against in-flight vibration and landing shock.

When special mounts (figure 13–23) are used to isolate radio equipment from vibrating structure, such mounts should provide adequate isolation over the entire range of expected vibration frequencies. When installing shock mounts, assure that the equipment weight does not exceed the weight-carrying capabilities of the mounts.

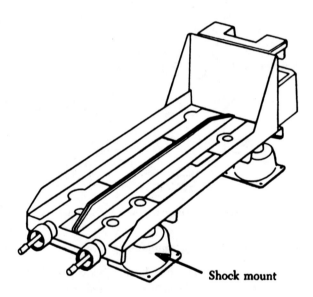

Shock mount

FIGURE 13–23. Typical shock-mounted base.

Radios installed in instrument panels do not ordinarily require vibration protection, since the panel itself is usually shock-mounted. However, make certain that the added weight can be safely carried by the existing mounts. In some cases it may be necessary to install larger capacity mounts or to increase the number of mounting points.

Radio equipment installed on shock mounts must have sufficient clearance from surrounding equipment and structure to allow for normal swaying of the equipment.

Periodic inspection of the shock mounts is required, and defective mounts should be replaced with the proper type. The factors to observe during the inspection are: (1) Deterioration of the shock-absorbing material, (2) stiffness and resiliency of the material, and (3) overall rigidity of the mount. If the mount is too stiff, it may not provide adequate protection against the shock of landing. If the mount is not stiff enough, it may allow prolonged vibration following an initial shock.

Shock-absorbing materials commonly used in shock mounts are usually electrical insulators. For this reason, each electronic unit mounted with shock mounts must be electrically bonded to a structural member of the aircraft, similar to that shown in figure 13–24. This may also be accomplished by using sheets of high-conductivity metal (copper or aluminum) where it is impossible to use a short bond strap.

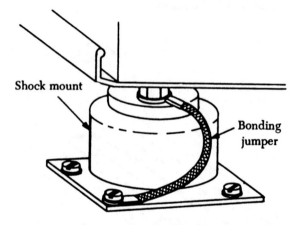

FIGURE 13–24. Typical shock mount bonding jumper.

REDUCING RADIO INTERFERENCE

Suppression of radio interference is a task of first importance. The problem has increased in proportion to the complexity of both the electrical system and the electronic equipment. Almost every component of the aircraft is a possible source of radio interference. Radio interference of any kind deteriorates the performance and reliability of the radio and electronic systems.

Isolation is the easiest and most practical method of radio noise suppression. This involves separating the source of radio noise from the input circuits of the affected equipment. In many cases, the noise in a receiver may be entirely eliminated simply by moving the antenna lead-in wire just a few inches away from the noise source. Some of the sources of radio interference in aircraft are rotating electrical devices, switching devices, ignition systems, propeller control systems, a.c. powerlines, and voltage regulators.

An aircraft can become highly charged with static electricity while in flight. If the aircraft is improperly bonded, all metal parts will not have the same amount of charge. A difference of potential will exist between various metal surfaces. The neutralization of the charges flowing in paths of variable resistance, due to such causes as intermittent contact from vibration or the movement of the control surfaces, will produce electrical disturbances (noise) in the radio receiver.

Bonding provides the necessary electrical connection between metallic parts of an aircraft. Bonding jumpers and bonding clamps are examples of bonding connectors. Bonding also provides the low-resistance return path for single-wire electrical systems.

Bonding radio equipment to the airframe will provide a low-impedance ground return and minimize radio interference from static electricity charges. Bonding jumpers should be as short as possible and installed in such a manner that the resistance does not exceed 0.003 ohm. When a jumper is used only to reduce radio noise and is not for current-carrying purposes, a resistance of 0.01 ohm is satisfactory.

The aircraft structure is also the ground for the radio. For the radio to function correctly, a proper balance must be maintained between the aircraft structure and the antenna. This means the surface area of the ground must be constant. Control surfaces, for example, may at times become partially insulated from the remaining structure. This would affect radio operation if the condition was not alleviated by bonding.

Shielding is one of the most effective methods of suppressing radio noise. The primary object of shielding is to electrically contain the radio frequency noise energy. In practical applications, the noise energy is kept flowing along the inner surface of the shield to ground instead of radiating into space. The use of shielding is particularly effective in situations where filters cannot be used. A good example of this is where noise energy radiates from a source and is picked up by the various circuits that eventually connect to the receiver input circuits. It would be impractical to filter all of the leads or units that are affected by the radiated noise energy; thus the application of effective shielding at the noise source itself is preferred, for it eliminates the radiated portion of the noise energy by confining it within the shield at its source.

Ignition wiring and spark plugs are usually shielded to minimize radio interference. If an intolerable radio noise level is present despite shielding, it may be necessary to provide a filter between the magneto and magneto switch to reduce the noise. This may consist of a single bypass capacitor or a combination of capacitors and choke coils. When this is done, the shielding between the filter and magneto switch can usually be eliminated.

The size of a filter may vary widely, depending on the voltage and current requirements as well as the degree of attenuation desired. Filters are usually incorporated in equipment known to generate radio interference, but since these filters are often inadequate it is frequently necessary to add external filters.

Static Discharger Wicks

Static dischargers are installed on aircraft to reduce radio receiver interference. This interference is caused by corona discharge emitted from the aircraft as a result of precipitation static. Corona occurs in short pulses which produce noise at the radio frequency spectrum. Static dischargers, normally mounted on the trailing edges of the control surfaces, wing tips, and vertical stabilizer, discharge the precipitation static at points a critical length away from the wing and tail extremities where there is little or no coupling of the static into the radio antenna.

Three major types of static dischargers are in use:

(1) Flexible vinyl-covered, silver- or carbon-impregnated braid.
(2) Semiflexible metallic braid.
(3) Null-field.

Flexible and semiflexible dischargers are attached to the aircraft by metal screws and should be periodically checked for tightness. At least 1 in. of the

inner braid of vinyl covered dischargers should extend beyond the vinyl covering. Null-field dischargers (figure 13–25) are riveted and epoxy bonded to the aircraft structure. A resistance measurement from the mount to the airframe should not exceed 0.1 ohm.

INSTALLATION OF AIRCRAFT ANTENNA SYSTEMS

An introductory knowledge of radio equipment is a valuable asset to the aviation mechanic, especially a knowledge of antenna installation and maintenance, since these tasks are often performed by the mechanic.

Antennas take many forms and sizes dependent upon the job they are to perform. Airborne antennas should be mechanically secure, mounted in interference-free locations, have the same polarization as the ground station, and be electrically matched to the receiver or transmitter which they serve.

The following procedures describe the installation of a typical rigid antenna:

(1) Place a template similar to that shown in figure 13–26 on the fore-and-aft center line at the desired location. Drill the mounting holes and correct diameter hole for the transmission line cable in the fuselage skin.

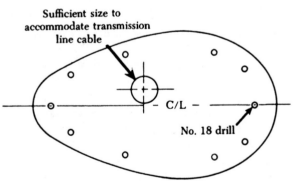

Sufficient size to accommodate transmission line cable

C/L

No. 18 drill

FIGURE 13–26. Typical antenna mounting template.

(2) Install a reinforcing doubler of sufficient thickness to reinforce the aircraft skin. The length and width of the reinforcing plate should approximate the example shown in figure 13–27.
(3) Install the antenna on the fuselage, making sure that the mounting bolts are tightened firmly against the reinforcing doubler and

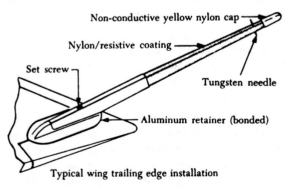

Non-conductive yellow nylon cap

Nylon/resistive coating

Set screw

Tungsten needle

Aluminum retainer (bonded)

Typical wing trailing edge installation

FIGURE 13–25. Null-field static discharger.

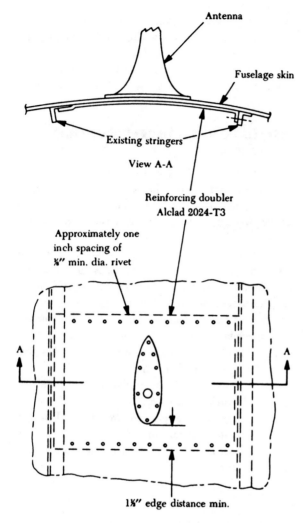

Antenna

Fuselage skin

Existing stringers

View A-A

Reinforcing doubler
Alclad 2024-T3

Approximately one
inch spacing of
⅛″ min. dia. rivet

A

A

1¼″ edge distance min.

FIGURE 13–27. Typical antenna installation on a skin panel.

the mast is drawn tight against the gasket. If a gasket is not used, seal between the mast and the fuselage with a suitable sealer, such as zinc chromate paste, or equal.

The mounting bases of antennas vary in shape and sizes; however, the aforementioned installation procedure is typical and may be used for mast-type antenna installations.

Transmission Lines

A transmitting or receiving antenna is connected directly to its associated transmitter or receiver by wire(s) which are shielded. The interconnecting shielded wire(s) are called a coaxial cable which connects the antenna to the receiver or transmitter. The job of the transmission line (coaxial cable) is to get the energy to the place where it is to be used and to accomplish this with minimum energy loss.

A transmission line connects the final power amplifier of a transmitter to the transmitting antenna. The transmission line for a receiver connects the antenna to the first tuned circuit of the receiver. Transmission lines may vary from only a few feet to several feet in length.

Transponders, DME, and other pulse type transceivers require transmission lines that are precise in length. The critical length of the transmission lines provides minimum attenuation of the transmitted or received signal. Refer to the equipment manufacture installation manual for type and allowable length of transmission lines.

Coaxial cable is utilized in most airborne installations for transmission lines. Coaxial cable is an unbalanced line that functions with a balanced antenna. To provide the proper impedance matching and the most efficient power transfer a balun is used. The balun is an integral part of the antenna and is not visible without disassembling the antenna.

When installing coaxial cable (transmission lines) secure the cables firmly along their entire length at intervals of approximately 2 ft. To assure optimum operation, coaxial cables should not be routed or tied to other wire bundles. When bending coaxial cable, be sure that the bend is at least 10 times the size of the cable diameter.

Maintenance Procedures

Detailed instructions, procedures, and specifications for the servicing of radio equipment are contained in the manufacturer's operation and service manuals. In addition, instructions for removal and installation of the units are contained in the maintenance manual for the aircraft in which the equipment is installed.

Although installation appears to be a simple procedure, many radio troubles can be attributed to carelessness or oversight when replacing radio equipment. Specific instances are loose cable connections, switched cable terminations, improper bonding, lack of or improper safety wiring, or failure to perform an operational check after installation.

Two additional points concerning installation of equipment needs emphasis. Prior to re-installing any unit, inspect its mounting for proper condition of shock mounts and bonding straps. After installation, safety wire as appropriate.

CHAPTER 14

CABIN ATMOSPHERE CONTROL SYSTEM

NEED FOR OXYGEN

Oxygen is essential for most living processes. Without oxygen, men and other animals die very rapidly. But before this extreme state is reached, a reduction in normal oxygen supplies to the tissues of the body can produce important changes in body functions, thought processes, and degree of consciousness. The sluggish condition of mind and body caused by a deficiency or lack of oxygen is called hypoxia. There are several causes of hypoxia, but the one which concerns aircraft operations is the decrease in partial pressure of the oxygen in the lungs.

The rate at which the lungs absorb oxygen depends upon the oxygen pressure. The pressure that oxygen exerts is about one-fifth of the total air pressure at any one given level. At sea level, this pressure value (3 p.s.i.) is sufficient to saturate the blood. However, if the oxygen pressure is reduced, either from the reduced atmospheric pressure at altitude or because the percentage of oxygen in the air breathed decreases, then the quantity of oxygen in the blood leaving the lungs drops and hypoxia follows.

From sea level to 7,000 ft. above sea level, the oxygen content and pressure in the atmosphere remain sufficiently high to maintain almost full saturation of the blood with oxygen and thus ensure normal body and mental functions.

At high altitude there is decreased barometric pressure, resulting in decreased oxygen content of the inhaled air. Consequently, the oxygen content of the blood is reduced.

At 10,000 ft. above sea level oxygen saturation of the blood is about 90%. Long exposure at this altitude will result in headache and fatigue. Oxygen saturation drops to 81% at 15,000 ft. above sea level. This decrease results in sleepiness, headache, blue lips and fingernails, impaired vision and judgment, increased pulse and respiration, and certain personality changes.

At 22,000 ft. above sea level the blood saturation is 68% and convulsions are likely to occur. Remaining without an oxygen supply at 25,000 ft. for 5 minutes where the blood saturation is down to 55 to 50% will cause unconsciousness.

COMPOSITION OF THE ATMOSPHERE

The mixture of gases commonly called air but more technically termed atmosphere is composed principally of nitrogen and oxygen, but there are smaller quantities of other important gases, notable carbon dioxide, water vapor, and ozone. Figure 14–1 indicates the respective percentage of the quantity of each gas in its relation to the total mixture.

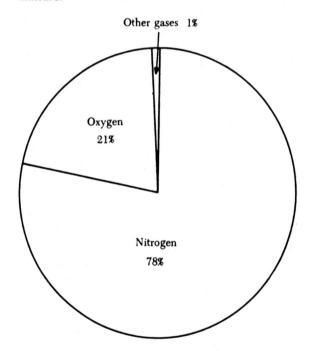

FIGURE 14–1. The gases of the atmosphere.

As the altitude increases, the total quantity of all the atmospheric gases reduces rapidly, and, except for water vapor and ozone, the relative proportions of the gaseous mixture remain unchanged up to about 50 miles altitude, or slightly above. Above 50 miles altitude, changes do take place, and different gases and new forms of the gases present at lower altitudes appear.

Nitrogen is the most common gas and comprises 78% of the total mixture of atmospheric gases. However, insofar as man is concerned nitrogen is an inert gas which cannot be used directly for his own life processes. But biologically it is of immense importance because many compounds containing nitrogen are essential to all living matter.

Oxygen and its importance cannot be overestimated. Without oxygen, life as we know it cannot exist. Oxygen occupies 21% of the total mixture of atmospheric gases.

Carbon dioxide is of biological interest. The small quantity in the atmosphere is utilized by the plant world to manufacture the complex substance which animals use as food. Carbon dioxide also helps in the control of breathing in man and other animals.

Water vapor in the atmosphere is variable, but, even under the moist conditions at sea level, it rarely exceeds 5%; yet this gas absorbs far more energy from the sun than do the other gases. Vapor is not the only form in which water occurs in the atmosphere; water and ice particles are nearly always present. These ice particles also absorb energy and, with water vapor, play an important part in the formation of atmospheric and weather conditions.

Ozone is a variety of oxygen which contains three atoms of oxygen per molecule rather than the usual two. The major portion of the ozone in the atmosphere is formed by the interaction of oxygen and the sun's rays near the top of the ozone layer.

Ozone is also produced by electrical discharges, and the peculiar odor of ozone, which is somewhat like that of weak chlorine, can be detected after lightning storms. Auroras and cosmic rays may also produce ozone. Ozone is of great consequence to both living creatures on earth and to the circulation of the upper atmosphere. Ozone is important to living organisms because it filters out most of the sun's ultraviolet radiation.

Pressure of the Atmosphere

The gases of the atmosphere (air), although invisible, have weight just like that of solid matter. The weight of a column of air stretching from the surface of the earth out into space is called the atmospheric pressure. If this column is 1 sq. in., the weight of air at sea level is approximately 14.7 lbs.; and the atmospheric pressure, therefore, can be

stated as 14.7 p.s.i. at sea level. Another common way of stating the atmospheric pressure is to give the height of a column of mercury which weighs the same as a column of the atmosphere of the same cross sectional area. When measured this way, the atmospheric pressure at sea level is normally 1013.2 millibars, or 29.92 in. Hg.

The atmospheric pressure decreases with increasing altitude. The reason for this is quite simple: the column of air that is weighed is shorter. How the pressure changes for a given altitude is shown in figure 14-2. The decrease in pressure is a rapid one, and at 50,000 feet the atmospheric pressure has dropped to almost one-tenth of the sea level value. At a few hundred miles above the earth, the air has become so rarefied (thin) that the atmosphere can be considered nonexistent, but the line of demarcation with space is very vague.

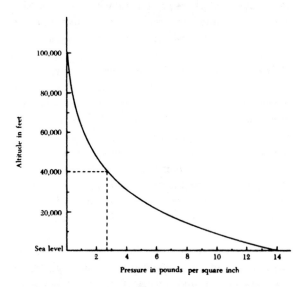

FIGURE 14-2. How the atmospheric pressure decreases with altitude. For example, at sea level the pressure is 14.7 p.s.i; while at 40,000 ft., as the dotted lines show, the pressure is only 2.72 p.s.i.

Temperature and Altitude

The variations in atmospheric temperature near the earth are well known and need no discussion. However, at high altitudes the atmospheric temperature is not so variable but tends to have a more set pattern.

The meteorologist finds it convenient to define, somewhat arbitrarily, the atmosphere as being made up of several layers. The lowest of these is called the troposphere. The air temperature decreases with increasing altitude in the troposphere and reaches a definite minimum at the top of the layer. The top of the troposphere is called the tropopause. The tropopause reaches its greatest height over the equator (about 60,000 ft.) and its lowest height over the poles (about 30,000 ft.). The tropopause marks the point at which air temperature stops decreasing with increasing altitude, and remains essentially constant.

The atmospheric layer above the tropopause is called the stratosphere. The lower stratosphere is an isothermal (constant temperature) region in which the temperature does not vary with altitude. The isothermal region continues up to about 82,000 to 115,000 ft. altitude. Above this level, the temperature increases sharply at the rate of about 1.5° C. per 1,000 ft. The temperature reaches a peak at about 164,000 to 197,000 ft. altitude. Above the 197,000 ft. altitude (approximately), the temperature decreases again, reaching a minimum of —10° F. to —100° F. at about 230,000 to 262,000 ft. altitude. Above this level, the temperature again increases and apparently continues to increase until the edge of space.

The foregoing paragraphs have presented a general knowledge of the atmosphere. It is obvious that a means of preventing hypoxia and its ill effects must be provided. When the atmospheric pressure falls below 3 p.s.i. (approximately 40,000 ft.), even breathing pure oxygen is not sufficient.

The low partial pressure of oxygen, low ambient air pressure, and temperature at high altitude make it necessary to create the proper environment for passenger and crew comfort. The most difficult problem is maintaining the correct partial pressure of oxygen in the inhaled air. This can be achieved by using oxygen, pressurized cabins, or pressure suits. The first and second methods are used extensively in civil aviation.

Pressurization of the aircraft cabin is now the accepted method of protecting persons against the effects of hypoxia. Within a pressurized cabin, people can be transported comfortably and safely for long periods of time, particularly if the cabin altitude is maintained at 8,000 ft., or below, where the use of oxygen equipment is not required. However, the flight crew in this type of aircraft must be aware of the danger of accidental loss of cabin pressure and must be prepared to meet such an emergency whenever it occurs.

PRESSURIZATION

When an aircraft is flown at a high altitude, it burns less fuel for a given airspeed than it does for the same speed at a lower altitude. In other words, the airplane is more efficient at a high altitude. In addition, bad weather and turbulence can be avoided by flying in the relatively smooth air above the storms. Aircraft which do not have pressurization and air conditioning systems are usually limited to the lower altitudes.

A cabin pressurization system must accomplish several functions if it is to assure adequate passenger comfort and safety. It must be capable of maintaining a cabin pressure altitude of approximately 8,000 ft. at the maximum designed cruising altitude of the aircraft. The system must also be designed to prevent rapid changes of cabin altitude which may be uncomfortable or injurious to passengers and crew. In addition, the pressurization system should permit a reasonably fast exchange of air from inside to outside the cabin. This is necessary to eliminate odors and to remove stale air.

In the typical pressurization system, the cabin, flight compartment, and baggage compartments are incorporated into a sealed unit which is capable of containing air under a pressure higher than outside atmospheric pressure. Pressurized air is pumped into this sealed fuselage by cabin superchargers which deliver a relatively constant volume of air at all altitudes up to a designed maximum. Air is released from the fuselage by a device called an outflow valve. Since the superchargers provide a constant inflow of air to the pressurized area, the outflow valve, by regulating the air exit, is the major controlling element in the pressurization system.

The flow of air through an outflow valve is determined by the degree of valve opening. This valve is ordinarily controlled by an automatic system which can be set by the flight crewmembers. A few simple minor adjustments are required on the average flight, but most of the time automatic controls need only to be monitored. In the event of a malfunction of the automatic controls, manual controls are also provided. A schematic of a basic pressurization system is shown in figure 14–3.

The degree of pressurization and, therefore, the operating altitude of the aircraft are limited by several critical design factors. Primarily the fuselage is designed to withstand a particular maximum cabin differential pressure. Cabin differential pres-

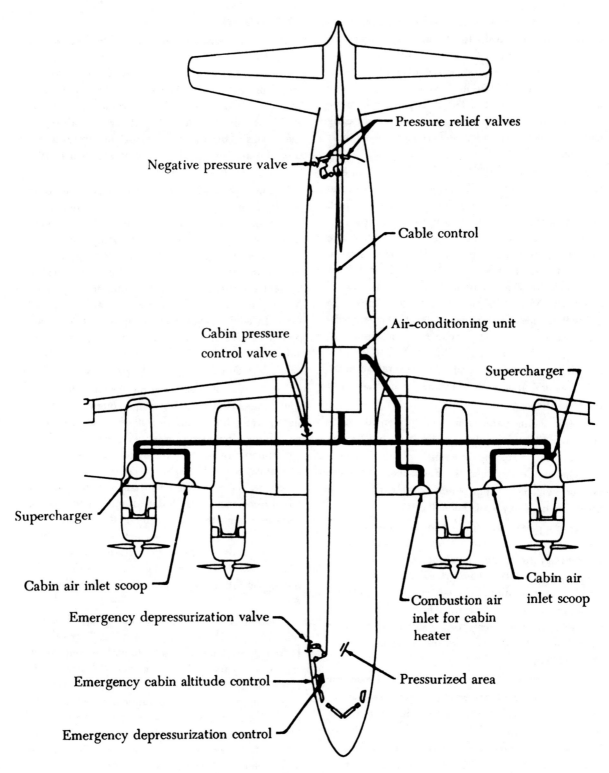

Negative pressure valve

Pressure relief valves

Cable control

Cabin pressure
control valve

Air-conditioning unit

Supercharger

Supercharger

Cabin air inlet scoop

Emergency depressurization valve

Emergency cabin altitude control

Emergency depressurization control

Combustion air
inlet for cabin
heater

Cabin air
inlet scoop

Pressurized area

FIGURE 14–3. Basic pressurization system.

sure is the ratio between inside and outside air pressures and is a measure of the internal stress on the fuselage skin. If the differential pressure be-comes too great, structural damage to the fuselage may occur. In addition, pressurization is limited by the capacity of the superchargers to maintain a constant volume of airflow to the fuselage. As altitude is increased, the pressure of the air entering the supercharger becomes less; consequently, the superchargers have to work harder to accomplish their part of the job. Eventually at some high altitude the superchargers will reach their designed limit of speed, power absorbed, or some other operating factor. The aircraft will normally not be flown higher than these limits allow.

Pressurization Problems

There are many complex technical problems associated with pressurized aircraft. Perhaps the most difficult problems are in the design, manufacturing, and selection of structural materials which will withstand the great differential in pressure that exists between the inside and outside of a pressurized aircraft when flying at high altitudes. If the weight of the aircraft structure were of no concern, it would be a relatively simple matter to construct a fuselage which could withstand tremendous pressures.

It is necessary to construct a fuselage capable of containing air under pressure, yet be light enough to allow profitable loading. As a general rule pressurized aircraft are built to provide a cabin pressure altitude of not more than 8,000 ft. at maximum operating altitude. If an aircraft is designed for operation at altitudes over 25,000 ft., it must be capable of maintaining a cabin pressure altitude of 15,000 ft., in the event of any reasonably likely failure.

The atmospheric pressure at 8,000 ft. is approximately 10.92 p.s.i., and at 40,000 ft. it is nearly 2.72 p.s.i. If a cabin altitude of 8,000 ft. is maintained in an aircraft flying at 40,000 ft. the differential pressure which the structure will have to withstand is 8.20 p.s.i. (10.92 p.s.i. minus 2.72 p.s.i.). If the pressurized area of this aircraft contains 10,000 sq. in., the structure will be subjected to a bursting force of 82,000 lbs., or approximately 41 tons. In addition to designing the fuselage to withstand this force, a safety factor of 1.33 must be added. The pressurized portion of the fuselage will have to be constructed to have an ultimate strength of 109,060 lbs. (82,000 times 1.33), or 54.5 tons.

From the foregoing example it is not difficult to grasp an idea of the difficulties encountered in designing and building a fuselage structure which will be light enough and strong enough at the same time.

AIR CONDITIONING AND PRESSURIZATION SYSTEMS

The cabin air conditioning and pressurization system supplies conditioned air for heating and cooling the cockpit and cabin spaces. This air also provides pressurization to maintain a safe, comfortable cabin environment. In addition to cabin air conditioning, some aircraft equipment and equipment compartments require air conditioning to prevent heat buildup and consequent damage to the equipment.

Some of the air conditioning systems installed in modern aircraft utilize air turbine refrigerating units to supply cooled air. These are called air cycle systems. Other model aircraft utilize a compressed gas cooling system. The refrigerating unit is a freon type, quite similar in operation to a common household refrigerator. Systems utilizing this refrigeration principle are called vapor cycle systems.

Terms and Definitions

The system which maintains cabin air temperatures is the air conditioning system. The sources of heat which make cabin air conditioning necessary are: (1) Ram-air temperature, (2) engine heat, (3) solar heat, (4) electrical heat, and (4) body heat.

It is necessary to become familiar with some terms and definitions to understand the operating principles of pressurization and air conditioning systems. These are:

(1) *Absolute pressure.* Pressure measured along a scale which has zero value at a complete vacuum.

(2) *Absolute temperature.* Temperature measured along a scale which has zero value at that point where there is no molecular motion (—273.1° C. or —459.6° F.).

(3) *Adiabatic.* A word meaning no transfer of heat. The adiabatic process is one in which no heat is transferred between the working substance and any outside source.

(4) *Aircraft altitude.* The actual height above sea level at which an aircraft is flying.

(5) *Ambient temperature.* The temperature in the area immediately surrounding the object under discussion.

(6) *Ambient pressure.* The pressure in the area immediately surrounding the object under discussion.

(7) *Standard barometric pressure.* The weight of gases in the atmosphere sufficient to

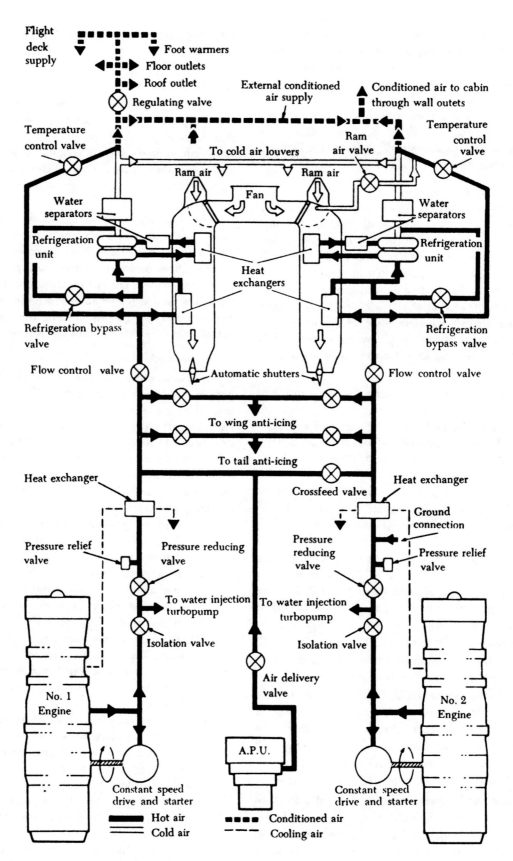

Flight deck supply

▼ Foot warmers

◀▶ Floor outlets

▶ Roof outlet

⊗ Regulating valve

External conditioned air supply

Conditioned air to cabin through wall outlets

Temperature control valve

Temperature control valve

To cold air louvers

Ram air valve

Ram air

Ram air

Fan

Water separators

Water separators

Refrigeration unit

Refrigeration unit

Heat exchangers

Refrigeration bypass valve

Refrigeration bypass valve

Flow control valve

Automatic shutters

Flow control valve

To wing anti-icing

To tail anti-icing

Heat exchanger

Crossfeed valve

Heat exchanger

Pressure relief valve

Pressure reducing valve

Pressure reducing valve

Ground connection

Pressure relief valve

To water injection turbopump

To water injection turbopump

Isolation valve

Isolation valve

Air delivery valve

No. 1 Engine

No. 2 Engine

Constant speed drive and starter

A.P.U.

Constant speed drive and starter

▬▬ Hot air
═══ Cold air

▄▄▄ Conditioned air
---- Cooling air

FIGURE 14–4. Typical pressurization and air conditioning system.

544

hold up a column of mercury 760 millimeters high (approximately 30 in.) at sea level (14.7 p.s.i.). This pressure decreases with altitude.

(8) *Cabin altitude.* Used to express cabin pressure in terms of equivalent altitude above sea level.

(9) *Differential pressure.* The difference in pressure between the pressure acting on one side of a wall and the pressure acting on the other side of the wall. In aircraft air conditioning and pressurizing systems, it is the difference between cabin pressure and atmospheric pressure.

(10) *Gage pressure.* A measure of the pressure in a vessel, container, or line, as compared to ambient pressure.

(11) *Ram-air temperature rise.* The increase in temperature created by the ram compression on the surface of an aircraft traveling at a high rate of speed through the atmosphere. The rate of increase is proportional to the square of the speed of the object.

(12) *Temperature scales.*

    (a) *Centigrade.* A scale on which 0° C. represents the freezing point of water, and 100° C. is equivalent to the boiling point of water at sea level.

    (b) *Fahrenheit.* A scale on which 32° F. represents the freezing point of water, and 212° F. is equivalent to the boiling point of water at sea level.

BASIC REQUIREMENTS

Five basic requirements for the successful functioning of a cabin pressurization and air conditioning system are:

(1) A source of compressed air for pressurization and ventilation. Cabin pressurization sources can be either engine-driven compressors, independent cabin superchargers, or air bled directly from the engine.

(2) A means of controlling cabin pressure by regulating the outflow of air from the cabin. This is accomplished by a cabin pressure regulator and an outflow valve.

(3) A method of limiting the maximum pressure differential to which the cabin pressurized area will be subjected. Pressure

relief valves, negative (vacuum) relief valves, and dump valves are used to accomplish this.

(4) A means of regulating (in most cases cooling) the temperature of the air being distributed to the pressurized section of the airplane. This is accomplished by the refrigeration system, heat exchangers, control valves, electrical heating elements, and a cabin temperature control system.

(5) The sections of the aircraft which are to be pressurized must be sealed to reduce inadvertent leakage of air to a minimum. This area must also be capable of safely withstanding the maximum pressure differential between cabin and atmosphere to which it will be subjected.

Designing the cabin to withstand the pressure differential and hold leakage of air within the limits of the pressurization system is primarily an airframe engineering and manufacturing problem.

In addition to the components just discussed, various valves, controls, and allied units are necessary to complete a cabin pressurizing and air conditioning system. When auxiliary systems such as windshield rain-clearing devices, pressurized fuel tanks, and pressurized hydraulic tanks are required, additional shutoff valves and control units are necessary.

Figure 14-4 shows a schematic diagram of a pressurization and air conditioning system. The exact details of this system are peculiar to only one model of aircraft, but the general concept is similar to that found in the majority of aircraft.

SOURCES OF CABIN PRESSURE

Reciprocating engine internal superchargers provide the simplest means of cabin pressurization. This is accomplished by ducting air from a manifold which supplies compressed air from a supercharger to the pistons. This arrangement can be used only when the engine carburetor is downstream of the supercharger. When the carburetor is upstream of the supercharger, as is often the case, this method cannot be used since the compressed air contains fuel. Air for cabin pressurization can also be ducted from a turbocharger used with a reciprocating engine.

There are several disadvantages in using these two methods. The cabin air becomes contaminated with fumes from lubricating oil, exhaust gases, and fuel. Also, cabin pressurization at high altitude becomes impossible as the discharge pressure of the supercharger decreases to nearly ambient. A third disadvantage is the decrease in engine performance near its design ceiling due to the air loss for cabin pressurization.

With gas turbine engines the cabin can be pressurized by bleeding air from the engine compressor. Usually the air bled from an engine compressor is sufficiently free from contamination and can be used safely for cabin pressurization. Even so, there are several disadvantages when using bleed air from turbine engine compressors. Among these disadvantages are: (1) The possibility of contamination of the air from lubricants or fuel in the event of leakage, and (2) dependence of the air supply on the engine performance.

Because of the many disadvantages associated with the pressurizing sources previously described, independent cabin compressors have been designed. These compressors can be engine driven through accessory drive gearing or can be powered by bleed air from a turbine engine compressor.

Generally, the compressors can be separated into two groups, (1) positive-displacement compressors and (2) centrifugal compressors.

Positive-Displacement Cabin Compressors (Superchargers)

Included in this group are reciprocating compressors, vane-type compressors, and Roots blowers. The first two are not very suitable for aircraft cabin pressurization because of the large quantity of oil present in the air delivered to the cabin.

The action of a Roots-type blower (figure 14–5) is based on the intake of a predetermined volume of air, which is subsequently compressed and delivered to the cabin duct.

The rotors are mounted in an airtight casing on two parallel shafts. The lobes do not touch each other or the casing, and both rotors turn at the same speed. Air enters the spaces between the lobes, is compressed, and is delivered to the cabin air duct.

A cutaway view of a cabin supercharger is shown in figure 14–6. The supercharger housing is usually finned on the external surface to increase its cooling area. The cooling effect is sometimes further increased by shrouding the supercharger housing and passing a stream of air through it. Air cooling is also used to reduce the temperature of internal parts. The cooling air is ducted through drilled

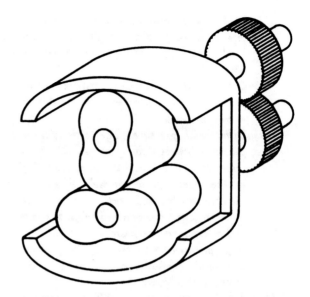

FIGURE 14–5. Schematic Roots-type cabin compressor.

passageways into the rotor cavities and is expelled at the inlet side of the supercharger cover.

To achieve an oil-free delivery of air, the supercharger bearings are contained in separate chambers. The rotor shafts can be fitted with seals made of oil-resistant rubber which prevents any lubricant from entering the compressor casing. The use of labyrinth seals permits a small amount of air leakage to ambient. Any drops of oil which may have passed the rubber seal are thus blown back.

Positive displacement compressors emit a shrill noise during their operation, because of the air pulsations caused by the rotors. Silencers are used with this type compressor to reduce the noise level.

Centrifugal Cabin Compressors

The operating principle of a centrifugal compressor is based on increasing the kinetic energy of the air passing through the impeller. With compressor impeller rotation, the induced air is not only accelerated, but it is also compressed because of the action of centrifugal force. The kinetic energy in the air is then converted into pressure in the diffuser. There are two basic types of diffusers: (1) Vaneless, where the air enters the diffuser space directly on leaving the impeller, and (2) those having guide vanes. A schematic of a centrifugal cabin compressor is shown in figure 14–7.

The cabin supercharger shown in figure 14–8 is essentially an air pump. It incorporates a centrifugal impeller similar to the supercharger in the induction system of a reciprocating engine. Outside

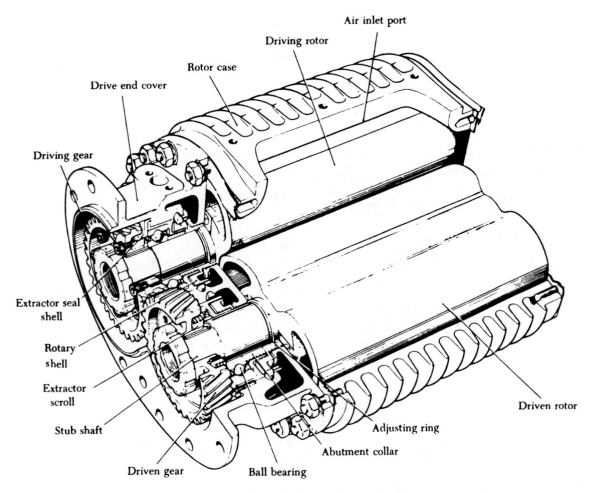

FIGURE 14–6. Cutaway view of a Roots-type cabin supercharger.

FIGURE 14–7. Centrifugal cabin compressor.

air at atmospheric pressure is admitted to the supercharger by suitable scoops and ducts. This air is then compressed by the high-speed impeller and delivered to the fuselage. The superchargers are usually driven by the engine through appropriate gearing; however, turbojet aircraft utilize superchargers (turbocompressors) which are pneumatically driven.

Engine-driven cabin superchargers are generally mounted in the engine nacelle. The supercharger is either splined directly to the engine accessory drive or is connected to an accessory drive by a suitable drive shaft. A mechanical disengaging mechanism is usually incorporated in the drive system to permit disconnecting the supercharger if it malfunctions. The disengaging mechanism can be operated from the flight deck by the crew. In most aircraft it is not possible or permissible to re-engage the supercharger in flight once it has been disconnected.

Engine-driven superchargers used on reciprocat-

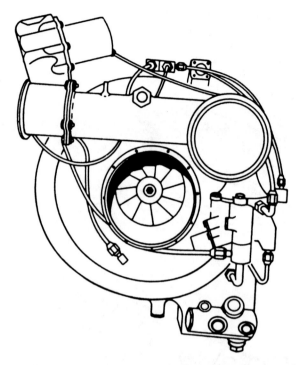

FIGURE 14–8. Pictorial view of a centrifugal cabin supercharger.

ing engine aircraft require a variable-ratio drive mechanism. The gear ratio of these superchargers is automatically adjusted to compensate for changes of engine r.p.m. or outside atmospheric pressure. Normally, the gear ratio is eight to 10 times engine speed when operating under cruising conditions. The drive ratio is at a maximum when operating at high altitude with low engine r.p.m.

Turbocompressors used on turbojet aircraft may be located in the engine nacelles or in the fuselage. There may be as many as four turbocompressors in an aircraft. Turbocompressors consist of a turbine rotated by air pressure which in turn rotates an impeller. The compressed air supply used to operate the turbocompressor is taken from the aircraft's pneumatic system. Speed of the turbocompressor is controlled by varying the supply of compressed air to its turbine.

Cabin superchargers of all types contain their own lubrication system. The lubricant used may be the same type oil used for the engine, or it may be a special oil similar to hydraulic fluid. Supercharger bearings and gears are lubricated by pressure and spray. Oil pressure is also used to operate the control system for the supercharger. The typical lubrication system incorporates a pump, relief valve, sump, cooling system, and sometimes a separate oil tank.

High impeller speed is an important limitation in all superchargers. When the tip speed of the impeller approaches the speed of sound, the impeller rapidly loses its efficiency as an air pump. An equally important limitation involves the back pressure created in the outlet air ducts. If the back pressure is excessive, the impeller may stall or surge.

Supercharger Control

The function of the supercharger control system is to maintain a fairly constant volume of air output from the supercharger. This is accomplished in the system used on reciprocating engine aircraft by varying the drive ratio of the supercharger. The drive ratio between the supercharger impeller and the engine is varied to compensate for changes in engine r.p.m. or atmospheric pressure. This is achieved by an automatic mechanism which samples the airflow output of the supercharger and, through a variable-speed drive gearbox, adjusts the impeller speed whenever the airflow output varies from a preset value.

The amount of f.hp. (friction horsepower) taken from the engine to drive the supercharger is dependent upon the drive ratio. Losses are lowest during low-ratio operation when the energy required to rotate the impeller is at a minimum. Losses are approximately 75 f.hp. in high ratio and 25 f.hp. in low ratio. This loss is indicated at high altitudes, where, the engines which drive the cabin superchargers may require 3 or 4 in. Hg additional manifold pressure to produce the same b.hp. (brake horsepower) as that of other engines.

The speed of the supercharger impeller is therefore adjusted by the control system to maintain a constant mass airflow output. If variables such as altitude tend to increase or decrease the output, the control mechanism causes a correction of the drive ratio. Changes of drive ratio are furthermore dampened by various system refinements to prevent rapid acceleration or deceleration which may result in uncomfortable surges of pressurization.

Serious consequences may occur if the impeller speed becomes higher than its designed maximum. To protect the supercharger against such an occurrence, the typical system has an overspeed governor. This unit is similar to a propeller flyweight governor. The overspeed governor actuates a valve to position the control mechanism to the low-ratio position. It works automatically to reduce impeller r.p.m. when an overspeed occurs.

Some installations also have an electrically operated valve which positions the control mechanism to the low-speed position. This minimum speed valve may be operated manually from the flight deck or automatically by a landing gear strut switch. It is used primarily to reduce supercharger drive ratio when pressurization is not being used or when emergencies occur.

SUPERCHARGER INSTRUMENTS

The principal instrument associated with the supercharger is an airflow gage. This gage usually measures the differential air pressure between the input and the output of the supercharger. In some cases there are two needles on the gage to indicate input and output pressures on the same scale. The airflow (or input and output pressure) gage indicates the proper operation of the supercharger. High readings, low readings, or fluctuating readings indicate various types of malfunctions.

Oil pressure and oil temperature indications are also made available by suitable instruments on the flight deck. In some cases warning lights may be used instead of, or in addition to, the actual gages.

Engine-driven cabin compressors are used on turboprop aircraft. These compressors do not have a variable-speed drive because the turboprop engine operates at a relatively constant speed. The output of this type compressor is controlled by automatically varying the inlet airflow through an airflow-sensing mechanism and a suitable inlet valve which maintains a constant compressor airflow output.

Ordinarily a surge and dump valve is used at the outlet of the compressor. In some systems this is the only type of control employed for the compressor. The surge and dump valve prevents surging of the compressor by partially reducing output pressure when system demands are heavy. The valve can also completely dump output pressure when the engine-driven compressor output is not needed. This valve can be operated from the flight deck and is also operated by various automatic control systems. When the surge and dump valve is opened, the engine-driven cabin compressor output is dumped overboard through suitable ducts.

Instruments used in conjunction with the engine-driven compressor are similar to those used with the variable-speed supercharger. An inlet and discharge pressure gage measures compressor pressures. Com-

pressor high oil temperature and low oil pressure are usually indicated by warning lights.

Turbocompressors used on turbojet aircraft are similar in operation to the exhaust-driven turbochargers used with some reciprocating engines. Power derived from the aircraft's pneumatic system is used to drive the turbine of the unit. Since the turbocompressors do not rely upon direct engine drive shafts, they can be placed either in the engine nacelles or in the fuselage. Ordinarily multiple turbocompressor units are used to provide the high airflow needed by the large turbojet aircraft. The output of the turbocompressor units is usually controlled by varying the pneumatic supply to the turbine.

The pneumatic air supply is obtained from the compressor section of the turbojet engine. This air supply is regulated to a constant pressure of approximately 45 p.s.i. to 75 p.s.i. Pneumatic system air pressure is also used to operate anti-icing and other aircraft systems; therefore, various shutoff valves and check valves are used to isolate inoperative units of the turbocompressor system.

The turbocompressor output is controlled automatically by an airflow control valve and servo-operated inlet vanes. The inlet vanes control the pneumatic system air supply to the turbocompressor turbine. The vanes open or close according to the air pressure signal sensed at the airflow control valve, and turbocompressor speed is increased or decreased to maintain a relatively constant output air volume. Turbocompressor speed will therefore increase with altitude.

The principal turbocompressor control is a simple "on/off" valve. This valve is located in the pneumatic air duct. In the "off" position it completely closes off the pneumatic supply to the turbine. Various special circuits may also actuate this shutoff valve when operation of the turbocompressor is not desired.

Most turbocompressor units incorporate an overspeed control. A typical overspeed control unit is a simple flyweight governor which causes the turbocompressor to completely shut down when a certain limiting r.p.m. is reached. Usually the pneumatic duct shutoff valve is closed by the overspeed control. The turbocompressor system also uses a surge and dump valve similar to that used for engine-driven compressors.

The flight deck instruments are the same as those used on engine-driven systems with the addition of a tachometer which measures turbocompressor speed. Turbocompressor speed on a typical aircraft varies from approximately 20,000 r.p.m. at sea level to 50,000 r.p.m. at 40,000 ft. The overspeed control may be set at about 55,000 r.p.m.

PRESSURIZATION VALVES

The principal control of the pressurization system is the outflow valve. This valve is placed in a pressurized portion of the fuselage, usually underneath the lower compartments. The purpose of the valve is to vent cabin air overboard through suitable openings in the wing fillet or the fuselage skin. Small aircraft use one outflow valve; large aircraft may use as many as three valves which work in unison to provide the required volume outflow.

One type of outflow valve is a simple butterfly which is opened or closed by an electric motor. The motor receives amplified electrical signals from the pressurization controller to vary the valve position required for pressurized flight.

Some aircraft use a pneumatic outflow valve (figure 14–9). This valve receives signals from the pressurization controller in the form of controlled air pressures. The air pressures which operate the valve are obtained from the high pressure inside the cabin, with assistance from the pneumatic system pressure in turbine-powered aircraft.

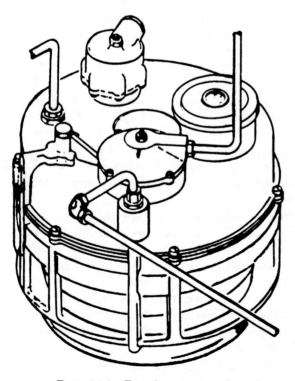

FIGURE 14–9. Typical pneumatic outflow valve.

In many aircraft, the outflow valve(s) will be held fully open on the ground by a landing gear operated switch. During flight, as altitude is gained, the valve(s) close(s) gradually to make a greater restriction to the outflow of cabin air. The cabin rate of climb or descent is determined by the rate of closing or opening of the outflow valve(s). During cruising flight the cabin altitude is directly related to the degree of outflow valve opening.

In addition to the controllable outflow valve(s), an automatic cabin pressure relief valve is used on all pressurized aircraft. This valve may actually be built into the outflow valve or may be an entirely separate unit. The pressure relief valve automatically opens when the cabin differential pressure reaches a preset value.

All pressurized aircraft require some form of a negative pressure relief valve. This valve may also be incorporated into the outflow valve or may be an individual unit. A common form of negative pressure relief valve is a simple hinged flap on the rear wall (pressure dome) of the cabin. This valve opens when outside air pressure is greater than cabin pressure. During pressurized flight the internal cabin pressure holds the flap closed. The negative pressure relief valve prevents accidentally obtaining a cabin altitude which is higher than the aircraft altitude.

The outflow of air from the cabin can also be accomplished through a manually operated valve. This valve may be called a safety relief valve, a manual depressurization valve, or some other similar term. The manual valve is used to control pressurization when all other means of control fail. It is primarily intended to permit rapid depressurization during fires or emergency descent.

Pressurization Controls

The pressurization controller (figure 14–10) is the source of control signals for the pressurization system. The controller provides adjustments to obtain the desired type of pressurized condition. Most operators specify standard operating procedures for the controller which have proven best for their particular type of operation.

The controller looks very much like an altimeter which has several added adjustment knobs. The dial is graduated in cabin altitude increments up to approximately 10,000 ft. Usually there is one pointer which can be adjusted to the desired cabin altitude by the cabin altitude set knob. In some cases there is another pointer or a rotating scale which also

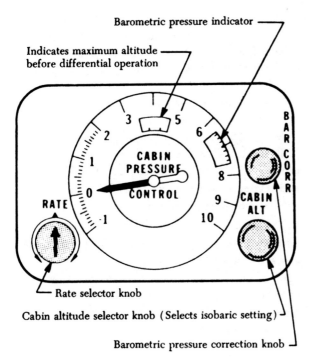

Barometric pressure indicator

Indicates maximum altitude
before differential operation

Rate selector knob

Cabin altitude selector knob (Selects isobaric setting)

Barometric pressure correction knob

FIGURE 14–10. Pressurization controller.

indicates the corresponding aircraft pressure altitude. A separate knob adjusts the controller to the existing altimeter setting (or sea level barometric pressure). The barometric setting selected is indicated on a separate dial segment. The third knob on the controller adjusts the cabin rate of altitude change. This adjustment can be made on a separate control in some installations.

When the controller knobs are set, adjustments are made on either an electric or a pneumatic signaling device inside the controller. The settings are compared to the existing cabin pressure by an aneroid or evacuated bellows. if the cabin altitude does not correspond to that which is set by the knobs, the bellows causes the appropriate signal to go to the outflow valve. When the bellows determines that the cabin altitude has reached that which has been set, the signals to the outflow valve are stopped. As long as other factors do not change, the outflow valve is held at the setting to maintain desired cabin pressure. The controller can sense any change, such as variance of aircraft altitude or loss of one supercharger, and re-adjust the outflow valve as necessary.

The rate control determines how fast the controller sends signals to the outflow valve. In some controllers the rate signal is partially automatic. The

barometric setting compensates the controller for the normal errors in altimetry which are encountered on most flights. This setting improves the accuracy of the controller and, as an example, protects the cabin from being partially pressurized while a landing is being made.

The signals which originate in the controller are very weak. This is because it is a delicate instrument and cannot handle high electric voltages or pneumatic forces. These weak signals are amplified, either electrically or pneumatically, to operate the outflow valve.

Several instruments are used in conjunction with the pressurization controller. The cabin differential pressure gage indicates the difference between inside and outside pressure. This gage should be monitored to assure that the cabin is not approaching the maximum allowable differential pressure. A cabin altimeter is also provided as a check on the performance of the system. In some cases, these two instruments are combined into one. A third instrument indicates the cabin rate of climb or descent. A cabin rate of climb instrument and a cabin altimeter are illustrated in figure 14–11.

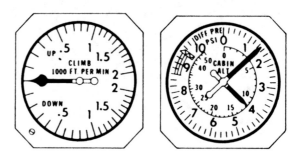

FIGURE 14–11. Instruments for pressurization control.

CABIN PRESSURE CONTROL SYSTEM

The cabin pressure control system is designed to provide cabin pressure regulation, pressure relief, vacuum relief, and the means for selecting the desired cabin altitude in the isobaric and differential range. In addition, dumping of the cabin pressure is a function of the pressure control system. A cabin pressure regulator, an outflow valve, and a safety valve are used to accomplish these functions.

Cabin Pressure Regulator

The cabin pressure regulator controls cabin pressure to a selected value in the isobaric range and limits cabin pressure to a preset differential value in the differential range. The isobaric range maintains the cabin at constant-pressure altitude during flight at various levels. It is used until the aircraft reaches the altitude at which the difference between the pressure inside and outside the cabin is equal to the highest differential pressure for which the fuselage structure is designed. Differential control is used to prevent the maximum differential pressure, for which the fuselage was designed, from being exceeded. This differential pressure is determined by the structural strength of the cabin and often by the relationship of the cabin size to the probable areas of rupture, such as window areas and doors.

The cabin pressure regulator is designed to control cabin pressure by regulating the position of the outflow valve. The regulator usually provides either fully automatic or manual control of pressure within the aircraft. Normal operation is automatic, requiring only the selection of the desired cabin altitude and rate-of-change of cabin pressure.

The cabin pressure regulator may be constructed integral with the outflow valve or may be mounted remote from the outflow valve and connected to it by external plumbing. In either instance the principle of operation is similar.

The regulator illustrated in figure 14–12 is integral with the outflow valve. This regulator is a differential pressure type, normally closed, pneumatically controlled and operated. This type regulator consists of two principal sections: (1) The head and reference chamber section, and (2) the outflow valve and diaphragm section.

The outflow valve and diaphragm section contains a base, a spring-loaded outflow valve, an actuator diaphragm, a balance diaphragm, and a baffle plate. The baffle plate is attached to the end of a pilot which extends from the center of the cover assembly. The outflow valve rides on the pilot

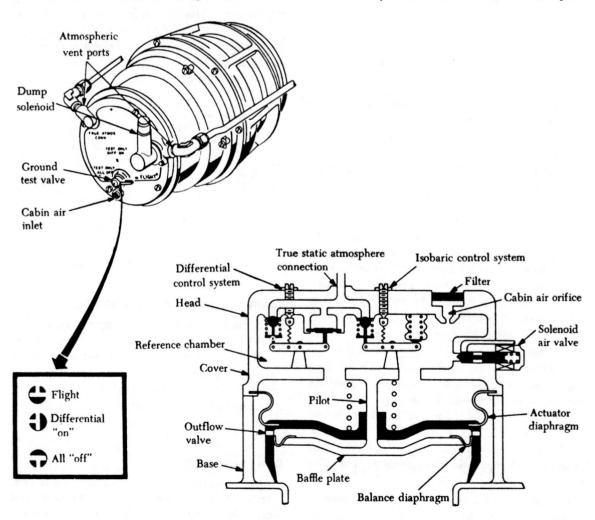

FIGURE 14–12.   Cabin air pressure regulator.

between the cover and the baffle plate and is spring loaded to a closed position against the base.

The balance diaphragm extends outward from the baffle plate to the outflow valve, creating a pneumatic chamber between the fixed baffle plate and the inner face of the outflow valve. Cabin air flows into this chamber through holes in the side of the outflow valve to exert a force against the inner face, opposing spring tension, to open the valve. The actuator diaphragm extends outward from the outflow valve to the cover assembly, creating a pneumatic chamber between the cover and the outer face of the outflow valve. Air from the head and reference chamber section flows through holes in the cover, filling this chamber and exerting a force against the outflow valve's outer face to aid the spring tension in holding the valve closed. The position of the outflow valve controls the flow of cabin air to atmosphere for cabin pressure control. The action of components in the head and reference chamber section controls the movements of the outflow valve by varying the pressure of reference chamber air being exerted against the outer face of the valve.

The head and reference chamber section includes an isobaric control system, a differential control system, a filter, a ground test valve, a true static atmosphere connection, and a solenoid air valve. The area inside the head is called the reference chamber.

The isobaric control system incorporates an evacuated bellows, a rocker arm, a follower spring, and an isobaric metering valve. One end of the rocker arm is connected to the head by the evacuated bellows. The other end of the arm positions the metering valve to a normally closed position against a passage in the head. A follower spring between the metering valve seat and a retainer on the valve causes the valve to move away from its seat as the rocker arm permits.

Whenever the reference chamber air pressure is great enough to compress the bellows the rocker arm pivots about its fulcrum. This allows the metering valve to move from its seat an amount proportionate to the amount of compression in the bellows. When the metering valve is open, reference chamber air flows to atmosphere through the true static atmosphere connection.

The differential control system incorporates a diaphragm, a rocker arm, a metering valve, and a follower spring. One end of the rocker arm is attached to the head by the diaphragm. The diaphragm forms a pressure-sensitive face between the reference chamber and a small chamber in the head. This small chamber is opened to atmosphere through a passage to the true static atmosphere connection. Atmospheric pressure acts on one side of the diaphragm and reference chamber pressure acts on the other. The opposite end of the rocker arm positions the metering valve to a normally closed position against a passage in the head. A follower spring between the metering valve seat and a retainer on the valve causes the valve to move away from its seat as the rocker arm permits.

When reference chamber pressure exceeds atmospheric pressure sufficiently to move the diaphragm, the metering valve is allowed to move from its seat an amount proportionate to the movement of the diaphragm. When the metering valve is open, reference chamber air flows to atmosphere through the true static atmosphere connection.

By regulating reference chamber air pressure, the isobaric and differential control systems control the actions of the outflow valve to provide for three modes of operation called unpressurized, isobaric, and differential.

During unpressurized operation, figure 14–13, reference chamber pressure is sufficient to compress the isobaric bellows and open the metering valve. Cabin air entering the reference chamber through the cabin air orifice flows to the atmosphere through the isobaric metering valve. Since the cabin air orifice is smaller than the orifice formed by the metering valve, reference chamber pressure is maintained at a value slightly less than cabin pressure. As pressure increases in the cabin, the differential pressure between the outflow valve inner and outer face increases. This unseats the outflow valve and allows cabin air to flow to the atmosphere.

As the isobaric range (figure 14–14) is approached reference chamber pressure, which has been decreasing at the same rate as atmospheric pressure, will have decreased enough to allow the isobaric bellows to expand and move the metering valve toward its seat. As a result, the flow of reference chamber air through the metering valve is reduced, preventing further decrease in reference pressure. In response to slight changes in reference chamber pressure, the isobaric control system modulates to maintain a substantially constant reference pressure in the chamber throughout the isobaric range of operation. Responding to the differential

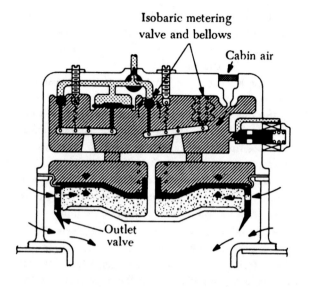

Isobaric metering
valve and bellows

Cabin air

Outlet
valve

░░░░░ Cabin air pressure
▨▨▨▨ Control pressure
▒▒▒▒ Atmospheric pressure

FIGURE 14–13. Cabin air pressure regulator in
the unpressurized mode.

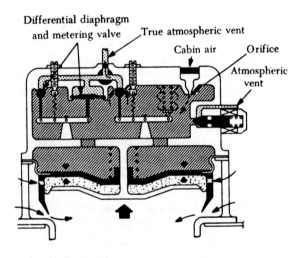

Differential diaphragm
and metering valve

True atmospheric vent

Cabin air   Orifice

Atmospheric
vent

░░░░░ Cabin air pressure
▨▨▨▨ Control pressure
▒▒▒▒ Atmospheric pressure

FIGURE 14–14. Cabin air pressure regulator in
the isobaric range.

between the constant reference chamber pressure and the variable cabin pressure, the outflow valve opens or closes, metering air from the cabin, as required, to maintain a constant cabin pressure.

As the differential range is approached, the pressure differential between the constant reference pressure and the decreasing atmospheric pressure becomes sufficient to move the diaphragm and open the differential metering valve. As a result, reference chamber air flows to atmosphere through the differential metering valve, reducing the reference pressure. Responding to the decreased reference pressure, the isobaric bellows expands and closes the isobaric metering valve completely. Reference chamber pressure is now controlled, through the differential metering valve, by atmospheric pressure being reflected against the differential diaphragm. As atmospheric pressure decreases, the metering valve opens more and allows reference pressure to decrease proportionately. Responding to the pressure differential between cabin and reference pressures, the outflow valve opens or closes as required to meter air from the cabin and maintain a predetermined differential pressure value.

In addition to the automatic control features just described, the regulator incorporates a ground test valve and a solenoid air valve, both of which are located in the head and reference chamber section. The solenoid air valve is an electrically activated valve spring-loaded to a normally closed position against a passage through the head that opens the reference chamber to atmosphere. When the cockpit pressure switch is positioned to "ram" the regulator solenoid opens, causing the regulator to dump cabin air to the atmosphere.

The ground test valve (see figure 14–12) is a three-position, manually operated control that allows for performance checks of the regulator and cabin pressurization system. In the "test only—all off" position the valve renders the regulator completely inoperative. In the "test only—differential on" position, the valve renders the isobaric control system inoperative so that the operation of the differential control system can be checked. In the "flight" position, the valve allows the regulator to function normally. The ground test valve should always be lockwired in the "flight" position unless being tested.

Cabin Air Pressure Safety Valve

The cabin air pressure safety valve (figure 14–15) is a combination pressure relief, vacuum relief, and dump valve. The pressure relief valve prevents cabin pressure from exceeding a predetermined differential pressure above ambient pressure. The vacuum relief prevents ambient pressure from

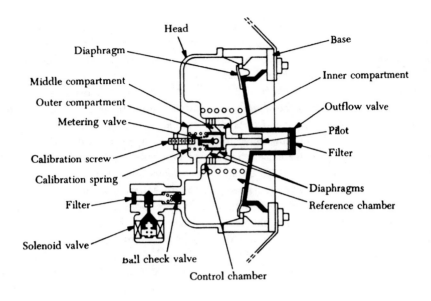

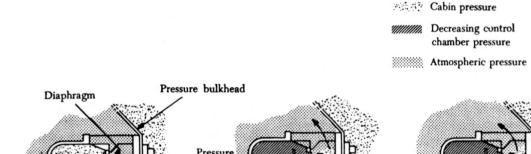

Cabin pressure

Decreasing control
chamber pressure

Atmospheric pressure

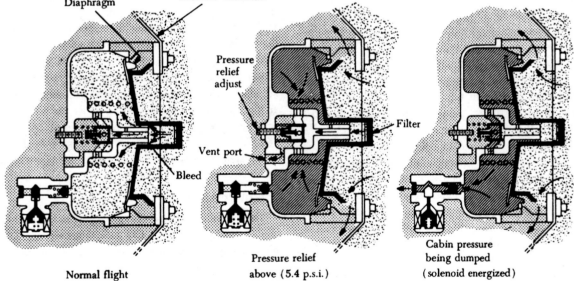

Normal flight

Pressure relief
above (5.4 p.s.i.)

Cabin pressure
being dumped
(solenoid energized)

FIGURE 14–15. Cabin air pressure safety valve.

exceeding cabin pressure by allowing external air to enter the cabin when ambient pressure exceeds cabin pressure. The dump valve illustrated is actuated by the cockpit control switch. When this switch is positioned to "ram," a solenoid valve opens, causing the valve to dump cabin air to atmosphere. On some installations a manual system,

using cables and bellcranks, is provided to actuate the dump valve.

The safety valve consists of an outflow valve section and a control chamber. The outflow valve section and the control chamber are separated by a flexible pressure-tight diaphragm. The diaphragm is exposed to cabin pressure on the outflow valve side and control chamber pressure on the opposite side. Movement of the diaphragm causes the outflow valve to open or close. A filtered opening in the outflow valve allows cabin air to enter the reference chamber. The outflow valve pilot extends into this opening to limit the flow of air into the chamber. Air pressure inside the reference chamber exerts a force against the inner face of the outflow valve to aid spring tension in holding the valve closed. The pressure of cabin air against the outer face of the outflow valve provides a force opposing spring tension to open the valve. Under normal conditions, the combined forces within the reference chamber are able to hold the outflow valve in the "closed" position. The movement of the outflow valve from closed to open allows cabin air to escape to atmosphere.

The head incorporates an inner chamber, called the pressure relief control chamber. Within the control chamber are located the two pressure relief diaphragms, the calibration spring, the calibration screw, and the spring-loaded metering valve. The action of these components within the chamber controls the movement of the outflow valve during normal operation.

The two diaphragms form three pneumatic compartments within the control chamber. The inner compartment is open to cabin pressure through a passage in the outflow valve pilot. The middle compartment is open to the reference chamber and is vented to the outer compartment through a bleed hole in the metering valve. The flow of reference chamber air from the middle compartment to the outer compartment is controlled by the position of the metering valve, which is spring-loaded to a normally closed position. The outer compartment, in which the calibration spring and screw are located, is opened to atmosphere through a passage in the head. Atmospheric pressure, reflected against the diaphragms, aids the calibration spring in keeping the metering valve closed. Cabin pressure, acting on the diaphragms through the inner compartment, tries to open the metering valve by moving it back against the calibration screw. Under normal conditions, the combined forces of atmospheric pressure and the calibration spring hold the metering valve away from the calibration screw, keeping it closed.

Pressure relief occurs when the cabin pressure exceeds atmospheric pressure by a predetermined value. At this point, cabin pressure overcomes the combined forces of atmospheric pressure and spring tension in the control chamber, moving the metering valve back against the calibration screw, opening the metering valve. With the valve open, reference chamber air can escape through the outer compartment to the atmosphere. As the reference chamber air pressure is reduced, the force of cabin pressure against the outflow valve overcomes spring tension and opens the valve, allowing cabin air to flow to atmosphere. The rate-of-flow of cabin air to atmosphere is determined by the amount the cabin-to-atmosphere pressure differential exceeded the calibration point. As cabin pressure is reduced, the forces opening the valve will be proportionately reduced, allowing the valve to return to the normally closed position as the forces become balanced.

In addition to the automatic operating provisions just described, the valve includes provisions for electrical activation to the dump position. This is accomplished by a passage in the head that allows reference chamber air to vent directly to atmosphere. The flow of air through the passage is controlled by a ball-check valve and an air solenoid valve. The solenoid valve is spring-loaded to a normally closed position. When the solenoid valve is opened by positioning the cockpit pressure switch to "ram," air flows from the reference chamber, decreasing the reference pressure and allowing the outflow valve to open and dump cabin air.

It should be remembered that the foregoing description of a pressure control system is for illustrative purposes and should not be construed to represent any particular make or model aircraft. Always refer to the applicable manufacturer's manual for the system details and limitations for the aircraft with which you are concerned.

AIR DISTRIBUTION

The cabin air distribution system includes: (1) Air ducts, (2) filters, (3) heat exchangers, (4) silencers, (5) nonreturn (check) valves, (6) humidifiers, (7) mass flow control sensors, and (8) mass flow meters. The distribution system shown in figure 14–16 is typical of the system used on small turboprop aircraft.

Air enters the cabin supercharger through a screen-covered opening in the left engine oil cooler airscoop. If the air inlet screen ices over, a spring-loaded door beside the screen opens allowing air to bypass the screen. From the cabin supercharger, the air passes through a firewall shutoff valve, a pressure relief valve, and a silencer which dampens the supercharger noise and pulsations. The air then

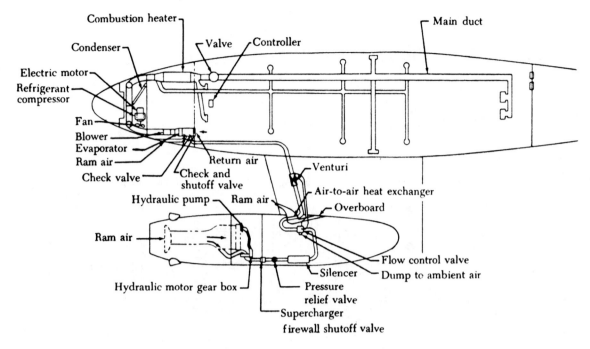

FIGURE 14-16. Typical air distribution system.

passes through a flow control valve which governs
the airflow rate to maintain the maximum pounds
per minute airflow.

Air Ducts

Ducts having circular or rectangular cross sections are most frequently used in air distribution
systems. Circular ducts are used wherever possible.
Rectangular ducts are generally used where circular
ducts cannot be used because of installation or
space limitations. Rectangular ducts may be used in
the cabin where a more pleasing appearance is desired.

Distribution ducts for various cabin zones, individual air outlets for passengers, and window de-
misters can have various shapes. Examples of circular, rectangular, elliptical and profiled ducts are
illustrated in figure 14-17.

Cabin air supply ducts are usually made from
aluminum alloys, stainless steel, or plastic. Main
ducts for air temperatures over 200° C. are made
from stainless steel. Those parts of the ducting
where the air temperature does not exceed 100° C.
are usually constructed from soft aluminum. Plastic
ducts, both rigid and flexible are used as outlet
ducts to distribute the conditioned air.

Since heated air is routed throughout the duct
system, it is important that the ducts be permitted
to grow (expand through heating) and to shrink

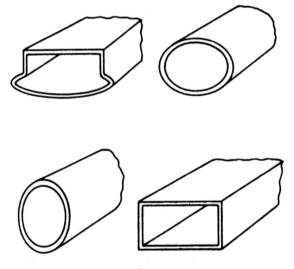

FIGURE 14-17.   Cross sections of air distribution ducts.

again when the air cools down. This expansion and
contraction must take place without loss of the pressure-tight integrity of the ducts. Expansion bellows
(figure 14-18) are incorporated at various places
throughout the duct system to permit the ducts to
expand or contract.

In general, supports are necessary on both sides
of a connecting bellows, a fixed support on one side
to prevent duct movement, and a sliding support
plus a fixed support on the other side. The sliding

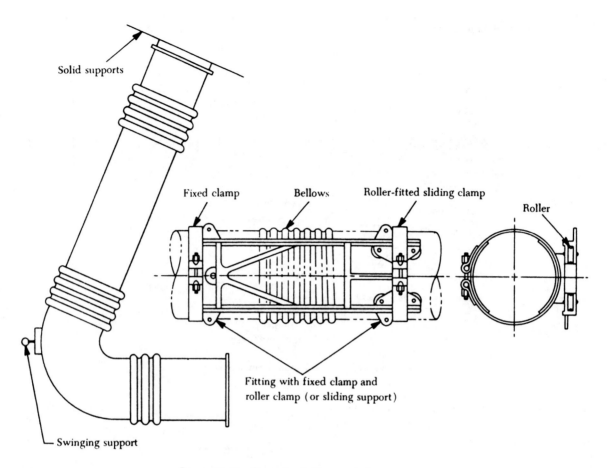

FIGURE 14-18. Expansion bellows and duct supports.

support permits movement of the bellows while the duct section is under pressure. Typical duct support systems are illustrated in figure 14-18.

Whenever a duct is angled, means are provided to take care of the end forces which tend to push the duct sections apart. This can be accomplished with external swinging supports which attach the duct to rigid airframe structure (figure 14-19).

In some instances a connecting link is incorporated within the duct itself to transmit end loads. The tension link within the bellows resembles a single link of chain that joins two segments of ducts. Figure 14-20 illustrates one such connecting link.

Filters

The air delivered to a pressurized cabin from a supercharger or engine compressor may contain dust particles, oil mist, or other impurities. Unfiltered air which contains a considerable amount of impurities usually has an offensive odor and causes

headache and nausea. Filters are generally incorporated into the ducting to clean the air.

AIR CONDITIONING SYSTEM

The function of an air conditioning system is to maintain a comfortable air temperature within the aircraft fuselage. The system will increase or decrease the temperature of the air as needed to obtain the desired value. Most systems are capable of producing an air temperature of 70° to 80° F. with normally anticipated outside air temperatures. This temperature-conditioned air is then distributed so that there is a minimum of stratification (hot and cold layers). The system, in addition, must provide for the control of humidity, it must prevent the fogging of windows, and it must maintain the temperature of wall panels and floors at a comfortable level.

In a typical system the air temperature is measured and compared to the desired setting of the temperature controls. Then, if the temperature is

558

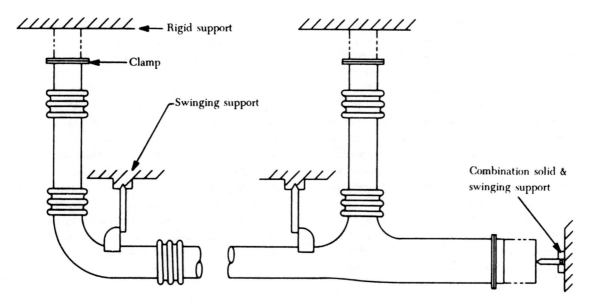

FIGURE 14–19. Typical supports for angled ducts.

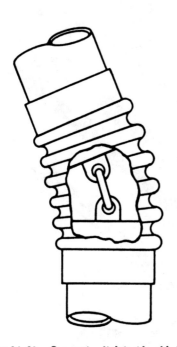

FIGURE 14–20. Connecting link inside of bellows.

not correct, heaters or coolers are set into operation to change the air temperature, and the air is mixed together to create a uniform temperature in the cabin. In summary, an air conditioning system is designed to perform any or all of the following functions: (1) Supply ventilation air, (2) supply heated air, and (3) supply cooling air.

Ventilation Air

Ventilation air is obtained through ram air ducts installed in the leading, lower, or upper surfaces of the aircraft or through other vents in the aircraft skin. Air entering these openings usually passes into and through the same duct system that is used for heating and cooling. On some aircraft, recirculating fans or blowers are present in the system to assist in circulating the air. Many aircraft have ground connections for receiving heated, cooled, or ventilating air from ground servicing equipment.

HEATING SYSTEMS

A large part of the heating requirements for the conditioned air is accomplished automatically when the air is compressed by the cabin superchargers. In many cases additional heat need not be added. Compression of the air often provides more than the necessary heating. Consequently, cooling, to some degree, is required even when the outside air temperature is not high.

When a degree of heating in addition to that obtained from the "heat of compression" is needed, one of the following types of systems is put into operation: (1) Gasoline combustion heaters, (2) electric heaters, (3) re-cycling of compressed air, and (4) exhaust gas air-to-air heat exchanger.

Combustion Heater

Combustion heaters operate similarly to the burner section of a turbojet engine. Gasoline is injected into the burner area under a pressure which breaks up the fuel into a fine mist. Combustion air is supplied to the burner by means of a ram air scoop or an electric motor driven fan. Ignition is supplied by continuous sparking of a special spark plug. The combustion of fuel and air takes place continuously. The temperature output of the heater is controlled by a cycling process whereby combustion is turned on and off for short periods of time depending upon the heating required. The air which eventually mixes with the cabin air is routed around the burner section in a separate air passage. This ventilating air picks up heat from the burner by convection through the metal walls of the burner. The burner combustion gasses are exhausted overboard to prevent carbon monoxide contamination of the cabin.

Various automatic combustion heater controls prevent operation of the heater when dangerous conditions exist. As examples, the flow of fuel is cut off if there is insufficient combustion air, insuffi-

cient ventilating air, and in some cases if the ignition system is not operating. Other controls prevent too rapid heating of the combustion chamber and prevent exceeding a maximum output temperature.

Electric heaters may be in the form of air duct heaters or electric radiant panels. The duct heater incorporates a series of high-resistant wire coils located in an air supply duct. When electric power is applied to the coils, they become hot. The air flowing through the duct carries the heat to the area where it is needed. Most duct heaters require the use of a fan to ensure sufficient airflow over the coils. Without the aid of the fan-produced airflow, the coils would fail due to overheating. Usually an electrical circuit is used which prevents heater operation unless the fan is in operation.

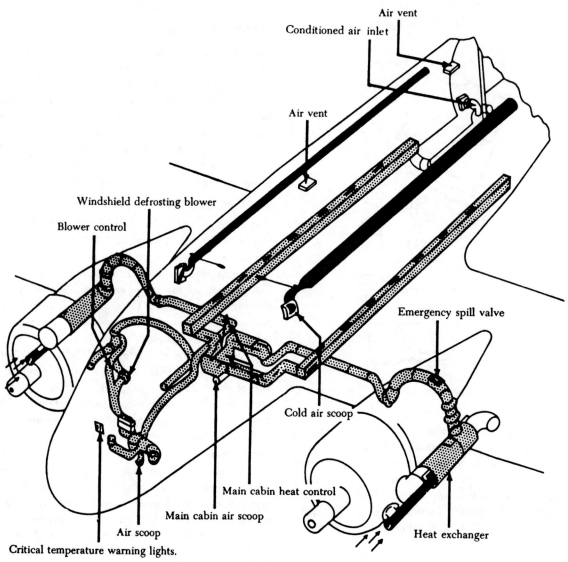

FIGURE 14–21. Engine exhaust heater system.

560

Radiant Panels

Radiant panels consist of wall and floor surfaces which have electric wires embedded in the panel material. When electric power is applied to the wires, the wires and panel surface become hot. This type panel supplies heat to the cabin air principally by radiation.

Electric Heaters

Electric heating systems require the expenditure of large amounts of electrical power. They cannot be used if the electrical system has limited capacity. Electric heating systems, however, are quick acting and can be used to preheat the aircraft on the ground before the engines are started if an adequate electric ground power source is available.

Compressed Air Heating

Some turbojet aircraft use a heating system in which the hot compressed air output of the cabin compressor is re-routed back into the compressor inlet. This double compression raises the temperature of the air to a sufficiently high degree so that other types of heating are usually not necessary.

Exhaust Gas Heaters

A relatively simple heating system used on a few large aircraft utilizes the engine exhaust gases, figure 14–21, as a heat source. This system is particularly effective on aircraft where the engine exhaust is ejected through a long tailpipe. A hot air muff or jacket is installed around the tailpipe. Air routed through the hot air muff picks up heat by convection through the tailpipe material. This heated air is then routed to an air-to-air heat exchanger, where its heat is given up to the air going to the cabin. By using the air-to-air heat exchanger in addition to the hot air muff, the danger of carbon monoxide entering the cabin is minimized.

Regardless of the type, heating systems provide heated air for comfort and furnish heat for defrosting, deicing, and anti-icing of aircraft components and equipment. Nearly all types of heating systems use the forward motion of the aircraft to force conditioned air to various points on the aircraft. A heating system consists of a heating unit and the necessary ducting and controls. The units, ducts, and controls used will vary considerably from system to system.

COMBUSTION HEATERS

The number and size of combustion heaters used in a particular aircraft depend upon its size and its heating demands. These heaters are installed singly or in combination to fit the heating requirements of specific aircraft. A large single heater or several smaller heaters may be used. Regardless of size, every combustion heater needs four things for operation: (1) Fuel to burn, (2) ignition to ignite the fuel, (3) combustion air to provide the oxygen required to support the flame, and (4) ventilating air to carry the heat to the places where it is needed.

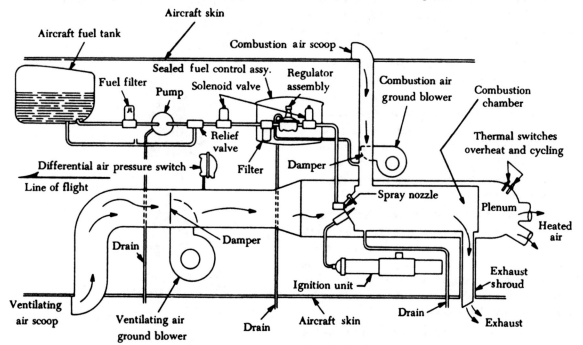

FIGURE. 14–22. Heater installation schematic.

All combustion heaters are similar in operation and construction features. The major differences are in the methods of introducing fuel and in the location of units and accessories. The various components that comprise a complete aircraft heating system are shown schematically in figure 14–22.

Heater Fuel System

The fuel used in the heaters is supplied in most cases from the same fuel tanks which supply the engines. Fuel flows from the tank to the heater by gravity or is pumped there by a fuel pump. Like the fuel which flows to the aircraft engine, heater fuel must first pass through a filter to remove impurities. If foreign particles are not removed, they may eventually clog heater system units and prevent heater operation.

After the fuel is filtered, it flows through a fuel solenoid valve and metering nozzle. There are several types of these valves and metering nozzles. Regardless of type, they usually have the same purpose, to maintain a constant volume at the fuel outlet to the combustion chamber. This uniform volume, in combination with a fixed combustion airflow, ensures a relatively constant fuel/air ratio to the heater. The result is a steady heater output.

To increase or decrease the cabin temperature, the heaters are permitted to operate longer when more heat is needed and for shorter periods of time when less heat is desired. On most heater systems this is accomplished automatically by an amplifier connected to temperature-sensing devices or by cycling switches which open and close the circuit to the fuel solenoid valve. Thus the heater cycles on and off to maintain the temperature selected on a temperature control rheostat located in the cabin.

Most heater systems also include overheat switches in each heater outlet to automatically turn off the heater fuel supply when the temperature reaches about 350° F. It can be seen that control of the heater fuel supply is necessary, not only for normal operation of the heater but also for stopping it when overheated.

Another essential unit of the heater fuel system is the one that feeds fuel into the combustion chamber. Depending on the installation, it may be either a spray nozzle or a vapor wick. The spray nozzle (figure 14–23) is shaped so as to inject a fine, steady spray into the stream of combustion air, where it is ignited by a spark plug.

The vapor wick is made of asbestos contained in a circular flanged casting or of stainless steel contained in a vertical standpipe. The latter type is illustrated in figure 14–24.

A preheater, in the form of an electrical wire coiled around the fuel line, is used with some heaters having a vapor wick. It warms the fuel to speed vaporization and aids ignition when the outside temperature is below zero. Its use is usually limited to 2 min., because longer operation would damage the wire coil.

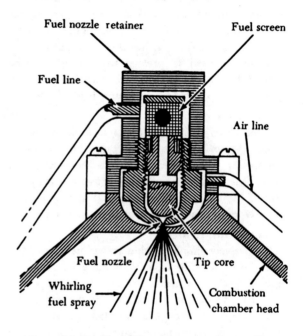

FIGURE 14–23. Typical heater spray nozzle assembly.

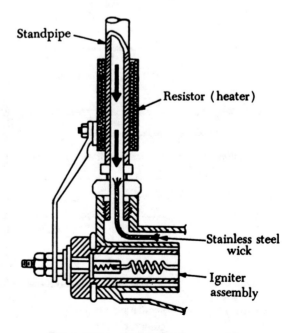

FIGURE 14–24. Stainless steel vapor wick.

Ignition System

High voltage for heaters using spark plug igniters is supplied either by a high-potential ignition unit operating from the 28-v. d.c. aircraft supply or by ignition transformers operating from a 115-v. a.c. aircraft source. The 28-v. d.c. ignition unit consists chiefly of a vibrator and step-up coil which produces a high-voltage spark at high frequency. A shielded lead is used to connect the step-up coil to the spark plug. The spark is produced between the center spark plug electrode and a ground electrode. About the same result is obtained where ignition transformers are used. Here, however, power is supplied by the 115-v., 400-Hz main inverter a.c. system. This is routed to the transformers, where it is stepped up to the very high voltage required to jump the spark gap at the spark plug. But whether a d.c. or an a.c. source is used to fire the spark plug, ignition is continuous during heater operation. This continuous operation prevents fouling of the spark plug electrodes.

It is the arrangement of the electrodes that makes the difference in the types of spark plugs used in aircraft combustion heaters. One type of spark plug is shown in figure 14–25A. This is known as a dual-electrode spark plug. Another type of plug to be found on combustion heaters is the shielded electrode plug (figure 14–25B). In this plug, the ground electrode forms a shield around the center electrode.

Although spark plug igniters differ somewhat in appearance, most glow coil igniters look similar to that shown in figure 14–25C. They consist of a resistance wire wound into a coil around a pin extending from the body of the igniter. The outer end of the coil is connected to the pin, thus providing both support and electrical continuity. The body of the igniter is fitted with two terminals, which are connected across the coil, and is threaded to provide for installation. The glow coil operates from the 24- or 28-v. d.c. power supply on the aircraft. The direct current causes the coil to become red hot, thereby igniting the fuel/air mixture until the heater is operating at a temperature sufficient to maintain the flame after the glow coil is turned off. A thermal cutout switch breaks the circuit to the glow coil when this temperature is reached. This prolongs the life of the igniter.

Another type of plug used is a single electrode type (not shown). The ground electrode used with this type of plug is a separate installation attached to the heater at an angle that will provide an airgap between the plug's electrode and ground.

Combustion Air System

Combustion air for each cabin heater is received through either the main air intake or through a separate outside scoop. On both pressurized and unpressurized aircraft it is provided by ram pressure during flight and by ground blowers during ground operation. To prevent too much air from entering the heaters as air pressure increases, either a combustion air relief valve or a differential pressure regulator is provided. The air relief valve is located in the line leading from the ram-air intake duct and is spring-loaded to dump excess air into the heater exhaust gas stream. The differential pressure regulator is also located in the combustion air intake line, but it controls the amount of air reaching the combustion chamber in a slightly different manner.

While the relief valve takes a large volume of air and bypasses the amount not needed, the pressure regulator allows only the needed amount of air to enter its intake in the first place. It does this by the use of a diaphragm and spring type control mechanism. One side of the diaphragm is vented to the heater intake air line and the other side to the heater exhaust gas line. Any change in the pressure

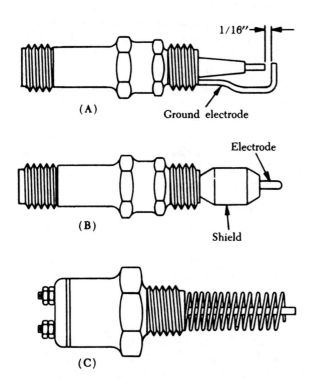

1/16"

(A)

Ground electrode

Electrode

(B)

Shield

(C)

FIGURE 14–25. Heater ignition plugs.

drop between these points is corrected at the regulator by letting in more or less air as required. Thus, a constant combustion air pressure is provided to the heater. Coupled with a steady fuel flow, this constant air pressure makes possible a regulated flow of combustion gases through the combustion chamber and the connecting radiator. If a fire breaks out near the heater, a fire valve automatically stops the supply of combustion air to prevent spreading of the fire to the heating system.

A damper-type combustion air fire valve (figure 14-26) is located in the combustion air inlet of some heaters. It has two semicircular, spring-loaded segments soldered together to permit maximum airflow through the combustion air duct. The segments will release to seal the duct when the solder melts at about 400° F.

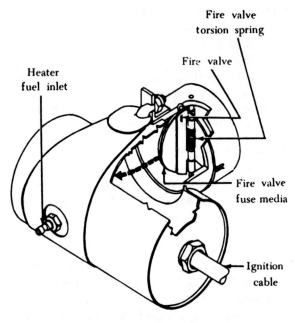

FIGURE 14-26. Cabin heater combustion air fire valve.

Ventilating Air

Ventilating air may come from one of three sources: (1) A blower for air circulation and heater operation on the ground, (2) a ram-air inlet, or (3) the cabin compressors on pressurized aircraft.

Ventilating air, ram or blower, enters at the burner head end of the heater and, passing over the heated radiator surfaces, becomes heated and passes through the outlet end into the plenum assembly and into the distribution system ductwork.

MAINTENANCE OF COMBUSTION HEATER SYSTEMS

Combustion heater components are subject to wear and damage which can result in system failure. When this occurs, troubleshooting procedures must be followed to isolate component failures. Then all damaged or excessively worn components must be replaced. During the replacement of the components, adjustments must be made to assure proper operation of the combustion heater system. Always follow the manufacturer's instructions when making any adjustments to a heater or heating system.

In this section, heater system adjustments which are representative of those performed by the aircraft mechanic are discussed. Keep in mind that the components of the system vary with the types of aircraft, and so do the adjustment procedures.

On some aircraft careful adjustment of cabin heat outlets is necessary to obtain uniform heat distribution. Some of the factors which cause the variation in distribution are: (1) The distance of the outlet from the source of heated air, (2) the cross sectional area of the outlet, (3) the space serviced by the outlet, and (4) any restrictions to airflow caused by duct size and routing.

Air mixing valves are installed in airborne heating systems so that hot and cold air can be mixed in the required proportions to maintain adequate heat. Some air mixing valves are preset on the ground and cannot be actuated during flight. External adjustments are provided on these valves to permit seasonal adjustment. During adjustment, the valves are set to a specified number of degrees from the fully closed position.

To assure the proper mixing of hot and cold air in motorized air mixing valves, adjustments are provided on each valve. The adjustments regulate the opening and closing positions of the valves.

Heater System Inspection

The inspection of combustion heater systems includes checking the air openings and outlets for obstructions. All controls are checked for freedom of operation. Turn on the fuel pump so that the fuel lines, solenoids, and valves can be checked for leakage. The heater unit is inspected for proper operation by turning it on and observing whether or not hot air comes out of the outlets. The outside of the heater unit is checked for signs of overheating. Any burned or darkened areas usually indicate a burned-through combustion chamber. Heaters damaged by overheating should be replaced. When replacing a heater due to overheating, always determine the cause of the trouble. The faulty operation of some part of the system, such as stopped-up heater air inlet ducts or improperly operating switches, regulators, valves, or other units, is the most likely cause of damage. The automatic and overheat control devices should be operationally checked. The cabin heating ducts should be examined for tears, breaks, and ballooning. To guarantee fuel flow, the heater fuel filter element should be inspected for cleanliness and the fuel injection nozzle or glow coil for freedom from carbon deposits.

To obtain proper operation of combustion heaters under freezing conditions, a special winterization inspection should be performed. Check heater drain lines regularly for restrictions caused by ice formation. During low temperature operation below $0°$ C. ($32°$ F.), water vapor in the combustion gases flowing through drain lines may condense and form ice. Under changing temperature conditions, water condenses and freezes in the ram and heater combustion sensing lines. Water produced during combustion may collect on the fuel nozzles and spark plug and form ice after the heater is turned off. This ice may be sufficient to make it difficult, if not impossible, to start the heater without preheating.

COOLING SYSTEMS

Air cooling systems are installed to provide a comfortable atmosphere within the aircraft both on the ground and at all altitudes. These systems keep the correct amount of air flowing through the interior of the aircraft at the right temperature and moisture content. Since the fuselage is a huge cavity, the capacity of the cooling system must be quite large. Several types of systems can be used to meet these requirements. Two of the more common types, air cycle and vapor cycle, are discussed in this section.

AIR CYCLE COOLING SYSTEM

An air cycle cooling system consists of an expansion turbine (cooling turbine), an air-to-air heat exchanger, and various valves which control airflow through the system. The expansion turbine incorporates an impeller and a turbine on a common shaft. High-pressure air from the cabin compressor is routed through the turbine section. As the air passes through the turbine, it rotates the turbine and the impeller. When the compressed air performs the work of turning the turbine, it undergoes a pressure and temperature drop. It is this temperature drop which produces the cold air used for air conditioning.

Before entering the expansion turbine, the pressurized air is directed through an air-to-air heat exchanger. This unit utilizes outside air at ambient temperature to cool the compressed air. It should be evident that the heat exchanger can only cool the compressed air to the temperature of the ambient air temperature. The primary purpose of the heat exchanger is to remove the heat of compression so that the expansion turbine receives relatively cool air on which to start its own cooling process.

The impeller part of the expansion turbine can perform several functions. In some installations the impeller is used to force ambient air through the heat exchanger. In this manner, the efficiency of the heat exchanger is increased whenever the speed of the expansion turbine is increased. Other installations use the impeller to further compress the cabin supercharger air as an aid to forcing it through the heat exchanger and the turbine.

A valve controls the compressed airflow through the expansion turbine. To increase cooling, the valve is opened to direct a greater amount of the compressed air to the turbine. When no cooling is required, the turbine air is shut off. Other valves, operated in conjunction with the turbine air valve, control the flow of ambient air through the heat exchanger. The overall control effect of these valves is to increase the heat exchanger cooling airflow at the same time increased cooling is obtained at the turbine.

The power required to drive the air cycle system is derived entirely from the cabin supercharger compressed air. Use of the air cycle system, therefore, imposes an increased load on the superchargers. As more cooling is demanded from the turbine, a greater back pressure is placed on the superchargers, which must work harder to supply the air demands. It is often necessary to make a choice between the desired amount of cooling and the desired degree of cabin pressurization, and a compromise is made by reducing the demand for one or the other. Maximum cooling and maximum pressurization cannot be obtained at the same time. Attempts to obtain both will cause the supercharger to surge or operate in an otherwise unsatisfactory manner.

System Operation

This description of the operation of an air conditioning system is intended to provide an under-

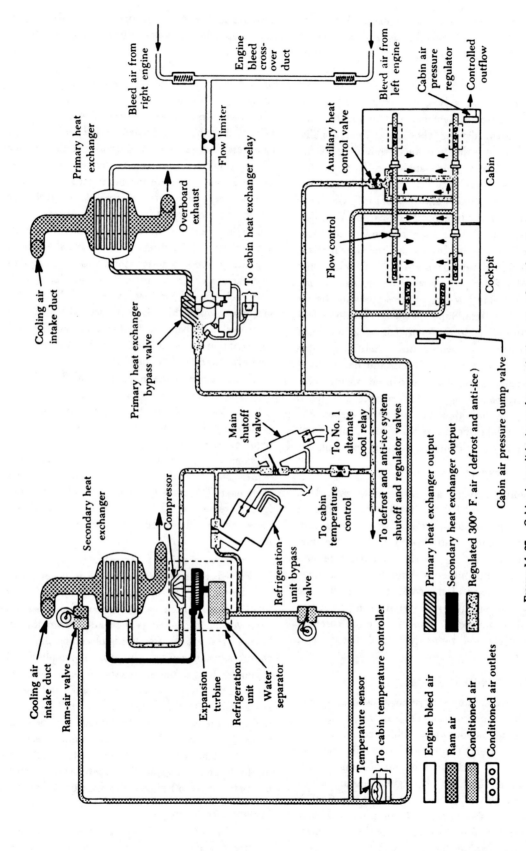

Bleed air from right engine

Engine bleed cross-over duct

Bleed air from left engine

Cabin air pressure regulator

Controlled outflow

Primary heat exchanger

Flow limiter

To cabin heat exchanger relay

Auxiliary heat control valve

Overboard exhaust

Flow control

Cabin

Cooling air intake duct

Primary heat exchanger bypass valve

Cockpit

Cabin air pressure dump valve

Secondary heat exchanger

Compressor

Main shutoff valve

To No. 1 alternate cool relay

To cabin temperature control

To defrost and anti-ice system shutoff and regulator valves

Cooling air intake duct

Ram-air valve

Expansion turbine

Refrigeration unit

Water separator

Refrigeration unit bypass valve

Temperature sensor

To cabin temperature controller

☐ Engine bleed air

░ Ram air

▒ Conditioned air

○○○ Conditioned air outlets

▨ Primary heat exchanger output

■ Secondary heat exchanger output

▨ Regulated 300° F. air (defrost and anti-ice)

Figure 14–27. Cabin air conditioning and pressurization system flow schematic.

566

standing of the manner in which the system is controlled, the functions of the various components and subassemblies, and their effect on total system operation. Figure 14-27 is a schematic of a typical system. Frequent reference to the schematic should be made during study of the following operational description.

The system is composed of a primary heat exchanger, primary heat exchanger bypass valve, flow limiters, refrigeration unit, main shutoff valve, secondary heat exchanger, refrigeration unit bypass valve, ram-air shutoff valve, and an air temperature control system. A cabin pressure regulator and a dump valve are included in the pressurization system.

Air for the cabin air conditioning and pressurization system is bled from the compressors of both engines. The engine bleed lines are cross-connected and equipped with check valves to ensure a supply of air from either engine.

A flow-limiting nozzle is incorporated in each supply line to prevent the complete loss of pressure in the remaining system if a line ruptures, and to prevent excessive hot air bleed through the rupture.

In reading the schematic, in figure 14-27, the initial input of hot air is indicated on the right-hand side. The flow is depicted across the page through each unit, in turn, and back to the squares on the lower right-hand side which represent the cockpit and cabin.

Air from the engine manifolds is ducted through a flow limiter to the primary heat exchanger and its bypass valve simultaneously. Cooling air for the heat exchanger is obtained from an inlet duct and is exhausted overboard.

The air supply from the primary heat exchanger is controlled to a constant temperature of 300° F. by the heat exchanger bypass valve. The bypass valve is automatically controlled by upstream air pressure and a downstream temperature-sensing element. These provide temperature data to cause the valve to maintain the constant temperature by mixing hot engine bleed air with the cooled air from the heat exchanger.

The cabin air is next routed through another flow limiter and a shutoff valve. The shutoff valve is the main shutoff valve for the system and is controlled from the cockpit.

From the shutoff valve, the air is routed to the refrigeration unit bypass valve, to the compressor section of the refrigeration unit, and to the second-ary heat exchanger. The bypass valve automatically maintains compartment air at any preselected temperature between 60° F. and 125° F. by controlling the amount of hot air which bypasses the refrigeration unit and mixes with the refrigeration unit output.

Cooling air for the secondary heat exchanger core is obtained from an inlet duct. Some installations use a turbine-driven fan to draw air through the heat exchanger; others use a hydraulically driven blower. After cooling the cabin air, the cooling air is exhausted overboard.

As the cabin air leaves the secondary heat exchanger, it is routed to the expansion turbine, which is rotated by the air pressure exerted on it. In performing this function, the air is further cooled before entering the water separator, where the moisture content of the air is reduced. From the water separator the air is routed through the temperature sensor to the cabin.

Air enters the cabin spaces through a network of ducts and diffusers and is distributed evenly throughout the spaces. Some systems incorporate directional vents that can be rotated by the cabin occupants to provide additional comfort.

An alternate ram-air system is provided to supply the cabin with ventilating air if the normal system is inoperative or to rid the cabin areas of smoke, foul odors, or fumes which might threaten comfort, visibility, or safety.

The air conditioning and the ram-air systems are controlled from a single switch in the cockpit. This switch is a three-position switch for selecting off, normal, and ram. In the "off" position (under normal conditions) all cabin air conditioning, pressurization, and ventilating equipment is off. In the "normal" position (under normal conditions) the air conditioning and pressurization equipment is functioning normally and ram air is off. In the "ram" position (under normal conditions) the main shutoff valve closes, and the cabin air pressure regulator and the cabin safety dump valve are opened. This allows ram air from the secondary heat exchanger cooling air inlet duct to be routed into the cabin air supply duct for cabin cooling and ventilation.

With the air pressure regulator and the safety dump valve energized open, existing cabin air and incoming ram air are constantly being dumped overboard, ensuring a steady flow of fresh air through the cabin.

A duct incorporated in the air conditioning system between the constant-temperature line downstream from the primary heat exchanger bypass valve and the cabin compartment supplies hot air for supplemental heating. Control of this air is provided by the auxiliary heat control valve, which is a butterfly type valve. The heat control valve is controlled by a manually operated heat control handle, which is connected by cable to a control arm mounted on the valve.

The temperature control system consists of a cabin temperature controller, a temperature selector knob, a two-position temperature control switch, a modulating bypass valve, and a control network. When the temperature control switch is in the "auto" position, the bypass valve will seek a valve gate position which will result in a duct temperature corresponding to the temperature controller setting. This is accomplished through the control network, which transmits signals from the sensing element to the cabin temperature controller, which then electrically positions the valve in relation to the settings of the temperature control knob. With the temperature control switch in the "man." position, the controller will control the bypass valve directly, without reference to the duct temperature. In this mode of operation the desired temperatures are maintained by monitoring the air temperature knob as varying conditions alter cabin temperature.

AIR CYCLE SYSTEM COMPONENT OPERATION

Primary Heat Exchanger

This unit, illustrated in figure 14–28, reduces the temperature of engine bleed air or supercharger discharge air by routing it through the veins in the core of the heat exchanger. During flight the core is cooled by ram air. The amount of air to be cooled in the primary heat exchanger is controlled by the primary heat exchanger bypass valve.

Primary Heat Exchanger Bypass Valve

The primary heat exchanger bypass valve (figure 14–29) is located in the high-pressure duct at the primary heat exchanger outlet. As previously stated, it modulates and controls the flow of primary heat exchanger air and primary heat exchanger bypass air to maintain a constant output air temperature of 300° F. The unit consists essentially of a regulator body assembly which contains a pressure regulator, a temperature control actuator, a solenoid valve, and a pneumatic thermostat. The body assembly of the unit contains two inlet ports marked "hot" and

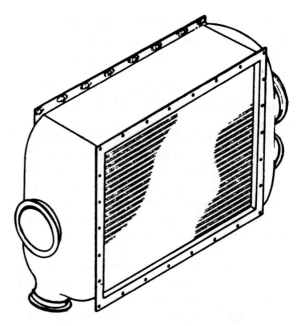

FIGURE 14–28. Primary heat exchanger.

"cold" and one outlet port. The two inlet ports incorporate butterfly valves, which are mounted on serrated shafts that extend across the width of the housing assembly and are attached to a common actuator control arm. The butterflies are positioned at 90° to each other and operate in such a manner that when one moves toward the "open" position the other moves toward the "closed" position. The actuator shaft contains an adjustable stop screw, which limits actuator travel, and pointers for indicating the position of the butterflies.

The temperature control actuator is mounted on the bypass valve body and consists of a housing and cover containing a spring-loaded diaphragm assembly. The diaphragm assembly is attached to the butterfly control arm and divides the actuator into a control pressure chamber and an ambient sensing chamber. The ambient chamber contains the diaphragm spring and the actuator rod.

As shown in the schematic in figure 14–29, pressure from the primary heat exchanger is routed through a filter and on through the pressure regulator into the control pressure chamber of the temperature control actuator. This internal pressure is called reference pressure. The reference pressure applied against the actuator diaphragm controls the position of the butterflies, which in turn controls the proportion of hot air from the bypass line and cooled air from the heat exchanger. The entire operation of the bypass valve is centered about the

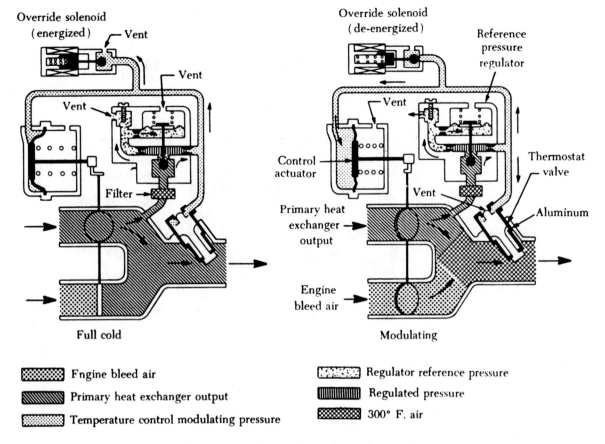

▨▨▨▨ Engine bleed air

▨▨▨▨ Primary heat exchanger output

▨▨▨▨ Temperature control modulating pressure

▨▨▨▨ Regulator reference pressure

▥▥▥▥ Regulated pressure

▨▨▨▨ 300° F. air

FIGURE 14–29. Primary heat exchanger bypass valve.

proportion of reference air pressure to heat. The greater the reference pressure that is supplied to the control actuator, the higher, will be the temperature of the output air.

A pressure regulator is provided in the bypass valve to ensure a supply of reference air pressure to the control actuator on a schedule that eliminates the effect of altitude on the controlled temperature. As the aircraft altitude increases, the constant reference pressure in the control actuator tends to move the control actuator diaphragm even further toward the ambient side. This moves the butterflies in a direction that causes the outlet temperature to rise. The pressure regulator offsets this condition with the help of the pneumatic thermostat.

The variable-orifice type thermostat consists of a spring-loaded, ball-type valve and seat assembly housed in a core assembly. The core assembly is composed of a high-expansion element (aluminum) and a low-expansion element (Invar). As can be seen in the diagram (figure 14–29), the aluminum housing and the end of the Invar core extend into the outlet duct. Linear expansion of the aluminum

housing moves the Invar core and ball-type valve assembly from the valve seat. This movement vents reference air pressure to the atmosphere. The resulting pressure applied against the temperature control actuator diaphragm controls the position of the butterflies.

The bypass valve regulating mechanism may be set to deliver cold air only by energizing the electromagnetic valve (override solenoid valve). The electromagnetic valve vents all reference air pressure to atmosphere when energized. With no reference air pressure, the spring-loaded diaphragm in the temperature control actuator returns the butterflies to the "full cold" position.

The electrical circuitry is so arranged that the solenoid can be energized only if the windshield anti-ice control switch is in the "off" position. This ensures a supply of hot air for anti-ice operation.

Shutoff Valve

The shutoff valve (figure 14–30), located in the air supply duct to the refrigeration unit, controls the air pressure to that unit. It is also the main

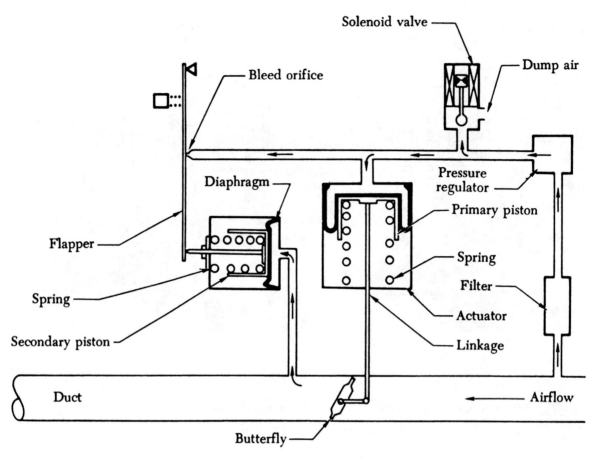

FIGURE 14–30. Shutoff valve.

shutoff valve for the cabin air conditioning and pressurization systems.

The valve requires electrical power and a minimum of 15 p.s.i. upstream pressure to function. It will regulate the downstream pressure to 115 p.s.i. Although this is an open/close valve, its major open function is to regulate. This is accomplished by a spring-loaded valve in the airflow line which is controlled by a primary piston. Upstream air pressure (if above 15 p.s.i.) bleeds through a filter and regulating mechanism to act on the primary piston, thereby opening the valve. After the downstream pressure rises to 115 p.s.i., it acts on a secondary piston which, through mechanical linkage, opens a bleed orifice to limit the amount of air acting on the primary piston. Since the primary piston is spring loaded to the "closed" position, it will then partially close, limiting the downstream pressure to 115 p.s.i.

The shutoff valve is operated by a solenoid valve that is spring loaded to off. In the "off" position, the control air from upstream is vented to atmosphere before it can operate the primary piston.

When the cockpit switch is actuated, the solenoid is energized and the vent closes, allowing pressure to build up to operate the primary piston.

Refrigeration Bypass Valve

The refrigeration bypass valve (figure 14–31) operates in conjunction with the temperature control system to modulate and control the flow of bypass air to the refrigeration unit. This action automatically maintains the cabin air at the temperature selected through the temperature controller. The valve is electrically controlled and pneumatically operated. Its operation relies on a signal from the downstream temperature-sensing element, which is controlled through the temperature control system, for an "open" position, but utilizes upstream pneumatic pressure to open the valve.

When electrical power is applied, a current-carrying coil and armature (transducer) is energized, closing a bleed port in the pressure chamber of the valve. The resulting pressure buildup in the chamber forces a piston to rotate a butterfly valve in the

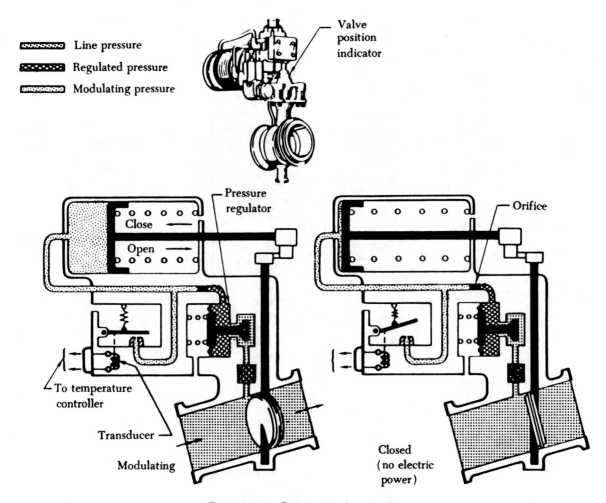

Line pressure

Regulated pressure

Modulating pressure

Valve position indicator

Pressure regulator

Close →

Open →

Orifice

To temperature controller

Transducer

Modulating

Closed (no electric power)

FIGURE 14–31. Refrigeration bypass valve.

cabin air duct to an "open" position. As the temperature varies or a new temperature is selected, the valve is re-positioned accordingly. Re-positioning is accomplished by action of the transducer in varying the amount of pressure allowed to bleed from the pressure chamber. Failure of the bypass valve or its components will cause the valve to move to the fail-safe (closed) position.

Secondary Heat Exchanger

The function of the secondary heat exchanger is to partially cool the air for cabin pressurization and air conditioning to a temperature which makes possible the efficient operation of the refrigeration unit.

The heat exchanger assembly consists mainly of dimpled aluminum alloy tubes. The tubes are arranged so that pressurized cabin air can flow through them and cooling air can flow across them.

The secondary heat exchanger operates in essentially the same manner as the primary heat exchanger. Cabin air that is to be further cooled is routed through the tubes in the heat exchanger core. Cooling air is forced through the secondary heat exchanger and returned to an engine air inlet or can be exhausted directly to the atmosphere.

Cabin air is regulated by the refrigeration bypass valve where it is directed to the secondary heat exchanger or to the refrigeration unit bypass line in metered quantities as required to meet the demands of the temperature control system.

Refrigeration Unit

The refrigeration unit or turbine is used in the air conditioning system to cool the pressurized air for the cabin. Operation of the unit is entirely automatic, the power being derived from the pressure and temperature of the compressed air passing through the turbine wheel. The refrigeration cycle is modulated to meet varying cabin cooling demands by a refrigeration bypass valve which pro-

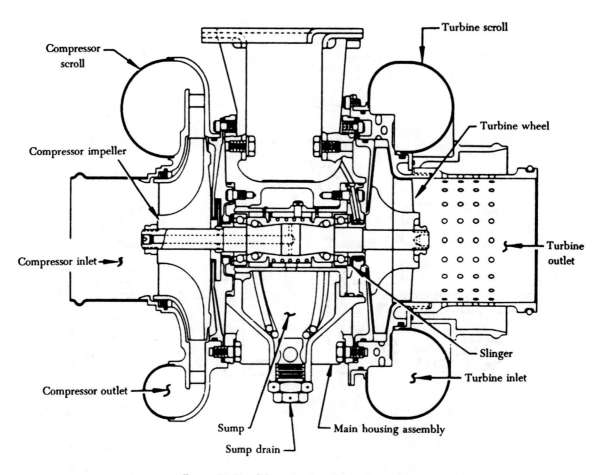

FIGURE 14–32. Schematic of a refrigeration turbine.

vides for a bypass of the entire refrigeration unit. Thus, cabin temperature is regulated by mixing bypassed air with that which has passed through the refrigeration unit.

The refrigeration turbine (figure 14–32) consists of three major sections: (1) Main housing assembly, (2) turbine scroll assembly, and (3) compressor scroll assembly.

The main housing assembly provides mounting for the two scroll assemblies and provides support for the two shaft bearings. It also serves as the oil reservoir from which the oil is supplied to the bearings by wicks. A dipstick for checking oil level is attached to the filler cap. The turbine scroll assembly is composed of two halves which confine the turbine nozzle within which the turbine wheel rotates. The compressor scroll assembly is composed of two halves which confine the diffuser within which the compressor wheel rotates.

A common shaft carries both assemblies and is supported by bearings in the housing assembly. An

oil slinger is mounted outboard of each of the bearings which carry the shaft. These slingers pump an oil/air mist through the bearings to provide for lubrication. Air/oil seals are provided between each slinger and the adjacent wheel.

The supply air which is being cooled drives the refrigeration turbine. An impeller, driven by this turbine, forces the cooling air through the refrigeration unit.

The refrigeration process takes place as the hot compressed air expands through the turbine wheel of the air expansion turbine. This results in a reduction in the temperature and pressure of the air. As this hot compressed air expands, it releases energy to the turbine wheel causing it to rotate at high speed.

Since the turbine wheel and compressor wheel are on opposite ends of a common shaft, the turbine wheel rotation results in a corresponding rotation of the compressor wheel. Thus, the energy released from the high temperature compressed air to the turbine wheel provides the energy required by the compressor wheel to further compress the incoming air. The load imposed on the turbine by the compressor holds the speed of rotation within the range of maximum efficiency. Reduction of the air temperature assists in maintaining the cabin temperature within desired limits.

Water Separators

Water separators (figure 14–33) are used in the cabin air conditioning system to remove excessive moisture from the air. In most refrigeration systems a water separator is installed in the discharge duct of the cooling turbine.

The water separator removes excess moisture from the conditioned air by passing the air through a coalescent bag or condenser. Very small water particles in the form of fog or mist in the air are formed into larger particles in passing through the condenser. As the moisture laden air passes through the vanes of the coalescent support, the water particles are carried with the swirling air and are thrown outward against the walls of the collector.

The water then drains into a collector sump and is drained overboard.

Some water separators also contain a pressure-relief and altitude-sensitive bypass valve. Since very little moisture is present in the air at high altitudes, the bypass valve in the water separator opens at a predetermined altitude, generally 20,000 ft., to permit cold air to pass directly through the water separator, bypassing the coalescent bag and reducing system back pressure. The bypass valve will also open if, for some reason, the coalescent bag should become obstructed.

A coalescent bag condition indicator is provided on some water separators to indicate when the bag is dirty. The indicator senses a pressure drop across the bag and indicates when the pressure drop is excessive. Since the indicator is pressure sensitive, the condition of the bag can be determined only while the system is in operation.

Ram-Air Valve

The ram-air valve is always closed during normal operations. It is energized to open when the cockpit

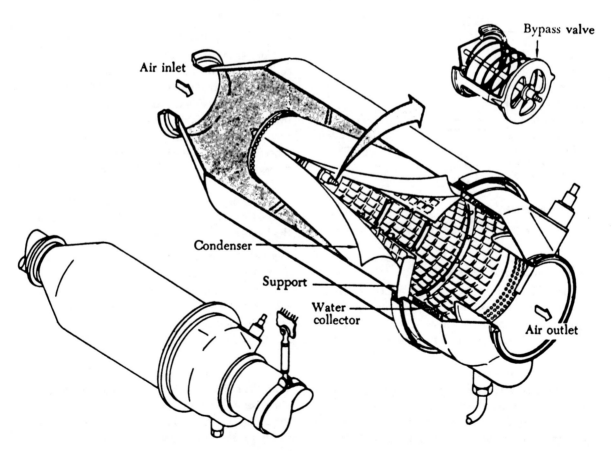

FIGURE 14–33. Water separator.

switch is placed in the "ram" position. With the ram-air valve open, air from the air inlet duct is admitted through the valve and directly into the cabin air supply duct.

ELECTRONIC CABIN TEMPERATURE CONTROL SYSTEM

The operation of the electronic temperature control system is based primarily on the balanced-bridge circuit principle. When any of the units which compose the "legs" of the bridge circuit change resistance value because of a temperature change, the bridge circuit becomes unbalanced. An electronic regulator receives an electrical signal as a result of this unbalance and amplifies this signal to control the mixing valve actuator.

In a typical application of the electronic temperature control system, three units are utilized: (1) Cabin temperature pickup (thermistor), (2) manual temperature selector, and (3) electronic regulator. Figure 14–34 shows a simplified schematic diagram of an electronic temperature control system.

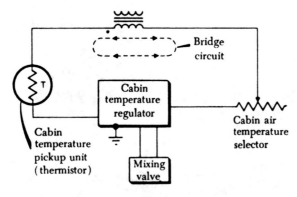

FIGURE 14–34. Electronic air temperature control system (simplified).

Cabin Temperature Pickup Unit

The cabin temperature pickup unit (temperature sensing unit) consists of a resistor that is highly sensitive to temperature changes. The temperature pickup unit is usually located in the cabin or cabin air supply duct. As the temperature of the air supply changes, the resistance value of the pickup unit also changes, thus causing the voltage drop across the pickup to change. The cabin temperature pickup is a thermistor type unit (figure 14–35). As the ambient temperature of the resistance bulb increases, the resistance of the bulb decreases.

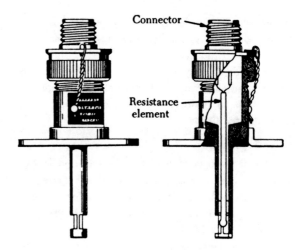

FIGURE 14–35. Thermistor.

Cabin Air Temperature Selector

The air temperature selector (see figure 14–34) is a rheostat located in the cabin. It permits selective temperature control by varying the effective temperature control point of the cabin air temperature pickup unit. The rheostat causes the cabin temperature pickup unit to demand a specific temperature of the supply air.

Cabin Air Temperature Control Regulator

The cabin air temperature control regulator, in conjunction with the air temperature selector rheostat and the air duct temperature pickup unit, automatically maintains the temperature of the air entering the cabin at a preselected value. The temperature regulator is an electronic device with a temperature regulating range. In some installations, this range may extend from as low as 32° F. to as high as 117° F.

The output of the regulator controls the position of the butterfly in the mixing valve, thus controlling the temperature of the inlet air to the cabin.

Typical System Operation

Figure 14–36 shows an electrical schematic of a typical air temperature control system. In most air temperature control systems, there is one switch to select the mode of temperature control. Usually, this switch will have four positions: "off," "auto," "man. hot," and "man. cold." In the "off" position, the air temperature control system is inoperative. With the switch in the "auto" range, the air temperature control system is in the automatic mode. With the switch in either the "man. hot" or "man. cold" position, the air temperature control system is in the manual mode.

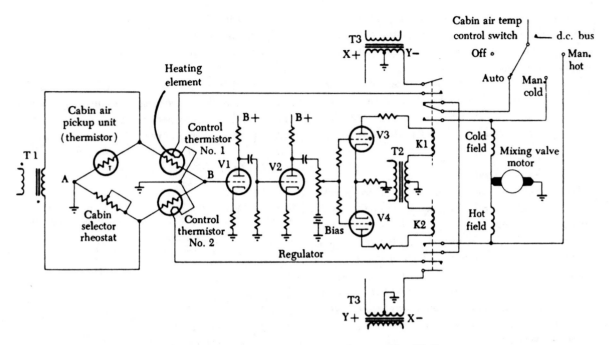

FIGURE 14–36. Air temperature control system (simplified).

ELECTRONIC TEMPERATURE CONTROL REGULATOR

The cabin selector rheostat and the cabin air pickup unit (thermistor) determine the direction and amount of rotation of the mixing valve motor. This function is controlled in the cabin air temperature regulator. The cabin selector rheostat and the cabin air pickup unit (see figure 14–36) are connected into a bridge circuit which also includes two thermistors that are located in the regulator.

The bridge circuit is energized by an a.c. source (T1). If the resistance of the cabin air pickup unit and the cabin selector rheostat were equal, then points A and B would have no potential difference.

Note that points A and B are the signal reference points for V1 (grid and cathode). If the cabin air temperature increases, the resistance value of the cabin air temperature pickup unit decreases, since the flow of the air passes over the pickup unit. This decrease in resistance of the pickup unit causes the voltage developed across the pickup unit to decrease, resulting in a potential difference between points A and B.

This signal, which is impressed on the grid of V1, goes through two stages of voltage amplification (V1 and V2). The amplified signal is applied to the grids of the two thyratron tubes (V3 and V4). The thyratron tubes (gas filled triode or te-

trode) are used for signal phase detection. For example, if the signal on the grid of V3 is in phase with the signal on the plate, V3 will conduct, causing current to flow through the coil of relay K1 and close its contacts.

One set of contacts completes a circuit for direct current flow to the cold-field coil of the mixing valve motor. This directs more hot air into the refrigeration unit, thereby cooling the cabin air.

At the same time, the remaining set of contacts of K1 completes a source of a.c. power (T3) to the heating element of thermistor No. 1 of the bridge circuit, causing the resistance of thermistor No. 1 to decrease. (Remember that a thermistor's resistance decreases as the temperature rises.) The resultant change in the voltage drop across thermistor No. 1 results in a balanced bridge across points A and B. This, in turn, causes relay K1 to become de-energized, stopping the rotation of the mixing valve motor.

At this point, heater voltage is removed from thermistor No. 1 and it cools, again unbalancing the bridge. This causes the mixing valve motor to drive farther towards the cool position, allowing still more refrigerated air to enter the cabin. Cycling continues until the drops in voltage across the pickup unit and the selector rheostat are equal.

Had the cabin air temperature been colder than the selected setting, the bridge would have become unbalanced in the opposite direction. This would have caused relay K2 on the regulator to become energized, thus energizing the hot-field coil of the mixing valve motor.

The bridge may also be unbalanced by another method, *i.e.*, by re-positioning the cabin selector rheostat. Again the mixing valve moves to regulate the temperature of the air until the bridge is re-balanced.

VAPOR CYCLE SYSTEM (FREON)

Vapor cycle cooling systems are used on several large transport aircraft. This system usually has a greater cooling capacity than an air cycle system, and in addition, can usually be used for cooling on the ground when the engines are not operating.

An aircraft freon system is basically similar in principle to the kitchen refrigerator or the home air conditioner. It uses similar components and operating principles and in most cases depends upon the electrical system for power.

Vapor cycle systems make use of the scientific fact that a liquid can be vaporized at any temperature by changing the pressure acting on it. Water, at sea level barometric pressure of 14.7 p.s.i.a. will boil if its temperature is raised to 212° F. The same water in a closed tank under a pressure of 90 p.s.i.a. will not boil at less than 320° F. If the pressure is reduced to 0.95 p.s.i.a. by a vacuum pump, the water will boil at 100° F. If the pressure is reduced further, the water will boil at a still lower temperature; for instance, at 0.12 p.s.i.a., water will boil at 40° F. Water can be made to boil at any temperature if the pressure corresponding to the desired boiling temperature can be maintained.

Refrigeration Cycle

The basic laws of thermodynamics state that heat will flow from a point of higher temperature to a point of lower temperature. If heat is to be made to flow in the opposite direction, some energy must be supplied. The method used to accomplish this in an air conditioner is based on the fact that when a gas is compressed, its temperature is raised, and, similarly, when a compressed gas is allowed to expand, its temperature is lowered.

To achieve the required "reverse" flow of heat, a gas is compressed to a pressure high enough so that its temperature is raised above that of the outside air. Heat will now flow from the higher temperature gas to the lower temperature surrounding air (heat sink), thus lowering the heat content of the gas.

The gas is now allowed to expand to a lower pressure; this causes a drop in temperature that makes it cooler than the air in the space to be cooled (heat source).

Heat will now flow from the heat source to the gas, which is then compressed again, beginning a new cycle. The mechanical energy required to cause this apparent reverse flow of heat is supplied by a compressor. A typical refrigeration cycle is illustrated in figure 14–37.

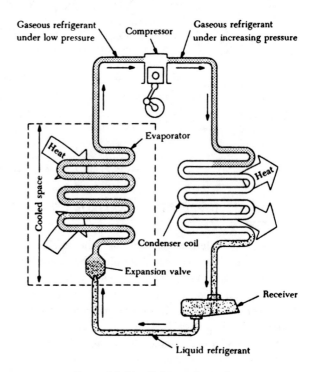

FIGURE 14–37. Refrigeration cycle.

This refrigeration cycle is based on the principle that the boiling point of a liquid is raised when the pressure of the vapor around the liquid is raised. The cycle operates as follows: A liquid refrigerant confined in the receiver at a high pressure is allowed to flow through the expansion valve into the evaporator. The pressure in the evaporator is low enough so that the boiling point of the liquid refrigerant is below the temperature of the air to be cooled, or heat source. Heat flows from the space to be cooled to the liquid refrigerant, causing it to boil (to be converted from liquid to a vapor). Cold vapor from the evaporator enters the compressor, where its pressure is raised, thereby raising the boiling point. The refrigerant at a high pressure and high temperature flows into the condenser.

576

Here heat flows from the refrigerant to the outside air, condensing the vapor into a liquid. The cycle is repeated to maintain the cooled space at the selected temperature.

Liquids that will boil at low temperatures are the most desirable for use as refrigerants. Comparatively large quantities of heat are absorbed when liquids are changed to a vapor. For this reason, liquid freon is used in most vapor cycle refrigeration units whether used in aircraft or in home air conditioners and refrigerators.

Freon is a fluid which boils at a temperature of approximately 39° F. under atmospheric pressure. Similar to other fluids, the boiling point may be raised to approximately 150° F. under a pressure of 96 p.s.i.g. These pressures and temperatures are representative of one type of freon. Actual values will vary slightly with different types of freon. The type of freon selected for a particular aircraft will depend upon the design of the freon system components installed.

Freon, similar to other fluids, has the characteristic of absorbing heat when it changes from a liquid to a vapor. Conversely, the fluid releases heat when it changes from a vapor to a liquid. In the freon cooling system, the change from liquid to vapor (evaporation or boiling) takes place at a location where heat can be absorbed from the cabin air, and the change from vapor to liquid (condensation) takes place at a point where the released heat can be ejected to the outside of the aircraft. The pressure of the vapor is raised prior to the condensation process so that the condensation temperature is relatively high. Therefore, the freon, condensing at approximately 150° F., will lose heat to the outside air which may be as hot as 100° F.

The quantity of heat that each pound of refrigerant liquid absorbs while flowing through the evaporator is known as the "refrigeration effect." Each pound flowing through the evaporator is able to absorb only the heat needed to vaporize it, if no superheating (raising the temperature of a gas above that of the boiling point of its liquid state) takes place. If the liquid approaching the expansion valve were at exactly the temperature at which it was vaporizing in the evaporator, the quantity of heat that the refrigerant could absorb would be equal to its latent heat. That is the amount of heat required to change the state of a liquid, at the boiling point, to a gas at the same temperature.

When liquid refrigerant is admitted to the evaporator, it is completely vaporized before it reaches the outlet. Since the liquid is vaporized at a low temperature, the vapor is still cold after the liquid has completely evaporated. As the cold vapor flows through the balance of the evaporator, it continues to absorb heat and becomes superheated.

The vapor absorbs sensible heat (heat which causes a temperature change when added to, or removed from matter) in the evaporator as it becomes superheated. This, in effect, increases the refrigerating effect of each pound of refrigerant. This means that each pound of refrigerant absorbs not only the heat required to vaporize it, but also an additional amount of sensible heat which superheats it.

FREON SYSTEM COMPONENTS

The major components of a typical freon system are the evaporator, compressor, condenser, and expansion valve (figure 14–38). Other minor items may include the condenser fan, receiver (freon storage), dryer, surge valve, and temperature controls. These items are interconnected by appropriate tubing to form a closed loop in which the freon is circulated during operation.

Freon System Operational Cycle Compressor

The principle of operation of the system can be explained by starting with the function of the compressor. The compressor increases the pressure of the freon when it is in vapor form. This high pressure raises the condensation temperature of the freon and produces the force necessary to circulate the freon through the system.

The compressor is driven either by an electric motor or by an air turbine drive mechanism. The compressor may be a centrifugal type or a piston type. The compressor is designed to act upon freon in a gaseous state and in conjunction with the expansion valve, maintains a difference in pressure between the evaporator and the condenser. If the liquid refrigerant were to enter the compressor, improper operation would occur. This type of malfunction is called "slugging." Automatic controls and proper operating procedures must be used to prevent slugging.

Condenser

The freon gas is pumped to the condenser for the next step in the cycle. At the condenser the gas passes through a heat exchanger where outside (ambient) air removes heat from the freon. When heat is removed from the high-pressure freon gas, a change of state takes place and the freon condenses to a liquid. It is this condensation process which releases the heat the freon picks up from

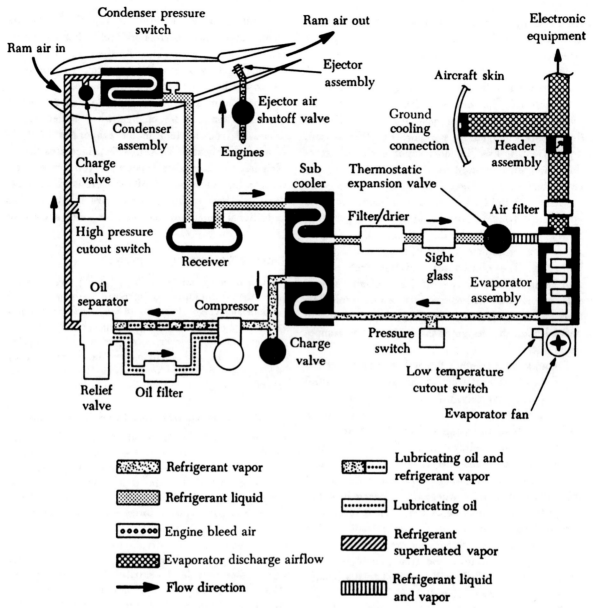

Refrigerant vapor

Refrigerant liquid

Engine bleed air

Evaporator discharge airflow

→ **Flow direction**

Lubricating oil and refrigerant vapor

Lubricating oil

Refrigerant superheated vapor

Refrigerant liquid and vapor

FIGURE 14–38. Vapor cycle system flow schematic.

the cabin air. The flow of ambient air through the condenser unit is ordinarily modulated by controllable inlet or outlet doors according to cooling requirements. A condenser cooling air fan or air ejector is often used to help force the ambient air through the condenser; this item is important for operation of the system on the ground.

Receiver

From the condenser the liquid freon flows to the receiver which acts as a reservoir for the liquid refrigerant. The fluid level in the receiver varies with system demands. During peak cooling periods, there will be less liquid than when the load is light. The prime function of the receiver

is to ensure that the thermostatic expansion valve is not starved for refrigerant under heavy cooling load conditions.

Subcooler

Some vapor cycle systems use a subcooler to reduce the temperature of the liquid refrigerant after it leaves the receiver. By cooling the refrigerant premature vaporization (flash-off) can be prevented. Maximum cooling takes place when the refrigerant changes from a liquid to a gaseous state. For efficient system operation this must occur in the evaporator. If the refrigerant vaporizes before it reaches the evaporator the cooling efficiency of the system is reduced.

The subcooler is a heat exchanger containing passages for liquid freon from the receiver on its way to the evaporator and cold freon gas leaving the evaporator on its way to the compressor. The liquid on the way to the evaporator is relatively warm in comparison to the cold gas leaving the evaporator. Although the gas leaving the evaporator has absorbed heat from the air being circulated through the evaporator, its temperature is still in the vicinity of 40° F. This cool gas is fed through the subcooler where it picks up additional heat from the relatively warm liquid freon that is flowing from the receiver. This heat exchange subcools the liquid freon to a level that ensures little or no flash-off (premature vaporization) on its way to the evaporator.

Subcooling is a term used to describe the cooling of a liquid refrigerant at constant pressure to a point below the temperature at which it was condensed. At 117 p.s.i.g., freon vapor condenses at a temperature of 100° F. If, after the vapor has been completely condensed, the liquid is cooled still further to a temperature of 76° F., it will have been subcooled 24°. Through subcooling, the liquid delivered to the expansion valve is cool enough to prevent most of the flash-off that would normally result, thereby making the system more efficient.

Filter/Drier

The system illustrated in figure 14–38 has a filter/drier unit installed between the subcooler and the sight glass. The filter/drier is essentially a sheet-metal housing with inlet and outlet connections and contains alumina desiccant, a filter screen, and a filter pad. The alumina dseiccant acts as a moisture absorbent so that dry freon flows to the expansion valve. A conical screen and fiber glass pad act as a filtering device, removing contaminants.

Scrupulously clean refrigerant at the expansion valve is a must because of the critical clearances involved. Moisture may freeze at the expansion valve, causing it to hang up with a resulting starvation or flooding of the evaporator.

Sight Glass

To aid in determining whether servicing of the refrigerating unit is required, a liquid line sight glass or liquid level gage is installed in the line between the filter/drier and the thermostatic expansion valve. The sight glass consists of a fitting with windows on both sides, permitting a view of fluid passage through the line. In some systems the sight glass is constructed as a part of the filter/drier.

During refrigeration unit operation, a steady flow of freon refrigerant observed through the sight glass indicates that sufficient charge is present. If the unit requires additional refrigerant, bubbles will be present in the sight glass.

Expansion Valve

The liquid freon flows to the expansion valve for the next step in the operation. The freon coming out of the condenser is high-pressure liquid refrigerant. The expansion valve lowers the freon pressure and thus lowers the temperature of the liquid freon. The cooler liquid freon makes it possible to cool cabin air passing through the evaporator.

The expansion valve, mounted close to the evaporator, meters the flow of refrigerant into the evaporator. Efficient evaporator operation depends upon the precise metering of liquid refrigerant into the heat exchanger for evaporation. If heat loads on the evaporator were constant, an orifice size could be calculated and used to regulate the refrigerant supply. A practical system, however, encounters varying heat loads and, therefore, requires a refrigerant throttling device to prevent starvation or flooding of the evaporator, which would affect the evaporator and system efficiency. This variable-orifice effect is accomplished by the thermostatic expansion valve, which senses evaporator conditions and meters refrigerant to satisfy them. By sensing the temperature and the pressure of the gas leaving the evaporator, the expansion valve precludes the possibility of flooding the evaporator and returning liquid refrigerant to the compressor.

The expansion valve, schematically portrayed in figure 14–39, consists of a housing containing inlet and outlet ports. The flow of refrigerant to the outlet ports is controlled by positioning a metering valve pin. Valve pin positioning is controlled by the pressure created by the remote sensing bulb, the superheat spring setting, and the evaporator discharge pressure supplied through the external equalizer port.

The remote sensing bulb is a closed system filled with refrigerant and the bulb is attached to the evaporator. Pressure within the bulb corresponds to the refrigerant pressure leaving the evaporator. This force is felt on top of the diaphragm in the power head section of the valve, and any increase in pressure will cause the valve to move towards an "open" position. The bottom side of the diaphragm has the forces of the superheat spring and evaporator discharge pressure acting in a direction to close the valve pin. The valve position at any instant is the result of these three forces.

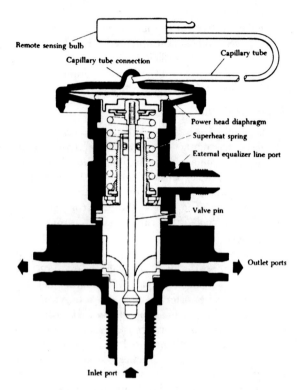

FIGURE 14–39. Schematic of thermostatic expansion valve.

If the temperature of the gas leaving the evaporator increases above the desired superheat valve, it will be sensed by the remote bulb. The pressure generated in the bulb is transmitted to the diaphragm in the power section of the valve, causing the valve pin to open. A decrease in the temperature of the gas leaving the evaporator will cause the pressure in the remote bulb to decrease, and the valve pin will move toward the "closed" position.

The superheat spring is designed to control the amount of superheat in the gas leaving the evaporator. A vapor is said to be superheated when its temperature is higher than that necessary to change it from a liquid to a gas at a certain pressure. This ensures that the freon returning to the compressor is in the gaseous state.

The equalizer port is provided to compensate for the effect the inherent evaporator pressure drop has on the superheat setting. The equalizer senses evaporator discharge pressure and reflects it back to the power head diaphragm, adjusting the expansion valve pin position to hold the desired superheat value.

Evaporator

The next unit in the line of cooling flow after the expansion valve is the evaporator, which is a heat exchanger forming passages for cooling air

flow and for freon refrigerant. Air to be cooled flows through the evaporator.

The freon changes from a liquid to a vapor at the evaporator. In effect, the freon boils in the evaporator, and the pressure of the freon is controlled to the point where the boiling (evaporation) takes place at a temperature which is lower than the cabin air temperature. The pressure (saturated pressure) necessary to produce the correct boiling temperature must not be too low; otherwise, freezing of the moisture in the cabin air will block the air passages of the evaporator. As the freon passes through the evaporator, it is entirely converted to the gaseous state.

This is essential to obtain the maximum cooling and also to prevent liquid freon from reaching the compressor. The evaporator is designed so that heat is taken from the cabin air; therefore, the cabin air is cooled. All the other components in the freon system are designed to support the evaporator, where the actual cooling is done.

After leaving the evaporator, the vaporized refrigerant flows to the compressor and is compressed. Heat is being withdrawn through the walls of the condenser and carried away by air circulating around the outside of the condenser. As the vapor condenses to a liquid it gives up the heat which was absorbed when the liquid changed to a vapor in the evaporator. From the condenser the liquid refrigerant flows back to the receiver, and the cycle is repeated.

DESCRIPTION OF A TYPICAL SYSTEM

Since the vapor cycle system used in Boeing aircraft models 707 and 720 are typical of most vapor cycle systems, they are used here to describe the operation of such systems.

The major components of the vapor cycle air conditioning system are the: (1) Air turbine centrifugal compressors, (2) primary heat exchangers, (3) refrigeration units, (4) heaters, and (5) necessary valves to control the airflow.

The vapor cycle system shown schematically in figure 14–40 is divided into a left-hand and a right-hand installation. Both installations are functionally identical.

Air Turbine Compressor

The cabin and flight compartments are pressurized by using two air turbine centrifugal compressors (turbo-compressors). Each compressor consists of a turbine section and a compressor section as shown in figure 14–41.

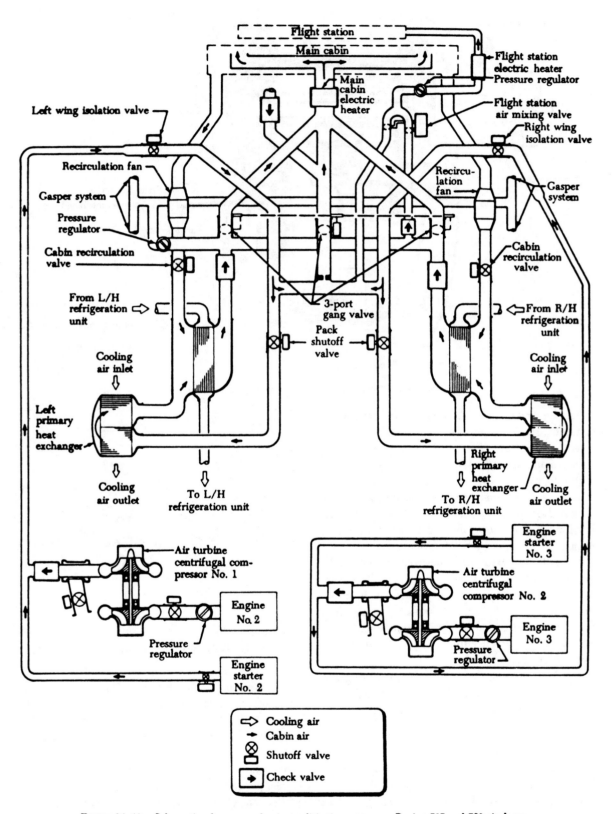

FIGURE 14-40. Schematic of vapor cycle air conditioning system on Boeing 707 and 720 airplanes.

581

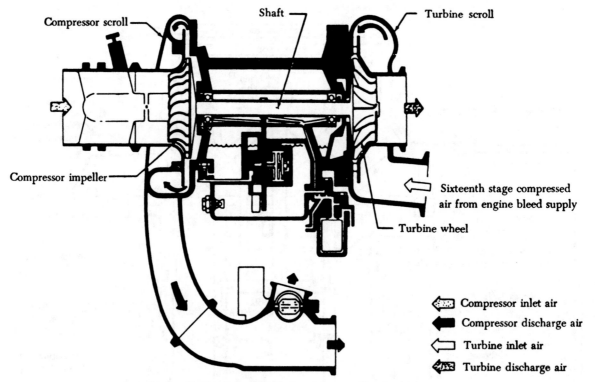

FIGURE 14–41. Schematic of air turbine centrifugal compressor.

The turbine section inlet duct is connected to the sixteenth stage compressed air from the engine bleed air manifold. The bleed air is under a pressure of approximately 170 p.s.i. This high-pressure, high-velocity air is reduced to approximately 76 p.s.i. by a differential pressure regulator located in the air duct leading to the turbine inlet. This regulated air pressure turns the turbine at about 49,000 r.p.m.

Since the compressor is connected directly to the turbine, it also turns at the same r.p.m. The compressor output is approximately 1,070 cu. ft. of air per minute at a maximum of 50 p.s.i. The compressor section inlet is connected to a ram-air scoop and the outlet is connected through ducts into the air conditioning system.

Air flows through the ducts, through a wing isolation valve, past the shutoff valve, and through the primary heat exchanger.

Primary Heat Exchangers

The two primary (air-to-air) heat exchangers are located in the left- and right-hand installation of the vapor cycle system as shown in figure 14–40.

Each primary heat exchanger consists of a duct assembly, a core assembly, and a pan assembly. The welded duct assembly contains both the inlet and outlet passages. The tube-type core assembly forms

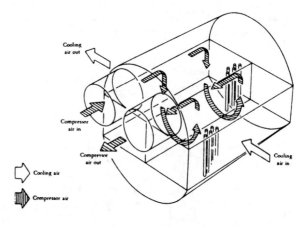

FIGURE 14–42. Schematic of primary heat exchanger.

the center portion of the unit. The pan assembly completes the enclosure of the tubes. Cabin ventilating air flows through the inside of the tubes of the core assembly. Ram air is forced around and between the outside of the tubes. Figure 14–42 shows a schematic diagram of the primary heat exchanger.

The primary heat exchangers remove about 10% of the heat of compression from the cabin ventilating air as it comes from the turbo-compressors, thus cooling the air to about 10° to 25° above outside air temperature.

582

Refrigeration Units

From the primary heat exchangers the ventilating air is ducted to the refrigeration units. The two refrigeration units are located in the left- and right-hand installation of the vapor cycle system as shown in figure 14–40. Each refrigeration unit consists of an electric motor driven freon compressor, an air-cooled refrigerant condenser, a receiver (freon container), an evaporator heat exchanger, a dual control valve, a heat exchanger (liquid-to-gas), and the necessary electrical components to assure proper operation of the unit. The refrigerant used in the system is freon 114. Lubricating oil is added to the freon each time the refrigeration unit is charged to provide lubrication for the compressor bearings.

After the air is cooled to the desired temperature, it is ducted into the cabin and flight deck.

Electric Heaters

The main cabin ventilating air and the flight compartment ventilating air are heated separately and independently by two electric heaters, one heater for each.

The flight compartment heater consists of a core which is made up of nine electrical heater elements mounted in a rectangular aluminum shell assembly, three protectors, a.c. power connection to the elements, and a control circuit to the thermal protectors.

The main cabin heater is similar but has a greater output capacity since it provides heat for a larger compartment and a greater volume of air.

Air Routing/Valves

The solid black arrows in figure 14–40 indicate the flow route of the ventilating air from the turbocompressor, through the air conditioning units to the cabin and flight compartment. The three-port gang valve regulates the flow of hot or cold air to the cabin in response to the selected temperature.

AIR CONDITIONING AND PRESSURIZATION SYSTEM MAINTENANCE

The maintenance required on air conditioning and pressurization systems varies with each aircraft model. This maintenance follows procedures given in the appropriate aircraft manufacturer or equipment manufacturer's maintenance manuals. It usually consists of inspections, servicing, removing, and installing components, performing operational checks, and troubleshooting for the isolation and correction of troubles within the system.

Inspections

Periodically inspect the system for component security and visible defects. Particular attention should be paid to the heat exchangers for signs of structural fatigue adjacent to welds. The ducting should be securely attached and adequately supported. Insulating blankets must be in good repair and secured around the ducting.

Servicing

Each refrigeration unit contains freon for absorbing heat, plus oil mixed with the freon for lubricating the compressor motor bearings. If there is insufficient freon in the unit, it is incapable of absorbing heat from the air going to the cabin. If there is insufficient oil, the motor bearings will overheat and eventually cause unsatisfactory compressor operation. It is important that sufficient amounts of freon and oil be in the unit at all times.

In contrast to a hydraulic system where the circuits consist of closed loops containing fluid at all times, a freon loop contains quantities of both liquid and vapor. This, in addition to the fact that it is unpredictable exactly where in the system the liquid will be at any one instant, makes it difficult to check the quantity of freon in the system.

Regardless of the amount of freon in the complete system, the liquid level can vary significantly, depending on the operating conditions. For this reason, a standard set of conditions should be obtained when checking the freon level. These conditions are specified by the manufacturer and, as mentioned previously, vary from aircraft to aircraft.

To check the freon level, it is necessary to operate the refrigeration unit for approximately 5 min. to reach a stable condition. If the system uses a sight glass, observe the flow of freon through the sight glass. A steady flow indicates that a sufficient charge is present. If the freon charge is low, bubbles will appear in the sight glass.

When adding freon to a system, add as much oil as is felt was lost with the freon being replaced. It is impossible to determine accurately the amount of oil left in a freon system after partial or complete loss of the freon charge. However, based on experience, most manufacturers have established procedures for adding oil. The amount of oil to be added is governed by: (1) The amount of freon to be added, (2) whether the system has lost all of its charge and has been purged and evacuated, (3) whether a topping charge is to be added, or (4) whether major components of the system have been changed.

Usually one-fourth-ounce of oil is added for each pound of freon added to the system. When changing a component, an additional amount of oil is added to replace that which is trapped in the replaced component.

Oil for lubrication of the compressor expansion valve, and associated seals must be sealed in the system. The oil used is a special highly refined mineral oil free from wax, water and sulfur. Always use the oil specified in the manufacturer's maintenance manual for a specific system.

Freon–12

Freon–12 is the most commonly used refrigerant. It is a fluorinated hydrocarbon similar to carbon tetrachloride with 2 of the chlorine atoms replaced by 2 fluorine atoms. It is stable at low or high temperatures, does not react with any of the materials or seals used in an air conditioning system, and is non-flammable.

Freon–12 will boil at sea level pressure at $-21.6°$ F. Any freon–12 dropped on the skin will result in frostbite. Even a trace in the eye can cause damage. If this should occur, treat the eye with clean mineral oil, or petroleum jelly followed with a boric acid wash. *GET TO A PHYSICIAN OR HOSPITAL AS SOON AS POSSIBLE.*

Freon is colorless, odorless, and non-toxic; however, being heavier than air, it will displace oxygen and can cause suffocation. When heated over an open flame, it converts to phosgene gas which is deadly! Avoid inhaling freon or contact with this gas.

Wear a face shield, gloves and protective clothing when working with freon.

Manifold Set

Whenever a freon system is opened for maintenance, a portion of the freon and oil will be lost. Replenishment of the freon and oil is a must for efficient system operation. This requires the use of a special set of gages and interconnected hoses.

The manifold set (figure 14–43) consists of a manifold with three fittings to which refrigerant service hoses are attached: two hand valves with "O" ring type seals, two gages, one for the low pressure side of the system and one for the high pressure side of the system.

The low pressure gage is a compound gage, meaning it will read pressures either side of atmospheric. It will indicate to about 30 inches of mercury, gage pressure (below atmospheric) to about 60 psi gage pressure above atmospheric.

The high pressure gages usually have a range from zero up to about 600 p.s.i. gage pressure.

The low pressure gage is connected on the manifold directly to the low side fitting. The high pressure gage likewise connects directly to the high side fitting. The center fitting of the manifold can be isolated from either of the gages or the high and low service fittings by the hand valves. When these valves are turned fully clockwise, the center fitting is isolated. If the low pressure valve is opened (turned counter-clockwise), the center fitting is opened to the low pressure gage and the low side service line. The same is true for the high side when the high pressure valve is opened.

Special hoses are attached to the fittings of the manifold valve for servicing the system. The high pressure charging hose attaches to the service valve in the high side, either at the compressor discharge, the receiver dryer, or on the inlet side of the expansion valve. The low pressure hose attaches to

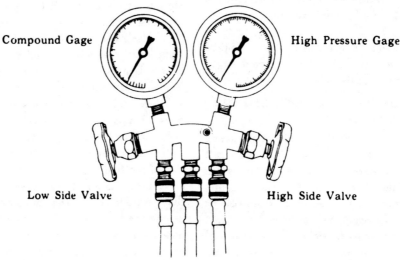

Compound Gage   High Pressure Gage

Low Side Valve   High Side Valve

FIGURE 14–43. Freon manifold set.

the service valve at the compressor inlet, or at the discharge side of the expansion valve. The center hose attaches to the vacuum pump for evacuating the system, or to the refrigerant supply for charging the system. Charging hoses used with Schrader valves must have a pin to depress the valve.

When not using the manifold set, be sure the hoses are capped to prevent moisture contaminating the valves.

Purging the System

Whenever a freon system is to be opened for maintenance it is necessary to purge the system. The manifold set is connected as previously described except the center hose is not yet connected to the vacuum pump. Cover the center hose with a clean shop towel, open both valves slowly. This will permit the gas to escape but will not blow the oil from the system. When both gages read zero, the system may be opened.

Evacuating the System

Only a few drops of moisture will contaminate and completely block an air conditioning system. If this moisture freezes in the expansion valve, the action stops. Water is removed from the system by evacuation. Anytime the system has been opened it must be evacuated before recharging.

The manifold set is connected into the system with the center hose connected to a vacuum pump. The pump reduces the pressure, moisture vaporizes and is drawn from the system. A typical pump used for evacuating air conditioning systems will pump 0.8 cubic feet of air per minute and will evacuate the system to about 29.62 inches of mercury (gage pressure). At this pressure water will boil at 45° F. Pumping down or evacuating a system usually requires about 60 minutes pumping time.

Recharging

With the system under vacuum from evacuation, close all valves, connect the center hose to a refrigerant supply. Open the container valve, loosen the high side hose at its connection to the system, and allow some freon to escape. This purges the manifold set. Tighten the hose.

Open the high pressure valve, this will permit freon to flow into the system. The low pressure gage should begin to indicate that the system is coming out of the vacuum. Close both valves. Start the engine, set the rpm at about 1250. Set the controls for full cooling. With the freon container upright to allow vapor to come out, open

the low pressure valve to allow vapors to enter the system. Put as many pounds of freon into the system as called for by the specifications. Close all valves, remove manifold set and perform an operational check.

Checking Compressor Oil

The compressor is a sealed unit in the refrigeration system. Any time the system is evacuated the oil quantity must be checked. Remove the filler plug and using the proper type dip stick, check the oil quantity. It should be maintained in the proper range using oil recommended by the manufacturer. After adding oil, replace the filler plug and recharge the system.

CABIN PRESSURIZATION OPERATIONAL CHECKS

Two operational checks can be performed on a cabin air conditioning and pressurization system. The first is a general operational check of the complete system, designed to ensure the proper operation of each major system component as well as the complete system. The second is a cabin pressurization check designed to check the cabin for airtightness.

To operationally check the air conditioning system, either operate the engines or provide the necessary ground support equipment recommended by the aircraft manufacturer.

With system controls positioned to provide cold air, ensure that cold air is flowing from the cabin distribution outlets. Position the system controls to provide heated air and check to see that there is an increase in the temperature of the airflow from the distribution outlets.

Checking the cabin pressurization system consists of the following: (1) A check of pressure regulator operation, (2) a check of pressure relief and dump valve operation, (3) a cabin static pressure test, and (4) a cabin dynamic pressure test.

To check the pressure regulator, connect an air test stand and a monometer (a gage for measuring pressure, usually in inches of Hg) to the appropriate test adapter fittings. With an external source of electrical power connected, position the system controls as required. Then pressurize the cabin to 7. 3 in. Hg, which is equivalent to 3.5 p.s.i. The pressurization settings and tolerances presented here are for illustrative purposes only. Consult the applicable maintenance manual for the settings for a particular make and model aircraft. Continue to pressurize the cabin, checking to see that the cabin pressure regulator maintains this pressure.

The complete check of this pressure relief and dump valve consists of three individual checks. First, with the air test stand connected to pressurize the cabin, position the cabin pressure selector switch to dump the cabin air. If cabin pressure decreases to less than 0.3 in. Hg (0.15 p.s.i.) through both the pressure relief and dump valves, the valves are dumping pressure properly. Second, using the air test stand, re-pressurize the cabin. Then position the manual dump valve to "dump." A lowering of the cabin pressure to 0.3 in. Hg (0.15 p.s.i.) and an airflow through the pressure relief and dump valve indicate that the manual dumping function of this valve is satisfactory. Third, position the master pressure regulator shutoff valve to "all off." (This position is used for ground testing only.) Then, using the air test stand, pressurize the cabin to 7.64 in. Hg (3.75 p.s.i.). Operation of the pressure relief and dump valves to maintain this pressure indicates that the relief function of the cabin pressure relief and dump valves is satisfactory.

The cabin static pressure test checks the fuselage for structural integrity. To perform this test, connect the air test stand and pressurize the fuselage to 10.20 in. Hg (5.0 p.s.i.). Check the aircraft skin exterior for cracks, distortion, bulging, and rivet condition.

Pressure checking the fuselage for air leakage is called a cabin pressure dynamic pressure test. This check consists of pressurizing the cabin to a specific pressure using an air test stand. Then with a monometer, determine the rate of air pressure leakage within a certain time limit specified in the aircraft maintenance manual. If leakage is excessive, large leaks can be located by sound or by feel. Small leaks can be detected using a bubble solution or a cabin leakage tester.

A careful observation of the fuselage exterior, prior to its being washed, may reveal small leaks around rivets, seams, or minute skin cracks. A telltale stain will be visible at the leak area.

CABIN PRESSURIZATION TROUBLESHOOTING

Troubleshooting consists of three steps: (1) Establishing the existence of trouble, (2) determining all possible causes of the trouble, and (3) identifying or isolating the specific cause of the trouble.

Troubleshooting charts are frequently provided in aircraft maintenance manuals for use in determining the cause, isolation procedure, and remedy for the more common malfunctions which cause the cabin air conditioning and pressurization systems to become inoperative or uncontrollable. These charts usually list the most common system failures. Troubleshooting charts are organized in a definite sequence under each trouble, according to the probability of failure and ease of investigation. To obtain maximum value, the following procedures are

Possible Cause	Isolation Procedure	Correction
(1) Trouble: Cabin temperature too high or too low (will not respond to control during "auto" operation).		
Defective temperature sensor.	Place system in manual operation and rotate air temperature control knob manually.	If system operates correctly, replace temperature sensor with one known to be operative and check system again in "auto" operation.
(2) Trouble: Cabin temperature too high or too low (will not respond to control during "auto" or manual operation).		
Defective temperature controller or refrigeration bypass valve inoperative.	With system being operated in manual position and the cabin air temperature control knob being cycled between "cold" and "hot" observe the valve position indicator (located on the valve).	If the valve is not opening and closing according to control settings, disconnect electrical plug from valve solenoid and check the power source. If the valve position indicates that the valve is opening and closing according to control settings, continue with the next troubleshooting item.

FIGURE 14–44. Troubleshooting an air cycle system.

586

recommended when applying a troubleshooting chart to system failures:

(1) Determine which trouble listed in the table most closely resembles the actual failure being experienced in the system.

(2) Eliminate the possible causes listed under the trouble selected, in the order in which they are listed, by performing the isolation procedure for each until the malfunction is discovered.

(3) Correct the malfunction by following the instructions listed in the correction column of the troubleshooting chart.

Figure 14–44 is an example of the type of troubleshooting chart provided in the maintenance manual for an aircraft that uses an air cycle system.

OXYGEN SYSTEMS GENERAL

The atmosphere is made up of about 21% oxygen, 78% nitrogen, and 1% other gases by volume. Of these gases, oxygen is the most important. As altitude increases, the air thins out and air pressure decreases. As a result, the amount of oxygen available to support life functions decreases.

Aircraft oxygen systems are provided to supply the required amount of oxygen to keep a sufficient concentration of oxygen in the lungs to permit normal activity up to indicated altitudes of about 40,000 ft.

Modern transport aircraft cruise at altitudes where cabin pressurization is necessary to maintain the cabin pressure altitude between 8,000 and 15,000 ft. regardless of the actual altitude of the aircraft. Under such conditions, oxygen is not needed for the comfort of the passengers and crew. However, as a precaution, oxygen equipment is installed for use if cabin pressurization fails. Portable oxygen equipment may also be aboard for first-aid purposes.

With some of the smaller and medium size aircraft designed without cabin pressurization, oxygen equipment may be installed for use by passengers and crew when the aircraft is flown at high altitudes. In other instances where there is no installed oxygen system, passengers and crew depend on portable oxygen equipment stowed in convenient positions.

The design of the various oxygen systems used in aircraft depends largely on the type of aircraft, its operational requirements, and, where applicable, the pressurization system. In some aircraft a continuous-flow oxygen system is installed for both passengers and crew. The pressure demand system is widely used as a crew system, especially on the larger transport aircraft. Many aircraft have a combination of both systems which may be augmented by portable equipment.

Continuous Flow System

In simple form a basic continuous-flow oxygen system is illustrated in figure 14–45. As shown in the illustration, with the line valve turned "on", oxygen will flow from the charged cylinder through the high-pressure line to the pressure-reducing valve, which reduces the pressure to that required at the mask outlets. A calibrated orifice in the outlets will control the amount of oxygen delivered to the mask.

The passenger system may consist of a series of plug-in supply sockets fitted to the cabin walls adjacent to the passenger seats to which oxygen masks can be connected, or it may be the "drop out" mask arrangement where individual masks are presented automatically to each passenger if pressurization fails. In both cases oxygen is supplied, often automatically, from a manifold. Any automatic control (e.g. barometric control valve) in the system can be overridden manually by a member of the crew.

Pressure-Demand System

A simple pressure-demand oxygen system is illustrated in figure 14–46. Note that there is a pressure-demand regulator for each crewmember, who can adjust the regulator according to his requirements.

PORTABLE OXYGEN EQUIPMENT

Typical portable oxygen equipment consists of a lightweight steel alloy oxygen cylinder fitted with a combined flow control/reducing valve and a pressure gage. A breathing mask, with connecting flexible tube and a carrying bag with the necessary straps for attachment to the wearer, completes the set.

The charged cylinder pressure is usually 1,800 p.s.i.; however, the cylinder capacities vary. A popular size for portable equipment is the 120-liter capacity cylinder.

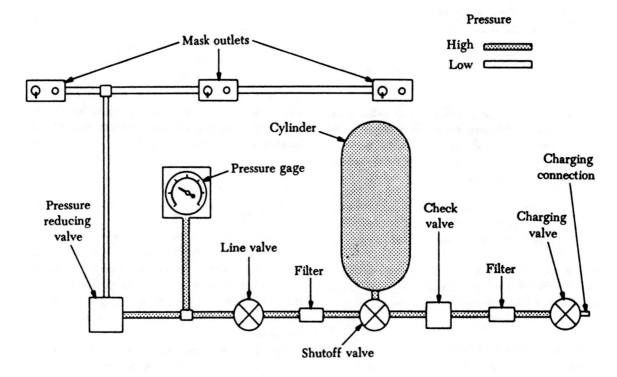

FIGURE 14–45. Continuous-flow oxygen system.

Depending on the type of equipment used, it is normally possible to select at least two rates of flow, normal or high. With some equipment three flow rate selections are possible, i.e., normal, high, and emergency, which would correspond to 2, 4, and 10 liters per minute. With these flow rates a 120-liter cylinder would last for 60, 30, and 12 min., respectively.

SMOKE PROTECTION EQUIPMENT

In some instances there is a requirement to carry smoke protection equipment for use by a member of the crew in a smoke or fume-laden atmosphere. Smoke protection equipment consists of a special smoke protection facial mask with eye protection in the form of a clear-vision visor, together with the necessary oxygen supply hose and head straps. Some are designed for use with oxygen from the aircraft oxygen system, and others are self-contained portable equipment.

OXYGEN CYLINDERS

The oxygen supply is contained in either high-pressure or low-pressure oxygen cylinders. The high-pressure cylinders are manufactured from heat-treated alloy, or are wire wrapped on the out-side surface, to provide resistance to shattering. All high-pressure cylinders are identified by their green color and have the words "AVIATORS' BREATH-ING OXYGEN" stenciled lengthwise in white, 1-in. letters.

High-pressure cylinders are manufactured in a variety of capacities and shapes. These cylinders can carry a maximum charge of 2,000 p.s.i., but are normally filled to a pressure of 1,800 to 1,850 p.s.i.

There are two basic types of low-pressure oxygen cylinders. One is made of stainless steel; the other, of heat-treated, low-alloy steel. Stainless steel cylinders are made nonshatterable by the addition of narrow stainless-steel bands that are seam-welded to the body of the cylinder. Low-alloy steel cylinders do not have the reinforcing bands but are subjected to a heat treatment process to make them nonshatterable. They have a smooth body with the word "NONSHATTERABLE" stenciled on them.

Both types of low-pressure cylinders come in different sizes and are painted light yellow. This color indicates that they are used for low-pressure oxygen only. The cylinders may carry a maximum charge of 450 p.s.i., but are normally filled to a pressure of

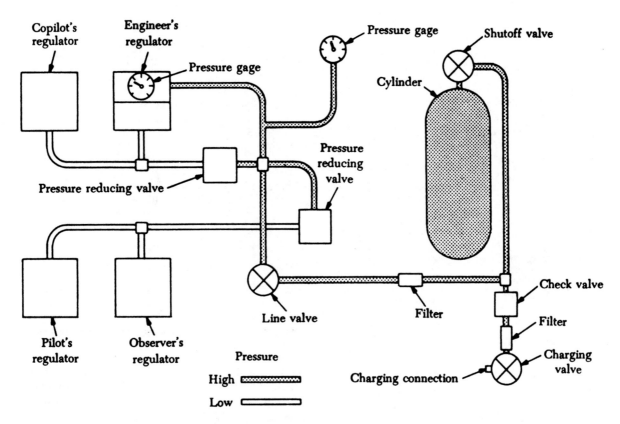

FIGURE 14–46. Typical pressure-demand oxygen system.

from 400 to 425 p.s.i. When the pressure drops to 50 p.s.i., the cylinders are considered empty.

The cylinders may be equipped with either of two types of valves. One type used is a self-opening valve which is automatically opened when the self-opening valve coupling assembly attached to the oxygen tubing is connected to the valve outlet. This coupling unseats a check valve, allowing oxygen from the cylinder to fill the oxygen system under high pressure. the other type is a hand-wheel, manually operated valve. This valve should be safety wired in the "full on" position when the cylinder is installed in the aircraft. This valve should be closed when removing or replacing parts of the oxygen system and when the cylinder is to be removed from the aircraft. Cylinders are often provided with a disk designed to rupture if cylinder pressure rises to an unsafe value. The disk is usually fitted in the valve body and vents the cylinder contents to the outside of the aircraft in the event of a dangerous pressure rise.

SOLID STATE OXYGEN SYSTEMS

Emergency supplemented oxygen is a necessity in any pressurized aircraft flying above 25,000 ft.

Chemical oxygen generators can be used to fulfill the new requirements. The chemical oxygen generator differs from the compressed oxygen cylinder and the liquid oxygen conveter in that the oxygen is actually produced at the time of delivery.

Solid-state oxygen generators have been in use for a number of years, dating back to 1920, when it was first used in mine rescues. During World War II the Japanese, British and Americans, all worked to develop oxygen generators for aircraft and submarines.

In figure 14–47, 120 standard cubic feet of oxygen (10 lbs.) is shown schematically in the number of cubic inches of space it would occupy as a gas, a liquid or a solid. In figure 14–48, the necessary hardware to install and operate the system has been included in the size and weight measurements. A close comparison of these values makes it apparent that the solid state oxygen generator system is the most efficient space wise. Likewise less equipment and maintenance is required for solid state oxygen converters. Integrity inspection is the only requirement until actual use is implemented.

Solid state describes the chemical source, sodium chlorate, formula $NaClO_3$. When heated to 478° F., sodium chlorate releases up to 45% of its weight as gaseous oxygen. The necessary heat for de-

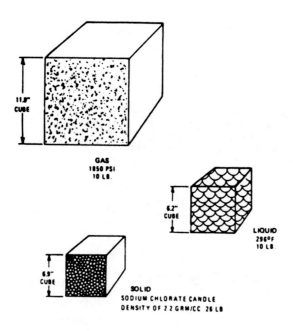

FIGURE 14–47. Volume comparison.

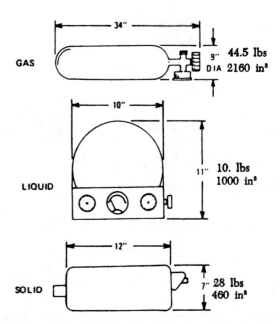

FIGURE 14–48. Weight and volume comparison—gas, liquid and solid oxygen storage.

composition of the sodium chlorate is supplied by iron which is mixed with the chlorate.

The Oxygen Generator

Figure 14–49 illustrates a schematic representation of a basic oxygen generator. The center axial position is occupied by a core of sodium chlorate, iron, and some other ingredients mixed together and either pressed or cast into a cylindrical shape. This item has been popularly referred to as an oxygen candle, because when it is ignited at one end, it burns progressively in much the same manner as a candle or flare. Surrounding the core is porous packing. It supports the core and filters salt particles from the gas as it flows toward the outlet. A chemical filter and particulate filter at the outlet end of the container provide final clean up of the gas so that the oxygen delivered is medically pure breathing oxygen. An initiation device is an integral part of the package. This may be either a mechanical percussion device or an electric squib. The choice depends on the application. The entire assembly is housed in a thin shelled vessel. Often included is a layer of thermal insulation on the inside shell, a check valve seal on the outlet, and a relief valve to protect against an inadvertent overpressure condition.

In operation, the burning is initiated at one end of the core by activating the squib or percussion device. Oxygen evolution rate is proportional to the cross sectional area of the core and the burn rate. The burn rate is determined by the concentration of fuel in the chlorate. In certain cases, one end of the core is larger than the other. The purpose of this is to program a high oxygen evolution rate during the initial few minutes of burning such as is required for an emergency descent supply. Burning continues until the core is expended.

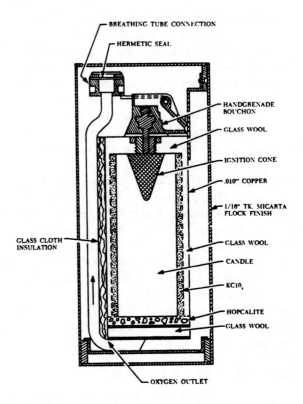

FIGURE 14–49. Apparatus for burning chlorate candles.

The simplicity of the process should be readily apparent; likewise, the limitations. There are no on/off valves and no mechanical controllers. Refill is accomplished by simply replacing, in total, the entire device. A limitation is that once the generator is initiated, flow is delivered at a predetermined rate, thus demand use is not very efficient.

In order to keep the process from consuming a great quantity of the oxygen, the quantity of iron is kept to a minimum. There is a tendency toward liberation of small amounts of chlorine. Barium peroxide, or barium dioxide, may be added by the manufacturer to provide an alkaline medium for removing the trace amounts of chlorine that may be present.

On a volume basis, which is extremely important in the aircraft installation, the storage capacity of oxygen in candles is about three times that of compressed gas.

A typical three outlet module for a 15 minute decompression emergency descent supply for a supersonic transport (25000 foot max. cabin altitude) weighs less than 0.9 pound and consists simply of a 2.1 inch diameter by 3.55 inch long stainless steel cylinder attached to three manifolded hose nipples. The cylinder contains the generator, initiator, salt, fume filter, enough insulation to keep the cylinder surface below 250° F. during burning, a pressure relief plug, and a temperature indicating paint spot for generator status visual inspection. The nipples contain orifices just small enough to assure essentially equal flow to all three masks.

The generators are inert below 400° F. even under severe impact. While reaction temperature is high and considerable heat is produced, the generators are insulated so that the outer surface of the cylinder is cool enough to avoid any fire hazard. The portable units may be held comfortably throughout the entire operation, as the heat generated is dissipated steadily over a long period of time. The same insulation works in reverse to delay initiation should a unit be subjected to an external fire. If such a fire is sufficiently prolonged to ignite the chlorate generator, oxygen production will be at a relatively low and continuous rate. In the simple continuous flow systems no pressure would be generated as all outlets would permit unrestricted flow of the oxygen, eliminating the intense jet torch effect of pressurized oxygen in fire.

Solid State vs. High Pressure Gaseous Oxygen

- Elimination of high pressure storage containers—saves weight.

- Elimination of distribution and regulation components—saves weight and maintenance.

- Simplification of individual distribution manifolds and drop-out mechanisms by the use of modular chlorate candle units.

- Improved reliability, hence safety by design of initiation circuitry, such that, an individual malfunction would not make other units inoperative (comparison here would be to ruptured lines, or high leakage in gaseous distribution systems).

- Simple, visual surveillance of each unit for condition of chlorate candle within the sealed container, by use of inspection window.

- Simple replacement of any unit, should it show any sign of deterioration, by plug-in cartridge, by relatively unskilled services crew; easily checked for installation and readiness for functioning from flight deck.

- Programmed oxygen release rates irrespective of the type of emergency.

OXYGEN PLUMBING

Tubing and fittings make up most of the oxygen system plumbing and connect the various components. All lines are metal except where flexibility is required. Where flexibility is needed, rubber hose is used.

There are several different sizes and types of oxygen tubing. The one most frequently used in low-pressure gaseous systems is made of aluminum alloy. Tubing made of this material resists corrosion and fatigue, is light in weight, and is easily formed. High-pressure gaseous supply lines are made from copper alloys.

Installed oxygen tubing is usually identified with color-coded tape applied to each end of the tubing, and at specified intervals along its length. The tape coding consists of a green band overprinted with the words "BREATHING OXYGEN" and a black rectangular symbol overprinted on a white background.

Oxygen System Fittings

Tube segments are interconnected or connected to system components by fittings. Tubing-to-tubing fittings are designed with straight threads to receive flared tube connections. Tubing-to-component (cylinder, regulator, and indicator) fittings have straight threads on the tubing end and external pipe threads (tapered) on the other end for attachment to the component, as shown in figure 14–50.

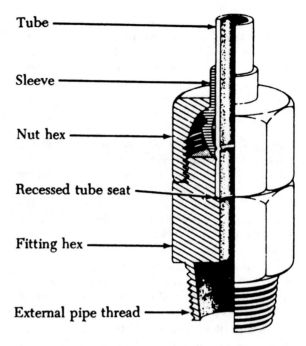

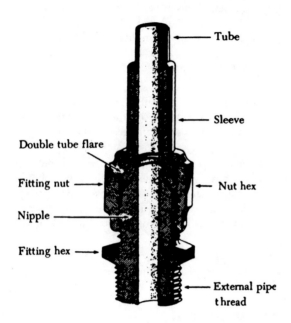

FIGURE 14–50. Sectional view of a typical oxygen system fitting.

FIGURE 14–51. A typical flareless fitting.

Oxygen system fittings may be made of aluminum alloy, steel, or brass. These fittings may be either of two types, flared or flareless. A typical flared fitting is shown in figure 14–50. A flareless fitting is shown in figure 14–51.

The sleeve in a flareless fitting must be preset before final installation in a flareless seat. Presetting causes the cutting edge of the sleeve to grip the tube sufficiently to form a seal between the sleeve and the tubing. The end of the tubing bottoms on the seat of the flareless fitting to provide tube end support after installation.

To seal oxygen system tapered pipe thread connections and to prevent thread seizure, use only an approved thread compound. Never use a mixture containing oil, grease, or any other hydrocarbon on any fittings used in oxygen systems.

Replacement Lines

The same cutting and bending methods described in Chapter 5, "Fluid Lines and Fittings," of AC 65–9A, Airframe and Powerplant Mechanics General Handbook, also apply to oxygen lines. As a general rule oxygen lines are double flared. The double flare makes the connection stronger and able to withstand more torque.

When installing a line, make sure there is proper clearance. The minimum clearance between oxygen plumbing and all moving parts should be 2 in. It is desirable to maintain a 6-in. clearance between oxygen tubing and electrical wires. When this is not possible, fasten all electrical wires securely with clips so that they cannot come to within 2 in. of the oxygen tubing.

OXYGEN VALVES

Five types of valves are commonly found in high-pressure gaseous oxygen systems. These are filler valves, check valves, shutoff valves, pressure reducer valves, and pressure relief valves. Low-pressure systems will normally contain only a filler valve and check valves.

592

Filler Valves

The oxygen system filler valve is located on most aircraft close to the edge of an access hatch or directly behind a cover plate in the skin. In either location, the valve is readily accessible for servicing. It is usually marked by a placard or a sign stenciled on the exterior, reading: "OXYGEN FILLER VALVE." There are two types of oxygen filler valves in use, a low-pressure filler valve and a high-pressure filler valve.

The low-pressure filler valve, figure 14–52, is used on systems equipped with low-pressure cylinders. When servicing a low-pressure oxygen system, push the recharging adapter into the filler valve casing. This unseats the filler valve and permits oxygen to flow from the servicing cart into the aircraft oxygen cylinders. The filler valve contains a spring-loaded locking device which holds the recharging adapter in place until it is released. When the adapter is removed from the filler valve, reverse flow of oxygen is automatically stopped by a check valve. A cap is provided to cover the filler opening and prevent contamination.

The high-pressure valve has a threaded fitting to receive the oxygen supply connector and a manual valve to control the flow of oxygen. To service an oxygen system that uses a high-pressure filler valve, screw the recharging adapter onto the aircraft filler valve. Open the manual valve on the filler valve and the servicing bottle. When recharging is completed, close the valves, remove the recharging adapter, and screw a valve cap on the valve to prevent contamination.

Check Valves

Check valves are installed in the lines between cylinders in all aircraft that have more than one storage cylinder. They are provided to prevent a reverse flow of oxygen or to prevent the loss of all oxygen as the result of a leak in one of the storage cylinders. Check valves permit the rapid flow of oxygen in only one direction. The direction of unrestricted flow is indicated by an arrow on the valves.

Of the two basic types of check valves commonly used, one type consists of a housing containing a spring-loaded ball. When pressure is applied to the

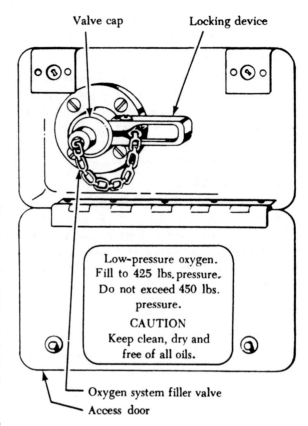

Valve cap   Locking device

Low-pressure oxygen.
Fill to 425 lbs. pressure.
Do not exceed 450 lbs.
pressure.
CAUTION
Keep clean, dry and
free of all oils.

Oxygen system filler valve
Access door

FIGURE 14–52. Low-pressure gaseous oxygen filler valve.

inlet side, the ball is forced against the spring, thus breaking the seal and allowing oxygen to flow. When pressure is equalized, the spring re-seats the ball, preventing any reverse flow of oxygen. The other type is a bell-mouthed hollow cylinder fitted with a captive ball in its bore. When pressure is applied at the bell-mouthed end (inlet), the ball will permit oxygen to flow. Any tendency of reverse flow causes the vall to move onto its seat, covering the inlet and preventing a reverse flow.

Shutoff Valves

Manually controlled two-position (on, off) shutoff valves are installed to control the flow of oxygen being emitted from a cylinder or a bank of cylinders. For normal operation, the knobs which control the valves are safetied in the "on" position. When necessary, such as changing a component, the appropriate valve can be closed. As a precaution, when opening a valve, the knob should be turned slowly to the "on" position. Otherwise, the sudden rush of highly pressurized oxygen into a depleted system·could rupture a line.

Pressure-Reducer Valves

In high-pressure oxygen systems, pressure-reducing valves are installed between the supply cylinders and the cockpit and cabin equipment. These valves reduce the high pressure of the oxygen supply cylinders down to approximately 300 to 400 p.s.i. required in the low-pressure part of the system.

Pressure-Relief Valves

A pressure-relief valve is incorporated in the main supply line of a high-pressure system. The relief valve prevents high-pressure oxygen from entering the system downstream of the pressure reducers if the reducer fails. The relief valve is vented to a blowout plug in the fuselage skin.

REGULATORS

Diluter-Demand Regulators

The diluter-demand regulator gets its name from the fact that it delivers oxygen to the user's lungs in response to the suction of his own breath. To prolong the duration of the oxygen supply, the oxygen is automatically diluted in the regulator with suitable amounts of atmospheric air. This dilution takes place at all altitudes below 34,000 ft.

The essential feature of a diluter-demand regulator is a diaphragm-operated valve called the demand valve (figure 14–53), which opens by slight suction on the diaphragm during inhalation and which closes during exhalation. A reducing valve upstream from the demand valve provides a controlled working pressure. Downstream from the demand valve is the diluter control closing mechanism. This consists of an aneroid assembly (a sealed, evacuated bellows) which controls the air inlet valve. When the diluter lever is set in the position marked "normal oxygen," atmospheric air at ground level is supplied with very little oxygen added. As altitude increases, the air inlet is gradually closed by the bellows to give a higher concen-

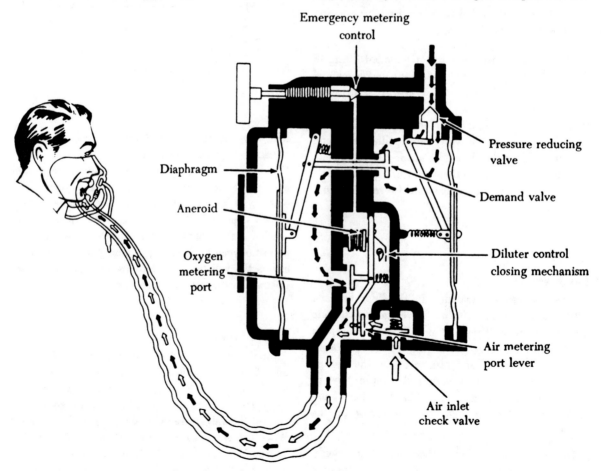

FIGURE 14–53. Schematic of a diluter-demand regulator.

594

tration of oxygen until at about 34,000 ft. the air inlet is completely closed and 100% oxygen is supplied. As altitude decreases, this process is reversed.

The diluter control as shown in figure 14–54, can be set by turning the lever to give 100% oxygen at any altitude. At moderate altitudes, however, this causes the oxygen supply to be consumed much more rapidly than normal. The diluter control should be set at "normal oxygen" for all routine operations. It can be set at "100 percent oxygen" for the following purposes: (1) Protection against exhaust gases or other poisonous or harmful gases in the aircraft, (2) to avoid the bends and chokes, and (3) to correct a feeling of lack of oxygen.

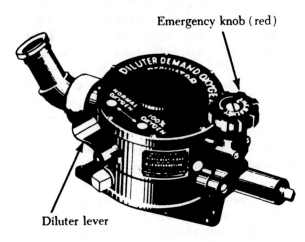

Figure 14–54. Diluter-demand regulator control.

The diluter-demand regulator is provided with an emergency valve, operated by a red knob (figure 14–54) on the front of the regulator. Opening this valve directs a steady stream of pure oxygen to the mask, regardless of altitude.

The following paragraphs illustrate a typical procedure for checking the operation of a diluter-demand regulator. First, check the oxygen system pressure gage, which should indicate between 425 and 450 p.s.i.; then check out the system using the following steps:

(1) Connect an oxygen mask to each diluter-demand regulator.

(2) Turn the auto-mix lever on the diluter-demand regulator to the "100 percent oxygen" position and listen carefully to make certain that no oxygen is escaping.

(3) Breathe oxygen normally from the mask. The oxygen flowmeter should blink once for each breath. (Figure 14–55 shows a

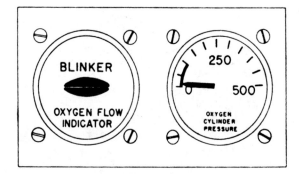

Figure 14–55. Flow indicator and pressure gage.

typical oxygen flowmeter and pressure gage.)

(4) With auto-mix lever in "100 percent oxygen" position, place the open end of the mask-to-regulator hose against the mouth and blow gently into the hose. Do not blow hard, as the relief valve in the regulator will vent. There should be positive and continued resistance, if not, the diaphragm or some part of the air-metering system may be leaking.

(5) Turn the auto-mix lever to "normal oxygen" position.

(6) Turn the emergency valve on the diluter-demand regulator to the "on" position for a few seconds. A steady flow of oxygen should result, ceasing when the emergency valve is turned off.

(7) Safety wire the emergency valve in "off" position with Federal Specification QQ–W–341, or equal, annealed copper wire, 0.0179-in. diameter.

Another type of diluter-demand regulator is the narrow panel type. This type regulator face (figure 14–56) displays a float-type flow indicator which signals oxygen flow through the regulator to the mask.

The regulator face also displays three manual control levers. A supply lever opens or closes the oxygen supply valve. An emergency lever is used to obtain oxygen under pressure. An oxygen selector lever is used for selecting an air/oxygen mixture or oxygen only.

Figure 14–57 illustrates how the narrow panel oxygen regulator operates. With the supply lever in the "on" position, the oxygen selection lever in the "normal" position, and the emergency lever in the "off" position, oxygen enters the regulator inlet.

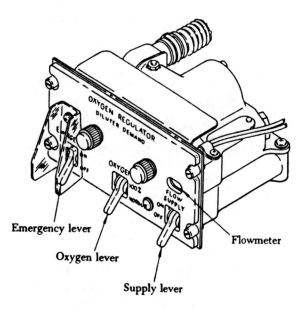

Figure 14–56. Typical narrow panel oxygen regulator.

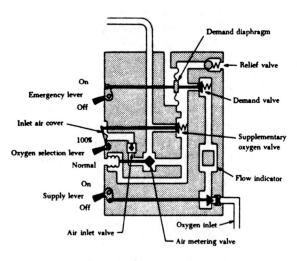

Figure 14–57. Schematic of a narrow panel.
oxygen regulator.

When there is sufficient differential pressure across the demand diaphragm, the demand valve opens to supply oxygen to the mask. This pressure differential exists during the user's inhalation cycle. After passing through the demand valve, the oxygen is mixed with air that enters through the air inlet port. The mixture ratio is determined by an aneroid-controlled air metering valve. A high oxygen ratio is provided at high altitudes and a high air ratio at lower altitudes. The air inlet valve is set to permit the airflow to begin at the same time as the oxygen flow.

The addition of air may be cut by turning the oxygen selection lever to "100%." When this lever is in "normal," air enters through the air inlet port, and the required amount is added to the oxygen to form the correct air/oxygen mixture.

Positive pressure at the regulator outlet may be obtained by turning the emergency lever to "on." This mechanically loads the demand diaphragm to provide positive outlet pressure.

Continuous-Flow Regulator

Continuous-flow regulators of the hand-adjustable and the automatic type are installed for the crew and passenger oxygen supply respectively.

The hand-adjustable, continuous-flow regulator delivers to the user's mask a continuous stream of oxygen at a rate that can be controlled. The system usually contains a pressure gage, a flow indicator, and a manual control knob for adjusting the oxygen flow. The pressure gage indicates the p.s.i. of oxygen in the cylinder. The flow indicator is calibrated in terms of altitude. The manual control knob adjusts the oxygen flow. The user adjusts the manual control knob until the altitude of the flow indicator corresponds to the cabin altimeter reading.

The automatic continuous-flow regulator is used in transport aircraft to supply oxygen automatically to each passenger when cabin pressure is equivalent to an altitude of approximately 15,000 ft. Operation of the system is initiated automatically by means of an electrically actuated device. The system can also be actuated electrically or manually should the automatic regulator malfunction.

Upon actuation, oxygen flows from the supply cylinders to the service units. A typical passenger service unit is shown in figure 14–58. During the

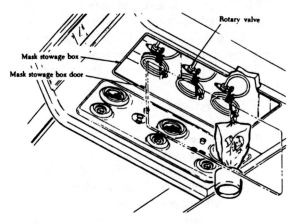

Figure 14–58. Typical passenger service unit.

first few seconds of oxygen flow, a pressure surge of 50 to 100 p.s.i. causes the oxygen mask box doors to open.

Each mask assembly then falls out and is suspended by the actuating attachment on the flexible tubing. The action of pulling the mask down to a usable position withdraws the outlet valve actuation pin, opening the rotary valve, allowing oxygen to flow to the mask.

OXYGEN SYSTEM FLOW INDICATORS

Flow indicators are used in oxygen systems to give visual indications that oxygen is flowing through the regulator. They do not show how much oxygen is flowing. Furthermore, their operation does not indicate that the user is getting enough oxygen.

In the blinker type indicator, figure 14–59, the eye opens and closes each time the user inhales and exhales. To check the flow indicator, set the diluter lever to "100% oxygen" and take several normal breaths from the mask-to-regulator hose. If the blinker opens and closes easily with each breath, it is in operating condition.

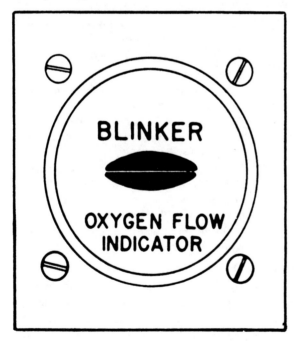

FIGURE 14–59. Oxygen flow indicator.

PRESSURE GAGES

Pressure gages are usually of the Bourdon tube type. Figure 14–60 shows the faces of two oxygen gages: (1) A low-pressure gage and (2) a high-pressure gage.

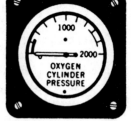

Low-pressure gage   High-pressure gage

FIGURE 14–60. Oxygen pressure gages.

Because of their connection into a system, the gages do not necessarily show the pressure in each cylinder. If the system contains only one supply cylinder, the pressure gage will indicate cylinder pressure. In systems where several cylinders are interconnected through check valves, the gage will indicate the pressure of the cylinder having the highest pressure.

Immediately after the system has been filled, pressure gage accuracy can be checked by comparing the aircraft pressure gage with the gage on the servicing cart. On low-pressure systems, the aircraft gage should read within 35 p.s.i. of the 425 p.s.i. servicing cart pressure. The same check can be used for high-pressure systems, but servicing pressure is 1,850 p.s.i. and a tolerance of 100 p.s.i. is allowed. The tolerances shown for pressure gage accuracy are typical and should not be construed as applying to all oxygen systems. Always consult the applicable aircraft maintenance manual for the tolerances of a particular system.

OXYGEN MASKS

There are numerous types of oxygen masks in use which vary widely in design detail. It would be impractical to discuss all of the types in this handbook. It is important that the masks used be compatible with the particular oxygen system involved. In general, crew masks are fitted to the user's face with a minimum of leakage. Crew masks usually contain a microphone. Most masks are the oronasal type, which covers only the mouth and nose.

Large transport aircraft are usually fitted with smoke masks for each crew position. The smoke masks are installed in stowage containers within easy grasp of the individual. These masks provide crew protection in an emergency and are not used frequently like the demand and continuous-flow masks. Smoke mask equipment consists of a full-face mask, a flexible breathing tube, and a coupling. The coupling connects to a demand regulator. A microphone is permanently installed in the mask.

Passenger masks, figure 14–61, may be simple, cup-shaped rubber moldings sufficiently flexible to obviate individual fitting. They may have a simple elastic head strap or they may be held to the face by the passenger.

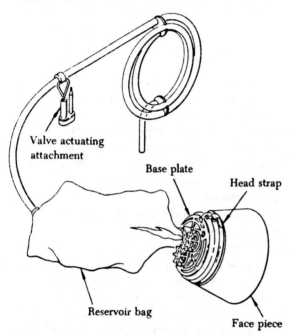

Valve actuating attachment

Base plate

Head strap

Reservoir bag

Face piece

FIGURE 14–61. Passenger oxygen mask.

All oxygen masks must be kept clean. This reduces the danger of infection and prolongs the life of the mask. To clean the mask, wash it with a mild soap and water solution and rinse it with clear water. If a microphone is installed, use a clean swab, instead of running water, to wipe off the soapy solution.

The mask must also be disinfected. A gauze pad, which has been soaked in a water solution of merthiolate can be used to swab out the mask. This solution should contain one-fifth teaspoon of merthiolate per quart of water. Wipe the mask with a clean cloth and air dry.

SERVICING GASEOUS OXYGEN SYSTEMS

The servicing procedures for a gaseous oxygen system depend upon the type of system. Before charging an aircraft system, consult the aircraft manufacturer's maintenance manual. Precautions such as purging the connecting hose before coupling to the aircraft filler valve, avoiding overheating caused by too rapid filling, opening cylinder

valves slowly, and checking pressures frequently during charging should be considered.

The type of oxygen to be used, the safety precautions, the equipment to be used, and the procedures for filling and testing the system must be observed.

Gaseous breathing oxygen used in aircraft is a special type of oxygen containing practically no water vapor and is at least 99.5% pure. While other types of oxygen (welder, hospital) may be pure enough, they usually contain water, which might freeze and block the oxygen system plumbing especially at high altitudes.

Gaseous breathing oxygen is generally supplied in 220- to 250-cu. ft. high-pressure cylinders. The cylinders are identified by their dark green color with a white band painted around the upper part of the cylinder. The words "OXYGEN AVIATORS' BREATHING" are also stenciled in white letters, lengthwise along the cylinders.

Oxygen Service Safety

Gaseous oxygen is dangerous and must be handled properly. It causes flammable materials to burn violently or even to explode. Listed below are several precautionary measures to follow:

(1) Tag all reparable cylinders that have leaky valves or plugs.

(2) Don't use gaseous oxygen to dust off clothing, etc.

(3) Keep oil and grease away from oxygen equipment.

(4) Don't service oxygen systems in a hangar because of the increased chances for fire.

(5) Valves of an oxygen system or cylinder should not be opened when a flame, electrical arc, or any other source of ignition is in the immediate area.

(6) Properly secure all oxygen cylinders when they are in use.

Gaseous Oxygen Servicing Trailers

Even though several types of servicing trailers are in use, each recharging system contains supply cylinders, various types of valves, and a manifold that connects the high-pressure cylinders to a purifier assembly. In the purifier assembly, moisture is removed from the oxygen. Coarse particles are trapped in the filter before reaching a reducing valve. The reducing valve has two gages which are used to monitor inlet and outlet pressures respectively. The reducing valve also has an adjusting screw for regulating the outlet pressure. This pressure is discharged into a flexible hose which connects to the charging valve and the adapter. The charging valve controls oxygen flowing away from the servicing trailer, and the adapter connects the recharging equipment to the aircraft filler valve.

On many aircraft a chart is located adjacent to the filler valve which shows the safe maximum charging pressure for the ambient temperature. This must be observed when charging the system.

It is common practice to have a warning placard cautioning against using oil or grease on the filler connections. Oxygen ground equipment should be maintained to a standard of cleanliness comparable to that of the aircraft system.

Leak Testing Gaseous Oxygen Systems

This test is performed at different times, depending on the inspection requirements for the particular type of aircraft. The system is allowed to cool, usually 1 hr., after filling before the pressures and temperatures are recorded. After several hours have elapsed, they are recorded again. Some manufacturers recommend a 6-hr. wait, others a 24-hr. wait. The recorded pressures are then corrected for any change in temperature since filling. Figure 14–62 is typical of the graphs provided in the aircraft main-

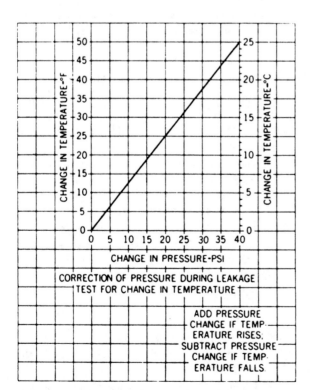

FIGURE 14-62. Pressure/temperature correction chart.

tenance manual to aid in making pressure/temperature corrections.

As an example for using the chart, assume that the oxygen system was recently charged. An hour later, the oxygen pressure gage read 425 p.s.i. at a temperature of 72° F. When the pressure gage was read 6 hrs. later, the pressure was 430 p.s.i. with a temperature of 79° F. By referring to figure 14–57, we see that the 7° temperature rise should have caused the pressure to increase 5 p.s.i., making the pressure gage read 430 p.s.i.

When oxygen is being lost from a system through leakage, the gage reading will be less than shown on the pressure/temperature correction chart. Leakage can often be detected by listening for the distinct hissing sound of escaping gas. If the leak cannot be located by listening, it will be necessary to soap-test all lines and connections with a castile soap and water solution or specially compounded leak-test material. To make this check, apply the test solution to areas suspected of leakage. Watch for bubbles. Make the solution thick enough to adhere to the contours of the fittings.

Any leak no matter how small must be found and repaired. A small leak may not cause trouble but if leak continues over a period of time, the surroundings and atmosphere become saturated. Such conditions are especially dangerous because personnel may not be aware that oxygen-enrichment exists. Oxygen-enriched conditions are almost always present in poorly ventilated areas. NO ATTEMPT SHOULD BE MADE TO TIGHTEN A LEAKING FITTING WHILE THE SYSTEM IS CHARGED.

Draining the Oxygen System

When it is necessary to drain the system, it can be done by inserting a filler adapter into the filler valve and opening the shutoff valves. Do not drain the system too rapidly as this will cause condensation within the system. An alternate method of draining the system is opening the emergency valve on the demand oxygen regulator. Perform this job in a well-ventilated area and observe all fire precautions.

Cleaning the Oxygen System

Always keep the external surfaces of the components of the oxygen system, such as lines, connections, and mounting brackets, clean and free of corrosion and contamination with oil or grease. As a cleaning agent, use anhydrous (waterless) ethyl alcohol, isopropyl alcohol (anti-icing fluid), or any other approved cleaner. If mask-to-regulator hoses are contaminated with oil or grease, the hoses should be replaced.

Cleaning Compound, Oxygen System

An approved cleaning formula for use on oxygen systems is available. This mixture of chlorinated, fluorinated hydrocarbons (freon) and isopropyl alcohol is safe for cleaning oxygen system components in aircraft, and for rinsing, flushing, and cleaning oxygen lines. Skin contact and prolonged inhalation of vapors should be avoided.

Purging the Oxygen System

An oxygen system needs to be purged if: (1) It has been depleted and not re-charged within 2 hrs., (2) if any line or component is replaced, requiring the draining or opening of the system for more than 2 hrs., or (3) it is suspected that the system has been contaminated.

The main cause of contamination in the system is moisture. Moisture in the system may be due to damp charging equipment. In very cold weather the small amount of moisture contained in breathing oxygen can cause contamination, due to repeated charging.

Although the introduction of moisture into the aircraft oxygen system can be considerably reduced by using the correct charging procedure, cumulative condensation in the system cannot be entirely avoided. There have been instances where oxygen systems, unused for long periods, have developed an unpleasant odor which necessitated purging to clear the system of moisture.

The procedure for purging may vary somewhat with each aircraft model. Generally speaking, on aircraft having the filler lines and the distribution lines commonly connected to one end of the storage cylinder, the system can be purged by filling the system with oxygen and then draining it at least three times. On aircraft that have the filler lines connected on one end of the cylinder and distribution lines connected to the opposite end of the cylinder, purge the system as follows: With all the regulator emergency palves open, pass oxygen at a pressure of 50 p.s.i. at the filler valve through the system for at least 30 min. Perform this job in a well-ventilated area and observe all fire precautions.

Dry nitrogen and/or dry air may also be used to purge oxygen systems. All open lines must be capped after use, also the system lines must be purged of the nitrogen by use of oxygen.

PREVENTION OF OXYGEN FIRES OR EXPLOSIONS

Many materials, particularly oils, grease, and non-metallic materials, are likely to burn when exposed to oxygen under pressure. To avoid fire or an explosion it is essential that all oxygen equipment be kept clean and free from oil or grease.

An oxygen fire or explosion depends on a combination of oxygen, a combustible material, and heat. The danger of ignition is in direct ratio to the concentration of oxygen, the combustible nature of the material exposed to the oxygen, and the temperature of the oxygen and material. Oxygen itself does not burn but it supports and intensifies a fire with any combustible material.

When working on an oxygen system it is essential that the warnings and precautions given in the aircraft maintenance manual be carefully observed. In general, before any work is attempted on an oxygen system the following fire precautions should be taken:

(1) Provide adequate fire-fighting equipment.

(2) Display "NO SMOKING" placards.

(3) Avoid checking aircraft radio or electrical systems.

(4) Keep all tools and oxygen servicing equipment free from oil or grease.

Oxygen System Inspection and Maintenance

Oxygen system inspection and maintenance should be accomplished according to these precautionary measures and any in addition to the manufacturer's instructions.

1. Never attempt maintenance until oxygen supply is turned off.

2. Fittings should be unscrewed slowly to allow residual pressure to dissipate.

3. Plug or cap all open lines immediately.

4. Do not use masking tape to seal openings; use caps or plugs designed for that purpose.

5. Maintain at least 2 inches clearance between oxygen lines and all moving equipment/parts within the aircraft to prevent the possibility of wearing oxygen lines.

6. Maintain at least 2 inches clearance between oxygen lines and all electrical wiring in the aircraft.

7. Provide adequate clearance between oxygen lines and all hot ducts, conduits and equip-

ment to prevent heating of the oxygen system.

8. Maintain at least 2 inches clearance between oxygen lines and all oil, fuel, hydraulic, or other fluid lines to prevent contamination.

9. Do not use lubricants unless specifically approved for oxygen system use.

10. A pressure and leak check must be performed each time the system is opened for maintenance.

Index

A

Absolute
  pressure, 543
  temperature, 543
Acceleration, 29,30
Access and inspection doors, 24
Accumulator, 316
  bladder type, 325
  diaphragm, 325
  maintenance type, 326
  piston type, 325
Acetone, 113
Acetylene gas and cylinders, 248
Acrylate, 214
Acrylic, 214
Acrylic nitrocellulose lacquer, 114,115
Acrylic nitro lacquer finishes, 144
  replacement of, 119
  touch up, 120
Actuating cylinders
  double-action actuating, 328,329
  single-action actuating, 328,329
ADF (automatic direction finder), 529
Adhesives, 205,208,209
Adiabatic, 543
Aerodynamic balance for control, 47
Aerodynamic center of horizontal tail, 38
Aerodynamic heating, 64
Aerodynamic high speed
  compressibility, 56
  expansion wave, 58,60-62
  hypersonic, 58
  mach number, 61,62
  oblique shock wave, 58
  shock wave, 59,62
  subsonic air flow, 59
  supersonic flow patterns, 56,58
  supersonic, 58,60-62
  transonic, 58,64
  vortex generator, 63
Aerodynamics, 27
Aileron, 3,4,11,19,36,40,41
  actuating horn, 19
  balance panel, 21
  control system, 42
  control tab, 21
  differential control, 42
  hinge, 11
  hinge pin fitting, 19
  hinge points, 19,21
  horn, 19
  inboard, 4,20
  outboard, 4,20
  spar, 11,19
  station, 6
  vent gap, 21
Airborne navigation equipment, 524
Airborne weather radar, 532
  diagram of system, 532
Air condition system (see Pressure system), 543-558
  air cycle system (see Air cycle), 543
  terms and definations, 543
  vapor cycle system (see Air cycle), 543
Aircraft lighting system (see Lightning system), 459-467
Aircraft steel structure welding (see Welding A/C), 276
Aircraft structures wooden (see A/C structure), 224
Aircraft Welding (see Welding), 276-283
Air cycle cooling system, 565-575,585,586
  cabin temperature control regulator, 574
  cabin temperature pick-up unit, 574
  cabin temperature selector, 574
  component operation, 568-574
  flow school, 566
  operation, 565-568

primary height exchanger, 568
primary height exchanger by-pass valve, 568,569
ram-air valve, 573
refrigeration by-pass valve, 570,571
refrigeration turbine, 572
refrigeration unit, 571-573
secondary height exchanger, 571
shutoff valve, 569,570
temperature control, electric cabin, 574,575
temperature control, regulator electronic, 575
thermistor, 574
water separators, 573
Air filter (see Vacuum system), 503
Airflow over wing section, 30
Airfoils, 30
  shape, 32
Air-oil separator, 502,503
Airspeed indicators, 479,480
Airspeed/mach indicators, 480,481
Airstream direct, detection, 488
Alcohol, 113
Alcohol de-icing windshield and carburetor system, 302
Alcohol, isopropyl, 600
Alternate source (see Static system), 475
Altimeter, 476-478
Altimeter, radio, 533
Altitude A/C (see Pressure), 543
Altitude, indicator, electric, 507,508
Aluminum washer, 453
Aluminum welding (see Oxyacetylene), 260,261
AM (Amplitude modulation), 521
Ambient
  pressure, 543
  temperature, 543
Angle of attack, 31
Angle of attack indicator, 488
  airstream direction detection, 488
  transmitter, 488
Angle of incidence, 32
Angular type piston pump, 322
Antenna, 520,521,529,530,533,538
  communication, 520,521
  distance-measuring-equipment, 521,528
  glidescope, 521,526
  loop, 529
  marker beacon, 521,527,529
  VHF, 521,523
Antenna system installation of, 537,538
  coaxial cable, 538
  impedance valve, 538
  maintenance procedures, 538
  mounting template, 537
  skin panel, 538
  transmission lines, 538
Anti-collision light, 461,462
Anti-drag wire, 11
Anti-skid system, 399
  fail safe protection, 400
  locked wheel skip control, 400
  normal skid control, 399,400
  pilot control, 400
  skid control box, 400
  skid control generator, 400
  skid control valves, 400
  touchdown protection, 400
Anti-tear strips, 93
Anti-icing system (see Thermal anti-icing system), 293
Anti-torque helicopter, 51
Arc welding
  gas shielded, 266,267
  metalic arc, 266,524,525
  metalic inert gas welding (MIG), 269
  tungsten inert gas welding (TIG), 267,268,269
Artificial horizon, 514

Assembly and rigging, 27-84
Atmosphere, 27
Atmosphere composition of, 539,540
  carbon dioxide, 540
  nitrogen, 539,540
  other gases, 539
  oxygen, 539,540
  ozone, 539,540
  water vapor, 539,540
Atmosphere pressure, 540
Attack, angle of, 31
Attitude, 36
Attitude indicator, 501,502,505,507,514
Autopilot system, 511-518
  annunciator system, 516,517
  aural warning, 517,518
  basic system, 512
  block diagram, 513
  command elements, 513
  components, 513
  computer or amplifier, 513
  controller, 513
  ILS, 515
  maintenance, 516
  output elements, 513-515
  principles of operation, 511
  sensing elements, 513
  servo's servomoters, 513,514
  transmitters, 513
  trim actuator, 514
  vor, 515,524
Automatic direction finder, 529
Autorotation, 54
Axis of aircraft, 35,36
  lateral (pitch), 35,36,45,55
  longitudinal (roll), 35,36,55
  vertical (yaw), 35,36,41,43,45,55
Axis of flight helicopter, 55

B

Balance pressure torch, 249
Balance, principals of
  balance set-up, 82-84
  dynamic balance, 83,84
  moveable surface, 82
  re-balance procedures, 83
  re-balancing, 80,81
  static, 82
Balance tab, 20
Bar folder, 141
Bayonet thermocouple, 494
Beads, 159
Bead welds (see Weld procedures), 271-273
Bearings, 376,377
Bellcrank, 44
Bench plate (see Sheet metal tools), 133
Bend allowance, 146
Bend allowance chart, 149
Bending, 1,2,132
Bends, making straight line
  bend allowance, 146
  bend allowance chart, 149
  brake line, 149
  duplication of pattern, 153,154
  layouts making, 151-154
  lighten holes, 154,155
  locating bend line, 151
  minimum bend radii, 151
  neutral axis, 147
  relief holes, 153,154,159
  set back, 148,149
  setback (K) chart, 150
  sight line, 149
  term, 149-151

Benzene, 113

Bernoulli's principle, 29

Bias, 85,88

Blade flapping helicopter, 52

Bleeding brakes, 372,373

Bleeding shock struts, 347

Blinker oxygen flow indicator, 597

Bogie gear (Landing gear) 341,342

Bolted sleeve repair (see Welding A/C steel structure), 277

Bonding and grounding, 452,455
  aluminum jumper connection to tubular structure, 453,454
  bonding conduit to structure, 453
  copper jumper connection to tubular structure, 453
  general bonding and grounding procedures, 45
  hardware combinations used in making bond, 453
  nut plate bonding and grounding to a flat surface, 454
  stud bonding or ground to flat surface, 454
  testing, 455

Bonding (Electric), 536

Boundary layer control devices, 48,49

Brace struts repair (see Welding A/C structure), 283

Brake assemblies
  dual-disk, 368
  expander tube, 371,372
  multiple-disk, 368,369
  segmented rotor, 369-371
  single disk, 366-368

Brake cornice, 142,143

Brake inspection and maintenance
  gravity method of bleeding brakes, 372,373
  pressure method of bleeding brakes, 373

Brake line, 149

Brake shuttle valve, 337

Brake system
  debooster cylinders, 364,365
  independent, 360-362
  master brake cylinder, 361
  nose wheel brakes, 366
  power boost, 365,366
  power brake control valve, 362
  pressure ball brake control valve, 363
  sliding spool type power brake control valve, 363,364
  typical system, 360
  Warner master brake cylinder, 362

Brakes, speed, 40

Brazing methods, 264

Break, determining length of, 194

Brinelling, 130

Bucking, 173

Bucking bars, 167,173

Bulkhead, 4,17

Bulkhead repair, 194

Bumping, 145,158-161

Burnishing, 130

Burr, 130

Butt joints, 255,273

Butt line, 6

Butt welds (see Weld procedures), 273,275,276

Buttock line, 6

Butyl-alcohol (Dope retarder), 110

C

Cabin atmosphere control system, 539-601
  air condition and pressure system, 543
  air condition system, 558
  air condition, maintenance of, 583
  air cycle cooling system, 565
  air cycle system component operation, 568
  air distribution, 556
  atmosphere composition of, 539
  basic requirements, 545
  cabin pressure control system, 551
  cabin pressure sources of, 545

combustion heaters, 561
  cooling system, 565
  electronic cabin temperature control system, 574
  electronic temperature control regulator, 575
  freon system component, 577
  heating system, 559
  maintenance of combustion heat system, 564
  pressurization, 541
  pressurization cabin operation check, 585
  pressurization cabin troubleshooting, 586
  pressurization gages, 597
  pressurization maintenance of, 583
  pressurization valves, 550
  regulator, 594
  servicing gages oxygen system, 59
  smoke protection equipment, 588
  solid-state oxygen system, 587
  supercharger instruments, 549
  typical system, description of, 580
  vapor cycle system (Freon), 576

Cabin-cockpit interior fire protection, 429

Cable assembly, 64

Cable (see Wire and cable), 433

Cadmium plated
  locknut, 453
  lockwasher, 453
  screws, 453

Calendering, 85

Cam-type pump, 322

Capacitor-type fuel quantity, 485-487
  indicator and probe, 486
  self-balancing bridge circuit, 487
  simplified bridge circuit, 486
  tank circuit, 486

Carbon dioxide, 540

Carburetor alcohol de-icing system, 302

Cases, instruments, 469

Category II, 527

CDI (Course deviation indicator), 526,527

Cellulose acetate base, 221

Cellulose nitrate dope, 113

Cellulose plastic, 213

Cementing plastic, 217,218

Center of gravity, 33,38

Center of pressure, 31

Centering nose wheel, 353,354

Centigrade, 545

Chamfering, 192

Chattering, 130

Check valves, 316,326,327,335,503,504

Chemical stability of hydraulic fluid, 310

Cherry and cherrylock (see Self-plugging), 182

Chlorinated, flourinated hydrocarbons, (Freon), 600

Chord line, 32

Chrome molybdenum welding, 259

Circuit breakers (see Circuit protection devices), 458

Circuit protection devices, 458

Clamps, 453

Cleaning solvents, 205

Cleaning (Honeycomb), 209

Cleco fastener, 134

Coaxial cable, 538

Cockpit-cabin interior fire protection, 429
  CO_2, 429
  dry chemical, 429,430
  extinguisher, types, 429
  freons, 430
  halon 1301 ,430
  unsuitable fire extinguisher, 430

Collective pitch control, 56

Color, 116

Combustion heaters, 294,561
  air system, 563
  fuel system, 562
  ignition plug, 563
  ignition system, 563
  inspection, 564,565
  maintenance of, 564,565
  nozzle assembly, 562

schematic, 561
  vapor wick, 562
  ventilating air, 564

Communication and navigation equipment
  installation of, 534-536
  bonding jumper, 536
  cooling & moisture, 535
  interference, radio reduction, 536
  shock mount base, 535
  type installation in panel, 535
  vibration isolation, 535

Compass magnetic, 507-511

Compass rose, 509

Compensator, 355,356

Compressibility (see Aerodynamic high-speed), 56

Compression, 1,2,132

Conduit, 457

Coning, 54

Connectors (Electrical), 455
  AN connector marking, 456
  AN 3100 wall receptacle, 456
  AN 3101 cable receptacle, 456
  AN 3102 box receptacle, 456
  AN 3106 angle plug, 456
  AN 3106 straight plug, 456
  AN 3107 MCK disconnect plug, 456
  AN 3108 angle plug, 456
  identification, 455,456
  installation of, 456,457
  types of, 455

Constant delivery pump, 318

Continuous-flow oxygen system, 587,588

Contraction and expansion of metal (Weld), 257

Control, 35,39
  lateral axis, 44,45
  longintudinal axis, 32,41
  vertical axis, 44

Control surface, adjustment of, 75

Control system helicopter, 57

Control tab, 20-23

Control valve (Pneumatic system), 334,335

Cooling system, 565-575

Copper terminal (Electrical), 453

Cord rib lacing, 89

Core material (Honeycomb), 205,206,209,210

Corner joints, 256,257

Cornice brake, 142,143

Correction card, compass, 511

Corrosion, 130

Corrosion preventives (Honeycomb), 206

Cotton fabric, 85

Count, 85,86

Countersinking, 170

Countersinks, 167

Course deviation indicator (CDI), 526,527

Covering fabric fuselage, 95

Covering (see Fabric covering, wings), 93

Cowling, 14-17

CO_2 cylinder installation, 417

CO_2 poisoning, human reaction, 432

Crack, 130

Cracks (Welding), 256

Crimping, 145

Cut, 130

Cutters, rivet, 166,167

Cutting plastic, 216

Cutting welding, 263

Cyclic, 55

Cylinders oxygen (Breathing), 588,589

D

Dacron, 88

Damage honeycomb
  causes, 201

evaluation, 201
inspection, 201
removal, 206,208
Damage inspection and classification of
brinelling, 130
burnishing, 130
burr, 319,130
chattering, 130
classification of, 131
corrosion, 130
crack, 130
cut, 130
definition of defects, 130
dent, 130
erosion, 130
galling, 130
gouge, 130
inclusion, 131
inspection of, 130
necessitating replacement of parts, 131
negligible, 131
nick, 131
pitting, 131
repairable by insertion, 131
repairable by patching, 131
score, 131
scratch, 131
stain, 131
upsetting, 131
Dampers shimmy (see Shimmy dampers), 356
D.C. selsyn instrument, 483,484
Debooster cylinder, 364,365
Decalcomanias, 125,126
Decals
metal with cellophane backing, 125
metal with no adhesive, 126
metal with paper backing, 126
paper, 125
removal, 126
vinyl film, 126
Defrost system for windshield, 302
De-icing of A/C on ground, 299
frost removal, 299
removing ice and snow deposits, 299
De-icing (see Pneumatic or thermal system), 292,293
Delta wing, 7
Density, 28
Dent, 130
Detection system, smoke (see Smoke detection system), 430
Deterioration of fabric (see Fabric deterioration), 104
Deutsch, 187
De-icing gages, 472,473
Deviation, 508
D.G. (Directional gyro), 504,514
Diagonal web member, 4
Dials, instruments, 469,470
Diaphragm-type gages, 473
Dihedral, 8,74
Dihedral angle, 39
Dihedral to lateral stability, 39
Dill lok-skrus, 186
Dimpling, 170
coin, 172
dies, 167
radius, 172
thermo, 171
Directional gyro, 504,514
Disk brakes (see Brake assemblies), 366
Dissymmetry of lift helicopter, 51
Distance-measuring-equipment, 528,529,538
antenna, 528
indicator, 528
type control, 528
vortac, 528
DME (distance-measuring-equipment), 528,529,538
Dollies, 133
Dope, 114

Dope materials, 108
aluminum-pigmented, 104
cellulose-acetate-butyrate, 108
cellulose-nitrate, 108
Dope proofing, 91
Dopes-doping, 107
applying reinforced patches, 110
applying surface tape, 110
blushing, 110
brittleness, 110
bubbles and blisters, 109
cold effects, 109
common troubles, 109
fungicidal dopes, 111
humidity effects, 109
inconsistent coloring, 110
installation of drain grommets, 110
number of coats, 11
peeling, 100,110
pinholes, 110
runs and sags, 110
slack panels, 109
techniques of application, 110
temperature effects, 109
Doppler navigation system, 530,531
doppler effect, 531
Double acting hand pump, 318
Double action actuating cylinder, 328,329
Drag, 33,34
induces, 35
parasite, 35
profile, 35
resultant, 34
Drag link (Brace), 342,348
Drag wire, 11
Drains for fabric, 95,110
Draw sets, rivet, 167
Drills, 136
air 360 degrees, 139
right-angle, 139
right-angle electric, 139
straight air, 139
straight electric, 139
Drill sizes pilot and reaming, 169
Drilling, 169
Drilling plastic, 216
Driving rivets, 173
Dual disk brake, 368
Dual main (Landing gear), 341

E

Edge joints, 256
E.G.T., exhaust gas temperature, 494-497
Electric arc welding (see Arc welding), 266
Electric altitude indicator, 507
Electrical landing gear retraction system, 349
Electrical systems
bonding (see Bonding and grounding), 452-455
conduit, 457
connectors (see Connectors), 455-457
connecting terminal lugs to terminal blocks, 451,452
emergency splicing repairs, 450,451
equipment installation, 457-459
general, 433
grounding (see Bonding and grounding), 452-455
lacing and tying wire bundles, 444,446
lighting system maintenance and inspection, 464-467
lighting system, 459-464
wire and cable cutting (see Wire-cable), 446-450
Electrical equipment installation, 457-459
circuit protection devices, 458
load limits, 457,458
load limits, controlling and monitoring, 458
relays, 459
switch derating factors, 459
switches, 459
wire and circuit protection chart, 458
Elevator, 3,4,20

Elevator action (Movement), 44
Elevator control tab, 20
Elongated octagonal patch, 190,191
ELT, Emergency locator transmitter, 533,534
batteries, 534
false alarms, 534
test equipment, 534
testing, 534
transmitter, 534
Emergency splicing repairs, 450,451
Emergency extension system
cut-away view of centering cam, 354
gear indicators, 353
ground locks, 351,353
landing gear safety devices, 351
nose wheel centering, 353-355
safety switch, 351
Empennage, 3,4,16
Enamel, 115
Enamel finishes, 121,122
Engine alignment, 75
Engine gages, 471
Engine mount repair (see Welding A/C steel structure), 279,280
Engine synchroscope, 491,492
Epoxy finishes, 120,121
Epoxy remover, 116
Erosion, 130
Erosion preventives (Honeycomb), 206
Exhaust gas temperature (EGT), 494-497
Exhaust heaters, 294,296
Expander tube (Brake), 371,372
Expansion and contraction of metal (Welding), 257
Expansion wave (see Aerodynamic high-speed), 58
Expansion, contraction allowances (Plastics), 220
Extinguishing agent characteristics
bromochlorodifluoromethane, 413
bromochloromethane, 413
carbon dioxide, 413
characteristics chart, 416
dibromodifluoromethane, 413
extinguishing agent comparison, 413
halogenated hydrocarbons, 412
inert cold gas agents, 415
methyl bromide, 413
nitrogen, 413

F

Fabric covering
aluminum pigmented dopes, 109
application techniques of, 110
causes of fabric deterioration, 104
checking condition of doped fabric, 106
common trouble in dope application, 109
covering fuselages, 95
covering wings, 93
dope materials, 108
dopes and doping, 107
fabrics, 85
miscellaneous textile materials, 88
number of coats, 111
recover-aircraft surface with glass cloth, 104
repair of fabric covers, 99
replacing panels in wing covers, 103
seams, 89
strength criteria for fabric, 107
temperature and humidity effects on dope, 109
testing fabric covering, 106
ventilation, drain, inspection openings, 95
Fabric covering, applying
chafe points, 91
dope proofing, 91
general, 91
glass cloth, 104
inter-rib bracing, 91
practices, 92
preparation of structure, 91
preparing plywood surfaces for doping, 92
taping, 93

Fabric covering fuselage, 95
  lacing, 95
Fabric covering wings
  anti-tear strips, 93
  double loop wing lacing, 95,98
  glass cloth, 104
  grommets, 99
  replacing panels, 103
  rib stitch spacing chart, 94,96-98
  single loop wing lacing, 94,96-98
  splice knot, 95
  tie off knots, 95
Fabric covers glass cloth, 104
Fabric covers repair
  doped-in panel, 103
  general, 99
  insewed (Doped-on), 102
  methods, 104
  replacing panels, 103
  sewed-in panel, 101,102
  sewed patch, 100,101
  tears, 100
  trailing edge, 102
Fabric deterioration, 104
  acid dope and thinners, 105
  checking, 106
  insufficient dope film, 105
  mildew, 105
  punch tester, 106
  storage conditions, 105
  strength criteria, 107
  tensile tester, 106,107
  testing, 106,107
Fabrics
  bias, 85,88
  calendering, 85
  cotton, 85
  count, 85,86
  dacron, 88
  filling, 85
  linen, 87
  mercerization, 85
  pinked edge, 85,88
  ply, 85
  quality, 85
  reinforcing tape, 89
  selvage edge, 85,88
  sizing, 85
  special fasteners, 89
  strength requrements, 85
  surface tape, 88
  tearing strength, 86
  tensile strength, 86
  warp, 85,88
  warp ends, 85
  weft, 85
  woof, 85,88
Fahrenheit, 545
Failures rivets (see Rivet failure), 175
Fairings, 24
Fairleads solid split, 66
FCC (Federal Communication Commission), 521,523
Ferrous metals oxyacetylene welding of, 258
Fiberglass cloth overlay (Honeycomb), 208,210
Fiberglass components, 221,222
  mat molded parts, 222
  repair procedures, 222,223
Fiberglass fabrics, 206
Fibestos, 214
Fillet flap, 20
Filling, 85
Filter, 558
Filters hydraulic, 313,314,316
  maintenance of, 314,315
  micronic type, 314,336
  schematic of micronic filter, 314,336
Filters pneumatic system, 335
  micronic, 336
  screen type, 337
Filter-air (Vacuum system), 503
Finishes applying
  brushing, 122
  dipping, 122
  spray painting, 122,123
  spray patterns at different settings, 123

Finishing material, storage of, 117
Fire detection system, 408
  connector joint, 424
  continuous element system, 411
  continuous loop system, 410
  fenwal sensing element, 410
  fenwal spot detection, 410
  fenwal spot detection circuit, 409
  fire switches, 423
  kiddle sensing element, 410
  Lindberg System School, 411
  loop clamp, 425
  maintenance, 423,425
  methods, 407
  overheat warning system, 411
  requirements, 407
  rubbing interference, 424
  sensing element defects, 424
  thermal switch circuit, 409
  thermal switch system, 408
  thermocouple circuit, 409
  thermocouple system, 408
  troubleshooting, 425
Fire detectors types of
  carbon monoxide, 407
  combustible mixture, 407
  fiber optic, 407
  maintenance, 423,425
  observation, 407
  overheat, 407
  radiation sensing, 407
  rate of temperature rise, 407
  smoke, 407
  troubleshooting, 425
Fire detection system maintenance practices, 423
Fire detection system troubleshooting, 425
Fire extinguishing agents (see Extinguishing agent characteristics), 412
Fire extinguishing system, 417
  conventional, 417
  high rate discharge, 417
Fire extinguishing system maintenance practice, 425
  bonnet sphere assembly, 427
  carbon dioxide cycle, 428
  container pressure check, 426
  container pressure/temperature curve, 426
  CO_2 cylinder construction, 428
  freon containers, 427
  freon discharge cartridges, 426
Fire point, 310
Fire precautions honeycomb, 204
Fire prevention and protection, 429
Fire protection systems, 407
  cockpit-cabin interiors, 429
  extinguishing agents characteristics, 412
  fire detection system, 408
  fire extinguishing system, 417
  fires, types of, 411
  fire zone classifications, 412
  general, 407
  maintenance practices, 425
  prevention and protection, 429
  reciprocating engine conventional CO_2 system, 417
  smoke detection system, 430
  troubleshooting, 425
  turbine engine extinguishing system, 420
  turbine engine ground fire protection, 422
  turbojet protection system, 419
Fires, types of, 411
  class A, 411
  class B, 411
  class C, 411
Fire zone classification
  class A zone, 412
  class B zone, 412
  class C zone, 412
  class D zone, 412
  class X zone, 412
Fittings (oxy), 591,592
Fixed wing A/C, 2
Flame adjustment (welding), 252
  neutral, 252
  oxidizing, 252
  reducing, 252
Flange wheels fixed and removable, 375,376

Flanging form blocks, 155
Flaps, 4
  aft flap, 22
  augmentation, 49
  fillet, 20
  fore flap, 22
  fowler, 22,47,48
  inboard, 20
  leading edge, 20,22,23,40
  mid flap, 22
  outboard, 20
  plain, 22,47,48
  slotted flaps, 47,48
  split, 22,47,48
  triple slotted flaps, 20
Flap station, 6
Flash point, 310
Flat position (see Weld procedures), 271
Flexible vinyl-covered static wicks, 537
Flight control surfaces, 18-23
  aerodynamic balance, 47
  aileron balance panel, 21
  aileron hinge, 19
  aileron location on wing tips, 18
  aileron rib end view, 19
  ailerons, 18,40
  auxiliary group, 40
  auxiliary wing flight surfaces, 21
  boundary layer control devices, 48,49
  elevator, 40
  fowler flap, 22
  large turbo jet aircraft, 20
  leading edge flaps, 20,22,23
  plain flap, 22
  primary group, 40
  rudder, 3,4,20,36,40
  rudder control tab location of, 23
  secondary group, 40
  split flap, 22
  supersonic flight, 64
  tabs, 23
  triple slotted flaps, 22
  wing flaps, 22
Flight control systems, 64
  bellcrank, 67
  cable assembly, 64
  cable connectors, 65
  cable drum, 67
  cable guides, 66
  gust lock, 66
  hardware, 64
  hydraulic operated control system, 65
  manual control, 66
  mechanical linkage, 64
  mechanism, 64
  push-pull rod, 67
  rig pin hole, 67
  solid fairleads, 66
  split fairleads, 66
  spring connector, 65
  torque tube, 67
  turnbuckles, 65
Flight theory of, 27
Flowmeter fuel system, 497-499
Fluorescent finishes, 121
Fluorescent paint remover, 116
Flush patch (wooden), 235
FM (frequency modulation), 521
F = ma, 30
Folded fell, 90
Folding, 146
Force, 30
Force equals mass acceleration (F = ma), 30
Forces acting in flight, 33
Form block flanging, 155
Form blocks hardwood, 133,158,159,161
Former repair, 194
Formers, 4
Forming of a weld, 258
Forming operations and terms
  bumping, 145,158-161
  crimping, 145
  folding, 146

606

processes, 144,156-164
shrinking, 145-147,158
stretching, 145,147,157,158
Forming plastic, 215
Four-way closed center selector valve, 330,331
Fowler flap, 22
Frame, 4,17
French fell, 90
Freon, 576,577,579,580,583,600
Freon system (see Vapor cycle), 12, 582
Frequency (radio), 520
Friction lock rivets, (see self-plugging), 178,179
Friese aerodynamic balance, 47
Fuel flowmeter system, 497-499
   indicator, 499
   turbine type, 499
   vane-type, 498
Fuel press remote indicator, 485
Fuselage, 2,34
Fuselage fabric covering, 95
Fuselage fittings repair (see Welding A/C system), 280,281
Fuselage stations, 28
Fuses (see Circuit protection devices), 458

G

Gage, pressure, 316,32
Galling, 130
Gas shielded welding (see Arc welding), 266,267
Gasket thermocouple, 494
Gate check valve, 503
Gear indicators, 353
Gear type pump, 319
Generator (oxy), 590
Gerotor type pump, 319
Glass cloth, 104
Glass cloth overlay (Honeycomb), 208,210
Glass fabrics, 206
Glide slope information, 526
Glueing, 232
   glued joints inspection, 225,226
   glued joints testing, 233
Glues, 230,231
Goggles (welding), 248,251
Gouge, 130
Gravity, 33,34
Grinders, 140
Grinders wheels, 140,141
Grip, 165
Grommets, 110
   metal, 99
   plastic, 99
   streamlined plastic, 99
Groove welds (see Weld procedures), 273,275,276
Ground effect helicopter, 54
Ground locks, 351,353
Gull wing, 8
Gun rivet (see Pneumatic rivet guns), 168
Gusset, 4
Gust lock, 66
Gyroscopic precession principle, 52
Gyroscopic instruments and system, 499-507
   air filter, 503
   air-oil separator, 502,503
   angular velocity, 500
   attitude gyros, 505
   attitude indicator, 501,502
   bearing friction, 500
   check valve, 504
   erecting mechanism, 505
   gate check valve, 502,503
   heading indicator, 501
   horizon indicator, 507
   inertia, 500

maintenance of system, 506
multi-engine, 504
needle valve, 504
operation of typical system, 504
pendulous vanes, 505
power sources, 501
precession, 500
pressure gage, 502
pressure operation gyros, 506
pressure relief valve, 502-504
pump, engine driven, 502
pump, typical, 502
radius, 500
regulator valve, 503
restrictor valve, 503
selector valve, 502-504
suction, 503
suction gage, 503
suction relief valve, 502-504
troubleshooting chart, 506
vacuum system, 501
venturi-tube system, 501
weight, 500

H

Hand driving rivets, 173
Hand forming, 155
   bumping, 160,161
   concave curve, 159
   convex curve, 160
   curved flanged parts, 158
   extruded angles, 156
   flaged angles, 157
   joggling, 162,163
   magnesium, 163
   nose ribs, 159
   stainless steel, 163
   straight line bends, 155
   stretching, 157
Hand sets, rivets, 167
Handley page aerodynamic balance, 47
Head pitot, 475
Heading indicator, 501,502,514
Heaters toilet drain, 303,304
Heat exchangers, 556
Heat lamps infared, 204
Heat treatment effect of (plastic), 218
Heating electric (pitot head), 475,476
Heating system, 559
   compressed air, 561
   electric heater, 561
   exhaust gas heater exchanger, 559,561
   exhaust heater system diagram, 560
   gasoline combustion heater, 559,560
   radiant panels, 561
   re-cycling compressed air, 559
Helicopter, 24
   anti-torque rotor, 51
   autorotation, 54
   axis of flight, 55
   blade flapping, 52,53
   collective pitch control, 56
   coning, 54
   control system, 57
   cyclic pitch, 55
   differential tilt of rotor thrusts, 51
   differential torque, 51
   dissymmetry of lift, 51
   ground effect, 54
   gyroscopic precession, 52
   location of major components, 25
   principle, 52
   rotor disk, 52
   structural components, 25
   structures, 24
   torque, 50
Helicopter controls, function of
   collective pitch stick, 77
   cyclic, 77
   pedals, 77
   throttle, 77
Helicopter rigging, 77
Helicopter structures, 24

Helicopter tachometer and synchroscope, 490
Helicopter tracking, 78,80
Helium, 27
Hertz (Hz), 525
H/F (high frequency), 520,522-524
   system diagram, 523
Hg mercury, 473
Hi-shear, 187
   inspection, 189
   removal, 188,190
High frequency (H/F), 520,522-524
   system diagram, 523
High lift devices, 47,48
High wing, 8
Hinge line, 23
Holding devices (Sheet metal tools), 134
Hole duplication, 166
Honeycomb, 14,15
   aileron, 14
   aileron tab, 14
   constant-thickness core, 14
   inboard flap, 14
   spoiler sandwich panels, 14
   tapered core, cord wedge, 14
   tapered core, phenalic wedge, 14
   tapered core, solid wedge, 14
   trailing edge sandwich panels, 14
Honeycomb (sandwich) bonded to metal wing, 15
Honeycomb damage (see Damage honeycomb), 201
Honeycomb metal bonded, 200
   construction features, 201
   damage, 201
   glass cloth overlay repairs, 208
   one skin core repair procedures, 209
   potted compoundeds, 206
   repair materials, 203
   repairs, 202
   section, 201
Honeycomb sections, 13,14
   constant-thickness, 13,14
   core, 13,14
   skin, 13,14
   tapered core, 13,14
Honeycomb repairs (see Repairs honeycomb), 201
Horizon indicating gyro, 507
Horizontal stabilizer moveable, 44
Horizontal stabilizer, 3,4
Huck riveting (see Self-plugging), 181
Human reactions to CO_2 poisoning, 432
Humidifiers, 556
Humidity, 28
Hydraulic and pneumatic power system
   actuating cylinders, 328,329
   A/C hydraulic system, 309
   A/C pneumatic system, 331-334
   basic hydraulic system, 315,316
   filters, 313,314
   hydraulic fluid, 309,310
   phospate ester base fluid, 311-313
   pneumatic power system, typical, 338-340
   pneumatic system components, 334-391
   pressure regulating, 323-328
   reservoirs, 316-323
   selector valve, 329,330
   types of hydraulic fluid, 310,311
Hydraulic fluids
   chemical stability, 310
   compatibility with A/C materials, 311
   contamination, 311,312
   contamination check, 312,213
   contamination control, 313
   filters, 313-315
   fire point, 310
   flash point, 310
   health and handling, 311
   intermixing fluids, 311
   mineral base, 311
   phosphate ester base fluid, 311
   types of, 310
   vegetable base, 310
   viscosity, 309
Hydraulic gages, 472
Hydraulic landing gear retraction system, 349, 351

Hydraulic system, basic, 315
   accumulator, 316,324-326
   actuating unit or cylinder, 316,328,329
   check valves, 316,326,327
   hand pump, 315,316
   power driven pump, 315,316
   pressure gage, 316
   pressure regulator, 316
   pumps, 318-322
   relief valve, 316,323
   reservoirs components, 317
   reservoirs, 315-317
   schematic hand pump system, 315
   schematic power pump, 316
   selector valve, 316,329
Hydroplaning, 395
Hydraulic pumps (see Pumps), 318
Hypersonic (see Aerodynamic high-speed), 58
Hz (Hertz), 525

I

Ice and rain protection
   de-icer boot construction, 287,288
   de-icing system components, 288-291
   general, 285,286
   ground de-icing of A/C, 299,300
   pneumatic de-icing system maintenance, 291-293
   pneumatic de-icing systems, 286
   pneumatic system ducting, 296-299
   rain eliminating system maintenance, 308
   rain eliminating systems, 303-308
   thermal anti-icing system, 293,296
   water and toilet drain heaters, 303
   windshield icing control system, 300-303
Icing control system, windshield, 300
Icing structural
   effects, 285
   prevention, 286
Icing (see Pneumatic thermal system), 293
Identification of plastics, 214
ILS (Instrument landing system), 515,525-528
Inboard flap, 20
Incidence angle of, 32
Incidence, 74
Inclusion, 131
Indicators, gear, 353
Inertial navigation system, 531,532
   schematic, 531
Infared heat lamps, 204
Injector type torch, 249
Inner sleeve splicing (see Welding A/C steel structure), 277,279,280
Inspection doors, 24
Inspection openings, 95
Inspection rivet, 176
Inspection lights (see Lighting systems), 464
Installation rivet, 166
Instrument cases, 469
Instrument landing system (ILS), 515,525-528
   CDI, 526
   diagram of system, 525
Instrument panels, 470,471
Instrument systems
   angle of attack indicator, 487,488
   autopilot system maintenance, 516,517
   autopilot system, 511-513
   bank and turn indicator, 482,483
   basic autopilot components, 513-515
   capacitor-type fuel quantity system, 485-487
   dials, 469,470
   electric attitude indicator, 507-511
   flight director systems, 515,516
   fuel flowmeter systems, 497-499
   fuel pressure gages, remote-indicating, 485
   general, 469
   gyro operation, sources of power, 501-504
   gyroscopic instruments, 499-501
   instrument cases, 469
   instrument panel, 470

maintenance of pitot static system, 481,482
oil pressure gages, remote-indicating, 485
pitot static system, 474-481
pressure gages, 471-474
pressure operated gyros, 506
range markings, 470
ratiometer electrical resistance thermometer, 497
repair of instruments, 470
synchro-type remote indicating instrument, 483-485
synchroscope, 491
tachometers, 488-491
temperature indicator, 491-497
turn and bank indicator, 482,483
vacuum-driven attitude gyros, 504-506
vacuum system maintenance practices, 506,507
Interference, radio reduction, 536,537
Inverted gull, 8
In-line check valve, 326,327
Isobaric range, 553,554
Isopropyl alcohol, 600

J

Jigs pressure, 204
Joggling, 162,163
Joints welded (see Welded joints), 255

K

Ketts saw, 138
kHz (kilohertz), 520,529
Kilohertz (kHz), 520,529
Knots
   modified seine, 91-93,96-98
   rib stitching, 90-92,96-98
   seine, 90-92,96-98
   splice, 95
   tie off, 95-98

L

Laminated joints, 226
Laminated plastic, 214,221
Landing gear system maintenance, 400
   adjustable door clearances, 403
   adjustable landing gear latches, 402,403
   disc and brace adjustable, 403,404
   landing gear rigging and adjustment, 402
   main gear door latch mechanism, 401
   schematic of overcenter adjustable, 404
Landing gear, 3,4,23
   main, 4
   nose, 4
Landing gear arrangement
   bogie gear, 341,342
   dual main, 341
   tricycle gear, 341
Landing gear repair (see Welding A/C steel structure), 281,282
Landing gear system
   air pressure, loss of in tubeless, 389,390
   anti-skid system, 399,400
   brake assemblies, 366-372
   brake systems, 360-366
   emergency extension system, 351
   general, 341-348
   inspection and maintenance of brake system, 372,373
   landing gear safety devices, 351-354
   landing wheels A/C, 373-377
   main gear alignment, support, retraction, 348-351
   maintenance, 400-405
   mounting and demounting, 385-389
   nose wheel steering system, 354-356
   operating and handling tips, 394-396
   pressure gage practice, good, 391,392
   repairing, 392-394
   shimmy dampers, 356-360

side wall inflated A/C tires, 396,397
tire inspection, mounted on wheel, 381,382
tire inspection summary, 397,398
tire inspection, tire demounted, 382,383
tire maintenance A/C, 379-381
tires, A/C, 377-379
tube inspection, 384,385
tube repair, 396
wheel, the, 390,391
Landing lights (see Light systems),462-464
Lap joints, 257
Lateral axis (see Axis of A/C), 35,36,45,55
Lateral control (see Control), 44,45
Lateral stability (see Stability), 38
Layout, 164
Layouts making, 151-154
Leading edge flaps, 20,22,23,40
Leading edge strip, 11
Leading edge repair, 197
Leak testing pitot static system, 482
Left wing light (red), 461
Length rivets, 165
L/F (low frequency), 520
Lift, 33,34
Lift resultant, 33,34
Lift tail, 38
Lift-drag ratio, 32
Lightening holes, 154,155
Lighting systems, aircraft, 459-467
   anti-collision installation typical unpressurized, 461
   anti-collision light installation vertical stabilizer, 462
   anti-collision lights, 461,462
   continuity testing, 465
   continuity testing with voltmeter, 467
   exterior lights, 459
   inspection lights wing, 464
   landing lights fixed, 463
   landing lights mechanism and circuitry, 463
   landing lights retractable, 462
   landing lights, 462-464
   left wing red light, 460
   maintenance and inspection of, 464-467
   ohmmeter internal circuitry, typical, 466
   position lights circuitry, 460
   position lights, 459-461
   right wing green light, 460
   single circuit position light circuitry without flasher, 461
   tail light clear (white), 460
   tail position light unit, 460
   taxi light fixed, 463,464
   taxi light mounted on non-steerable position, 464
   wing tip position light unit, 460
Linen, 87
Lines and tubing, 337,338
Line-disconnect valve, 326,328
Linseed oil, 115
Load limits (electrical), 457,458
   controlling monitoring, 458
L.O.C., 526,527
Location numbering system, 5
   aileron station, 6
   butt line, 6
   buttock line, 6
   flap station, 6
   fuselage stations, 6
   nacelle station, 6
   water line, 6
Locking devices, 69
Lok, 186
Longeron, 4,17
Longeron repair, 531,194
Longitudinal axis (see Axis of A/C), 35,36,55
Longitudinal control (see Control), 32, 41
Longitudinal stability (see Stability), 37
Loop antenna, 529
   enclosed, 529
   internal, 529

Low frequency (L/F), 520
Low wing, 8
Lubber line, 508
Lucite, 214
Lumarith, 214

M

Mach number (see Aerodynamic hi-speed), 61,62
Mach/airspeed indicators, 480,481
Magnesium welding (oxyacetylene), 261,262
Magnesyn system, 484,485
Magnetic compass, 507-511
    correction card, 511
    deviation, 508
    geographic north, 508
    lubber line, 508
    magnetic north, 508
    rose, 509
    swinging, 509
    variation, 508
Main landing gear alignment, support, retraction, 348
Major structural stresses, 1
Manifold gages, 473,474
Marker beacon, 527
    category II, 527
    diagram of system, 527
Masking material, 116
Masks, oxygen, 597,598
Mass, 30
Master brake cylinder, 361
Mechanical linkage, 68
    external surface locks, 70
    internal locking devices, 69
    locking devices, 69
    snubbers, 69
    stops, 68
    tension regulators, 70
    torque tubes, 68
Mechanical lock rivet (see Self-plugging), 181
Mechanical tachometers, 488-490
Medium frequency (M/F), 520
megahertz (MHz), 520,522,524,525,533,534
Mek, 205,206,210
Mercerization, 85
Mercury Hg, 473
Mercury, 28
M/F (medium frequency), 520
Metal structure laminated, 15
Metallic arc welding (see Arc welding), 266,524,525
Metalworking machines
    bar folder, 141
    cornice brake, 142,143
    drill press, 140
    drills, 139
    grinders, 140
    grinding wheels, 140,141
    ketts-saw, 138
    nibblers, 138
    reciprocating saw, 138
    rotary punch, 137
    scroll shears, 136
    slip roll former, 143
    throatless shears, 137
Metering pin-type shock strut, 343
Metering tube-type shock strut, 343
Metering valve, 355,356
Methyl-ethyl-ketone (see Mek), 205,206,210
MHz (megahertz), 520,522,524,525,533,534
Micro shaving, 175
Micronic type filters, 314,336
Microphones, 520-522
Mid wing, 8
Mig welding (see Arc welding), 269
Mineral base hydraulic fluid, 311
Mineral spirits volatile, 114
Modified seine knot, 90-92,96-98

Monocoque construction, 4
Monocoque type fuselage, 3
Motion, 29
Motion, Newton's laws of, 30
Multipass welding (see Welding procedures), 271
Multiple disk brake, 368,369

N

Nacelles, 13
Nacelle station, 6
Natural resins (plastic), 213
Navigation light, 8
Navigation and communication systems, 519-538
    airborne navigation equipment, 524
    airborne weather radar system, 532,533
    automatic direction finder, 529, 530
    basic equipment components, 520-522
    basic radio principles, 519,520
    communication system, 522-524
    distance-measuring equipment (DME), 528,529
    Doppler navigation system, 530,531
    emergency locator transmitter (ELT), 533,534
    general, 519
    inertial navigation system, 531,532
    installation of navigation and communication
        equipment, 534-536
    installation of antenna system, 537,538
    instrument landing system 525-529
    power supply, 522
    radio altimeter, 533
    radio beacon transponder, 530
    reducing radio interference, 536,537
    VHF omnirange system, 524,525
Navigation equipment, airborne, 524
Navigation equipment installation of, 534-536
Needle valve, 504
Neutral flame, 252
Newton's Laws of Motion, 30
Nibblers, 138
Nick, 131
Nitrocellulose lacquer, 114
Nitrocellulose lacquer finishes, 144
    replacement of existing finish, 118,119
Nitrogen, 539
Nixonite, 214
North magnetic pole, 507,508
North pole
    geographic, 508
    magnetic, 508
    variation,508
Nose shock strut, 344
Nose wheel brakes, 366
Nose wheel centering, 353,354
Nose wheel steering system
    compensator, 355,356
    followup linkage, 355,356
    heavy aircraft, 354,355
    hydraulic flow diagram, 356
    light aircraft, 354
    metering valve, 355,356
    operation, 355
Null-field static wicks, 537
Numbering system location, 5

O

Oblique shock (see Aerodynamic hi-speed), 58
OBS (Omnibearing selector), 525
Octagonal elongated patch, 191
Ohmmeter usage, 466,467
Oil pressure remote indicator, 485
    schematic, 485
Oil stain, 116
Omnibearing selector (OBS), 525
Omnirange system (Vor), 515,524,525

Opens, 466,467
Outboard flap, 20
Outflow valves, 550
Oval patch (wooden), 238
Overhang aerodynamic balance, 47
Overhead position welds (see Weld procedure), 269-271
Oxidizing, 252
Oxyacetylene cutting, 263
Oxyacetylene welding equipment
    acetylene cylinder, 248
    acetylene gas, 248
    balance pressure type torch, 249,250
    flame adjustment, 252
    goggles, 248,251
    injection-type torch, 249,250
    oxygen cylinder, 248,249
    pressure requlators, 249
    rods, 251
    setting up, 251
    tips, 250,251
    torch, 249
Oxyacetylene welding of ferrous metals, 258
    chrome molybdenum, 259
    stainless steel, 259
    steel, 258,259
Oxyacetylene welding of nonferrous metal
    aluminum 260,261
    magnesium, 261,262
    titanium, 262,263
Oxyacetylene welding process, 252
    backhand welding, 254
    extinguishing torch, 253
    flame adjustment, 252
    forehand welding, 254
    position, 255
    welding techniques, 254
Oxygen, 27
Oxygen, 248,249,587-600
    cylinder, 588
    equipment, portable, 587
    fires or explosion, prevention of, 600
    masks, 597
    need for, 539
    plumbing, 591
    system flow indicator, 597
    system generator, 587
    valves, 592
Oxygen system, 539
    check valve, 593
    chlorinated, fluorinated, hydronated (freon
        cleaning), 600
    cleaning, 599,600
    cleaning compound, 600
    continuous flow, 587-588
    continuous-flowing, 596-597
    cylinders, 588,589
    dilute-demand regulator, 594-596
    draining system, 599
    explosion prevention, 600
    filler valve, 593
    fires, prevention, 600
    fittings, 591,592
    flow indicator, 595,597
    general, 587
    generator, 590
    inspection, 600,601
    isopropyl alcohol, 600
    leak test system 599
    maintenance, 600,601
    masks, 597,598
    need for, 539
    plumbing, 591
    pressure gage, 595,597
    pressure demand, 587,589,591
    pressure reduction valve, 594
    pressure relief valve, 594
    pressure, temperature correction chart, 599
    purging, 600
    regulators, 594
    replacement lines, 592
    servicing gas system, 598
    servicing safety, 598
    servicing trailers, 598-599
    solid state system, 589,591
    valves, 592
    volume comparisons (liquid, gas, solid), 590
Ozone, 539,540

P

Paint, 116
  removal, 117
  remover, 116
  touchup, 117
Paint drier, 115
Paint finishes applying (see Finish application),
  122,123
Painting and finishing
  acrylic nitrocellulose lacquer finish, 114,119,120
  common paint troubles, 124,125
  decalcomanias (decals), 125,126
  enamel finishes, 121,122
  epoxy finishes, 120,121
  finishing materials, 113-116
  fluorescent finishes, 121
  general, 113
  identification numbers, 125
  methods of applying finishes, 122,123
  nitrocellulose lacquer finishes, 114,118,119
  paint and removal, 117
  paint finishes identification of, 117
  paint finishes resoration of, 118
  paint system compatibility, 122
  paint touchup, 117
  painting trim, 125
  preparation of paint, 123,124
Paint finishes identification of, 117
Paint troubles
  blushing, 124
  poor adhesion, 124
  sags and runs, 124
  spray dust, 124
  spray mottle, 124
Panel repair, 192
Panels, instrument, 470,471
Patch repair welding (see Welding A/C steel structure),
  276
Perspex, 214
Phospate ester base hydraulic fluid, 311
Pinked edge, 85,88
Piston type pump, 320-322
Piston type shimmy dampers, 356-358
Pitch (see Axis of A/C), 35,36,45,55
Pitch, rivet, 166
Pitot-static system, 474-484
  airspeed indicators, 479,480
  airspeed indicators maximum allowance, 480
  airspeed/mach indicators, 480,481
  alternate source, 475
  altimeter and errors, 476,477
  head, 475
  heating element, 475,476
  leak testing, 482
  mach indicator, 480,481
  maintenance of, 481
  press multi-engine system, 475,476
  rate-of-climb indicators, 478,479
  schematic, 475-477
  sources of pressure, separate, 476
  static pressure, 475
  tube head, 475
  turn and bank indicators, 482,483,502,504,514
Pitot system, 474-482
Pitting, 131
Plain flap, 22
Plain overlap seam, 90
Plastacele, 214
Plastics, 213
  acetate, 213
  acrylate, 214
  acrylic, 214
  bolt mounting, 221
  care, 218,219,220
  cellulose, 214
  cellulose acetate base, 221
  cellulose plastics, 213
  cementing, 217,218
  characteristics of, 214
  cutting, 216
  drillings, 216,217
  expansion and allowances, 220

  fabricating, 215
  fibestos, 214
  forming, 215,536
  heat treatment, effects of, 218
  identification, 214
  installation procedures, 220
  laminated, 214,221
  lucite, 214
  lumarith, 214
  maintenance, 218-220
  natural resin, 213
  nixonite, 214
  optical considerations, 213
  perspex, 214
  plastacele, 214
  plexiglass, 214
  protection, 214,215
  protein plastics, 213
  protein resin, 213
  rivet mounting, 221
  solvent cement, 217
  storage, 214,215
  synthetic fiberedge attachment, 221
  syrup/cement, 218
  thermosetting, 214
  transparent, 213
Plexiglass, 214
Plug patch (wooden), 235
Ply, 85
Plywood skin repairs, 235
Pneumatic driving, 174
Pneumatic de-icing systems
  adjustment, 292
  boot construction, 287
  combination regulator unloading valve, oil
    separator, 290
  components, 288-291
  de-icier boot maintenance, 292,293
  ducting, 296-299
  electronic timer, 291
  engine driven air pump, 288
  inspection, 292
  maintenance, 291,292
  oil-separator, 290
  operating checks, 291
  safety valve, 290
  solenoid distributor valve, 291
  suction regulating valve, 291
  system operation, 287-289
  systems, 286-293
  trouble shooting, 292,293
Pneumatic power system (see Hydraulic pneumatic
  power), 338-340
Pneumatic rivet guns, 168
  fast hitting, 168
  long stroke, 168
  nomenclature, 169
  offset handle, 168
  pistol grip, 168
  push button, 168
  slow hitting, 168
Pneumatic systems
  brake shuttle valve, 337
  brake system emergency, 337
  check valves, 334,335
  components, 334-340
  control, 334, 335
  filters, 335,336
  high pressure, 331,332
  jet engine compressor, 333
  line and tubing, 337
  low pressure, 333,334
  maintenance, 340
  medium pressure, 333,334
  micronic filter, 336
  relief valve, 334
  restrictors, 335,336
  schematic of two stage air compressor, 333
  screen type filter, 337
  steel cylinder high pressure, 333
  typical power system, 338
  variable restrictors, 335,336
Pods, 13
Portable oxygen equipment, 587,588
Position lights (see Lighting systems), 459-461
Potted compound repairs, 206-208,210
Potting compound (for electrical), 451,452

Power boost, 365,366
Power driven pump, 318
Power supply, 522
Precession, 500
Pressure
  absolute, 543
  ambient, 543
Pressure (atmospheric), 27,28,540
Pressure demand oxygen system, 587,589,591
Pressure gages, 471
  de-icing, 472,473
  diaphram type, 473
  engine, 471,472
  hydraulic, 472
  manifold, 473,474
  shock mounting, 470
  suction, 473
Pressurization (see Cabin atmosphere), 541-548
  absolute pressure, 543
  absolute temperature, 543
  aircraft altitude, 543
  adiabatic, 543
  air-conditioning, 543
  air distribution, 556
  ambient pressure, 543
  ambient temperature, 543
  basic requirements, 545
  basic system, 542
  cabin altitude, 545
  centigrade, 545
  centrifugal compression, 546-548,582
  compressors (positive displacement), 546
  controller, 551
  controls, 550
  control system, 551
  differential pressure, 545
  ducts, 557
  fahrenheit, 545
  filter (air system), 548
  flow sch., 556
  gage pressure, 545
  isobaric range, 553,554
  maintenance, 583
  operational checks, 585
  out flow valve, 550
  pressure regulator, cabin, 551,552
  problems, 543
  ram-air temperature rise, 545
  roots type compressor, 546,547
  safety valve, cabin, 554,555
  sources of pressure, 545,546
  standing barometric pressure, 543,545
  superchargers and instruments, 546-550
  temperature scales, 545
  terms and definitions, 543,544
  troubleshooting, 586
  typical system, 544
  valves, 550
Pressure gage, 316,324
Pressure regulation, 323-328
Primer
  acrylic cellulose nitrate modified, 115
  standard wash, 115
  zinc chromate, 115
Primers, 205
Procedures and technique in welding, 269,271
Proper synchroscope, 491,492
Protein plastic, 213
Power driven pump, 318
Pumping mechanisms, 319
Pumps, hydraulic
  angular type piston, 322
  cam type, 322
  constant delivery pump, 318
  double action hand, 318
  gear type, 319
  gerotor type, 319
  piston type, 320-322
  power driven, 318
  pumping mechanisms, 319
  vane type, 319,320,334
  variable-delivery pump, 319
Pylon, 4

Q

Quick-disconnect valve, 327,328

R

Radar, airborne weather, 532
Radio altimeter, 533
  diagram of system, 533
Radio basic components, 520
  AM, 520,521
  amplitude, 521,522
  antenna, 520,521
  FCC, 521
  FM, 521
  headset, loudspeaker, 520
  microphone, 520,521
  receiver, 520,521
  RF, 520,521
  transmitter, 520,521
  VHF, 521
Radio beacon transponder, 530
Radio frequency (RF), 520,521
Radio interference, reduction, 536,537
  bonding, 536
  shielding, 536,537
  static discharge wicks, 537
Radio principle, basic, 519,520
  frequency bands, 520
  high frequency (H/F), 520,522-524
  kilohertz, 520
  low frequency (L/F), 520
  medium frequency (MF), 520
  megahertz, 520,522
  super high frequency (SHF), 520
  transformer circuit, simple, 519
  ultra high frequency (UHF), 520
  very high frequency (VHF), 520-524,528
Radomes, 223
  detection, removal of oil and moisture, 223
  handling, installation, storage, 223
  inspection of damage, 223
  repairs, 224
  testing of repairs, 224
Rain and ice protection (see Ice-rain protection), 285
Rain eliminating system, 303
  adjustment of wiper system, 308
  circuit diagram, 306
  electric windshield wipers system, 303,304
  hydraulic shield wiper system, 304-306
  maintenance of, 308
  pneumatic rain removal system, 307,308
  typical system, 307
  windshield rain repellant, 307,308
  windshield wiper actuator, 307
Range markings, 470
Rate-of-climb indicators, 478,479
Receivers, 520,521
Reciprocating engine conventional CO₂ system, 417
  CO₂ cylinder installation, 417
  maintenance, 423,425
  system in twin engine, 418,419
  troubleshooting, 425
Reciprocating saw, 138
Reducing flame, 252
Reducing radio interference, 536,537
Reflector rod, 8
Regulation, pressure, 323-328
Regulator valve (vacuum system), 503
Regulators (welding), 249
Relative wind, 31,38,39
Relief holes, 153,154
Relief valve, 316,323,334
  pressure, 503
  suction, 503,504
Remote indicating instrument, 480-488
  angle of attack indicator, 487,488
  capacitor type fuel quantity system, 485-487
  D.C. selsyn system, 483,484
  fuel pressure, 485

magnesyn system, 484,485
oil pressure, 485
schematic diagram of D.C. selsyn, 483
synchro-type, 480-485
transmitters, 488
Removing rivets, 177
Repair of aircraft instruments, 470
Repairs (see Sheet metal, structural repair), 261
Repairs, emergency splicing (electrical), 450,451
Repairs honeycomb, 201
  adhesives, 205,208,209
  cleaning, 209
  cleaning solvents, 205
  core material, 205,206
  core plug, 209,210
  corrosion preventives, 206
  erosion preventives, 206
  fire precautions, 204
  glass cloth overlay, 208,210
  glass fabrics, 206
  infared heat lamps, 204
  materials, 205
  potted compound repair, 206-208,210
  pressure jigs, 204
  primers, 205
  procedures, 212,213
  removal of damage, 206,207,210
  resins, 205
  router, 202
  routing damaged areas, 203
  skin and core repairs procedures, 209,213
  techniques, 206
  tools and equipment, 202,204
  transition area, 211
Repair spar wooden (see Spar repair wood), 195
Repair specific type (see Sheet metal repair), 189-529
Reservoirs, 315-317
Resins, 205
Resultant lift, 31,34
RF (radio frequency), 520,521
Rib and spar structure, 11
Rib lacing cord, 89
Ribs, 11,17
  bulkhead, 11
  butt, 11
  compression, 11
  false, 11
  leading edge, 8
  plain, 11
  wing, 11
  wooden, 10
Rib repair, 196,197
Rib stitching knot, 90-92,96-98
Rib stitch spacing chart, 94
Rigging, 70
  cable rigging chart, 71
  checks, 72
  contour templates, 72
  dihedral, 74
  engine alignment, 75
  incidence, 74
  incidence board, 75
  measuring cable tension, 70
  rigging fixtures, 72
  rulers, 72
  structural alignment, 72
  surface travel measurement, 71
  symmetry, 75,76
  tensiometer, 70
  universal propeller protractor, 73
  verticality, fin, 75
Rigging and assembly, 27-84
Rigging helicopter, 77
Right wing light (green), 460
Rivet failures, 175
  bearing, 176
  head, 176
  shear, 176
Rivet gun (see Pneumatic rivet guns), 168
Rivets and riveting
  bucking, 173
  bucking bars, 167,173
  cherrylock rivets, 181
  coin dimpling, 172

countersinking, 170
countersinks, 167
cutters, 166,167
determining length of, 165
deutsch rivets, 187
dill lok-skrus, 186
dimpling dies, 167
draw sets, 167
drilling, 169
drill sizes, pilot and reaming, 167
driving rivets, 173
failure, 175
grip, 165
hand driving, 173
hand sets, 167
hi-shear, 187-189
hole duplicators, 166
huck rivets, 181
inspection, 176
installation, 166
layout, 164
length, 165
lok-rivets, 186
micro shaving, 175
pitch, 166
pneumatic driving, 174
pneumatic rivet guns, 168
preparation of rivet holes, 168
pull thru, 181
radius dimpling, 172
removing, 177
rivet gun nomenclature, 169
Rivnuts, 184
rivnut tool, 184,185
self-plugging (friction lock), 178,179
self-plugging (mechanical lock), 181
shear, bearing, head failure, 176
shophead, 165
spacing, 167
special rivets, 177
squeeze riveting, 174
squeeze rivets, 168
thermo dimpling, 171
transverse pitch, 166
Rivets special (see Special rivets), 177
Rivnuts, 184
Rods (welding), 251
Roll (see Axis of A/C), 35,36,55
Roots compression, 546,547
Rose, compass, 509
Rotary punch, 137
Round patch, 191
  three row, 191
  two row, 191,529
Round plug patch (wooden), 237
Router, 202
Rudder, 17,20,36,40,43,45
Rudder construction, 17
Rudder control tab, 20,23
Ruddervator, 45

S

Safety devices for landing gear, 351
Safety switch (landing gear), 351
Sand bags, 134
Sandwich panels (see Honeycomb), 14,15
Saw
  ketts, 138
  reciprocating, 138
Scarf joint, 234,235
Scarf patch (wooden), 239
Score, 131
Scratch, 131
Scroll shears, 136
Sealing (see Structural sealing), 198
Seams (sewing)
  doped, 91
  durability, 90
  elasticity, 90
  folded fell, 90
  French fell, 90

good appearance, 90
machine sewed, 90
plain overlap, 90
sewed, 90
strength, 89
Segment rotor, 369-371
Seine knot modified, 90-92,96-98
Selector valves, 316,329,503,504
four way closed-center, 426
spool type, 330,332
Self-plugging
cherrylock hydroshift tooling, 182
friction, 178,179
guns, 179
huck, 181
inspection, 180
installation procedures, 182
installation tools, 178
mechanical, 181
removal, 180.184
Selsyn D.C. instrument, 483,484
Selvage edge, 85,88
Semi-flex static wicks, 537
Semimonocoque construction, 4
Semimonocoque type fuselage, 3
Semimonoque steel engine mounts, 15
Sensitive altimeter, 478
Servo tab, 46
Set back (k) chart, 150
Setback, 148,149
Sewed seams, 90
Sewing (see Seams), 89-91
Shear, 1,2,132
Shears
scroll, 136
throatless, 137
Shear squaring, 135,136
Sheet metel forming (see Forming operation), 145
Sheet metel hand forming (see Hand forming), 155
Sheet metal machines (see Metalworking machines), 136
Sheet metal repairs
bulkhead, 194
chamfering, 192
determining length of break, 194
elongated octagonal patch, 190.191
example, 128former, 194
keeping weight to minimum, 129
leading edge, 197
longeron, 194
maintain original contour, 129
maintain original strength. 127,128
panel repair, 192
rib, 196,197
round patch, 191
smooth skin repair, 190
spar, 195
stringer repair, 193
trailing edge, 198,199
web, 196,197
Sheet metal special tools and devices
bench plate and stakes, 133
cleco fastener, 134
dollies and stakes, 133
hardwood form blocks, 133,158,159,161
holding devices, 131
machinist vise, 134
sand bags, 134
shrinking blocks, 134,159,161
squaring shear, 135,136
utility bench vise, 134
v-block, 133,156
vises, 134
SHF, 520
Shielding, 536,537
Shimmy dampers
piston type, 356-358
steer type, 359,360
vane type, 358,359
Shock struts, 341
bleeding, 347
drag link, 348

hose, 344
metering pin type, 343
metering tube type, 343
operation, 345
servicing, 346
torque link, 348
trunnion and brake arrangement, 348
Shock wave (see Aerodynamic hi-speed), 59,62
Shophead, 165
Shorts, 466,467
Shrinking, 145-147,158
Shrinking blocks, 134
Shuttle valve, brake, 337
Sight line, 149
Silver solder, 264,265
Single action actuating cylinder, 328,329
Single disk brake, 366-368
Sizing, 185
Skin, 4,17,24
Skin and fairing structure, 24
Skin leading edge, 8
Skin patch panel, 192
Skin smooth repair, 529,530
Slats, 40
Sleeve repair weld (see Welding A/C steel structure), 276
Slots, 40
Slotted flaps, 47,48
automatic, 49
fixed, 49
Smoke detection system
carbon monoxide detection, 430,431
CO_2 poisoning, human reaction, 432
photoelectric smoke detector, 430
test circuit, 431
visual smoke detector, 431
Smoke protection equipment, 588
Smooth skin repair, 190
Snubbers, 169
Soft soldering, 265
Soldering
iron, 265
silver, 264,265
soft, 265
Solvent-emulsion compounds, 114
Spacing, rivet, 167
Spar aileron, 11
Spar and rib structure, 11
Spar front, 11
Spar nomenclature, 10
diagonal tube, 10
lower cap strip, 10
rib attach angle, 10
splice, 10
stiffener, 10
upper spar cap, 10
vertical tube, 10
web upper lower, 10
Spar patch (wooden), 242,243,246
Spar repair, 195
Spars, 8,9,12,17
cross-sectional configuration, typical, 9
fail safe, 10
metal spar shapes, 9
plate web, 10
structure, 11
truss, 10
Special rivets
cherrylock, 181
deutsch, 187
dill lok-skrus, 186
hi-shear, 187-189
huck, 181,188,190
lok, 186
pull thru, 184
Rivnuts, 184
self-plugging (friction lock), 178,179
self-plugging (mechanical lock), 181
Speed brakes, 21,40
Splayed patch (wooden), 239

Splice knot, 95
Spliced joints (wooden), 234
Splicing repairs emergency, 450,451
Split flap, 22
Slit wheels, 373-375
Spoilers, 4,20,40,41,43
inboard, 20
outboard, 20
Spool type selector valve, 330,332
Spray gun, 123,124
Spring tab, 46
Squaring shear, 135,136
Squeeze riveters, 166
Squeeze riveting, 174
Stability, 37
dihedral, 39
directional, 37,38
dynamic, 37
lateral, 38
longintudinal, 37
static, 37
Stability and control, 35
Stabilizer actuated elevator tab (horizontal), 20
Stabilizer
horizontal, 34
vertical, 3,4
Stabilizers, 45
Stain, 131
Stainless steel, 259
Stakes (sheet metal tools), 133
Standing barometric pressure, 543,545
Static discharge wicks, 537
flexible vinyl, silver or carbon, 537
null-field, 537
semi-flex metallic braid, 537
Static pressure, 475
Stations (see Location numbering system), 5
Steel welding, 258,259
Steer dampers, 359,360
Straight leading and trailing edges, 7
Straight leading, tapered trailing edge, 7
Stress in structural members, 131,132
bending, 132
compression, 132
shear, 132
tension, 132
torsion, 132
Stresses, 1,2,132
acting on aircraft, 2,214
bending, 1,2,132
compression, 1,2,132
shear, 1,2,132
tension, 1,2,132
torsion, 1,2,132
Stretching, 145,147,157,158
Stringer repair, 193
Stringers, 4,8,17
Structural repair, general, 129
Structural repairs
basic principles of sheet metal repairs, 127-129
bends straight line, 146-148
bolt and bushing holes, 244
bonded honeycomb metal, 200
damage, classification of, 131
damage, inspection of, 130,131
deutsch rivets, 187
dill lok-skus, 186,187
fiberglass components, 221,222
forming, hand, 155-164
forming machines, 141-145
forming operations and terms, 145,146
general structural repairs, 129
glass fabric cloth overlay repairs, 208
glues, 230,232
gluing, 232,233
hi-shear rivets, 187-189
honeycomb construction features, 201
honeycomb damage, 201,202
honeycomb repair materials, 205,206
honeycomb repair, 202,205
layouts, marking, 151-155

lok-rivets, 186
metal working machines, 136-141
one skin and core repairs procedures, 209,213
plastics, 213
plastics forming, 215-220
plastics installation procedures, 220,221
plastics laminated, 221
plastics storage and protection, 214,215
plastics transparent, 213
plywood skin repairs, 235-242
potted compound repair, 206-208
radomes, 223,224
rib repairs, 245,246
rivet failures, 175-177
rivet holes, preparation of, 168-172
rivet installation, 166-168
rivet layout, 164-166
rivets, driving, 173-175
rivet self-plugging (friction lock), 178-180
rivet self-plugging (mechanical lock), 181-184
rivets pull thru, 184
rivets removing, 177
rivets special, 177,178
Rivnuts, 184-186
sealing, structural, 198-200
setback, 148-151
spar and rib repairs, 242,244
special repair, types, 189-198
special tools, devices for sheet metal, 133-135
spliced joints, 234,235
stress in structural repairs, 131-133
wooden aircraft structures, 224
wooden structures, inspection of, 224,228
wooden structures, service and repair, 228-230

Structural members, stresses in, 131

Structural components, 3-5
aileron, 3
elevator, 3
empennage, 3
fuselage, 3
horizontal stabilizer, 3
landing gear, 3
nacelle, 3
propeller, 3
rudder, 3
vertical stabilizer, 3
wing, 3

Structural icing, 285

Structural sealing, 198
determing defects, 199
sealant repair, 199

Structure, helicopter (see Helicopter), 24

Structures, aircraft
empennage, 16
fairings, 24
fixed wing, 2
flight control surfaces, 18
fuselage, 2,3
general, 1
helicopter, 24
landing gear, 23
major structural stresses, 1
monocoque, 3
nacelles, 13
pods, 13
semimonocoque, 3
skin, 24
truss type, 2
wing, 6

Strut, 4

Subsonic air flow (see Aerodynamic hi-speed), 59

Subsonic flow, 29

Suction, 503

Suction gages, 573

Suction gage (vacuum), 503,504

Super high frequency (SHF), 520

Superchargers, 546,547
centrifugal, 546,547
control, 548
instruments, 549,550
pictorial view, centrifugal, 548
roots, 546,547

Supersonic air (see Aerodynamic hi-speed), 59

Supersonic (see Aerodynamic hi-speed), 58, 60-62

Surfaces patch (wooden), 236

Sweepback, 39

Sweptback wing, 7

Swinging a compass, 509

Symmetry, 39,75,76

Synchroscope, 491,492

Synchro-type remote indicating instrument, 483-485

Synethic resins (plastic), 213

T

Tabs, 20
balance, 20,46
control, 20,23
elevator control, 20
general, 45
rudder control, 27
servo, 46
spring, 46
trim, 40,46

Tachometers, 488-491
electric, 489,490
generators, 490
helicopter, 490
maintenance, 490
mechanical, 489
reciprocating type, 489
schematic diagram (electric), 490
true, 480
turbine type, 489,490

Tail cone, 4

Tail light clear (white), 460

Tail surface spar repair (see Welding A/C), 276

Tapered leading and trailing edge, 7

Tapered leading, straight trailing edge, 7

Taxi lights (see Lighting system), 463,464

Tearing strength of fabric, 86

Techniques and procedures in welding, 269-271

Tee joints, 256

Temperature
absolute, 543
ambient, 543
centigrade, 545
fahreheit, 545

Temperature indicator, 491-497
bayonet type, 494
cylinder head temperature, 494
EGT (exhaust gas temperature), 494-497
electric resistor thermocoupler, 492
gasket type, 494
gas temperature indicator, 494-497
ratiometer thermocoupler, 497
thermocouple thermometer indicator, 493,494
T.I.T. (turbine inlet temperature), 495,496
types of thermocouple indicators, 495
wheatstone-bridge metering circuit, 493

Temperature/altitude of atmosphere, 540,541

Templates, 161

Tensiometer, 70

Tension, 1,2,132

Tension regulators, 70

Terminal copper, 453

Thermal anti-icing system, 293
combustion heaters, 294
ducting, 296-299
exhaust heaters, 294-296
schematic heat source, 296
schematic typical system, 298
schematic wing & tail, 297

Thermister, 574

Thermocouples (see Temperature indicator), 491-497

Thermometers (see Temperature indicator), 491-497

Thermoplastic, 214

Thermosetting, 214

Thinner, 113

Throatless shears, 137

Thrust, 33,34

Tie off knot, 95-98

Tie rod, 11

TIG welding (see Arc welding), 267-269

Tips (welding), 250,251

Tire inspection demounted
basic tire maintenance, 382-384
bead area, tubeless tires, 383
bead damage, 382,383
bulges, 383
cords broken, 383
good pressure gage practice, 391
injuries, probe, 382
injuries repairable, 382
inspection, 386
liner blisters, 383
sidewall conditions, 382
thermal fuze, 383

Tire inspection mounted on wheel
damage at wheel, 381
good pressure gage practice, 391
leaks, 381
recapping, when to remove, 381
sidewall injuries, 381
tread injuries, 381
wear, uneven, 381
wheel damage, 381

Tire inspection summary
blister and tread separation, 398
brake heat damage, 399
chafe damage, 399
chevron cutting, 398
circumferential cracks, 398
contamination, 399
cut, 398
fabric and chipping, 398
flaking and chipping, 398
groove and cracking rib undercutting, 398
impact break, 399
liner breakdown, 399
normal, 397
open tread splice, 397
over inflation, 397
peeled rib, 397
skip, 398
thrown tread, 397
tread chunking, 397
tread rubber reversion, 398
types of tire damage, 397-399
under inflation, 397
weather checking, 398
wornout, 397

Tire maintenance
air diffusion loss, tubeless, 379
equalize pressure, for duals, 379
good pressure gage practices, 391
inflation, effects of under, 380
load recommendations, observe, 380
new mountings, 379
nylon flap spotting, 380
nylon stretch, allow for, 379
pressure data source of, 379,380
preventive maintenance summary, 380
proper inflation, 379
satisfactory service, 379
sidewall inflated, 396

Tire mounting and demounting
balance, 385,386
demounting, 388
demounting safety, 387
drop center wheels removable side flange, 389
flat base wheels, 389
good pressure gage practice, 391
handle beads and wheels with care, 387
inflation safety, 386
inspection tube installation, 385
let stand, then recheck, 387
lubrication, 385
mounting, 387
one piece drop center wheels, 389
seating tube type tires, 387
smooth contour tail wheel tires, 389
tires, 389
tube type tires, 388

Tire operating and handling tips
braking and pivoting, 394
condition of landing field, 395
hydroplaning, 395
takeoff and landings, 395

Tires and tubes, storage of
    fuel and solvent hazards, 392
    moisture ozone, avoid, 392
    permissible tire stacking, 392
    racks, 392
    safe tube storage, 392
    store in dark, 392

Tires aircraft, 377
    apex strip, 377
    bead heel, 377
    beads, 377
    bead toe, 377
    breakers, 377
    care, 378,379
    casing plies, 377
    chafers, 377
    construction, 377,378
    cord body, 377
    flippers, 377
    fuel hazards, 398
    good pressure gage practice, 391
    innerliner, 377
    moisture, avoid, 392
    ozone, avoid, 392
    racks, 392
    sidewall, 377
    sidewall inflated, 396
    solvent hazards, 392
    storage, 391
    store in dark, 392
    tread, 377
    tread reinforcement, 377
    tread reinforcing ply, 377
    tube storage, 392

Tires repairing
    bead-to-bead retread, 393
    damage, types of, 397-399
    full recapping, 383
    nonrecappable, 393
    operational damage, 392
    recapping, 393
    repairable, 393
    spot repairs, 394
    three-quarter retread, 393
    tires that may be recapped, 393
    top recapping, 393

Tires tubeless, air pressure loss, 389
    air temperature, 390
    beads damaged, 390
    beads improperly seated, 390
    cut, 390
    good pressure gage practice, 391
    puncture, 390
    split wheels, 389
    stretch period, initial, 390
    venting, 390

T.I.T. (turbine inlet temperature), 495,496
Titanium welding (Oxyacetylene), 262,263
Toilet drain heaters, 303,304
Toluene, 114
Torch (welding), 249
    balance pressure type, 249
    injector type, 249
Torque helicopter, 50
Torque link, 342,343,348
Torque tube, 67,68
Torsion, 1,2,132
Tracking, 78
Trailing edge repair, 197,198
Transmitters, 520,521
Transonic (see Aerodynamic hi-speed), 58,64
Transparent plastic, 213
Transponder, radio beacon, 530,538
Transverse pitch, 166
Tricycle gear (landing gear), 341
Trim tab, 40,46
Triple slotted flap, 22
Troubleshooting
    fire detection system, 425
    fire detectors, 425
    fire protection system 425
    gyro instrument, 506
    pneumatic deicing system, 292,293

    pressurization, 586
    vapor cycle (freon), 586
True airspeed indicator, 480
Trunnion and bracket arrangement, 348
Truss type fuselage, 2
Truss warren, 4
Tube head pitot, 475
Tube inspection
    chaffing, 384
    fabric base tubes, 385
    good pressure gage practice, 391
    inner tube inspection, 384
    inspection, 386
    natural contour, 385
    proper size, 384
    repair, 396
    storing, 391,392
    taking a "set", 385
    thinned out edges, 385
    thinning, 384
    valve stems, 384
    wrinkles, 384
Tubing and lines, 337,338
Tubular repair welding (see Welding A/C steel structure), 276
Tubular steel fuselage, 4
Tubular wing repair (see Welding A/C steel structure), 276
Turbine engine ground fire protection, 422
Turbine engine fire extinguishing system, 420-422
    detection system and switches, schematic drawing, 423
    discharge nozzle locator, schematic drawing, 422
    discharge tube, schematic drawing, 422
    oval containers, 422
    schematic drawing of multi-engine system, 421
    t-handle switch, 420
Turbine inlet temperature (TIT), 495,496
Turbine tachometer, 489,490
Turbojet fire protection system, 419,420
    schematic drawing of system, 419
Turn and bank indicator, 482,483,502,504,514
    four minute turn indicator, 482
    maintenance practices, 483
    two minute turn indicator, 482
Turnbuckles, 65
Turpentine, 114

U

Ultra high frequency (UHF), 520,528
Universal propeller protractor, 73
Upsetting, 131

V

Vacuum system (see Gyroscoptic instrument system), 499-507
Valves
    brake shuttle, 337
    check, 316,326,327,334
    control, 334
    in-line check, 326,327
    line-disconnect, 327,328
    orifice type check, 326,327
    quick-disconnect, 327,328
    relief, 316,323,334
    selector, 316,329,330
Valves (oxygen), 592
    check, 593
    filler, 593
    outflow, 550
    pressure reducer, 594
    pressure relief, 594
    shut off, 593,594
Vane type pump, 319,320,334
Vane type shimmy dampers, 358,359
Vane type fuel flow system, 498

Vapor cycle system (freon), 576-587
    air routing valve, 583
    air turbine compressor, 580-582
    centrifugal compressor, 582
    components, 577
    compressor, oil checking, 585
    condenser, 577,578
    cycle compressor, 577
    description of typical system, 580-583
    electric heaters, 583
    evacuating system, 585
    evaporator, 580
    expansion valve, 579,580
    filter/drier, 579
    freon-12, 584
    inspection, 583
    maintenance, 583
    manifold set, 584
    operational checks, 585,586
    primary heat exchanger, 582
    purging system, 585
    receiver, 578
    recharging, 585
    refrigeration cycle, 576,577
    refrigeration units, 583
    schematic, 578
    servicing, 583,585
    sight glass, 579
    subcooler, 578,579
    thermostatic expansion valve, 580
    troubleshooting, 586
    707, 720 vapor cycle system, 581
Variable delivery pump, 319
Variation, 508
Varnish, 116
V-blocks, 133,156
Vegetable base hydraulic fluid, 310
Velocity, 29
Ventilation, 95,110
Venturi-tube system (vacuum), 501
Vertical axis (see Axis of A/C), 35,36,41,43,45,55
Vertical control (see Control), 44
Vertical fin, 4
Vertical position welds (see Weld procedures), 269-271
Vertical stability (see Stability), 37
Vertical stabilizer, 3,17
Vertical stabilizer and rudder construction. 17
Vertical stabilizer, 464
Very high frequency (VHF), 520-524,528
    system diagram, 523
VHF omnirange system, 524
Viscosity, 30
Vises
    machinist, 134
    utility bench, 134
Voltmeter, 467
VOR (omnidirectional range), 515,524,525
Vortex generator (see Aerodynamic hi-speed), 63
Vortices wing tip, 35

W

Warner master brake cylinder, 362
Warp, 85,88
Warren truss, 4
Washer, aluminum, 453
Water and toilet drain heaters, 303,304
Water line, 6
Water penetration (wooden structures), 227
Weather radar, airborne, 532
Web repair, 196,197
Weft, 85
Weight, 34
Weld
    characteristics of good, 258
    correct forming of, 258

Welded joints
  butt, 255,273
  corner, 256,257
  edge, 256
  lap, 257
  tee, 256

Welding
  aircraft steel structures, 276-283
  brazing methods, 264
  correct forming of a weld, 256
  cutting, 263
  electric arc welding, 266-269
  expansion and contraction of metals, 257,258
  general, 247,248
  joints welded, 255
  nonferrous metals using oxyacetylene, 260,261
  oxyacetylene welding equipment, 248-254
  oxyacetylene welding of ferrous metal, 258
  positions, 255
  procedures and techniques arc, 269-276
  soft soldering, 265
  titanium, 262,263

Welding arc (see Arc welding), 266

Welding A/C steel structure, 276
  bolted sleeve, 277
  brace struts, 233
  built-in fuselage fitting, 280,281
  built-up tubular wing repair, 283
  engine mount repairs, 279,280
  inner sleeve method of splicing, 277,279,280
  landing gear repair, 281,282
  part not to be welded, 276,277
  patch repair welded, 277,278
  tail surface spar repair, 283
  tubular repair, 277
  welded sleeve repair, 277
  wing struts, 283

Welding nonferrous metals oxyacetylene, 260-263

Welding of ferrous metals, 258

Welding procedures and techniques (arc), 269-271
  bead welds, 271-273,275
  butt joint, 273,275,276
  fillet welds, 273,275,276
  flat position, 271
  groove welds, 273,275,276
  multipass welding, 271,272
  overhead position, 274
  techniques of position, 271
  vertical position, 274

Wheatstone-bridge meter circuit, 493

Wheel centering, nose, 353,354

Wheels, landing
  bearings, 376,377
  corrosion, bead ledge area, 390
  cracks, bead ledge flange area, 390
  damage, 381
  flange wheels fixed, 376
  flange wheels removable, 375
  good pressure gage practice, 391
  knurls, 390
  O-rings, improper installation of, 390
  porous wheel assemblies, 390
  scratches in ledge and flange areas, 390
  sealing surfaces, 390
  seating poor, in bead area, 390
  split wheels, 373-375
  thermal fuze installation, 391
  tie bolts, 391
  tubeless valve holes, 391
  wear bead ledge area, 390

Windshield icing control system (see Wind), 300
  alcohol de-icing windshield and carburetor, 302
  circuit, 303
  defrost system, 302
  pitot tube anti-icing, 303

Wing area, 33,34

Wing box-beam milled, 13

Wing configuration, 8
  dihedral, 8
  gull wing, 8
  high wing, 8
  inverted gull, 8
  low wing, 8
  mid wing, 8

Wing landing gear operation, 351,352

Wing lights
  left (red), 460
  right (green), 460

Wing nomemclature, 11
  aileron, 11
  aileron hinge, 11
  anti-drag wire, 11
  attach fillings, 11
  bulk head rib, 11
  butt rib, 11
  compression rib, 11
  drag wire, 11
  false rib, 11
  false spar, 11
  front spar, 11
  leading edge strip, 11
  nose rib false rib, 11
  plain rib, 11
  rear spar, 11
  tie rod, 11
  wing rib, 11
  wing tip, 11

Wing ribs, 10,11

Wing shapes, 7
  delta wing, 7
  straight leading and trailing edges, 7
  straight leading, tapered trailing edge, 7
  swept back wing, 7
  tapered leading and trailing edges, 7
  tapered leading, straight trailing edge, 7

Wing spars (see Spars), 8,9,12

Wing structure, 6-12

Wing struts repair (see Welding A/C steel structure), 276-282

Wing tips, 8,11,12
  access door, 12
  anti-icing exhaust air outlet, 12
  cap, 12
  corrugated inner skin, 12
  heat duct, 12
  leading edge outer skin, 12
  navigation light, 12
  reflector rod, 12
  upper skin, 12

Wire and cable
  aluminum wire into aluminum terminal lugs, 451
  aluminum wire terminal, 450
  bend radii, 441
  cable clamps, 444,445
  chafing, protection against, 442
  characteristics of copper and aluminum, 433
  clamp and grommet in bulkhead, 443
  conductor insulation, 439
  copper wire terminal, 449
  crimping tools, 449,450
  current carrying capacity of wire, 436
  cutting, 447
  drain hole intubing, 444
  factors affecting the selection of wire size, 433,435
  gage chart, 34
  groups and bundles, 441
  indentifying, 439-441
  installation, 441
  insulation sleeve, 450
  lacing, 444
  lacing branch offs, 446,447
  reducing wire size with permanent splice, 451
  replacing broken wire by soldering and potting, 452
  routing and installation, 442
  routing precautions, 443,444
  selection of conductor material, 435
  single cord lacing, 445
  size, 433
  slack in bundles, 441,443
  solvents and fluids protection against, 442,443
  solderless terminals and splices, 448,449
  spliced connection in bundles, 441
  splicing copper with pre-insulated splice, 450
  splicing repairs, emergency, 450,451
  splicing with solder and potting compound, 451
  staggered splice, 442
  stripping, 447-449
  temperature, high protection against, 442
  terminal lugs to terminal blocks, connection, 451,452
  twisting, 441
  tying, 44

  use of wire chart, 436-439
  voltage drop, 436
  wheel well areas, protection, 443
  wire and circuit protection chart, 458
  wire chart, 437,438

Wooden aircraft structure
  bolt and bushing holes, 244
  flush patch, 235
  glue joint inspection, 225,226
  glued joints testing, 233
  glues, 230
  gluing, 232
  inspection of, 224
  laminated joint, 226
  oval patch, 238
  plug patch, 235
  plywood skin repairs, 235
  rib repair, 242,245
  round plug patch, 237
  scarf joint, 234,235
  scarf patch, 239
  service and repair, 228
  single ply structure, 225
  spar repair, 242,243,246
  splayed patch, 239
  splice joints, 234
  service patch, 236
  water penetration, 227
  wooden condition, 226-228

Woof, 85,88

Y

Yaw (see Axis of A/C), 35,36,41,43,45,55

Z

Zinc chromate primer, 115